Physics, Optics, and Spectroscopy of Materials

Physics, Optics, and Spectroscopy of Materials

Zeev Burshtein
Ness Ziona, Israel

WILEY

This edition first published 2022

Registered Office
John Wiley & Sons, Inc., 111 River Street, Hoboken, NJ 07030, USA

Editorial Office
John Wiley & Sons, Inc., 111 River Street, Hoboken, NJ 07030, USA

For details of our global editorial offices, customer services, and more information about Wiley products visit us at www.wiley.com.

Wiley also publishes its books in a variety of electronic formats and by print-on-demand. Some content that appears in standard print versions of this book may not be available in other formats.

Library of Congress Cataloging-in-Publication Data Applied for:

Hardback ISBN: 9781119768739

Cover image: © yuanyuan yan/Getty Images
Cover design by Wiley

Set in 11/14pt STIXTwoText by Straive, Pondicherry, India

Contents

Introduction

This book, titled "Physics, Optics, and Spectroscopy of Materials", is an English translation of my book "Light and Matter", which was originally written in Hebrew (see Appendix D1, reference [37]). It is an outgrowth of lecture notes I have prepared for my students at the Ben-Gurion University in Beer-Sheva, Israel, and of notes I have myself used in my research efforts. The latter research efforts were mostly performed at the Nuclear Research Center, Negev (NRCN) in Israel, although a considerable portion were also done during my sabbaticals at the NRC in Ottawa, Canada, at Fisk University in Nashville, Tennessee, USA, and at the Toronto University in Toronto, Canada. My central research area over the last 30 years has involved laser science and technology. Accordingly, I refrained from extending the presented scope of this book to extreme regions of the electromagnetic radiation, like X-rays and gamma rays, or nuclear processes.

The intended audience are undergraduate and graduate students in the physics, chemistry, and electro-optics departments, as well as their teachers. I wrote the book in a form that supports a methodically organized lecturing, and not solely as a textbook. Thus, I have attempted to present the subjects in a deductive order, and grade the mathematical sophistications and difficulties. Figures, graphs, and data tables represent only a small part of the scientifically known information. They are principally aimed at demonstrating the physical effects and their relations. The used mathematics rest somewhere between practices used in "theoretical" and "experimental" science. The mathematical proofs and developments used to present the physical issues appear where they are first needed, and are not concentrated at the book's beginning or end, as is often the practice in textbooks. I have cut short some long mathematical developments, particularly where their continuation is just technical; I have pointed out, however, the path for completing them. I refrained from presenting proofs and developments related to advanced subjects in relativity, group theory, or angular momentum algebra. The background subjects needed to comprehend the material are usually addressed in undergraduate introductory lectures on quantum physics. Familiarity with these subjects is helpful but not imperative for understanding the material in this book.

Several exercises are provided at the end of each chapter. They are aimed at demonstrating a desired mix of what I consider proper to ensure an understanding of the chapter content. I consider the carrying out of numerical calculations to be very important; it helps to clarify the meaning of the different physical terms, their relations, and the effects' magnitudes. I recommend using the cgs units system. The number of presented exercises is quite small; many more collections of exercise are available for the student and lecturer at any university library.

A list of literature references for further reading is provided at the end of the book. To a large or small extent, I have consulted most of them in selecting the subjects and their presentation methods. I am thus highly indebted to all of their authors. I am also indebted to several colleagues who read parts of the manuscript and provided useful comments, especially Prof. G. Kimmel, Dr. Y. Shimony, Dr. P. Blau, and Mr. S. Kamin. I am further indebted to my son, architect Joseph Burshtein, who spent a considerable amount of time preparing various figures, especially where it was important to demonstrate depth in basically three-dimensional structures.

Zeev Burshtein
Beer-Sheva
2022

1 Electromagnetic Radiation/Matter Interaction – A Classical Approach

To date, quantum mechanical theory provides the most complete and accurate description of the interaction between electromagnetic radiation and matter. It also provides a clear description of internal nuclear processes. However, it is still possible to account for a host of effects in classical terms. Moreover, presentation of the classical approach accentuates its failure points, those that eventually led to the development of quantum mechanics. Therefore, we start by presenting the subject via the classical approach.

1.1 Electromagnetic Radiation by Atoms and Molecules

Atoms and molecules emit electromagnetic radiation because of the persistent motion experienced by their electrons about the nucleus. These motions generate vibrating electromagnetic fields propagating through space. Let's consider first the electric charge distribution in an atom. Any charge distribution may be presented as a sum of multipole moments. The charge distribution expansion into multipole moments is described in detail in Section 1.5. Here we indicate that the zero-order one, the monopole, is just the total charge located at the mass center. Among the others, most important is the first-order one, the dipole moment. The relative influence of higher-order moments is greatly reduced when observed at a large distance from the mass center.

Even without specifying the electrons motion about the nuclei, we may assume that they form a system of normal vibrational modes. In a normal mode, all system parts vibrate at a common frequency and phase. Classical theory predicts a continuum of possible frequencies for normal modes, much like the frequencies continuum allowed for the motion of stars around the sun. Quantum mechanics, on the other hand, predicts the existence of only specific normal modes, those that exhibit an internal constructive interference of the electronic wave function. Thus, a discrete set of possible frequencies is obtained for the normal modes of bound electrons, like the characteristic frequencies of a both-ends-fixed violin cord. The existence of discrete sets of frequencies for electrons bound to nuclei is demonstrated experimentally in countless cases. We thus artificially impose this fact into all our classical calculations. We may assume that a proper choice of initial conditions forces the charge multipoles to vibrate at a single angular frequency ω_0.

We are not concerned with the monopole addend. It vanishes in non-ionized atoms, and ions emit electromagnetic radiation only under acceleration or deceleration. A constant-velocity movement won't cause emission of any radiation. Thus, we consider the radiation emitted by the vibrating dipole moment; higher-order moments need to be considered only if the dipole moment diminishes.

Notably, we neglected all vibrating magnetic multipoles. They are formed by angular motion of the electric charge. It is possible to show that except for special cases, the magnetic dipole moment contribution to the electromagnetic radiation is negligibly small compared to that of the electric dipole one. The units of the electric dipole moment are charge times length units; in the cgs system they are $\text{esu} \cdot \text{cm}$, or alternatively $(\text{erg} \cdot \text{cm}^3)^{1/2}$.

Physics, Optics, and Spectroscopy of Materials, First Edition. Zeev Burshtein.

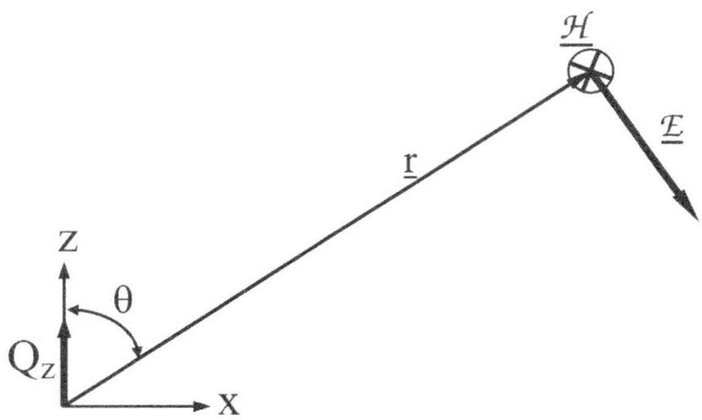

Figure 1.1 Orientations of electric and magnetic fields induced by a vibrating electric dipole Q_Z. The X axis represents a direction perpendicular to Z in the $(Z,\underline{r})$ plane. $\underline{\mathcal{E}}$ will always reside on the $(Z,\underline{r})$ plane, and $\underline{\mathcal{H}}$ is always perpendicular to it.

We consider the radiation emitted by an electric dipole Q_Z, vibrating harmonically at an ω_0 angular frequency

$$Q_Z(t) = Q_0 \cos\omega_0 t \quad . \tag{1.1}$$

For simplicity, we begin by assuming that the maximal dipole strength Q_0 is a time-independent constant. We will show that this assumption requires correction to be quantitatively assessed. The issue of radiation emitted by the dipole has been classically solved. In free space, at a distance r which is large compared to the wavelength λ ($\lambda = 2\pi c/\omega_0$, where c is the vacuum speed-of-light constant), the electric and magnetic fields are given by

$$\boxed{|\underline{\mathcal{E}}| = |\underline{\mathcal{H}}| = \frac{1}{c^2 r}\ddot{Q}_Z\left(t - \frac{r}{c}\right)\sin\theta} \quad . \tag{1.2}$$

The electric field $\underline{\mathcal{E}}$ and magnetic field $\underline{\mathcal{H}}$ vectors are mutually perpendicular (see Figure 1.1), and θ is the angle between the Z-axis direction and the $\underline{r}$ vector direction. Both are perpendicular to $\underline{r}$. The situation addressed in the figure is for the part of the vibrational period where the magnetic field $\underline{\mathcal{H}}$ points into the page. When the magnetic field $\underline{\mathcal{H}}$ points outward, the electric field $\underline{\mathcal{E}}$ reverses orientation as well. The phase common to both $\underline{\mathcal{E}}$ and $\underline{\mathcal{H}}$ lags behind the vibrating dipole phase according to the radiation propagation time-delay r/c. Notably, the vibrating fields amplitudes diminish when the $\underline{r}$ vector direction coincides with the Z-axis direction ($\theta = 0$), and attain maximal value when $\underline{r}$ is perpendicular to Z. An equation identical to Eq. (1.2) will be obtained for any $\underline{r}$ orientation if the dipole in Figure 1.1 is exchanged by its projection on a plane perpendicular to $\underline{r}$. The radiation is linearly polarized in the $(\underline{r},\underline{Q})$ plane.

Eq. (1.2) describes an electromagnetic radiation propagating away from the vibrating dipole. To calculate the radiated power, it is necessary to integrate the Poynting vector $\underline{S} = (c/4\pi)\underline{\mathcal{E}} \times \underline{\mathcal{H}}$ over a surface σ enclosing the vibrating dipole. Recall that the Poynting vector describes the radiated power density; its orientation indicates the direction of energy propagation. In most cases, the time resolution of interest in the power variation is not during the $T = 2\pi/\omega_0$ oscillation period, but over many such periods. Thus, for simplicity, we calculate the average power over a single full period, and consider that value as representative of the temporal power.

The average power $\overline{P}$ over a full period (in units of energy per unit of time) will thus be given by

$$\overline{P} = \frac{1}{T}\int_0^T dt \int_\sigma \underline{S}\cdot d\underline{\sigma} \quad . \tag{1.3}$$

We perform integration over the solid angle Ω using a variable change; $d\Omega = \frac{d\underline{\sigma}}{r^2}\cdot\frac{\underline{r}}{r}$.

$$\overline{P} = \frac{1}{T}\int_0^T dt \int_\Omega \frac{1}{4\pi c^3}\frac{|\ddot{Q}_Z|^2}{r^2}\sin^2\theta \cdot r^2 d\Omega \quad . \tag{1.4}$$

We use Eq. (1.1) to calculate $\ddot{Q}_z$ and set it in Eq. (1.4). A further variable change is by setting $d\Omega = 2\pi\,\mathbf{sin}\,\theta \cdot d\theta$. We get

$$\overline{P} = \frac{\omega_0{}^4 Q_0{}^2}{2c^3}\int_0^{\pi} d\theta\,\mathbf{sin}^3\theta \cdot \frac{1}{T}\int_0^{T} dt\,\mathbf{cos}^2\omega_0 t = \frac{\omega_0{}^4 Q_0{}^2}{2c^3}\cdot\left[\frac{\mathbf{cos}^3\theta}{3} - \mathbf{cos}\,\theta\right]_0^{\pi}\cdot\frac{1}{T}\cdot\frac{1}{\omega_0}\left[\frac{1}{2}\omega_0 t + \frac{1}{4}\,\mathbf{sin}\,2\omega_0 t\right]_0^{T} \quad ;$$

$$\boxed{\overline{P} = \frac{1}{3}\cdot\frac{\omega_0{}^4}{c^3}Q_0{}^2} \quad . \tag{1.5}$$

Let's consider a model oscillator, of a single electron vibrating harmonically and one-dimensionally against an equally charged positive center. The attractive force exerted by a positive center on the negative electron is obviously nonharmonic. However, any periodic motion may be expanded into a sum of harmonic components, and we consider the dominant one. For a vibrational amplitude x_0, the maximal dipole moment is $Q_0 = ex_0$, where e is the elementary charge. Because we are interested in radiation emitted by atoms and molecules, x_0 would exhibit atomic dimensions, namely $\sim 1\,\text{Å}$. The mechanical energy W (kinetic + potential) is then

$$W = \frac{1}{2}m\omega_0{}^2 x_0{}^2 \quad , \tag{1.6}$$

where m is the electron mass. Thus,

$$\overline{P} = \frac{2}{3}\cdot\frac{e^2}{mc^3}\omega_0{}^2 W = \gamma W \quad , \tag{1.7}$$

where $\gamma \equiv 2e^2\omega_0{}^2/3mc^3$. The value of $2\pi\gamma/\omega_0$ expresses the fraction of radiation energy emitted during a vibrational period relative to the oscillator mechanical energy. This value is always small compared to a unity. For demonstration, let's use a $\nu = 10^{15}$ Hz frequency, namely a $\lambda = 3000\,\text{Å}$ wavelength, belonging to the near ultraviolet. We now calculate the quantity $2\pi\gamma/\omega_0$:

<u>Given data:</u>

$$e = 4.8 \times 10^{-10}\ \text{esu} \ ; \ m = 9.11 \times 10^{-28}\ \text{g} \ ; \ c = 3 \times 10^{10}\ \text{cm/s} \ ; \ \nu = 10^{15}\ \text{s}^{-1}$$

<u>Developing:</u>

$$2\pi\gamma/\omega_0 = \frac{4\pi}{3}\cdot\frac{e^2\omega_0}{mc^3} = \frac{8\pi^2}{3}\,\frac{e^2\nu}{mc^3} \quad ;$$

<u>Setting numerical values:</u>

$$2\pi\gamma/\omega_0 = \frac{8 \times 3.14^2}{3}\,\frac{4.8^2 \times 10^{-20} \times 10^{15}}{9.11 \times 10^{-28} \times 3^3 \times 10^{30}} = 2.45 \times 10^{-7} \quad .$$

The latter result clearly demonstrates that the oscillator mechanical energy change within a single oscillation is so tiny that averaging over a single oscillation while assuming constant amplitude for that duration is fully justified.

As a result of the radiation losses, the oscillator energy gradually decays, hence also the maximal dipole moment Q_0. The radiation power $\overline{P}$ expresses the energy loss of the mechanical oscillator which forms the oscillating dipole. Thus, we may write a simple rate equation describing the mechanical energy stored by the oscillator at any moment:

$$\frac{dW}{dt} = -\gamma W \quad , \tag{1.8}$$

the solution of which is

$$W(t) = W_0 e^{-\gamma t} \quad , \tag{1.9}$$

where W_0 is the initial oscillator mechanical energy. The solution describes an exponential decay. The time duration $\tau = 1/\gamma$ for the energy to assume an e^{-1} fraction of its initial value is called **lifetime**. The oscillator forming the vibrating dipole practically loses all its energy within several lifetimes, and practically ceases to radiate. In the optical region, lifetimes are of the order of ~10^{-8} s. Occurrence of that order of fluorescence emission lifetimes of exciting atoms and molecules is an indication of its type being of an **electric dipole radiation** nature.

It is customary to denote $\gamma = \tau_0 {\omega_0}^2$, where $\tau_0 = \frac{2}{3}\frac{e^2}{mc^3} = \frac{4}{3} r_0/c$. The quantity $r_0 = e^2/2mc^2$ is considered the **electron electrostatic radius**, whose magnitude is $r_0 = 1.409 \times 10^{-13}$ cm. This "electrostatic" radius expresses the idea that the electron rest energy mc^2 stems from the energy $e^2/2r_0$ generated by total electron charge e distributed uniformly over a spherical surface of r_0 radius♣. The quantity $\tau_0 = 6.26 \times 10^{-24}$ s expresses the time required for electromagnetic radiation to traverse 4/3 times the distance equal to the electrostatic electron radius.

Because the emitted electromagnetic radiation causes a continual damping of the oscillator energy, its equation of motion should be corrected to include a factor expressing this effect. We shall see shortly that an addend to the equation of motion resulting in Eq. (1.9)'s temporal behavior is $\gamma m\dot{x}$; the equation of motion will be

$$\boxed{m\ddot{x} + \gamma m\dot{x} + m{\omega_0}^2 x = 0} \quad . \tag{1.10}$$

We assume a solution of the form $x = x_0 e^{i\varepsilon t}$. Setting in Eq. (1.10) and solving, yields $\varepsilon = \frac{1}{2}i\gamma \pm \sqrt{{\omega_0}^2 - \frac{1}{4}\gamma^2}$, namely $x(t) = x_0 e^{-\frac{1}{2}\gamma t} e^{\pm i\omega_1 t}$, where ${\omega_1}^2 = {\omega_0}^2 - \frac{1}{4}\gamma^2$. The two solutions for ε differ by only their phase,

♣ This classical model is inspired by the fact that many multiparticle quantum systems converge into near-spherical-shell configurations like shell structures of electronic and nuclear orbits. In relation to electron orbits, see Sections 5.4.2 and 5.5.3 in this book. However, it is unclear whether it is at all possible to attribute any spatial distribution to the electronic charge. In most scientific literature, the "electron electrostatic radius" is defined simplistically as $r_0 \equiv e^2/mc^2$, twice the above defined value. This book adheres to our definition and value. Surprisingly, this value is also *exactly* the attractive potential range between *nucleons*. Its value being finite is also expressed by nonzero rest-masses of its quanta, the π-mesons. As an aside, the rest masses of the two oppositely charged π-mesons are *exactly* $2/\alpha = 274$ times heavier than the electron rest mass m, where α is the fine-structure constant of atomic electron states spectra; see Eq. (4.20) and related text.

which may be chosen arbitrarily. The corrected time dependence of the electric dipole moment becomes (see Eq. [1.1])

$$Q_z(t) = Q_0 e^{-\frac{1}{2}\gamma t} \cos \omega_1 t \quad . \tag{1.11}$$

Because $\gamma \ll \omega_0$, it may be written with excellent approximation that $\omega_1 \cong \omega_0 - \frac{1}{4}\left(\frac{\gamma}{\omega_0}\right) \cdot \left(\frac{1}{2}\gamma\right)$. However, as we shall see shortly, the frequency change $\Delta\omega_L \equiv \omega_0 - \omega_1 \cong \frac{1}{4}\left(\frac{\gamma}{\omega_0}\right) \cdot \left(\frac{1}{2}\gamma\right)$ is tiny compared to the **natural width** of the emitted frequency band (a concept discussed in Section 1.2.1). Therefore, we shall always replace ω_1 with ω_0. The electric field at any point in space attains the form

$$\mathcal{E}(t) = \mathcal{E}_0 e^{-\frac{1}{2}\gamma t} \cos(\omega_0 t - \varphi) \quad , \tag{1.12}$$

where $\mathcal{E}_0$ is the electric field amplitude at the chosen point, and $\varphi = \omega_0 r/c$ is the phase lag relative to the oscillator's phase.

Quantum mechanically, the interaction between the electron oscillator and the electromagnetic field vacuum vibrations (**zero-point oscillations** of the radiation field) induces a frequency shift called **Lamb –Shift**. In some cases it may become measurably larger than the natural width. This subject will be discussed in detail in Section 4.7.

1.2 Spectral Line Widths

1.2.1 Natural Width

Section 1.1 shows that the radiation emitted by the oscillating electric dipole is not purely harmonic, but rather exhibits a damped nature (Eq. [1.12]). In complex numbers algebra it reads

$$\mathcal{E}(t) = \begin{cases} \mathcal{E}_0 e^{i\left(\omega_0 t + \frac{1}{2} i\gamma t\right)} & t > 0 \\ 0 & t \le 0 \end{cases} \quad . \tag{1.13}$$

In the present case, the physical meaning of the use of complex algebra is that at any moment, the electric field is the real part of the complex quantity. For simplicity, we avoided including the phase φ in the expression because it is independent of time for any chosen point in space. Such wave may be presented as a group of pure harmonic components: by **Fourier** expression, the function $\mathcal{E}(t)$ may be presented in the form

$$\mathcal{E}(t) = \frac{1}{\sqrt{2\pi}} \int_{-\infty}^{+\infty} \mathbf{E}(\omega) e^{i\omega t} d\omega \quad . \tag{1.14}$$

The function $\mathbf{E}(\omega)$ is called **Fourier transform** of the $\mathcal{E}(t)$ function, and is given by

$$\mathbf{E}(\omega) = \frac{1}{\sqrt{2\pi}} \int_{-\infty}^{+\infty} \mathcal{E}(t) e^{-i\omega t} dt \quad . \tag{1.15}$$

The $\boldsymbol{E}(\omega)$ function provides information on the wave spectral distribution: its absolute value expresses the amplitude of the ω angular frequency component, and its phase describes the phase distribution among the various frequency components. The latter information is crucial when dealing with interference among different beams. We now calculate the Fourier transform of a wave of the form given in Eq. (1.13):

$$\boldsymbol{E}(\omega) = \frac{1}{\sqrt{2\pi}} \int_0^\infty e^{-i\left[(\omega-\omega_0)-\frac{1}{2}i\gamma\right]t} dt = \frac{1}{\sqrt{2\pi}} \cdot \frac{-1}{i(\omega-\omega_0)+\frac{1}{2}\gamma} e^{i\left[(\omega_0-\omega)+\frac{1}{2}i\gamma\right]t} \Bigg]_0^\infty =$$

$$= \frac{1}{\sqrt{2\pi}} \cdot \frac{1}{i(\omega-\omega_0)+\frac{1}{2}\gamma} = \frac{1}{\sqrt{2\pi}} \cdot \frac{\frac{1}{2}\gamma - i(\omega-\omega_0)}{\frac{1}{4}\gamma^2 + (\omega-\omega_0)^2} .$$

In most cases, the measurable quantity is not the spectral distribution of the wave amplitude, but rather the distribution of energy $I(\omega)$, which is proportional to $|\boldsymbol{E}(\omega)|^2$. Calculation yields

$$\boxed{I(\omega) = I_0 \frac{(\gamma/2)}{\pi\left[(\omega-\omega_0)^2 + (\gamma/2)^2\right]}} . \tag{1.16}$$

$I(\omega)$ is normalized such that $\int_{-\infty}^{+\infty} I(\omega)d\omega = I_0$, where $I_0 = W_0$ is the total emitted energy.

Figure 1.2 displays the $I(\omega)$ function. Its shape is called a **Lorentzian** by the physicists, and **Cauchy distribution** (when normalized to a unity integral) by the mathematicians. The energy distribution is symmetric around the central angular frequency ω_0, and the full width at half maximum is γ. This width is called the **natural width** of the emission spectrum, or sometimes the **natural emission band**. It is often symbolized by $\Delta\omega_{1/2}$ or $\Delta\omega_N$:

$$\Delta\omega_N = \gamma = 1/\tau , \tag{1.17}$$

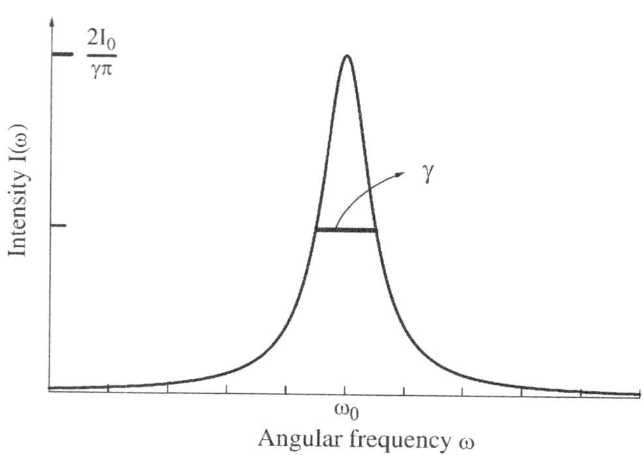

Figure 1.2 A Lorentzian function describing the spectral distribution of the light intensity emitted during decay of an excited atom. The distribution is symmetric about the central ω_0 angular frequency, and its full width at half maximum is equal to the damping parameter γ.

We consider now how the natural frequency spectral width is expressed in wavelength spectral shape. We use the relation $\omega = 2\pi c/\lambda$. Differentiation yields $\Delta\omega_N \cong (2\pi c/\lambda^2)\cdot|\Delta\lambda_N|$. Thus, $|\Delta\lambda_N| \cong \Delta\omega_N/(2\pi c/\lambda^2)$. Setting former expressions for $\Delta\omega_N$ and λ yields

$$|\Delta\lambda_N| \cong \frac{4\pi e^2}{3mc^2} = \frac{8\pi}{3} r_0 \ , \tag{1.18}$$

where $\Delta\lambda_N$ is the natural wavelength spectral width. *The natural wavelength spectral width is a fundamental constant, independent of the central emission wavelength.* The natural frequency width, on the other hand, is proportional to the central angular frequency (Eq. [1.7]). Setting the magnitude of r_0 in Eq. (1.18) comes to $|\Delta\lambda_N| = 1.18\times10^{-12}$ cm $= 1.18\times10^{-12}$ Å. In comparison with, for example, a green light of 5,000 Å wavelength, the natural width is negligible, particularly in comparison with other common spectral broadening mechanisms.

By quantum theory, excitation of the electromagnetic field occurs by energy portions (quanta) called **photons**, each given by $E = \hbar\omega$, where $\hbar$ is the reduced **Planck constant**. This is another fact that classical theory cannot deduce; we thus impose it artificially into our former classical derivations. The smallest amount of energy that may be emitted by an atom or a molecule at an ω angular frequency is a single photon. In view of the existing broadening in the emitted frequencies, there will be an uncertainty in the energy $\Delta E = \hbar\,\Delta\omega_N = \hbar\gamma = \hbar/\tau$ of the emitted photon energy. This quantity is called the **energetic width** of the spectral line. It is a special case of **Heisenberg's uncertainty principle**, usually presented in the form (see also Eq. [4.12] and the related text)

$$\boxed{\Delta E\cdot\tau \geq \hbar} \ . \tag{1.19}$$

1.2.2 Doppler Broadening

One of the means to induce electromagnetic emission of atoms is to heat their gas phase to a high temperature. In such a state, the emitting atoms exhibit a high velocity; some have a velocity component directed toward the observer and others have a velocity component directed away from the observer. Radiation at angular frequency ω_0 emitted by an atom of velocity component v_x relative to the observer will be seen as if its frequency is $\omega \cong \omega_0 + \omega_0(v_x/c)$. That frequency shift is called **Doppler shift**. The approximation provided in the above expression is valid only for $v_x << c$, that is, as long as relativity effects may be neglected. For the said gases, this condition is satisfied up to high temperatures. Because the gases exhibit a distribution of velocities, the observer will see a distribution of Doppler shifts, namely a frequency broadening.

The velocity distribution of gas molecules is given by **Maxwell's equation**

$$dN(v_x) = dv_x\cdot\frac{N}{\sqrt{2\pi k_B T/M}}\,\boldsymbol{exp}\left(-\frac{Mv_x{}^2}{2k_B T}\right) \ , \tag{1.20}$$

where $dN(v_x)$ is the concentration of molecules whose x-velocity component resides between v_x and $v_x + dv_x$, N is the total concentration of gas molecules, M is the molecular mass, k_B is Boltzmann constant, and T is the absolute temperature. Ignoring natural broadening, the relative emitted radiation intensity measured at ω frequency will be proportional to the number of molecules emitting at that frequency. The emission spectral line will thus exhibit the form

$$I(\omega) = I_0 \frac{1}{\sqrt{2\pi}\omega_0 (k_B T/Mc^2)^{1/2}} \boldsymbol{exp}\left[-\frac{(\omega-\omega_0)^2}{2\omega_0{}^2(k_B T/Mc^2)}\right] \quad . \tag{1.21}$$

An assumption underlying Eq. (1.21) is that there is no intermolecular interaction that causes frequency changes in their emissions, a reasonably valid assumption for low-pressured gases. Eq. (1.21), like Eq. (1.16), is normalized such that $\int_{-\infty}^{+\infty} I(\omega)d\omega = I_0$, where I_0 is the total emitted power. The shape expressed by Eq. (1.21) is called **Gauss Bell-shaped function**, or **normal distribution**; it is shown in Figure 1.3, normalized to a unity integral. The function is symmetric about ω_0, and its width at half maximum γ_D is given by

$$\gamma_D = 2\sqrt{2\,\boldsymbol{ln}\,2}\,\omega_0 \left(\frac{k_B T}{Mc^2}\right)^{1/2} \quad . \tag{1.22}$$

Thus, the Doppler width of the angular frequency at half maximum is proportional to the square root of the absolute temperature, and inversely proportional to the square root of the molecular mass. As administered in calculations related to the natural width, the Doppler width of the wavelength spectrum $\Delta\lambda_D$ would be

$$\Delta\lambda_D = 2\sqrt{2\,\boldsymbol{ln}\,2}\,\lambda_0 \left(\frac{k_B T}{Mc^2}\right)^{1/2} \quad . \tag{1.23}$$

For demonstration of the Doppler broadening, we calculate the relative Doppler broadening $\Delta\lambda_D/\lambda_0$ for an atomic low-pressured hydrogen H gas at 1 degree Kelvin (1 K):

Given data:

$k_B = 1.38 \times 10^{-16}\,\text{erg}\cdot\mathbf{deg}^{-1}$; $T = 1\,\text{K}$; $M = M_p = 1.673 \times 10^{-24}\,\text{g}$; $c = 3 \times 10^{10}\,\text{cm}\cdot\text{s}^{-1}$.

Developing:

$$\frac{\Delta\lambda_D}{\lambda_0} \cong 2\sqrt{2\,\boldsymbol{ln}\,2} \quad \left(\frac{k_B T}{Mc^2}\right)^{1/2} \quad .$$

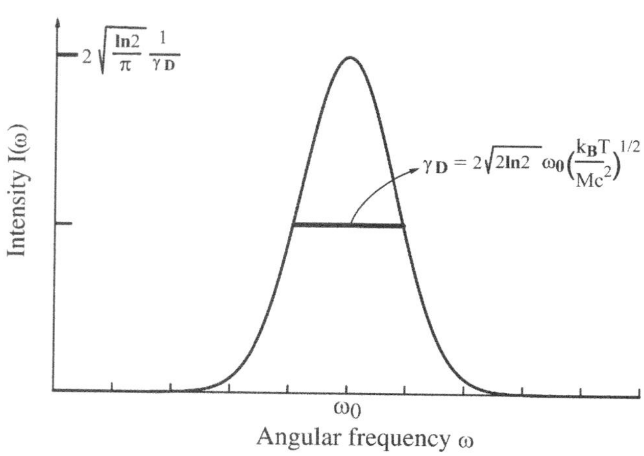

Figure 1.3 Emission spectral line-shape of low-pressured atomic or molecular gas at high temperature T caused by Doppler broadening.

Setting numerical values:

$$\frac{\Delta\lambda_D}{\lambda_0} \cong 2\sqrt{2\,\boldsymbol{ln}\,2}\frac{1}{3\times10^{10}}\left(\frac{1.38\times10^{-16}\times1}{1.673\times10^{-24}}\right)^{1/2} = 7.13\times10^{-7} \quad .$$

This latter result suggests expressing a general technical equation for any gas at large

$$\boxed{\Delta\lambda_D[\text{Å}] = 7.13\times10^{-7}\left(\frac{T[K]}{A}\right)^{1/2}\lambda_0[\text{Å}]} \quad , \tag{1.24}$$

where $A \equiv M/M_p$ is the atomic weight of the emitting element in the gas (or molecular weight in the case of a molecular gas). If we consider as an example the atomic hydrogen H_α emission at $\lambda_0 = 5467.47$ Å, the Doppler broadening at 1 K temperature would be $\Delta\lambda_D = 3.9\times10^{-3}$ Å. It is 33 times larger than the natural broadening (1.18×10^{-4}Å ; see section 1.2.1). At the sun surface temperature (6000 K), the emission will exhibit a $\Delta\lambda_D = 0.3$ Å Doppler broadening. In most practical cases, the Doppler broadening is much greater than the natural broadening.

Under extreme conditions, the natural and Doppler broadenings exhibit the same order of magnitude (heavy molecules at low temperatures). The spectral shape then exhibits a more complex analytic form. Writing $f_1(\xi)$ for the Lorenz function, and $f_2(\omega)$ for the Gauss function, namely

$$\begin{cases} f_1(\xi) = \dfrac{(\gamma/2\pi)}{(\xi-\omega_0)^2 + (\gamma/2)^2} \\ f_2(\omega) = 2\sqrt{\dfrac{\boldsymbol{ln}\,2}{\pi}}\dfrac{1}{\gamma_D}\,\boldsymbol{exp}\left(-\dfrac{\omega^2}{2(\gamma_D{}^2/4\,\boldsymbol{ln}\,2)}\right) \end{cases} , \tag{1.25}$$

the frequency spectrum exhibits the following shape:

$$I(\omega) = I_0\int_{-\infty}^{+\infty} f_2(\omega-\xi)f_1(\xi)d\xi \quad . \tag{1.26}$$

The latter equation expresses a convolution between the two shaped functions. We won't elaborate on its precise analytical shape. We just indicate that in this case as well, the shape is symmetric relative to ω_0, for which it exhibits a maximum. To an excellent approximation, the convoluted function width is given by

$$\boxed{\gamma_{eff}{}^2 \cong \gamma_D{}^2 + \gamma^2} \quad . \tag{1.27}$$

Note that Eq. (1.27) is exact for two distribution functions if the second order moment exists for both. In our case, the second order moment exists for $f_2(\omega)$; however, it diverges for $f_1(\xi)$. Eq. (1.27), however provides an excellent fit in most cases.

1.2.3 Additional Broadening Mechanisms

A further emission line broadening is observed for atomic and molecular gases under high pressure, and interpreted as resulting from two different processes. One is interruption of the emitting electric dipole oscillation by collision with a different species, whereby a part of the oscillator energy is dissipated non-radiatively. This expresses itself by a shorter lifetime, hence an additional broadening contribution given by $\gamma_P = 1/\tau_c$, where τ_c is the average duration between collisions. Another process is the interaction potential between approaching atoms/molecules, which changes to some extent their excited oscillatory frequencies. Because there is a distribution of distances among the atoms/molecules, this process expresses itself by distribution in their emission ω_0 frequencies. The two effects are termed **pressure broadening**. An extreme case of the pressure broadening is the **excimer broadening**. In that case, an excited gas molecule bonds with another atom/molecule, forming an **excimer**. The oscillation frequency of the excited electron depends on the distance between the two bonded species. The distribution in emission frequencies reflects the distribution of distance between the bonded atoms. It is indeed used for the study of the excimer bonding effect.

Additional broadening occurs for atoms (especially for ions) embedded in a solid. For a disordered solid like a glass, the emitting ions reside in a variety of environments, yielding variations in the emission frequencies. Thus, the ions emission spectrum reflects the distribution in the influence of the environmental differences on the residing ions. Spectral line broadening of that nature is termed **heterogeneous broadening**, because the same-type atoms exhibit different emission frequencies. The Doppler broadening is also heterogeneous.

Line broadening occurs also in an ordered crystalline solid due to the embedded ion interaction with the lattice vibrations. During lattice vibration, the shape of the ion vicinity changes, hence the frequency of the oscillating dipole. A typical lattice vibrational period is of the order of 10^{-12} s. On one hand, this duration is extremely long compared to the 10^{-15} s typical period of the electron dipole oscillation. On the other hand, it is extremely short compared to the 10^{-8} s shortest possible lifetime of the decaying dipole energy. Thus, the frequency of the oscillating dipole changes periodically over the lifetime. This results in broadening of the emission line; its value depends on the lattice vibrational frequency and the dipole emission frequency amplitude. This broadening is common to all emitting ions, and is thus addressed as **homogeneous broadening**. The natural broadening is also homogeneous.

The development of theoretical expressions to all broadening mechanisms is a broad subject that requires the construction of a specific model for each case, and even separate models per each gas, solid, or ion residing in a solid environment. This chapter does not elaborate on this subject.

1.3 Electromagnetic Radiation Absorption by Atoms and Molecules

This section discusses the energy acquired by the electron oscillator when exposed to electromagnetic radiation. We examine a vibrating electron whose resonance frequency is ω_0 and damping constant γ. The equation of motion when the electron is limited to a motion along the X axis is (see also Eq. [1.10] above)

$$\boxed{m\ddot{x} + m\gamma\dot{x} + m\omega_0^{\,2}x = e\,\mathcal{E}(t)} \quad , \tag{1.28}$$

where $\mathcal{E}(t)$ is the electric field component in the X - axis orientation. We don't limit $\mathcal{E}(t)$ to a sinusoidal time dependence. Still, it is possible to perform a Fourier expansion of $\mathcal{E}(t)$ and $x(t)$ according to

$$\left.\begin{aligned} \mathcal{E}(t) &= \frac{1}{\sqrt{2\pi}}\int_{-\infty}^{\infty} \mathbf{E}(\omega)e^{-i\omega t}d\omega\ ; \\ x(t) &= \frac{1}{\sqrt{2\pi}}\int_{-\infty}^{\infty} \mathbf{X}(\omega)e^{-i\omega t}d\omega\ . \end{aligned}\right\} \tag{1.29}$$

If $\mathcal{E}(t)$ is presented by a real function, we may utilize the following important relation♣

$$\mathbf{\Phi} \equiv \frac{c}{4\pi}\int_{-\infty}^{\infty} \mathcal{E}^*(t)\mathcal{E}(t)dt = \frac{c}{2\pi}\int_{0}^{\infty} \boldsymbol{E}^*(\omega)\boldsymbol{E}(\omega)d\omega \quad . \tag{1.30}$$

The function $\mathbf{\Phi}(\omega)$ describes the incident electromagnetic energy flux density (energy per unit area). Its meaning is that the incident electromagnetic energy flux density per angular frequency unit is given by

$$\mathbf{F}(\omega) \equiv \frac{d\mathbf{\Phi}(\omega)}{d\omega} = \left(\frac{c}{2\pi}\right) |\boldsymbol{E}(\omega)|^2 \quad . \tag{1.31}$$

The relation between the Fourier transformed $\boldsymbol{E}(\omega)$ and $\boldsymbol{X}(\omega)$ is obtained by setting them in Eq. (1.28). The result is

$$\boldsymbol{X}(\omega) = \frac{(e/m)}{\omega_0{}^2 - \omega^2 - i\gamma\omega} \cdot \boldsymbol{E}(\omega) \quad . \tag{1.32}$$

Correspondingly, by taking the time derivative of the expression for $x(t)$ in Eq. (1.29), it is easy to see that the Fourier transform of the velocity $\dot{x}(t)$ is given by

$$\boldsymbol{V}(\omega) = -i\omega\boldsymbol{X}(\omega) = \frac{-i\omega(e/m)}{\omega_0{}^2 - \omega^2 - i\gamma\omega} \cdot \boldsymbol{E}(\omega) \quad . \tag{1.33}$$

The total energy ΔW absorbed by the electron oscillator is given by the algebraic real part of the time integral over the time-dependent power of the electric field action on the oscillator

$$\Delta W = \boldsymbol{Re}\left\{\int_{-\infty}^{\infty} e\,\mathcal{E}(t)\cdot\dot{x}(t)dt\right\} = \frac{2e^2}{m}\int_{0}^{\infty} \frac{\gamma\omega^2|\boldsymbol{E}(\omega)|^2}{(\omega_0{}^2 - \omega^2)^2 + \gamma^2\omega^2}d\omega \quad . \tag{1.34}$$

The energy absorbed per angular frequency unit will thus be

$$\boxed{\frac{d(\Delta W)}{d\omega} = \left(\frac{2e^2}{m}\right)\cdot|\boldsymbol{E}(\omega)|^2\cdot\frac{\gamma\omega^2}{(\omega_0{}^2 - \omega^2)^2 + \gamma^2\omega^2}} \quad . \tag{1.35}$$

Now we define the absorption cross-section quantity

$$\sigma_{abs} \equiv \frac{\text{energy absorbed per angular frequency unit}}{\text{incident radiation flux density per angular frequency unit}} \quad . \tag{1.36}$$

♣ The expression on the right follows **Parseval**'s equality $\int_{-\infty}^{\infty} |\boldsymbol{F}(k)|^2dk = \int_{-\infty}^{\infty} |f(x)|^2dx$, where $\boldsymbol{F}(k)$ is the Fourier transform of $f(x)$, and that $\boldsymbol{E}^*(\omega) = \boldsymbol{E}(-\omega)$ for real $\mathcal{E}(t)$.

The absorption cross section has area units. In setting the proper equations (see Eqs. [1.35] and [1.31]) we get the expression

$$\boxed{\sigma_{abs} = \frac{4\pi e^2}{mc} \cdot \frac{\gamma\omega^2}{\left(\omega_0{}^2 - \omega^2\right)^2 + \gamma^2\omega^2}} \quad . \tag{1.37}$$

We now analyze the absorption cross-section dependence on the angular frequency:

{1} In the limit of frequencies considerably higher than resonance, that is, $\omega^2 >> \omega_0{}^2$, we get

$$\sigma_{abs} \approx \frac{4\pi e^2}{mc}\tau_0\left(\frac{\omega_0}{\omega}\right)^2 \quad . \tag{1.38}$$

For obtaining Eq. (1.38), we set the relation $\gamma = \tau_0\omega_0{}^2$ (see Section 1.1). Thus, we found that in the high-frequency region, the absorption cross section is inversely proportional to the squared angular frequency. Remember, however, that we are restricted to frequencies below the ionization frequency, that is, to photon energies $\hbar\omega$ smaller than the ionization energy.

{2} In the limit of frequencies considerably lower than resonance, namely $\omega^2 << \omega_0{}^2$ we get

$$\sigma_{abs} \approx \frac{4\pi e^2}{mc}\tau_0\left(\frac{\omega}{\omega_0}\right)^2 \quad . \tag{1.39}$$

In that region the absorption cross section is rather directly proportional to the squared frequency.

{3} In the limit of frequencies close to resonance, namely $\omega^2 \sim \omega_0{}^2$ we get

$$\boxed{\sigma_{abs} \approx \left(\frac{2\pi^2 e^2}{mc}\right) \cdot \frac{\gamma/2\pi}{(\omega - \omega_0)^2 + (\gamma/2)^2}} \quad , \tag{1.40}$$

which exhibits the known Lorentzian shape (Eq. [1.14] and Figure 1.2), of identical width as the emission's shape. It is interesting to note the magnitude of the cross-section integral

$$\int_{-\infty}^{\infty} d\omega \cdot \sigma_{abs}(\omega - \omega_0) \cong \left(\frac{2\pi^2 e^2}{mc}\right) = 0.166\ \text{cm}^2 \cdot \text{rad/s} \quad ,$$

which is entirely independent of the resonance frequency ω_0.

When an electromagnetic beam passes through a gas, its intensity reduces gradually because of absorption by the different atoms/molecules constituting the gas. The decay process of a beam propagating in the Z direction is characterized by an absorption coefficient parameter α_{abs} defined as follows:

$$\alpha_{abs}(\omega) \equiv -\frac{1}{\mathbf{F}(\omega)}\frac{d\mathbf{F}(\omega)}{dz} \quad . \tag{1.41}$$

In the cgs units system, the absorption coefficient unit is cm^{-1}.

Let's now consider a slice of space perpendicular to the Z direction, of S area and an infinitesimal thickness dz. We further assume that the said slice contains a single atom. The absorbed radiation flux density per unit of angular frequency will thus be given by

$$\frac{1}{S}\frac{d(\Delta W)}{d\omega} = -d\mathbf{F}(\omega) = \frac{1}{S}\sigma_{abs}(\omega)\mathbf{F}(\omega) \quad .$$

If, on the other hand, the absorbing atoms density in that volume is N_0, the said slice will contain $N_0 \cdot S \cdot dz$ atoms. The absorbed radiation flux density per unit angular frequency will accordingly grow:

$$-d\mathbf{F}(\omega) = \frac{1}{S}\sigma_{abs}(\omega)\mathbf{F}(\omega) \cdot N_0 \cdot S \cdot dz = \sigma_{abs}(\omega)\mathbf{F}(\omega) \cdot N_0 \cdot dz \ .$$

This relation is valid as long as all absorbing atoms are identical, and absorption occurring by one atom has no influence on any other atom. One thus obtains that

$$\alpha_{abs}(\omega) = \sigma_{abs}(\omega)N_0 \ .$$

If there is a distribution of ω_0 resonance frequencies of the constituent atoms, each group of atoms of same ω_0 provides its own contribution to the absorption coefficient:

$$d\alpha(\omega) = \sigma_{abs}(\omega - \omega_0) \cdot N(\omega_0)d\omega_0 \ ,$$

or

$$\alpha(\omega) = \int_0^{\infty} \sigma_{abs}(\omega - \omega_0) \cdot N(\omega_0)d\omega_0 \ ,$$

where $N(\omega_0)$ is the atoms' distribution function by their resonance frequencies.

For demonstration, we consider a gas of same atoms (or same molecules) whose resonance frequency ω_p is identical at rest. However, at a finite temperature, the atoms' distribution function by their resonance frequencies relates to the Doppler broadening per Eq. (1.21).

$$N(\omega_0) = \frac{N}{\sqrt{2\pi}\omega_p\left(k_BT/Mc^2\right)^{1/2}} \boldsymbol{exp}\left[-\frac{(\omega_0 - \omega_p)^2}{2\omega_p{}^2\left(k_BT/Mc^2\right)}\right] \ , \tag{1.42}$$

where N is the total volume concentration of the gas atoms. We have already seen that the Doppler width is almost always large compared to the natural width. The $N(\omega_0)$ function in Eq. (1.42) changes slowly in the $|\omega_0 - \omega_p| \lesssim (2-3)\gamma_D$ range, compared to the change in the above $\sigma_{abs}(\omega - \omega_0)$ function (Eq. [1.40]), whose dependence on ω exhibits a narrow peak at the $\omega = \omega_0$ point. Thus, it is possible to set $N(\omega)$ instead of $N(\omega_0)$ in the above expression for $\alpha(\omega)$, and place it in front of the integration. The resultant expression for $\alpha_{abs}(\omega)$ would be (see also Eq. [1.41])

$$\boxed{\alpha_{abs}(\omega) = N\frac{\left(2\pi^2e^2/mc\right)}{\sqrt{2\pi}\omega_p\left(k_BT/Mc^2\right)^{1/2}} \boldsymbol{exp}\left[-\frac{(\omega - \omega_p)^2}{2\omega_p{}^2\left(k_BT/Mc^2\right)}\right]} \ , \tag{1.43}$$

for the $|\omega - \omega_p| \lesssim (2-3)\gamma_D$ range. The absorption spectrum in this range accurately reflects the Doppler broadening, and would equally reflect any other type of spectral absorption broadening. On the other hand, for frequencies far from the central ω_p frequency, one gets

$$\boxed{\alpha_{abs}(\omega) \cong N \cdot \sigma_{abs}(\omega - \omega_p)} \ , \tag{1.44}$$

for $|\omega - \omega_p| >> \gamma_D$. The absorption spectrum in that range reflects the spectral dependence of the absorption cross section.

1.4 Radiation Scattering by Atoms and Molecules

Section 1.3 described a process of energy subtraction from a radiation field propagating in a medium containing atoms or molecules. The physical reason for the energy subtraction was a "friction-like process" in the equations of motion of the electron dipole moment. However, this same friction is also responsible for electromagnetic radiation emission, that is, subtraction of energy from the vibrating dipole. Radiation **scattering** occurs when the two processes are executed simultaneously, although not necessarily at the same propagation direction.

We turn now to an analytical study of the scattered radiation. If the photon energy is large compared to the atom ionization energy, the involved electrons should be attributed as free electrons, and linear momentum exchange between each electron and the radiation field should be included. This scattering process is termed **Compton scattering**, and requires a more advanced treatment. We thus limit ourselves to cases of bound electrons, where linear momentum exchanges involve the entire heavy atom or molecule. The energy involved in these momentum exchanges is then small compared to the photon energy, and is thus neglected. In addition, we assume that the radiation wavelength λ is long compared to the atomic dimensions, which allows the assumption that the electric field is constant across the electron motion range at any point in time. This approximation is valid for the case of our study, because the atomic and molecular dimensions never exceed several angstroms, while the wavelengths are of the order of several thousand angstroms.

We consider a plane-parallel, linearly polarized electromagnetic wave, whose electric field is given by

$$\underline{\mathcal{E}} = \underline{u}\mathcal{E}_0 e^{i(\underline{k}\cdot\underline{r} - \omega t)} , \tag{1.45}$$

where $\underline{u}$ is a unit vector in the electric field direction. We choose the Z axis to describe the beam propagation direction. Therefore, the polarization orientation lies in the (x, y) plane. Figure 1.4 shows the wave incidence on an electron residing at the origin of a Cartesian axis system. We consider the scattered wave propagating in the $\underline{n} \equiv (\boldsymbol{\sin\theta\cos\varphi}, \boldsymbol{\sin\theta\sin\varphi}, \boldsymbol{\cos\varphi})$ direction. The angle between $\underline{n}$ and $\underline{u}$ is marked as υ. The electron is bonded to the axes' origin, exhibiting an ω_0 resonance frequency and a $\gamma \equiv 2e^2\omega_0^2/3mc^3$ damping factor.

The electron equation of motion under the influence of a driving force (see Eq. [1.10]) would be

$$\ddot{\underline{x}} + \gamma\dot{\underline{x}} + \omega_0^2\underline{x} = \underline{u}\frac{e}{m}\mathcal{E}_0\, e^{-i\omega t} . \tag{1.46}$$

We ignore the interaction between the oscillating electron and the magnetic field component. It is important to emphasize the physical meaning of this neglect, though. When the electron acquires a velocity

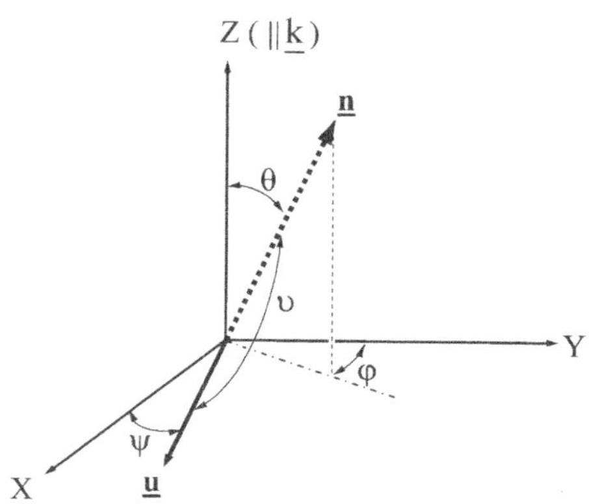

Figure 1.4 Definition of orientations and angles for the incident and scattered radiation description. The incident light propagates along the Z- axis with the electric field oriented parallel to the $\underline{u}$ vector direction in the (x, y) plane. The text discussion relates to a scattered beam propagating in the $\underline{n}$ vector direction.

$\dot{x}$ in the electric field direction, it is correspondingly subject to an ***additional force*** emanating from the associated magnetic field. The orientation of this force is *always directed in the propagation direction of the incident electromagnetic field*. This is the classical account for the linear momentum exchange between photons and electrons. The second paragraph of this section mentioned that this effect is negligibly small for bound electrons.

The steady-state solution to Eq. (1.47) is a harmonic motion about the origin of the form

$$\underline{x} = \underline{x}_0 e^{-i\omega t} \quad , \tag{1.47}$$

where

$$\underline{x}_0 = \frac{\mathcal{E}_0 e/m}{\omega_0{}^2 - \omega^2 - i\gamma\omega} \cdot \underline{u} \quad .$$

Such motion results in a vibrating electric dipole, emitting electromagnetic radiation in all directions as described in Figure 1.1. By Eq. (1.4), the scattered radiation power per unit solid angle is given by

$$\frac{dP(t)}{d\Omega} = \frac{e^2}{4\pi c^3}\left|\ddot{x}\right|^2 \boldsymbol{\sin}^2 \upsilon \quad . \tag{1.48}$$

Averaged over a full vibrational period, it is given by

$$\frac{d\bar{P}}{d\Omega} = \left(\frac{c\,\mathcal{E}_0{}^2}{8\pi}\right)\left(\frac{e^2}{mc^2}\right)^2 \frac{\omega^4}{(\omega_0{}^2 - \omega^2)^2 + \gamma^2\omega^2} \boldsymbol{\sin}^2 \upsilon \quad . \tag{1.49}$$

The first parentheses-enclosed factor on the right is just the incident electromagnetic intensity (in units of energy per unit area, per unit of time). The second factor is the squared twice the electron electrostatic radius (see Section 1.1). The third factor, now denoted as $\phi(\omega)$, provides the angular frequency dependence of the scattered energy:

$$\phi(\omega) \equiv \frac{\omega^4}{(\omega_0{}^2 - \omega^2)^2 + \gamma^2\omega^2} \quad .$$

We now define the term **differential scattering cross section** as

$$\frac{d\sigma}{d\Omega} \equiv \frac{\text{scattered energy per unit time per unit spatial angle}}{\text{incident energy per unit time per unit area}} \quad . \tag{1.50}$$

The differential scattering cross section has area units. For our case one obtains

$$\boxed{\frac{d\sigma}{d\Omega} = 4r_0{}^2 \frac{\omega^4}{(\omega_0{}^2 - \omega^2)^2 + \gamma^2\omega^2} \boldsymbol{\sin}^2 \upsilon = 4r_0{}^2 \phi(\omega)\, \boldsymbol{\sin}^2 \upsilon} \quad . \tag{1.51}$$

Eq. (1.51) expresses the differential scattering cross section by the angle υ between the incident beam electric field orientation and the scattered radiation propagation direction. This expression is useful for cases when the incident beam is linearly polarized. It is also useful as a starting point for analysis of other cases. For example, for an elliptically polarized incident beam, weighted averaging must be performed over all ψ angles (Figure 1.4). The calculation is relatively simple for circularly polarized or unpolarized incident beams, for which all ψ angles exhibit equal weights. To perform the calculation for those cases, we use the following trigonometric identity

$$\sin^2 \upsilon = 1 - \sin^2 \theta \cos^2 (\varphi - \psi) \quad . \tag{1.52}$$

Averaging over ψ results in

$$\overline{\sin^2 \upsilon} = 1 - \frac{1}{2} \sin^2 \theta = \frac{1}{2}\left(1 + \cos^2 \theta\right) \quad . \tag{1.53}$$

The differential scattering cross section for circularly or unpolarized light will thus be

$$\frac{d\sigma}{d\Omega} = 2r_0^2 \phi(\omega)\left(1 + \cos^2 \theta\right) \quad , \tag{1.54}$$

and it expresses the scattering intensity per unit solid angle according to the angle between the light propagation direction and that of the scattered light.

To visualize the scattering cross section intensity distribution according to the scattering angle, it is customary to present it as **polar plots**. The primitive plot is planar, involving a plane crossing the origin perpendicularly to the "equator's" $X-Y$ plane, and forming an angle φ with, say, the $Y-$ axis (Figure 1.4). The edges of vectors whose lengths equal the **differential scattering intensity** $d\sigma/d\Omega$, starting at the origin and inclined by θ from the $Z-$ axis, are plotted on the right side of the plane. Such primitive plot per Eq. (1.54) is provided in Figure 1.5a. Further visualization is obtained by the surface of a solid body

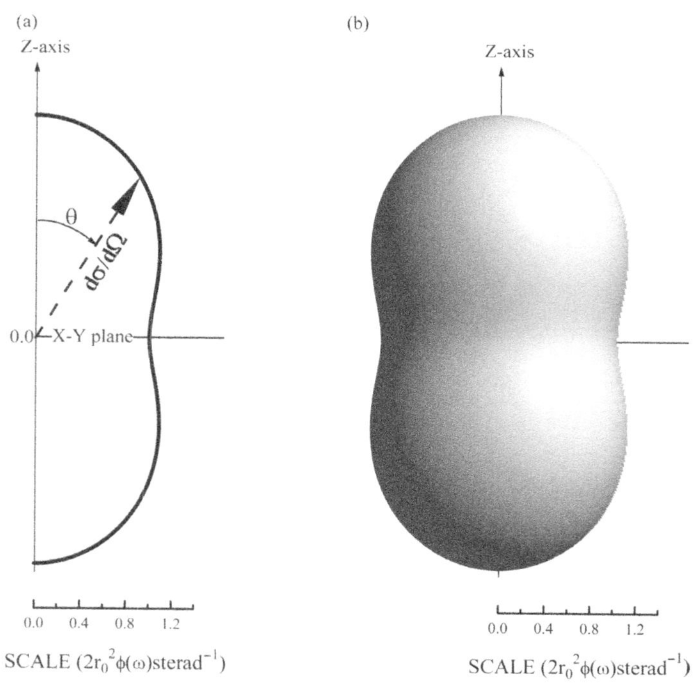

Figure 1.5 Polar plots of the differential scattering intensity per unit solid angle for beams circularly polarized or unpolarized. (a) In primitive form, and (b) by a solid spinning top of Figure 1.5a; the top surface describes the spatial orientation distribution of scattering cross section per unit solid angle.

generated by rotating the primitive polar plot around the $Z-$ axis. In our present case, there is no dependence on the φ azimuth angle. The image obtained resembles a simple, vertically symmetric peanut, as demonstrated in Figure 1.5b.

The total scattering cross section $\sigma_T(\omega)$ is obtained by setting $d\Omega = 2\pi\ \boldsymbol{\sin}\,\theta\ d\theta$ and integrating $d\sigma/d\theta$ over all θ angles:

$$\boxed{\sigma_T(\omega) = \frac{32\pi}{3} r_0{}^2 \phi(\omega)} \quad , \tag{1.55}$$

a result known as a **Thomson equation**.

We turn now to analyze the angular frequency dependence of the scattering cross section provided by the $\phi(\omega)$ function.

{1} At high frequencies compared to the resonance frequency, satisfying $\omega^2 >> \omega_0{}^2$, we obtain that $\phi(\omega) \cong 1$. However, recall that this relation is valid only when the photon energy $\hbar\omega$ is small compared to the ionization energy. On the other hand, this scattering mechanism (**Thomson scattering**) still plays an important role at higher frequencies as well. This book does not discuss that frequencies range.

{2} At frequencies that are small compared to the resonance frequency, namely $\omega^2 << \omega_0{}^2$, we obtain that $\phi(\omega) \cong (\omega/\omega_0)^4$. Scattering in this range is termed **Rayleigh scattering**. Proportionality of scattering to the fourth power of frequency accounts for the blue color of the sky. The origin of the sky's color is the sunlight scattering by constituent atmospheric molecules. Visible light ranges from 4,000 Å (blue) to 7,000 Å (red). The blue color frequency, 7.5×10^{14} Hz, is thus larger than the 4.28×10^{14} Hz red color frequency. Their fourth power ratio is about 9.4. Thus, blue light scattering of the sunlight is over nine times more efficient than the red edge of the visible spectrum.

{3} At frequencies close to resonance, that is, $\omega{\sim}\omega_0$, the $\phi(\omega)$ function attains the form

$$\phi(\omega) \cong \frac{\omega_0{}^2/4}{(\omega-\omega_0)^2 + (\gamma/2)^2} \quad . \tag{1.56}$$

The frequency dependence exhibits a Lorentzian shape of a γ full width at half-maximum (Eq. [1.16] and Figure 1.2). The total scattering cross section σ_T at $\omega{\sim}\omega_0$ may thus be written as

$$\boxed{\sigma_T(\omega) \cong \left(\frac{2\pi^2 e^2}{mc}\right) \cdot \frac{\gamma/2\pi}{(\omega-\omega_0)^2 + (\gamma/2)^2}} \quad . \tag{1.57}$$

It is not at all surprising that the total scattering cross section at ω angular frequency turns out to be identical to the absorption cross section at the same angular frequency (Eq. [1.40]). As with absorption, the spectral width of the total atomic/molecular gas scattering at a finite temperature depends on the various broadening mechanisms, such as Doppler broadening.

In addition, we note that because the motions of the oscillating electron are limited to the plane perpendicular to the incident beam propagation direction (the (X, Y) plane; see Figure 1.4), the scattered electromagnetic field will always be partially polarized, depending on the θ angle. More details are provided in Section 8.5.1.

1.5 Reminder: Multipole Moments Expansion

We consider an arbitrary spatial charge density $\rho(x_0, y_0, z_0)$ confined to a sphere of radius a (Figure 1.6). The potential at a point $\underline{r}(x, y, z)$ is given by

$$\psi(\underline{r}) = \iiint\limits_V \frac{\rho(x_0, y_0, z_0)}{R} dx_0 dy_0 dz_0 \quad , \tag{1.58}$$

where V is the sphere volume confining the charge distribution and R is the point distance from the charge element, satisfying $R^2 = (x - x_0)^2 + (y - y_0)^2 + (z - z_0)^2$. For $r > a$, it is possible to expand $(1/R)$ as a Taylor series in x_0, y_0, z_0 around $1/r$, and obtain a converging series:

$$\frac{1}{R} = \sum_{n=0}^{\infty} \frac{(-1)^n}{n!} \left[x_0 \frac{\partial}{\partial x} + y_0 \frac{\partial}{\partial y} + z_0 \frac{\partial}{\partial z} \right]^n \left(\frac{1}{r} \right) \quad , \tag{1.59}$$

where $r = (x^2 + y^2 + z^2)^{1/2}$. We use the multinomial expansion law:

$$(\alpha + \beta + \gamma)^n = \sum_{\lambda + \mu + \nu = n} \frac{n!}{\lambda! \mu! \nu!} \alpha^\lambda \beta^\mu \gamma^\nu \quad , \tag{1.60}$$

getting for $r > a$:

$$\psi(\underline{r}) = \sum_{n=0}^{\infty} \sum_{\ell=0}^{n} \sum_{k=0}^{n-\ell} \frac{(-1)^n}{\ell! k! (n-\ell-k)!} \left\{ \iiint x_0^{\ell} y_0^{k} z_0^{n-\ell-k} \cdot \rho(x_0, y_0, z_0) dx_0 dy_0 dz_0 \right\} \cdot \frac{\partial^n}{\partial x^\ell \partial y^k \partial z^{n-\ell-k}} \left(\frac{1}{r} \right) \quad . \tag{1.61}$$

Observation of the latter equation may seemingly suggest that measuring a potential ψ and *all* its derivatives outside the confining sphere of a radius, may provide the quantities enclosed within the braces of Eq. (1.61), but not on the actual charge distribution. Thus, it is seemingly possible to replace the actual charge distribution by any “artificial” distribution that provides the same values for the quantities within the said braces. Mathematically that is not true. If *all* potential derivatives exhibit the said identity, the “artificial” distribution equals the true one. Still, for practical approximation purposes, there is some latitude in selecting the artificial distribution, particularly because for large distances compared to a, the higher-order derivatives of $1/r$ become small, practically unmeasurable.

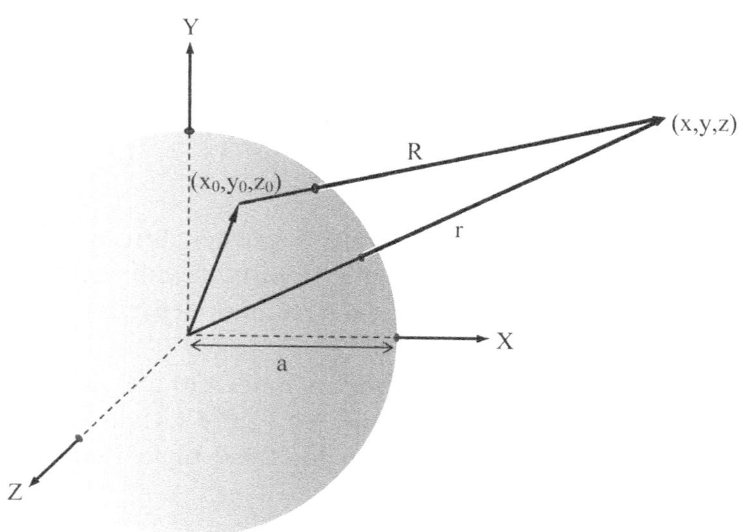

Figure 1.6 Demonstration of variables used for development of the charge distribution multipoles.

At large r distances compared to a, it would be possible to measure only the $n = 0$ coefficient of $1/r$. It is termed **Monopole moment** and is given by $Q \equiv \int\int\int \rho(x_0, y_0, z_0) dx_0dy_0dz_0$, namely, it is the total electric charge. The monopole won't change if we replace any distribution with an artificial charge Q (of esu units) concentrated at the origin. In the δ - functions language, it is written as $\rho_0 = Q \cdot \delta(x)\delta(y)\delta(z)$. At shorter distances it is possible to observe higher-order moments as well. The first-order moments, the dipole moments, are of the form

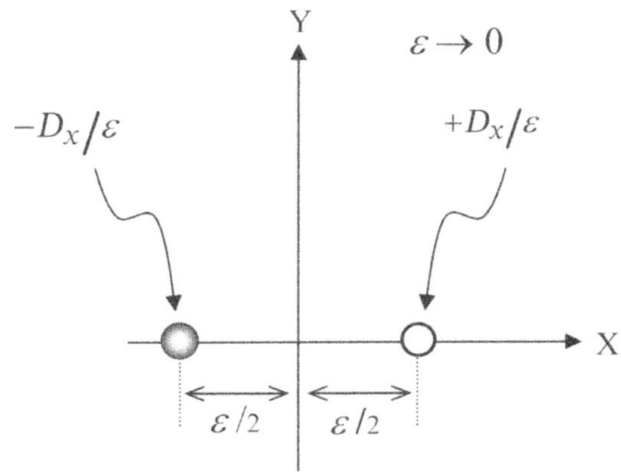

Figure 1.7 Artificial charge distribution providing the same D_x dipole moment as the real charge distribution.

$$D_x = \iiint x_0\rho(x_0, y_0, z_0)dx_0dy_0dz_0$$
$$D_y = \iiint y_0\rho(x_0, y_0, z_0)dx_0dy_0dz_0 \quad .$$
$$D_z = \iiint z_0\rho(x_0, y_0, z_0)dx_0dy_0dz_0$$

The same dipole moments would be obtained by an artificial charge distribution ρ_1 as follows:

$$\rho_1 = D_x\delta'(x)\delta(y)\delta(z) + D_y\delta(x)\delta'(y)\delta(z) + D_z\delta(x)\delta(y)\delta'(z) \quad .$$

The D_x, D_y, and D_z quantities are vectors of esu·cm units. For example, the first addend in ρ_1 may then be described geometrically by a combination of a point charge $+D_x/\varepsilon$ positioned at $\varepsilon\left(\frac{1}{2}, \ 0, \ 0\right)$, and an opposite point charge $-D_x/\varepsilon$ positioned at the symmetrically opposite point $\varepsilon\left(-\frac{1}{2}, \ 0, \ 0\right)$; that for an infinitely small ε parameter (practically, for ε being very small compared to the distance from the origin where this moment is measured). Figure 1.7 demonstrates such an artificial D_x moment. The total ρ_1 is a vector sum of such components along X, Y, and Z axes.

The second-order moments, the quadrupole moments, are six expressions of the form

$$Q_{xy} = \iiint x_0y_0\rho(x_0, y_0, z_0)dx_0dy_0dz_0 \quad .$$

Three expressions are "mixed," namely of the form Q_{xy}, Q_{xz}, and Q_{yz}, and the other three are "diagonal," namely of the form Q_{xx}, Q_{yy}, and Q_{zz}. It is possible to obtain identical quadrupole moments from an artificial charge distribution ρ_2 as follows:

$$\rho_2 = Q_{xy}\delta'(x)\delta'(y)\delta(z) + Q_{xz}\delta'(x)\delta(y)\delta'(z) + Q_{yz}\delta(x)\delta'(y)\delta'(z) +$$
$$+ Q_{xx}\delta''(x)\delta(y)\delta(z) + Q_{yy}\delta(x)\delta''(y)\delta(z) + Q_{zz}\delta(x)\delta(y)\delta''(z) .$$

The different $Q_{\alpha\beta}$ quadrupole quantities are of esu·cm^2 units. It is possible to artificially obtain the first mixed addend of ρ_2 (usually symbolized as ρ_2^{xy}) by a combination of two $+Q_{xy}/\varepsilon^2$ point charges located at the $\left(\pm\frac{1}{2}\varepsilon, \pm\frac{1}{2}\varepsilon, 0\right)$ points, and two $-Q_{xy}/\varepsilon^2$ point charges located at the $\left(\pm\frac{1}{2}\varepsilon, \mp\frac{1}{2}\varepsilon, 0\right)$; see Figure 1.8a. The first diagonal addend of ρ_2 (usually symbolized as ρ_2^{xx}) may correspondingly be artificially described by a combination of a point charge $-2Q_{xx}/\varepsilon^2$ located at the origin, a point charge $+Q_{xx}/\varepsilon^2$ located at $(\varepsilon, \ 0, \ 0)$, and another $+Q_{xx}/\varepsilon^2$ point charge located at $(-\varepsilon, \ 0, \ 0)$; see Figure 1.8b. In the two cases, ε is infinitely small. Each such addend may be pictured as a "dipole of dipoles." The entire set of second-order moments is a combination of all individual quadrupole moments.

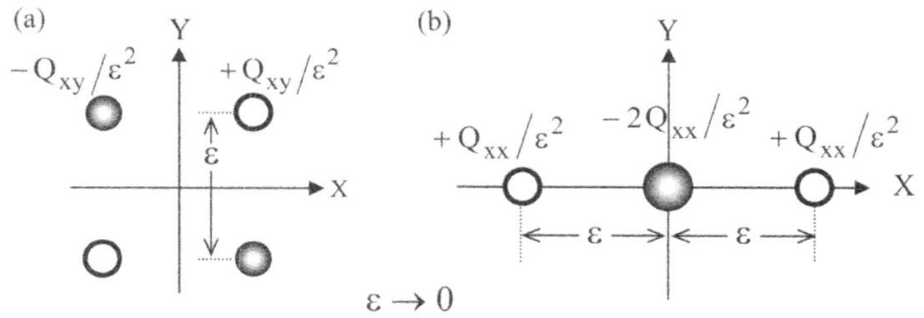

Figure 1.8 Artificial charge distribution providing the same Q_{xy} and Q_{xx} quadrupole moments, (a) and (b), respectively, as provided by the real charge distribution.

Exercises for Chapter 1

1.1 A two-plate capacitor is constructed of two metallic plates of 400 cm^2 area and 1 mm distance, and is connected to an alternating voltage source of 40 MHz frequency and 300 V amplitude.
(a) Consider the capacitor as an oscillating point dipole and calculate the electric field amplitude perpendicularly to the dipole direction at 5 m distance. Compare it with the electric field amplitude between the capacitor plates.
(b) Calculate the power dissipated by the capacitor resulting from the electromagnetic radiation by the capacitor. What would the dissipated power be for only 50 Hz source frequency?

1.2 Estimate the high-frequency limit, for which neglecting the power dissipation within a single period of an oscillating electron dipole is no longer justified. What is the spectral region it belongs to? Calculate the energy of a single photon of that frequency. Express it also in electron-volt (eV) units.

1.3 Assuming that the entire energy of an oscillating electron dipole is dissipated as a single photon, calculate the initial oscillation amplitude x_0 for a $\nu = 10^{15}$ Hz frequency.

1.4 (a) Calculate the Fourier transform of the function $\mathcal{E}(t) = \begin{cases} \mathcal{E}_0 e^{i\left(\omega_0 t + \frac{1}{2} i\gamma t\right)} & t > 0 \\ 0 & t \leq 0 \end{cases}$.
(b) Calculate and plot the amplitude and phase spectral dependence of the Fourier components in the vicinity of the center ω_0 frequency. Specifically, indicate the amplitude and phase at the $\omega_0 \pm \frac{1}{2}\gamma$ frequencies.

1.5 (a) Calculate the Doppler width of the two D emission lines of sodium Na vapor (5889.9 and 5895.5 Å) at 1,300°C.
(b) Calculate the temperature that would bring the Doppler width to equal the separation between the two emission lines.
(c) Calculate the absorption cross section at the peak of each of the sodium D-lines at 1,300°C.

1.6 Calculate the scattering cross section of an unpolarized light of $\nu = 10^{15}$ Hz frequency by an atom of the same transition frequency. What would be the scattering cross section if the light frequency changed into $\nu = 2 \times 10^{14}$ Hz or $\nu = 3 \times 10^{15}$ Hz?

2 Electromagnetic Radiation/Matter Interaction – A Semi-Quantum Approach

The basic assumption underlying the semi-quantum approach for describing the interaction between electromagnetic radiation and matter is that the atomic or molecular system is quantum, whereas the electromagnetic radiation field is simultaneously classic. The radiation field causes perturbation that induces transitions between allowed atomic or molecular quantum states.

2.1 A Reminder of Perturbation Theory

2.1.1 Static Perturbation Theory

We consider a quantum system of a known Hamiltonian $\mathbf{H}_0$, whose steady-state solutions $\{\psi_\ell\}$ satisfy the relation

$$\mathbf{H}_0\psi_\ell = \varepsilon_\ell\psi_\ell \quad , \tag{2.1}$$

where $\{\psi_\ell\}$ is a complete family of orthonormal functions (eigenfunctions). For simplicity, we assume first that the set of $\{\varepsilon_\ell\}$ eigenenergies is non-degenerate; namely, any two different solutions ψ_{ℓ_1} and ψ_{ℓ_2} belong to different eigenenergies:- $\varepsilon_{\ell_1} \neq \varepsilon_{\ell_2}$.

The central problem in perturbation theory is the finding of eigenenergies and eigenfunctions of the perturbed Hamiltonian $\mathbf{H}_0 + \mathbf{V}$, where $\mathbf{V}$ is the perturbation Hamiltonian. We assume a static perturbation Hamiltonian, namely independent of time. We must solve the equation

$$(\mathbf{H}_0 + \mathbf{V})\varphi_n = E_n\varphi_n \quad . \tag{2.2}$$

We take advantage of the fact that the $\{\psi_\ell\}$ set of solutions to the unperturbed Hamiltonian is a **complete family**. It is thus possible to represent every solution φ_n to Eq. (2.2) as a linear combination of $\{\psi_\ell\}$ functions. Namely,

$$\varphi_n = \sum_\ell a_{n\ell}\psi_\ell \quad , \tag{2.3}$$

where the expansion coefficients $a_{n\ell}$ are constants. Setting this expansion in Eq. (2.2) yields the following equation:

$$\sum_\ell a_{n\ell}(E_n - \varepsilon_\ell)\psi_\ell = \sum_\ell a_{n\ell}\mathbf{V}\psi_\ell \quad . \tag{2.4}$$

We multiply this equation from the left by ψ_m^* and integrate over the entire space. Using the orthonormality quality among the $\{\psi_\ell\}$ functions, we get

Physics, Optics, and Spectroscopy of Materials, First Edition. Zeev Burshtein.

$$a_{nm}(E_n - \varepsilon_m) = \sum_{\ell} a_{n\ell} V_{m\ell} \ , \tag{2.5}$$

where $V_{m\ell} = \langle m|\mathbf{V}|\ell\rangle \equiv \int \psi_m^*(\underline{r})V(\underline{r})\psi_\ell(\underline{r})d^3\underline{r}$.

Eq. (2.5) is general and accurate, but difficult to solve. The cases of usual interest are those where the perturbation Hamiltonian **V** is relatively small, and we limit our discussion to such cases. The following conditions express the "being relatively small" nature of the perturbation:

1) $V_{m\ell} << \varepsilon_\ell$ for all ℓ .
2) $V_{m\ell} << |\varepsilon_m - \varepsilon_\ell|$ for all ℓ ; (this condition also implies that there may not be any degeneracy in the solutions system of $\mathbf{H}_0$ anyway).
3) $a_{n\ell} \sim 1$ for $\ell = n$, while $a_{n\ell} << 1$ for all $\ell \neq n$. Beyond the aspect of the perturbation being small, this condition also means that the ψ_n function always provides the major contribution in the expansion of any φ_n function.

Using the above conditions, Eq. (2.5) reduces to

$$\boxed{\left.\begin{array}{ll} E_n - \varepsilon_n \cong V_{nn} & (m = n) \\ a_{nm} \cong \dfrac{V_{mn}}{E_n - \varepsilon_m} \cong \dfrac{V_{mn}}{\varepsilon_n - \varepsilon_m} & m \neq n \end{array}\right\}} \ . \tag{2.6}$$

These relations constitute the first-order approximation to the eigenenergies and expansion coefficients of $\{\varphi_n\}$.

We turn to discuss the case where solution of the unperturbed Hamiltonian exhibits some degeneracy. Namely, there exists a set of eigenenergies such that $\varepsilon_{n\ell} = \varepsilon_{n1}$ for a specific n , and for specific ℓ numbers $\ell = 1...k$, but their eigenfunctions $\psi_{n1}...\psi_{nk}$ are different and orthonormal. Eq. (2.5) is accurate for this case as well. We assume a perturbation where all expansion coefficients $a_{n\ell}$ are of the same order, while expansion coefficients of other n or ℓ are negligibly small. It means that *mixing* with other, far energy states, is negligible. Eq. (2.5) then reduces to

$$a_{nm}(E_{nm} - \varepsilon_{n1}) = \sum_{\ell=1}^{k} a_{n\ell} V_{m\ell} \quad \text{for} \quad m = 1...k \ . \tag{2.7}$$

We denote $\Delta_m \equiv E_{nm} - E_{n1}$. This quantity expresses the correction to the state eigenenergy caused by the perturbation. We further denote by $\left(\widetilde{\Delta}_m\right)$ the line-vector of the Δ_m values. The equations described in Eq. (2.7) may be written in a matrix form as

$$\boxed{\left[(V_{m\ell}) - \left(\widetilde{\Delta}_m\right) \cdot I\right] (a_{n\ell}) = 0} \ , \tag{2.8}$$

where $(a_{n\ell})$ is a column vector of the expansion coefficients, $(V_{m\ell})$ is a $k \times k$ matrix of the perturbation elements, and I is a unity matrix of order k. The energy corrections are thus the elements of the $(V_{m\ell})$ matrix, and the expansion coefficients are the eigenvectors $(a_{n\ell})$. If the eigenenergy corrections Δ_m of $(V_{m\ell})$ are different, this effect is termed *degeneracy removal* by the **V** perturbation. Sometimes, only a partial removal of degeneracy is obtained.

Lack of degeneracy is just a special case of $k = 1$ degeneracy. It may be summarized that by first-order approximation, the corrections to the eigenenergy caused by the perturbation are the eigenenergies of the perturbation matrix $(V_{m\ell})$.

2.1.2 Time-Dependent Perturbation Theory

As in the static perturbation case, we consider a system of a known Hamiltonian, whose solutions $\{\breve{\psi}(\underline{r},t)\}$ satisfy Schrödinger's equation

$$\mathbf{H}\breve{\psi}(\underline{r},t) = i\hbar\frac{\partial\breve{\psi}(\underline{r},t)}{\partial t} \quad . \tag{2.9}$$

The steady-state solutions of this equation are of the form

$$\breve{\psi}(\underline{r},t) = \psi_\ell(\underline{r})\cdot e^{-i(\varepsilon_\ell/\hbar)t} \quad , \tag{2.10}$$

where ε_ℓ are the eigenenergies of the $\psi_\ell(\underline{r})$ functions. For simplicity we assume no degeneracy, then introduce a given time-dependent perturbation of the form $\mathbf{H}' = V(\underline{r})\cdot \mathrm{f}(t)$ for $t > 0$. We are seeking solutions for the equation

$$\left(\mathbf{H} + \mathbf{H}'\right)\varphi_n = i\hbar\frac{\partial}{\partial t}\varphi_n \quad . \tag{2.11}$$

Because the unperturbed Hamiltonian solutions $\{\breve{\psi}(\underline{r},t)\}$ are a complete set, it is possible to represent every φ_n solution as a linear combination of $\{\breve{\psi}(\underline{r},t)\}$ functions. Namely,

$$\varphi_n(\underline{r},t) = \sum_\ell a_{n\ell}(t)\breve{\psi}_\ell(\underline{r},t) \quad . \tag{2.12}$$

Recall that we consider in fact a perturbation starting at $t = 0$. For continuousness considerations, at sufficiently short times after the start of perturbation, it should be possible to describe any φ_n descendent-function of E_n eigenenergy as an evolution of a particular eigenfunction $\breve{\psi}_n$ belonging to the unperturbed Hamiltonian. We set Eq. (2.12) 's expression into Eq. (2.11), multiply on the left by $\breve{\psi}_n^*$, and integrate over the entire space. As a result, we get

$$i\hbar\dot{a}_{nn}(t) = \sum_\ell e^{\frac{i}{\hbar}(E_n-\varepsilon_n)t}\cdot \mathrm{f}(t)\cdot a_{n\ell}(t)\cdot V_{n\ell} \quad , \tag{2.13}$$

where $V_{n\ell} = \langle n|\mathbf{V}|\ell\rangle \equiv \int \psi_n^*(\underline{r})V(\underline{r})\psi_\ell(\underline{r})d^3\underline{r}$.

We assume that for times $t \le 0$, namely before the perturbation "ignition" moment, the system existed at a certain quantum $|k\rangle$ state of ε_k energy for $k \ne n$. The initial condition for the expansion coefficients $a_{n\ell}(t)$ would then be

$$\left.\begin{aligned} a_{nk} &= 1; \\ a_{n\ell} &= 0 \quad for \quad \ell \ne k \end{aligned}\right\} \quad . \tag{2.14}$$

For sufficiently short durations past the perturbation ignition, we may assume that the addend providing the major contribution at the right side of Eq. (2.13) is the one containing $a_{nk} = 1$, whereas other addends are still negligible. Eq. (2.13) then reduces to

$$i\hbar\dot{a}_{nn}(t) = e^{\frac{i}{\hbar}(E_n - \varepsilon_n)t} \cdot \mathrm{f}(t) \cdot V_{nk} \quad . \tag{2.15}$$

We choose to express f(t) as a real function. Its Fourier transform then satisfies the relation $\boldsymbol{F}(-\omega) = \boldsymbol{F}^*(\omega)$. In such case, the Fourier transform satisfies the further relation

$$\mathrm{f}(t) = \frac{1}{\sqrt{2\pi}} \int_0^\infty d\omega \left[\boldsymbol{F}(\omega) e^{-i\omega t} + \boldsymbol{F}^*(\omega) e^{i\omega t} \right] \quad . \tag{2.16}$$

We set this decomposition in the (2.15) differential equation, set the approximation $E_n - \varepsilon_k \cong \varepsilon_n - \varepsilon_k$, integrate over time, and get

$$a_{nn}(t) = \frac{V_{nk}}{\sqrt{2\pi}} \int_0^\infty d\omega \left[\boldsymbol{F}(\omega) \frac{1 - e^{\frac{i}{\hbar}(\varepsilon_n - \varepsilon_k + \hbar\omega)t}}{\varepsilon_n - \varepsilon_k + \hbar\omega} + \boldsymbol{F}^*(\omega) \frac{1 - e^{\frac{i}{\hbar}(\varepsilon_n - \varepsilon_k - \hbar\omega)t}}{\varepsilon_n - \varepsilon_k - \hbar\omega} \right] \quad . \tag{2.17}$$

The integration constant is obviously zero, in compliance with the initial condition $a_{nn}(0) = 0$. Practically, only one of the addends in the square brackets is important: the one exhibiting a pole. It is the second addend if $\varepsilon_n > \varepsilon_k$; it is the first addend if $\varepsilon_k > \varepsilon_n$. For the following, we assume the first case, namely that $\varepsilon_n > \varepsilon_k$, and thus entirely neglect the first addend. The $|k\rangle$ to the $|n\rangle$ state transition probability $P_{nk}(t)$ would be given by $P_{nk}(t) = |a_{nn}(t)|^2 = a_{nn}^*(t) \cdot a_{nn}(t)$; namely

$$P_{nk}(t) = \frac{|V_{nk}|^2}{2\pi} \int_0^\infty \int_0^\infty d\omega' d\omega \boldsymbol{F}\left(\omega'\right) \boldsymbol{F}^*(\omega) \frac{1 - e^{-\frac{i}{\hbar}(\varepsilon_n - \varepsilon_k - \hbar\omega')t}}{\varepsilon_n - \varepsilon_k - \hbar\omega'} \cdot \frac{1 - e^{\frac{i}{\hbar}(\varepsilon_n - \varepsilon_k - \hbar\omega)t}}{\varepsilon_n - \varepsilon_k - \hbar\omega} \quad . \tag{2.18}$$

As an interim calculation, we treat the following expression

$$\mathcal{L}(\omega, t) \equiv \frac{1 - e^{-\frac{i}{\hbar}(\varepsilon_n - \varepsilon_k - \hbar\omega)t}}{\varepsilon_n - \varepsilon_k - \hbar\omega} \quad . \tag{2.19}$$

We denote $z = \dfrac{\varepsilon_n - \varepsilon_k - \hbar\omega}{\hbar}$; The z variable expresses the angular frequency departure from the resonance one $\omega_0 \equiv (\varepsilon_n - \varepsilon_k)/\hbar$. Obviously $d\omega = -dz$. We set this variable change in Eq. (2.19) and obtain $\mathcal{L}(z, t) = \dfrac{2\sin\frac{1}{2}zt}{\hbar z} \cdot e^{i\left(\frac{1}{2}zt - \pi/2\right)}$. The close-to-resonance region where the $\mathcal{L}$ function exhibits a pole is the one important for integration. This region covers both positive and negative z values. The integration may thus be described as within the $(-\infty, \infty)$ limits. We denote $\breve{\boldsymbol{F}}(z) \equiv \boldsymbol{F}\left(\frac{\varepsilon_n - \varepsilon_k}{\hbar} - z\right)$, set all in Eq. (2.18), and get

$$\begin{aligned} P_{nk}(t) &= \frac{2|V_{nk}|^2}{\pi} \int_{-\infty}^{\infty} \int_{-\infty}^{\infty} dz' dz\, \breve{\boldsymbol{F}}(z') \breve{\boldsymbol{F}}^*(z) \mathcal{L}(z', t) \cdot \mathcal{L}^*(z, t) = \\ &= \frac{2|V_{nk}|^2}{\pi} \int_{-\infty}^{\infty} \int_{-\infty}^{\infty} dz'\, dz\, \breve{\boldsymbol{F}}(z') \breve{\boldsymbol{F}}^*(z) \left(\frac{\sin\frac{1}{2}z't \cdot \sin\frac{1}{2}z\, t}{\hbar^2 z' z} \right) \cdot e^{\frac{1}{2}i(z' - z)\, t} \quad . \end{aligned} \tag{2.20}$$

The last two factors of the integrand oscillate as functions of time. Therefore, the transition probability $P_{nk}(t)$ also contains a time-oscillating component. After sufficiently long times to be estimated shortly, the oscillations moderate, and calculation of the end probability becomes possible.

To understand the approximations that we undertake hereafter, we estimate the range of the z parameter in terms of the perturbation spectral width. We assume first that the $\breve{\boldsymbol{F}}(z)$ function is nonzero in a typical δz region surrounding $z = 0$. By the uncertainty principle, the perturbation f(t) existence time duration is at least $1/\delta z$. No option exists to interrupt the perturbation and test the system's status earlier than $t = 1/\delta z$, because it will increase the perturbation's spectral width. Thus, we limit ourselves to sufficiently long times, satisfying $t >> 1/\delta z$. Therefore, the condition $|z' - z| >> 1$ is satisfied through most of the integration range. We then perform a gross approximation in calculating Eq. (2.20)'s integral. Specifically, *we replace the last factor by its average.* Such assumption may be justified if the number of oscillations performed over time by the last factor is large compared to those performed by the preceding factor. Such behavior occurs only in a particular half of the two-dimensional integration space (z', z) , where z' and z exhibit opposite signs. In the other half space, this condition is not satisfied. Nevertheless, we employ this approximation throughout the entire (z', z) space. We still obtain the correct result, as explained in the bottom footnote♣

$$e^{i(z'-z)t} \underset{\text{exchange by}}{\Rightarrow} \frac{1}{2t}\int_{-t}^{+t} e^{i(z'-z)\xi}d\xi \underset{t >> 1/\delta z}{\cong} \frac{\pi}{t}\delta\left(z'-z\right) . \qquad (2.21)$$

We set the latter result in Eq. (2.20) and get

$$P_{nk}^{av}(t) \cong \frac{|V_{nk}|^2}{\hbar^2}\int_{-\infty}^{\infty} d\left(\frac{1}{2}z\,t\right)\breve{\boldsymbol{F}}(z)\breve{\boldsymbol{F}}^*(z)\left(\frac{\boldsymbol{sin}^2\frac{1}{2}z\,t}{\left(\frac{1}{2}z\,t\right)^2}\right) . \qquad (2.22)$$

We denote $x = \frac{1}{2}z\,t$. Figure 2.1 demonstrates the $y = \boldsymbol{sin}^2 x/x^2$ function. It exhibits a high central peak at $x = 0$, of an approximately π width. Beyond the central peak, the function turns negligibly small. Its spectral width is thus π/t , which is small compared to the perturbation's spectral width, for which we have demanded to satisfy $\delta z >> 1/t$. We now introduce a further assumption that the perturbation $\left|\breve{\boldsymbol{F}}(z)\right|^2$ spectrum is also *smooth* over the integration region. Thus, we may pull out the quantity $\left|\breve{\boldsymbol{F}}(0)\right|^2$ from inside the integration expression, and place it at its front. We then get

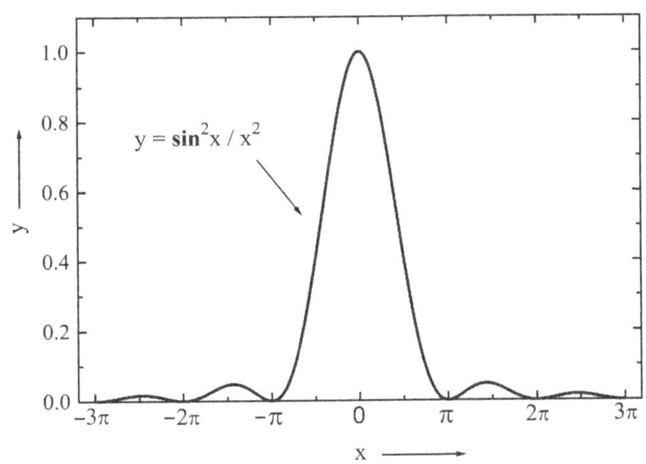

Figure 2.1 $y = sin^2\,x/x^2$ as function of x.

♣ For mathematic punctuality, the (z', z) integration plane in Eq. (2.20) should be divided in two. Then perform on the second-to-last factor, which oscillates many more times than the last factor, an approximation analogous to that performed in Eq. (2.21) on the last factor. The result is identical, because each of the factors is an (indeed different) representation of Dirac's δ function (see Appendix C).

$$P_{nk}^{av}(t) \cong \frac{|V_{nk}|^2}{\hbar^2} \left|\breve{\boldsymbol{F}}(0)\right|^2 \int_{-\infty}^{\infty} d\left(\frac{1}{2} z\, t\right) \left(\frac{\sin^2 \frac{1}{2} z\, t}{\left(\frac{1}{2} z\, t\right)^2} \right) \quad ;$$

The value of the integral in the latter equation is π . We then get

$$\boxed{P_{nk}^{av}(t) \cong \frac{\pi}{\hbar^2} \cdot \left|V_{nk}\right|^2 \cdot \left|\boldsymbol{F}\left(\omega_0 \equiv \frac{\varepsilon_n - \varepsilon_k}{\hbar}\right)\right|^2} \quad . \tag{2.23}$$

Importantly, Eq. (2.23) is valid only at a time long after the end of perturbation, and only for a spectrally smooth perturbation near the resonance frequency. Under such conditions, *the transition probability is independent of time*!

2.2 A Reminder of Planck's Black-Body Radiation

Planck's equation relates to the electromagnetic radiation emitted by a **black body**. An Ideal black body is a substance that **absorbs all incident electromagnetic radiation**. However, a black body must also radiate, otherwise its temperature will rise to infinity. A system that approximates most closely the ideal black body is a small hole in a large container, whose side walls are at thermal equilibrium with a heat reservoir; the reemission probability of light entering the container through the small hole practically vanishes, due to the large number of the inside involved reflections and diffuse scattering at the walls. The electromagnetic emission out of the hole then depends only on the electromagnetic field existing inside the container. This is the subject of our concern.

By quantum mechanics, the energy related to an oscillating electromagnetic field may attain only a discrete set of values $E_n = \left(n + \frac{1}{2}\right)\hbar\omega$ where $n = 0, 1, 2, \cdots \infty$. The energy corpuscle $\hbar\omega$ is the **photon**, and n is the number of photons, also addressed as **photons population**. The discreteness of energies implies that the electric field amplitudes (and correspondingly the magnetic field amplitudes) may also attain only discrete values determined by the given set of energies. According to **Boltzmann**, the photon population probability P_n of n ω-frequency mode photons at thermal equilibrium with the container walls at temperature T satisfies

$$P_n \propto e^{-n\hbar\omega/k_B T} \quad . \tag{2.24}$$

This is known as the **Boltzmann Equation**. Using it, one obtains

$$P_n = \frac{e^{-n\hbar\omega/k_B T}}{\sum_{s=0}^{\infty} e^{-s\hbar\omega/k_B T}} \quad . \tag{2.25}$$

The average number of photons occupying a vibrational (oscillational) mode of ω angular frequency $\langle n(\omega)\rangle_T$ is given by

$$\langle n(\omega)\rangle_T = \sum_{n=0}^{\infty} nP_n = \frac{\sum_{n=0}^{\infty} ne^{-n\hbar\omega/k_B T}}{\sum_{s=0}^{\infty} e^{-s\hbar\omega/k_B T}} \quad ;$$

$$\boxed{\langle n(\omega)\rangle_T = \frac{1}{e^{-\hbar\omega/k_B T} - 1}} \quad . \tag{2.26}$$

For the latter development, we have used the following identities:

$$\sum_s sx^s = x\frac{d}{dx}\sum_s x^s = x\frac{d}{dx}\left(\frac{1}{1-x}\right) = x\frac{1}{(1-x)^2} \quad ; \tag{i}$$

$$\sum_s e^{-s\hbar\omega/k_B T} = \frac{1}{1-e^{-\hbar\omega/k_B T}} \quad . \tag{ii}$$

To obtain the vibrational electromagnetic modes in a medium at thermal equilibrium, we consider a homogeneous transparent cube-shaped medium of a refractive index n♠ (not to be confused with the photons population number of that previously employed same letter). The medium has a macroscopic side-length L and side walls that are in intimate thermal contact with a heat reservoir of absolute temperature T. For demonstration, we imagine metallic walls, such that the inner electromagnetic waves form junctions on each. The wavelengths of standing waves in each dimension satisfy the relation $\lambda_j = 2L/j$ for $j = 1,\ 2,\ 3\ \ldots\infty$. Their corresponding wave numbers are $k_j = 2\pi/\lambda_j = (\pi/L)j$. The gap between two adjacent wave numbers is $\Delta k = \pi/L$. For positive k-vectors, the density of states in each single dimension of the k-space is thus $\mathcal{D}_k = 1/\Delta k = (L/\pi)$. Noting that a standing wave may be represented by a sum of two oppositely propagating waves (positive and negative k vectors), we reduce the density by a factor of 2, but include negative k-vectors as entities separate from their positive k-vectors counterparts, namely $\mathcal{D}_k = (L/2\pi)$. In three dimensions, the density of states per unit volume would be

$$\mathcal{D}(\underline{k}) = 2\frac{1}{(2\pi)^3} \quad . \tag{2.27}$$

The factor of 2 multiplication for the three-dimensional states density is added to account for the two possible independent polarization states belonging to each wave vector $\underline{k}$.

Notably, it is possible to present different conditions for the electromagnetic amplitudes at the medium walls, resulting, however, with the same macroscopic density of states. We further assume that the same volume density of $\underline{k}$ vectors is valid for any macroscopic shape of the medium.

We turn now to calculate the density of states along the ω angular frequency axis. The density of ω states per unit medium volume is defined by the relation $D(\omega)d\omega \equiv \mathcal{D}(\underline{k})d^3\underline{k}$. The angular frequency dispersion relation is given by $\omega(\underline{k}) = (c/n)k$; namely, the angular frequency depends on the size of the $\underline{k}$ vector, but not on its orientation. This means that the equi-frequency surfaces $\omega(\underline{k}) = Const.$ are spherical. Thus,

$$D(\omega) = \mathcal{D}(\underline{k})\cdot 4\pi k^2\frac{dk}{d\omega} = \frac{2}{(2\pi)^3}\cdot\frac{4\pi\omega^2 n^2}{c^2}\cdot\frac{n}{c} \quad ;$$

$$\boxed{D(\omega) = \frac{n^3\omega^2}{\pi^2 c^3}} \quad . \tag{2.28}$$

♠ The medium transparency requirement implies that the frequencies related to Planck's equation are lower than all resonance frequencies $\{\omega_0^j\}$ of the medium constituent atoms/molecules. The medium effective response then expresses itself as a real macroscopic refractive index n; namely the effective electromagnetic radiation phase velocity c/n replaces the speed of light constant c. In-depth discussion of this issue is provided in Chapter 10, especially in Section 10.4.

The volume electromagnetic radiation energy density per unit angular frequency is then given by $\rho_{\omega}^{BB}(\omega) = \hbar\omega \cdot D(\omega)\langle n(\omega)\rangle_T$ (note! $\langle n(\omega)\rangle_T$ here marks the thermal average number of photons per Eq. (2.26)). Namely,

$$\boxed{\rho_{\omega}^{BB}(\omega) = \frac{n^3\hbar\omega^3}{\pi^2 c^3} \cdot \frac{1}{\exp\left(\hbar\omega/k_B T\right) - 1}} \quad . \tag{2.29}$$

This result is called the **Planck equation**. The "*BB*" superscript is an abbreviation of the **Black Body** term; it is included to note that this equation forms the basis for calculating the black body electromagnetic radiation characteristics. Eq. (2.29) is sufficient for continuation of the current discussion. Development completion of the black body radiation characteristics, and subsequent discussion, follows in Section 2.6.

2.3 An Atom or Molecule in an Electromagnetic Radiation Field

This section discusses the specific case of a single bound electron in an atom or a molecule under perturbation of an electromagnetic field. Classically, the oscillating electron forms an electric dipole $e\underline{r}$, where $\underline{r}$ is the electron departure vector from the origin (where the positive binding nucleus resides). The radiation field expresses itself by an electric field $\underline{\mathcal{E}}(t) = \underline{u}\mathcal{E}(t)$, where $\underline{u}$ is a unit vector along the field direction. Accordingly, the $\underline{r}$ vector must oscillate under the influence of the oscillating electric field. There is also a phase difference between the field and $\underline{r}$ oscillations, and equally important, $\underline{r}$ and $\underline{\mathcal{E}}$ are not necessarily co-oriented (Chapter 1). Still, in view of the occurrence of steady orbital states in both atoms and molecules, we interpret $\underline{r}$ as rather *the temporal displacement vector from the steady unperturbed orbital.* For simplicity, we assume that in calculating the interaction with the electromagnetic field, we may replace the oscillating, time-dependent $\underline{r}$ with an average, time-independent displacement, which we also denote as $\underline{r}$. The perturbation Hamiltonian then assumes the form

$$\boxed{\mathbf{H}' = e(\underline{r}\cdot\underline{u})\mathcal{E}(t)} \quad . \tag{2.30}$$

We denote $V(\underline{r}) = e(\underline{r}\cdot\underline{u})$ and $\mathrm{f}(t) = \mathcal{E}(t)$. Then we use Eq. (2.23) to get the transition probability between the $|k\rangle$ and $|n\rangle$ states

$$P_{nk}^{av}(t) \cong \frac{\pi e^2}{\hbar^2} |\langle n|\underline{r}|k\rangle\cdot\underline{u}|^2 \cdot |\boldsymbol{E}(\omega_0)|^2 \quad . \tag{2.31}$$

Recall (Eq. [2.23]) that Eq. (2.31) is valid only long past termination of the perturbation, and only for a smoothly varying spectrum, and that the end probability is independent of time! However, in contrast to the last feature, we consider that the transition probability grows gradually and smoothly with the perturbation time duration from zero to its final value given by Eq. (2.31). For justification, consider the following view: if the perturbing wave group is interrupted at any time τ past its start, the quantity $|\boldsymbol{E}(\omega_0)|^2$ correspondingly lowers, and so does the end transition probability. Thus, the gradual probability growth assumption as a function of time correlates with the *perturbation duration*, and in any case, it is the value it assumes only a long time *after* its termination. With that understanding, we calculate the probability growth rate as a function of the perturbation τ duration:

$$W_{nk} \equiv \frac{dP_{nk}^{av}}{d\tau} \cong \frac{\pi e^2}{\hbar^2}|\langle n|\underline{r}|k\rangle\cdot\underline{u}|^2\cdot\frac{d|\boldsymbol{E}(\omega_0)|^2}{d\tau} \quad . \tag{2.32}$$

We have seen in Eqs. (1.30) and (1.31) that the expression $\boldsymbol{F}(\omega) = \frac{c}{2\pi n}|\boldsymbol{E}(\omega)|^2$ describes the radiation energy flux density per unit angular frequency incident on the atom/molecule. We set $d\boldsymbol{F}(\omega)/d\tau \equiv (c/n)\rho_\omega(\omega_0)$, where ρ_ω is the volume density of the electromagnetic energy per unit angular frequency in the atom/molecule vicinity. Then we get

$$W_{nk} \cong \frac{2\pi^2 e^2}{\hbar^2}|\langle n|\underline{r}|k\rangle\cdot\underline{u}|^2\cdot\rho_\omega(\omega_0) \quad . \tag{2.33}$$

This last equation is an application of **Fermi's golden rule** to the case of an atom or molecule interaction with the electromagnetic radiation field.

As previously noted, there is a random angular distribution between the electric field orientation $\underline{u}$ and the transition vector $\langle n|\underline{r}|k\rangle$. We thus calculate their average scalar product.

$$\overline{|\langle n|\underline{r}|k\rangle\cdot\underline{u}|^2} = |\langle n|\underline{r}|k\rangle|^2\cdot\overline{\boldsymbol{cos}^2\theta_u} = |\langle n|\underline{r}|k\rangle|^2\cdot\frac{1}{4\pi}\int_\Omega \boldsymbol{cos}^2\theta_u d\Omega$$
$$= \frac{1}{4\pi}|\langle n|\underline{r}|k\rangle|^2\int_0^\pi \boldsymbol{cos}^2\theta_u\cdot 2\pi\,\boldsymbol{sin}\theta_u d\theta_u = |\langle n|\underline{r}|k\rangle|^2\left[-\frac{1}{2}\cdot\frac{\boldsymbol{cos}^3\theta_u}{3}\right]_0^\pi = \frac{1}{3}|\langle n|\underline{r}|k\rangle|^2 \quad , \tag{2.34}$$

where θ_u is the angle between the $\underline{u}$ and $\langle n|\underline{r}|k\rangle$ vectors, and Ω is the solid angle over which the averaging is performed. To correct the transition probability rate accordingly, we set the above result in Eq. (2.33) and obtain

$$\boxed{W_{nk} \cong \frac{2\pi^2 e^2}{3\hbar^2}|\langle n|\underline{r}|k\rangle|^2\rho_\omega\left(\omega_0 = \frac{\varepsilon_n - \varepsilon_k}{\hbar}\right)} \quad . \tag{2.35}$$

The $\underline{\mu} \equiv e\langle n|\underline{r}|k\rangle$ vector is termed **transition dipole** between the $|k\rangle$ and $|n\rangle$ states. In Chapter 1 we symbolized this very quantity by the letter Q (see Eq. [1.1] and the following text).

Often, the expression received in Eq. (2.35) is granted a different interpretation. The atom plus the electromagnetic radiation field are assumed to form a closed thermodynamic system. An initial state includes the atom at a low energy state $|k\rangle$ and several $\hbar\omega$ photons in the radiation field. The final state includes an atom at a higher state $|n\rangle$ yet with one less photon in the radiation field. The said thermodynamic system exhibits no change of energy, that is, the initial and final states are degenerate. The interaction between the atom and radiation field is considered as a **coupling**, causing **exchange** between the two system states.

A result identical with that given by Eq. (2.35) would have been obtained also for an inverse starting condition $\varepsilon_n < \varepsilon_k$, with $\omega_0 = (\varepsilon_k - \varepsilon_n)/\hbar$. The initial system state is an atom at a high-energy state $|k\rangle$ residing in the radiation field; the final state is an atom at a low-energy state $|n\rangle$ and the radiation field is enriched with an additional $\hbar\omega$ photon.

Our analysis suggests that absorption and emission of electromagnetic radiation are symmetric, depending on the presence of energy in the radiation field. The absorption is termed **induced absorption**; the emission is termed **stimulated emission**. Our treatment fails in accounting for the spontaneous emission by excited atoms, a common process established by numerous experimental tests. This issue is the subject of our next section.

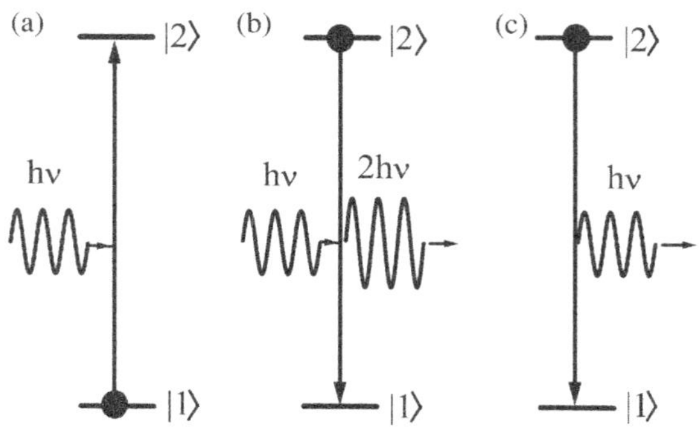

Figure 2.2 Schematic illustration of radiative transitions between two energy states of matter. (a) Absorption; a photon in the radiation field is absorbed by an atom of a lower energy state $|1\rangle$, resulting in lowering of the radiation field energy. (b) Stimulated emission; the radiation field stimulates transition from a higher atomic energy state $|2\rangle$ to a lower energy state $|1\rangle$, and the radiation field energy increases by one photon, co-phased with the stimulating field. (c) Spontaneous emission; an atom suddenly changes from a higher state $|2\rangle$ to a lower state $|1\rangle$, and enriches the radiation field energy with one photon.

2.4 Stimulated Emission and Einstein's Coefficients

The semi-quantum calculation carried out in the previous section suggested entire symmetry of energy exchange between an atom and an electromagnetic radiation field: absorption and emission are both stimulated by the radiation field. However, experiment suggests the existence of an additional emission mechanism: **spontaneous radiation emission**, where a higher energy state decays to a lower energy state even in the absence of any radiation, or in other words, in total darkness. Figure 2.2 summarizes the possible processes for energy exchange between an atom or molecule and the electromagnetic radiation field.

It is important to emphasize that the spontaneous emission is a mechanism consistent with the second law of thermodynamics: an atom at a high energy state is a highly ordered system, hence of a low entropy; it transforms to a low energy state atom plus a photon propagating in space at a *random orientation*. The two systems exhibit the same total energy. The latter, however, is less ordered, hence of a higher entropy.

To account for the spontaneous emission, Einstein proposed to extend Eq. (2.35) for the emission case (namely for $\varepsilon_n < \varepsilon_k$) by including an addend A_{nk} that describes the spontaneous emission occurrence probability per unit of time, as follows:

$$\boxed{W_{nk}^{em} = B_{nk}\rho_\omega(\omega_0) + A_{nk}} \quad , \tag{2.36}$$

where $B_{nk} = \frac{2\pi^2 e^2}{3\hbar^2}\left|\langle n|\underline{r}|k\rangle\right|^2$. If $|n\rangle$ is the ground-state, the A_{nk} addend describes the inverse lifetime of the excited $|k\rangle$ state ; that is, $A_{nk} = \tau^{-1} = \gamma$ (see Chapter 1).

The A_{nk} and B_{nk} quantities are called **Einstein radiative transition coefficients**. A_{nk} is the **Einstein spontaneous emission coefficient** and B_{nk} is the **Einstein stimulated emission coefficient**.

Thermodynamic considerations require a linkage between the two Einstein coefficients. Let's consider the analyzed atom situated together with many other identical atoms, at thermal equilibrium with a heat reservoir of T temperature. The number of atoms emitting radiation at ω_0 frequency must equal the number of those absorbing radiation of the same frequency over the same duration. Let N_k atoms have the higher energy $|k\rangle$ state, and N_n have the lower energy $|n\rangle$ state, $\varepsilon_n < \varepsilon_k$. Equilibrium conditions require that

$$N_k\left(B_{nk}\rho_\omega^{BB}(\omega_0) + A_{nk}\right) = N_n B_{kn}\rho_\omega^{BB}(\omega_0) \quad . \tag{2.37}$$

Simultaneously, the Boltzmann population ratio must also be satisfied:

$$\frac{N_k}{N_n} = \boldsymbol{exp}\left(-\frac{\varepsilon_k - \varepsilon_n}{k_B T}\right) \quad , \tag{2.38}$$

where $k_B = 1.38 \times 10^{-16}$ erg · deg^{-1} is the Boltzmann constant. Using the above two equations, and setting $B_{nk} = B_{kn}$, we may extract the following expression for $\rho_\omega^{BB}(\omega_0)$:

$$\rho_\omega^{BB}(\omega_0) = \frac{A_{nk}/B_{nk}}{\mathbf{exp}(\hbar\omega_0/k_BT) - 1} \quad . \tag{2.39}$$

Recall that $\rho_\omega^{BB}(\omega_0)$, the volume energy concentration per unit of angular frequency of electromagnetic radiation at thermal equilibrium, is also given by Planck's equation (Eq. (2.29))

$$\rho_\omega^{BB}(\omega_0) = \frac{n^3\hbar\omega_0{}^3}{\pi^2c^3}\frac{1}{\mathbf{exp}(\hbar\omega_0/k_BT) - 1} \quad . \tag{2.40}$$

Comparing the above two expressions, one gets

$$\boxed{\frac{A_{nk}}{B_{nk}} = \frac{n^3\hbar\omega_0{}^3}{\pi^2c^3};\ A_{nk} = \frac{2n^3e^2\omega_0{}^3}{3\hbar c^3}|\langle n|\underline{r}|k\rangle|^2} \quad . \tag{2.41}$$

When there are several energy states lower than the excited $|k\rangle$ state, the total transition probability to lower energy states is a sum of transition probabilities to all lower $|n\rangle$ states; namely

$$\tau_{k,rad}{}^{-1} = \sum\nolimits_n A_{nk} \quad , \tag{2.42}$$

where $\tau_{k,rad}$ is the effective radiative lifetime of the excited $|k\rangle$ state.

Recall the classically developed Eq. (1.7). It provides an expression for the spontaneous emission probability per unit time as $\gamma = 2e^2\omega_0{}^2/3mc^3$. Let's set in the above expression for γ the relation $\hbar\omega_0 = \frac{1}{2}m\omega_0{}^2x_0{}^2$ (the mechanical energy of the classical oscillator). Then assume $n > 1$ as the medium refractive index (namely set c/n instead of c) , and get $\gamma = \frac{2n^3e^2\omega_0{}^3}{3\hbar c^3}\cdot\frac{1}{2}x_0{}^2$. This result is consistent with the semi-quantum approach result for A_{nk} , by identifying the $|\langle n|\underline{r}|k\rangle|^2$ quantity as half the squared classical amplitude of the vibrating electron $\frac{1}{2}x_0{}^2$. This is not surprising, because we have started by defining $\underline{r}$ as the average displacement (see Section 2.3). Indeed, in a harmonic oscillation, the averaged squared displacement satisfies $\overline{x^2} = \frac{1}{2}x_0{}^2$.

In view of the last results, the total emission probability from a higher to a lower state (spontaneous plus stimulated) assumes the form (see Eq. [2.36])

$$W_{nk}^{em} = \tau_{rad}{}^{-1}\left[\frac{\pi^2c^3}{n^3\hbar\omega_0{}^3}\rho_\omega(\omega_0) + 1\right] \quad . \tag{2.43}$$

Eq. (2.43) indicates that the total transition probability per unit time increases with an increase in radiation energy concentration at resonance transition frequency in the atom-containing volume. In other words, the effective $|k\rangle$ excited state lifetime shortens. Correspondingly, the transition spectrum broadens. That kind of broadening mechanism is termed **power broadening**. This issue won't be discussed further.

In the specific case where the atom is located in a thermal radiation field, Eq. (2.43) reduces to the form (see also Eq. [2.26])

$$\boxed{W_{nk}^{em} = \tau_{rad}{}^{-1}[\langle n(\omega_0)\rangle_T + 1]} \quad . \tag{2.44}$$

Eq. (2.43) suggests that the total emission probability per unit time increases with temperature (namely the effective $|k\rangle$ excited state lifetime shortens); it is a result of enhanced thermal photon population in the medium. A similar expression is used to explain excited state lifetime shortening of atoms residing in a material medium due to coupling with the material vibrations. Emission of energy to the material vibrational field, namely to the phonon field, is a *non-radiative mechanism*, competing with the radiative photon emission, quite often causing its practical quenching. This subject will be discussed in detail in Chapter 13.

In quantum mechanics, spontaneous emission is also interpreted as stimulated by the vacuum electromagnetic radiation (**vacuum fluctuations**). The vacuum radiation is the $\frac{1}{2}\hbar\omega$ addend in the oscillator energy – see the discussion at the beginning of section 2.2 – emanating from solution of the harmonic oscillator wave equation (see also Section 5.3). The mere existence of spontaneous emission is in fact an experimental verification of the physical validity of the solution. We discuss the subject further in Section 4.7, in relation to the **Lamb shift** in atomic spectra.

2.5 Radiation Absorption and Amplification in Matter

Eq. (2.41) indicates that the radiative stimulated transition probability per unit time is given by

$$W^s_{nk} = A_{nk}\frac{\pi^2 c^3}{n^3\hbar\omega_0{}^3}\rho_\omega(\omega_0) = \frac{\pi^2 c^3}{n^3\hbar\omega_0{}^3\tau_{rad}}\rho_\omega(\omega_0) \quad . \tag{2.45}$$

Often in laser spectroscopy, the frequency ν , or alternatively the vacuum wavelength λ, rather than the angular frequency, are used as variables. Recall that $\rho_\nu(\nu_0)d\nu_0 = \rho_\lambda(\lambda_0)(-d\lambda_0) = \rho_\omega(\omega_0)\,d\omega_0$, where $\rho_\nu(\nu_0)$ is the radiation field volume energy density at ν_0 frequency per unit frequency, and $\rho_\lambda(\lambda_0)$ is the radiation field volume energy density at λ_0 per unit wavelength. The "0" subscript in the frequency or in the wavelength expresses their attribution to resonance. Thus, we get

$$\left.\begin{aligned} W^s_{nk}(\nu_0) &= \frac{c^3}{8\pi n^3 h\nu_0{}^3\tau_{rad}}\rho_\nu(\nu_0) \\ W^s_{nk}(\lambda_0) &= \frac{n\lambda_0{}^5}{8\pi hc\tau_{rad}}\rho_\lambda(\lambda_0) \end{aligned}\right\} \quad . \tag{2.46}$$

In the following, we use the presentation of the stimulated transition probability as function of frequency, because most physical measuring equipment operates under approximately standard atmospheric conditions, where the refractive index is very close to a unity. The frequency of a beam crossing a medium, where absorption and emission take place, is virtually identical to its frequency in the measuring equipment, independent of the difference in refractive indices. On the other hand, the radiation wavelength inside the measuring equipment is different than inside the medium. The wavelength appearing in the lower line of Eq. (2.46) is that of the absorbing/amplifying medium; therefore, it would be necessary to convert the equation to match the wavelength in the measuring equipment. Anyway, the reader should practice the conversion between the two alternative representations.

We represent now the $g(\nu)$ function that describes the *shape* of an absorption or emission line, normalized to a unity integral ♣:

$$\boxed{\int_0^\infty g(\nu)d\nu = 1} \quad . \tag{2.47}$$

If the transition shape exhibits a *single peak*, as in the natural broadening case (see Figure 1.2) or in different homogeneous broadening cases, it is possible to write $g(\nu_0) = 1/\Delta\nu$, where $\Delta\nu$ is the effective line width. In a Lorentzian line shape, this parameter is close to the line *Full Width at Half Maximum* γ – see Figure 1.2 and related text. Then, Eq. (2.46) may be written as

$$W_{nk}^{S}(\nu_0) = \frac{c^3}{8\pi n^3 h\nu_0{}^3 \tau_{rad}}[\rho_\nu(\nu_0)\Delta\nu]\cdot g(\nu_0) \quad . \tag{2.48}$$

The quantity $[\rho_\nu(\nu_0)\Delta\nu]$ is the volume concentration of the electromagnetic radiation within a $\Delta\nu$ frequency range around the atomic resonance frequency ν_0 . We denote $I(\nu_0) = (c/n)[\rho_\nu(\nu_0)\Delta\nu]$. This quantity (in units of energy per unit area, per unit time) is the *monochromatic* **radiation beam intensity** at ν_0 frequency that exhibits the same volume concentration of the electromagnetic radiation. It is also sometimes addressed as a **beam flux density**. In other words,

$$W_{nk}^{S}(\nu_0) = \frac{c^2 I(\nu_0)}{8\pi n^2 h\nu_0{}^3 \tau_{rad}}\cdot g(\nu_0) \quad . \tag{2.49}$$

If the radiation volume contains $N_2 = N_k$ identical atoms per unit volume at the excited state level, and $N_1 = N_n$ identical atoms per unit volume at the lower state level, such that $N_2 > N_1$, the radiation field intensity increases. Such a situation exists in the amplifying medium of a laser system. The increase rate of volume energy density is given by $(N_2 - N_1)W_{12}^{S}\cdot h\nu_0$. The $(N_2 - N_1)$ quantity is termed **population inversion**. "Inversion" is used in the name because the population condition described is inversed compared to thermal equilibrium conditions, under which the density of lower states is rather larger than the density of higher states. Let's denote by Z the axis representing the electromagnetic beam propagation direction. Then

$$\frac{dI(\nu_0)}{dz} = \left[\frac{(N_2 - N_1)c^2}{8\pi n^2 \nu_0{}^2 \tau_{rad}}\cdot g(\nu_0)\right]\cdot I(\nu_0) \quad .$$

The energy amplification coefficient of a progressing beam at resonance frequency is then given by

$$\Gamma(\nu_0) = \frac{(N_2 - N_1)\, c^2}{8\pi n^2 \nu_0{}^2 \tau_{rad}}\cdot g(\nu_0) \quad . \tag{2.50}$$

The amplification coefficient units are cm^{-1} in the cgs units system. In Section 1.2, we derived the natural broadening of absorption and emission atomic lines as a result of the finite radiation emission duration which forms a wave group. In quantum mechanics, the lines spectral frequency dependence is interpreted as frequency dependence of the emission or absorption *probability*. By this interpretation, Eq. (2.50)

♣ If by a random choice of scale for the $g(\nu)$ function, its integral is not a unity, it is possible to normalize it simply by dividing $g(\nu)$ by the value of the said integral. The newly obtained function satisfies the normalization condition required by equation (2.47).

provides a calibration of the energy amplification at the peak frequency ν_0. Hence, in homogeneous broadening cases, and frequencies ν around the peak vicinity, we get

$$\Gamma(\nu) = \frac{(N_2 - N_1)\, c^2}{8\pi n^2 {\nu_0}^2 \tau_{rad}} \cdot g(\nu) \quad . \tag{2.51}$$

For heterogeneous broadening cases, namely where there exists a *distribution of peak frequencies* described by a distribution function $G(\nu_0)$ (for example, in the Doppler broadening), we have to replace the population inversion density $(N_2 - N_1)$ of Eq. (2.50) by $(N_2 - N_1)G(\nu_0)\Delta\nu$. Specifically,

$$\Gamma(\nu_0) = \frac{(N_2 - N_1)G(\nu_0)\Delta\nu \cdot c^2}{8\pi n^2 {\nu_0}^2 \tau_{rad}} \cdot g(\nu_0) \quad ;$$

setting $g(\nu_0)\Delta\nu = 1$ (Eq. [2.47]), and following $\Delta\nu$ ' s definition, one obtains

$$\Gamma(\nu_0) = \frac{(N_2 - N_1) \cdot c^2}{8\pi n^2 {\nu_0}^2 \tau_{rad}} \cdot G(\nu_0) \quad . \tag{2.52}$$

In the heterogeneous broadening case (Eq. [2.52]), it is customary to avoid the "0" subscript used for noting the resonance frequencies. In addition, the letter g replaces G for noting the resonance frequencies distribution function. Thus, in heterogeneous broadening cases we have

$$\Gamma(\nu) = \frac{(N_2 - N_1)c^2}{8\pi n^2 \nu^2 \tau_{rad}} \cdot g(\nu) \quad . \tag{2.53}$$

In summary:

$$\left.\boxed{\begin{array}{ll} \Gamma(\nu) = \dfrac{(N_2 - N_1)c^2}{8\pi n^2 {\nu_0}^2 \tau_{rad}} \cdot g(\nu) & \textit{for homogeneous broadening} \\ \Gamma(\nu) = \dfrac{(N_2 - N_1)c^2}{8\pi n^2 \nu^2 \tau_{rad}} \cdot g(\nu) & \textit{for heterogeneous broadening} \end{array}}\right\} \quad . \tag{2.54}$$

The difference between the expressions for heterogeneous and homogeneous broadening is usually small, especially for narrow absorption and emission lines. The dependence on the squared frequency appearing in the denominator of the heterogeneous broadening then has little impact, and the gain frequency dependence reflects quite accurately the spectral shape of the spontaneous emission $g(\nu)$. In most cases, the contributions to broadening are partially homogeneous and partially heterogeneous; the practically used expression is that of the heterogeneous broadening.

In the special $N_2 = 0$ case, the gain turns negative; that is, absorption of beam energy occurs. The absorption coefficient would be given by

$$\alpha(\nu) = \frac{N_1 c^2}{8\pi n^2 {\nu_0}^2 \tau_{rad}} g(\nu) \text{ or } \alpha(\nu) = \frac{N_1 c^2}{8\pi n^2 \nu^2 \tau_{rad}} g(\nu) \quad , \tag{2.55}$$

for homogeneous or heterogeneous broadening, respectively.

Absorption spectra measurements of atoms/molecules embedded in dielectric media is relatively easy compared to their emission spectra measurements. Recalling the (2.47) normalization condition on the g(ν) function, we obtain

$$\left.\begin{aligned}\tau_{rad}{}^{-1} &= \frac{8\pi n^2\nu_0{}^2}{N_1c^2}\int_0^{\infty}\alpha(\nu)d\nu \quad \textit{for homogeneous broadening}\\ \tau_{rad}{}^{-1} &= \frac{8\pi n^2}{N_1c^2}\int_0^{\infty}\nu^2\alpha(\nu)d\nu \quad \textit{for heterogeneous broadening}\end{aligned}\right\} . \tag{2.56}$$

Thus, measurement of the absorption coefficient spectrum of atoms/molecules embedded in a dielectric transparent medium establishes a direct means for calculation of the excited-state radiative lifetime. Usually, fluorescence emission decay rate measurement provides shorter effective lifetimes. The close proximity of the embedded atom/molecule to the medium constituents promotes parallel non-radiative decay processes, as mentioned in relation to Eq. (2.44).

In spectroscopic studies, it is customary to characterize electron transitions by the **transition dipole** defined by $\underline{\mu} \equiv e\langle n|\underline{r}|k\rangle$. It relates to the radiative lifetime by $\tau_{rad}{}^{-1} = \frac{2n^3\omega_0{}^3}{3\hbar c^3}\mu^2$ (see Eqs. [2.35] and [2.41]). Setting the relation in Eq. (2.55) yields

$$\mu^2 = \frac{3hc}{4\pi^3 n\nu_0}\cdot\frac{1}{N_1}\int_0^{\infty}\alpha(\nu)d\nu \quad \text{or} \quad \mu^2 = \frac{3hc}{4\pi^3 n}\cdot\frac{1}{N_1}\int_0^{\infty}\frac{\alpha(\nu)}{\nu}d\nu \ , \tag{2.57}$$

for homogeneous and heterogeneous broadenings, respectively.

For practical usage of Eqs. (2.57) and (2.56), limited integration over frequency regions of strong absorption coefficient is quite satisfactory, usually done around the central peak.

Electric dipole transition is given by $\mu_d{}^2 = e^2\cdot\frac{1}{2}x_0{}^2 = \frac{e^2\hbar}{m\omega_0} = \frac{e^2h}{(2\pi)^2m\nu_0}$. If an electric dipole transition exhibits a heterogeneous frequency distribution given by g(ν), the average dipole transition would be given by

$$\overline{\mu_d{}^2} = \frac{e^2h}{(2\pi)^2m}\int_0^{\infty}\frac{g(\nu)}{\nu}d\nu \ . \tag{2.58}$$

Spectroscopic absorption measurements analyzed by Eq. (2.57) quite often provide dipole moments that are several orders of magnitude *smaller* than the ideal value provided by Eq. (2.58). This "discrepancy" is interpreted as indicating that electric moments involved in the electron oscillation during transition doesn't necessarily produce large dipole moments. It obviously relates to the $\langle n|\underline{r}|k\rangle$ integral value (see Eq. [2.35] and the following definition of $\underline{\mu}$). To characterize such important effects, the **oscillator strength** is defined as the ratio between the squared measured dipole moment and the squared ideal dipole moment: $\boldsymbol{f} \equiv \mu^2/\overline{\mu_d{}^2}$. Recalling that the g(ν) function is the experimental $\alpha(\nu)$ function normalized to a unity integral, we get

$$\boxed{\boldsymbol{f} = \frac{3mc}{\pi ne^2}\cdot\frac{1}{N_1}\int_0^{\infty}\alpha(\nu)d\nu} \ ; \tag{2.59}$$

an expression identical for both homogeneous and heterogeneous broadenings.

Because of the practical importance of this equation, it is worthwhile to present it in a technical form, which is suitable for spectroscopic studies:

$$\boxed{f = \frac{113}{n}\frac{1}{N_1[\mathrm{cm}^{-3}]}\int_0^{\infty}\alpha(\nu)\left[\mathrm{cm}^{-1}\right]\cdot d\nu[\mathrm{Hz}]} \quad . \tag{2.60}$$

2.6 Black Body Radiation – Continuation and Completion

The starting point for calculation of a black body radiation is Eq. (2.29), which provides the volume concentration of the electromagnetic radiation per unit angular frequency inside a closed macroscopic container that maintains thermal equilibrium with its side-walls. For calculation, we perform several changes and conversions in the equation. First, we note that Eq. (2.29) could not be valid at all frequency regions, as the refractive index n is frequency dependent. In different frequency regions, the inner medium is no more transparent, that is, the complex refractive index has a dominant imaginary component. Therefore, the black body is described as a small hole in an *empty* large container, satisfying $n = 1$ at all frequencies (see Figure 2.3).

We further convert the frequency dependence into a wavelength dependence, by presenting the volume concentration of the electromagnetic radiation per *unit wavelength*! The wavelength dependence is the form practiced in spectroscopic measurements related to black body radiation. The relation between the quantities is $\rho_{\omega}^{BB}(\omega)d\omega = \rho_{\lambda}^{BB}(\lambda)(-d\lambda)$, or alternatively, $\rho_{\lambda}^{BB}(\lambda) = \rho_{\omega}^{BB}(\omega)(-d\omega/d\lambda)$. Setting $n = 1$ and performing the said conversion, we get

$$\rho_{\lambda}^{BB}(\lambda, T) = \frac{8\pi hc}{\lambda^5}\cdot\frac{1}{\boldsymbol{exp}\,(hc/\lambda k_B T) - 1} \quad , \tag{2.61}$$

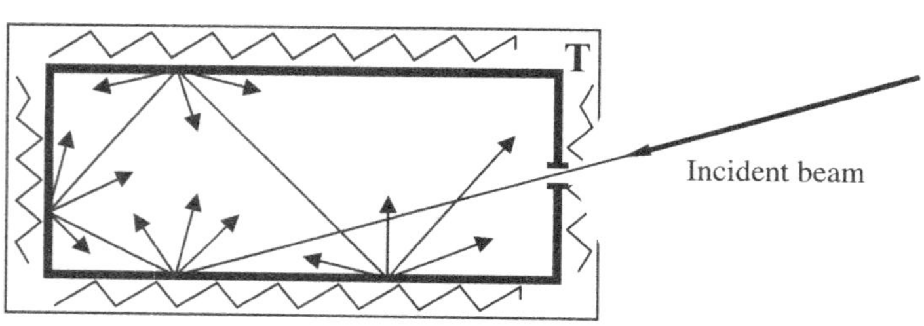

Figure 2.3 Demonstration of a black body realization by a small hole in a container surrounded by a furnace that keeps the container side-walls at a constant T temperature. The odds that an electromagnetic radiation beam entering the container through the hole will be re-emitted through the same hole are extremely small due to multiple diffuse scattering events on the container side-walls. The present section considers the thermal electromagnetic radiation *emitted* through the hole, which is independent of the presence or absence of any entering radiation.

where $\rho_{\lambda}^{BB}(\lambda, T)$ is the volume concentration of the electromagnetic energy inside the container volume per unit wavelength. On dividing Eq. (2.61) by the photon energy hc/λ, we obtain the volume density of photons per unit wavelength. Thus,

$$N_{\lambda}^{BB}(\lambda, T)d\lambda = \frac{8\pi}{\lambda^4}\cdot\frac{1}{\boldsymbol{exp}\,(hc/\lambda k_B T) - 1}d\lambda \quad . \tag{2.62}$$

For brevity and convenience in the following development, we denote $N \equiv N_{\lambda}^{BB}(\lambda, T)d\lambda$, a quantity that provides the number of photons per unit volume in the container between λ and $\lambda + d\lambda$. All these photons exhibit a common velocity – c

(the speed-of-light constant), equally distributed about all directions. In other words, if we describe a sphere of radius c in the three-dimensional velocity space, the velocity vectors' edges will be evenly distributed on that sphere surface at a $N/4\pi c^2$ density (Figure 2.4). We denote by $N(v_x)dv_x$ the number of photons in the wavelength range, whose velocity component in the x orientation ranges between v_x and $v_x + dv_x$.

Therefore, $N(v_x)dv_x = \frac{N}{4\pi c^2}\cdot c d\theta \cdot 2\pi c \cos\theta = \frac{1}{2} N \cos\theta \cdot d\theta$ (see Figure 2.4). Also, $v_x = c\cdot \sin\theta \Rightarrow dv_x = c\cdot\cos\theta\cdot d\theta$.

Therefore, $N(v_x)dv_x = \frac{1}{2}\frac{N}{c}dv_x$. The photon flux density (photons per unit time per unit area) propagating in the x direction (chosen to be the orientation toward the exit hole) would be given by summing all *positive x* orientation components, namely by the integral $\int_0^c v_x \cdot N(v_x)dv_x = \frac{1}{2}\frac{N}{c}\int_0^c v_x dv_x = \frac{1}{2}\frac{N}{c}\cdot\left.\frac{{v_x}^2}{2}\right]_0^c = \frac{1}{4}c\cdot N$. The energy flux density (emitted energy per unit time, per unit area, per unit wavelength) is given by multiplying the latter result by the photon energy hc/λ and dividing by the $d\lambda$ wavelength section; namely

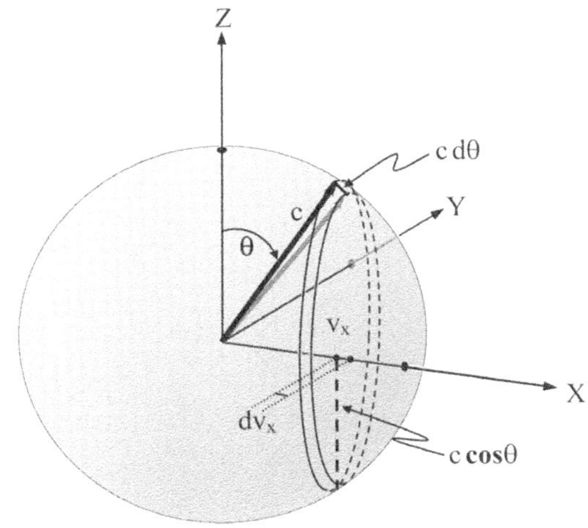

Figure 2.4 Geometric plot aimed at clarifying the various quantities used in calculating the photon flux density toward a given orientation in a medium, where the photons propagate equally in all directions.

$$\boxed{F_{BB}(\lambda, T) = \frac{2\pi hc^2}{\lambda^5}\cdot\frac{1}{\exp(hc/\lambda k_B T) - 1}} \quad . \tag{2.63}$$

We turn now to analyze the spectral features of the black body electromagnetic radiation per Eq. (2.63). As a starting step, we consider two extreme limits: very long wavelengths satisfying $hc/\lambda k_B T << 1$, and very short wavelengths satisfying $hc/\lambda k_B T >> 1$. In those limits, Eq. (2.63) exhibits the forms

$$F_{BB}(\lambda, T) \cong \begin{cases} \frac{2\pi c}{\lambda^4} k_B T & hc/\lambda k_B T \ll 1 \\ \frac{2\pi hc^2}{\lambda^5}\exp(-hc/\lambda k_B T) & hc/\lambda k_B T \gg 1 \end{cases} \quad . \tag{2.64}$$

Thus, in the very long wavelengths range, the energy density of the black body emission reduces with increasing wavelength. On the other hand, in the very short wavelengths range, the black body emission energy density increases with increasing wavelength. This behavior suggests that the energy density exhibits a maximum at some wavelength, to be denoted λ_{peak}. The energy density increases with temperature for all wavelengths (see Figure 2.5).

To find the peak position, we consider the wavelength derivative $dF_{BB}(\lambda, T)/d\lambda$ of Eq. (2.63). It vanishes at a wavelength satisfying $e^{-hc/\lambda k_B T} = 1 - \frac{1}{5}hc/\lambda k_B T$. Its approximate solution for λ satisfies $hc/\lambda_{peak} k_B T \cong 4.965$. Then

$$\boxed{\lambda_{peak}\cdot T = 0.2014\frac{hc}{k_B}} \quad . \tag{2.65}$$

This result indicates that the wavelength at which the maximal energy density is obtained for the black body radiation is inversely proportional to the absolute temperature. This very result was experimentally

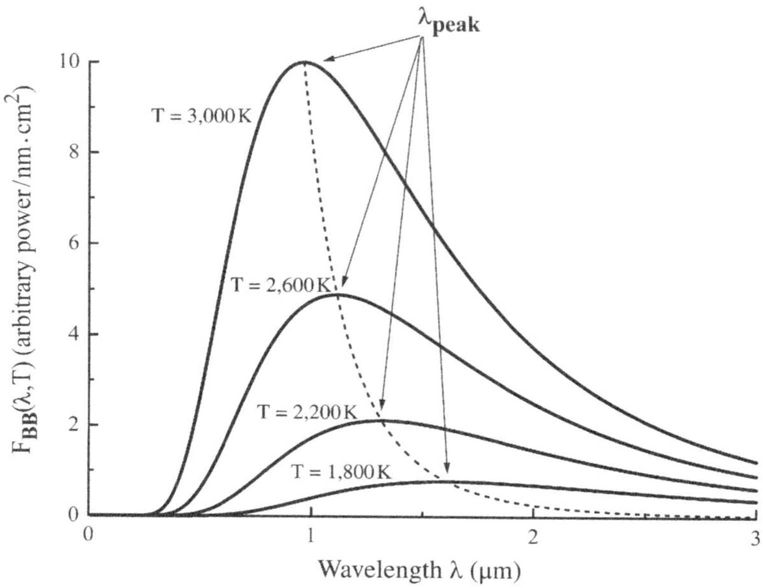

Figure 2.5 Black body radiation spectra at different temperatures. The spectra demonstrate both peak displacement toward shorter wavelengths, and total emission increase with temperature.

known prior to Planck's derivation, and is termed **Wien's displacement law**. The numerical value of the right side of Eq. (2.66) is 0.2898 cm·K.

In addition, we calculate the total black body flux density emission (energy per unit area, per unit time) by integrating over the entire wavelength dimension:

$$\Phi_{BB}(T) = \int_0^\infty \frac{2\pi hc^2}{\lambda^5} \cdot \frac{1}{\mathbf{exp}\,(hc/\lambda k_B T) - 1} d\lambda \quad . \tag{2.66}$$

We perform a variable change by defining $\xi = hc/\lambda k_B T$. Then

$$\Phi_{BB}(T) = \frac{2\pi k_B{}^4}{c^2 h^3} T^4 \int_0^\infty \frac{\xi^3}{e^\xi - 1} d\xi = \frac{2\pi k_B{}^4}{c^2 h^3} T^4 \cdot \frac{\pi^4}{15} \quad ;$$

$$\boxed{\Phi_{BB}(T) = \left[\frac{2\pi^5 k_B{}^4}{15 c^2 h^3}\right] \cdot T^4 = \sigma_{SB} T^4} \quad . \tag{2.67}$$

The relation provided in Eq. (2.67) is called **Stefan-Boltzmann law**, and the constant $\sigma_{SB} \equiv \frac{2\pi^5 k_B{}^4}{15 c^2 h^3} = 5.67 \times 10^{-5}$ erg·cm^{-1} s^{-1}K^{-4} is the **Stefan-Boltzmann constant**. The law expresses the fact (empirically known before the advent of Planck's equation [2.29]) that the total black body radiation emission is proportional to the fourth power of the absolute temperature.

Real bodies could be considered "black" only approximately, and they are never ideally black. Thermal emission of a real body is always smaller than of an ideal black body, yet it is quite similar in many aspects. It may be often expressed as

$$F(\lambda, T) = \varepsilon(\lambda, T) F_{BB}(\lambda, T) \quad . \tag{2.68}$$

The $\varepsilon(\lambda, T)$ quantity is termed **emissivity**, and in principle depends on the wavelength and temperature. It is always smaller than a unity; it is a characteristic material parameter and, to a large extent, characteristic of the material surface conditions. Surface polishing of a solid body always reduces the emissivity, and surface roughening increases the emissivity. Experimental evidence shows that for most solid materials, the emissivity spectral and temperature dependences are quite moderate. For many practical needs it may be safely assumed to be constant over a broad range of wavelengths.

Exercises for Chapter 2

2.1 An electron resides in a one-dimensional potential well, zero in the $-\frac{1}{2}a < x < +\frac{1}{2}a$ range, and infinitely high outside. The known solution to Schrödinger's equation is

$$\psi_n(x) = \sqrt{\frac{2}{a}}\,\boldsymbol{sin}\left(\frac{n\pi}{a}x + n\frac{\pi}{2}\right); \quad \varepsilon_n = \frac{\pi^2\hbar^2}{2ma^2}n^2; \quad n = 1, 2, 3....$$

The system is afterwards perturbed by a potential of the form $V(x) = V_0x^2$ in the $-\frac{1}{2}a < x < +\frac{1}{2}a$ range. Calculate the corrections to the electron eigenenergies using the static perturbation theory. What limits should be imposed on the perturbation potential "spring constant" V_0, to be considered as "small"? Calculate the numerical limits for $a = 50$ Å in dyn/cm and in eV/cm^2 units.

2.2 Calculate the oscillator strengths between different states of the system described in Exercise 2.1. Plot an energy scheme for up to $n = 7$, and indicate allowed and forbidden transitions using solid and dashed arrows, respectively. Summarize the selection rules.

2.3 Effective losses in a laser resonance cavity that includes a $Nd : YAG$ rod are 0.02 cm^{-1}. Calculate the density of excited Nd^{3+} ions needed (population inversion) to overcome the losses. The excited Nd^{3+} ions fluoresce is centered at 1064 nm, with a line width of 0.45 nm and a radiative lifetime of 550 μs. The host YAG refractive index is 1.823.

2.4 The absorption constant spectrum of a glass, embedded with triply ionized ytterbium atoms (Yb^{3+}) may be resolved as a sum of three adjacent Gaussian peaks as follows:
(a) Peak position: 10,260 cm^{-1}, standard deviation width: 50 cm^{-1}, peak height: 9.2 cm^{-1};
(b) Peak position: 10,430 cm^{-1}, standard deviation width: 400 cm^{-1}, peak height: 1.55 cm^{-1};
(c) Peak position: 10,900 cm^{-1}, standard deviation width: 300 cm^{-1}, peak height: 1.9 cm^{-1}.
The glass refractive index is 1.45 and the excited states radiative lifetime is 2.0 ms. Calculate the ytterbium ions concentration in the glass and the oscillator strength of each transition band.

2.5 A melting furnace is formed as a container, whose side walls are kept at 1,500°C. The furnace has a small door of 400 cm^2 area used for inserting and pulling out of manufactured raw materials.
(a) Calculate the peak emission wavelength when the door opens.
(b) Calculate the radiation power losses when the door is open.

3 The Hydrogen Atom – Electrostatic Attraction Approximation

3.1 De Broglie Waves and Schrödinger's Equation

Employing reciprocity considerations, **de Broglie** conjectured in 1923 that because electromagnetic radiation exhibits some particle behavior, matter particles should exhibit some wave behavior. In other words, particles' propagation through space must abide by some wave propagation rules, and would exhibit all interference and diffraction effects known for waves at large, and electromagnetic radiation specifically. The conjecture, quite revolutionary at the time, was validated by experiment, and de Broglie was awarded the 1929 Nobel Prize in physics. The wave group velocity is the quantity that describes the propagation of a time- and space-limited disturbance in a wave-carrying medium. De Broglie correspondingly assumed that the group velocity of a wave assigned to a massive particle provides the particle's translational speed through space. In deducing the wave characteristics of matter, he used the fact that propagation of a light beam through space exhibits a mechanical-like behavior when considering an area segment of the wave front that is sufficiently small compared to the entire wave-front area, yet still large compared to the wavelength. The segment wave-front propagation is the **light ray**. Propagation analysis of a broad wave-front area via the distributed rays is termed **geometrical optics**. In that range of physics, it is possible to interpret every light ray as a kind of a mechanical arrow, subject to special laws of motion: the known **Snell** law of refraction, or its equivalent **Fermat** transition time minimization principle. The Fermat principle states that the ray path between any two A and B points in space is the one of the shortest time possible. In other words, propagation by any different path requires more time. Namely,

$$\delta \int_A^B \frac{ds}{\mathrm{v}} = \delta \int_A^B \frac{n}{c} ds = 0 \quad , \tag{3.1}$$

where s is the path length from the starting A point, v is the wave propagation velocity, and n is the medium refractive index. Figure 3.1 demonstrates the geometrical optics law applied to propagating electromagnetic radiation crossing between two media of different refractive indices.

De Broglie equated Snell's law with Maupertuis's minimal action principle that refers to mechanical systems dynamics

$$\delta_{\boldsymbol{H}} \int_A^B \boldsymbol{T} dt = 0 \quad , \tag{3.2}$$

where $\boldsymbol{T} = \frac{1}{2}\sum_j \frac{\partial \boldsymbol{L}}{\partial \dot{q}_j} \dot{q}_j$ is the system kinetic energy, $\boldsymbol{L}\left(q_j, \dot{q}_j, t\right)$ is the Lagrangian, q_j and $p_j \equiv \partial \mathbf{L}/\partial \dot{q}_j$ are the system generalized coordinate and momentum, respectively, in the $j-$ number dimension, and A and B are the generalized system coordinates at arbitrary start and end points, respectively. The added $\boldsymbol{H}$

Physics, Optics, and Spectroscopy of Materials, First Edition. Zeev Burshtein.

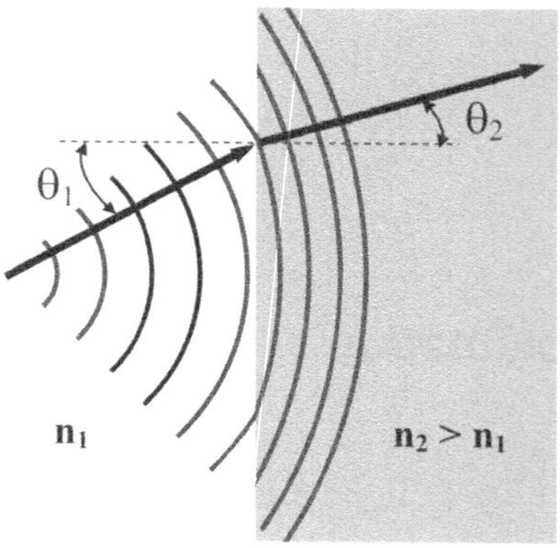

Figure 3.1 "Mechanical" model of light wave propagation including transition between two different media. Bold arrows display a representative ray belonging to the beam. By Snell's law, the ray refracts upon transition between the media such that $\frac{\sin\theta_1}{\sin\theta_2} = n \equiv \frac{n_2}{n_1}$.

subscript to the variation operator δ is aimed at emphasizing that the concerned (infinitely close) paths are Hamiltonian invariant. Maupertuis principle is equivalent to the mechanical system equations of motion. We change the form of Maupertuis' principle as follows:

$$\delta_{\boldsymbol{H}} \int_A^B \boldsymbol{T} dt = \delta_{\boldsymbol{H}} \int_A^B \boldsymbol{T} \frac{dt}{ds} ds = \delta_{\boldsymbol{H}} \int_A^B \frac{\boldsymbol{T}}{\mathrm{v}} ds = 0 \ , \tag{3.3}$$

where s is the generalized path length of the mechanical system in the q_j space, and v is the velocity of motion along that path. In the special case of a single m mass particle, Maupertuis' principle reduces to the form

$$\delta_{\boldsymbol{H}} \int_A^B \frac{1}{2} \frac{m\mathrm{v}^2}{\mathrm{v}} ds = \delta_{\boldsymbol{H}} \int_A^B \frac{1}{2} (m\mathrm{v}) ds = \delta_{\boldsymbol{H}} \int_A^B \frac{1}{2} p \, ds = 0 \ , \tag{3.4}$$

where $p \equiv m\mathrm{v}$ is the particle momentum. Comparing Eqs. (3.1) and (3.4) suggests that in identifying between the particle mechanical motion and its propagating assigned wave, simple proportionality relation should exist between its momentum and the inverse wave propagation velocity

$$p = C\left(\frac{1}{\mathrm{v}}\right) = \frac{2\pi \, C}{\lambda\omega} = \frac{2\pi\overline{C}}{\lambda} \ , \tag{3.5}$$

where λ is the central wavelength of the particle group of waves, and ω the central group angular frequency related by $\lambda\omega = 2\pi\mathrm{v}$. The quantities C and $\overline{C} \equiv C/\omega$ are proportionality constants between p and $1/\mathrm{v}$ or p and $1/\lambda$, respectively. *We assume* that the proportionality constant $\overline{C}$ is not explicitly position dependent. To analyze the impact of the latter assumption, we use the original assumption that the particle velocity is in fact the group velocity of its assigned wave; thus

$$\frac{1}{\mathrm{v}} = \frac{dk}{d\omega} = \frac{d(2\pi/\lambda)}{d\omega} = \frac{d\left(p/\overline{C}\right)}{d\omega} = \frac{1}{\overline{C}} \frac{dp}{d\omega} - \frac{p}{\overline{C}^2} \frac{d\overline{C}}{d\omega} = \frac{1}{\overline{C}} \frac{dp}{dE} \frac{dE}{d\omega} - \frac{p}{\overline{C}^2} \frac{d\overline{C}}{d\omega} \ , \tag{3.6}$$

where E is the particle energy and k is the wave vector. For a point particle, the energy is given by $E = \left(p^2/2m\right) + V(\underline{r})$, where $V(\underline{r})$ is the potential energy. Notably, $V(\underline{r})$ is explicitly and solely rather position dependent. Thus,

$$\frac{dE}{dp} = \frac{p}{m} \quad \Rightarrow \quad \frac{dp}{dE} = \frac{m}{p} = \frac{1}{\mathrm{v}} \ . \tag{3.7}$$

We set the latter result in the right side of Eq. (3.6), multiply the equation by $\overline{C}^2\mathrm{v}$, and get

$$\overline{C}^2 = \overline{C}\frac{dE}{d\omega} - p\mathrm{v}\frac{d\overline{C}}{d\omega} = \overline{C}\frac{dE}{d\omega} - 2\boldsymbol{T}\frac{d\overline{C}}{d\omega} \ . \tag{3.8}$$

By assumption, the left side is position independent, and that must therefore apply to each addend on the right side. However, the kinetic energy $\boldsymbol{T}$ factor in the second addend on the right side is position dependent; therefore the second addend on the right side must be identically zero. Namely,

$$\frac{d\overline{C}}{d\omega} \equiv 0 \quad \Rightarrow \quad \overline{C} = Const. \ , \tag{3.9}$$

Thus, $\overline{C}$ is a constant, independent of both angular frequency and position. Eq. (3.8) then reduces to the form

$$\frac{dE}{d\omega} = \overline{C} \quad \Rightarrow \quad E = \overline{C}\omega + Const1 \ . \tag{3.10}$$

Because energy is in principle determined up to an additive constant, the integration constant $Const1$ in Eq. (3.10) is chosen as zero, hence $E = \overline{C}\omega$. Recalling Planck's relation for electromagnetic radiation quantum $E = \hbar\omega$ (photons), we conclude that $\overline{C} \equiv \hbar$. Returning to Eq. (3.5), we conclude the validity of the two following relations:

$$\boxed{\begin{array}{l} p = \dfrac{2\pi\hbar}{\lambda} = \hbar k \\ E = \hbar\omega \end{array}} \ . \tag{3.11}$$

The first line in Eq. (3.11) is the new result based on de Broglie's conjecture, and as indicated, has been verified thoroughly by experiment. The two above relations establish the starting point for all further theoretical studies in quantum mechanics.

Schrödinger undertook a further step to specifically express de Broglie's particle wave equation. Recall that the electromagnetic wave equation is of the form

$$\Delta\boldsymbol{\Psi} = \frac{1}{\mathrm{v}^2}\frac{\partial^2\boldsymbol{\Psi}}{\partial t^2} \ , \tag{3.12}$$

where $\Delta \equiv \dfrac{\partial^2}{\partial x^2} + \dfrac{\partial^2}{\partial y^2} + \dfrac{\partial^2}{\partial z^2}$ is the Laplace operator, $\boldsymbol{\Psi} = \boldsymbol{\Psi}(\underline{r}, t)$ is the wave function, which is actually the electric field $\mathcal{E}(\underline{r}, t)$ (or alternatively the magnetic field $\mathcal{H}(\underline{r}, t)$), and v is the wave propagation velocity. Schrödinger assumed that a wave equation of the same form, with the addition of de Broglie relations, applies equally to a wave function describing the particle motion. Later on, Born proposed an interpretation to the particle wave function, namely that its squared absolute value $|\boldsymbol{\Psi}(\underline{r}, t)|^2$ provides the probability of the particle to be present at the specific position $\underline{r}$ at each moment t. It was a generalization of the meaning of the expression $\dfrac{1}{4\pi}|\mathcal{E}(\underline{r}, t)|^2$ that provides the volume density of electromagnetic energy at $\underline{r}$ position at each moment t.

Eq. (3.12) allows the separation between space and time variables in the form $\boldsymbol{\Psi}(\underline{r}, t) = \psi(\underline{r})\cdot e^{-i\omega t}$ such as to obtain eigensolutions that describe the vibrational modes of any system. Setting the time and space separation assumption in Eq. (3.12) yields

$$\Delta\psi(\underline{r}) + \frac{\omega^2}{\mathrm{v}^2}\psi(\underline{r}) = 0 \ . \tag{3.13}$$

We then use de Broglie's relations (Eq. [3.11]) for the following development:

$$\frac{\omega^2}{\mathrm{v}^2} = \frac{4\pi^2}{\lambda^2} \underset{\substack{de\ Broglie \\ relation}}{=} \frac{p^2}{\hbar^2} = \frac{2m\boldsymbol{T}}{\hbar^2} = \frac{2m(E-V)}{\hbar^2} \ .$$

We set the result in Eq. (3.13), multiply by $\hbar^2/2m$, and get

$$\boxed{-\frac{\hbar^2}{2m}\Delta\psi(\underline{r}) + V\psi(\underline{r}) = E\psi(\underline{r})} \ . \tag{3.14}$$

This equation is called **time independent Schrödinger equation**, or **steady-state Schrödinger equation.**

We consider now the time derivative of $\Psi(\underline{r}, t)$:

$$\frac{\partial \Psi(\underline{r}, t)}{\partial t} = -i\omega\cdot\Psi(\underline{r}, t) = \frac{-iE}{\hbar}\Psi(\underline{r}, t) \quad \Rightarrow \quad E\Psi(\underline{r}, t) = i\hbar\frac{\partial \Psi(\underline{r}, t)}{\partial t} \quad . \tag{3.15}$$

Setting the last result in Eq. (3.14) yields

$$\boxed{-\frac{\hbar^2}{2m}\Delta\Psi(\underline{r}, t) + V\Psi(\underline{r}, t) = i\hbar\frac{\partial \Psi(\underline{r}, t)}{\partial t}} \quad . \tag{3.16}$$

This equation is termed **time-dependent Schrödinger equation**.

As expected, setting the time and space separation in Eq. (3.16), and dividing by the common $e^{-i\omega t}$ factor , the time-independent part is again obtained (Eq. [3.14]).

3.2 Differential Operators and Physical Quantities

Observation of Schrödinger equations suggests the definition of various differential operators on the Ψ functions space. The Laplace operator may be written in the form $\Delta = \underline{\nabla}^2$, where the "Del" vector is $\underline{\nabla} \equiv \left(\frac{\partial}{\partial x}, \frac{\partial}{\partial y}, \frac{\partial}{\partial z}\right)$. One defines a "vector" operator $\mathbf{p} = -i\hbar\underline{\nabla}$. It is easily seen that the time-independent Schrödinger equation may be written in the form

$$\frac{\mathbf{p}^2}{2m}\psi(\underline{r}) + V\psi(\underline{r}) = E\psi(\underline{r}) \quad . \tag{3.17}$$

This form suggests an obvious correlation between the $\mathbf{p}^2$ operator and the squared particle momentum $p^2(\underline{r})$ (that is, it will always be true that $\mathbf{p}^2\psi(\underline{r}) = p^2(\underline{r})\cdot\psi(\underline{r})$) . In view of the probability interpretation of the particle wave function, the mean value of the squared momentum will be given by

$$\langle p^2 \rangle = \int_{all\ space} p^2(\underline{r})|\psi(\underline{r})|^2\cdot d^3\underline{r} = \int \psi(\underline{r})^* \mathbf{p}^2 \psi(\underline{r})\cdot d^3\underline{r} \quad ; \tag{3.18}$$

that, on condition that the wave function is properly normalized, that is, that its squared absolute represents a legitimate probability density; in other words, that it satisfies $\int_{all\ space} |\psi(\underline{r})|^2\cdot d^3\underline{r} = 1$. In the following, we assume that the dealt-with wave functions are all normalized to satisfy this condition.

In analogy to the correlation between the $\mathbf{p}^2$ operator and the squared linear momentum p^2 , it is further *conjectured* that there is a similar correlation between the $\mathbf{p}$ operator and the $\underline{p}$ linear momentum vector itself, such that its average value (termed **expectation value**) would be given by

$$\langle \underline{p} \rangle = \int_{all\ space} \psi(\underline{r})^* \mathbf{p}\psi(\underline{r})\cdot d^3\underline{r} \quad . \tag{3.19}$$

This is an example of the fundamental assumptions in quantum mechanics, whose predictions have been verified by countless experiments. As a generalization, it is assumed that Hermitian differential operators

Table 3.1 List of quantum operators

Physical quantity	Classical definition	Quantum operator
Position	$\underline{r}$	$\underline{r}$
Momentum	$\underline{p}$	$\mathbf{p} = -i\hbar\underline{\nabla}$
Angular momentum	$\underline{r} \times \underline{p}$	$\mathbf{L} \equiv \underline{r} \times \mathbf{p} = -i\hbar\underline{r} \times \underline{\nabla}$
Kinetic energy	$\frac{p^2}{2m}$	$-\frac{\hbar^2}{2m}\underline{\nabla}^2$
Total energy	$\frac{p^2}{2m} + V(\underline{r})$	$-\frac{\hbar^2}{2m}\underline{\nabla}^2 + V(\underline{r})$
Energy*	E	$i\hbar\frac{\partial}{\partial t}$

* Relating only to the time-dependent wave function $\Psi(\underline{r}, t)$.

$\mathbf{O}(\underline{r}, \mathbf{p})$, which correlate to classical mechanics quantities, describe the actual classical quantities by their average (expectation value)

$$\langle O \rangle \equiv \int_{all\ space} \psi(\underline{r})^* \mathbf{O}(\underline{r}, \mathbf{p}) \psi(\underline{r}) \cdot d^3\underline{r} \ . \tag{3.20}$$

Hermitian operator $\mathbf{O}(\underline{r}, \mathbf{p})$ is one that satisfies the equality $\int \Phi_1^* [\mathbf{O}\Phi_2] d^3\underline{r} = \int [\mathbf{O}\Phi_1]^* \Phi_2 d^3\underline{r}$ for all Φ functions belonging to the functions space spanned by the eigenfunctions of Schrödinger's equation (namely those solving Schrödinger's equation). The special trait of Hermitian operators in representing physical quantities stems from the fact that their eigenvalues are always real (eigenvalue "o" of an operator "$\mathbf{O}$" is one that solves the equation $\mathbf{O}\Phi = o\Phi$ for a certain Φ function). Table 3.1 lists such operators correlated to physical quantities. All are Hermitian. In a special case where a Φ function is also an eigenfunction of the operator, its expectation value coincides with its eigenvalue. We won't elaborate further on quantum operators' algebra, or in formal proof of the above statements. In our future discussions we will sometimes use conclusions of that algebra, and in special cases also include proofs.

3.3 Schrödinger Equation Solution for Hydrogen and Hydrogen-Like Atoms

The hydrogen atom consists of a single proton of positive charge and a single electron of negative charge, orbiting around the proton. All atoms and ions having a single orbiting electron are considered "hydrogen-like". They include protonium H, whose nucleus consists of a single proton; deuterium ^2D, whose nucleus consists of a single proton and a single neutron, and tritium ^3T, whose nucleus consists of a single proton and two neutrons. Also included are a singly ionized helium He^+, doubly ionized lithium Li^{2+}, triply ionized beryllium Be^{3+}, and so on. Generally, the hydrogen-like atoms consist of a nucleus of mass M and electric charge $+Ze$, where Z is the atomic number, and a single electron of mass m and electric charge $-e$.

As a starting point we consider the nucleus and electron as point mass and point charge particles. In addition, we assume that the electrostatic attraction continues to act during their mutual motions, with no

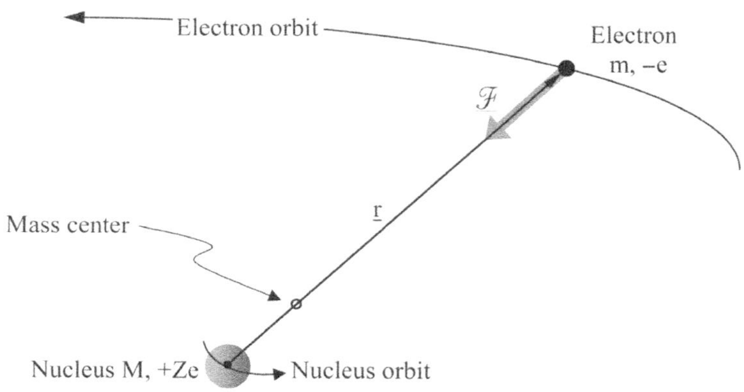

Figure 3.2 Illustration of the coordinated motions of the atomic nucleus and electron around their mass center. The nucleus orbit size and its distance from the mass center have been grossly exaggerated; the nucleus mass is very large compared to the electron mass. Therefore, its orbit is actually tiny compared to the electron orbit.

time delays, and entirely neglect gravitational forces and magnetic effects. The electrostatic attractive force $\underline{\mathscr{F}}$ is always central. Therefore, the equations of motion of these two bodies may be described as an effective motion of a single body, whereby the effective attractive force is always oriented toward their mass center. The effective mass is the reduced mass $\mu \equiv \frac{mM}{m+M}$, where m and M are the electron and nuclear masses, respectively. The potential energy is given by $V(\underline{r}) = -\frac{Ze^2}{r}$, where $\underline{r}$ is the distance vector from the nucleus to the electron. In the systems of our interest, the nuclear mass is very large compared to the electron mass. Therefore, the reduced mass is very close to the electron mass, and the electron-nucleus distance is very close to the electron distance from the mass center (see Figure 3.2).

A classical solution of the nuclear and electron motion is well established: the electron would form an elliptical orbit, of which the mass center is one of the two focuses. A similar motion would be performed by the nucleus, such that the two-system mass center remains still, and the nucleus, electron, and mass center always portray a common straight line. The solution provides no restrictions on the system binding energy. As to be shown, solution of Schrödinger's equation, on the other hand, poses limitations on the "allowed" binding energies, and accordingly on the allowed orbits. The "orbit" concept itself will be fundamentally revised.

The steady-state Schrödinger equation for hydrogen-like atoms is (Eq. [3.14])

$$-\frac{\hbar^2}{2\mu}\left(\frac{\partial^2\psi}{\partial x^2}+\frac{\partial^2\psi}{\partial y^2}+\frac{\partial^2\psi}{\partial z^2}\right)-\frac{Ze^2}{r}\psi = E\psi \quad . \tag{3.21}$$

The current system under study exhibits a full rotational symmetry. It means that the Hamiltonian functional form is invariant under any "stiff" axis rotation whose origin is preserved at the mass center. To allow a more convenient expression of that property, it is customary to use the spherical coordinate system (r, θ, φ) defined in Figure 3.3.

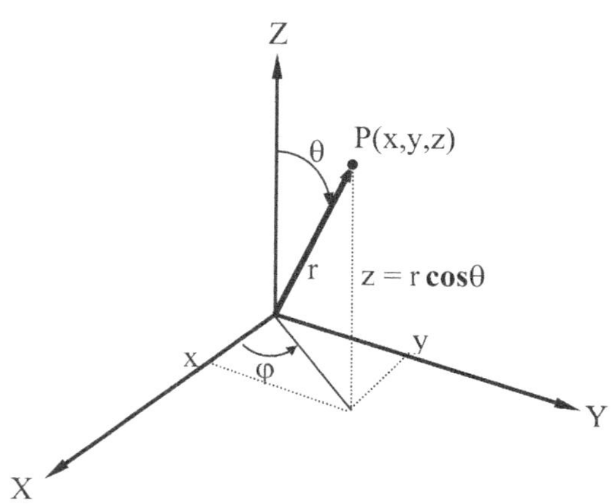

Figure 3.3 Cartesian and spherical coordinates: The Cartesian P point (x, y, z) coordinates are converted into (r, θ, φ) in the spherical coordinates.

Mathematically, the spherical coordinates are given by

$$r = \sqrt{x^2+y^2+z^2}\ ; \quad \theta = \boldsymbol{tan}^{-1}\left[\frac{\sqrt{x^2+y^2}}{z}\right]\ ; \text{ and}$$

$\varphi = \tan^{-1}(y/x)$. Conversion of Laplace operator into spherical variables is a long (and tiring) algebraic procedure. To demonstrate the conversion principle, we detail only its start, skip its continuation, and provide the result. We start with the partial derivative operator $\partial/\partial x$:

$$\frac{\partial}{\partial x} = \frac{\partial r}{\partial x} \cdot \frac{\partial}{\partial r} + \frac{\partial \theta}{\partial x} \cdot \frac{\partial}{\partial \theta} + \frac{\partial \varphi}{\partial x} \cdot \frac{\partial}{\partial \varphi} \quad .$$

Accordingly, we calculate

$$r = \left(x^2 + y^2 + z^2\right)^{1/2} \Rightarrow \frac{\partial r}{\partial x} = x\left(x^2 + y^2 + z^2\right)^{-1/2} = \frac{x}{r} = \frac{r \sin\theta \cos\varphi}{r} = \sin\theta\cos\varphi \quad ;$$

$$\theta = \tan^{-1}\left[\frac{\sqrt{x^2+y^2}}{z}\right] \Rightarrow \frac{\partial \theta}{\partial x} = \cos^2\theta \cdot \frac{x}{z\sqrt{x^2+y^2}} = \frac{1}{r}\cos\theta \sin\varphi \quad ;$$

$$\varphi = \tan^{-1}(y/x) \Rightarrow \frac{\partial \varphi}{\partial x} = \cos^2\varphi \cdot \frac{-y}{x^2} = -\frac{1}{r} \cdot \frac{\cos^2\theta \sin\varphi}{\cos^2\varphi \sin\theta} \quad ,$$

and get

$$\boxed{\frac{\partial}{\partial x} = \sin\theta\cos\varphi \cdot \frac{\partial}{\partial r} + \frac{1}{r}\cos\theta\sin\varphi \cdot \frac{\partial}{\partial \theta} - \frac{1}{r}\frac{\cos^2\theta \sin\varphi}{\cos^2\varphi\sin\theta} \cdot \frac{\partial}{\partial\varphi}} \quad . \tag{3.22}$$

Continuing using the same recipe, we calculate the $\partial^2/\partial x^2$ operator, and similarly the x and y second order partial derivatives. In summary, Schrödinger's equation 3.21 attains the form

$$\boxed{-\frac{\hbar^2}{2\mu}\left[\frac{1}{r^2}\frac{\partial}{\partial r}\left(r^2\frac{\partial\psi}{\partial r}\right) + \frac{1}{r^2 \sin\theta}\frac{\partial}{\partial\theta}\left(\sin\theta\frac{\partial\psi}{\partial\theta}\right) + \frac{1}{r^2\sin^2\theta}\frac{\partial^2\psi}{\partial\varphi^2}\right] - \frac{Ze^2}{r}\psi = E\psi} \quad . \tag{3.23}$$

All quantum operators exhibit polar presentations. We will present them for use when needed.

We turn now to solving Schrödinger's equation 3.23. In the solving procedure course we emphasize, present, and discuss in detail, those stages that lead to the conclusion that the solutions form a discrete set, related to a discrete set of "allowed" energies and "allowed" quantum numbers. In other stages, various results will be shown without detailed proof.

We assume the solutions to exhibit variables separation as follows:

$$\psi(r,\theta,\varphi) = \mathbf{R}(r)\mathbf{Y}(\theta,\varphi) \quad . \tag{3.24}$$

We set this assumption in Eq. (3.23) and get

$$-\frac{\hbar^2}{2\mu}\left[\frac{1}{r^2}\frac{d}{dr}\left(r^2\frac{d\mathbf{R}}{dr}\right)\mathbf{Y} + \frac{1}{r^2\sin\theta}\frac{\partial}{\partial\theta}\left(\sin\theta\frac{\partial\mathbf{Y}}{\partial\theta}\right)\mathbf{R} + \frac{1}{r^2\sin^2\theta}\frac{\partial^2\mathbf{Y}}{\partial\varphi^2}\mathbf{R}\right] - \frac{Ze^2}{r}\mathbf{RY} = E\mathbf{RY} \quad ;$$

we divide by $\mathbf{RY}$, multiply by r^2 , and get

$$-\frac{\hbar^2}{2\mu}\left[\frac{1}{\mathbf{R}}\frac{d}{dr}\left(r^2\frac{d\mathbf{R}}{dr}\right)\right] - r^2\left(\frac{Ze^2}{r} + E\right)$$
$$-\frac{\hbar^2}{2\mu}\left[\frac{1}{\mathbf{Y}\sin\theta}\frac{\partial}{\partial\theta}\left(\sin\theta\frac{\partial\mathbf{Y}}{\partial\theta}\right) + \frac{1}{\mathbf{Y}\sin^2\theta}\frac{\partial^2\mathbf{Y}}{\partial\varphi^2}\right] = 0 \quad .$$

The first line in the last equation's left side depends only on the r variable. The second line depends only on θ and φ. Because they are functionally independent, they must be each independently constant, with their sum zero. Thus, we get two equations. The first one is

$$-\frac{\hbar^2}{2\mu}\left[\frac{1}{r^2}\frac{d}{dr}\left(r^2\frac{d\boldsymbol{R}}{dr}\right)\right]-\left(\frac{Ze^2}{r}+\frac{\Gamma_1}{r^2}\right)\boldsymbol{R}=E\boldsymbol{R}\quad, \tag{3.24-i}$$

with Γ_1 a constant.

Regarding the second part, we perform further variables separation of the form $\boldsymbol{Y}(\theta,\varphi)=\boldsymbol{\Theta}(\theta)\boldsymbol{\Phi}(\varphi)$. We get $-\frac{\hbar^2}{2\mu}\left[\frac{1}{\boldsymbol{\Theta}\,\boldsymbol{\sin}\,\theta}\frac{d}{d\theta}\left(\boldsymbol{\sin}\,\theta\frac{d\boldsymbol{\Theta}}{d\theta}\right)+\frac{1}{\boldsymbol{\Phi}\,\boldsymbol{\sin}^2\theta}\frac{d^2\boldsymbol{\Phi}}{d\varphi^2}\right]=\Gamma_2$, where Γ_2 is a constant satisfying $\Gamma_1+\Gamma_2=0$. We multiply both sides by $\boldsymbol{\sin}^2\theta$ and get

$$\left[\frac{\boldsymbol{\sin}\,\theta}{\boldsymbol{\Theta}}\frac{d}{d\theta}\left(\boldsymbol{\sin}\,\theta\frac{d\boldsymbol{\Theta}}{d\theta}\right)+\frac{2\mu\Gamma_2}{\hbar^2}\boldsymbol{\sin}^2\theta\right]+\left[\frac{1}{\boldsymbol{\Phi}}\frac{d^2\boldsymbol{\Phi}}{d\varphi^2}\right]=0\quad.$$

Here as well, we obtain a sum of two independent addends: the first is only a function of θ, the second only a function of φ. Each addend thus must be a constant, and their sum zero. Then we get the following equations:

$$\frac{\boldsymbol{\sin}\,\theta}{\boldsymbol{\Theta}}\frac{d}{d\theta}\left(\boldsymbol{\sin}\,\theta\frac{d\boldsymbol{\Theta}}{d\theta}\right)+\frac{2\mu\Gamma_2}{\hbar^2}\boldsymbol{\sin}^2\theta=\Gamma_3\quad; \tag{3.24-ii}$$

$$\left[\frac{1}{\boldsymbol{\Phi}}\frac{d^2\boldsymbol{\Phi}}{d\varphi^2}\right]=\Gamma_4\quad; \tag{3.24-iii}$$

where $\Gamma_3+\Gamma_4=0$.

We start by solving equation (3.24-iii). A non-divergent solution obtains for $\Gamma_4<0$. We define $\Gamma_4=-m^2$ for a real m. The solution is $\boldsymbol{\Phi}=e^{\pm im\varphi}$. We demand the solution to be single-valued, namely that the function must exhibit a 2π period; we get that m must be an integer; we choose the positive integer as representative. On adding a normalization condition, requiring that $\int_0^{2\pi}|\boldsymbol{\Phi}|^2d\varphi=1$, we get

$$\boldsymbol{\Phi}_{\pm m}(\varphi)=\frac{1}{\sqrt{2\pi}}e^{\pm im\varphi}\quad. \tag{3.25}$$

We turn to solving equation (3.24-ii). We set $\Gamma_3=-\Gamma_4=m^2$, and change to a new variable $\xi=\boldsymbol{\cos}\theta$. Because $\frac{d\boldsymbol{\Theta}}{d\theta}=\frac{d\boldsymbol{\Theta}}{d\xi}(-\boldsymbol{\sin}\,\theta)$, we get for the first addend $\frac{d}{d\theta}\left(\boldsymbol{\sin}^2\theta\frac{d\boldsymbol{\Theta}}{d\xi}\right)=\frac{d}{d\xi}\left\{\left(\xi^2-1\right)\frac{d\boldsymbol{\Theta}}{d\xi}\right\}(-\boldsymbol{\sin}\,\theta)$. We further define $\lambda\equiv\frac{2\mu\Gamma_2}{\hbar^2}$ and get

$$\left(1-\xi^2\right)\frac{d^2\boldsymbol{\Theta}}{d\xi^2}-2\xi\frac{d\boldsymbol{\Theta}}{d\xi}+\left(\lambda-\frac{m^2}{1-\xi^2}\right)\boldsymbol{\Theta}=0\quad. \tag{3.24-ii-a}$$

We attempt a solution of the form $\boldsymbol{\Theta}=(1-\xi^2)^{m/2}\mathcal{V}$, set in equation (3.24-ii-a), and get the following differential equation for $\mathcal{V}$:

$$\left(1-\xi^2\right)\mathcal{V}''-2\xi(m+1)\mathcal{V}'+[\lambda-m(m+1)]\mathcal{V}=0\quad. \tag{3.24-ii-b}$$

For brevity, in the latter equation we used the superscript prime signs to indicate a derivative. To solve the equation, we represent $\mathcal{V}$ as a power series:

$$\mathcal{V} = \sum_{\nu=0}^{\infty} a_\nu \xi^\nu \Rightarrow \mathcal{V}' = \sum_{\nu=0}^{\infty} \nu a_\nu \xi^{\nu-1} \Rightarrow \mathcal{V}'' = \sum_{\nu=0}^{\infty} \nu(\nu-1) a_\nu \xi^{\nu-2} .$$

We set the above result in Equation (24-ii-b) and obtain

$$\sum \nu(\nu-1)a_\nu\xi^{\nu-2} - \sum \nu(\nu-1)a_\nu\xi^\nu - 2(m+1)\sum \nu\ a_\nu\xi^\nu + [\lambda - m(m+1)]\sum a_\nu\xi^\nu = 0 .$$

We now set the relation $\sum_{\nu=0}^{\infty} \nu(\nu-1)a_\nu\xi^{\nu-2} = \sum_{\nu=0}^{\infty} (\nu+2)(\nu+1)a_{\nu+2}\xi^\nu$ in the above, and extract and divide by the ξ^ν, the common factor to all summed addends. The reduced summation would be identically zero if and only if each of its addends, independently, is identically zero. Thus, we get

$$a_\nu[\lambda - m(m+1) - 2\nu(m+1) - \nu(\nu-1)] + (\nu+2)(\nu+1)a_{\nu+2} = 0 ,$$

or

$$\frac{a_{\nu+2}}{a_\nu} = \frac{1 + \dfrac{2m+1}{\nu} + \dfrac{m(m+1)-\lambda}{\nu^2}}{1 + \dfrac{3}{\nu} + \dfrac{2}{\nu^2}} . \tag{3.26}$$

We are interested in a non-diverging solution for $\mathcal{V}$. Under some conditions, the a_ν coefficients diverge: if $m > 1$ and ν large enough, the third addends in the numerator and denominator (having ν^2 in their denominator) exhibit very little impact on the $a_{\nu+2}/a_\nu$ ratio. In comparing the addends of the second numerator and denominator, when $2m + 1 > 3$ the summed numerator is larger than the denominator, their ratio is then larger than a unity. The power series describing $\mathcal{V}$ would then diverge for any m at $\xi = \pm 1$. It is possible to prevent the power series divergence if all a_ν coefficients vanish for ν values exceeding a certain ν_0. Obviously, If $a_{\nu_0} \neq 0$ and $a_{\nu_0+1} = 0$, then also $a_{\nu_0+2}/a_{\nu_0} = 0$, and so on. In accordance, the λ quantity must satisfy $\lambda = (m+\nu_0)(m+\nu_0+1) \equiv \ell(\ell+1)$. Because $\nu_0 \geq 0$, then $\ell \geq m$. The solution of Eq. (3.24-ii-b) would then be $\boldsymbol{\Theta}_\ell^m(\xi) = \left(1-\xi^2\right)^{m/2}\mathcal{V}_\ell^m(\xi)$, where $\mathcal{V}_\ell^m(\xi)$ is a polynomial of order $\nu_0 = \ell - m$. The thus-defined $\boldsymbol{P}_\ell^m(\xi) \equiv \left(1-\xi^2\right)^{m/2}\mathcal{V}_\ell^m(\xi)$ functions are called **associated Legendre polynomials**.

We now devote some space to analyze the properties of Legendre associated polynomials. We start by setting $m = 0$, and get $\boldsymbol{P}_\ell^0(\xi) = \mathcal{V}_\ell^0(\xi)$. Then we differentiate equation (3.24-ii-b) and get

$$\begin{gathered}\left(1-\xi^2\right)\mathcal{V}''' - 2\xi\mathcal{V}'' - 2(m+1)\mathcal{V}' - 2\xi(m+1)\mathcal{V}'' + [\lambda - m(m+1)]\mathcal{V}' = 0 ; \\ \left(1-\xi^2\right)\mathcal{V}''' - 2\xi(m+2)\mathcal{V}'' + [\lambda - (m+1)(m+2)]\mathcal{V}' = 0 .\end{gathered} \tag{3.24-ii-c}$$

We got a second-order differential equation for $\mathcal{V}'$, which is of identical form as equation (3.24-ii-b) for $\mathcal{V}$, except for replacing m by $m + 1$. This means that up to a multiplying constant factor, $\left(\mathcal{V}_\ell^m\right)' = \mathcal{V}_\ell^{m+1}$. Introducing mathematical induction considerations, we get

$$\boldsymbol{P}_\ell^m(\xi) = \left(1-\xi^2\right)^{m/2}\frac{d^m}{d\xi^m}\boldsymbol{P}_\ell^0(\xi) . \tag{3.27}$$

We further provide without proof a solution to equation (3.24-ii-b) for $m = 0$:

$$\boldsymbol{P}_\ell^0(\xi) = \frac{1}{2^\ell \ell!}\frac{d^\ell}{d\xi^\ell}\left(\xi^2 - 1\right)^\ell ,$$

known as **Rodrigues equation**. We then get

$$\boxed{\boldsymbol{P}_\ell^m(\xi) \equiv \frac{1}{2^\ell \ell!}\left(1-\xi^2\right)^{m/2}\frac{d^{\ell+m}}{d\xi^{\ell+m}}\left(\xi^2-1\right)^\ell} . \tag{3.28}$$

It is possible to generalize the definition of Legendre associated polynomials to include negative m values (**Hobson-Bateman**'s generalization), by adopting the formal structure of Eq. (3.28). The generalized function is a valid solution to equation (3.24-ii-a) for any integer m which satisfies $|m| \le \ell$, namely for $m = -\ell, -\ell+1, -\ell+2, ...0, +1, +2, ..., +\ell$. In using that generalized definition, we get the relation $P_\ell^{-m}(\xi) = (-1)^m \frac{(\ell-m)!}{(\ell+m)!} P_\ell^m(\xi)$; that is, on reversing the sign of m , the $\boldsymbol{P}$-function reverses sign for an odd m, but preserves its sign for an even m. Such properties of reversing or preserving signs, are termed "**negative parity**" or "**positive parity**", respectively.

The $\boldsymbol{P}_\ell^m(\xi)$ functions are subject to several recurrence relations. Three of them, aimed at being used later, are quoted below without proof:

$$\xi \cdot \boldsymbol{P}_\ell^m(\xi) = \frac{\ell+1-m}{2\ell+1}\boldsymbol{P}_{\ell+1}^m(\xi) + \frac{\ell+m}{2\ell+1}\boldsymbol{P}_{\ell-1}^m(\xi) ; \tag{3.29-i}$$

$$\left.\begin{aligned}\left(1-\xi^2\right)^{1/2} \cdot \boldsymbol{P}_\ell^m(\xi) &= \frac{1}{2\ell+1}\boldsymbol{P}_{\ell+1}^{m+1}(\xi) - \frac{1}{2\ell+1}\boldsymbol{P}_{\ell-1}^{m+1}(\xi) \\ &= \frac{(\ell+m)(\ell+m-1)}{2\ell+1}\boldsymbol{P}_{\ell-1}^{m-1}(\xi) - \frac{(\ell-m+1)(\ell-m+2)}{2\ell+1}\boldsymbol{P}_{\ell+1}^{m-1}(\xi)\end{aligned}\right\} ; \tag{3.29-ii}$$

$$\left.\begin{aligned}\left(1-\xi^2\right)\frac{d}{d\xi}\boldsymbol{P}_\ell^m(\xi) &= (\ell+1)\xi\boldsymbol{P}_\ell^m(\xi) - (\ell-m+1)\boldsymbol{P}_{\ell+1}^m(\xi) \\ &= (\ell+m)\boldsymbol{P}_{\ell-1}^m(\xi) - \ell\xi\boldsymbol{P}_\ell^m(\xi)\end{aligned}\right\} . \tag{3.29-iii}$$

In addition, the $\boldsymbol{P}_\ell^m(\xi)$ functions exhibit the following normalization relations:

$$\int_{-1}^{+1} \boldsymbol{P}_{\ell'}^m(\xi)\boldsymbol{P}_\ell^m(\xi)d\xi = \frac{2}{2\ell+1}\frac{(\ell+m)!}{(\ell-m)!}\delta_{\ell'\ell} \equiv N_{\ell m}\delta_{\ell'\ell} ; \tag{3.30-i}$$

$$\int_{-1}^{+1} \frac{\boldsymbol{P}_\ell^{m'}(\xi)\boldsymbol{P}_\ell^m(\xi)}{1-\xi^2}d\xi = \frac{1}{m}\frac{(\ell+m)!}{(\ell-m)!}\delta_{m'm} , \tag{3.30-ii}$$

where in Eq. (3.30-i), $N_{\ell m} \equiv \frac{2}{2\ell+1} \cdot \frac{(\ell+m)!}{(\ell-m)!}$.

Proper normalization properties are also required of the $\boldsymbol{\Theta}_\ell^m(\xi)$ functions. Pursuant to Eq. (3.30-i), the following normalization is selected to indicate their sign selection property per the m-parity:

$$\boxed{\boldsymbol{\Theta}_\ell^m(\theta) = (-1)^m\left\{\frac{2\ell+1}{2}\frac{(\ell-m)!}{(\ell+m)!}\right\}^{1/2}\boldsymbol{P}_\ell^m(\cos\theta)} . \tag{3.31}$$

With that normalization, the $\boldsymbol{\Theta}_\ell^m(\xi)$ functions exhibit the property:

$$\boxed{\int_0^\pi \left|\boldsymbol{\Theta}_\ell^m(\theta)\right|^2 \boldsymbol{\sin}\,\theta d\theta = 1} \quad . \tag{3.32}$$

Notably, all $\boldsymbol{\Theta}_\ell^m(\xi)$ functions are independent; however, they are not fully orthonormal.

The previous generalization of $\boldsymbol{P}_\ell^m(\xi)$ functions to negative m values allows rewriting the different solutions of the $\boldsymbol{\Phi}$ functions. As a straightforward continuation to Eq. (3.25), we rewrite the $\boldsymbol{\Phi}$ solutions simply as

$$\boxed{\boldsymbol{\Phi}_m(\varphi) = \frac{1}{\sqrt{2\pi}} e^{im\varphi}} \quad , \tag{3.33}$$

where $m = -\ell, -\ell+1, -\ell+2, \ldots 0, +1, +2, \ldots, +\ell$; a total of $2\ell+1$ functions. They are all orthonormal, namely satisfying

$$\boxed{\int_0^{2\pi} \boldsymbol{\Phi}_{m'}^* \boldsymbol{\Phi}_m \, d\varphi = \delta_{mm'}} \quad . \tag{3.34}$$

Solutions of the $\boldsymbol{Y}$ function defined in Eq. (3.24) are $\boldsymbol{Y}_{\ell m}(\theta,\varphi) = \boldsymbol{\Theta}_\ell^m(\theta)\boldsymbol{\Phi}_m(\varphi)$. They are called **spherical harmonics** and are fully orthonormal, satisfying

$$\int_0^\pi \int_0^{2\pi} \boldsymbol{Y}_{\ell' m'}^*(\theta,\varphi)\boldsymbol{Y}_{\ell m}(\theta,\varphi)\,\boldsymbol{\sin}\,\theta \cdot d\theta \cdot d\varphi = \delta_{\ell'\ell}\delta_{m'm} \quad . \tag{3.35}$$

Table 3.2 provides a list of specifically expressed $\boldsymbol{Y}_{\ell m}(\theta,\varphi)$ spherical harmonics for $\ell = 0, 1, 2, 3$, and the different m values belonging to each ℓ. Extension up to $\ell = 6$ is provided in Table 12.1.

We turn now to finding the $\boldsymbol{R}(r)$ function, namely to solving Eq. (3.24-i). In view of the specific expressions obtained for Eqs. (3.24-ii) and (3.24-iii) solutions, the constant in Eq. (3.24-i) should be expressed as $\Gamma_1 = -\frac{\hbar^2}{2\mu}\ell(\ell+1)$. Thus, in relation to Eq. (3.24-i), we get

$$\frac{d^2\boldsymbol{R}}{dr^2} + \frac{2}{r}\frac{d\boldsymbol{R}}{dr} + \left(\frac{2\mu Ze^2}{\hbar^2 r} - \frac{\ell(\ell+1)}{r^2} + \frac{2\mu E}{\hbar^2}\right)\boldsymbol{R} = 0 \quad . \tag{3.24-i-a}$$

In the following we limit ourselves to consider cases of negative energy $E < 0$, namely only bound electron states. We perform changes of constants names, and redefine the equation's variable: We define $\beta^2 \equiv -\frac{2\mu E}{\hbar^2}$; $\gamma \equiv \frac{\mu Ze^2}{\hbar^2\beta}$, and $\rho \equiv 2\beta r$. The differential equation for $\boldsymbol{R}$ then exhibits the form

$$\frac{d^2\boldsymbol{R}}{d\rho^2} + \frac{2}{\rho}\frac{d\boldsymbol{R}}{d\rho} + \left(\frac{\gamma}{\rho} - \frac{\ell(\ell+1)}{\rho^2} - \frac{1}{4}\right)\boldsymbol{R} = 0 \quad . \tag{3.24-i-b}$$

Table 3.2 $Y_{\ell m}(\theta, \varphi)$ functions for ℓ = 0, 1, 2, 3 and their possible m values per each ℓ. In spectroscopy, the ℓ = 0, 1, 2, 3, 4, 5, 6, 7 states are correspondingly symbolized by the letters s, p, d, f, g, h, i, j. In the table they are accordingly indicated in parentheses.

ℓ	m	$Y_{\ell m}(\theta, \varphi)$
0 (s)	0	$Y_{00} = 1/\sqrt{4\pi}$
1 (p)	0	$Y_{10} = \sqrt{3/4\pi}\ \cos\theta$
	±1	$Y_{1\pm1} = \mp\sqrt{3/8\pi}\ \sin\theta\cdot e^{\pm i\varphi}$
2 (d)	0	$Y_{20} = \frac{1}{2}\sqrt{5/4\pi}\left(3\cos^2\theta - 1\right)$
	±1	$Y_{2\pm1} = \mp\sqrt{15/8\pi}\ \sin\theta\ \cos\theta\cdot e^{\pm i\varphi}$
	±2	$Y_{2\pm2} = \frac{1}{4}\sqrt{15/2\pi}\ \sin^2\theta\cdot e^{\pm i2\varphi}$
3 (f)	0	$Y_{30} = \frac{1}{2}\sqrt{7/4\pi}\left(5\cos^3\theta - 3\cos\theta\right)$
	±1	$Y_{3\pm1} = \mp\frac{1}{4}\sqrt{21/4\pi}\ \sin\theta\cdot\left(5\cos^2\theta - 1\right)\cdot e^{\pm i\varphi}$
	±2	$Y_{3\pm2} = \frac{1}{4}\sqrt{105/2\pi}\cdot\cos\theta\ \sin^2\theta\cdot e^{\pm i2\varphi}$
	±3	$Y_{3\pm3} = \mp\frac{1}{4}\sqrt{35/4\pi}\cdot\sin^3\theta\cdot e^{\pm i3\varphi}$

We attempt a solution of the form $\boldsymbol{R}(\rho) = e^{-\rho/2}\cdot\rho^{\ell}\,\boldsymbol{G}(\rho)$. Under certain conditions about the $\boldsymbol{G}$ function, the attempted form assures that $\boldsymbol{R}$ diminishes at both $\rho = 0$ and $\rho \to \infty$ ends. Setting this form in Eq. (3.24-i-b), the following differential equation is obtained for the $\boldsymbol{G}(\rho)$ function:

$$G'' + \left[\frac{2(\ell+1)}{\rho} - 1\right] G' + \left[\frac{\gamma - (\ell+1)}{\rho}\right] G = 0 \ . \tag{3.24-i-c}$$

For solution we present $\boldsymbol{G}(\rho)$ as a power series; namely

$$G = \sum_{\nu=0}^{\infty} c_\nu \rho^\nu \quad \Rightarrow \quad G' = \sum_{\nu=0}^{\infty} \nu c_\nu \rho^{\nu-1} \quad \Rightarrow \quad G'' = \sum_{\nu=0}^{\infty} \nu(\nu-1) c_\nu \rho^{\nu-2} \ .$$

Setting in (3.24-i-c) above, we get

$$\sum \nu(\nu-1)c_\nu\rho^{\nu-2} + 2(\ell+1)\sum \nu c_\nu \rho^{\nu-2} - \sum \nu c_\nu \rho^{\nu-1} + [\gamma - (\ell+1)]\sum c_\nu \rho^{\nu-1} = 0 \ .$$

We set

$$\sum_{\nu=0}^{\infty} \nu(\nu-1)c_\nu\rho^{\nu-2} = \sum_{\nu=0}^{\infty}(\nu+1)\nu c_{\nu+1}\rho^{\nu-1} \quad \text{and} \quad \sum_{\nu=0}^{\infty} \nu c_\nu \rho^{\nu-2} = \sum_{\nu=0}^{\infty}(\nu+1)\,c_{\nu+1}\rho^{\nu-1}$$

in the above equation, then extract $\rho^{\nu-1}$ as a common factor for all addends. The summed remaining power series, which multiplies $\rho^{\nu-1}$, vanishes if and only if each of its addends independently vanishes. Thus, we get

$$\frac{c_{\nu+1}}{c_\nu} = \frac{(\ell+1) + \nu - \gamma}{(\nu+1)[2(\ell+1) + \nu]} \ . \tag{3.36}$$

Because any solution to the (3.24-i-c) differential equation is unique up to a constant factor, we choose $c_0 = 1$. Then we get

$$c_\nu = \frac{1}{\nu!}\prod_{\nu=0}^{\infty}\frac{(\ell+1)+\nu-\gamma}{2(\ell+1)+\nu} \quad . \tag{3.37}$$

Generally, the power series describing $\boldsymbol{G}$ is infinite. At large ν values, the c_ν constants behave like $c_\nu \sim 1/\nu!$, therefore $\boldsymbol{G} \sim e^\rho$, and as a result the $\boldsymbol{R}$ function will diverge at infinite ρ . On the other hand, if the power series is finite, the $\boldsymbol{G}$ function at infinite ρ behaves according to its highest power, and the exponential factor will dominate the general $\boldsymbol{R}$ behavior there. The meaning of a "finite power series" is that the expansion coefficients c_ν are zero for ν values exceeding a certain ν_0 . Then $\gamma = (\ell + 1) + \nu_0 \equiv n$. That means that n is an integer, satisfying $n \geq (\ell + 1)$. In other words, for a given n , the possible ℓ values are $\ell = 0, 1, 2, 3, \cdots, n-1$. Returning to the definition of the γ constant, we get

$$\boxed{E_n = -\frac{Z^2e^4\mu}{2\hbar^2}\frac{1}{n^2} = -\frac{Z^2e^2}{2a_0}\frac{1}{n^2}} \quad , \tag{3.38}$$

where $a_0 = \hbar^2/\mu e^2$ is called the **Bohr radius**. Later we will see that there is a strong linkage between the Bohr radius and the average electron distance from the nucleus. Interestingly, the work needed to collect an electric $-Ze$ charge at a uniform electrostatic distribution over a sphere of a_0 radius is given by $-E_1$. Thus, the ground-state electron binding energy (the $n = 1$ case) is offsetting that work. The Bohr radius for the $\mu = m$ case and the electron binding energy for the $\mu = m$ and $Z = 1$ cases are universal constants, whose values are $a_0 = \hbar^2/me^2 = 0.52917 \times 10^{-8}$ cm , and $E_1 = -me^4/2\hbar^2 = -2.1797 \times 10^{-11}$ erg.

The different solutions to the $\boldsymbol{G}$ function are marked $\boldsymbol{G}_{n\ell}$. We now claim without proof that the $\boldsymbol{G}_{n\ell}(\rho)$ functions are proportional to the $\boldsymbol{L}_{n-\ell-1}^{2\ell+1}(\rho)$ **Laguerre associated polynomials**. The general definition of the Laguerre associated polynomials for a ζ variable and integral s and t is given by

$$\boldsymbol{L}_s^t(\zeta) \equiv (-1)^t\frac{d^t}{d\zeta^t}\left[e^\zeta\frac{d^{s+t}}{d\zeta^{s+t}}\left(\zeta^{s+t}e^{-\zeta}\right)\right] \quad . \tag{3.39}$$

We quote without proof two recurrence and orthogonality relations of the $\boldsymbol{L}_s^t(\zeta)$ functions, which will be later used for demonstration purposes

$$\zeta\boldsymbol{L}_s^t(\zeta) = (2s+t+1)\boldsymbol{L}_s^t(\zeta) - \frac{s+1}{s+t+1}\boldsymbol{L}_{s+1}^t(\zeta) - (s+t)^2\boldsymbol{L}_{s-1}^t(\zeta) \quad ; \tag{3.40-a}$$

$$\int_0^\infty \zeta^t e^{-\zeta}\boldsymbol{L}_s^t(\zeta)\boldsymbol{L}_{s'}^t(\zeta)d\zeta = \delta_{ss'}\frac{[(s+t)!]^3}{s!} \quad . \tag{3.40-b}$$

Using Eq. (3.39) we get

$$\boldsymbol{L}_{n-\ell-1}^{2\ell+1}(\rho) = -\frac{d^{2\ell+1}}{d\rho^{2\ell+1}}\left\{e^\rho\frac{d^{n+\ell}}{d\rho^{n+\ell}}\left[\rho^{n+\ell}e^{-\rho}\right]\right\} \quad , \tag{3.41}$$

And the (3.24-i) differential equation solution would thus be

$$\boxed{\boldsymbol{R}_{n\ell}(\rho) = -\left[\left(\frac{2Z}{na_0}\right)^3\frac{(n-\ell-1)!}{2n[(n+\ell)!]^3}\right]^{1/2}e^{-\rho/2}\rho^\ell\frac{d^{2\ell+1}}{d\rho^{2\ell+1}}\left\{e^\rho\frac{d^{n+\ell}}{d\rho^{n+\ell}}\left[\rho^{n+\ell}e^{-\rho}\right]\right\}} \quad , \tag{3.42}$$

where $\rho = \dfrac{2Zr}{na_0}$ and $\ell = 0, 1, 2, \cdots, n-1$. The $\boldsymbol{R}_{n\ell}(r)$ functions are independent, but not orthogonal. However, they are normalized to satisfy the relations (see Eq. [3.40-b])

$$\boxed{\int_0^\infty \boldsymbol{R}_{n\ell}^*(r)\boldsymbol{R}_{n\ell}(r)\cdot r^2 dr = 1} \quad , \tag{3.43}$$

Under the above normalization, the $\boldsymbol{R}_{n\ell}(r)$ functions satisfy the following orthogonality relations

$$\int_0^\infty \boldsymbol{R}_{n'\ell'}^*(r)\boldsymbol{R}_{n\ell}(r)\cdot r^2 dr = \left(1 + \delta_{n'n} - \delta_{\ell'\ell}\right)B_{n'n\ell'\ell} \quad , \tag{3.44}$$

where $B_{n'n\ell'\ell}$ are real constants satisfying $0 < \left|B_{n'n\ell'\ell}\right| \le 1$. For brevity we don't present their specific values, except for noting that $B_{nn\ell\ell} = 1$. We present Eq. (3.44) without proof. The meaning of this relation is that the integral on the left side of Eq. (3.44) is identically zero if and only if the two following condition are satisfied: the equality $\ell' = \ell$ and the inequality $n' \neq n$. In Table 3.3 we present the specific expression of $\boldsymbol{R}_{n\ell}(r)$ functions for $n = 1, 2, 3, 4$, and the ℓ values permitted per each n.

Table 3.3 $R_{n\ell}(\rho)$ functions for n = 1, 2, 3, 4, and ℓ values permitted per each n.

n	ℓ	$\boldsymbol{R}_{n\ell}(\rho)$; $(\rho = 2Zr/na_0)$
1	0	$\boldsymbol{R}_{10}(\rho) = 2\left(\frac{Z}{a_0}\right)^{3/2} e^{-\rho/2}$
2	0	$\boldsymbol{R}_{20}(\rho) = \frac{1}{2\sqrt{2}}\left(\frac{Z}{a_0}\right)^{3/2}(2-\rho)e^{-\rho/2}$
	1	$\boldsymbol{R}_{21}(\rho) = \frac{1}{2\sqrt{6}}\left(\frac{Z}{a_0}\right)^{3/2}\rho e^{-\rho/2}$
3	0	$\boldsymbol{R}_{30}(\rho) = \frac{1}{9\sqrt{3}}\left(\frac{Z}{a_0}\right)^{3/2}\left(6-6\rho+\rho^2\right)e^{-\rho/2}$
	1	$\boldsymbol{R}_{31}(\rho) = \frac{1}{9\sqrt{6}}\left(\frac{Z}{a_0}\right)^{3/2}(4-\rho)\rho e^{-\rho/2}$
	2	$\boldsymbol{R}_{32}(\rho) = \frac{1}{9\sqrt{30}}\left(\frac{Z}{a_0}\right)^{3/2}\rho^2 e^{-\rho/2}$
4	0	$\boldsymbol{R}_{40}(\rho) = \frac{1}{96}\left(\frac{Z}{a_0}\right)^{3/2}\left(-24+36\rho-12\rho^2+\rho^3\right)e^{-\rho/2}$
	1	$\boldsymbol{R}_{41}(\rho) = \frac{1}{32\sqrt{15}}\left(\frac{Z}{a_0}\right)^{3/2}\left(-20+10\rho-\rho^2\right)\rho e^{-\rho/2}$
	2	$\boldsymbol{R}_{42}(\rho) = \frac{1}{96\sqrt{5}}\left(\frac{Z}{a_0}\right)^{3/2}(-6+\rho)\rho^2 e^{-\rho/2}$
	3	$\boldsymbol{R}_{43}(\rho) = \frac{1}{96\sqrt{35}}\left(\frac{Z}{a_0}\right)^{3/2}\rho^3 e^{-\rho/2}$

Considering the "scalar product" properties of Eqs. (3.40-b) and (3.44), and scalar product properties of other functions (Eqs. [3.30-i] or [3.32] and [3.34]), it may be readily verified that the Schrödinger equation solutions $\psi_{n\ell m}(\underline{r}) = \boldsymbol{R}_{n\ell}(r)\boldsymbol{\Theta}_\ell^m(\theta)\boldsymbol{\Phi}_m(\varphi)$ are fully orthonormal, satisfying

$$\int\limits_{all\ space} \psi^*_{n'\ell'm'}(\underline{r})\cdot\psi_{n\ell m}(\underline{r})d^3\underline{r} = \delta_{n'n}\delta_{\ell'\ell}\delta_{m'm} \quad , \tag{3.45}$$

The integers n, ℓ, and m are called **quantum numbers**. n is the **principal quantum number** because of its role at determining the state energy. ℓ is called the **orbital angular-momentum quantum number**, and m is called the **magnetic quantum number**. The reason for the latter naming will be clarified in the next chapter.

An important characteristic property of the solutions is their parity. A function's **parity** relates to its behavior under **inversion**; namely when the $\underline{r}$ variable is replaced by $-\underline{r}$. If the function maintains its algebraic sign, it is granted the adjective **even**; if it reverses sign, it is granted the adjective **odd**. The importance of knowing the parity property is accentuated when the function is integrated over its entire definition space. The integral vanishes for an odd function because its integral over any half space exhibits the same absolute value as over the other half, but an opposite sign! In the spherical coordinate system (Figure 3.3) an inversion operation means replacing the (r, θ, φ) variables by $(r, \pi - \theta, \pi + \varphi)$. In Schrödinger's equation solutions for hydrogen-like atoms, the function's parity is determined by that of $\boldsymbol{Y}_{\ell m}(\theta, \varphi) = \boldsymbol{\Theta}_\ell^m(\theta)\boldsymbol{\Phi}_m(\varphi)$. The inversion operation performed on $\boldsymbol{\Phi}_m(\varphi)$ changes only the φ angle, causing a parity switching determined by $(-1)^m$. As to the $\boldsymbol{\Theta}_\ell^m(\theta)$ function, a change in θ is such that $\cos\theta \rightarrow (-\cos\theta)$. Recall that the $\boldsymbol{\Theta}_\ell^m(\theta)$ function is obtained through differentiating $\ell + m$ times the expression $(\xi^2 - 1)^\ell$, where $\xi = \cos\theta$ (Eq. [3.28]). Each differentiation step thus involves a single parity switch with respect to the former, and the function's parity thus switches according to $(-1)^{\ell + m}$. In conclusion, the sign of the $\boldsymbol{Y}_{\ell m}(\theta, \varphi)$ product function switches according to $(-1)^\ell$.

3.4 Physical Meanings of Schrödinger Equation Solutions for Hydrogen-Like Atoms

The physical meanings of the $\boldsymbol{Y}_{\ell m}(\theta, \varphi)$ and $\boldsymbol{R}_{n\ell}(r)$ functions are derived from the fundamental meaning of an electron wave function. Recall that a wave function square is interpreted to describe the electron probability density to reside within a certain segment in space. In other words, the $|\psi(\underline{r})|^2 d^3\underline{r}$ is the probability of the electron to be in the $d^3\underline{r}$ segment of space. We may write $|\psi(\underline{r})|^2 d^3\underline{r} = |\boldsymbol{R}_{n\ell}(r)|^2|\boldsymbol{Y}_{\ell m}(\theta, \varphi)|^2 r^2 d\Omega\cdot dr$, where $d\Omega$ is a solid angular segment about the origin. Integration over the r variable provides the probability for the electron to be anywhere in the $d\Omega$ solid angular segment:

$$\int_0^\infty |\boldsymbol{R}_{n\ell}(r)|^2|\boldsymbol{Y}_{\ell m}(\theta, \varphi)|^2 r^2 d\Omega\cdot dr = |\boldsymbol{Y}_{\ell m}(\theta, \varphi)|^2 d\Omega \int_0^\infty |\boldsymbol{R}_{n\ell}(r)|^2 r^2\cdot dr = |\boldsymbol{Y}_{\ell m}(\theta, \varphi)|^2 d\Omega \quad .$$

In other words, $|\boldsymbol{Y}_{\ell m}(\theta, \varphi)|^2$ is the probability density per unit solid angle. In a similar way, integration over the solid angle variable shows that $|\boldsymbol{R}_{n\ell}(r)|^2 \cdot r^2$ is the probability density per unit distance from the origin (the nucleus).

Figures 3.4 and 3.5 show the probability distribution by the $|\boldsymbol{Y}_{\ell m}(\theta, \varphi)|^2$ function. Consider first Figure 3.4. All $|\boldsymbol{Y}_{\ell m}|^2$ functions are independent of the φ azimuth angle. The $|Y_{00}|^2$ function is constant, independent of both φ and θ (Fig. 3.4a). This means that the electron probability is identical at all orientations.

In this case, the angular part of the wave function has a perfect spherical distribution. It is not so for the other probability distributions; for example, the $|Y_{10}|^2$ function (Fig. 3.4b) exhibits peaks at the **"poles"** ($\theta = 0$ and $\theta = \pi$), and it vanishes on the **equator plane** (the $X - Y$ plane for $\theta = \pi/2$). On the other hand, the $|Y_{1\,\pm\,1}|^2$ functions vanish at the poles, and peak on the equator plane (Fig. 3.4b). Each maximum in the probability density on the sphere surface located between two zeros is addressed as **lobe**. It is possible to prove that the number of probability density lobes is given by $(\ell + 1) - |m|$.

There is a way to emphasize the probability distribution structure by using a **polar figure**. A polar figure displays the distribution in a plane perpendicular to the equator $X - Y$ plane that includes the Z polar axis, and forms a φ angle with the X axis (see Figure 3.3). The edges of origin-initiated vectors that form a θ angle with the Z axis, of having a length of $|\mathbf{Y}_{\ell m}(\theta, \varphi)|^2$ probability density (Figure 3.4), are plotted on the right side of the plane. Polar figures of that type for the $\ell = 0, 1, 2$ cases described in Figure 3.4 are displayed in Figure 3.5. In Figure 3.5-1a, the θ angle and its belonging probability density vector are specifically indicated. To maintain clarity, the number of presented cases is limited, and does not include, for example, the $\ell = 3$ case of Table 3.2, as well as various other cases. It is impressively demonstrated that all distributions are symmetric in relation to the equator plane. Probability maxima are obtained on the poles for $m = 0$. On the other hand, the probabilities on the poles vanish in all $|m| \neq 0$ cases. Obviously, these properties may be deduced directly by analysis of the original functions.

Identical polar figures are obtained for any φ angle, since $|\mathbf{Y}_{\ell m}|^2$ is φ independent. This property is often illustrated by drawing of solid spinning tops obtained by rotating the polar figures around their polar axis. The cases drawn in Figure 3.5 (1) are also drawn as solid spinning tops in Figure 3.5 (2). The $|\mathbf{Y}_{00}|^2$ case

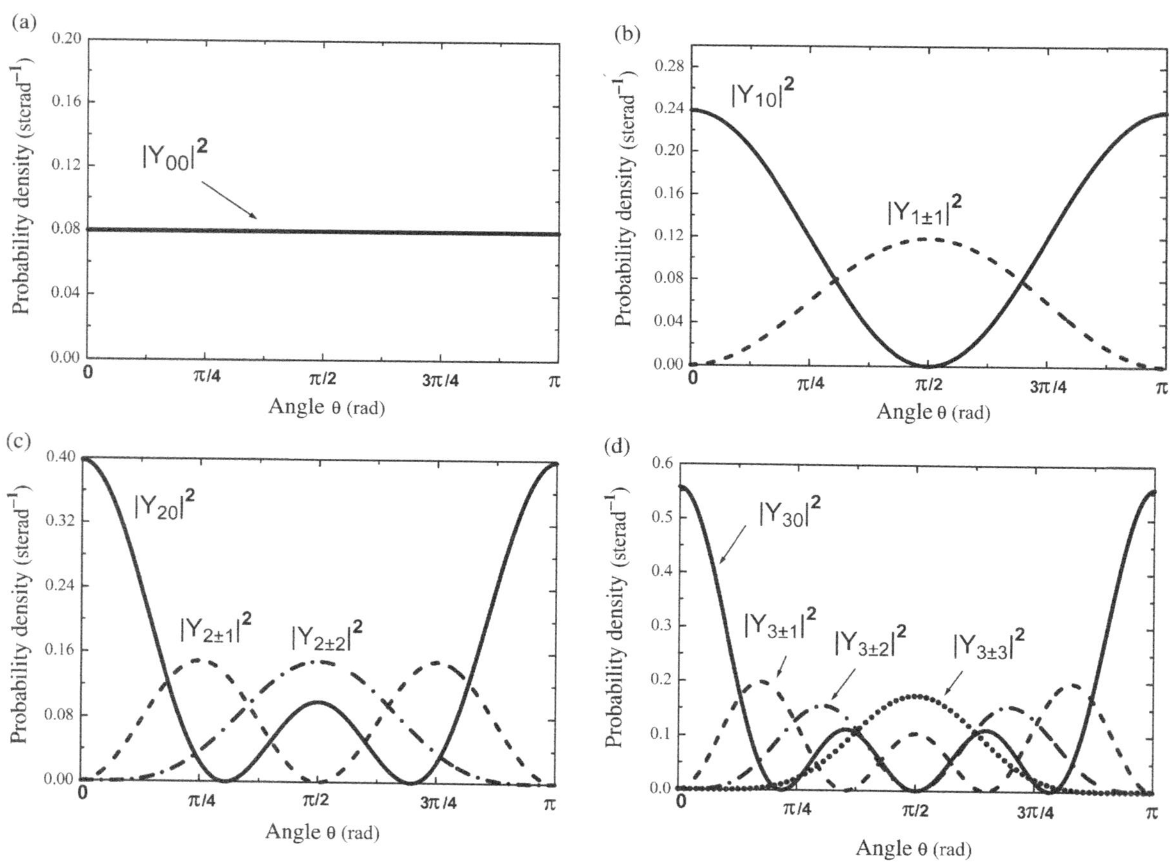

Figure 3.4 Probability density distribution per solid angle unit for ℓ = 1, 2, 3, and their allowed m values. Mathematical expressions for the $\boldsymbol{Y}_{\ell m}(\theta, \varphi)$ functions are provided in Table 3.2.

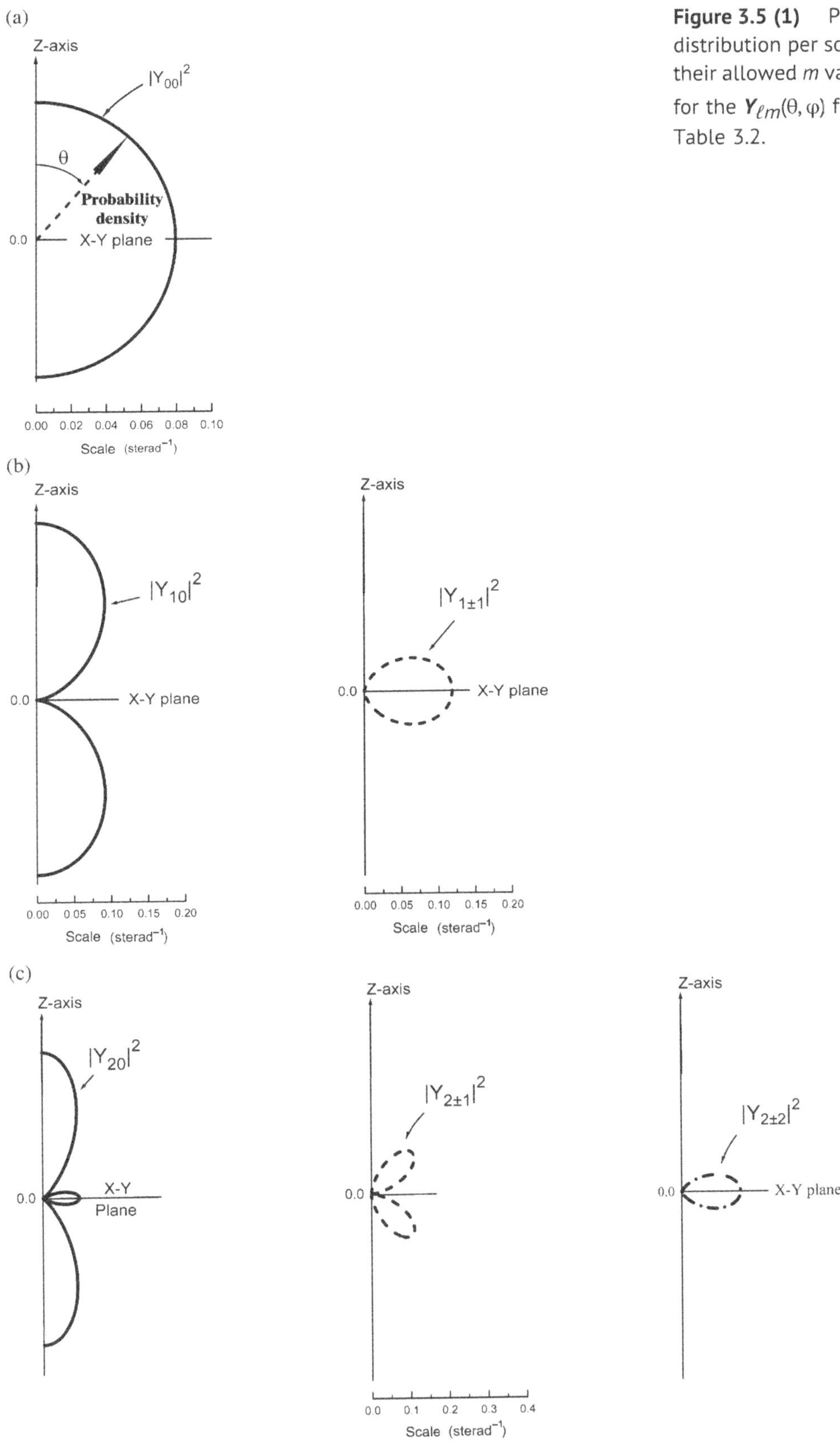

Figure 3.5 (1) Polar figure of probability density distribution per solid angle unit for $\ell = 0, 1, 2$ and their allowed *m* values. Mathematical expressions for the $\boldsymbol{Y}_{\ell m}(\theta, \varphi)$ functions are provided in Table 3.2.

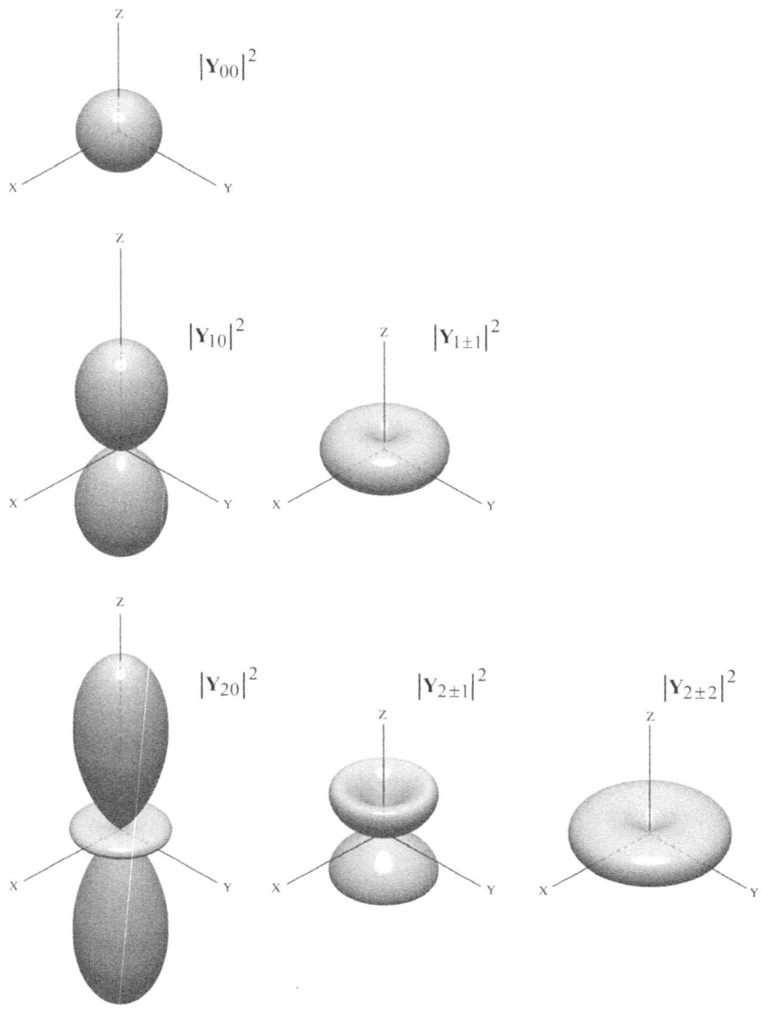

Figure 3.5 (2) Solid spinning tops obtained by rotating each polar figure of Figure 3.5(1) around its polar axis. The top surfaces describe the spatial orientation distribution of probability density per unit solid angle for the ℓ and m quantum numbers they belong to.

displays a spherical surface. In the other cases, the lobes exhibiting nonzero values on the poles, display spinning top surfaces pointed toward the origin. Those that zero on the poles, display bagel-like surfaces.

Table 3.3 provided a list of $\boldsymbol{R}_{n\ell}(r)$ functions in their specific form for $n = 1, 2, 3, 4$, and their various possible ℓ values. Following this table, Figure 3.6 shows the radial probability density $|\boldsymbol{R}_{n\ell}(r)|^2 \cdot r^2$ as a function of the distance from the nucleus r. The distance is provided in a_0/Z units. The simplest case is obtained for $n = 1$ and $\ell = 0$ (Fig. 3.6a). The probability density portrays a single peak in the vicinity of a_0/Z , yet the distribution is broad, and the electron exhibits considerable probabilities of being found at distances as large as $(3 - 4)a_0/Z$. On considering the behavior at excited, $n > 1$ states, the distance increases with increasing n. For example, the peak position of the probability density $|\boldsymbol{R}_{40}(r)|^2 \cdot r^2$ is as high as ~$25a_0/Z$.

A further important property of the probability density functions is that for a constant n, the peak positions occur at shorter r distances for growing ℓ. For example, the peak in the $|\boldsymbol{R}_{43}(r)|^2 \cdot r^2$ probability distribution is obtained at ~$16a_0/Z$. The physical meaning of that behavior will attain a further significance in the next chapter, when the linkage between the ℓ quantum number and the electron orbital angular momentum will be presented. For all n values, when $\ell = 0$, the probability density exhibits lobes

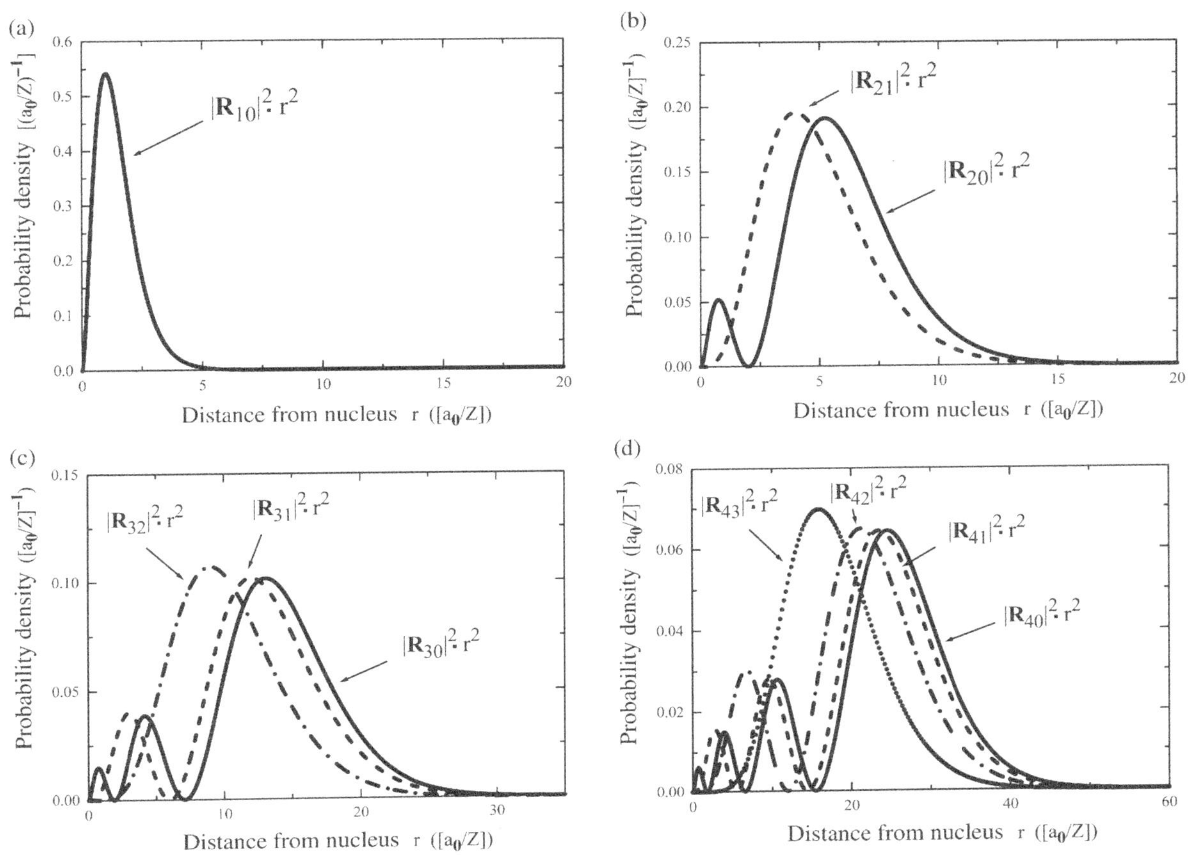

Figure 3.6 Radial probability density distribution for $n = 1,\ 2,\ 3,\ 4$, and their allowed ℓ values. Mathematical expressions for the $R_{n\ell}(r)$ functions forming the calculation basis are provided in Table 3.3.

approaching the nucleus. This means that the electrons spend some time *inside the nucleus*. Often the wording used to describe this behavior is to say that "the *s* electron orbits *penetrate* the nucleus". The interaction between the penetrating electron and the nucleus exhibits a measurable effect on the state energy scheme structure, establishing a spectroscopic means to study the inner nuclear structure. This issue will be addressed shortly in Section 4.10.1.

It is possible to show that the average electron distance from the nucleus is given by

$$r_{ave} = \langle r \rangle \equiv \int_0^{\infty} r \cdot |\boldsymbol{R}_{n\ell}(r)|^2 r^2 dr = n^2 \left(\frac{a_0}{Z}\right) \left[\frac{3}{2} - \frac{\ell(\ell+1)}{2n^2}\right] \quad . \tag{3.46}$$

This expression is presented without detailed calculation. Mean values of other powers of r relating to the same $|n\ell m\rangle$ eigenenergy state would be

$$\langle r^2 \rangle = \frac{1}{2} n^4 \left(\frac{a_0}{Z}\right)^2 \left[5 + \frac{1 - 3\ell(\ell+1)}{n^2}\right] \quad ; \tag{3.47-a}$$

$$\left\langle \frac{1}{r} \right\rangle = \frac{1}{n^2} \cdot \left(\frac{Z}{a_0}\right); \quad \left\langle \frac{1}{r^2} \right\rangle = \frac{1}{n^3} \cdot \frac{2}{2\ell + 1} \cdot \left(\frac{Z}{a_0}\right)^2 \quad . \tag{3.47-b}$$

An important, quite useful recurrence relation related to mean values of r powers is given by

$$\frac{q+1}{n^2}\left\langle r^q\right\rangle-(2q+1)\left(\frac{a_0}{Z}\right)\left\langle r^{q-1}\right\rangle+\frac{q}{4}\left[(2\ell+1)^2-q^2\right]\left(\frac{a_0}{Z}\right)^2\left\langle r^{q-2}\right\rangle=0\ , \tag{3.48}$$

for an integer q satisfying $q > -(2\ell+1)$. Eq. (3.48) may be derived using the radial part of Schrödinger's differential equation (Eq. [3.24-i-a]). Using the recurrence relation given by Eq. (3.48) above, and the relations provided by Eqs. (3.46) and (3.47), allows obtaining the mean (average) values of other r powers as well.

3.5 Spectroscopy of Hydrogen and Hydrogen-Like Atoms

We have seen in the previous section that the eigen wave functions of the Hamiltonian are given by the $\psi_{n\ell m}(r,\theta,\varphi)=\boldsymbol{R}_{n\ell}(r)\boldsymbol{Y}_{\ell m}(\theta,\varphi)$ functions. Recall that wave functions are often shorthand written as "ket" vectors; specifically $|n\ell m\rangle$ in the present case. The eigenenergies are given by $E_n=-\frac{Z^2e^2}{2a_0}\frac{1}{n^2}$, and depend solely on the principal quantum number n. A phenomenon where states belonging to independent (actually orthogonal) wave functions have identical energies, is termed **degeneracy**. Because there are n different ℓ values per each n quantum number $(\ell=0,1,2\cdots n-1)$, and $2\ell+1$ different m values per each ℓ quantum number $(m=-\ell+1,-\ell+2,\cdots,0,+1,+2,\cdots,+\ell)$, the total degeneracy of each n - number is $\sum_{\ell=0}^{n-1}(2\ell+1)=n^2$. A quantity $R\equiv e^2/2a_0hc$ is called **Rydberg constant** and is presented in cm^{-1} units. Its actual value obviously depends on the Bohr radius a_0, which depends on the atomic reduced mass (see Eq. [3.38] and following text). Different nuclei have different masses, thus their atomic Rydberg constants are slightly different. In the limit of an infinitely heavy nucleus, the atomic reduced mass equals the electron mass. The numerical value of the Rydberg constant then equals $R_\infty=109{,}737.3\ \text{cm}^{-1}$. The eigenenergies in erg units are then given by

$$\boxed{E_n=-\frac{R_\infty hcZ^2}{n^2}}\ . \tag{3.49}$$

Figure 3.7 provides an energy scheme of hydrogen-like atoms. The states are classified vertically by the principal quantum number n, and horizontally by the orbital quantum number ℓ (up to $\ell=4$). It remains to split and classify each $|n\ell\rangle$ level into its magnetic m quantum numbers. It is possible to do so along an axis perpendicular to the page plane, but for obvious reasons, the image will be more complicated. The degeneracy among the different states belonging to the same n will be fully or partially removed when corrections related to various effects and interactions not yet accounted for will be presented. Marked arrows on the scheme describe dipole-allowed transitions for photon emission, a subject to be detailed in Section 3.6. The opposite transitions related to photon absorptions are equally dipole-allowed. The states energies turn denser as n grows, converging to an infinite density toward the zero-energy limit (the free electron energy). The discussion we had so far relates to bound electron states. As to unbound electrons, namely having *positive* total energy, a continuum of state energies exists, and that property is displayed by the gray rectangles above the zero-energy line. We won't address that region in this book.

Absorption and emission spectra of hydrogen-like atoms are interpreted as radiative transitions between two states of different n values. Thus, the emitted/absorbed photon energies are given by

$$h\nu=E_{n_2}-E_{n_1}=-\frac{Z^2e^2}{2a_0}\left(\frac{1}{{n_2}^2}-\frac{1}{{n_1}^2}\right)\ , \tag{3.50}$$

where n_1 and n_2 are, respectively, the initial and final principal quantum numbers of the states involved. The various emission lines obtained are classified by the lowest-energy n value, and named as follows:

Lyman series	$n_1 = 1;\ n_2 = 2, 3, 4, \cdots\infty$
Balmer series	$n_1 = 2;\ n_2 = 3, 4, 5, \cdots\infty$
Paschen series	$n_1 = 3;\ n_2 = 4, 5, 6, \cdots\infty$
Brackett series	$n_1 = 4;\ n_2 = 5, 6, 7, \cdots\infty$
Pfund series	$n_1 = 5;\ n_2 = 6, 7, 8, \cdots\infty$

In the hydrogen atom, Lyman series wavelengths belong to the far ultraviolet, the Balmer series ranges in the vicinity of the visible region, and the others in the infrared and far infrared.

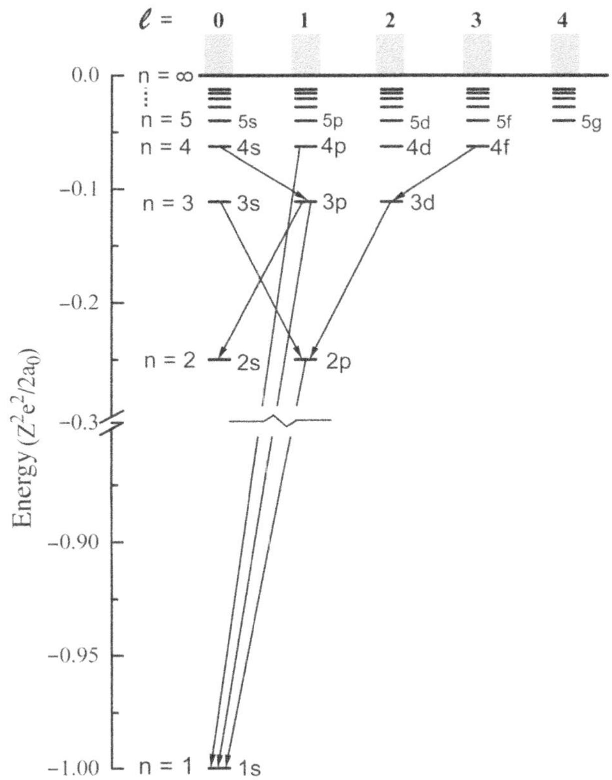

Figure 3.7 Energy levels scheme in hydrogen-like atoms. Several dipole-allowed radiative transmissions are marked by solid arrows. For details see text.

3.6 Selection Rules

The simplistic presentation of Eq. (3.50) fails in bringing forward selection rules that allow electric dipole transitions between certain ℓ and m states, and forbid others. To find out the selection rules one needs to study the wave functions characteristics of the electron states, as well as the photon characteristics. Following Eq. (2.33), the $W_{n'\ell'm',n\ell m}$ transition probability rate (the transition probability per unit time) from the initial $|n\ell m\rangle$ state to the final $|n'\ell'm'\rangle$ one is given by

$$W_{n'\ell'm',n\ell m} \cong \frac{2\pi^2e^2}{\hbar^2}\left|\left\langle n'\ell'm'|\underline{r}|n\ell m\right\rangle\cdot\underline{u}\right|^2\cdot\rho_\omega\left(\omega_0 = \frac{E_{n'\ell'm'} - E_{n\ell m}}{\hbar}\right), \tag{3.51}$$

where $\underline{u}$ is a unit vector oriented with the electric field component of the electromagnetic radiation, and $\underline{r}$ is the average displacement vector of the electron from its initial state distance from the nucleus (Section 2.3), and ρ_ω is the volume concentration of electromagnetic radiation per unit of angular frequency. A primary "selection rule" stems from the last factor on the right side of Eq. (3.51), and is obviously related to the transition energy. For this factor to be nonzero, it is imperative that the photons involved should exhibit an energy $\hbar\omega_0 = E_{n'\ell'm'} - E_{n\ell m} = E_{n'} - E_n$. This condition expresses the trivial energy conservation requirement. Additional selection rules result by analysis of the central, right side factor of Eq. (3.51).

Let θ_u be the angle between $\underline{u}$ and the $\langle n'\ell'm'|\underline{r}|n\ell m\rangle$ vector. Then

$$\left|\left\langle n'\ell'm'|\underline{r}|n\ell m\right\rangle\cdot\underline{u}\right|^2 = \left|\left\langle n'\ell'm'|\underline{r}|n\ell m\right\rangle\right|^2\cdot\cos^2\theta_u\ . \tag{3.52}$$

A **forbidden** transition is such that the $\langle n'\ell'm'|\underline{r}|n\ell m\rangle$ vector vanishes; in an **allowed** transition, $\langle n'\ell'm'|\underline{r}|n\ell m\rangle$ is nonzero. We consider first the Z component of the $\langle n'\ell'm'|\underline{r}|n\ell m\rangle$ vector.

$$\left.\begin{aligned}&\langle n'\ell'm'|\underline{r}|n\ell m\rangle_Z = \langle n'\ell'm'|r\cos\theta|n\ell m\rangle\\ &= \int_0^\infty r\cdot\boldsymbol{R}^*_{n'\ell'}(r)\boldsymbol{R}_{n\ell}(r)r^2dr \times \int_0^\pi \cos\theta\cdot\boldsymbol{\Theta}^{*m'}_{\ell'}(\theta)\boldsymbol{\Theta}^m_\ell(\theta)\sin\theta\cdot d\theta \times \int_0^{2\pi}\boldsymbol{\Phi}^*_{m'}(\varphi)\boldsymbol{\Phi}_m(\varphi)d\varphi\end{aligned}\right\}\ . \tag{3.53}$$

To be nonzero, it is obviously necessary that the last factor on the right side of Eq. (3.53) is nonzero. As the $\boldsymbol{\Phi}_m(\varphi)$ functions are orthonormal, to be nonzero it is sufficient and necessary that the transition satisfies $m' = m$, or in other words, $\Delta m \equiv m' - m = 0$. This selection rule is thus a necessary condition to emission or absorption of a Z-polarized photon.

The projection of an $\underline{r}$ vector on an (X, Y) plane is $r\boldsymbol{sin}\theta$. Its components on the X and Y axes are $r\boldsymbol{sin}\theta\boldsymbol{cos}\varphi$ and $r\boldsymbol{sin}\theta\,\boldsymbol{sin}\varphi$, respectively (see Figure 3.3). We manufacture, however, a different axes system: $Q^+ \equiv X + iY$ and $Q^- \equiv X - iY$. The $\underline{r}$ vector components on those axes would be $q^+ = r\,\boldsymbol{sin}\,\theta\,(\boldsymbol{cos}\,\varphi + i\,\boldsymbol{sin}\,\varphi) = r\,\boldsymbol{sin}\,\theta \cdot e^{i\varphi}$, and $q^- = r\,\boldsymbol{sin}\,\theta\,(\boldsymbol{cos}\,\varphi - i\,\boldsymbol{sin}\,\varphi) = r\,\boldsymbol{sin}\,\theta \cdot e^{-i\varphi}$, respectively. For example, the Q^+ component of the $\langle n'\ell'm'|\underline{r}|n\ell m\rangle$ vector would be

$$\left.\begin{aligned} &\langle n'\ell'm'|\underline{r}|n\ell m\rangle_{Q^+} = \langle n'\ell'm'|r\,\boldsymbol{sin}\,\theta{\cdot}e^{i\varphi}|n\ell m\rangle \\ &= \int_0^\infty r{\cdot}\boldsymbol{R}^*_{n'\ell'}(r)\boldsymbol{R}_{n\ell}(r)r^2dr \times \int_0^\pi \boldsymbol{sin}\,\theta{\cdot}\boldsymbol{\Theta}^{*m'}_{\ell'}(\theta)\boldsymbol{\Theta}^m_\ell(\theta)\,\boldsymbol{sin}\,\theta{\cdot}d\theta \times \int_0^{2\pi} e^{i\varphi}\boldsymbol{\Phi}^*_{m'}(\varphi)\boldsymbol{\Phi}_m(\varphi)d\varphi \end{aligned}\right\}. \tag{3.54}$$

Explicitly, the last factor on the right side is $\frac{1}{2\pi}\int_0^{2\pi} \boldsymbol{exp}\left[i\varphi\left(m - m' + 1\right)\right]$. A sufficient and necessary condition for it to be nonzero is $\Delta m = +1$. It is thus a necessary condition for a right-handed *circularly polarized* photon to be absorbed or emitted. By an analogous calculation using the Q^- component, we get that $\Delta m = -1$ is a necessary condition for a left-handed *circularly polarized* photon to be absorbed or emitted. In summary, the necessary selection rules for optical transitions related to the magnetic quantum number are

$$\Delta m = 0,\ \pm 1. \tag{3.55}$$

Necessary conditions related to the orbital quantum number ℓ are obtained by analysis of each of the integrals acting as second factors on the right side of Eqs. (3.53) and (3.54). They involve the $\boldsymbol{\Theta}^m_\ell(\theta)$ functions. We consider first the factor appearing in Eq. (3.53); returning to the $\xi = \boldsymbol{cos}\,\theta$ variable, and setting $\Delta m = 0$ (namely $m' = m$), we get

$$\begin{aligned} &\int_0^\pi \boldsymbol{cos}\,\theta{\cdot}\boldsymbol{\Theta}^{*m}_{\ell'}(\theta)\boldsymbol{\Theta}^m_\ell(\theta)\,\boldsymbol{sin}\,\theta{\cdot}d\theta = \frac{1}{\sqrt{N_{\ell'm}N_{\ell m}}}\int_{-1}^{+1} \xi{\cdot}\boldsymbol{P}^m_{\ell'}(\xi)\boldsymbol{P}^m_\ell(\xi){\cdot}d\xi \\ &= \frac{1}{(2\ell+1)\sqrt{N_{\ell'm}N_{\ell m}}}\left\{\int_{-1}^{+1} \boldsymbol{P}^m_{\ell'}(\xi){\cdot}(\ell+1-m)\boldsymbol{P}^m_{\ell+1}(\xi){\cdot}d\xi + \int_{-1}^{+1} \boldsymbol{P}^m_{\ell'}(\xi){\cdot}(\ell+m)\boldsymbol{P}^m_{\ell-1}(\xi){\cdot}d\xi\right\} \\ &= \frac{(\ell+1-m)N_{\ell+1,m}}{(2\ell+1)\sqrt{N_{\ell'm}N_{\ell m}}}\delta_{\ell',\ell+1} + \frac{(\ell+m)N_{\ell-1,m}}{(2\ell+1)\sqrt{N_{\ell'm}N_{\ell m}}}\delta_{\ell',\ell-1} \\ &= \sqrt{\frac{(\ell+1)^2-m^2}{(2\ell+3)(2\ell+1)}}\,\delta_{\ell',\ell+1} + \sqrt{\frac{\ell^2-m^2}{(2\ell+1)(2\ell-1)}}\,\delta_{\ell',\ell-1}\ . \end{aligned} \tag{3.56}$$

Transition from the first line to the second one relies on Eq. (3.29-a); from the second to the third line it relies on Eq. (3.30-a). The result yields, that a sufficient and necessary condition for the second factor in Eq. (3.54) to be nonzero, is that either $\ell' = \ell + 1$ or $\ell' = \ell - 1$, namely $\Delta\ell = \pm 1$.

We turn now to consider the second factor on the right side of Eq. (3.54). Returning to the $\xi = \boldsymbol{cos}\theta$ variable, and setting $\Delta m = +1$ (namely, $m' = m + 1$), we get

$$
\begin{aligned}
\int_0^{\pi} \boldsymbol{sin}\,\theta\cdot\boldsymbol{\Theta}_{\ell'}^{*m+1}(\theta)\boldsymbol{\Theta}_{\ell}^{m}(\theta)\,\boldsymbol{sin}\,\theta\cdot d\theta &= \frac{1}{\sqrt{N_{\ell',m+1}N_{\ell m}}}\int_{-1}^{+1}\left(1-\xi^2\right)^{1/2}\cdot\boldsymbol{P}_{\ell'}^{m+1}(\xi)\boldsymbol{P}_{\ell}^{m}(\xi)\cdot d\xi \\
&= \frac{1}{(2\ell+1)\sqrt{N_{\ell',m+1}N_{\ell m}}}\left\{\int_{-1}^{+1}\boldsymbol{P}_{\ell'}^{m+1}(\xi)\cdot\boldsymbol{P}_{\ell+1}^{m+1}(\xi)\cdot d\xi - \int_{-1}^{+1}\boldsymbol{P}_{\ell'}^{m+1}(\xi)\cdot\boldsymbol{P}_{\ell-1}^{m+1}(\xi)\cdot d\xi\right\} \\
&= \frac{N_{\ell+1,m+1}}{(2\ell+1)\sqrt{N_{\ell',m+1}N_{\ell m}}}\delta_{\ell',\ell+1} - \frac{N_{\ell-1,m+1}}{(2\ell+1)\sqrt{N_{\ell',m+1}N_{\ell m}}}\delta_{\ell',\ell-1} \\
&= \sqrt{\frac{(\ell+m+1)(\ell+m+2)}{(2\ell+3)(2\ell+1)}}\,\delta_{\ell',\ell+1} - \sqrt{\frac{(\ell-m-1)(\ell-m)}{(2\ell-1)(2\ell+1)}}\,\delta_{\ell',\ell-1}\ .
\end{aligned}
\tag{3.57}
$$

Transition from the first line to the second one relies on Eq. (3.29-b); from the second to the third line relies on Eq. (3.30-a). The result yields that a sufficient and necessary condition for the second factor in Eq. (3.54) to be nonzero is either $\ell' = \ell + 1$ or $\ell' = \ell - 1$, namely $\Delta\ell = \pm 1$. An identical conclusion is obtained for the $\Delta m = -1$ case (namely $m' = m - 1$).

Properties of the $\boldsymbol{R}_{n\ell}(r)$ functions add no further selection rules. Therefore, the necessary conditions formulated so far are also sufficient. In other words, to declare allowed a radiative electric dipole transition, it is sufficient and necessary that the following selection rules are satisfied:

$$
\boxed{\Delta\ell = \pm 1 \textit{ together with } \Delta m = 0, \pm 1}\ . \tag{3.58}
$$

For deductive brevity of addressing the selection rules issue, several teaching texts argue that they all stem from parity properties of the $\boldsymbol{Y}_{\ell m}(\theta, \varphi)$ function. It is however erroneous, because by parity consideration alone, it may only be concluded that necessary conditions for radiative transitions are $\Delta\ell = \pm 1, \pm 3, \pm 5, \cdots$. It may not be concluded that they are sufficient. Especially, the above presented accurate proof of Eq. (3.58) shows that the radiative transition probabilities for $\Delta\ell = \pm 3, \pm 5, \cdots$ rather vanish!

The $\Delta\ell = \pm 1$ selection rule for electric dipole transitions has been demonstrated in the states energy scheme of Figure 3.7 by solid arrows connecting states belonging to allowed transitions.

As to the transition probabilities quantitative aspects, we may write

$$
\left.
\begin{aligned}
\langle n', \ell+1, m|\underline{r}|n\ell m\rangle_Z &= \sqrt{\frac{(\ell+1)^2 - m^2}{(2\ell+3)(2\ell+1)}}\left\langle \boldsymbol{R}_{n\ell}^{n',\ell+1}\right\rangle \\
\langle n', \ell-1, m|\underline{r}|n\ell m\rangle_Z &= \sqrt{\frac{\ell^2 - m^2}{(2\ell+1)(2\ell-1)}}\left\langle \boldsymbol{R}_{n\ell}^{n',\ell-1}\right\rangle
\end{aligned}
\right\}\ , \tag{3.59}
$$

and

$$
\left.
\begin{aligned}
\langle n', \ell+1, m+1|\underline{r}|n\ell m\rangle_{Q^+} &= \sqrt{\frac{(\ell+m+2)(\ell+m+1)}{(2\ell+3)(2\ell+1)}}\left\langle \boldsymbol{R}_{n\ell}^{n',\ell+1}\right\rangle \\
\langle n', \ell-1, m+1|\underline{r}|n\ell m\rangle_{Q^+} &= -\sqrt{\frac{(\ell-m)(\ell-m-1)}{(2\ell+1)(2\ell-1)}}\left\langle \boldsymbol{R}_{n\ell}^{n',\ell-1}\right\rangle \\
\langle n', \ell+1, m-1|\underline{r}|n\ell m\rangle_{Q^-} &= -\sqrt{\frac{(\ell-m+2)(\ell-m+1)}{(2\ell+3)(2\ell+1)}}\left\langle \boldsymbol{R}_{n\ell}^{n',\ell+1}\right\rangle \\
\langle n', \ell-1, m-1|\underline{r}|n\ell m\rangle_{Q^-} &= \sqrt{\frac{(\ell+m)(\ell+m-1)}{(2\ell+1)(2\ell-1)}}\left\langle \boldsymbol{R}_{n\ell}^{n',\ell-1}\right\rangle
\end{aligned}
\right\}\ , \tag{3.60}
$$

where $\left\langle \boldsymbol{R}_{n\ell}^{n'\ell'} \right\rangle \equiv \int_0^\infty \boldsymbol{R}_{n'\ell'}(r)\boldsymbol{R}_{n\ell}(r)\cdot r^3 dr$; they vanish for ℓ' and ℓ not satisfying the (3.58) selection rule.

Eqs. (3.59) and (3.60) provide recipes for direct calculation of probabilities and oscillator strengths in transitions among various states of hydrogen-like atoms. There are general expressions for the radial transition components $\left\langle \boldsymbol{R}_{n\ell}^{n'\ell'} \right\rangle$ appearing in Eqs. (3.59) and (5.60); they are, however, long and complicated. It is often easier to perform a direct (even numerical) calculation of needed integration. There are also summing rules for transitions between different groups of states, which are important in spectroscopic studies of hydrogen-like atoms. An example of such summing rule is given as an exercise; this aspect won't be elaborated on further in this book.

A fundamental issue worth emphasizing concerns the determination of the "preferred" Z-axis in an atom that supposedly exhibits a full rotational symmetry. It is important to appreciate that the said symmetry exists only in the *absence* of interaction with the surrounding medium. When a linearly polarized photon is either absorbed or emitted, the atomic Z-axis orientation coincides with the said polarization orientation. When a circularly polarized photon is absorbed or emitted, the atomic Z-axis orientation coincides with the photon propagation orientation. Under the influence of a static external magnetic or electric field, the atomic Z-axis orientation coincides with the orientation of the applied fields. The action of the said fields is not limited to defining the Z-axis. It also causes splitting of the atomic states; they are called **Zeeman** splitting by the magnetic field and **Stark** splitting by the electric field. These effects will be discussed in detail in the next chapter, Sections 4.9.1 and 4.9.2, respectively.

An intuitive, physically correct interpretation of the $\langle n'\ell'm'|\underline{r}|n\ell m\rangle$ role in determining the selection rules (Eq. [3.51]), is that transitions between two states are dipole-allowed only if they involve a change in the atomic dipole moment; namely that the end state dipole moment is different than the dipole moment of the initial state.

Exercises for Chapter 3

3.1 Diffraction of an X-ray whose wavelength is shorter than a typical inter-atomic distance (about 1.0 Å) is used for the study of crystalline structure. Calculate a typical velocity of neutrons that can be used for the same purpose. Compare their kinetic energy with the typical room-temperature thermal energy k_BT at $T = 300$K.

3.2 Solve Schrödinger's equation for a particle of mass M residing in a one-dimensional square potential well, whose potential energy is $V = 0$ in the $0 \le x \le a$ range and is $V = \infty$ for $x > a$ and $x < 0$. Specify the particle state energy spectrum and the normal wave function of each state.

3.3 Calculate the difference between the Bohr radiuses of a regular hydrogen atom (protonium) and a tritium atom (nucleus consisting of one proton and two neutrons). Calculate the transition energies between the $n = 1$ and $n = 2$ states of each atom, and indicate the spectroscopic splitting cased by the difference in the nuclear masses of the different isotopes in cm^{-1} units.

3.4 Perform a direct calculation of the integral $\int_0^\infty \boldsymbol{R}_{20}^*(r)\boldsymbol{R}_{21}(r)\cdot r^2 dr$ of the hydrogen atom. Specific expressions of the $\boldsymbol{R}_{n\ell}(r)$ functions are provided in Table 3.3.

3.5 (a) Plot a graph describing the average electron-nucleus distance as a function of the principal quantum number n for $n = 1, 2, 3, \cdots, 10$ and $\ell = 0$ in (a_0/Z) units.

(b) Plot a graph describing the average electron-nucleus distance as function of the allowed orbital quantum number ℓ for $n = 10$ in (a_0/Z) units.
(c) Calculate a general expression for the width (mean deviation) of the electron-nucleus distance distribution. Plot two figures: one, a graph describing the width as a function of the principal quantum number n for $n = 1, 2, 3, \cdots, 10$ and $\ell = 0$ in (a_0/Z) units; and a second graph describing the width as a function of the allowed orbital quantum number ℓ for $n = 10$ in (a_0/Z) units.

3.6 Calculate the squared transition dipole moment between the $|321\rangle$ and $|211\rangle$ states of a hydrogen atom. Indicate the polarization state of the emitted photon.

3.7 Prove that by adding the squared transition radiuses $\left|\langle n'\ell' m|\underline{r}|n\ell m\rangle\right|^2$ from a certain $|n\ell m\rangle$ state to all allowed m' sub-states of an $|n'\ell'\rangle$ hydrogen atom (namely including all involved photon polarization states) one gets: $\frac{\ell+1}{2\ell+1}\left\langle \boldsymbol{R}_{n\ell}^{n',\ell+1}\right\rangle^2$ for $\ell' = \ell + 1$ and $\frac{1}{2\ell+1}\left\langle \boldsymbol{R}_{n\ell}^{n',\ell-1}\right\rangle^2$ for $\ell' = \ell - 1$.
Readers working to solve this exercise are advised to consider for use the following general identity: $|\underline{r}|^2 \equiv x^2 + y^2 + z^2 = z^2 + \frac{1}{2}\left\{|x+iy|^2 + |x-iy|^2\right\}$.

4 Hydrogen Atom – Corrections to the Electrostatic Attraction Approximation

4.1 Angular Momentum and the Orbital Quantum Number

In continuation of Table 3.1, the angular momentum operator is given by

$$\mathbf{L} = -i\hbar \underline{r} \times \underline{\nabla} = -i\hbar \begin{vmatrix} \mathbf{i} & \mathbf{j} & \mathbf{k} \\ x & y & z \\ \partial/\partial x & \partial/\partial y & \partial/\partial z \end{vmatrix}, \tag{4.1}$$

where **i**, **j**, and **k** are unit vectors oriented along the *X*, *Y*, and *Z* axes, respectively, and *x*, *y*, and *z* are the Cartesian coordinates of the $\underline{r}$ vector. The Cartesian components of the angular momentum operator would be

$$\mathbf{L}_X = -i\hbar\left(y\frac{\partial}{\partial z} - z\frac{\partial}{\partial y}\right) \; ; \; \mathbf{L}_Y = -i\hbar\left(z\frac{\partial}{\partial x} - x\frac{\partial}{\partial z}\right) \; ; \; \mathbf{L}_Z = -i\hbar\left(x\frac{\partial}{\partial y} - y\frac{\partial}{\partial x}\right) \; . \tag{4.2}$$

We now express the **L** vector components in spherical coordinates, as defined in Figure 3.3. In Eq. (3.22) we have obtained that

$$\frac{\partial}{\partial x} = \sin\theta\cos\varphi \cdot \frac{\partial}{\partial r} + \frac{1}{r}\cos\theta\sin\varphi \cdot \frac{\partial}{\partial \theta} - \frac{1}{r}\frac{\cos^2\theta\sin\varphi}{\cos^2\varphi\sin\theta} \cdot \frac{\partial}{\partial \varphi} \; .$$

We may similarly progress in finding the other partial derivatives appearing in Eq. (4.2):

$$\frac{\partial}{\partial y} = \sin\theta\sin\varphi \cdot \frac{\partial}{\partial r} + \frac{1}{r}\cos\theta\sin\varphi \cdot \frac{\partial}{\partial \theta} + \frac{1}{r}\frac{\cos\varphi}{\sin\theta} \cdot \frac{\partial}{\partial \varphi} \; ;$$

$$\frac{\partial}{\partial z} = \cos\theta\frac{\partial}{\partial r} - \frac{\sin\theta}{r} \cdot \frac{\partial}{\partial \theta} \; .$$

On setting the relations $x = r\sin\theta\cos\varphi$, $y = r\sin\theta\sin\varphi$, and $z = r\cos\theta$ in Eq. (4.2), we get

$$\left.\begin{aligned} \mathbf{L}_X &= -i\hbar\left(-\sin\varphi\frac{\partial}{\partial\theta} - \cot\theta\cos\varphi\frac{\partial}{\partial\varphi}\right) \\ \mathbf{L}_Y &= -i\hbar\left(\cos\varphi\frac{\partial}{\partial\theta} - \cot\theta\sin\varphi\frac{\partial}{\partial\varphi}\right) \\ \mathbf{L}_Z &= -i\hbar\frac{\partial}{\partial\varphi} \end{aligned}\right\} \; . \tag{4.3}$$

The squared angular momentum operator is defined as

$$\mathbf{L}^2 = \mathbf{L}_X{}^2 + \mathbf{L}_Y{}^2 + \mathbf{L}_Z{}^2 \quad , \tag{4.4}$$

which, in spherical coordinates, is expressed as

$$\boxed{\mathbf{L}^2 = \mathbf{L}_X{}^2 + \mathbf{L}_Y{}^2 + \mathbf{L}_Z{}^2 = -\hbar^2\left(\frac{1}{\sin\theta}\frac{\partial}{\partial\theta}\left(\sin\theta\frac{\partial}{\partial\theta}\right) + \frac{1}{\sin^2\theta}\frac{\partial^2}{\partial\varphi^2}\right)} \quad . \tag{4.5}$$

A detailed calculation of Eq. (4.5) is quite long (and tiring) and won't be presented here. Importantly, all components of the angular momentum (Eq. [4.3]) and squared angular momentum (Eq. [4.5]) are independent of the radius, and therefore don't operate on the radial part of the wave function.

We consider the result of the $\mathbf{L}_Z$ quantum operation on a certain eigenfunction of the hydrogen atom Hamiltonian:

$$\begin{aligned}\mathbf{L}_Z|n\ell m\rangle &= -i\hbar\frac{\partial}{\partial\varphi}\boldsymbol{R}_{n\ell}(r)\boldsymbol{\Theta}_\ell^m(\theta)\boldsymbol{\Phi}_m(\varphi) = -i\hbar\boldsymbol{R}_{n\ell}(r)\boldsymbol{\Theta}_\ell^m(\theta)\frac{d\boldsymbol{\Phi}_m(\varphi)}{d\varphi} \\ &= -i\hbar\boldsymbol{R}_{n\ell}(r)\boldsymbol{\Theta}_\ell^m(\theta)\cdot im\boldsymbol{\Phi}_m(\varphi) = m\hbar|n\ell m\rangle\end{aligned} \quad . \tag{4.6}$$

The $|n\ell m\rangle$ function is therefore also an eigenfunction of the Z angular momentum operator, with $m\hbar$ being its eigenvalue. This eigenvalue is thus the expectation value of that angular momentum component; namely $\langle\mathbf{L}_Z\rangle = m\hbar$, where $m = -\ell, -\ell+1, -\ell+2, \ldots, 0, +1, +2, \ldots, +\ell$. The largest projection of an angular momentum on the Z axis is therefore $\langle\mathbf{L}_Z\rangle_{max} = \ell\hbar$

Following the same procedure we used with the $\mathbf{L}_Z$ operator, we consider the result of the $\mathbf{L}^2$ operation on an eigenfunction of the hydrogen atom Hamiltonian:

$$\left.\begin{aligned}\mathbf{L}^2|n\ell m\rangle &= -\hbar^2\boldsymbol{R}_{n\ell}(r)\left(\frac{1}{\sin\theta}\frac{\partial}{\partial\theta}\left(\sin\theta\frac{\partial}{\partial\theta}\right) + \frac{1}{\sin^2\theta}\frac{\partial^2}{\partial\varphi^2}\right)\boldsymbol{\Theta}_\ell^m(\theta)\boldsymbol{\Phi}_m(\varphi) \\ &= -\hbar^2\boldsymbol{R}_{n\ell}(r)\left(\frac{1}{\sin\theta}\frac{\partial}{\partial\theta}\left(\sin\theta\frac{\partial}{\partial\theta}\right) + \frac{1}{\sin^2\theta}\frac{\partial^2}{\partial\varphi^2}\right)\boldsymbol{Y}_{\ell m}(\theta,\varphi) \\ &= \hbar^2\boldsymbol{R}_{n\ell}(r)\cdot\ell(\ell+1)\boldsymbol{Y}_{\ell m}(\theta,\varphi) = \ell(\ell+1)\hbar^2\boldsymbol{R}_{n\ell}(r)\boldsymbol{Y}_{\ell m}(\theta,\varphi) = \ell(\ell+1)\hbar^2|n\ell m\rangle\end{aligned}\right\} \quad . \tag{4.7}$$

The $|n\ell m\rangle$ eigenfunction of the Hamiltonian is therefore an eigenfunction of the squared angular momentum operator, with $\ell(\ell+1)\hbar^2$ being its eigenvalue. This eigenvalue is therefore the expectation value of the squared angular momentum operator; namely $\langle\mathbf{L}^2\rangle = \ell(\ell+1)\hbar^2$.

The last results connect the electron angular momentum with the orbital quantum number. The size (absolute value) of the angular momentum is a time-independent quantity: $\langle\underline{\boldsymbol{L}}^2\rangle^{1/2} = \hbar\sqrt{\ell(\ell+1)}$, which is an invariant property of the state. However, it is obvious that it could not be uniquely oriented; if it were uniquely oriented, we could assign that orientation to the Z axis, and its projection on that axis would be $\hbar\sqrt{\ell(\ell+1)}$, in contrast to the previous calculation that it should be $\ell\hbar$. Therefore, one must conclude that the angular momentum vector belongs equally to all φ azimuth angles. To demonstrate the conclusion in classical terms, the angular momentum vector $\underline{\boldsymbol{L}}$ is imagined to form some precession along a circular circumference about the Z axis. Recall that a somewhat similar motion occurs in classical mechanics, when a spinning top connected at a single point is made to rotate around an axis that differs from its symmetry axis; the freely rotating time-dependent angular rotation frequency $\underline{\boldsymbol{\omega}}$ then displays a circular circumference around the top's symmetry axis. Such motion is called **precession**. This idea is displayed in Figure 4.1. In contrast to the classical example, the time variation of the X and Y components of the quantum angular momentum may not follow an ordered pattern, because then it would be possible to assign a temporal orientation to the angular momentum vector.

In contrast to the Z component of the angular momentum, its X and Y components expectation values are each identically zero. Namely, $\langle \underline{L}_X \rangle \equiv \langle n\ell m | \mathbf{L}_X | n\ell m \rangle = 0$ and $\langle \underline{L}_Y \rangle \equiv \langle n\ell m | \mathbf{L}_Y | n\ell m \rangle = 0$. These properties may be proved using the $\boldsymbol{\Phi}_m(\varphi)$ functions properties and recurrence and orthogonality relations of the $\boldsymbol{P}_\ell^m(\xi)$ functions provided in Eqs. (3.29) and (3.30). However, the expectation values of their squared operators are nonzero; it must obviously be so because, by definition, the expectations sum of all component squared operators provide the expectation value of the squared angular momentum operator (Eqs [4.5] and [4.7]).

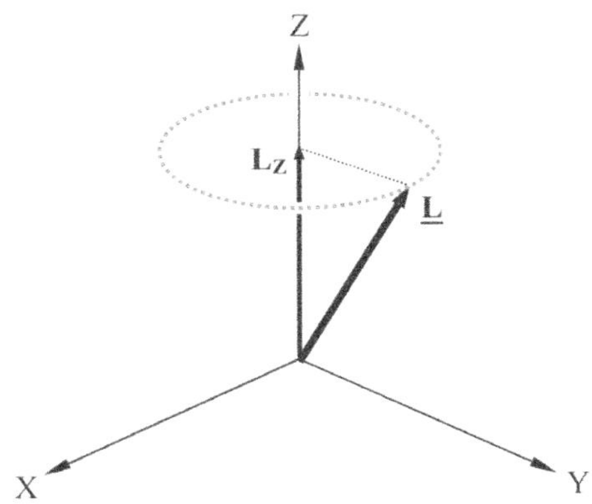

Figure 4.1 Imagined precession-like motion of the electron angular momentum vector, emphasizing the fact that it could not exhibit a unique orientation in space.

The latter conclusion is linked directly to the uncertainty principle. Qualitatively, it is obvious that there should be some distribution in the expectation values of squared angular velocity X and Y components. For example, let $\Delta\underline{L}_X$ be the standard mean of $\underline{L}_X$. It satisfies the relation $(\Delta\underline{L}_X)^2 = \left\langle (\underline{L}_X - \langle \underline{L}_X \rangle)^2 \right\rangle = \langle \underline{L}_X{}^2 \rangle$. Similarly, $(\Delta\underline{L}_Y)^2 = \langle \underline{L}_Y{}^2 \rangle$. The squared angular momentum expectation value is given by

$$\left\langle \underline{L}^2 \right\rangle = \ell(\ell+1)\hbar^2 = \left\langle \underline{L}_X{}^2 \right\rangle + \left\langle \underline{L}_Y{}^2 \right\rangle + \left\langle \underline{L}_Z{}^2 \right\rangle = \left\langle \underline{L}_X{}^2 \right\rangle + \left\langle \underline{L}_Y{}^2 \right\rangle + m^2\hbar^2 \quad ,$$

therefore

$$(\Delta\underline{L}_X)^2 + (\Delta\underline{L}_Y)^2 = \left(\ell^2 + \ell - m^2\right)\hbar^2 \quad . \tag{4.8}$$

This result clearly shows that the minimal uncertainty in the X and Y angular momentum components obtains for $|m| = \ell$, namely when the angular momentum vector points closest to the Z axis. There is a unique case for which a complete certainty exists for the X and Y angular momentum components; it is when $\ell = 0$.

It is possible to express an additional relation among the standard deviations of the angular momentum components, emanating from commutation relations among the angular momentum operator components. A commutation relation operator between two **A** and **B** operators is defined by

$$[\mathbf{A}, \mathbf{B}] = \mathbf{AB} - \mathbf{BA} \quad . \tag{4.9}$$

By direct calculation, it is possible to prove that

$$[\underline{r}_s, \mathbf{p}_t] = i\hbar\delta_{st} \quad ; \quad [\underline{r}_s, \underline{r}_t] = 0 \quad ; \quad [\mathbf{p}_s, \mathbf{p}_t] = 0 \quad , \tag{4.10}$$

where the numeral subscripts s, $t = 1, 2, 3$ indicate, respectively, the X, Y, Z components of the $\underline{r}$ position vector and the **p** linear momentum operator (Table 3.1). These relations also yield the following relations:

$$[\mathbf{L}_X, \mathbf{L}_Y] = i\hbar\mathbf{L}_Z \quad ; \quad [\mathbf{L}_Y, \mathbf{L}_Z] = i\hbar\mathbf{L}_X \quad ; \quad [\mathbf{L}_Z, \mathbf{L}_X] = i\hbar\mathbf{L}_Y \quad . \tag{4.11}$$

Commutation relations between quantum operators **A** and **B** connect between their related physical A and B expectations uncertainties (mean deviations) by the inequality

$$\boxed{(\Delta A)^2(\Delta B)^2 \geq \left\langle \left\{ \frac{i}{2}[\mathbf{A},\mathbf{B}] \right\}^2 \right\rangle} \quad . \tag{4.12}$$

We are quoting this important relation without proof. Setting $\mathbf{A} = \mathbf{L}_X$ and $\mathbf{B} = \mathbf{L}_Y$, and the first commutation relation of Eq. (4.11), we get

$$\boxed{(\Delta \underline{\mathbf{L}}_X)^2(\Delta \underline{\mathbf{L}}_Y)^2 \geq \frac{\hbar^2}{4}\left\langle \underline{\mathbf{L}}_Z{}^2 \right\rangle = \frac{m^2\hbar^4}{4}} \quad . \tag{4.13}$$

The above properties of the angular momentum allow expression of various operators in terms of the angular momentum. As an example, let's consider the spherical coordinates presentation of the kinetic energy:

$$\left.\begin{aligned} \mathbf{E}_k \equiv \frac{\mathbf{p}^2}{2\mu} &= -\frac{\hbar^2}{2\mu}\left[\frac{1}{r^2}\frac{\partial}{\partial r}\left(r^2\frac{\partial}{\partial r}\right) + \frac{1}{r^2\,\mathbf{sin}\,\theta}\frac{\partial}{\partial\theta}\left(\mathbf{sin}\,\theta\frac{\partial}{\partial\theta}\right) + \frac{1}{r^2\,\mathbf{sin}^2\theta}\frac{\partial^2}{\partial\varphi^2}\right] \\ &= -\frac{\hbar^2}{2\mu}\left[\frac{1}{r^2}\frac{\partial}{\partial r}\left(r^2\frac{\partial}{\partial r}\right)\right] + \frac{\mathbf{L}^2}{2\mu r^2} \end{aligned}\right\} \quad . \tag{4.14}$$

Transition from the first to second line is based on Eq. (4.5). Accordingly, the Hamiltonian operator may be written as

$$\boxed{\mathbf{H} = -\frac{\hbar^2}{2\mu}\left[\frac{1}{r^2}\frac{\partial}{\partial r}\left(r^2\frac{\partial}{\partial r}\right)\right] + \frac{\mathbf{L}^2}{2\mu r^2} - \frac{Ze^2}{r}} \quad . \tag{4.15}$$

Eq. (4.15) is useful as a starting step for expressing Schrödinger's equation in a form similar to a one-dimensional differential equation. We perform the usual variables separation $\psi = \boldsymbol{R}(r)\boldsymbol{Y}_{\ell m}(\theta,\varphi)$, and recall that $\mathbf{L}^2\boldsymbol{Y}_{\ell m}(\theta,\varphi) = \ell(\ell+1)\hbar^2\boldsymbol{Y}_{\ell m}(\theta,\varphi)$. Then, for the radial part $\boldsymbol{R}$, we get

$$\boxed{-\frac{\hbar^2}{2\mu}\left[\frac{1}{r^2}\frac{d}{dr}\left(r^2\frac{d}{dr}\right)\right]\boldsymbol{R}(r) + \left(\frac{\ell(\ell+1)\hbar^2}{2\mu r^2} + \boldsymbol{V}(r)\right)\boldsymbol{R}(r) = E\boldsymbol{R}(r)} \quad .$$

We define a new function $\boldsymbol{u}(r) \equiv \boldsymbol{R}(r)/r$, for which the radial part of Schrödinger's equation assumes the form

$$-\frac{\hbar^2}{2\mu}\frac{d^2\boldsymbol{u}(r)}{dr^2} + \left[\boldsymbol{V}(r) + \frac{\ell(\ell+1)\hbar^2}{2\mu r^2}\right]\boldsymbol{u}(r) = E\boldsymbol{u}(r) \quad . \tag{4.16}$$

This equation is often addressed as the **radial Schrödinger equation**. It exhibits the form of a one-dimensional Schrödinger's equation belonging to an effective potential

$$\boldsymbol{V}_{eff}(r) = \boldsymbol{V}(r) + \frac{\ell(\ell+1)\hbar^2}{2\mu r^2} \quad .$$

4.2 Mechanical Relativistic Correction to the Eigenenergies of the Hydrogen Atom

An underlying assumption of the initial Schrödinger equation (Eq. [3.16]) is that the particles' velocities are sufficiently small compared to the speed of light, such that it is possible to ignore relativity mass changes effects. It is thus possible to add corrections to the eigenenergies provided in Eq. (3.38) or Eq. (3.49) that account for those changes, and appropriate corrections that follow below. However, no such correction pretends to render Schrödinger's equations (3.15) and (3.16) consistent with fundamental relativistic principles. It would just be a correction, artificially imposed on our calculations, that are basically non-relativistic.

Relativistically, the total energy of a moving particle is given by

$$E = V + c\sqrt{m^2c^2 + p^2} \ , \tag{4.17}$$

where V is the body's potential energy, m is its rest mass, and p is its linear momentum. We expand Eq. (4.17) in powers of p^2 around the $p^2 = 0$ point to get

$$E = V + mc^2 + \frac{p^2}{2m} - \frac{1}{2mc^2}\left(\frac{p^2}{2m}\right)\left(\frac{p^2}{2m}\right) + \ldots \ . \tag{4.18}$$

The first addend on the right side is obviously the body's potential energy, and it is expressed in the $-Ze^2/r$ form in Eq. (3.21) for hydrogen-like atoms. The second addend provides the electron rest energy; being constant, it need not be expressed in the dynamic equations. The third addend is the kinetic energy. The kinetic energy operator appears in Schrödinger's equation as $(-\hbar^2/2\mu)\underline{\nabla}^2$, where μ is the reduced mass of the integrated electron plus nucleus system. The higher-order addends don't appear in the equations. In the following they are considered as perturbations to the states' energies (see Section 2.1.1). The most important is obviously the first. When transformed into a quantum operator, it assumes the form

$$\mathbf{E}_{rc} = -\frac{1}{2\mu c^2}\left(\frac{-\hbar^2\underline{\nabla}^2}{2\mu}\right)\left(\frac{-\hbar^2\underline{\nabla}^2}{2\mu}\right) \ . \tag{4.19}$$

The "rc" subscript on the equation's left side is the acronym for the "relativistic correction" term. Considering the now "zero-order" hydrogen Schrödinger equation, we may see that

$$\left(\frac{-\hbar^2\underline{\nabla}^2}{2\mu}\right)|n\ell m\rangle = \left(E_n + \frac{Ze^2}{r}\right)|n\ell m\rangle \ .$$

Therefore,

$$\mathbf{E}_{rc}|n\ell m\rangle = \frac{-1}{2\mu c^2}\left(E_n + \frac{Ze^2}{r}\right)^2|n\ell m\rangle = \frac{-1}{2\mu c^2}\left(E_n{}^2 + 2E_n\frac{Ze^2}{r} + \frac{Z^2e^4}{r^2}\right)|n\ell m\rangle \ .$$

Multiplying from the left by $\langle n\ell m|$ yields the diagonal elements of the perturbation matrix (see Eq. [2.8] and relevant text).

$$E_{rc} \equiv \langle \mathbf{E}_{rc} \rangle_{n\ell m\ n\ell m} \equiv \langle n\ell m | \mathbf{E}_{rc} | n\ell m \rangle = \frac{-1}{2\mu c^2}\left(E_n^{\ 2} + 2E_n Z e^2 \left\langle \frac{1}{r} \right\rangle + Z^2 e^4 \left\langle \frac{1}{r^2} \right\rangle \right) .$$

Expectation values for $1/r$ and $1/r^2$ are given in Eqs. (3.47-b). For demonstration we ignore this information and perform an independent evaluation of $(1/r)$'s expectation value. Following Eqs. (3.40) and (3.42), we get

$$\left\langle \frac{1}{r} \right\rangle = \int_0^\infty r^{-1} |\mathbf{R}_{n\ell}(r)|^2 r^2 dr = \left(\frac{2Z}{na_0} \right) \frac{(n-\ell-1)!}{2n[(n+\ell)!]^3} \int_0^\infty \rho^{2\ell+1} e^{-\rho} \mathbf{L}_{n-\ell-1}^{2\ell+1}(\rho) \mathbf{L}_{n-\ell-1}^{2\ell+1}(\rho) d\rho .$$

Then, using the orthogonality property provided in Eq. (3.40-b), we get

$$\left\langle \frac{1}{r} \right\rangle = \left(\frac{2Z}{na_0} \right) \frac{(n-\ell-1)!}{2n[(n+\ell)!]^3} \cdot \frac{[(n+\ell)!]^3}{(n-\ell-1)!} = \frac{1}{n^2} \left(\frac{Z}{a_0} \right) .$$

Further, copying from Eq. (3.47-b), we get

$$\left\langle \frac{1}{r^2} \right\rangle = \frac{2}{n^3(2\ell+1)} \left(\frac{Z}{a_0} \right)^2 .$$

Therefore,

$$E_{rc} = \frac{-1}{2\mu c^2}\left(E_n^{\ 2} + 2E_n Z e^2 \frac{1}{n^2}\left(\frac{Z}{a_0} \right) + Z^2 e^4 \frac{2}{n^3(2\ell+1)} \left(\frac{Z}{a_0} \right)^2 \right) .$$

Recall the relations $a_0 = \dfrac{\hbar^2}{\mu e^2}$ and $E_n = -\left(\dfrac{Z^2 e^2}{2a_0} \right) \dfrac{1}{n^2}$. We further define $\alpha \equiv e^2/c\hbar$ and get

$$\boxed{E_{rc} = \left(\frac{|E_n|}{n} \right) Z^2 \alpha^2 \left(\frac{3}{4n} - \frac{2}{2\ell+1} \right)} . \tag{4.20}$$

The quantity $\alpha \equiv e^2/c\hbar = 7.29736 \times 10^{-3} \cong 1/137$ is a universal constant termed **fine-structure constant**.

It may be directly verified that all off-diagonal elements of the perturbation matrix are identically zero. Thus, Eq. (4.20) provides the required energy correction for each state. This correction removes the normal degeneracy assumed previously between different ℓ states belonging to the same principal quantum number n; in Eq. (4.20), different ℓ states belonging to the same n will assume different corrections. The corrections will always be only a small fraction, of the order of $\alpha^2 \sim 5 \times 10^{-5}$ relative to the absolute state energy; the splitting between adjacent ℓ states (a possible situation for large n's), turns smaller for larger ℓ's.

4.3 Electron Spinning

So far, we have considered the electron as a point mass and electric charge. It has been conjectured that the electron could also be spinning about an axis. Because the electron has a rest mass, its spinning will be accompanied by an angular momentum, and because it is electrically charged, its spinning will be accompanied by a magnetic moment. Later in time, it has been realized that modeling the electron as a stiff particle spinning about an axis is too simplistic. However, as we shall see, this model serves well as a useful starting point for various approximate calculations. This section concerns the electron spinning effects.

4.3.1 Infinitesimal Rotations and the Angular Momentum Operator

By **Euler**'s theorem, the motion of a stiff body that keeps a single unique point stable in space can always be described as a combination of three consecutive rotations about the axes of a stiffly attached Cartesian coordinate system $X'Y'Z'$, termed **rest frame**: rotation by a $\Delta\varphi_1$ angle around the Z' axis; rotation by $\Delta\varphi_2$ angle about the Y' axis; an additional rotation by a $\Delta\varphi_3$ angle about the Z' axis. A particle steady-state wave function establishes a "stiff body" in the sense that its values don't change in the rest system even when this system rotates about the measuring laboratory coordinate system (**laboratory frame** or **Lab frame**). Let's consider a rotation component around the laboratory Z axis which coincides with the Z' axis, and assume the body to have performed a $\Delta\varphi$ rotation about the Z axis. The relation between the wave-function coordinates (x, y, z) in the lab frame and the wave-function coordinates (x', y', z') in the rest frame is given by

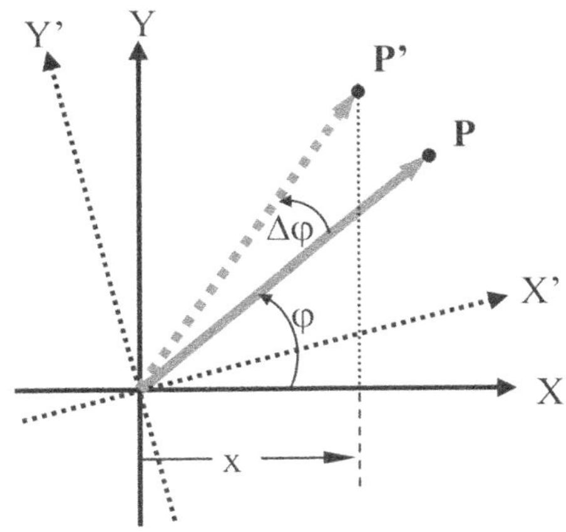

Figure 4.2 Geometric Drawing aimed at assisting the obtainment of the $\boldsymbol{A}_Z$ matrix expression in Eq. 4.21. The $Z = Z'$ axes emerge out of the page toward the viewer.

$$\begin{pmatrix} x \\ y \\ z \end{pmatrix} = \begin{pmatrix} \cos\Delta\varphi & -\sin\Delta\varphi & 0 \\ \sin\Delta\varphi & \cos\Delta\varphi & 0 \\ 0 & 0 & 1 \end{pmatrix} \begin{pmatrix} x' \\ y' \\ z' \end{pmatrix} \equiv \boldsymbol{A}_Z \begin{pmatrix} x' \\ y' \\ z' \end{pmatrix} . \tag{4.21}$$

The right side of the last equation defines the rotation matrix $\boldsymbol{A}_Z$.

To prove Eq. (4.21), we consider the drawing provided in Figure 4.2. (X, Y, Z) is the lab frame remaining unchanged during the stiff body rotation. (X', Y', Z') is the body rest frame that initially coincided with the lab frame. The P point is a projected body point on the (X, Y) plane. Following a body rotation by a $\Delta\varphi$ angle, the body rest frame assumes the (X', Y', Z') coordinates, and the P point moves to the P' position. The P' coordinates in the body rest frame experienced no change, but underwent a change in the lab frame. For example, let's calculate x:

$$x = \sqrt{x'^2 + y'^2} \cdot \cos(\varphi + \Delta\varphi) = \sqrt{x'^2 + y'^2}(\cos\varphi \cos\Delta\varphi - \sin\varphi \sin\Delta\varphi) = x' \cos\Delta\varphi - y' \sin\Delta\varphi \quad .$$

Similarly, one obtains $y = x' \sin\Delta\varphi + y' \cos\Delta\varphi$. And per definition, $z = z'$. Hence **QED**.

Any stiff body rotation may be represented by a matrix connecting the two coordinate frames. Two consecutive rotations (around same or different axes) would be represented by matrix multiplication. Namely, if $\boldsymbol{A}$ is a first rotation and $\boldsymbol{B}$ is a second rotation, the representative matrix would be $\boldsymbol{BA}$. Rotations are not commutative; namely changing the operations order might change the outcome. Mathematically it means that the $\boldsymbol{AB}$ matrix might be different from the $\boldsymbol{BA}$ one. On the other hand, *infinitesimal rotations* are commutative. In that limit, any $\boldsymbol{A}$ rotation matrix may be represented as

$\boldsymbol{A} = \boldsymbol{I} + \boldsymbol{\varepsilon}_A$, where $\boldsymbol{I}$ is the unity matrix and $\boldsymbol{\varepsilon}_A$ is a matrix all of whose elements are infinitesimal. Multiplication of two matrices representing infinitesimal rotations would then look as

$$\boldsymbol{BA} = (\boldsymbol{I} + \boldsymbol{\varepsilon}_B)(\boldsymbol{I} + \boldsymbol{\varepsilon}_A) = \boldsymbol{I} + \boldsymbol{\varepsilon}_B + \boldsymbol{\varepsilon}_A + \boldsymbol{\varepsilon}_B\boldsymbol{\varepsilon}_A \underset{\substack{\text{first-order}\\ \text{infinitesimal}\\ \text{limit}}}{=} \boldsymbol{I} + \boldsymbol{\varepsilon}_B + \boldsymbol{\varepsilon}_A \ . \tag{4.22}$$

Because matrix addition over a commutative field (numbers) is commutative, in the infinitesimal limit, matrix multiplication is also commutative. This property provides a mathematical convenience without reducing the generality of conclusions, because any finite rotation may be represented as the limit of a long series of consecutive infinitesimal rotations.

In the infinitesimal limit, where we exchange $\Delta\varphi$ by $d\varphi$, the $\boldsymbol{A}_Z$ matrix of Eq. (4.21) assumes the shape

$$\boldsymbol{A}_Z = \begin{pmatrix} \cos\Delta\varphi & -\sin\Delta\varphi & 0 \\ \sin\Delta\varphi & \cos\Delta\varphi & 0 \\ 0 & 0 & 1 \end{pmatrix} \underset{\text{limit}}{\overrightarrow{\text{infinitesimal}}} \begin{pmatrix} 1 & 0 & 0 \\ 0 & 1 & 0 \\ 0 & 0 & 1 \end{pmatrix} - d\varphi \begin{pmatrix} 0 & 1 & 0 \\ -1 & 0 & 0 \\ 0 & 0 & 0 \end{pmatrix} \equiv \boldsymbol{I} - d\varphi\boldsymbol{Q}_Z \ .$$

The right side is a definition of a $\boldsymbol{Q}_Z$ designated matrix.

We now perform an infinitesimal rotation of a stiff body around the Z axis. In the *lab frame*, the body wave function satisfies

$$\begin{aligned} \psi(\underline{r}) &= \psi\left(\boldsymbol{A}_Z\underline{r}'\right) \equiv \mathbf{A}_Z\psi\left(\underline{r}'\right) = \psi\left(\underline{r}' - d\varphi\boldsymbol{Q}_Z\underline{r}'\right) \underset{\substack{\text{Taylor}\\ \text{expansion}}}{\cong} \\ &= \psi\left(\underline{r}'\right) - \left(\underline{\nabla}'\psi\left(\underline{r}'\right)\cdot d\varphi\boldsymbol{Q}_Z\underline{r}'\right) = \psi\left(\underline{r}'\right) - d\varphi\left(\underline{\nabla}'\psi\left(\underline{r}'\right)\cdot\boldsymbol{Q}_Z\underline{r}'\right) \end{aligned} ,$$

where the $\mathbf{A}_Z$ quantum operator is in fact *defined* by the $\mathbf{A}_Z\psi(\underline{r}') \equiv \psi(\boldsymbol{A}_Z\underline{r}')$ relation. Therefore,

$$\frac{\psi(\underline{r}) - \psi(\underline{r}')}{d\varphi} = \left(\frac{\partial\psi}{\partial x'}, \frac{\partial\psi}{\partial y'}, \frac{\partial\psi}{\partial z'}\right)\begin{pmatrix} -y' \\ +x' \\ 0 \end{pmatrix} = -y'\frac{\partial\psi}{\partial x'} + x'\frac{\partial\psi}{\partial y'} = \frac{i}{\hbar}\mathbf{L}_Z\psi\left(\underline{r}'\right) \ . \tag{4.23}$$

For transition to the right side, see Eq. (4.2). The $\psi(\underline{r}')$ wave function remains constant in the body rest frame; its value is identical with the value that $\psi(\underline{r})$ *had* in the lab frame *before* the rotation operation. On the other hand, $\psi(\underline{r})$ assumes a new value as a result of rotation. Therefore, the left side of Eq. (4.23) is simply the derivative $\partial\psi(\underline{r})/\partial\varphi$. Then we may formulate an identity relation between the following two quantum operators:

$$\boxed{\mathbf{L}_Z \equiv -i\hbar\frac{\partial}{\partial\varphi}} \ . \tag{4.24}$$

Seemingly, this result appears identical with the third line in Eq. (4.3). However, the physical meaning is different. In the present result, the $\partial/\partial\varphi$ operator on the right side considers the change in the wave function as observed in the *lab frame* resulting from a stiff rotation of the particle. The $\mathbf{L}_Z$ operator on the left side operates on the wave function *in the particle rest frame*. The meaning of the identity is that the *result* of these two different operations is identical! Calculations performed in relation to infinitesimal rotations about other axes yield the same formal mathematical identities as in other lines in Eq. (4.3); the physical meaning of the two equations' sides should be understood in the same manner.

Another physical meaning relates to the wave function symmetry. A rotation operation performed to test the wave function symmetry in relation to that rotation could be represented by the angular momentum operator acting in the particle rest frame. For example, the $\mathbf{Y}_{\ell m}(\theta, \varphi)$ wave function exhibits a rotational symmetry of order m around the Z axis (namely, frame rotation by $2\pi/m$ exhibits no change in the wave function value; see Section 3.3). Notably, letting $\mathbf{L}_Z$ operate on the $\mathbf{Y}_{\ell m}(\theta, \varphi)$ function provides $m \cdot \hbar$ as the eigenvalue (namely, $\mathbf{L}_Z|\mathbf{Y}_{\ell m}(\theta, \varphi)\rangle = m\hbar \cdot |\mathbf{Y}_{\ell m}(\theta, \varphi)\rangle$; see Eq. [4.6]). The angular momentum operators thus reflect symmetry properties belonging to the full rotational symmetry group. Group theory deals with algebra of symmetry operations, and we shall occasionally use, directly or indirectly, terms and conclusions of that theory.

4.3.2 Generalization of the Angular Momentum Concept

In Section 4.1 we saw that the angular momentum operators exhibit some commutation relations summarized in Eq. (4.11). It is possible to generalize the angular momentum concept by an opposite procedure: it is assumed that the commutation relations are those that *define* the angular momentum operators, and look for their eigenvalues and eigenfunctions. As we shall see, new eigenfunctions emerge, with far-reaching and quite surprising physical conclusions. To emphasize the generalization fact, the generalized operators are symbolized by the letter $\mathbf{J}$ instead of $\mathbf{L}$. Namely, we assume the existence of Hermitian operators $\mathbf{J}_X$, $\mathbf{J}_Y$, and $\mathbf{J}_Z$, satisfying the following commutation relations (see Eq. [4.11]):

$$[\mathbf{J}_X, \mathbf{J}_Y] = i\hbar\mathbf{J}_Z \quad ; \quad [\mathbf{J}_Y, \mathbf{J}_Z] = i\hbar\mathbf{J}_X \quad ; \quad [\mathbf{J}_Z, \mathbf{J}_X] = i\hbar\mathbf{J}_Y \quad . \tag{4.25}$$

In analogy we further define

$$\mathbf{J}^2 = \mathbf{J}_X{}^2 + \mathbf{J}_Y{}^2 + \mathbf{J}_Z{}^2 \quad . \tag{4.26}$$

For calculation convenience, we define new (non-Hermitian) operators of the form

$$\mathbf{J}_+ \equiv \mathbf{J}_X + i\mathbf{J}_Y \quad ; \quad \mathbf{J}_- \equiv \mathbf{J}_X - i\mathbf{J}_Y \quad .$$

The commutation relations exhibited by the above $\mathbf{J}$ operators are

$$\left.\begin{array}{lll} [\mathbf{J}^2, \mathbf{J}_X] = 0 \ ; & [\mathbf{J}^2, \mathbf{J}_Y] = 0 \ ; & [\mathbf{J}^2, \mathbf{J}_Z] = 0 \\ [\mathbf{J}^2, \mathbf{J}_\pm] = 0 & & \\ [\mathbf{J}_Z, \mathbf{J}_+] = \hbar\mathbf{J}_+ \ ; & [\mathbf{J}_Z, \mathbf{J}_-] = -\hbar\mathbf{J}_- \ ; & [\mathbf{J}_+, \mathbf{J}_-] = 2\hbar\mathbf{J}_Z \end{array}\right\} \tag{4.27}$$

4.3.2.1 Basis Functions Properties

As basis for the wave functions space, we require that such functions be orthonormal and simultaneously eigenfunctions of both $\mathbf{J}^2$ and $\mathbf{J}_Z$. We symbolize the eigenfunctions by $|jm\rangle$, where the j and m symbols serve as labels for the λ_j and λ_m eigenvalues of $\mathbf{J}^2$ and $\mathbf{J}_Z$, respectively. Accordingly, the $\{|jm\rangle\}$ set of functions forms a complete family, and eigenfunctions belonging to different j or m are orthonormal. In a strict mathematical writing, the requirements are written as:

$$\left.\begin{array}{ll} \text{(a)Orthogonality :} & \langle j'm'|jm\rangle = \delta_{j'j}\delta_{m'm} \\ \text{(b)Completeness :} & \mathbf{O}|jm\rangle = \sum\limits_{j'm'} |j'm'\rangle\langle j'm'|\mathbf{O}|jm\rangle \end{array}\right\} , \tag{4.28}$$

for any **O** operator acting on the eigenfunctions space of the generalized angular momentum.

We further define $m \equiv \lambda_m/\hbar$; this definition will serve hereinafter also as a label for the $|jm\rangle$ function.

Now we consider the matrix elements of the $\mathbf{J}_Z\mathbf{J}_+ - \mathbf{J}_+\mathbf{J}_Z = \hbar\mathbf{J}_+$ commutation relation between $|jm'\rangle$ and $|jm\rangle$. Because the $\mathbf{J}_Z$ operator is Hermitian by assumption (see Hermitianity definition in relation to Table 3.1), we get

$$\left(m' - m\right)\hbar\left\langle jm'|\mathbf{J}_+|jm\right\rangle = \hbar\left\langle jm'|\mathbf{J}_+|jm\right\rangle \ . \tag{4.29}$$

Eq. (4.29) is trivially satisfied for any value of m if the $\langle jm'|\mathbf{J}_+|jm\rangle$ matrix element vanishes; for example, when $\mathbf{J}_+|jm\rangle = 0$. On the other hand, the only non-vanishing $\mathbf{J}_+$ matrix elements are those satisfying $(m' - m) = 1$. For such cases we may write

$$\mathbf{J}_+|jm\rangle = \xi_m\hbar|j, m+1\rangle \ , \tag{4.30}$$

where the ξ_m quantity is a number which may be complex. Using similar considerations, it may be obtained that

$$\mathbf{J}_-|jm\rangle = \xi'_m\hbar|j, m-1\rangle \ . \tag{4.31}$$

By the definition of the $\mathbf{J}_+$ and $\mathbf{J}_-$, and Hermitianity of $\mathbf{J}_X$ and $\mathbf{J}_y$, it may be readily verified that ξ_m and ξ'_{m+1} are complex conjugates of each other; namely that $\xi'_{m+1} = \xi^*_m$. As a result, operating the commutation operator $\mathbf{J}_+\mathbf{J}_- - \mathbf{J}_-\mathbf{J}_+ = 2\hbar\mathbf{J}_Z$ yields that $\xi_{m-1}\xi^*_{m-1} - \xi^*_m\xi_m = 2m$. We thus obtained a difference equation for the $|\xi_m|^2$ series of elements:

$$|\xi_{m-1}|^2 - |\xi_m|^2 = 2m \ . \tag{4.32}$$

The general solution for the $|\xi_m|^2$ series (easily proved by induction) is

$$|\xi_m|^2 = C - m(m+1) \ , \tag{4.33}$$

where C could arbitrarily be any real constant. Eq. (4.33) indicates that for any finite value of C, the right side becomes negative for sufficiently large positive or negative values of m. However, $|\xi_m|^2$ is necessarily non-negative. Recalling Eqs. (4.29) and (4.30), $|\xi_m|^2$ *could vanish* when $\mathbf{J}_+|jm\rangle = 0$. m-values related to such special state/s are given as solutions of the square equation $C - m(m+1) = 0$: a lower valued $\underline{m}$ and a higher valued $\overline{m}$ are given by

$$\underline{m} = -\frac{1}{2} - \frac{1}{2}(1+4C)^{1/2} \text{ and } \overline{m} = -\frac{1}{2} + \frac{1}{2}(1+4C)^{1/2} \ . \tag{4.34}$$

Obviously, $C = \overline{m}(\overline{m}+1)$, then $\underline{m} = -(\overline{m}+1)$, or $\overline{m} - \underline{m} = 2\overline{m} + 1$. We may correspondingly suppose that $|\xi_m|^2$ could all be *nonzero* for a restricted range of m values given by

$$m = \underline{m}+1, \underline{m}+2, \ldots, \overline{m}-2, \overline{m}-1 \ . \tag{4.35}$$

Because $\overline{m}$ and $\underline{m}$ differ by an integer, $2\overline{m}$ is also a nonnegative integer, and $\overline{m}$ may only take the values $0, \frac{1}{2}, 1, \frac{3}{2}, 2, \ldots$.

4.3.2.2 Eigenvalues of the $\mathbf{J}^2$ Operator

We express the $\mathbf{J}^2$ operator in the following manner:

$$\mathbf{J}^2 = \frac{1}{2}(\mathbf{J}_+\mathbf{J}_- + \mathbf{J}_-\mathbf{J}_+) + \mathbf{J}_Z{}^2 \ . \tag{4.36}$$

We consider the eigenvalue equation $\mathbf{J}^2|jm\rangle = \lambda_j|jm\rangle$. We get

$$\lambda_j = \frac{\hbar^2}{2}\left(|\xi_{m-1}|^2 + |\xi_m|^2\right) + m^2\hbar^2$$
$$= \frac{\hbar^2}{2}[\overline{m}(\overline{m}+1) - (m-1)m + \overline{m}(\overline{m}+1) - m(m+1)] + m^2\hbar^2 = \overline{m}(\overline{m}+1)\hbar^2 \; .$$

Notably, λ_j is independent of m. We may thus shorten the λ_j notation and use the symbol "j" instead, and further identify between $\overline{m}$ and j. In summary,

$$\boxed{\begin{aligned}\mathbf{J}^2|jm\rangle &= \hbar^2 j(j+1)|jm\rangle \\ \mathbf{J}_Z|jm\rangle &= \hbar m|jm\rangle\end{aligned}} \; , \tag{4.37}$$

where the possible j and m values are

$$\left.\begin{aligned} j &= 0,\ \frac{1}{2},\ 1,\ \frac{3}{2},\ 2,\ \frac{5}{2},\ \dots \\ m &= -j,\ -j+1,\ \dots,\ j-1, j \end{aligned}\right\} . \tag{4.38}$$

Thus, $2j+1$ values of m correspond to each j state value, hence $2j+1$ different eigenfunctions. **The extraordinary novelty of the latter results is the prospect for the existence of angular momentum operators having half-odd eigenvalues.**

4.3.2.3 Matrix Elements of Angular Momentum Operators

We start by considering the matrix elements of $\mathbf{J}_+$ and $\mathbf{J}_-$. Their calculation basis are Eqs. (4.30) to (4.35), defined, however, up to a phase factor. It is customary to select it as +1. Using this selection, we get

$$\begin{aligned}\mathbf{J}_+|jm\rangle &= \hbar[(j-m)(j+m+1)]^{1/2}|j,m+1\rangle \\ \mathbf{J}_-|jm\rangle &= \hbar[(j+m)(j-m+1)]^{1/2}|j,m-1\rangle\end{aligned} \; . \tag{4.39}$$

Nonzero $\mathbf{J}_X$ and $\mathbf{J}_Y$ elements are

$$\left.\begin{aligned}\langle j,m+1|\mathbf{J}_X|jm\rangle &= \frac{1}{2}\hbar[(j-m)(j+m+1)]^{1/2} \\ \langle j,m-1|\mathbf{J}_X|jm\rangle &= \frac{1}{2}\hbar[(j+m)(j-m+1)]^{1/2} \\ \langle j,m+1|\mathbf{J}_Y|jm\rangle &= -\frac{i}{2}\hbar[(j-m)(j+m+1)]^{1/2} \\ \langle j,m-1|\mathbf{J}_Y|jm\rangle &= \frac{i}{2}\hbar[(j+m)(j-m+1)]^{1/2}\end{aligned}\right\} . \tag{4.40}$$

4.3.2.4 Electron Spin

A case deserving special elaboration is $j = \frac{1}{2}$, for which $m = \pm\frac{1}{2}$. Such property is addressed as "**internal spinning**" or simply "**spin**". The angular momentum operator gains the special symbol $\mathbf{S}$ instead of $\mathbf{J}$, the eigenvalue $s = \frac{1}{2}$, and m values $m_s = \pm\frac{1}{2}$. The matrix elements are

$\left\langle\frac{1}{2},m'\lvert\mathbf{s}_X\rvert\frac{1}{2},m\right\rangle$			$\left\langle\frac{1}{2},m'\lvert\mathbf{s}_Y\rvert\frac{1}{2},m\right\rangle$			$\left\langle\frac{1}{2},m'\lvert\mathbf{s}_Z\rvert\frac{1}{2},m\right\rangle$		
m' \ m	$+\frac{1}{2}$	$-\frac{1}{2}$	m' \ m	$+\frac{1}{2}$	$-\frac{1}{2}$	m' \ m	$+\frac{1}{2}$	$-\frac{1}{2}$
$+\frac{1}{2}$	0	$\hbar/2$	$+\frac{1}{2}$	0	$-i\hbar/2$	$+\frac{1}{2}$	$\hbar/2$	0
$-\frac{1}{2}$	$\hbar/2$	0	$-\frac{1}{2}$	$+i\hbar/2$	0	$-\frac{1}{2}$	0	$-\hbar/2$

The obtained matrices are often written as

$$\sigma_X = \frac{\hbar}{2}\begin{pmatrix} 0 & 1 \\ 1 & 0 \end{pmatrix} \quad ; \quad \sigma_Y = \frac{\hbar}{2}\begin{pmatrix} 0 & -i \\ +i & 0 \end{pmatrix} \quad ; \quad \sigma_Z = \frac{\hbar}{2}\begin{pmatrix} 1 & 0 \\ 0 & -1 \end{pmatrix} \quad . \tag{4.41}$$

For brevity, the $(\hbar/2)$ factor in the matrices is often deleted, and they are then termed "**Pauli spin matrices**". It may be readily verified that the σ_X, σ_Y, and σ_Z matrices obey the same multiplication and commutation relations as the $\mathbf{s}_X$, $\mathbf{s}_Y$, and $\mathbf{s}_Z$ operators, respectively. It is impossible to find neither one-dimensional matrices (numbers) that would exhibit those relations, nor single-valued continuous functions of, for example, the (r, θ, φ) spherical coordinates that could serve as eigenfunctions of $\mathbf{s}^2$ and $\mathbf{s}_Z$. Suitable functions need to be a two-dimensional vector of single-valued functions

$$\chi(\underline{r}, t) = \begin{pmatrix} \psi_+(\underline{r}, t) \\ \psi_-(\underline{r}, t) \end{pmatrix} \quad . \tag{4.42}$$

The probability density of the particle to occur at a certain $\underline{r}$ position in space is given by

$$\left(\psi_+^*(\underline{r},t), \psi_-^*(\underline{r},t)\right)\begin{pmatrix} \psi_+(\underline{r}, t) \\ \psi_-(\underline{r},t) \end{pmatrix} = |\psi_+(\underline{r}, t)|^2 + |\psi_-(\underline{r}, t)|^2 \quad , \tag{4.43}$$

where the ψ_+ and ψ_- functions are normalized to satisfy $\int\left[|\psi_+(\underline{r}, t)|^2 + |\psi_-(\underline{r}, t)|^2\right]d^3\underline{r} = 1$. The $\mathbf{s}^2$ and $\mathbf{s}_Z$ steady-state eigenfunctions attain the form

$$|\chi_+\rangle \equiv \begin{pmatrix} \psi_+(\underline{r}) \\ 0 \end{pmatrix} \quad ; \quad |\chi_-\rangle \equiv \begin{pmatrix} 0 \\ \psi_-(\underline{r}) \end{pmatrix} \quad . \tag{4.44}$$

The $|\chi_+\rangle$ function is called **spin up**; the $|\chi_-\rangle$ function is called **spin down**. They satisfy the relations

$$\boxed{\mathbf{s}_Z|\chi_+\rangle = \frac{1}{2}\hbar|\chi_+\rangle \quad ; \quad \mathbf{s}_Z|\chi_-\rangle = -\frac{1}{2}\hbar|\chi_-\rangle \quad ; \quad \mathbf{s}^2|\chi_\pm\rangle = \frac{3}{4}\hbar^2|\chi_\pm\rangle} \quad . \tag{4.45}$$

To further explore these eigenfunctions properties we discuss their behavior under a finite rotation of the particle. We assume that Eq. (4.24) is valid for *all* angular momentum operators, including the generalized $\mathbf{J}_Z$ operators. We consider the azimuth angle dependence of a point in the steady state wave function, namely $\underline{r} = \underline{r}(\varphi)$. We get

$$\psi[\underline{r}(\varphi + d\varphi)] - \psi[\underline{r}(\varphi)] = d\varphi \cdot \frac{i}{\hbar}\mathbf{J}_Z\psi[\underline{r}(\varphi)] \quad ;$$

$$\psi[\underline{r}(\varphi + d\varphi)] = \left(1 + d\varphi \cdot \frac{i}{\hbar}\mathbf{J}_Z\right)\psi[\underline{r}(\varphi)] \quad .$$

If $\mathbf{J}_Z$ is represented by a matrix, and correspondingly ψ is represented by a column vector, the unity matrix $\boldsymbol{I}$ replaces the 1 number in the last equation. If we choose a finite rotation angle $\Delta\varphi$, we define a related infinitesimal rotation angle $d\varphi$ as $d\varphi = \Delta\varphi/N$, where N is very large to justify the "infinitesimal rotation" approximation. Then, for the finite rotation we get

$$\psi[\underline{r}(\varphi + \Delta\varphi)] = \left[1 + \frac{1}{N}\left(\Delta\varphi\cdot\frac{i}{\hbar}\mathbf{J}_Z\right)\right]^N \psi[\underline{r}(\varphi)] \ .$$

One then concludes that the finite rotation operator about the Z axis, $\mathbf{U}_Z(\Delta\varphi)$, is given by

$$\boxed{\mathbf{U}_Z(\Delta\varphi) = \lim_{N\to\infty}\left[1 + \frac{1}{N}\left(i\frac{\Delta\varphi}{\hbar}\mathbf{J}_Z\right)\right]^N \equiv e^{i\cdot\Delta\varphi\mathbf{J}_Z/\hbar}} \ . \tag{4.46}$$

The expression on the right side is symbolic, to remind the limit of the expression on the left when $\mathbf{J}_Z$ is a number. For demonstration, we choose the special (important) case of $\mathbf{J}_Z = \sigma_Z$. Because σ_Z is a diagonal matrix (Eq. [4.41]), the calculation is quite simple. We get

$$\mathbf{U}_Z(\Delta\varphi) = \begin{pmatrix} e^{i\Delta\varphi/2} & 0 \\ 0 & e^{-i\Delta\varphi/2} \end{pmatrix} \ . \tag{4.47}$$

We apply this operation to the interesting case of a full rotation, namely $\Delta\varphi = 2\pi$:

$$\mathbf{U}_Z(2\pi)\begin{pmatrix} \psi_+(\underline{r},t) \\ \psi_-(\underline{r},t) \end{pmatrix} = \begin{pmatrix} e^{\pi i} & 0 \\ 0 & e^{-\pi i} \end{pmatrix}\begin{pmatrix} \psi_+(\underline{r},t) \\ \psi_-(\underline{r},t) \end{pmatrix} = \begin{pmatrix} -1 & 0 \\ 0 & -1 \end{pmatrix}\begin{pmatrix} \psi_+(\underline{r},t) \\ \psi_-(\underline{r},t) \end{pmatrix} = -\begin{pmatrix} \psi_+(\underline{r},t) \\ \psi_-(\underline{r},t) \end{pmatrix} \ . \tag{4.48}$$

The result shows that when the particle performs a full rotation, **the wave function reverses sign**. In other words, the particle needs to perform *two full rotations* for the wave function to appear at its original phase. **Dirac** has shown that it is possible to conclude half-odd angular momentums for a particle, as solution of a properly relativity-modified Schrödinger equation.

It may be readily verified that the wave function changes its phase gradually upon rotation. Notably, the above features relate to the phase observation *in the lab frame*; the particle wave function itself is a "rigid body", undergoing no phase change in the particle frame while rotating.

The author of the present book is unaware of any straightforward graphical means of demonstrating the features of the spin-half wave function, because there should be two complex numbers adjoined to each point in space exhibiting proper phase relations. However, to remove some mystery feeling related to the half-odd spin concept, we show a geometric body exhibiting such topological property. Figure 4.3 describes a **Moebius stripe** wrapped around the Z axis. Arrows plotted continually across the stripe represent the wave function and its phase; in other words, an arrow projection on the Z axis provides, for example, the real part of the wave function, and its projection on the (X, Y) plane correspondingly describes the imaginary part. Each arrow is given a dark "back" and a bright "front", emphasizing the arrow rest phase. Rotation about the Z axis transfers the wave function, as observed in the lab frame, continuously

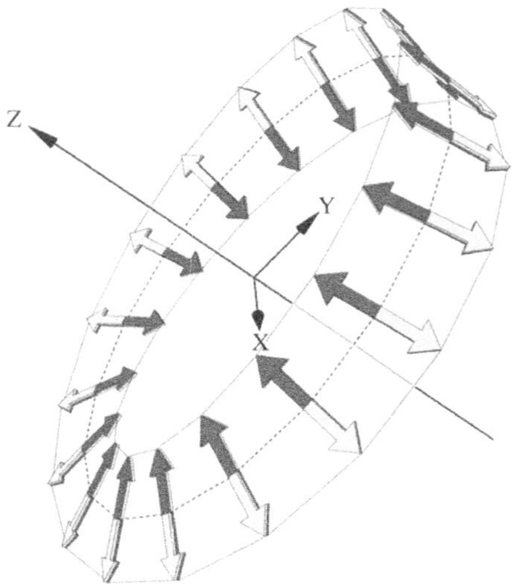

Figure 4.3 Demonstrated internal structure of a system whose rotational symmetry belongs to a spin 1/2 value, that is, it must perform two full rotations to restore its original phase, **as observed in the lab frame.**

between states described by adjacent arrows. Apparently, a full rotation presents an arrow with an inverted front and back relative to the stripe. Only an additional full rotation will present an arrow at the original posture. Further mathematical properties of half-odd spin states are provided in Section 5.3.2.

We are unable to add any information concerning the spatial dependence of the ψ_+ or ψ_- spin functions, because of our yet complete ignorance about the electron internal structure. We are enforced to be satisfied by knowing the eigenvalues and symmetry properties.

The possibility that the electron may exhibit a 1/2 angular momentum eigenstate ($s = 1/2$), usually addressed as **internal spin** or just **spin**, has been discovered experimentally by analysis of states splitting in atomic spectroscopy, even before Dirac's theoretical work. Spin 1/2 property exists for several elementary particles: protons, neutrons, neutrinos, μ-mesons, and their corresponding anti-particles.

4.4 Combining Orbital Angular Momentum and Spin

The occurrence of an internal spin is a relativistic effect. We include this property artificially into our calculations, which are basically non-relativistic. In view of the result summarized in Eq. (4.38), we could suppose that the spin 1/2 property ($s = 1/2$) is a ground-state of the electron angular momentum. However, as a zero-order approach, we assume that the spin property is a separate effect, independent of the non-relativistic orbital motion of the electron. This assumption supports the image of the electron as a stiff ball rotating about its axis in addition to rotating about the nucleus. Correspondingly, the electron wave function may be described as a product function

$$\psi = \psi_{n\ell m} \cdot \chi \quad . \tag{4.49}$$

However, in classical mechanics, the resultant angular momentum of two rotational motions is their vector sum. That sum is the invariant property under a central force field, and not each component separately. We thus assume that the resultant angular momentum is the invariant property in quantum mechanics as well. Correspondingly, we correct our calculations by performing a vector addition of the orbital angular momentum and the internal spin. The resultant angular momentum operator related to $\underline{J} = \underline{L} + \underline{s}$ will be $\mathbf{J} = \mathbf{L} + \mathbf{s}$, and also $\mathbf{J}_Z = \mathbf{L}_Z + \mathbf{s}_Z$. We further require that $\mathbf{J}$ would be a generalized angular momentum operator by its definition provided in Section 4.3.2. The mathematical meaning of the latter requirement is that the wave function, as designed by Eq. (4.49), will be an eigenfunction of both $\mathbf{J}_Z$ and $\mathbf{J}^2$.

For the first requirement we get

$$\mathbf{J}_Z\psi = (\mathbf{L}_Z + \mathbf{s}_Z)\psi_{n\ell m}\chi = m\hbar\psi_{n\ell m}\chi + \psi_{n\ell m}\mathbf{s}_Z\chi \quad .$$

The simple product function $\psi = \psi_{n\ell m} \cdot \chi$ would thus be an eigenfunction of $\mathbf{J}_Z$ if and only if χ is an eigenfunction of $\mathbf{s}_Z$, namely if $\chi = \chi_+$ (namely $m_s = +1/2$) or $\chi = \chi_-$ (namely $m_s = -1/2$). For given ℓ

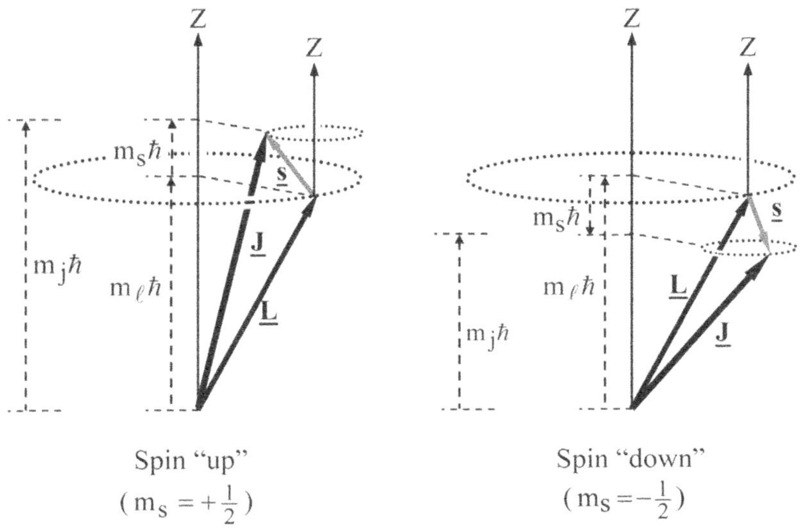

Figure 4.4 Vector addition of orbital and spin angular momentums. The resultant $\underline{J}$ angular momentum size (length) would not be constant if the $\underline{L}$ and $\underline{s}$ component vectors' Z axis projections are required to be invariant. Cases shown are "spin up" for which $j = \ell + s = \ell + \frac{1}{2}$, and "spin down", for which $j = \ell - s = \ell - \frac{1}{2}$. For detailed discussion – see text.

and m there are two eigenfunctions of $\mathbf{J}_Z$ of Eq. (4.49): $\psi_{n\ell m} \cdot \chi_+$ and $\psi_{n\ell m} \cdot \chi_-$. These functions are symbolized as $|n\ell m_\ell m_s\rangle$, and are called spin/orbit functions.

Figure 4.4 portrays two addition states between a spin and a given orbital angular momentum, each being an eigenfunction of $\mathbf{J}_Z$, where there is no linkage between their separate precession motions. The result is that *the total angular momentum size is not invariant*, although its projection on the Z axis remains constant. We now demand additionally, that an eigenfunction of $\mathbf{J}_Z$ is also an eigenfunction of $\mathbf{J}^2$. We write such eigenfunction as $|n\,\ell s\, j\, m_j\rangle$. Mathematically, it means that

$$\mathbf{J}^2 = (\mathbf{L} + \mathbf{s})^2 = \mathbf{L}^2 + \mathbf{s}^2 + \mathbf{Ls} + \mathbf{sL} = \mathbf{L}^2 + \mathbf{s}^2 + 2\mathbf{s}\cdot\mathbf{L} \quad . \tag{4.50}$$

Thus,

$$\begin{aligned}\mathbf{J}^2\left|n\ell sjm_j\right\rangle &= \mathbf{L}^2\left|n\ell sjm_j\right\rangle\mathbf{s} + \mathbf{s}^2\left|n\ell sjm_j\right\rangle + 2\mathbf{s}\cdot\mathbf{L}\left|n\ell sjm_j\right\rangle \\ &= \left(\ell(\ell+1)\hbar^2 + s(s+1)\hbar^2 + 2\mathbf{s}\cdot\mathbf{L}\right)\left|n\ell sjm_j\right\rangle\end{aligned} \quad .$$

On the other hand, we also require that $\mathbf{J}^2|n\,\ell s\, jm_j\rangle = j(j+1)\hbar^2|n\,\ell s\, jm_j\rangle$. Therefore, the $\mathbf{s}\cdot\mathbf{L}$ operator eigenvalue must satisfy the relation

$$\mathbf{s}\cdot\mathbf{L}\left|n\ell sjm_j\right\rangle = \tfrac{1}{2}\{\, j(\,j+1) - \ell(\ell+1) - s(s+1)\}\hbar^2\left|n\ell sjm_j\right\rangle .$$

In other words, the $\underline{s}\cdot\underline{L}$ expectation value, which also equals its eigenvalue, would be

$$\langle\underline{s}\cdot\underline{L}\rangle = \frac{1}{2}\{\, j(\,j+1) - \ell(\ell+1) - s(s+1)\}\hbar^2 \quad ; \tag{4.51}$$

$$\boxed{\langle\underline{s}\cdot\underline{L}\rangle = \begin{cases} \dfrac{1}{2}\ell\hbar^2, & \text{spin-up: } j = \ell + \dfrac{1}{2} \\ -\dfrac{1}{2}(\ell + h)\hbar^2 & \text{spin-down: } j = \ell - \dfrac{1}{2} \end{cases}} \quad . \tag{4.52}$$

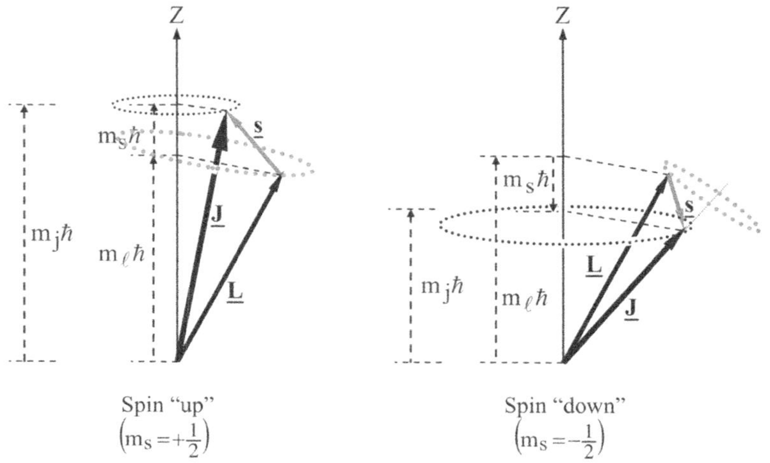

Figure 4.5 A vector addition scheme of given orbital angular momentum $\underline{\boldsymbol{L}}$ and spin $\underline{\boldsymbol{s}}$, where spin/orbit interaction keeps their resultant $\underline{\boldsymbol{J}}$ size constant, together with the sizes of each. It is clearly seen that while their separate Z projections vary, their sum projection is invariant. For detailed discussion, see text.

The relation obtained in Eq. (4.51) expresses the need to include as correction a coupling between the orbital angular momentum and spin states that were naively assumed independent. It is called **spin/orbit coupling**, or **L-S coupling**. Inclusion of that coupling assures that the $\underline{\boldsymbol{J}}$ vector absolute value is an invariant of motion, and its related angular momentum operator is indeed a generalized angular momentum described in Section 4.3.2. Figure 4.5 displays the same $\underline{\boldsymbol{L}}$ and $\underline{\boldsymbol{s}}$ vector additions as in Figure 4.4, however where their spin/orbit coupling binds them now in a manner that their resultant sum is an invariant of motion. In the figure, the coupling expresses itself by coordinated precessions of the $\underline{\boldsymbol{L}}$ and $\underline{\boldsymbol{s}}$ vectors about their $\underline{\boldsymbol{J}}$ resultant. The separate projections of $\underline{\boldsymbol{L}}$ and $\underline{\boldsymbol{s}}$ on the Z axis don't *remain* constant, but their sum does! Thus, the m_ℓ and m_s projections are no longer considered "good quantum numbers", but the sum $m_j \equiv m_\ell + m_s$ is considered a "good quantum number". We discuss this issue in more detail in Section 4.9. We shall see that any $|n\ell s\, jm_j\rangle$ function can be expressed as a linear combination of spin/orbit $|n\ell m_\ell m_s\rangle$ functions.

Correspondingly to the inclusion of the independent spin, the letters notation agreement of angular momentum is updated by addition of j as a subscript to the right of the ℓ letter. For example, the $\ell = 2$ and $j = \frac{5}{2}$ case will be noted as $d_{5/2}$ (see Table 3.2).

4.5 Gyromagnetic Ratio and Spin/Orbit Coupling

4.5.1 The Gyromagnetic Ratio

Being an electrically charged particle, an electron circulating in a closed loop generates an electric current loop. The current loop necessarily establishes a magnetic dipole. The current loop is obviously related to the angular momentum vector, as described in Figure 4.5. In a classical approximation, an electric charge $(-e)$ circulating at angular velocity ω generates a current "vector" given by $\underline{\boldsymbol{i}} = -e\underline{\boldsymbol{\omega}}/2\pi$. The angular momentum is $\underline{\boldsymbol{L}} = m\underline{\boldsymbol{\omega}}\cdot r^2$; thus $\underline{\boldsymbol{i}} = -e\underline{\mathbf{L}}/2\pi mr^2$. The generated magnetic dipole is given by $\underline{\boldsymbol{M}}^L = (\underline{\mathbf{i}}/c)A$, where A is the area enclosed by the moving charge. Setting $A = \pi r^2$ we get

$$\underline{\boldsymbol{M}}^L = -\frac{1}{\hbar}\left(\frac{e\hbar}{2mc}\right)\underline{\boldsymbol{L}} = -\frac{\mu_B}{\hbar}\underline{\boldsymbol{L}}\ , \tag{4.53}$$

where the quantity $\mu_B \equiv e\hbar/2mc = 9.2741 \times 10^{-21}\text{erg} \cdot \text{gauss}^{-1}$ is a fundamental constant called **Bohr magneton**. Converting Eq. (4.53) into a relation between operators, we get $\mathbf{M}^L \equiv -(\mu_B/\hbar)\mathbf{L}$. Thus, the expectation value of the magnetic dipole moment projection on the Z axis will be

$$\left\langle \mathbf{M}^L{}_Z \right\rangle = \langle -(\mu_B/\hbar)\mathbf{L}_Z \rangle = -\mu_B m_\ell \quad . \tag{4.54}$$

Namely, the expected projection on the Z axis of a magnetic dipole induced by an orbital angular momentum is an integer number of Bohr magnetons.

It would be just natural to assume a similar relation for the spin angular momentum, namely

$$\underline{M}^S \stackrel{?}{=} -\frac{\mu_B}{\hbar}\underline{s} \quad . \tag{4.55}$$

Had this relation been valid, the expectation value of a magnetic dipole Z-projection induced by the electron spin would be a half Bohr magneton. Experiment shows that the electron spin indeed exhibits a magnetic dipole; however, its value is a whole Bohr magneton, namely *twice* that expected by Eq. (4.55). This fact may be intuitively justified by considering that the 1/2 angular momentum quantum property of the internal spin emanates from the gradual phase change during spinning of the spin wavefunction (Eq. [4.48] and Figure 4.3). The spatial probability distribution, and apparently the electric charge distributions, are phase independent. Thus, the magnetic moment generated by spinning behaves as if the spin quantity is a unity. To correct Eq. (4.55) to match the experimental results, we write:

$$\boxed{\underline{M}^S = -g\frac{\mu_B}{\hbar}\underline{s}} \quad , \tag{4.56}$$

where $g \equiv 2$ is called **gyromagnetic ratio**, or **g-factor**. Experiments, whereby magnetic dipole moments generated by spin 1/2 electrons were measured, yielded a marginally different value: $g \equiv 2.0024$. For practical purposes, the approximation $g = 2.00$ is amply satisfactory.

4.5.2 Spin/Orbit Interaction

The energy correction termed **spin/orbit interaction** is attributed to two separate effects entailing the assumed approximation that the electron spinning and orbital motions are independent: electric dipole generation, and **Thomas precession**.

4.5.2.1 Electric Dipole of a Moving Magnetic Dipole

While the electron orbits in the atom, its internal magnetic moment is also orbiting and entails an electric dipole. This dipole is influenced by the electrostatic electric field generated by the nucleus, changing its energy. The entailed electric dipole is given by $\underline{Q} = \frac{1}{c}\underline{v} \times \underline{M}^S$, where $\underline{v}$ is the velocity of the moving magnetic dipole. We use this expression without proof. For some physical feeling note that the generated electric dipole diminishes if the velocity and magnetic dipole are co-oriented. The interaction energy between the electric dipole and the electrostatic field by the nucleus $\underline{\mathcal{E}}$ is $U_{SL} = -\underline{Q}\cdot\underline{\mathcal{E}} = -\frac{1}{c}\underline{v} \times \underline{M}^S\cdot\underline{\mathcal{E}}$. For a hydrogen-like atom we may set $\underline{\mathcal{E}} = (Ze/r^2)\underline{u}_r$, where $\underline{u}_r$ is a unity vector in the radial orientation. We further use the relation $\underline{L} = m\underline{r} \times \underline{v} = mr\underline{u}_r \times \underline{v}$ and get

$$U_{SL}^{ed} = g\frac{Ze^2}{2m^2c^2r^3}\underline{s}\cdot\underline{L} \quad . \tag{4.57}$$

The U_{SL}^{ed} expectation value will thus be

$$E_{SL}^{ed} \equiv \left\langle U_{SL}^{ed} \right\rangle = g\frac{Ze^2}{2m^2c^2}\langle \underline{s}\cdot\underline{L}\rangle\left\langle \frac{1}{r^3}\right\rangle \ .$$

Using Eq. (3.47) and the recurrence relation of Eq. (3.48), one readily obtains

$$\left\langle \frac{1}{r^3}\right\rangle = \frac{1}{n^3}\frac{2}{\ell(\ell+1)(2\ell+1)}\left(\frac{Z}{a_0}\right)^3 \ , \qquad (4.58)$$

for $\ell > 0$. Therefore,

$$\boxed{E_{SL}^{ed} = g\left(\frac{|E_n|}{n}\right)\frac{Z^2\alpha^2}{\hbar^2}\cdot\frac{2}{\ell(\ell+1)(2\ell+1)}\langle \underline{s}\cdot\underline{L}\rangle = g\cdot a_{SL}\langle \underline{s}\cdot\underline{L}\rangle} \ , \qquad (4.59)$$

where

$$\boxed{a_{SL} \equiv \left(\frac{|E_n|}{n}\right)\frac{Z^2\alpha^2}{\hbar^2}\frac{2}{\ell(\ell+1)(2\ell+1)}} \ . \qquad (4.60)$$

The a_{SL} quantity is called **spin/orbit interaction coefficient**.

4.5.2.2 Thomas Precession

An additional effect related to the circular motion of the electron is called **Thomas precession**, and sometimes **Thomas rotation**. An electron circulating in a closed orbit is constantly being accelerated (centrifugal acceleration). A generalization of special relativity suggests that any accelerating body will appear also as rigidly rotating in the lab frame, termed Thomas rotation. We will hereby use this effect without proof, and present only its conclusion. The angular velocity of Thomas rotation $\underline{\omega}_{Th}$ is given by

$$\underline{\omega}_{Th} = -\frac{1}{2c^2}\underline{v}\times\underline{a} \quad \text{for} \quad \frac{v}{c} << 1 \ , \qquad (4.61)$$

where $\underline{v}$ and $\underline{a}$ are the velocity and acceleration vectors, respectively. As a result, the body kinetic energy changes. Under conditions where the Thomas angular velocity induced momentum is small compared to the spin, the kinetic energy change is given by

$$U_{SL}^{Th} = \underline{s}\cdot\underline{\omega}_{Th} = +\frac{1}{2c^2}\underline{a}\times\underline{v}\cdot\underline{s} \ . \qquad (4.62)$$

We set now $\underline{a} = (-e\underline{E}/m) = -(Ze^2/mr^2)\underline{u}_r$ and get

$$U_{SL}^{Th} = \underline{s}\cdot\underline{\omega}_{Th} = -\frac{1}{2c^2}\frac{Ze^2}{m^2r^3}(r\underline{u}_r)\times(m\underline{v})\cdot\underline{s} = -\frac{Ze^2}{2m^2c^2r^3}\underline{s}\cdot\underline{L} \ .$$

Continued calculation follows the steps used after Eq. (4.57). We get

$$\boxed{E_{SL}^{Th} = -\left(\frac{|E_n|}{n}\right)\frac{Z^2\alpha^2}{\hbar^2}\frac{2}{\ell(\ell+1)(2\ell+1)}\langle \underline{s}\cdot\underline{L}\rangle = -a_{SL}\langle \underline{s}\cdot\underline{L}\rangle} \ . \qquad (4.63)$$

The above energy correction stemming from the Thomas precession is also often addressed as **Darwin component**.

4.5.2.3 Total Spin/Orbit Coupling

When the two above contributions to the spin/orbit coupling are added, one gets

$$E_{SL} = E_{SL}^{ed} + E_{SL}^{Th} = (g-1)\left(\frac{|E_n|}{n}\right)\frac{Z^2\alpha^2}{\hbar^2}\frac{2}{\ell(\ell+1)(2\ell+1)}\langle \underline{s}\cdot\underline{L}\rangle \ .$$

By setting g = 2:

$$\boxed{E_{SL} = \left(\frac{|E_n|}{n}\right)\frac{Z^2\alpha^2}{\hbar^2}\frac{2}{\ell(\ell+1)(2\ell+1)}\langle \underline{s}\cdot\underline{L}\rangle} \ . \tag{4.64}$$

Inclusion of the electron spinning in our calculations thus changes the energy of each state by the strength of the spin/orbit coupling $\langle \underline{s}\cdot\underline{L}\rangle$ (Eq. [4.52]). When including the coupling strength of the spin-up state $\left(j = \ell + \frac{1}{2}\right)$ compared to a spinless state of $j = \ell$, the energy addition would be

$$E_{SL}(\uparrow) = \left(\frac{|E_n|}{n}\right)Z^2\alpha^2\frac{1}{(\ell+1)(2\ell+1)} \ . \tag{4.65}$$

As to the spin-down state $\left(j = \ell - \frac{1}{2}\right)$, the energy addition would be

$$E_{SL}(\downarrow) = -\left(\frac{|E_n|}{n}\right)Z^2\alpha^2\frac{1}{\ell(2\ell+1)} \ . \tag{4.66}$$

The energy splitting of an $|n\,\ell m\rangle$ state caused by the spin/orbit coupling would thus be

$$\boxed{\Delta E_{SL} \equiv E_{SL}(\uparrow) - E_{SL}(\downarrow) = \left(\frac{|E_n|}{n}\right)Z^2\alpha^2\frac{1}{\ell(\ell+1)}} \ . \tag{4.67}$$

4.5.3 Summed Energy Spectrum Correction

On combining the mechanical relativistic correction (Eq. [4.20]) and the spin/orbit one for the two spin states (Eqs. [4.65] and [4.66]), the summed correction may be written as follows:

$$\boxed{E_{rc}^{(j)} = \left(\frac{|E_n|}{n}\right)Z^2\alpha^2\left(\frac{3}{4n} - \frac{2}{2j+1}\right)} \ . \tag{4.68}$$

We have obtained a correction that depends solely on the j quantum number for each principal quantum number n. This correction removes the degeneracy among the ℓ quantum numbers belonging to the same n. However, it does not remove the degeneracy among the different m_j numbers. The above spectral corrections emanating from the spin/orbit coupling are called **fine structure**. Figure 4.6 provides an energy scheme of several lower levels of the hydrogen atom and their fine structure. The states' splitting are about five orders of magnitude smaller than the energy separations among the states. The figure also indicates electric dipole allowed transitions between states. The $3d_{3/2} \rightarrow 2p_{3/2}$ transition is relatively weak, and thus plotted by a dashed line. In Section 4.9 we provide an extended discussion concerning the selection rules and transition probabilities.

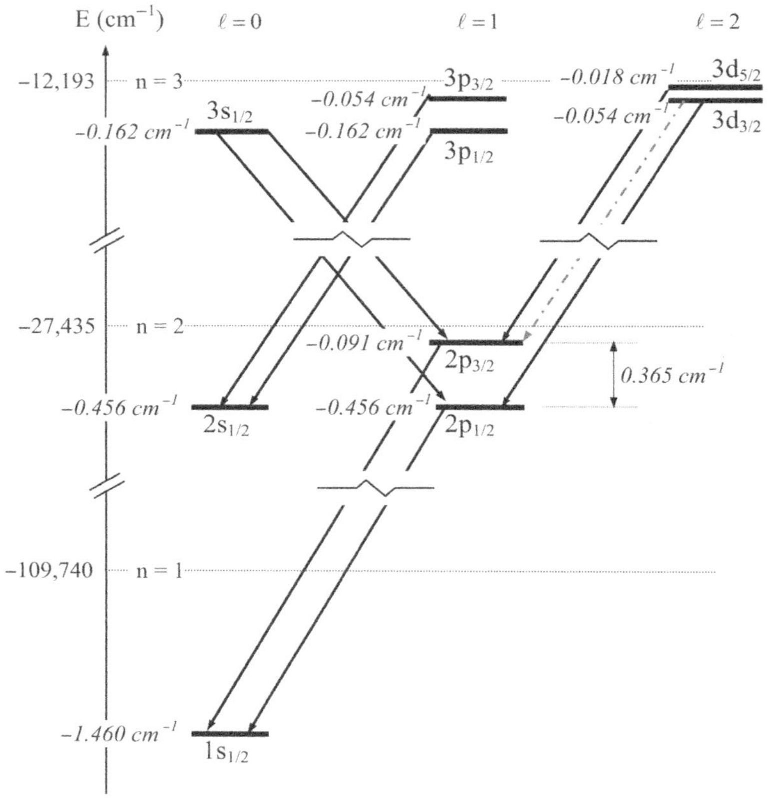

Figure 4.6 Energy scheme of lower states in the hydrogen atom that includes the fine structure shifts and splitting. The energies are indicated in cm^{-1} units. The original electrostatic energy approximation is indicated by dotted horizontal lines. The scale for the energy gaps belonging to the same principal quantum number *n* has been extended to emphasize the splitting cases. For a detailed discussion, see text.

Splitting of each hydrogen atomic state belonging to an $\ell = 0$, as expected by Eqs. (4.65) and (4.66), has been verified experimentally. Historically, it has been the need to account for the *optical splitting in two* of atomic sodium *Na* vapor fluorescence, which triggered the conjecture of an electron exhibiting an internal spinning of $s = 1/2$ eigenvalue (see Figure 5.11).

4.6 Landé Factor

An important effect of the non-unity gyromagnetic ratio is that the resultant magnetic dipole of the orbital with the internal electron spin is not necessarily co-oriented with the resultant angular momentum (except in the $\ell = 0$ case). This issue is demonstrated in Figure 4.7. Because the $\underline{s}$ and $\underline{L}$ orientations are not steady in relation to their resultant $\underline{J}$ (as demonstrated in Figure 4.5), the orientation of the magnetic moment $\underline{M}$ is not constant. Its projection $\underline{M}_{av}$ along the $\underline{J}$ orientation is constant, however. The orientations of both $\underline{J}$ and $\underline{M}_{av}$ also vary; only their absolute values (sizes) are constants of motion, as well as their projections along the *Z* axis.

The resultant magnetic moment $\underline{M}$ is given by

$$\underline{M} = -\frac{\mu_B}{\hbar}(\underline{L} + g\underline{s}) = -\frac{\mu_B}{\hbar}(\underline{J} + \underline{s}) \ . \qquad (4.69)$$

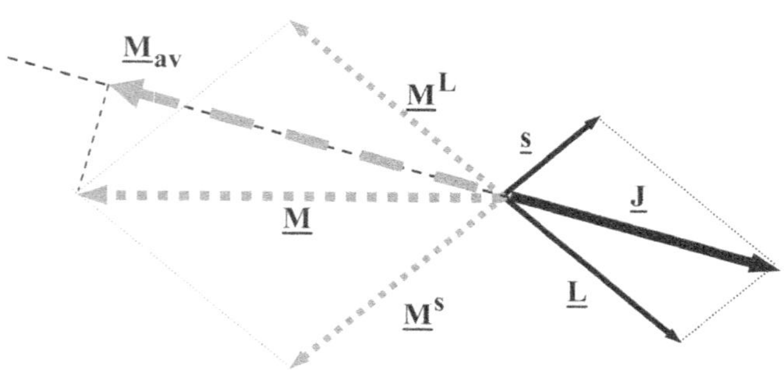

Figure 4.7 Example of combining $\underline{M}^L$ and $\underline{M}^S$ magnetic dipoles that entail the corresponding combination of $\underline{L}$ and $\underline{s}$ angular momentums. Being the spin $\underline{s}$ gyromagnetic ratio different than a unity, the resultant magnetic moment $\underline{M}$ is not co-oriented with the resultant angular momentum $\underline{J}$.

We turn to calculate the *average* magnetic moment $\underline{M}_{av}$. As indicated, this is the projection vector along the $\underline{J}$ vector orientation:

$$\underline{M}_{av} = (\underline{M}\cdot\underline{u}_J)\underline{u}_J = -\frac{\mu_B}{\hbar}\left[(\underline{J} + \underline{s})\cdot\frac{\underline{J}}{|\underline{J}|^2}\right]\underline{J}$$

$$= -\frac{\mu_B}{\hbar}\left(1 + \frac{\underline{s}\cdot\underline{J}}{|\underline{J}|^2}\right)\underline{J} \ . \qquad (4.70)$$

In addition, because $\underline{\boldsymbol{J}} = \underline{\boldsymbol{L}} + \underline{\boldsymbol{s}}$, we may write that $\underline{\boldsymbol{s}}\cdot\underline{\boldsymbol{J}} = \frac{1}{2}\left(\underline{\boldsymbol{J}}^2 + \underline{\boldsymbol{s}}^2 - \underline{\boldsymbol{L}}^2\right)$, set it in Eq. (4.70), and convert the relation into an operator expression. We get

$$\mathbf{M}_{av} = -\frac{\mu_B}{\hbar}\left(1 + \frac{j(j+1) + s(s+1) - \ell(\ell+1)}{2j(j+1)}\right)\mathbf{J} \equiv -\frac{\mu_B}{\hbar}g_L\mathbf{J} \quad . \tag{4.71}$$

The quantity

$$\boxed{g_L \equiv \left(1 + \frac{j(j+1) + s(s+1) - \ell(\ell+1)}{2j(j+1)}\right)} \quad , \tag{4.72}$$

is called **Landé factor**. In the hydrogen atom, Landé factor attains two different values for each ℓ, related to the two possible spin orientations, spin-up $\left(j = \ell + \frac{1}{2}\right)$ and spin-down $\left(j = \ell - \frac{1}{2}\right)$:

$$g_L = 1 \pm \frac{1}{2\ell + 1} \quad for \quad j = \ell \pm \frac{1}{2} \quad . \tag{4.73}$$

This relation becomes useful in analysis of the **Zeeman effect**, to be presented in Section 4.8.1.

4.7 Lamb Shift

A free (unbound) electron could not be completely at rest because of the influence of electromagnetic **vacuum fluctuations** (also addressed as **zero-point fluctuations**). The existence of that electromagnetic field type is expected by quantum electrodynamics theory. Each normal vibration of the electromagnetic field is like a harmonic oscillator, whose lowest vibrational energy is $\frac{1}{2}\hbar\omega$ (see also Section 5.3.1.2). We have already mentioned in Section 2.4 that the existence of spontaneous emission by excited atoms establishes a proof for the existence of the vacuum electromagnetic radiation. Following Eq. (2.28), the volume density per unit angular frequency of the vacuum electromagnetic radiation is given by

$$\rho_\omega^{vac}(\omega) = \frac{\omega^2}{\pi^2 c^3}\cdot\frac{1}{2}\hbar\omega = \frac{\hbar\omega^3}{2\pi^2 c^3} \quad . \tag{4.74}$$

This density is also given by (see Eq. [1.31])

$$\rho_\omega^{vac}(\omega) = \frac{1}{2\pi}|\boldsymbol{E}(\omega)|^2 = \frac{1}{2\pi}\left[|\boldsymbol{E}_X(\omega)|^2 + |\boldsymbol{E}_Y(\omega)|^2 + |\boldsymbol{E}_Z(\omega)|^2\right] \quad , \tag{4.75}$$

where $\boldsymbol{E}(\omega)$ is the Fourier component of the electric field at angular frequency ω, and $\boldsymbol{E}_X(\omega)$, $\boldsymbol{E}_Y(\omega)$, and $\boldsymbol{E}_Z(\omega)$ are its components along the X, Y, and Z orientations, respectively. It is assumed that over a sufficiently long time, the average squared Fourier components of the electric field are equal in the different orientations in space. The response of a free electron to action by an alternating electric field differs from that of a bound electron in that its resonance frequency ω_0 is identically zero, as well as its damping coefficient γ (Eq. [1.7] and related text). Setting the above in Eq. (1.32), we get

$$|\boldsymbol{X}(\omega)|^2 = \frac{(e/m)^2}{\omega^4}\frac{|\boldsymbol{E}(\omega)|^2}{3} = \frac{(e/m)^2}{\omega^4}\frac{\hbar\omega^3}{3\pi c^3} = \frac{(e/m)^2\hbar}{3\pi\omega c^3} \quad , \tag{4.76}$$

where $\boldsymbol{X}(\omega)$ is the Fourier transform of the electron deviation from its resting position $\delta x(t)$ along the X direction. Following Eqs. (1.29) and (1.30), we may write

$$\int_{-\infty}^{\infty} |\delta x(t)|^2 dt = 2\int_{0}^{\infty} |\boldsymbol{X}(\omega)|^2 d\omega = \frac{2(e/m)^2\hbar}{3\pi c^3}\cdot \lim_{\substack{\omega_{max}\to\infty \\ \omega_{min}\to 0}} \left(\boldsymbol{ln}\,\frac{\omega_{max}}{\omega_{min}}\right) .$$

Identical expressions are obtained for resting position deviations along the Y and Z axes. Thus, the expectation value of the squared position vector deviation from the rest position $|\delta \underline{r}(t)|^2 = |\delta x(t)|^2 + |\delta y(t)|^2 + |\delta z(t)|^2$ satisfies

$$\left\langle |\delta \underline{r}|^2 \right\rangle \equiv \int_{-\infty}^{\infty} |\delta \underline{r}(t)|^2 dt = \frac{6(e/m)^2\hbar}{\pi c^3}\cdot \lim_{\substack{\omega_{max}\to\infty \\ \omega_{min}\to 0}} \left(\boldsymbol{ln}\,\frac{\omega_{max}}{\omega_{min}}\right) . \tag{4.77}$$

The element on the right diverges logarithmically to infinity. In contrast to Eq. (4.77), we assume that *both* ω_{max} and ω_{min} are finite! The phenomenon of receiving infinite values in calculation of physical quantities that are definitely final, is a known and outstanding pitfall in the applications of quantum electrodynamics. To render the theory still useful, it is a custom to select mathematically finite segments of the integrals and relate them to the calculated physical effects. As to the remainder infinites, it is assumed that they will disappear in a future generalized theory that will relate them to the physical background, such as the particles' rest masses, their charges, or to the finite value of the speed of light. This physical/mathematical procedure is termed **renormalization.**♣ It is required to express special assumptions for each specific case.

The "Lamb shift" actually considers the case of a bound electron. The theory adopts Eq. (4.77) as a basis, yet selecting special finite, non-zero values for the ω_{max} and ω_{min} integration limits. ω_{max} is assumed to satisfy the relation $\hbar\omega_{max} = mc^2$, namely the electron rest mass. The underlying hidden assumption is that in a future theory, the energy provided to the electron by the higher-frequency electromagnetic vacuum fluctuations will converge, at least partially, to the electron rest mass. As to lower frequencies, no vacuum-fluctuating photon of frequency below ionization one has sufficient energy to have any impact. Thus, the lower-limit frequency is selected as the electron state energy $\hbar\omega_{max} = |E_n|$. With these choices, Eq. (4.77) reduces to

$$\left\langle |\delta \underline{r}|^2 \right\rangle = \frac{2(e/m)^2\hbar}{\pi c^3}\cdot \boldsymbol{ln}\left(\frac{2n^2}{Z^2\alpha^2}\right) . \tag{4.78}$$

To demonstrate the order of magnitude of the electron deviation effect from its resting position, we calculate $\left\langle |\delta \underline{r}|^2 \right\rangle$ for the ground electron state of the hydrogen atom. *The quantities' given data are*:

$$e = 4.8\times 10^{-10}\,\text{esu} \ ; \quad m = 9.11\times 10^{-28}\,\text{g} \ ; \quad c = 3\times 10^{10}\,\text{cm/s} \ ; \quad \hbar = 1.0546\times 10^{-27}\,\text{erg·s} \ ;$$

$$\alpha = 1/137 \ ; \quad n = 1 \ ; \quad Z = 1 \ .$$

♣ We have already encountered a similar consideration on defining the electron electrostatic radius r_0 (Eq. [1.9] related text and footnote). The electrostatic self energy of an electron would diverge to infinity for a point-like electron. It is not argued that r_0 is indeed the radius of the electron, but that perhaps, in a future theory, the calculation of the electrostatic energy will be revised for distances shorter than the order of r_0, such as to converge to a finite value. Or perhaps a different model of the electron will account for the coexistence of a rest mass and an electric charge, without internal inconsistency.

Setting in Eq. (4.78):

$$\left\langle |\delta \underline{r}|^2 \right\rangle = \frac{2 \times 4.8^2 \times 10^{-20} \times 1.0546 \times 10^{-27}}{3.1416 \times 9.11^2 \times 10^{-56} \times 27 \times 10^{30}} \boldsymbol{ln}\left(2 \times 137^2\right) = 7.27 \times 10^{-23}\ \mathrm{cm}^2 \ .$$

For comparison, the squared electrostatic radius of the electron is $r_0{}^2 = 1.96 \times 10^{-26} \mathrm{cm}^2$. Thus, because of the electromagnetic vacuum fluctuations, the squared average deviation of the orbiting electron from its "smooth" orbital shape is about 3,700 times larger than the squared electron electrostatic radius! The average deviation is obviously zero. Still, the electron potential energy changes while deviation occurs, and this deviation is not symmetric for increased or decreased distance from the expectation value of the radial distance. This entails some change in the electron energy. To quantify that effect, we expand the electron potential energy to a Taylor series around the equilibrium radius:

$$\boldsymbol{V}(\underline{r} + \delta\underline{r}(t)) = \boldsymbol{V}(\underline{r}) + (\delta\underline{r}(t)\cdot\underline{\nabla})\boldsymbol{V}(\underline{r}) + \frac{1}{2}(\delta\underline{r}(t)\cdot\underline{\nabla})^2\boldsymbol{V}(\underline{r}) + \dots \ . \tag{4.79}$$

The linear element in the Taylor expansion vanishes because the equilibrium radius remains unchanged. The other element to consider is the square one to be seen not to vanish, and we neglect all other higher-order ones. We get

$$\delta\boldsymbol{V}(\underline{r}(t)) \equiv \boldsymbol{V}(\underline{r} + \delta\underline{r}(t)) - \boldsymbol{V}(\underline{r}) =$$

$$\cong \frac{1}{2}\left[(\delta x(t))^2 \frac{\partial^2}{\partial x^2} + (\delta y(t))^2 \frac{\partial^2}{\partial y^2} + (\delta z(t))^2 \frac{\partial^2}{\partial z^2} + \right.$$

$$\left. + (\delta x(t)\cdot\delta y(t))\frac{\partial^2}{\partial x\partial y} + 2(\delta y(t)\cdot\delta z(t))\frac{\partial^2}{\partial y\partial z} + 2(\delta z(t)\cdot\delta x(t))\frac{\partial^2}{\partial z\partial x}\right]\boldsymbol{V}(\underline{r}) \ .$$

We average this expression by integrating over time. The $\delta x(t)$, $\delta y(t)$, and $\delta z(t)$ are independent. The average of each mixed product is a product of each average; because each average is zero, each mixed product is zero. We further assume spatial isotropy, namely that $\left\langle(\delta x(t))^2\right\rangle = \left\langle(\delta y(t))^2\right\rangle = \left\langle(\delta z(t))^2\right\rangle = \frac{1}{3}\left\langle|\delta\underline{r}|^2\right\rangle$. Then we get

$$\langle\delta\boldsymbol{V}(\underline{r})\rangle = \frac{1}{6}\left\langle|\delta\underline{r}|^2\right\rangle\underline{\nabla}^2\boldsymbol{V}(\underline{r}) = \frac{\alpha}{\pi}\left(\frac{\hbar}{\mu c}\right)^2 \boldsymbol{ln}\left(\frac{2n^2}{Z^2\alpha^2}\right)\cdot\underline{\nabla}^2\boldsymbol{V}(\underline{r}) \ . \tag{4.80}$$

In the last expression we replaced the electron mass m by the atom reduced mass μ, because we deal in fact with a bound electron-nucleus system. We consider the expression on the right as a product of quantum operators; the expectation value of the energy change will then be

$$\delta E_{Lamb} = \frac{\alpha}{3\pi}\left(\frac{\hbar}{\mu c}\right)^2 \boldsymbol{ln}\left(\frac{2n^2}{Z^2\alpha^2}\right)\cdot\left\langle n\ell m\middle|\underline{\nabla}^2\cdot\mathbf{V}\middle|n\ell m\right\rangle \ . \tag{4.81}$$

We set $\mathbf{V} = \boldsymbol{V}(\underline{r}) = -(Ze^2/r)$, and perform an intermediate calculation. By Schrödinger equation (Eq. [3.21]) we may write

$$\underline{\nabla}^2\cdot\boldsymbol{V}|n\ell m\rangle = \underline{\nabla}^2\left(\frac{-Ze^2}{r}\right)|n\ell m\rangle = \left(\frac{2\mu}{\hbar^2}\right)\left(-\frac{\hbar^2\underline{\nabla}^2}{2\mu}\right)\left\{-\frac{\hbar^2\underline{\nabla}^2}{2\mu} - E_n\right\}|n\ell m\rangle \ .$$

Because the two last operators on the right side (within the parenthesis and braces) are commutative, we may continue by writing

$$= \left(\frac{2\mu}{\hbar^2}\right)\left(\frac{Ze^2}{r}\right)\left(E_n + \frac{Ze^2}{r}\right)|n\ell m\rangle = \left(\frac{2\mu}{\hbar^2}\right)\left\{E_n Ze^2\left(\frac{1}{r}\right) + Z^2e^4\left(\frac{1}{r^2}\right)\right\}|n\ell m\rangle \ .$$

To receive the expectation value, we multiply from the left by $\langle n\ell m|$ and get

$$\delta E_{Lamb} = \frac{\alpha}{3\pi}\left(\frac{\hbar}{\mu c}\right)^2\left(\frac{2\mu}{\hbar^2}\right) \boldsymbol{ln}\left(\frac{2n^2}{Z^2\alpha^2}\right)\left\{E_n Ze^2\left\langle\frac{1}{r}\right\rangle + Z^2e^4\left\langle\frac{1}{r^2}\right\rangle\right\} \ . \tag{4.82}$$

We set the following results already obtained by the intermediate calculations preceding Eq. (4.20):

$$\left\langle\frac{1}{r^2}\right\rangle = \frac{2}{n^3(2\ell+1)}\left(\frac{Z}{a_0}\right)^2 \ ; \quad \left\langle\frac{1}{r}\right\rangle = \frac{1}{n^2}\left(\frac{Z}{a_0}\right) \ ,$$

And get

$$\boxed{\delta E_{Lamb} = \left(\frac{|E_n|}{n}\right)\left[\frac{2Z^2\alpha^3}{3\pi} \boldsymbol{ln}\left(\frac{2n^2}{Z^2\alpha^2}\right)\right]\left\{\frac{4}{2\ell+1} - \frac{1}{n}\right\}} \ . \tag{4.83}$$

For demonstration of the **Lamb shift**, we calculate the difference between the shifts of the $2s_{1/2}$ and $2p_{1/2}$ (Fig. 4.7). We use the equation

$$\delta E_{Lamb}(2s) - \delta E_{Lamb}(2p) = \left(\frac{|E_n|}{n}\right)\left[\frac{2Z^2\alpha^3}{3\pi} \boldsymbol{ln}\left(\frac{2n^2}{Z^2\alpha^2}\right)\right]\left\{\frac{4}{2\ell_1+1} - \frac{4}{2\ell_2+1}\right\}$$

Given data:

$$n = 2 \ ; \quad |E_2| = 27,435 \text{ cm}^{-1} \ ; \quad Z = 1 \ ; \quad \alpha = 1/137.04 \ ; \quad \ell_1 = 0 \ ; \quad \ell_2 = 1 \ .$$

Setting in the above equation, we get:

$$\delta E_{Lamb}(2s) - \delta E_{Lamb}(2p) =$$
$$= \left(\frac{27,435}{2}\right)\left[\frac{2}{3 \times 3.1416 \times 137.04^3} \boldsymbol{ln}\left(2 \times 4 \times 137.04^2\right)\right]\left\{\frac{4}{1} - \frac{4}{3}\right\} = 0.03595 \text{ cm}^{-1} \ .$$

The result may be expressed in frequency units by just multiplying the latter result by the speed of light c constant. One obtains

$$\boxed{\{\delta E_{Lamb}(2s) - \delta E_{Lamb}(2p)\} = 1077.9 \text{ MHz}} \ . \tag{4.84}$$

The splitting obtained is positive. It means that the $2s_{1/2}$ state is located higher than the $2p_{1/2}$ one by the amount obtained. Selection rules for the hydrogen atom summarized in Eq. (3.58) allow dipole transitions between states of same principal quantum number n and different orbital quantum numbers ℓ which satisfy the selection rule $\Delta\ell = \pm 1$. Therefore, the $2p_{1/2} \to 2s_{1/2}$ transition is electric dipole allowed. In 1947, **Lamb** measured that transition energy using radio-frequency radiation. The experimental result was 1057 MHz, very close to the theoretically calculated result.

4.8 Selection Rules and Transition Probabilities

In Section 3.6 we showed that a dipole transition between $|n\ell m\rangle$ and $|n'\ell' m'\rangle$ is allowed if and only if $\langle n'\ell' m'|\underline{r}|n\ell m\rangle \neq 0$ (Eq. [3.49] and related text). When considerations include the internal electron spin existence, the expression to study is $\left\langle n'\,\ell'\,s'\,j'\,m_j'|\underline{r}|n\,\ell\,s\,j\,m_j\right\rangle$. Spin/orbit coupling may be neglected when it is small compared to other perturbations that render the attractive potential acting on the electron essentially non-central. Prominent examples are strong external electric or magnetic fields. The total angular momentum may not be considered then as a constant of motion. We may then replace the wave function with a product function $|n\ell m_\ell m_s\rangle = |n\ell m_\ell\rangle|m_s\rangle$. Because spin coordinates are independent of space coordinates, we have to study the integral

$$\left\langle n'\ell' m_\ell' m_s'|\underline{r}|n\,\ell m_\ell m_s\right\rangle = \left\langle n'\ell' m_\ell'|\underline{r}|n\ell m_\ell\right\rangle\left\langle m_s'|m_s\right\rangle = \left\langle n'\ell' m_\ell'|\underline{r}|n\ell m_\ell\right\rangle\delta_{m_s' m_s} \ .$$

Thus, *on neglecting spin/orbit coupling*, the selection rules are

$$\boxed{\Delta m_\ell = 0,\ \pm 1 \ \ ; \quad \Delta\ell = \pm 1 \ \ ; \quad \Delta m_s = 0 \ \ ; \quad \Delta j = \Delta\ell = \pm 1} \ . \tag{4.85}$$

In Section 4.4, we showed that when the value of total angular momentum $\underline{\boldsymbol{J}}$ is a constant of motion, the separate Z-axis projections of $\underline{\boldsymbol{L}}$ and $\underline{\boldsymbol{s}}$ are not constant, but their sum is constant. The belonging wavefunction for that case, $|n\ell sjm_j\rangle$ may be expressed as a linear combination of the $|n\ell m_\ell m_s\rangle$ basis functions. Namely,

$$\left|n\ell sjm_j\right\rangle = \sum_{m_\ell + m_s = m_j} |n\ell m_\ell m_s\rangle\langle n\ell m_\ell m_s|n\ell sjm_j\rangle \ . \tag{4.86}$$

This property demonstrates the fact that the $|n\ell m_\ell m_s\rangle$ basis functions generate a sub-space invariant under the angular momentum operators. The principal quantum number n plays no role in the calculation. It is thus customary to delete it from the states' symbols in calculations of angular momentums. The $\langle n\ell m_\ell m_s|n\ell s\,jm_j\rangle$ expansion constants, or in their abbreviated symbols $\langle \ell m_\ell m_s|\ell s\,jm_j\rangle$, are called **Clebsch-Gordan coefficients**, or **Vector-coupling coefficients**.

Calculation of Clebsch-Gordan coefficients for addition of two general angular momentum vectors is a long, quite tiring labor. The present case where ℓ is integer and $s = 1/2$ is relatively simple. Because $m_j = m_\ell + m_s$, the summing in Eq. (4.86) includes only two addends:

$$\begin{aligned}\left|\ell s j m_j\right\rangle &= \left|\ell, m_j - \frac{1}{2}, \uparrow\right\rangle\left\langle \ell, m_j - \frac{1}{2}, \uparrow\right|\left.\ell s j m_j\right\rangle + \left|\ell, m_j + \frac{1}{2}, \downarrow\right\rangle\left\langle \ell, m_j + \frac{1}{2}, \downarrow\right|\left.\ell s j m_j\right\rangle \\ &\equiv a\left|\ell, m_j - \frac{1}{2}, \uparrow\right\rangle \ + \ b\left|\ell, m_j + \frac{1}{2}, \downarrow\right\rangle \ .\end{aligned} \tag{4.87}$$

In Eq. (4.87), and in several following expressions, we adopt the common "graphic" nomenclature for m_s: $\uparrow \ \equiv +\frac{1}{2}$, and $\downarrow \ \equiv -\frac{1}{2}$ ("spin-up" and "spin-down", respectively). For m_ℓ, on the other hand, we keep the numerical nomenclature.

The squared total angular momentum operator may be written as follows:

$$\begin{gathered}\mathbf{J}^2=\mathbf{L}^2+\mathbf{s}^2+(\mathbf{L}_X+i\mathbf{L}_Y)(\mathbf{s}_X-i\mathbf{s}_Y)+(\mathbf{L}_X-i\mathbf{L}_Y)(\mathbf{s}_X+i\mathbf{s}_Y)+2\mathbf{L}_Z\mathbf{s}_Z,\\ \mathbf{J}^2=\mathbf{L}^2+\mathbf{s}^2+\mathbf{L}_+\mathbf{s}_-+\mathbf{L}_-\mathbf{s}_++2\mathbf{L}_Z\mathbf{s}_Z\ .\end{gathered}\tag{4.88}$$

We apply the two sides of Eq. (4.88) on the $|\ell\, s\, j\, m_j\rangle$ wave function. Using Eq. (4.87), we get

$$\begin{aligned}\mathbf{J}^2\left|\ell sjm_j\right\rangle &= \hbar^2 j(j+1)\left|\ell sjm_j\right\rangle\\ &= \hbar^2\left[\ell(\ell+1)+\frac{3}{4}\right]\cdot\left\{a\left|\ell,m_j-\frac{1}{2}\uparrow\right\rangle+b\left|\ell,m_j+\frac{1}{2},\downarrow\right\rangle\right\}\\ &+\hbar^2\sqrt{\ell(\ell+1)-\left(m_j-\frac{1}{2}\right)\left(m_j+\frac{1}{2}\right)}\cdot a\left|\ell,m_j+\frac{1}{2},\downarrow\right\rangle\\ &+\hbar^2\sqrt{\ell(\ell+1)-\left(m_j+\frac{1}{2}\right)\left(m_j-\frac{1}{2}\right)}\cdot b\left|\ell,m_j-\frac{1}{2},\uparrow\right\rangle\\ &+\hbar^2\left(m_j-\frac{1}{2}\right)\cdot a\left|\ell,m_j-\frac{1}{2},\uparrow\right\rangle-\hbar^2\left(m_j+\frac{1}{2}\right)\cdot b\left|\ell,m_j+\frac{1}{2},\downarrow\right\rangle.\end{aligned}$$

Because $\left|\ell,m_j-\frac{1}{2},\uparrow\right\rangle$ and $\left|\ell,m_j+\frac{1}{2},\downarrow\right\rangle$ are independent wavefunctions, we get the following two equations:

$$\begin{aligned}&a\cdot\left[j(j+1)-\left(\ell+\frac{1}{2}\right)^2-m_j\right]-b\cdot\sqrt{\left(\ell+\frac{1}{2}\right)^2-m_j^2}=0\\ &-a\cdot\sqrt{\left(\ell+\frac{1}{2}\right)^2-m_j^2}+b\cdot\left[j(j+1)-\left(\ell+\frac{1}{2}\right)^2-m_j\right]=0\end{aligned}\ ;$$

For the $j=\ell+\frac{1}{2}$ case we get

$$a\sqrt{\ell+\frac{1}{2}-m_j}\ -b\sqrt{\ell+\frac{1}{2}+m_j}=0\ .$$

With addition of the $a^2+b^2=1$ normalization condition, the solutions are

$$a=\sqrt{\frac{\ell+\frac{1}{2}+m_j}{2\ell+1}}\ ;\ b=\sqrt{\frac{\ell+\frac{1}{2}-m_j}{2\ell+1}}\ .\tag{4.89}$$

For the $j=\ell-\frac{1}{2}$ case, and under the $a^2+b^2=1$ normalization condition, we get

$$a=\sqrt{\frac{\ell+\frac{1}{2}-m_j}{2\ell+1}}\ ;\ b=-\sqrt{\frac{\ell+\frac{1}{2}+m_j}{2\ell+1}}\ .\tag{4.90}$$

The above results are tabulated in Table 4.1. It is interesting to observe that the Clebsch-Gordan coefficients are independent of m_ℓ. An expression providing Clebsch-Gordan coefficients for coupling between two general angular momentums is given in Section 4.11.

Table 4.1 Clebsch-Gordan $\left\langle \ell m_\ell m_s | \ell \frac{1}{2} j m_j \right\rangle$ coefficients

j \ m_s	$[a]$; $m_s = +\frac{1}{2}(\uparrow)$	$[b]$; $m_s = -\frac{1}{2}(\downarrow)$
$j = \ell + \frac{1}{2}$	$\sqrt{\frac{\ell + \frac{1}{2} + m_j}{2\ell + 1}}$	$\sqrt{\frac{\ell + \frac{1}{2} - m_j}{2\ell + 1}}$
$j = \ell - \frac{1}{2}$	$\sqrt{\frac{\ell + \frac{1}{2} - m_j}{2\ell + 1}}$	$-\sqrt{\frac{\ell + \frac{1}{2} + m_j}{2\ell + 1}}$

We use the latter results to calculate the selection rules. Following Eq. (3.51), the criterion for being a dipole-forbidden or dipole-allowed optical transition is the corresponding vanishing or non-vanishing of the $\langle n'\ell' s\, j'\, m'_j | \underline{r} | n\, \ell\, s\, j\, m_j \rangle$ transition vector. In a calculation procedure administered in relation to Eq. (3.52), we calculate first the Z-component of the transition vector, related to absorption or emission of a Z-polarized light:

$$\left\langle n'\ell' \frac{1}{2}\, j'\, m'_j | \underline{r} | n\ell \frac{1}{2} j m_j \right\rangle_Z = \left\langle n'\ell' \frac{1}{2}\, j'\, m'_j | r\, \mathbf{cos}\,\theta | n\,\ell \frac{1}{2} j\, m_j \right\rangle$$

$$= \left\{ \int_0^\infty r \boldsymbol{R}^*_{n'\ell'}(r) \boldsymbol{R}_{n\ell}(r) r^2 dr \right\} \times \left\langle \ell' \frac{1}{2}\, j' m'_j | \,\mathbf{cos}\,\theta | \ell \frac{1}{2} j\, m_j \right\rangle \ .$$

Each state function appearing in the matrix element that constitutes the last factor on the second line can be developed to a sum of two orbital angular momentums per Clebsch-Gordan (see Eq. [4.87] and Table 4.1). Thus, in principle, the last term is a sum of four integrals:

$$\left\langle \ell' \frac{1}{2}\, j' m'_j | \,\mathbf{cos}\,\theta | \ell \frac{1}{2} j\, m_j \right\rangle$$

$$= \left\{ \left\langle \ell', m'_j - \frac{1}{2}, \uparrow \right| a' + \left\langle \ell', m'_j + \frac{1}{2}, \downarrow \right| b' \right\} \mathbf{cos}\,\theta \left\{ a \left| \ell, m_j - \frac{1}{2}, \uparrow \right\rangle + b \left| \ell, m_j + \frac{1}{2}, \downarrow \right\rangle \right\}$$

$$= \left\langle \ell', m'_j - \frac{1}{2}, \uparrow \middle| a' a\, \mathbf{cos}\theta \middle| \ell, m_j - \frac{1}{2}, \uparrow \right\rangle$$

$$+ \left\langle \ell', m'_j - \frac{1}{2}, \uparrow \middle| a' b\, \mathbf{cos}\theta \middle| \ell, m_j + \frac{1}{2}, \downarrow \right\rangle$$

$$+ \left\langle \ell', m'_j + \frac{1}{2}, \downarrow \middle| b' a\, \mathbf{cos}\theta \middle| \ell, m_j - \frac{1}{2}, \uparrow \right\rangle$$

$$+ \left\langle \ell', m'_j + \frac{1}{2}, \downarrow \middle| b' b\, \mathbf{cos}\theta \middle| \ell, m_j + \frac{1}{2}, \downarrow \right\rangle \ .$$

Integrals containing opposite spin orientations vanish identically. The remaining are the first and last ones. The scalar product over spin coordinates provides a unity. We are left with the following sum:

$$a'a\left\langle \ell', m'_j - \frac{1}{2} \middle| \cos\theta \middle| \ell, m_j - \frac{1}{2} \right\rangle + b'b\left\langle \ell', m'_j + \frac{1}{2} \middle| \cos\theta \middle| \ell, m_j + \frac{1}{2} \right\rangle \quad .$$

Each of the integrals vanishes identically unless the following conditions are simultaneously satisfied: $m'_j = m_j$ (namely $\Delta m_j = 0$), and $\ell' = \ell + 1$ or $\ell = \ell' + 1$ (namely $\Delta\ell = \pm 1$) – see Section 3.6, especially Eq. (3.53).

We now calculate the Q^+ component of the $\langle n'\ell' s' j' m'_j | \underline{r} | n\ell s j m_j \rangle$ vector:

$$\begin{aligned}
&\left\langle n'\ell'\frac{1}{2}j'm'_j|\underline{r}|n\ell\frac{1}{2}jm_j\right\rangle_{Q^+} \\
&= \left\langle n'\ell'\frac{1}{2}j'm'_j\middle| r\sin\theta\cdot e^{i\varphi}\middle| n\ell\frac{1}{2}jm_j\right\rangle_{Q^+} \\
&= \left\{\int_0^\infty rR^*_{n'\ell'}(r)R_{n\ell}(r)r^2dr\right\} \times \left\langle \ell'\frac{1}{2}\ j'm'_j\middle| \sin\theta\cdot e^{i\varphi}\middle| \ell\frac{1}{2}jm_j\right\rangle \quad .
\end{aligned}$$

Each state function appearing in the matrix element that constitutes the last factor on the third line can be developed to a sum of two orbital angular momentums per Clebsch-Gordan (see Eq. [4.87] and Table 4.1). Thus, in principle, the said last term is a sum of four integrals:

$$\begin{aligned}
&\left\langle \ell'\frac{1}{2}\ j'm'_j\middle| \sin\theta\cdot e^{i\varphi}\middle| \ell\frac{1}{2}jm_j\right\rangle \\
&= \left\{\left\langle \ell', m'_j - \frac{1}{2}, \uparrow\right| a' + \left\langle \ell', m'_j + \frac{1}{2}, \downarrow\right| b'\right\} \sin\theta\cdot e^{i\varphi}\left\{a\left|\ell, m_j - \frac{1}{2}, \uparrow\right\rangle + b\left|\ell, m_j + \frac{1}{2}, \downarrow\right\rangle\right\} \\
&= \left\langle \ell', m'_j - \frac{1}{2}, \uparrow\middle| a'a\sin\theta\cdot e^{i\varphi}\middle| \ell, m_j - \frac{1}{2}, \uparrow\right\rangle \\
&+ \left\langle \ell', m'_j - \frac{1}{2}, \uparrow\middle| a'b\sin\theta\cdot e^{i\varphi}\middle| \ell, m_j + \frac{1}{2}, \downarrow\right\rangle \\
&+ \left\langle \ell', m'_j + \frac{1}{2}, \downarrow\middle| b'a\sin\theta\cdot e^{i\varphi}\middle| \ell, m_j - \frac{1}{2}, \uparrow\right\rangle \\
&+ \left\langle \ell', m'_j + \frac{1}{2}, \downarrow\middle| b'b\sin\theta\cdot e^{i\varphi}\middle| \ell, m_j + \frac{1}{2}, \downarrow\right\rangle
\end{aligned}$$

Integrals containing opposite spin orientations vanish identically. The remaining are the first and last ones. Scalar product over spin coordinates provides a unity. We are left with the following sum:

$$a'a\left\langle \ell', m'_j - \frac{1}{2}\middle| \sin\theta\cdot e^{i\varphi}\middle| \ell, m_j - \frac{1}{2}\right\rangle + b'b\left\langle \ell', m'_j + \frac{1}{2}\middle| \sin\theta\cdot e^{i\varphi}\middle| \ell, m_j + \frac{1}{2}\right\rangle \quad .$$

Each of the integrals vanishes identically unless the following conditions are simultaneously satisfied: $m'_j = m_j + 1$ (namely $\Delta m_j = +1$), and $\ell' = \ell + 1$ or $\ell = \ell' + 1$ (namely $\Delta\ell = \pm 1$) – see Section 3.6, especially Eq. (3.54).

On similarly calculating the Q^- component of the $\langle n'\ell' s j' m'_j | \underline{r} | n \ell s j m_j \rangle$ vector, we obtain as a non-vanishing requirement that $\Delta m_j = -1$ simultaneously with $\Delta\ell = \pm 1$. In summary, a necessary condition for the transition probability to not vanish, is fulfillment of the requirement that $\Delta m_j = 0, \pm 1$ simultaneously with $\Delta\ell = \pm 1$. As in the present $j = \ell \pm \frac{1}{2}$ case, the above requirement may be formulized as "$\Delta m_j = 0, \pm 1$ simultaneously with $\Delta j = 0, \ \pm 1, \ \pm 2$". **However, in the case where $\Delta j = \pm 2$, the transition probabilities vanish, identically**. It is provable by setting the quantities belonging to that case in the relevant equations. For example, the $\Delta j = +2$ case obtains when $\ell' = \ell + 1$ along with $j' = \ell' + \frac{1}{2}$ and $j = \ell - \frac{1}{2}$. **The transition probability for $\Delta j = 0$ doesn't vanish**; however, it is possible to show that compared to the $\Delta j = \pm 1$ case, its value turns smaller for growing ℓ.

In summary, selection rules for electric dipole transitions between hydrogen electron states of well-defined angular momentums are

$$\boxed{\Delta m_j = 0, \pm 1 \ \textit{together with}\ \Delta j = 0, \pm 1} \quad . \tag{4.91}$$

4.9 Static External Magnetic and Electric Fields: Zeeman and Stark Effects

4.9.1 Zeeman Splitting

Zeeman splitting or **Zeeman effect** is a name used to describe the changes in atomic states energies when exposed to a static external magnetic field $\underline{\mathcal{B}}$. There are two aspects that require some emphasis. One aspect is that under application of a static external magnetic field, the force field acting on the electron is no more central, and the total angular momentum vector size is no longer a constant of motion. A second aspect is that under the external field, the Z axis is well defined for every atom, that is, it is codirected with the field. The Zeeman splitting is grossly classified into "weak-field" and "strong-field" effects. In the former, the induced splitting is small compared to the spin/orbit induced splitting, where the size of the total angular momentum $\underline{\boldsymbol{J}}$ may still be regarded a constant of motion. In the latter, the induced splitting is large compared to the spin/orbit splitting, hence the $\underline{\boldsymbol{J}}$-size may no longer be regarded as a constant of motion.

4.9.1.1 Weak Magnetic Field

An electron orbiting around the nucleus generates a magnetic dipole whose average value $\underline{\boldsymbol{M}}_{\text{av}}$ is provided in Eqs. (4.71) and (4.72) (see also Figure 4.7). The magnetic field changes the magnetic dipole energy, given by

$$U_{\mathcal{B}} = -\underline{\boldsymbol{M}}_{\text{av}} \cdot \underline{\mathcal{B}} = \frac{\mu_B}{\hbar} g_L \underline{\boldsymbol{J}} \cdot \underline{\mathcal{B}} = \frac{\mu_B}{\hbar} g_L \mathcal{B} \underline{\boldsymbol{J}}_Z \quad . \tag{4.92}$$

We now consider this relation as a relation between quantum operators. Thus, the expectation value of an energy change of each quantity is given by

$$E_{\mathcal{B}} \equiv \langle U_{\mathcal{B}} \rangle = g_L \frac{\mu_B}{\hbar} \mathcal{B} \langle \underline{\mathbf{J}}_Z \rangle = g_L \mu_B \mathcal{B} m_j \quad . \tag{4.93}$$

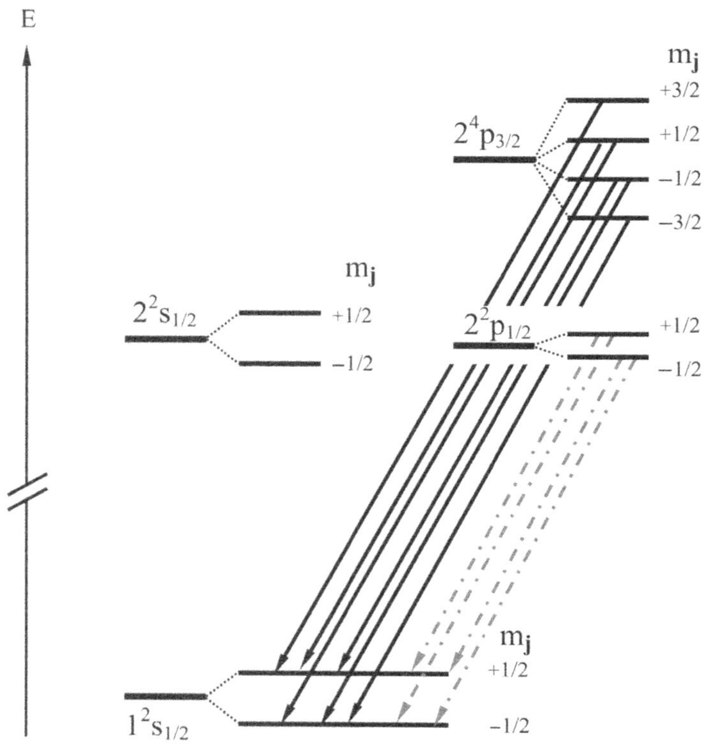

Figure 4.8 Lower energies scheme of a hydrogen atom, detailing the splitting caused by a weak magnetic field, and dipole-allowed optical transitions. Relatively weak transitions are indicated by gray dash-dotted arrows. The scale related to the energy separations among states belonging to the principal quantum number n is shrunk by several orders of magnitude to emphasize the splitting. The different splitting shown per different state expresses the different Landé factor of each.

The $|n\,\ell\, s\, j\, m_j\rangle$ state thus splits into $2j+1$ states, exhibiting equal separations between adjacent ones, where $j = \ell \pm \frac{1}{2}$ and $m_j = -j, -j+1, \ldots -\frac{1}{2}, +\frac{1}{2}, +\frac{3}{2}, \ldots +j$. The splitting size, namely the separation between adjacent states belonging to the same j, is identical, being proportional to the magnetic field strength $\mathcal{B}$. This splitting expresses itself in spectroscopic studies of absorption or emission lines, reflecting the $\Delta m_j = 0, \pm 1$ selection rule.

Figure 4.8 shows an example of the Zeeman weak-field splitting for the hydrogen $1s_{1/2}$ ground state, and the $2s_{1/2}$, $2p_{1/2}$, and $2p_{3/2}$ excited states, and their related dipole-allowed transitions. We have selected the following, more complex nomenclature, $n^{2j+1}\ell_j$ for specifying the different states. The added $2j+1$ superscript to the left of ℓ_j shows the number of split states generated per each unperturbed state upon application of the magnetic field. The split $2^2s_{1/2}$ state is meta-stable, that is, it exhibits no direct optical transitions to the ground (also split) $1^2s_{1/2}$ state. The energy splitting in the $2^2p_{1/2} \rightarrow 1^2s_{1/2}$ and $2^2p_{3/2} \rightarrow 1^2s_{1/2}$ transitions form quite a complex structure in the absorption spectrum. Spectroscopic study of the emission structure and its dependence on the magnetic field strength is a powerful tool to determine the Landé factor g_L, as well as the ℓ and j quantum numbers belonging to each of the involved states.

It is important to assess the range of magnetic field strengths that could be labeled “weak” for the considered transitions. The total splitting of any state is given by $2jg_L\mu_B\mathcal{B}$. The spin/orbit splitting, on the other hand, is given by Eqs. (4.65) and (4.66). Thus, we get the condition

$$\mathcal{B} \ll \left(\frac{|E_n|}{n}\right) Z^2\alpha^2 \frac{1}{\ell(\ell+1)} \frac{1}{2j\, g_L \mu_B} \quad . \tag{4.94}$$

For $j = \ell + \frac{1}{2}$ for example, the condition becomes

$$\mathcal{B} \ll \frac{1}{\mu_B} \left(\frac{|E_n|}{n}\right) Z^2\alpha^2 \frac{1}{2\ell(\ell+1)^2} \quad . \tag{4.95}$$

For the particular $n = 2;\ \ell = 1$ case we get $\mathcal{B} \ll 2 \times 10^3$ gauss.

4.9.1.2 Strong Magnetic Field

In a strong magnetic field, the quantum number j loses its stance as a good quantum number; the magnetic field effect on the electron orbit and spin are treated separately. Thus, we get that the corresponding energy corrections would be

$$\left.\begin{aligned} U_{\mathcal{B}}^{L} &= -\underline{\mathbf{M}}_{L}\cdot\underline{\mathcal{B}} = \frac{\mu_B}{\hbar}\underline{\mathbf{L}}\cdot\underline{\mathcal{B}} = \frac{\mu_B}{\hbar}\mathcal{B}\underline{\mathbf{L}}_Z \\ U_{\mathcal{B}}^{S} &= -\underline{\mathbf{M}}_{S}\cdot\underline{\mathcal{B}} = g\frac{\mu_B}{\hbar}\underline{\mathbf{s}}\cdot\underline{\mathcal{B}} = g\frac{\mu_B}{\hbar}\mathcal{B}\underline{\mathbf{s}}_Z \\ U_{\mathcal{B}} &= U_{\mathcal{B}}^{L} + U_{\mathcal{B}}^{S} \end{aligned}\right\} . \tag{4.96}$$

We then get

$$\left.\begin{aligned} E_{\mathcal{B}}^{L} &= \langle U_{\mathcal{B}}^{L}\rangle = \frac{\mu_B}{\hbar}\mathcal{B}\langle \mathbf{L}_Z\rangle = \mu_B\mathcal{B}m_\ell \\ E_{\mathcal{B}}^{S} &= \langle U_{\mathcal{B}}^{S}\rangle = g\frac{\mu_B}{\hbar}\mathcal{B}\langle \mathbf{s}_Z\rangle = g\mu_B\mathcal{B}m_s \\ E_{\mathcal{B}} &= E_{\mathcal{B}}^{L} + E_{\mathcal{B}}^{S} \end{aligned}\right\} . \tag{4.97}$$

The states splitting may be described as a two-stage process: in the first stage, each $|n\,\ell m_\ell m_s\rangle$ state splits into $2\ell + 1$, equally separated states ($\mu_B\mathcal{B}$) belonging to the $2\ell + 1$ possible values of the m_ℓ quantum numbers. Then, each such state splits in two belonging to the two possible values of the m_s quantum number. Because the gyromagnetic ratio is $g = 2$ and $|\delta m_s| = 1$, the splitting effected between the two spin states is *twice as large* as between two orbital projection states.

Figure 4.9 displays the states system of Figure 4.8 obtained under a strong external magnetic field ($\mathcal{B} \gg 2 \times 10^3$ gauss) such that the splitting it causes is large compared to the spin/orbit coupling. The $2s$ state is meta-stable, and no transitions take place to the ground $1s$ state. The $2p$ state splits in six, but two are congruent, so there are effectively five states. The $1s$ ground state splits in two. Abided by the $\Delta m_\ell = 0, \pm 1$ and $\Delta m_s = 0$ selection rules (Eq. [4.85]), six $2p \to 1s$ transitions are allowed. However, because all energy separations are identical, they may be organized in pairs of identical transition energies, thus only three lines are observed in the emission spectrum. If spin/orbit interaction effects are also included, some states would shift slightly and some split, such that eventually the number of observed transitions will be the same as for the weak magnetic field (Figure 4.8). Spin/orbit states splitting in this case are obviously very small compared to the others.

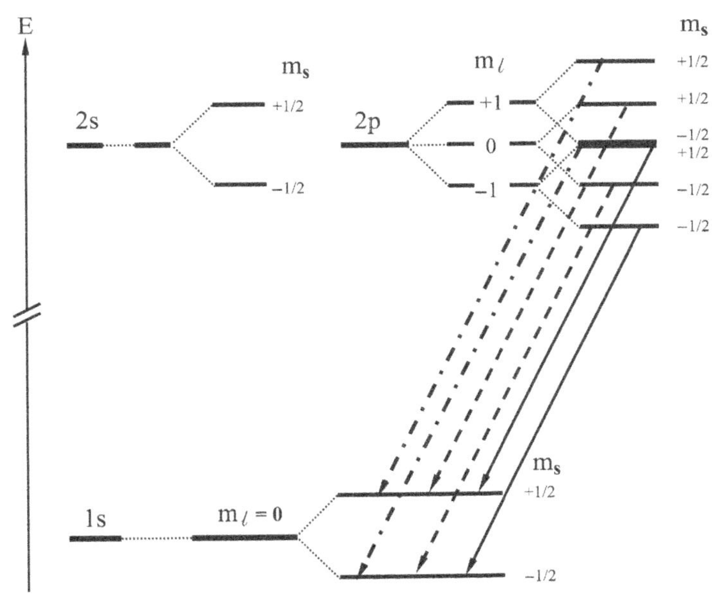

Figure 4.9 Lower energies scheme of a hydrogen atom detailing the splitting caused by a strong magnetic field, while neglecting spin/orbit splitting effects. Dipole-allowed optical transitions are indicated. The scale related to the energy separations among states belonging to the principal quantum number n is shrunk by several orders of magnitude to emphasize the splitting. Six allowed transitions are indicated, however organized in pairs of equal transition energy. Each identical pair member is plotted as arrows of same appearance: solid, dashed, and dash-dotted.

4.9.2 Stark Splitting

Stark splitting or **Stark effect** is the name given to changes in states' energy because of external static electric field application. As in the case of an external magnetic field (Section 4.9.1), the resultant force field acting on the electron is no more central, and the size of the total angular momentum is no longer a constant of motion. In addition, the Z-axis orientation is well defined for each atom, being codirectional with the electric field. The perturbation potential is $U_{\mathcal{E}} = e\underline{\mathcal{E}}\cdot\underline{r}$, and the same is the perturbation operator:

$$\mathbf{U}_{\mathcal{E}} = e\underline{\mathcal{E}}\cdot\underline{r} \quad . \tag{4.98}$$

According to perturbation theory, all diagonal elements of the perturbation matrix vanish. On the other hand, some of the off-diagonal elements are non-zero (see Section 2.1.1). As a result, in first order, there is no impact on the ground state. States mixing occurs for the excited states, as will be demonstrated.

4.9.2.1 Ground State; First-Order Perturbation Theory

We start by considering the Stark effect on the 1s ground state. Neglecting the electron spin, this is a singlet state, namely exhibiting no degeneracy. Neglecting mixing by energetically distant states, the first-order energy correction is $E_{\mathcal{E}} = e\langle 1s|\underline{\mathcal{E}}\cdot\underline{r}|1s\rangle = e\mathcal{E}\langle 1s|\underline{u}\cdot\underline{r}|1s\rangle = e\mathcal{E}\langle 1s|\underline{r}|1s\rangle\cdot \boldsymbol{cos}\,\theta_u$, where $\underline{u}$ is a unity vector in the electric field $\underline{\mathcal{E}}$ orientation, and θ_u is the angle between the $\underline{u}$ and $\langle 1s|\underline{r}|1s\rangle$ vectors (see also Eqs. [2.32] and [2.34]). In the present case, and in the Stark effect in general, the electric field orientation is constant; thus, we choose the atomic Z orientation to coincide with the electric field orientation. Then $\boldsymbol{cos}\theta_u = 1$. We have seen in our discussion on transition probabilities, that a necessary condition for the transition vector $\langle n'\ell' m'|\underline{r}|n\ell m\rangle$ to not vanish is that the orbital angular momentum quantum numbers satisfy $\ell' - \ell = \pm 1$ (Section 3.6, particularly Eq. [3.58]). In the present case, the transition vector vanishes, and in the first-order perturbation theory there would be no shift of the state energy. In the following section we consider the second order correction for the energy, and its physical implications.

4.9.2.2 Ground State; Second-Order Perturbation Theory

Because the first-order perturbation theory predicts a null shift of the state energy, we consider using the second-order correction. To that end, we develop first a reminder to the second-order theory. It would obviously include some repetition of Section 2.1, yet in a more compact and general form.

We consider the $\mathbf{H} = \mathbf{H}_0 + \mathbf{V}$ Hamiltonian operator, where $\mathbf{H}_0$ is the unperturbed Hamiltonian and $\mathbf{V}$ is the perturbation Hamiltonian $V(\underline{r})$. The unperturbed Hamiltonian satisfies $\mathbf{H}_0\psi_n = \varepsilon_n\psi_n$, where ψ_n is the eigenfunction of the unperturbed n state. We assume that the $\{\psi_n\}$ set of eigenfunctions establishes a full orthonormal system, which spans the entire space of the unperturbed Hamiltonian eigenfunctions. The perturbed Hamiltonian satisfies $\mathbf{H}\varphi_n = E\varphi_n$, where φ_n is the perturbed n state wavefunction, which is a "daughter" of the ψ_n "parent". Let λ be any number between zero and one, and let's observe the "partially perturbed" Hamiltonian $\mathbf{H}_\lambda = \mathbf{H}_0 + \lambda\mathbf{V}$. We expand the $\varphi_{n\lambda}$ eigenfunction of $\mathbf{H}_\lambda$ about ψ_n, and its eigenvalue $E_{n\lambda}$ about ε_n, both in powers of λ as follows:

$$\left.\begin{aligned}\varphi_{n\lambda} &= \psi_n + \lambda\varphi_n^{(1)} + \lambda^2\varphi_n^{(2)} + \ldots + \lambda^s\varphi_n^{(s)} + \ldots \\ E_{n\lambda} &= \varepsilon_n + \lambda E_n^{(1)} + \lambda^2 E_n^{(2)} + \ldots + \lambda^s E_n^{(s)} + \ldots\end{aligned}\right\} \quad . \tag{4.99}$$

Therefore,

$$(\mathbf{H}_0 + \lambda \mathbf{V})\left(\psi_n + \lambda \varphi_n^{(1)} + \lambda^2 \varphi_n^{(2)} + \dots\right)$$
$$= \left(\varepsilon_n + \lambda E_n^{(1)} + \lambda^2 E_n^{(2)} + \dots\right)\left(\psi_n + \lambda \varphi_n^{(1)} + \lambda^2 \varphi_n^{(2)} + \dots\right) .$$

We open the parentheses, equate elements belonging to the same power of λ between the two equation sides, set $\lambda = 1$, and finally get the following set of equations:

$$\left.\begin{array}{l} (\mathbf{H}_0 - \varepsilon_n)\psi_n = 0 \\ (\mathbf{H}_0 - \varepsilon_n)\varphi_n^{(1)} = \left(E_n^{(1)} - \mathbf{V}\right)\psi_n \\ (\mathbf{H}_0 - \varepsilon_n)\varphi_n^{(2)} = \left(E_n^{(1)} - \mathbf{V}\right)\varphi_n^{(1)} + E_n^{(2)}\psi_n \\ (\mathbf{H}_0 - \varepsilon_n)\varphi_n^{(3)} = \left(E_n^{(1)} - \mathbf{V}\right)\varphi_n^{(2)} + E_n^{(2)}\varphi_n^{(1)} + E_n^{(3)}\psi_n \\ \vdots \end{array}\right\} . \tag{4.100}$$

It is possible to select all $\varphi_n^{(s)}$ functions to be each orthogonal to its related ψ_n function, without influencing the validity of the separate equations constituting Eq. (4.100). (If it occurs that the scalar product $\left\langle \psi_n | \varphi_n^{(s)} \right\rangle = C \neq 0$ for some $\varphi_n^{(s)}$ function satisfying Eq. [4.100], that function may be replaced by $\varphi'^{(s)}_n \equiv \varphi_n^{(s)} - C\psi_n$; the replacing function is orthogonal to ψ_n, while still satisfying Eq. [4.100]). Another observation to be used later is that a scalar multiplication from the left of each left side of the (4.100) equations by ψ_n vanishes. This may be verified by recalling that $\mathbf{H}_0$ is Hermitian (see Hermitianity definition in relation to Table 3.1).

We now perform the scalar multiplication by $\langle \psi_n |$ on the left of each left side of the (4.100) equations. The first-line equation yields $\varepsilon_n = \langle \psi_n | \mathbf{H}_0 | \psi_n \rangle$; the second-line equation yields $E_n^{(1)} = \langle \psi_n | \mathbf{V} | \psi_n \rangle$; the third-line equation yields $E_n^{(2)} = \left\langle \psi_n | \mathbf{V} | \varphi_n^{(1)} \right\rangle$. Generally, we have

$$E_n^{(s)} = \left\langle \psi_n | \mathbf{V} | \varphi_n^{(s-1)} \right\rangle . \tag{4.101}$$

Notably, for calculation of the energy correction of s order, it is sufficient to know the corrected wavefunction of a lower order, namely of the $(s - 1)$ order.

We are interested in calculating the Stark second-order energy correction $E_n^{(2)}$; the first-order correction has already been detailed in Section 2.1.1. The $\varphi_n^{(1)}$ is developed by a linear combination of the zero-order Hamiltonian: $\varphi_n^{(1)} = \sum_{m \neq n} a_{nm}^{(1)} \psi_m$. In first-order correction (Eq. [2.6]), one gets

$$\begin{array}{l} E_n^{(1)} = \langle \psi_n | \mathbf{V} | \psi_n \rangle \\ a_{nm}^{(1)} = \dfrac{\langle \psi_m | \mathbf{V} | \psi_n \rangle}{\varepsilon_n - \varepsilon_m} \quad ; \quad m \neq n \end{array} . \tag{4.102}$$

Therefore,

$$E_n^{(2)} = \sum_{m \neq n} a_{nm}^{(1)} \langle \psi_n | \mathbf{V} | \psi_m \rangle = \sum_m \frac{\langle \psi_m | \mathbf{V} | \psi_n \rangle \langle \psi_n | \mathbf{V} | \psi_m \rangle}{\varepsilon_n - \varepsilon_m} \quad ;$$

$$\boxed{E_n^{(2)} = \sum_{m \neq n} \frac{|\langle \psi_m | \mathbf{V} | \psi_n \rangle|^2}{\varepsilon_n - \varepsilon_m}} \quad . \tag{4.103}$$

By the last result, the standard procedure for calculating Stark shift for the ground state ($n = 1$) is to perform all integrals specified in the numerator of Eq. (4.103) in relation to all excited states, and add up all specified ratios. This, indeed a clear procedure, is however long and tiring. Therefore, a shortcut procedure is selected based on the *principles* of the second-order perturbation theory.

Based on the remark that follows Eq. (4.101), we proceed to calculate the $\varphi_1^{(1)}$ for a hydrogen-like atom. We specify the equation appearing on the second line of Eq. (4.100):

$$\left(-\frac{\hbar^2}{2\mu} \nabla^2 - \frac{Ze^2}{r} + \frac{Z^2 e^2}{2a_0} \right) \varphi_1^{(1)} = -e\mathcal{E} r \cos\theta \cdot \frac{1}{\sqrt{\pi}} \left(\frac{Z}{a_0} \right)^{3/2} e^{-Zr/a_0} \quad . \tag{4.104}$$

(for setting of the specific functional form of $\psi_1 = |100\rangle$, see Tables 3.2 and 3.3). Eq. (4.104) is an inhomogeneous differential equation for $\varphi_1^{(1)}$. We attempt a solution of the form

$$\varphi_1^{(1)}(\underline{r}) = f(r) \cos\theta \quad . \tag{4.105}$$

Such function is odd; in that respect it is as required, orthogonal to the ground-state wavefunction, which is even. We convert Eq. (4.104) into spherical coordinates (see Eq. [3.23]), set $\varphi_1^{(1)}$ and get

$$\frac{d^2 f}{dr^2} + \frac{2}{r}\frac{df}{dr} - \frac{2}{r^2} f + \frac{2}{r}\left(\frac{Z}{a_0} \right) f - \left(\frac{Z}{a_0} \right)^2 f = \frac{2}{\sqrt{\pi}} \left(\frac{1}{e} \right) \left(\frac{\mathcal{E}}{Z} \right) \left(\frac{Z}{a_0} \right)^{5/2} e^{-(Z/a_0)\cdot r} \cdot r \quad . \tag{4.106}$$

We attempt a solution of the form $f(r) = \mathcal{V}(r) e^{-(Z/a_0)\cdot r}$. We then get the following differential equation for $\mathcal{V}$:

$$\mathcal{V}'' + \left\{ \frac{2}{r} - 2\left(\frac{Z}{a_0} \right) \right\} \mathcal{V}' - \frac{2}{r^2} \mathcal{V} = \frac{2}{\sqrt{\pi}} \left(\frac{1}{e} \right) \left(\frac{\mathcal{E}}{Z} \right) \left(\frac{Z}{a_0} \right)^{5/2} \cdot r \quad . \tag{4.107}$$

As a solution for $\mathcal{V}$ we attempt a power series of r. Physical considerations suggest that it would not contain a power lower than a unity (1) because otherwise the $\varphi_1^{(1)}$ function would diverge at the origin. The mathematical solution provides a second-order polynomial:

$$\mathcal{V} = -\frac{2}{\sqrt{4\pi}} \left(\frac{1}{e} \right) \left(\frac{\mathcal{E}}{Z} \right) \left(\frac{Z}{a_0} \right)^{1/2} \left\{ r + \frac{1}{2}\left(\frac{Z}{a_0} \right) \cdot r^2 \right\} \quad . \tag{4.108}$$

And in summary

$$\varphi_1^{(1)} = -\frac{2}{\sqrt{4\pi}}\left(\frac{1}{e}\right)\left(\frac{\mathcal{E}}{Z}\right)\left(\frac{Z}{a_0}\right)^{1/2}\left\{r + \frac{1}{2}\left(\frac{Z}{a_0}\right)\cdot r^2\right\}\cdot e^{-(Z/a_0)r}\cdot \boldsymbol{\cos\theta} \ . \tag{4.109}$$

As off this point, we can calculate directly the energy correction by the second-order perturbation theory, namely the Stark effect:

$$E_{\mathcal{E}} = E_1^{(2)} = \left\langle \psi_1 | \mathbf{V} | \varphi_1^{(1)} \right\rangle$$

$$= \frac{-1}{\pi}\left(\frac{Z}{a_0}\right)^2\left(\frac{\mathcal{E}^2}{Z}\right)\int_0^\infty dr \int_0^\pi d\theta \int_0^{2\pi} d\varphi \cdot e^{-2(Z/a_0)\cdot r}\cdot\left\{r + \frac{1}{2}\left(\frac{Z}{a_0}\right)\cdot r^2\right\} r^3 \boldsymbol{\cos}^2\theta\, \boldsymbol{\sin}\theta \ ;$$

$$\boxed{E_{\mathcal{E}} = -\frac{9}{4}\left(\frac{a_0}{Z}\right)^3\frac{\mathcal{E}^2}{Z}} \ . \tag{4.110}$$

The square dependence of the ground-state Stark shift on the electric field strength is interpreted in terms of an induced static electric dipole: assume that the atom exhibits some polarizability denoted by α (the polarizability employs the same letter as the fine structure constant, but is obviously an entirely different physical quantity). An applied static electric field induces a static electric dipole $\underline{Q} = \alpha\underline{\mathcal{E}}$. The energy invested in its generation is given by $|E_{dipole}| = \frac{1}{2}\alpha\mathcal{E}^2$. If the dipole energy is identified as the ground-state Stark shift, the atom polarizability is given by

$$\boxed{\alpha = \frac{9}{2Z}\left(\frac{a_0}{Z}\right)^3} \ . \tag{4.111}$$

Note that the polarizability has volume units (cm^3 in cgs units).

4.9.2.3 First Excited State; First-Order Perturbation Theory

We now turn to finding the energy corrections to the $|2\ell m\rangle$ first excited state, for which $\ell = 0$, 1, and $m = 0, \pm 1$. This system exhibits four-fold degenerate states (degeneracy of order 4; see Section 3.5, especially Figure 3.7). Because the electric field $\underline{\mathcal{E}}$ has a constant orientation, we limit our consideration to the Z component of $\langle 2\ell' m' | \underline{r} | 2\ell m\rangle$, namely to $\langle 2\ell' m' | \underline{r} | 2\ell m\rangle_Z$. In Section 3.6 we showed that a necessary condition for that component to not vanish is that $m' = m$. The perturbation elements form a 4×4 matrix. Using the above selection rules, the perturbation matrix is of the form

$$\begin{array}{c} \\ \langle 0,0| \\ \langle 1,0| \\ \langle 1,1| \\ \langle 1,-1| \end{array}\begin{array}{c} \begin{array}{cccc} |0,0\rangle & |1,0\rangle & |1,1\rangle & |1,-1\rangle \end{array} \\ \begin{pmatrix} 0 & \langle 0,0|\mathbf{U}_{\mathcal{E}}|1,0\rangle & 0 & 0 \\ \langle 0,0|\mathbf{U}_{\mathcal{E}}|1,0\rangle & 0 & 0 & 0 \\ 0 & 0 & 0 & 0 \\ 0 & 0 & 0 & 0 \end{pmatrix} \end{array} . \tag{4.112}$$

The non-vanishing elements are thus

$$\langle 2,0,0|\mathbf{U}_{\mathcal{E}}|2,1,0\rangle = \langle 2,1,0|\mathbf{U}_{\mathcal{E}}|2,0,0\rangle^{*} = e\mathcal{E}\langle 2,0,0|\underline{r}|2,1,0\rangle_{Z} = e\mathcal{E}\langle 2,0,0|r\,\boldsymbol{\cos}\,\theta|2,1,0\rangle$$

$$= e\mathcal{E}\int_0^{\infty} r\cdot \boldsymbol{R}_{20}^{*}(r)\boldsymbol{R}_{21}(r)r^2dr \times \int_0^{\pi} \boldsymbol{\cos}\,\theta\cdot\boldsymbol{\Theta}_0^{0*}(\theta)\boldsymbol{\Theta}_1^{0}(\theta)\,\boldsymbol{\sin}\,\theta\cdot d\theta \times \int_0^{2\pi}\boldsymbol{\Phi}_0^{*}(\varphi)\boldsymbol{\Phi}_0(\varphi)d\varphi,$$

$$= e\mathcal{E}\frac{1}{8\sqrt{3}}\int_0^{\infty}\left(\frac{r}{a_0}\right)^4\left(2-\frac{r}{a_0}\right)e^{-r/a_0}dr\cdot\frac{\sqrt{3}}{2}\int_0^{\pi}\boldsymbol{\cos}^2\theta\,\boldsymbol{\sin}\,\theta\cdot d\theta \quad .$$

(See Tables 3.2 and 3.3.) Performing the above integrals finally yields

$$\boxed{\langle 2,0,0|\mathbf{U}_{S}|2,1,0\rangle = 3e\mathcal{E}a_0} \quad . \tag{4.113}$$

The secular equation needing solution for obtaining the self energies of the perturbation matrix is then (see Eq. [2.8])

$$\begin{vmatrix} -E_{\mathcal{E}} & 3e\mathcal{E}a_0 & 0 & 0 \\ 3e\mathcal{E}a_0 & -E_{\mathcal{E}} & 0 & 0 \\ 0 & 0 & -E_{\mathcal{E}} & 0 \\ 0 & 0 & 0 & -E_{\mathcal{E}} \end{vmatrix} = 0 \quad . \tag{4.114}$$

The four roots of the secular equation are $E_{\mathcal{E}1} = 0$; $E_{\mathcal{E}2} = 0$; $E_{\mathcal{E}3} = +3e\mathcal{E}a_0$; $E_{\mathcal{E}4} = -3e\mathcal{E}a_0$. Thus, half of the four-fold degeneracy is removed so that only a two-fold degeneracy remains. The eigenvectors are of the form

$$(A_1) = \begin{pmatrix} 0 \\ 0 \\ a_{11} \\ a_{1,-1} \end{pmatrix} \; ; \; (A_2) = \begin{pmatrix} 0 \\ 0 \\ a'_{11} \\ a'_{1,-1} \end{pmatrix} \; ; \; (A_3) = \begin{pmatrix} 1/\sqrt{2} \\ -1/\sqrt{2} \\ 0 \\ 0 \end{pmatrix} \; ; \; (A_4) = \begin{pmatrix} 1/\sqrt{2} \\ 1/\sqrt{2} \\ 0 \\ 0 \end{pmatrix} \; , \tag{4.115}$$

where the "a" elements are required to satisfy $a_{11}{}^2 + a_{1,-1}{}^2 = 1$ and $a'_{11}{}^2 + a'_{1,-1}{}^2 = 1$. It is also required that the matrix $(a) \equiv \begin{pmatrix} a'_{11} & a_{11} \\ a'_{1,-1} & a_{1,-1} \end{pmatrix}$ be regular. Correspondingly, the new state functions are

$$\psi_1 = a_{11}\cdot|211\rangle + a_{1,-1}|2,1,-1\rangle \; ; \; \psi_2 = a'_{11}\cdot|211\rangle + a'_{1,-1}|2,1,-1\rangle$$
$$\psi_3 = \frac{|200\rangle - |210\rangle}{\sqrt{2}} \; ; \; \psi_4 = \frac{|200\rangle + |210\rangle}{\sqrt{2}} \quad . \tag{4.116}$$

Regularity of the (a) matrix assures that the ψ_1 and ψ_2 functions are independent.

The physical meaning of the last result indicated that application of the external static electric field $\underline{\mathcal{E}}$ induces mixing among the different $|\ell m\rangle$ states belonging to the principal $n = 2$ quantum number. In the new states, the atom behaves *as if it has a constant dipole moment*, whose size is $\boldsymbol{Q}_{ss} = 3ea_0$; in the ψ_3 state, the dipole is oriented oppositely to the electric field orientation; in the ψ_4 state it is co-oriented with the electric field; in the two other states, ψ_1 and ψ_2, its projection on the field orientation vanishes. Accordingly, a state splitting is obtained according to Figure 4.10.

To demonstrate the order of magnitude of the Stark effect, we now calculate the splitting obtained under an external electric field 3×10^5 V/cm strong:

Given data:

$$e = 4.8 \times 10^{-10}\ \text{esu} \ ; \ a_0 = 0.52917 \times 10^{-8}\ \text{cm} \ ;$$

$$\mathcal{E} = 3 \times 10^5\ \text{V/cm} = 10^3\ \text{statV/cm} \ .$$

Used equation:

$$E_{\mathcal{E}} = 3ea_0\mathcal{E} \ .$$

Setting:

$$E_{\mathcal{E}} = 3 \times 4.8 \times 10^{-10} \times 0.52917 \times 10^{-8} \times 10^3 = 7.62 \times 10^{-15}\text{erg} = \frac{7.62 \times 10^{-15}}{6.626 \times 10^{-27} \times 3 \times 10^{10}} = 38.35\ \text{cm}^{-1} \ .$$

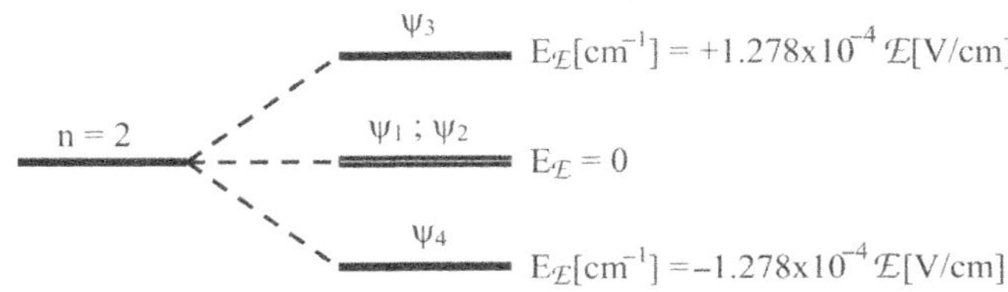

Figure 4.10 Stark splitting of the first excited state of the hydrogen atom. The splitting is proportional to the electric field strength. The written technical equation on the right provides the energy shift in cm^{-1} units, and the electric field strength in V/cm units.

Dopant atoms or ions residing in a solid material are usually exposed to strong static electric fields that cause Stark shifting and splitting, typically of the order of several hundreds cm^{-1}. This effect is called "crystal field splitting". The major difference between the basic Stark splitting and crystal field splitting is the *huge electric field gradients* involved in the latter, whereas in the former the field spatial gradient is zero by definition.

It should be emphasized, although we won't elaborate on proving, that fundamentally *the summed corrections of all orders diverges*! In other words, an atom under an external, even a tiny static electric field, exhibits no bonded steady-state solutions because there would always be a distance in space where the potential energy becomes smaller than the ground-state of the unperturbed atom. The atom then disintegrates, with probability growing exponentially with the electric field strength. The above calculations we have performed are valid as long as no such disintegration occurred.

4.10 The Fine Structure

In addition to shifting and splitting described so far, the hydrogen-like atomic spectra exhibit shifting and splitting originated by interactions between the electron and the nucleus. These effects are called **Hyperfine structure**, and they are considerably small compared to the spin/orbit interaction effects. In this section, we limit ourselves to pointing out the different effects and providing a qualitative explanation of their origin, with no quantitative analysis.

4.10.1 Isotope Shifting

Isotopes are atoms whose nuclei contain an identical number of protons, but different numbers of neutrons. There are two reasons for isotopes to exhibit spectral differences. One is a difference in their nuclear mass, hence their reduced mass μ appearing in the energy expression (Eq. [3.38]). A gas containing several isotopes will show corresponding splitting in the optical transition energies. The other reason is differences in the nuclear *size*. Up to now we have modeled the nucleus as a point charge and mass. Having finite sizes, different isotopic nuclei exhibit different charge distributions. This causes different internal electric potential when the electron crosses through the nucleus. This reality expresses itself by a different energy compared to that given by Eq. (3.38). It causes removal of degeneracy among different ℓ values of the same principal quantum number, because of different penetrations through the nucleus (Table 3.3 and Figure 3.6). Most penetrating wave functions are those belonging to S states (namely those of $\ell = 0$). Those of higher ℓ values penetrate less, and their effected shifts are much less important.

4.10.2 Nuclear Magnetic Shifting

The nucleus itself carries a magnetic dipole resulting from the orbital motion of its constituent protons, and the internal spinning of both protons and neutrons. The magnetic dipole obviously interacts with the orbiting electrons, causing spectral shifting and splitting.

4.10.3 Nuclear Quadrupole Shifting

At first-order approximation, the nucleus has a full spherical symmetry. A distorted sphere, apparently a shape assumed by many nuclei, has a quadrupole component. Therefore, an electron motion about the nucleus is changed, an effect expressing itself by spectral shifting and splitting.

4.11 Appendix: Clebsch-Gordan Coefficients for Coupling of Two Angular Momentums

A general equation for calculating the Clebsch-Gordan coefficients for two coupling angular momentums is given by

$$\begin{aligned}\langle j_1 m_1 j_2 m_2 | j_1 j_2 j m\rangle &= \delta(m, m_1 + m_2)\sqrt{\frac{(j_1 + j_2 - j)!(j + j_1 - j_2)!(j + j_2 - j_1)!(2j+1)!}{(j + j_1 + j_2 + 1)!}} \\ &\times \sum_k \frac{(-)^k \sqrt{(j_1 + m_1)!(j_1 - m_1)!(j_2 + m_2)!(j_2 - m_2)!(j + m)!(j - m)!}}{k!(j_1 + j_2 - j - k)!(j_1 - m_1 - k)!(j_2 + m_2 - k)!(j - j_2 + m_1 + k)!(j - j_1 - m_2 + k)!},\end{aligned} \tag{4.117}$$

where k runs over all integers between zero and the lower quantity among $(j_1 + j_2 - j)$, $(j_1 - m_1)$, and $(j_2 + m_2)$. We won't prove this equation in the present book.

Exercises for Chapter 4

4.1 (a) Use Eq. (4.12) for calculating the uncertainty relations between a particle position and its linear momentum. Calculate the uncertainty in the electron linear momentum if its position is limited to the typical dimensions of an atomic nucleus (~5×10^{-13} cm). Calculate the kinetic energy related to that linear momentum, and compare it with the rest masses of an electron and a proton.
(b) Use Eq. (4.12) to calculate the uncertainty relations between a particle lifetime and its energy.

4.2 Present an energy scheme of a hydrogen atom that includes the relativistic correction, still without electron spin (Eq. [4.20]) for the principal quantum numbers $n = 4$ and $n = 5$. Indicate by arrows the radiative dipole-allowed transitions.

4.3 Present an energy scheme of a hydrogen atom for the principal quantum number $n = 4$ that accounts for corrections emanating from the electron spinning (Eq. [4.68]).

4.4 Calculate numerically the average magnetic moment $\underline{\boldsymbol{M}}_{\mathrm{av}}$ of all states belonging to the principal quantum number $n = 4$ of the hydrogen atom.

4.5 Calculate the energy separation between the lamb-shifted $3s_{1/2}$ and $3p_{1/2}$ states of the hydrogen atom.

4.6 Calculate in cm^{-1} units the $1^2s_{1/2}$, $2^2p_{1/2}$, and $2^4p_{1/2}$ hydrogen atom states' splitting under an applied external magnetic field of 150 gauss.

4.7 Demonstrate that the electric dipole transition probability in a hydrogen-like atom vanishes identically when $\Delta j = \pm 2$ by testing the $\Delta m_j = 0$ case.

4.8 Plot an energy states' splitting scheme belonging to $n = 2$ and $n = 3$ of the hydrogen atom under application of a "strong" magnetic field. Use arrows to indicate the radiative allowed electric dipole transitions between states.

5 Many-Electron Atoms

5.1 Preamble

Except for hydrogen-like atoms that contain a single electron, all other atoms contain several – up to more than 100 – electrons. However, as we shall see, the insight attained concerning the hydrogen-like atoms plays a key role in achieving a solution, indeed approximate but mostly satisfactory, to the properties of many-electron atoms. The difficulty in providing accurate solutions to these atoms' properties relates to the large number of interparticle interactions. The potential energy of the entire electron aggregation in an atom is given by

$$\boxed{V(\underline{r}_1, \underline{r}_2, \cdots \underline{r}_i \cdots) = \sum_{\substack{all \\ electrons}} -\frac{Ze^2}{r_i} + \sum_{\substack{all \\ pairs}} \frac{e^2}{\left|\underline{r}_i - \underline{r}_j\right|}}, \tag{5.1}$$

where $\underline{r}_i$ is the distance vector of the electron relative to the nucleus. It would be mathematically hopeless to describe the motion of any particular electron independently of the motions of other electrons, because any displacement of one electron affects the potential energy of all others. The selected approach is to treat an atom as a unified entity. Accordingly, in many-electron atoms one speaks of "atomic states" and "atomic wave functions". Solution is still mathematically complicated, and requires simplifying approximations. The zero-order approximation describes the motion of each electron separately as if independent of the motions of all other electrons. This approach is called the **independent electrons model**.

5.2 Helium-Like Atoms

The simplest many-electron atom is one that contains two electrons, like the negative hydrogen ion H^- $(Z = 1)$, a helium atom He $(Z = 2)$ whose nucleus (of most common isotope) comprises two protons and two neutrons, a positive lithium ion Li^+ $(Z = 3)$ whose nucleus comprises three protons and four neutrons, and so on. The electrons Hamiltonian of each such system is given by

$$\boxed{\mathbf{H} = -\frac{\hbar^2}{2m}\left(\underline{\nabla}_1^{\,2} + \underline{\nabla}_2^{\,2}\right) - \frac{Ze^2}{r_1} - \frac{Ze^2}{r_2} + \frac{e^2}{r_{12}}}, \tag{5.2}$$

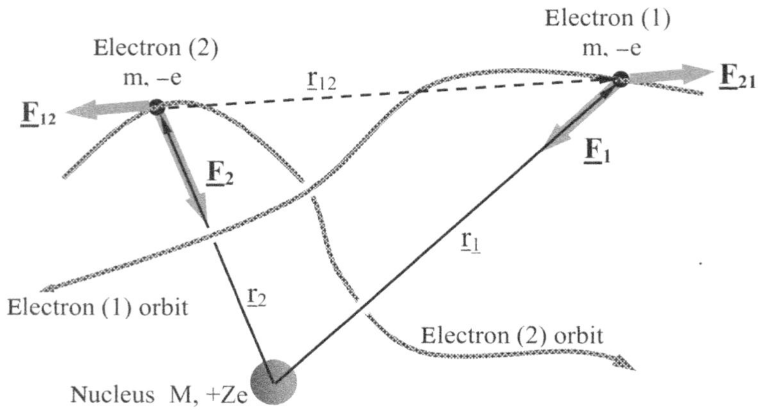

Figure 5.1 Schematic illustration of a helium-like atom comprising a nucleus and two electrons, and forces between each and the nucleus and between themselves. Two shown orbit sections of the electrons demonstrate that the electrons orbits are rather complex.

where $\underline{\nabla}_1{}^2$ and $\underline{\nabla}_2{}^2$ are Laplace operators acting, respectively, on the first and second electron coordinates, and $r_{12} \equiv |\underline{r}_1 - \underline{r}_2|$ is the distance between the two electrons. Figure 5.1 describes the fundamental structure of a helium-like atom.

5.2.1 Zero-Order Approximation under the Independent Electron Model

Under the zero-order approximation of the independent electron model we ignore the last addend in the Hamiltonian of Eq. (5.2), and write

$$\mathbf{H}_0 = \mathbf{H}_{01} + \mathbf{H}_{02} = \left[-\frac{\hbar^2}{2m}\underline{\nabla}_1{}^2 - \frac{Ze^2}{r_1} \right] + \left[-\frac{\hbar^2}{2m}\underline{\nabla}_2{}^2 - \frac{Ze^2}{r_2} \right]. \tag{5.3}$$

In such a manner, the Hamiltonian is divided into two independent contributions, where each separately establishes a Hamiltonian of a hydrogen-like atom. If we further assume a solution through a variable separation format $\psi(\underline{r}_1, \underline{r}_2) = \psi_1(\underline{r}_1)\psi_2(\underline{r}_2)$, we get the two independent equations:

$$\left.\begin{matrix} \left(-\frac{\hbar^2}{2m}\underline{\nabla}_1{}^2 - \frac{Ze^2}{r_1} \right)\psi_1 = E_1\psi_1 \text{ (i)} \\ \left(-\frac{\hbar^2}{2m}\underline{\nabla}_2{}^2 - \frac{Ze^2}{r_2} \right)\psi_2 = E_2\psi_1 \text{ (ii)} \end{matrix}\right\}. \tag{5.4}$$

Separately, each equation is a Schrödinger equation of a hydrogen-like atom. The solutions would be $\psi_1 = |n_1\ell_1m_1\rangle$ and $\psi_2 = |n_2\ell_2m_2\rangle$ of E_{n_1} and E_{n_2} eigenenergy solutions, as provided in Eq. (3.38). The total energy, that is, the energy required to ionize the atom entirely, would be $-E = E_{n_1} + E_{n_2}$.

When examined against experimental results, the latter simplistic prediction reveals non-negligible differences. The experimental states are considerably higher. For example, the calculated helium ground state ($n_1 = n_2 = 1$; $Z = 2$) suggests $E_{gs}(He) = 2 \times Z^2 E_{gs}(H) \underset{Z=2}{=} -8.789 \times 10^5\ \text{cm}^{-1}$. The measured one is lower - $E_{gs}^{(measured)}(He) = -6.365 \times 10^5\ \text{cm}^{-1}$ (about 30% difference). The overestimated ground-state ionization energy should be related to ignoring the rejection potential between the two equally negatively charged electrons.

5.2.2 First-Order Correction and the Effective Screening Idea

As first-order correction we address the $\mathbf{H}_{12} \equiv e^2/r_{12}$ coupling term between the electrons as a small perturbation, and calculate the correction to the ground-state energy in the first-order perturbation theory. It is given by

$$E_{gs}^{(1)} = \left\langle \frac{e^2}{r_{12}} \right\rangle = \langle 100|\langle 100\,|\frac{e^2}{r_{12}}|\,100\rangle|100\rangle\,. \tag{5.5}$$

We set the specific functional form of the $|100\rangle$ state function using Tables 3.2 and 3.3:

$$E_{gs}^{(1)} = \frac{1}{(8\pi)^2}\left(\frac{2Z}{a_0}\right)^6 \iint \frac{e^2}{r_{12}} e^{-(2Z/a_0)(r_1+r_2)} d^3\underline{r}_1 d^3\underline{r}_2\,. \tag{5.6}$$

The above integral expresses the electrostatic rejection energy between two identical charge distributions $\rho(r) = (-e/8\pi)(2Z/a_0)^3 \cdot e^{-(2Z/a_0)r}$ each, about the origin. We start by calculating the electrostatic energy required to accumulate a *single* such charge distribution. The related electric field $\mathcal{E}$ would be radial, and its size may be calculated as $\mathcal{E}(r) = q(r)/r^2$, where $q(r)$ is the partial charge enclosed by a sphere of radius r. A direct calculation yields

$$q(r) = (-e)\cdot\left[1 - e^{-(2Z/a_0)r}\left\{1 + \left(\frac{2Z}{a_0}\right)r + \frac{1}{2}\left(\frac{2Z}{a_0}\right)^2 r^2\right\}\right], \tag{5.7}$$

Thus,

$$\mathcal{E}(r) = \frac{(-e)}{r^2}\cdot\left[1 - e^{-(2Z/a_0)r}\left\{1 + \left(\frac{2Z}{a_0}\right)r + \frac{1}{2}\left(\frac{2Z}{a_0}\right)^2 r^2\right\}\right]. \tag{5.8}$$

The energy density is given by $\mathcal{E}^2/8\pi$; the total energy is given by integration over the entire space

$$E_{es} = \frac{1}{8\pi}\int_0^\infty \mathcal{E}^2(r)\cdot 4\pi r^2 dr\,. \tag{5.9}$$

Setting Eq. (5.8) and changing the variable from r to $\xi \equiv (2Z/a_0)r$, we get

$$E_{es} = \frac{Ze^2}{a_0}\int_0^\infty \left\{1 - e^{-\xi}\left(1 + \xi + \frac{1}{2}\xi^2\right)\right\}^2 \frac{1}{\xi^2} d\xi \tag{5.9-i}$$

$$= \frac{Ze^2}{a_0}\left[\frac{1}{\xi}\left(e^{-\xi}(\xi+2) - e^{-2\xi} - 1\right) - \frac{e^{-2\xi}}{16}\left(2\xi^2 + 10\xi + 21\right)\right]_0^\infty = \frac{5}{8}\cdot\frac{Ze^2}{2a_0}\,.$$

When the two identical distributions join, the electrostatic energy becomes fourfold larger. This growth may be viewed as the addition of the two distributions separately, plus the rejection energy between the two distributions under full overlap. The latter is the targeted calculation result. It is thus simply twice the calculated value by Eq. (5.9-i). Namely,

$$E_{gs}^{(1)} = \frac{5}{4} \cdot \frac{Ze^2}{2a_0}. \tag{5.10}$$

Finally, we get

$$E_{gs} \cong E_{gs}^{(0)} + E_{gs}^{(1)} = -2\frac{Z^2e^2}{2a_0} + \frac{5}{4}\frac{Ze^2}{2a_0} = -2\frac{e^2}{2a_0}\left[Z^2 - \frac{5}{8}Z + \left(\frac{5}{16}\right)^2 - \left(\frac{5}{16}\right)^2\right];$$

$$E_{gs} \cong -2\frac{e^2}{2a_0}\left[\left(Z - \frac{5}{16}\right)^2 - \left(\frac{5}{16}\right)^2\right]. \tag{5.11}$$

For simplicity we neglect the last (squared) addend in the square parentheses in Eq. (5.11), which means some reduction in the correction given by Eq. (5.10). We then get the approximation

$$\boxed{E_{gs} \cong -2\frac{(Z-5/16)^2e^2}{2a_0}}. \tag{5.12}$$

The experimental correction obtained for the helium atom ($Z = 2$) is only about 2% larger than that given by Eq. (5.12). The physical meaning is that the two electrons may indeed be approximated as independent, yet in a manner that each electron performs a partial screening of the nuclear charge to the motion of the other electron. The screening per Eq. (5.12) is $5/16 = 0.3125$. The experimental screening is 0.2959. This conclusion establishes a powerful guide for calculating the energy states of the helium atom itself, and the various states of many-electron atoms in general. The first-order approximation for these systems assumes the electrons to be independent, each being subject to an effective central electrostatic attraction that is indeed like a basic electrostatic attraction, yet with a partial screening of the nuclear charge effected by the other electrons.

In first-order approximation, the effective Hamiltonian expressing the said principles is of the form

$$\boxed{\mathbf{H} = \mathbf{H}_1 + \mathbf{H}_2 = \left\{-\frac{\hbar^2}{2m}\underline{\nabla}_1{}^2 - \frac{[Z-\varsigma(r_1)]e^2}{r_1}\right\} + \left\{-\frac{\hbar^2}{2m}\underline{\nabla}_2{}^2 - \frac{[Z-\varsigma(r_2)]e^2}{r_2}\right\}}, \tag{5.13}$$

where $\varsigma(r)$ is the screening fraction (**Hartree screening function**), which is a continually increasing function of the electron r distance from the nucleus, such that it diminishes for $r = 0$, assuming the value $(Z - N + 1)$ for $r \to \infty$, where N is the number of electrons. The underlying physical idea is that when any specific electron approaches the nucleus vicinity, it is likely that the other electrons are far from the nucleus, and provide no screening. On the other hand, when the specific electron is far from the nucleus, it is likely that the others are close to the nucleus, providing their full $(Z - N + 1)$ screening capacity. As a further conjecture, it is assumed that the wave function of each individual electron may be written in the form $\psi_i(\underline{r}_i) = \boldsymbol{R}_{n_i\ell_i}(r_i)\boldsymbol{Y}_{\ell_i m_i}(\theta_i, \varphi_i)$, where the $i = 1, 2, \cdots N$ numbers provide tagging of the electrons as individual entities. Under the assumption of the potential central character, the angular functions $\boldsymbol{Y}_{\ell_i m_i}(\theta_i, \varphi_i)$ would be identical with the ordinary $\boldsymbol{Y}_{\ell m}$ spherical harmonics described in Sections 3.3 and 3.4 (especially Table 3.2). This approach is supported by experimental results. As to the $\boldsymbol{R}_{n_i\ell_i}(r_i)$ radial wave functions, because the assumed potential is not proportional to the ordinary $1/r$ - shaped regular electrostatic potential, they won't exhibit simple similarity to the $\boldsymbol{R}_{n\ell}$ functions of the hydrogen atom (Table 3.3). As a result, the energy degeneracy among different ℓ states belonging to the same principal quantum number n is removed. The symbol $\boldsymbol{R}_{n\ell}$ is still used for the effective radial functions. The 5/16 screening quantity appearing in Eq. (5.12) is thus an average of the $\varsigma(r)$ function belonging to the helium ground state. There are usually no simple analytic expressions for these functions. Their calculation is long and tiring even for approximate results: the relevant task is actually the finding of $\varsigma(r)$ functions, or alternatively, $\boldsymbol{R}_{n\ell}$ functions that provide the best fit to experimental results. This book does not elaborate further on the quantitative aspect. However, we will extend our discussions to some fundamental issues.

5.2.3 Exchange Symmetry

For clarity and brevity, we symbolize the independent wave functions of each of the helium atom electrons by $\psi_a(1)$ and $\psi_b(2)$, where a and b symbolize each the *set* of quantum numbers belonging to its independent electron. The 1 and 2 indicate the tagging numbers given to each electron and its corresponding coordinates.

Now, even if the separate wave functions $\psi_a(1)$ and $\psi_b(2)$ are modified to exhibit the best fit to measured states energy, their product $\psi_a(1)\psi_b(2)$ fails in describing the total electronic state. The reason is that in principle, there is no actual means of discriminating between a combination of electron 1 being in state $|a\rangle$ and electron 2 in state $|b\rangle$, and a combination where electron 1 is in state $|b\rangle$ and electron 2 is in state $|a\rangle$. This phenomenon is called **exchange symmetry**. It is a necessary condition that the total wave function $\psi_{atom}(1,2)$ will be such that the total probability density $|\psi_{atom}(1,2)|^2$ won't change on exchanging the electrons between the $|a\rangle$ and $|b\rangle$ states. The following two linear combinations of $\psi_a(i)\psi_b(j)$ product functions conform to that condition:

$$\left.\begin{aligned}\psi_{atom}^{(S)}(1,2) = \psi_S(1,2) = \frac{\psi_a(1)\psi_b(2) + \psi_a(2)\psi_b(1)}{\sqrt{2}}\\ \psi_{atom}^{(A)}(1,2) = \psi_A(1,2) = \frac{\psi_a(1)\psi_b(2) - \psi_a(2)\psi_b(1)}{\sqrt{2}}\end{aligned}\right\}. \tag{5.14}$$

Each function is a valid eigenfunction for Schrödinger's equation in the independent electron model (Eq. [5.3]). Division by $\sqrt{2}$ of each combination is made for normalization, such that integration over space of their squared absolute value is a unity. The ψ_S function is symmetric, that is, it preserves its sign on exchange of roles between the electrons. The ψ_A function is antisymmetric, meaning it reverses its sign on exchange of roles between the electrons. The "S" and "A" subscripts and superscripts are given to indicate the symmetry property. The physical meaning related to each combination is that in any eigenstate, **each of the electrons can orbit equally in both $|a\rangle$ and $|b\rangle$ states**. The ψ_S and ψ_A functions are called **orbital** wave functions, or **spatial** wave functions, to indicate that they overlook the role of the spinning properties.

If the interelectron repulsion term is included (last addend in Eq. [5.2]), the energies belonging to the ψ_A and ψ_S wave functions must also be different. The ψ_A function is diminished if the two electrons come close or reside in the same place (that is, when $\underline{r}_1 = \underline{r}_2$). This means that the electrons rarely approach each other. The average repulsion energy is then small, hence they exhibit a lower energy state (higher ionization energy). In the symmetric case, on the other hand, the electrons will often approach each other; the average repulsion energy is then large, hence exhibiting a higher state energy (a smaller ionization energy). We thus expect to have two systems of energy states for the helium atom: one belonging to symmetric states and the other to antisymmetric states. In special cases, in which the quantum numbers belonging to the two independent electrons are identical, the antisymmetric wave function vanishes identically. The only possible state is then the symmetric one.

Consequently, seeking solutions for the Hamiltonian of Eq. (5.2) starts by introducing zero-order, unperturbed symmetric or antisymmetric functions as given in Eq. (5.14). Corrections to the states' energies would be given by (see also Eq. [5.5])

$$E_{ab}^{(1)} = \int \psi_{atom}^{*}(\underline{r}_1, \underline{r}_2)\mathbf{H}_{12}\psi_{atom}(\underline{r}_1, \underline{r}_2)d^3\underline{r}_1 d^3\underline{r}_2 = C_{ab} \pm K_{ab}, \tag{5.15}$$

where

$$C_{ab} = \int d^3\underline{r}_1 d^3\underline{r}_2 \frac{e^2}{|\underline{r}_1 - \underline{r}_2|} |\psi_a(\underline{r}_1)|^2 |\psi_b(\underline{r}_2)|^2;$$

$$K_{ab} = \int d^3\underline{r}_1 d^3\underline{r}_2 \frac{e^2}{|\underline{r}_1 - \underline{r}_2|} \psi_a^*(\underline{r}_1)\psi_b(\underline{r}_1) \cdot \psi_b^*(\underline{r}_2)\psi_a(\underline{r}_2).$$

In summary,

$$\boxed{E_{ab} = E_a + E_b + C_{ab} \pm K_{ab}}, \tag{5.16}$$

where the (+) and (−) signs in front of the K_{ab} addend refer to the symmetric ψ_S and antisymmetric ψ_A functions, respectively. It is possible to show that the K_{ab} integral is positive, as expected by the previous qualitative discussion, namely that symmetric states energies are higher than antisymmetric energies (the separation is $2K_{ab}$). The integral signed C_{ab} is called **Coulomb integral**; in classical terms its value expresses interaction between the first and second electrons at assumed $\rho_1 = -e|\psi_a(1)|^2$ and $\rho_2 = -e|\psi_b(2)|^2$ electric charge distributions, respectively. The K_{ab} integral is called **exchange integral**; its value expresses a presumed interaction existing between the first and second electrons at $\rho_1^{ex} = -e\psi_a^*(1)\psi_b(1)$ and $\rho_1^{ex} = -e\psi_a^*(1)\psi_b(2)$ electric charge distributions, respectively. This kind of interaction is called **exchange interaction**. This component in the calculated state energy *has no classical analog*. In other words, *no such fundamental interaction exists between electrons*. The need to include it as a correction in the calculated state energy stems from our incapability of providing a direct solution to Schrödinger's equation, which led us to assume an independent electron approximation. The "exchange interaction" term should be interpreted on a limited aspect, solely as a correction component, resulting from exchange symmetry between independent electrons. Energy correction calculation to different combinations of independent electron states is performed in a manner like that performed for the ground state (Eqs. [5.5] to [5.10]), including proper Coulomb and exchange integrals.

Full description of the atomic states requires consideration of the different possible combinations of the electrons' spin states. We start by ignoring the relation of these spin states to the orbital angular momentum states of their owning electrons♣. Separately, the projection of each independent electron spin on the Z-axis could be co-oriented with the axis (spin up) or oppositely oriented (spin down). On coupling the spins, their precessions may no longer be considered independent. Because there is no way of discriminating between different electrons, considerations identical to those employed in relation to their orbital motions lead to propose the existence of separate symmetric χ_S and antisymmetric χ_A atomic spin functions in relation to exchange between the two electrons. These functions would be

♣ This same approach is the starting point in a summing method of angular momentum vectors called **Russell-Saunders coupling**: initially, the orbital and spin angular momentums are summed separately. Only then, the two separate orbital and spin resultants are summed to represent a total angular momentum vector.

$$\chi_A = \frac{\chi_+(1)\chi_-(2) - \chi_+(2)\chi_-(1)}{\sqrt{2}}; \qquad M_S = 0$$
$$\chi_S = \begin{cases} \chi_+(1)\chi_+(2); & M_S = +1 \\ \dfrac{\chi_+(1)\chi_-(2) + \chi_+(2)\chi_-(1)}{\sqrt{2}}; & M_S = 0 \\ \chi_-(1)\chi_-(2); & M_S = -1 \end{cases} \tag{5.17}$$

The M_S quantity noted per each spin function marks the resultant spin angular momentum projection $\underline{S} \equiv \underline{s}_1 + \underline{s}_2$ on the Z axis. It may be verified by activating each function by the resultant spin operator $\mathbf{S}_Z \equiv \mathbf{s}_{Z1} + \mathbf{s}_{Z2}$ on each function. Figure 5.2 describes these functions in terms of vector summing and independent precession motions of the component spins (see Figures 4.4 and 4.5 and related text). For simplicity, the resultant spin is not plotted as should be done in a full presentation of the coupled states (Figure 4.5); the precession motion of this resultant spin about the Z axis is overlooked.

Eq. (5.14) and Figure 5.2 show that there is a single function which is antisymmetric to exchange between electrons. Its resultant angular momentum vanishes, and obviously its projection on the Z axis as well. On the other hand, there are three different symmetric spin functions belonging to a total $S = 1$ spin, of differing projections on the Z axis: $M_S = -1$, $M_S = 0$, and $M_S = +1$ (generally, the multiplicity of a state of a resultant J angular momentum is $2J + 1$). The combined spin states are named by their multiplicity; the $S = 0$ state is called **singlet**; the two possible spin states of a single electron for which $S = s = 1/2$ (spin up and spin down) are called **doublet**; the three possible states of $S = 1$ are called **triplet**; and the several possible states of a more complex case of $S > 1$ (that may arise by coupling of several electrons) are called **multiplet**.

The atom wave function is achieved by multiplying the orbital wave function with a spin wave function. By electron exchange symmetry, there are seemingly four possible combinations: $\psi_A\chi_A$, $\psi_A\chi_S$, $\psi_S\chi_A$, and $\psi_S\chi_S$. However, as we shall elaborate on, **Pauli exclusion principle** allows only those combinations that provide an antisymmetric wave function: $\psi_A\chi_S$ and $\psi_S\chi_A$. Pauli exclusion principle asserts that **independent electrons of same orbital quantum state must have different spin states**. As a result, helium atom states belonging to an antisymmetric orbital wave function exhibit a triplet state; those

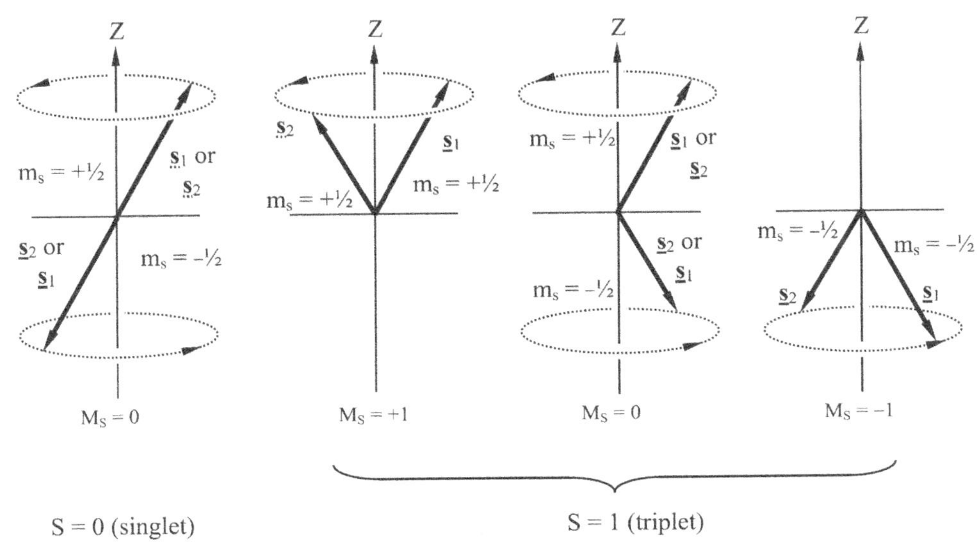

Figure 5.2 Schematic vector display of different spin states of two coupled electrons. See detailed discussion in text.

belonging to a symmetric orbital wave function exhibit a singlet state. Section 5.4 elaborates on Pauli's exclusion principle and its physical implications.

5.2.4 Helium Energy Level Scheme

We limit ourselves to discuss only those helium atom bonded states in which one independent electron is in its ground state, while the other is in an excited state. Theoretical calculations as well as experiment show that simultaneous excitation of the two electrons provides a positive total energy, causing immediate subsequent ionization. Thus, in a bonded state, one electron always belongs to the 1*s* state (Figure [3.7]), exhibiting $n = 1,\ \ell = 0,\ m_\ell = 0$ quantum numbers. The other electron belongs to one of the single-electron *ns, np, nd*, ... states. Those helium atoms forming a singlet ($S = 0$) state are called **parahelium**; those forming a triplet ($S = 1$) state are called **orthohelium**. Under certain conditions, these two-type states may be treated as two different gases, because radiative transitions between any para- and ortho-states are strongly forbidden, and the atoms stay for long periods in their "para" or "ortho" states.

We elaborate on the selection rules. The momentary atom electric dipole moment is given by a sum of the separate dipoles of the independent electrons: $e\underline{r}_1 + e\underline{r}_2 = e(\underline{r}_1 + \underline{r}_2)$. To calculate the radiative transitions between the ortho and para states, one must consider the size of the $\langle\psi_A|\underline{r}_1 + \underline{r}_2|\psi_S\rangle$ vector (see Section 3.6, particularly Eq. [3.52]). The $(\underline{r}_1 + \underline{r}_2)$ operator as well as the ψ_S function don't reverse sign on exchanging roles between the two electrons. However, the ψ_A wave function does reverse its sign. Because the $(\underline{r}_1 + \underline{r}_2)$ vector size should be independent of the order of the subscript's notation, it must be zero, hence also the radiative transition probability. Notably, even if the $\langle\psi_A|\underline{r}_1 + \underline{r}_2|\psi_S\rangle$ had not vanished, the scalar product between the corresponding spin states would provide the same result, because the triplet spin wave functions are all orthogonal to the singlet wave function. As to other radiative transitions in the helium atom, between two parastates or between two orthostates, they conform to the selection rules of the hydrogen-like atoms; namely, allowed transitions satisfy the $\Delta\ell = \pm 1;\ \Delta m_\ell = 0, \pm 1$ conditions (see Eq. [3.58]).

Transitions between singlet and triplet states are eventually allowed, though by a tiny probability, mainly because of spin-orbit interactions causing mixing between the states.

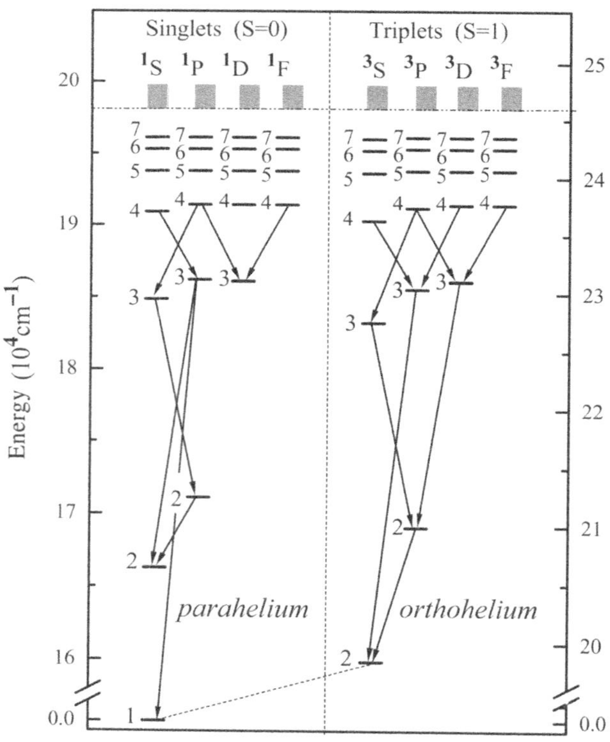

Figure 5.3 Experimental energy levels scheme of a helium atom, also showing several allowed optical transitions. See text for details.

Figure 5.3 shows the energy levels scheme of a single-electron ground and excited states in a helium atom. The ground state is assigned a zero energy. The single electron ionization energy is 198,305 cm^{-1} or 24.58 eV. A number to the side of each state marks the principal quantum number belonging to the single electron forming the level. The resultant orbital angular momentum (whose components are the orbital angular momentums of the separate independent electrons) is indicated by capital letters♣: *S, P, D, F*, ... for $L = 0, 1, 2, 3, ...,$

respectively. In this case, the resultant angular momentum equals that of the excited independent electron, because the other independent electron is in its 1*s* state. The $2S + 1$ multiplicity by the resultant spin is marked by a superscript number left to the letter marking the orbital angular momentum: 1 for singlet states and 3 for triplet states. As previously indicated, no more is there degeneracy among angular momentum states belonging to the same principal quantum number. Because of spin-orbit coupling, all triplet states are split into three adjacent states belonging to the three different orientations of the resultant spin relative to the total orbital angular momentum vector (see also Eq. [5.17] and Figure 5.2). This splitting is called **fine structure**, and is not included in the figure. Singlet states obviously exhibit no fine structure.

Except for the ground state, which is a singlet, and overlooking the fine structure, the energy schemes of the ortho- and parahelium are quite similar (Figure 5.3). Each of the separate schemes is granted the **manifold** adjective. That is, the parastates scheme is termed **singlet manifold**; the orthostates scheme is termed **triplet manifold**.

5.3 Bosons, Fermions, and Pauli Exclusion Principle

5.3.1 Harmonic Oscillator

Harmonic oscillations play roles in various physical phenomena such as atomic vibrations in crystals and electromagnetic oscillations, as well as in any vibrational quantization. Section 1.1 introduced classical harmonic oscillations of an electric dipole. The motivation to address the issue at this stage from a quantum aspect is to get acquainted with quantum particles called **bosons** in comparison to others called **fermions**.

5.3.1.1 Hamiltonian and Creation and Destruction Operators

Consider a particle of mass m under the influence of a one-dimensional potential of the form $V(x) = \frac{1}{2}m\omega_0{}^2x^2$. The system Hamiltonian would be

$$\boxed{\mathbf{H} = \frac{\mathbf{p}^2}{2m} + \frac{1}{2}m\omega_0{}^2x^2} \quad . \tag{5.18}$$

The $\mathbf{p}$ and x operators satisfy the following commutation relation $[x, \mathbf{p}] = i\hbar$. We define corresponding new operators $\mathbf{P} \equiv \frac{\mathbf{p}}{\sqrt{m\omega_0\hbar}}$; $\mathbf{Q} \equiv \sqrt{\frac{m\omega_0}{\hbar}}\cdot x$. Their commutation relation is $[\mathbf{Q}, \mathbf{P}] = i$. Expressing the Hamiltonian (Eq. [5.18]) by the new operators yields

$$\mathbf{H} = \left(\frac{1}{2}\mathbf{P}^2 + \frac{1}{2}\mathbf{Q}^2\right)\hbar\omega_0 \, . \tag{5.19}$$

It is possible to circumvent the need for a direct solution of the Hamiltonian differential equation expressed in Eq. (5.19), yet still deduce many properties of the solutions by analysis of the commutation relations among different Hamiltonian related operators. We have already used a similar approach in relation to the generalization of the angular momentum concept in Section 4.3.2.

♣ The resultant values of the orbital and spin angular momentums emanating from contribution of several electrons are indicated by capital letters. Their corresponding components belonging to the contributing lone electrons are marked by lowercase letters.

We define the following new operators:

$$\left.\begin{aligned}\mathbf{a} &= \frac{1}{\sqrt{2}}(\mathbf{Q} + i\mathbf{P}) \\ \mathbf{a}^{\dagger} &= \frac{1}{\sqrt{2}}(\mathbf{Q} - i\mathbf{P})\end{aligned}\right\}. \tag{5.20}$$

The $\mathbf{a}$ and $\mathbf{a}^{\dagger}$ are mutually conjugate Hermitian operators. Their commutation relation is

$$\left[\mathbf{a}, \mathbf{a}^{\dagger}\right] = 1\,. \tag{5.21}$$

Expressing $\mathbf{P}$ and $\mathbf{Q}$ using $\mathbf{a}$ and $\mathbf{a}^{\dagger}$, their setting in Eq. (5.19) yields

$$\mathbf{H} = \frac{1}{2}\left(\mathbf{a}^{\dagger}\mathbf{a} + \mathbf{a}\mathbf{a}^{\dagger}\right)\hbar\omega_0\,. \tag{5.22}$$

We further define another operator:

$$\mathbf{N} \equiv \mathbf{a}^{\dagger}\mathbf{a}\,. \tag{5.23}$$

Using this definition and the commutation relation (Eq. [5.21]), one gets

$$\boxed{\mathbf{H} = \left(\mathbf{N} + \frac{1}{2}\right)\hbar\omega_0}\,. \tag{5.24}$$

Because the Hamiltonian operator is Hermitian, $\mathbf{N}$ is also Hermitian. Using the above commutation relation and the definition by Eq. (5.23), we also get the following relations:

$$\mathbf{N}\mathbf{a} = \mathbf{a}(\mathbf{N} - 1); \quad \mathbf{N}\mathbf{a}^{\dagger} = \mathbf{a}^{\dagger}(\mathbf{N} + 1)\,.$$

We assume the existence of orthonormal eigenfunctions of the $\mathbf{N}$ operator, being also nondegenerate. **We denote these assumed functions by $|n\rangle$ and their corresponding eigenvalues by** n. These notations exhibit a one-to-one correspondence by virtue of the nondegeneracy assumption. By the Hermitianity of $\mathbf{N}$, its eigenvalue n must be real. We further assume their scales to be positive; their scalar product is obviously positive: $\langle n|n\rangle > 0$. We analyze the scales of the $\mathbf{a}|n\rangle$ and $\mathbf{a}^{\dagger}|n\rangle$ functions:

$$\left.\begin{aligned}\left\langle n\middle|\mathbf{a}^{\dagger}\mathbf{a}\middle|n\right\rangle &= \langle n|\mathbf{N}|n\rangle = n\langle n|n\rangle \\ \left\langle n\middle|\mathbf{a}^{\dagger}\mathbf{a}\middle|n\right\rangle &= \langle n|\mathbf{N} + 1|n\rangle = (n + 1)\langle n|n\rangle\end{aligned}\right\}. \tag{5.25}$$

Because the scale of any $|n\rangle$ function is non-negative, its eigenvalue always satisfies $n \geq 0$. If $n = 0$, we then conclude that $\mathbf{a}|n\rangle = 0$. If, on the other hand, $n \neq 0$, we shall see that $\mathbf{a}|n\rangle$ is an eigenfunction of $\mathbf{N}$, of $(n - 1)$ eigenvalue; $\mathbf{a}^{\dagger}|n\rangle$ is also an eigenfunction of $\mathbf{N}$, of $(n + 1)$ eigenvalue:

$$\left.\begin{aligned}\mathbf{N}\cdot\mathbf{a}|n\rangle &= \mathbf{a}\cdot(\mathbf{N} - 1)|n\rangle = \mathbf{a}\cdot(n - 1)|n\rangle = (n - 1)\cdot\mathbf{a}|n\rangle \\ \mathbf{N}\cdot\mathbf{a}^{\dagger}|n\rangle &= \mathbf{a}^{\dagger}(\mathbf{N} + 1)|n\rangle = \mathbf{a}^{\dagger}(n + 1)|n\rangle = (n + 1)\cdot\mathbf{a}^{\dagger}|n\rangle\end{aligned}\right\}. \tag{5.26}$$

The $\mathbf{a}$ and $\mathbf{a}^{\dagger}$ operators are correspondingly termed **annihilation** and **creation** operators, because by acting on the $\mathbf{N}$ operator eigenfunctions they form other eigenfunctions of eigenvalues that are lowered or increased by a unity, respectively. The issue of annihilation and creation will be emphasized, and attain additional implications in our discussions. We will use the terms "function" and "vector" interchangeably.

5.3.1.2 Energy Levels Scheme of the Harmonic Oscillator

If an eigenvalue of the harmonic oscillator satisfies $n > 0$, we may consider the $\mathbf{a}|n\rangle$ vector whose eigenvalue is $(n-1)$, and apply an identical consideration for $|n\rangle$. We then conclude that $(n-1) \geq 0$, hence $n \geq 1$. We may then consider the vector $\mathbf{a}^2|n\rangle$ whose eigenvalue is $(n-2)$ and conclude that $n \geq 2$. Thus, starting with any $|n\rangle$-vector, we may form a series of other eigenvectors

$$|n\rangle, \ \mathbf{a}|n\rangle, \ \mathbf{a}^2|n\rangle, \ \cdots, \ \mathbf{a}^p|n\rangle, \ \cdots,$$

of corresponding eigenvalues

$$n, \ (n-1), \ (n-2), \ \cdots, \ (n-p), \ \cdots.$$

Those series of vectors and eigenvalues of the **N** operator obtained that way must be final, because the eigenvalues are required to be all non-negative. In other words, for a certain p value, the $\mathbf{a}^p|n\rangle$ function becomes a null vector. As Eq. (5.26) shows, $n-p=0$, hence $n=p$ is a **positive integer**! Furthermore, $n=0$ is the lowest possible eigenvalue of the **N** operator.

It is possible to follow a similar procedure using the $\mathbf{a}^\dagger$ operator, starting from an eigenvector $|n\rangle$ (of an eigenvalue which is a positive integer), and form a series of eigenvectors

$$|n\rangle, \ \mathbf{a}^\dagger|n\rangle, \ \left(\mathbf{a}^\dagger\right)^2|n\rangle, \ \cdots, \ \left(\mathbf{a}^\dagger\right)^p|n\rangle, \ \cdots,$$

of corresponding eigenvalues

$$n, \ (n+1), \ (n+2), \cdots, \ (n+p), \ \cdots.$$

In contrast to the case of activating the annihilation operator, there is no limit for activating the creation operator any number of times. Thus, the eigenvalues spectrum of the **N** operator coincides with the set of non-negative integers. The Hamiltonian energy spectrum would be (Eq. [5.24])

$$\boxed{E_n = \left(n + \frac{1}{2}\right) \cdot \hbar\omega_0; \quad n = 0, 1, 2, \cdots, \infty} \ . \tag{5.27}$$

5.3.1.3 Eigenfunctions of the Harmonic Oscillator

As we have seen, knowing a single eigenfunction allows obtaining all other ones by a repetitive application of the **a** or $\mathbf{a}^\dagger$ operators, as needed. The entire set of eigenfunctions establishes a complete set. They are all orthogonal, yet not orthonormal; that is, their norm is not necessarily a unity. It is possible to find their norms relation using relations obtained from Eq. (5.25). Still, if we normalize the $|0\rangle$ function to a unity scalar product, it is possible to obtain an orthonormal set of eigenfunction by the equation

$$|n\rangle \equiv \frac{1}{\sqrt{n!}}\left(\mathbf{a}^\dagger\right)^n|0\rangle, \tag{5.28}$$

all satisfying the orthonormality condition

$$\left\langle n|n'\right\rangle = \delta_{nn'} . \tag{5.29}$$

To find the specific form of the $\psi_0 \equiv |0\rangle$ function, we express the wave equation for the eigenvalue $E = \frac{1}{2}\hbar\omega_0$: Schrödinger equation in terms of the new variables Q and P would be

$$\left(Q^2 - \frac{d^2}{dQ^2}\right)\psi_0(Q) = \psi_0(Q),$$

whose solution is

$$\psi_0 = \frac{1}{\pi^{1/4}}e^{-\frac{1}{2}Q^2}. \text{ Also,}$$

$$\mathbf{a}^\dagger = \frac{1}{\sqrt{2}}\left(Q - \frac{d}{dQ}\right).$$

Thus, we get (Eq. [5.28])

$$\boxed{\psi_n(Q) = \frac{1}{\sqrt{\pi^{1/2}2^n n!}}\left(Q - \frac{d}{dQ}\right)^n e^{-\frac{1}{2}Q^2}}. \tag{5.30}$$

The $\psi_n(Q)$ functions are even or odd for even or odd n, respectively. It is possible to show that they may be expressed in terms of polynomials called **Hermite polynomials** $\boldsymbol{H}_n(Q)$ as follows:

$$\psi_n(Q) = \frac{1}{\sqrt{\pi^{1/2}2^n n!}}e^{-\frac{1}{2}Q^2}\boldsymbol{H}_n(Q). \tag{5.31}$$

Table 5.1 expresses the first six wave functions of the harmonic oscillator written in a format related to Eq. (5.31). As to special properties of these functions, we note an important recurrence relation that will be used later:

$$Q\psi_n(Q) = \sqrt{\frac{n+1}{2}}\psi_{n+1}(Q) + \sqrt{\frac{n}{2}}\psi_{n-1}(Q). \tag{5.32}$$

Table 5.1 Normalized wave functions of the harmonic oscillator $\psi_n(Q)$ for $n = 0, 1, 2, 3, 4$.

n	E_n	$\psi_n(Q);\quad (Q = (m\omega_0/\hbar)^{1/2}x)$
0	$\frac{1}{2}\hbar\omega_0$	$\frac{1}{\sqrt{\pi^{1/2}}}e^{-\frac{1}{2}Q^2}$
1	$\frac{1}{2}\hbar\omega_0 + \hbar\omega_0$	$\frac{1}{\sqrt{2\pi^{1/2}}}e^{-\frac{1}{2}Q^2}\cdot 2Q$
2	$\frac{1}{2}\hbar\omega_0 + 2\hbar\omega_0$	$\frac{1}{2\sqrt{2\pi^{1/2}}}e^{-\frac{1}{2}Q^2}\left(4Q^2 - 2\right)$
3	$\frac{1}{2}\hbar\omega_0 + 3\hbar\omega_0$	$\frac{1}{4\sqrt{3\pi^{1/2}}}e^{-\frac{1}{2}Q^2}\left(8Q^3 - 12Q\right)$
4	$\frac{1}{2}\hbar\omega_0 + 4\hbar\omega_0$	$\frac{1}{8\sqrt{6\pi^{1/2}}}e^{-\frac{1}{2}Q^2}\left(16Q^4 - 48Q^2 + 12\right)$
5	$\frac{1}{2}\hbar\omega_0 + 5\hbar\omega_0$	$\frac{1}{16\sqrt{15\pi^{1/2}}}e^{-\frac{1}{2}Q^2}\left(32Q^5 - 160Q^3 + 120Q\right)$

5.3.1.4 Bosons

Results obtained in the previous section show that above the lowest energy of $\frac{1}{2}\hbar\omega_0$, the possible vibrational energies of a harmonic oscillator are an integer number of a basic $\hbar\omega_0$ "chunk" of energy. The physical meaning of the result is generalized, to consider each $\hbar\omega_0$ energy as a separate particle called phonon, identical to all others. The harmonic oscillator vibration is thus considered a combination of an integer number of phonons occupying the same quantum system. There is no limit to the number of phonons occupying the system, and it could be infinitely large. Particles of such nature are called **bosons**; they include the photons, and the different types of $\boldsymbol{\pi}$ (pi), $\boldsymbol{K}$ (Kappa), and $\boldsymbol{\eta}$ (etta) mesons. Their occupation in a complex thermodynamic system is called **Bose-Einstein statistics**, which, in addition to the possibility for any occupation number, includes the Boltzmann distribution relation as a weight function (Eq. [2.38] and on).

As a further generalization of the phonon concept introduction, the harmonic oscillator wave function is attributed to the separate phonons. Particularly, the $|n\rangle$ wave function is considered the wave function of the "n-th occupying phonon". The action of the $\mathbf{a}^\dagger$ creation operator is considered as a phonon absorption process by the harmonic oscillator; the action of the $\mathbf{a}$ annihilation operator is considered as a phonon emission by the harmonic oscillator; both without a need to specify the actual system coupled to the oscillator that gives up or accepts the phonon energy, respectively. The $|0\rangle$ ground state is called **vacuum state** because it can't emit a phonon. The $\mathbf{N}$ operator defined previously in Eq. (5.23) is called occupation operator, because its eigenfunction shows the number of phonons occupying the quantum system.

5.3.2 Angular Momentum

5.3.2.1 Annihilation, Creation, and Occupation Operators

We return to the to the generalized angular momentum operators provided in Eqs. (4.25) to (4.27). We redefine the same operators via their commutation relations, yet divided by $\hbar$ for brevity:

$$\mathbf{b}_X \equiv \mathbf{J}_X/\hbar; \quad \mathbf{b}_Y \equiv \mathbf{J}_Y/\hbar; \quad \mathbf{b}_Z \equiv \mathbf{J}_Z/\hbar\,, \tag{5.33}$$

which satisfy the following commutation relations:

$$[\mathbf{b}_X, \mathbf{b}_Y] = i\mathbf{b}_Z; \quad [\mathbf{b}_Y, \mathbf{b}_Z] = i\mathbf{b}_X; \quad [\mathbf{b}_Z, \mathbf{b}_X] = i\mathbf{b}_Y\,. \tag{5.34}$$

In analogy to Eq. (5.20), we define "creation" and "annihilation" operators by

$$\mathbf{b}^\dagger \equiv \mathbf{b}_X + i\mathbf{b}_Y; \quad \mathbf{b} \equiv \mathbf{b}_X - i\mathbf{b}_Y\,, \tag{5.35}$$

satisfying the following commutation relations:

$$\left[\mathbf{b}_Z, \mathbf{b}^\dagger\right] = \mathbf{b}^\dagger; \quad [\mathbf{b}_Z, \mathbf{b}] = -\mathbf{b}; \quad \left[\mathbf{b}^\dagger, \mathbf{b}\right] = 2\mathbf{b}_Z\,. \tag{5.36}$$

We will show that the action of the $\mathbf{b}$ and $\mathbf{b}^\dagger$ operators on the angular momentum $|jm\rangle$ vectors exhibits similarity to the action of the $\mathbf{a}$ and $\mathbf{a}^\dagger$ annihilation and creation operators on the harmonic oscillator $|n\rangle$ state vectors. Following Eq. (4.39), we get

$$\left.\begin{aligned} \mathbf{b}^\dagger|jm\rangle &= [(j-m)(j+m+1)]^{1/2}|j, m+1\rangle \\ \mathbf{b}|jm\rangle &= [(j+m)(j-m+1)]^{1/2}|j, m-1\rangle \end{aligned}\right\}. \tag{5.37}$$

The action of $\mathbf{b}$ on $|jm\rangle$ generates a vector whose projection on the Z-axis is lowered by one unit; the action of $\mathbf{b}^\dagger$ on $|jm\rangle$ generates a vector whose projection on the Z-axis is increased by one unit. The $\mathbf{b}$ and $\mathbf{b}^\dagger$

operators are thus annihilation and creation operators of angular momentum projections on the Z-axis, respectively.

Now we define a different general relation between general **A** and **B** quantum operators called anticommutation relation:

$$\{\mathbf{A}, \mathbf{B}\} \equiv \mathbf{AB} + \mathbf{BA}\,. \tag{5.38}$$

We emphasize, that the commutation and anticommutation relations are themselves quantum operators. The first is called **commutator** and the second is called **anticommutator**. Based on their definitions, $\mathbf{b}$ and $\mathbf{b}^\dagger$ satisfy the following anticommutation relation:

$$\left\{\mathbf{b}^\dagger, \mathbf{b}\right\} = 2\left(\mathbf{b}_X{}^2 + \mathbf{b}_Y{}^2\right).$$

We limit our discussion to the specific case of the electron spin. Considering the expressions for the spin matrices in Eq. (4.41), we get

$$\mathbf{b}_X = \frac{1}{2}\begin{pmatrix} 0 & 1 \\ 1 & 0 \end{pmatrix}; \quad \mathbf{b}_Y = \frac{1}{2}\begin{pmatrix} 0 & -i \\ i & 0 \end{pmatrix}; \quad \mathbf{b}_Z = \frac{1}{2}\begin{pmatrix} 1 & 0 \\ 0 & -1 \end{pmatrix}. \tag{5.39}$$

Then

$$2\left(\mathbf{b}_X{}^2 + \mathbf{b}_Y{}^2\right) = \begin{pmatrix} 1 & 0 \\ 0 & 1 \end{pmatrix} = \boldsymbol{I},$$

where $\boldsymbol{I}$ is the unity matrix of order 2. In other words, the anticommutation relation satisfies

$$\left\{\mathbf{b}, \mathbf{b}^\dagger\right\} = 1, \tag{5.40}$$

and in addition

$$\left\{\mathbf{b}^\dagger, \mathbf{b}^\dagger\right\} = \{\mathbf{b}, \mathbf{b}\} = 0\,. \tag{5.41}$$

Returning to the *spin* angular momentum operator per Eq. (4.41), it may be shown that

$$\mathbf{s}_Z = \frac{1}{2}\hbar\left(\mathbf{b}^\dagger\mathbf{b} + \mathbf{b}\mathbf{b}^\dagger\right). \tag{5.42}$$

Using the anticommutation relation given in Eq. (5.40), we get

$$\mathbf{s}_Z = \left(\mathbf{b}^\dagger\mathbf{b} - \frac{1}{2}\right)\cdot\hbar\,.$$

We now define a new operator

$$\mathbf{N}_Z^{\mathbf{S}} \equiv \mathbf{b}^\dagger\mathbf{b}, \tag{5.43}$$

whose action on an angular momentum vector yields $\mathbf{N}_Z^{\mathbf{S}}|jm_s\rangle = \left(m_{\mathbf{S}} + \frac{1}{2}\right)|jm_s\rangle$. Thus, the eigenvalue of the $\mathbf{N}_Z^{\mathbf{S}}$ operator reflects the number of $\hbar$ angular momentum chunks exhibited by angular momentum projection on the Z axis above the "possible minimum" $m_s = -\frac{1}{2}$. We now imagine an $\hbar$ angular momentum chunk as an **individual particle populating a quantum system**. For a moment, we consider the $m_s = -\frac{1}{2}$ state as a system that occupies no such particle, and symbolize it by $|0\rangle$; we symbolize $\mathbf{b}^{\dagger}|0\rangle \equiv |1\rangle$. Naturally, $|0\rangle = \left|j,\ -\frac{1}{2}\right\rangle = |j,\ \downarrow\rangle$ and $|1\rangle = \left|j,\ +\frac{1}{2}\right\rangle = |j,\ \uparrow\rangle$. The $\mathbf{N}_Z^{\mathbf{S}}$ operator is thus called **spin occupation operator** (see also Eq. [5.23]). Using all above operators' algebra, it may be readily seen that the system may not populate more than a single particle. We verify it in two ways. One is to act with the creation operator twice on the ground state. By Eq. (4.41) the result is a null state:-

$$\mathbf{b}^{\dagger}\mathbf{b}^{\dagger}|0\rangle = \frac{1}{2}\left\{\mathbf{b}^{\dagger},\mathbf{b}^{\dagger}\right\}|0\rangle = 0\,. \tag{5.44}$$

The second way is to explore the properties of the occupation operator:

$$\left(\mathbf{N}_Z^{\mathbf{S}}\right)^2 = \mathbf{b}^{\dagger}\mathbf{b}\cdot\mathbf{b}^{\dagger}\mathbf{b} = \mathbf{b}^{\dagger}\left(-\mathbf{b}^{\dagger}\mathbf{b} + 1\right)\mathbf{b} = \mathbf{b}^{\dagger}\mathbf{b} = \mathbf{N}_Z^{\mathbf{S}}\,;$$

$$\mathbf{N}_Z^{\mathbf{S}}\left(\mathbf{N}_Z^{\mathbf{S}} - 1\right) = 0\,. \tag{5.45}$$

Obviously, therefore, the eigenvalues of the occupation operator are zero or one.

To provide a qualitative understanding of this important result, we have seen several times that increase of the eigenvalue of a wave equation by a single quantum entails the appearance of an additional zero point in the related eigenfunction. This has a formal mathematical proof that we won't present here. A common demonstration is to count the number of zeros in the standing waves of a string tied on both sides. Other examples are the number of lobes in the $\boldsymbol{Y}_{\ell 0}$ functions for different ℓ's (Figure 3.4) and the number of zeros in the $\boldsymbol{R}_{n0}$ functions for different n's (Figure 3.6). The reader may verify it for the harmonic oscillator wave functions of Table 5.1. Accordingly, if we enforce a zero point in the spin wave function demonstrated in Figure 4.3, for example by modulating the amplitude, the wave described by the function will deliver both a positive and a negative amplitude to each point, namely induce a destructive interference throughout, resulting in a null wave function. This same conclusion may be obtained by imagining the system as a circular tense string; such a system could not support a standing wave with a single zero point.

5.3.2.2 Pauli Exclusion Principle

Thus far we have considered the states of a single independent electron, and have concluded (what we already know) that it may exhibit only one of two possible spin states. Pauli has extended the meaning of that conclusion to a supposed state of *two* independent electrons exhibiting the same orbital quantum state, yet of opposite spin states. He portrayed the transforming of the spin-down electron into a spin-up state, as equivalent to an operation of a $\mathbf{b}^{\dagger}$ creation operator on the spin-up electron. The result is a null wave function for both electrons (Eq. [5.44]). In other words, the above assumed state may not exist. A symmetric conclusion obtains for transforming the spin-up electron into a spin-down wave function. Correspondingly, Pauli exclusion principle declares that

Only a single independent electron can occupy a specified atomic quantum state . (5.46)

As a result, it may be concluded that the wave function of a many-electron system must be zero if the coordinates of two or more independent electrons belong to the same set of quantum numbers, spin included. A function antisymmetric to exchange between electrons exhibits that behavior: exchange between two electrons will then reverse the wave function sign, and if the two sets of quantum numbers are identical, the function value wouldn't change; only an identically zero function could satisfy such a condition. We will return to discuss that issue in relation to atoms containing more than two electrons.

Particles conforming to the condition that only a single one may occupy a given quantum state are called **fermions**. Their thermodynamic occupation in a complex quantum system is called **Fermi-Dirac statistics**. Electrons thus belong to the Fermionic family, along with all particles exhibiting half-odd-sized angular momentum like the positrons, protons, neutrons, neutrinos, muons, the various hyperons, and the omega-minus.

5.4 Electronic Structure of Many-Electron Atoms

5.4.1 Slater Determinant

Slater proposed to write the wave function of an atom containing N independent electrons in a form of a determinant:

$$\psi_{abc\ldots} = \frac{1}{\sqrt{N!}} \begin{vmatrix} \psi_a(1) & \psi_a(2) & \psi_a(3) & \cdots \\ \psi_b(1) & \psi_b(2) & \psi_b(3) & \cdots \\ \psi_c(1) & \psi_c(2) & \psi_c(3) & \cdots \\ \cdots & \cdots & \cdots & \cdots \end{vmatrix}, \tag{5.47}$$

where each of the $a, b, c, \cdots$ letter subscripts is a collective symbol for the independent electron n, ℓ, m_ℓ, m_s quantum numbers set. The $1, 2, 3, \cdots N$ numerals are numbering of the electrons, where each also collectively describes the correspondingly spatial and internal spin coordinates of its belonging electron: $\underline{r}_1, \underline{r}_1^{(i)}; \underline{r}_2, \underline{r}_2^{(i)}; \underline{r}_3, \underline{r}_3^{(i)}; \cdots; \underline{r}_N, \underline{r}_N^{(i)}$, where the (i) superscript refers to the relevant internal spin coordinate. The expression in Eq. (5.47) describes an equal participation of each electron in all states, in a manner that is consistent with the principle that there is no way of noticing an exchange between electrons. The function is in fact antisymmetric in respect to exchange between two electrons, because mathematically, exchange between electrons means exchange between two columns in the determinant, which inverses the determinant's sign. Setting two identical electrons (namely electrons of the same quantum numbers) means two identical columns in the determinant, making it identically zero. In physical terms, it means that the probability for two electrons to exhibit identical sets of quantum numbers is identically zero. The division by $\sqrt{N!}$ appears for normalization purpose.

For demonstration, we consider the wave function created of two spatial states, $|a\rangle$ and $|b\rangle$ of two independent electrons belonging to the same spin (say spin-up) state.

$$\psi = \frac{1}{\sqrt{2}} \begin{vmatrix} \psi_a(1)\chi_+(1) & \psi_a(2)\chi_+(2) \\ \psi_b(1)\chi_+(1) & \psi_b(2)\chi_+(2) \end{vmatrix}$$

$$= \frac{1}{\sqrt{2}}[\psi_a(1)\psi_b(2) - \psi_a(2)\psi_b(1)]\chi_+(1)\chi_+(2) = \psi_A\chi_S.$$

As a result, we obtained a product between an antisymmetric spatial function ψ_A as in Eq. (5.14), and a symmetric total-spin function χ_S (belonging to $M_S = 1$) as in Eq. (5.17).

5.4.2 Electron Configuration and the Shell Structure

Pauli exclusion principle establishes a key for achieving both qualitative and quantitative understanding of many-electron atoms. The electrons orbiting about the nucleus are treated as independent in a zero-order approximation. The atom ground state is formed when the electrons occupy independent single-electron states according to their energetic depth (indeed with a screened attractive potential; see Section 5.2, especially Eq. [5.13]). The $|n\ell\rangle$ state will inhabit up to $2(2\ell + 1)$ electrons because there are $2\ell + 1$ different m_ℓ values, and each may belong to two different possible values of the spin projection: $m_s = \pm\frac{1}{2}$. Usually, the occupancy increases for higher n values. A set of electrons belonging to a certain principal quantum number n forms a **shell**. This classification is not entirely strict, and states of a different n but close in energy to an adjacent shell are assigned to the latter. Shells belonging to $n = 1, 2, 3, 4, \cdots$ are named using the capital letters $K, L, M, N, \cdots$, respectively.

Independent electron energies occupying the same shell are relatively close, and exhibit large separations from electron energies belonging to the different shells. The separations are often termed **energy gaps**. Electrons of the same shell and same ℓ orbital quantum number form a **sub-shell**. Electrons belonging to different sub-shells exhibit larger energy gaps than between members of the same sub-shell. In all shells, *s*- electrons form the deepest sub-shell. Among other sub-shells, the deeper are those belonging to higher ℓ numbers. Figure 5.4 shows a scheme of the shell structure and their possible occupancy. A full shell belongs to atoms whose atomic numbers are $Z = 2$, $Z = 10$, $Z = 18$, $Z = 36$, $Z = 54$, and $Z = 86$; these are respectively the Noble gasses helium (*He*), neon (*Ne*), argon (*Ar*), krypton (*Kr*), xenon (*Xe*), and radon (*Rn*). The energetic reason why these atoms are considered "noble" will be discussed. An atom that would contain 118 electrons is also expected to be noble; however, no such atom has ever been observed in nature, neither has such been produced artificially. Undoubtedly, the reason is that a *nucleus* of an atom containing 118 protons would be highly unstable.

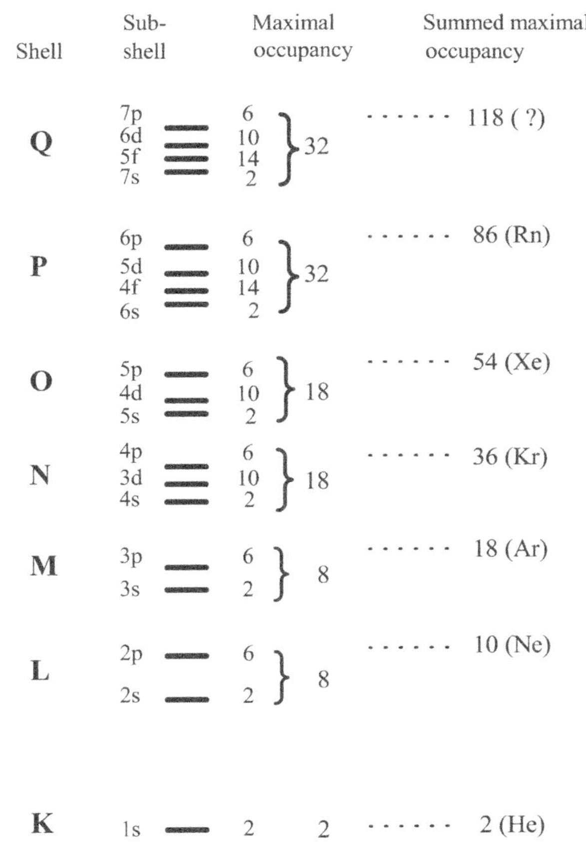

Figure 5.4 Schematic "zero-order" classification of independent electron states in atoms into shells and subshells structures. See text for details.

A combination of independent electron states that form an atomic state is called **electronic configuration**. The noting method is to write up a series of the form *Configuration* $= \cdots n\ell^q \cdots$, where q indicates the number of electrons belonging to the principal quantum number n and orbital quantum number ℓ (the $n\ell$ sub-shell). For example, the electronic configuration of the fluorine atom F in its ground state is $1s^2 2s^2 2p^5$. It means that the atom occupies two electrons in the $1s$ state, two in the $2s$ state, and five electrons in the $2p$ state. Sometimes, this configuration is written in the form $[He]2s^2 2p^5$, to emphasize that the configuration of the two first written electrons is identical with the configuration of the noble helium *He* gas.

The electronic configuration does not uniquely determine the electronic state, because there are usually various options of arranging the electrons by different combinations of their angular momentum and spin vectors. Experiments provide a following empiric rule, one of several such rules addressed as **Hund's rules**:

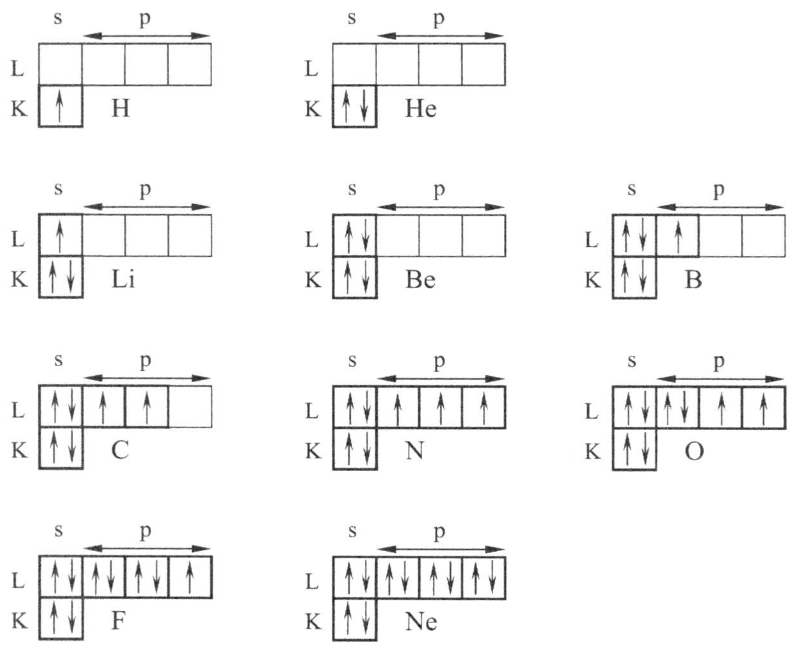

Figure 5.5 Ground-state electronic configuration of the first 10 elements in the periodic table according to Hund's rule (Eq. [5.48]).

In an atomic ground state, the resultant spin is the highest possible in consistency with Pauli exclusion principle . (5.48)

The ground state is obviously the one of largest binding energy. Indeed, by the Hund's rule, as many electrons as possible would exhibit parallel spin projections in the ground state. As a result, a maximal number of electron pairs will be spin-symmetric for exchange between electrons; as a result, the exchange operation will be antisymmetric for the same couples with respect to their spatial part of the wave functions. That way, the probability for those pairs to approach each other spatially reduces, the average repulsion is then smaller, hence the binding energy larger (see Eq. [5.14] and following text).

For demonstration, Figure 5.5 describes the construction procedure of the first 10 atoms of the periodic table of the elements according to Hund's rule. Specific principal and orbital quantum numbers of the independent electrons are represented by boxes. The boxes are classified by their K and L shells and their s- and p-subshells. An s-subshell obviously contains a single box, and a p-subshell contains three boxes (related to the $m_\ell = 0, \pm 1$ possible independent electron states). The spin-up and spin-down states are noted inside the boxes by up-, and down-oriented arrows, respectively. Submission to Hund's rule is demonstrated in the filling of the p-subshells, where as many as possible electrons of parallel spins have been introduced.

5.4.3 Electronic Configuration and Chemical Stability

The chemical properties of any atom are derived mostly by its ground state, but also by the presence or absence of higher, near-lying states. All noble gases exhibit closed shells and relatively large energy separations from higher lying shells. Formation of chemical bonding requires transition to a higher state; therefore, under normal conditions the noble gases are inert with respect to reaction with identical or other atoms. Many-electron atoms containing a single electron on top of a closed shell are usually metals

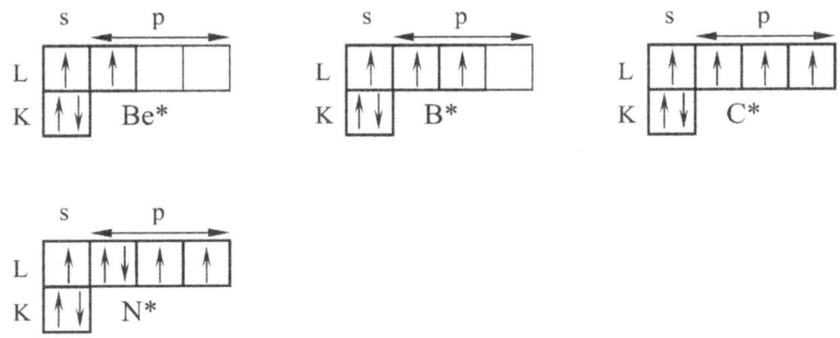

Figure 5.6 Electronic configurations of beryllium, boron, carbon, and nitrogen excited states involved in forming chemical bonding of these elements. The asterisk beside each element symbol indicates the configuration's relation to an excited state.

because that electron is only lightly bonded. The hydrogen *H* and lithium *Li* atoms are similar in the sense that both contain a single electron in an *s* state: 1*s* in hydrogen, and 2*s* in lithium. However, unlike lithium, hydrogen is not a metal. Lithium exhibits a nearby *p* state, and the excited atomic configuration may change into $1s^2 2p$ instead of $1s^2 2s$, with only a small invested energy. It is possible to account in a similar way to the difference between the chemical properties of helium *He* and beryllium *Be* atoms; like helium, the beryllium atom exhibits a filled *s* subshell. However, one of its electrons may be excited into an energetically close *p* state (to form a $1s^2 2s2p$ configuration instead of the $1s^2 2s^2$ one). Thus, beryllium will exhibit a doubled chemical valency (namely two electrons may participate in a chemical bonding). The energy required to excite beryllium to the $1s^2 2s2p$ configuration is approximately 2.7 eV. For comparison, the lowest energy required to excite a helium atom to a higher state is about 20 eV. In the case of boron *B*, the first excited state involves moving a 2*s* electron into a 2*p* state, requiring an energy investment of about 3.6 eV. In that state, the boron exhibits a threefold valency. The carbon *C*, nitrogen *N*, and oxygen *O* are more complex, because they all exhibit several excited states of energy close to the ground state. Figure 5.6 shows the electronic configurations of the beryllium, boron, carbon, and nitrogen excited states involved in forming their chemical reactions.

5.4.4 Spin/Orbit Coupling and Term Determination

Even after determination of the electronic configuration, there exist several different relations among the orbital and spin orientations of the independent electrons. The quantity that forms an atomic constant of motion is the resultant angular momentum, characterized by the *J* quantum number, and satisfying (see Eq. [4.37])

$$\left\langle \mathbf{J}^2 \right\rangle = J(J+1)\hbar^2; \quad \langle \mathbf{J}_Z \rangle = M_J \hbar; \quad M_J = \pm J, \pm (J-1), \cdots. \tag{5.49}$$

Recall that a resultant angular momentum contributed by several independent electrons is indicated by a capital letter, while that of a single electron is indicated by a lowercase letter.

A central issue in atomic spectroscopy is finding the possible allowed values of *J* per each configuration, and correspondingly the wave function and energy of each state. The reason is that different combinations within the same configuration result in a different angular momentum, hence a different energy. A widely accepted calculation scheme is called **Russell-Saunders coupling**. Initially, the orbital angular momentum contributions and spin angular momentum contributions are summed separately. Then the separate resultants are summed. Thus, the first performed summation is $\mathbf{L} = \sum_i \mathbf{L}_i$; $\mathbf{L}_Z = \sum_i \mathbf{L}_{Zi}$, where the i subscript is a numerator of an independent electron. The quantum numbers L and $M_L = \sum_i m_{\ell i}$ which characterize the $\mathbf{L}^2$ and $\mathbf{L}_Z$ operators, satisfy

$$\left\langle \mathbf{L}^2 \right\rangle = L(L+1)\hbar^2; \quad \langle \mathbf{L}_Z \rangle = M_L \hbar; \quad M_L = \pm L, \pm (L-1), \ldots . \tag{5.50}$$

The next summation is $\mathbf{S} = \sum_i \mathbf{s}_i$; $\mathbf{S}_Z = \sum_i \mathbf{s}_{Zi}$. The quantum numbers S and $M_S = \sum_i m_{si}$ characterizing the $\mathbf{S}^2$ and $\mathbf{s}_Z$ operators, satisfy

$$\left\langle \mathbf{S}^2 \right\rangle = S(S+1)\hbar^2; \quad \langle \mathbf{S}_Z \rangle = M_S \hbar; \quad M_S = \pm S, \pm (S-1), \cdots . \tag{5.51}$$

The third and final stage is the summation $\mathbf{J} = \mathbf{L} + \mathbf{S}$. It is possible to show, though we won't detail the proof here, that the possible allowed J quantum numbers are given by

$$J = L + S, \quad L + S - 1, \quad \cdots, \quad |L - S| . \tag{5.52}$$

The Russell-Saunders method is justified by several underlying assumptions. One is that each independent electron is indeed characterized by individual orbital angular momentum and spin. Second is that the resultant orbital angular momentum and resultant spin of several independent electrons are indeed "good quantum numbers". Namely, that coupling among the orbital angular momentum components, among the spin angular momentum components, and between the resultant orbital angular momentum $\underline{L}$ and the resultant spin $\underline{S}$, are "small perturbations". It is further assumed that the coupling between $\underline{L}$ and $\underline{S}$ is small compared to couplings among their components. Obviously, these assumptions require experimental validation.

Figure 5.7 schematically describes the vector addition procedure of orbital and spin angular momentum components.

The state of an atom is largely characterized by its L, S, and J quantum numbers. The set of L and S quantum numbers is called **Term**, and is the dominant factor in determining the state energy, because different states are characterized by different combinations of the independent electron orbits. Therefore, the average electrostatic repulsion between electron pairs is different. Different J-states of the same Term also exhibit different energies, but it stems from spin/orbit interaction, which is small compared to any electrostatic attraction or repulsion. A set of different J states belonging to the same Term are called **multiplet**. A relevant discussion is provided at the end of Section 5.2, especially Figure 5.3 and the related text. The number of possible different J values for each Term is $2S+1$ if $S \leq L$, or $2L + 1$ if $L < S$ (see Eq. [5.52]). This number describes the degree of spin/orbit Term splitting, and is often called **multiplicity**. A Term is symbolized by ^{2S+1}L, and each of its multiplet components by $^{2S+1}L_J$. According to the known convention, the resultant orbital angular momentum is indicated by the S, P, D, F, $\cdots$ letters for $L = 0, 1, 2, 3, \cdots$, respectively. The other (S and J) quantities are usually indicated by numbers.

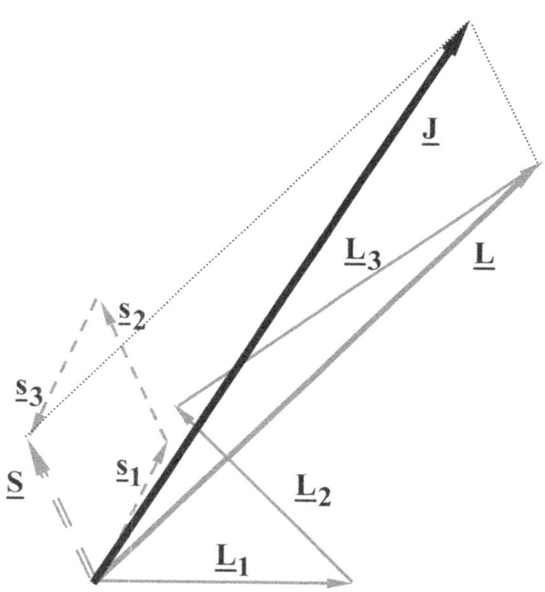

Figure 5.7 Demonstration of a vector addition scheme of orbital and spin angular momentums according to Russell-Saunders method.

We now turn to discuss the specific (simplest) case of coupling between only two independent electrons. The possible values of the resultant orbital quantum number L and its magnetic quantum number M_L are

$$\left.\begin{array}{l} L = \ell_1 + \ell_2, \quad \ell_1 + \ell_2 - 1, \quad \cdots, \quad |\ell_1 - \ell_2| \\ M_L = +L, \quad L-1, \quad \cdots, -L \end{array}\right\}. \tag{5.53}$$

Because $s = \frac{1}{2}$ for each electron, then

$$\left.\begin{array}{l} S = s_1 + s_2, \quad s_1 - s_2 = 1, \quad 0 \\ M_S = m_{s_1} + m_{s_2} = \begin{cases} 1, \quad 0, \quad -1 & if \;\; S = 1 \\ 0 & if \;\; S = 0 \end{cases} \end{array}\right\}. \tag{5.54}$$

In addition to these relations, conforming to Pauli exclusion principle is required, meaning that the joint electronic wave function must be antisymmetric with respect to exchange between the independent electron coordinates (both spin and space). The procedure in forming a Russell-Saunders wave function based on single independent spin/orbit wave functions is as follows:

i) Formation of $\mathbf{S}^2$ and $\mathbf{S}_Z$ eigenfunctions.
ii) Formation of $\mathbf{L}^2$ and $\mathbf{L}_Z$ eigenfunctions.
iii) Formation of a function antisymmetric to electrons exchange based on (i) and (ii).

Step (i) has already been presented in Section 5.2, Eq. (5.15). We present it again using the $|SM_S\rangle$ "ket" symbols for the wave functions:

$$\left.\begin{array}{ll} \chi_A = |00\rangle = \dfrac{\chi_+(1)\chi_-(2) - \chi_+(2)\chi_-(1)}{\sqrt{2}}; & M_S = 0 \\ \chi_S = \begin{cases} |11\rangle = \chi_+(1)\chi_+(2); \\ |10\rangle = \dfrac{\chi_-(1)\chi_-(2) + \chi_+(2)\chi_-(1)}{\sqrt{2}}; \\ |1-1\rangle = \chi_-(1)\chi_-(2); \end{cases} & \begin{array}{l} M_S = +1 \\ M_S = 0 \\ M_S = -1 \end{array} \end{array}\right\}. \tag{5.55}$$

Each $|LM_L\rangle$ wave function may be expanded using single electron orbital functions

$$|LM_L\rangle \equiv |\ell_1\ell_2LM_L\rangle = \sum_{m_1 + m_2 = M_L} |\ell_1 m_1 \ell_2 m_2\rangle\langle \ell_1 m_1 \ell_2 m_2 | \ell_1 \ell_2 L M_L\rangle\,. \tag{5.56}$$

This expansion is like the one presented in Sections 4.4 and 4.8 in relation to spin/orbit coupling, particularly Eq. (4.86) and following text. The $\langle \ell_1 m_1 \ell_2 m_2 | \ell_1 \ell_2 L M_L\rangle$ factors are Clebsch-Gordan coefficients (see Eq. [4.117]).

The $|LM_L\rangle$ wave function exhibits no specific symmetry with respect to electron exchange. However, it is possible to use it to form an either symmetric or antisymmetric wave function, as required, by subjecting it to either the $\hat{\mathbf{S}}$ or $\hat{\mathbf{A}}$ operators, respectively, that are defined by

$$\boxed{\hat{\mathbf{S}} \equiv \frac{1}{\sqrt{2}}(1 + \hat{\mathbf{P}}_{12}); \quad \hat{\mathbf{A}} \equiv \frac{1}{\sqrt{2}}(1 - \hat{\mathbf{P}}_{12})}\;, \tag{5.57}$$

where the $\hat{\mathbf{P}}_{12}$ operator exchanges the coordinates between electron 1 and 2. The end-target antisymmetric wave function is the product $|LM_LSM_S\rangle = |LM_L\rangle|SM_S\rangle$.

For demonstration, we calculate the Terms obtained for several combinations of electron configurations. In simple cases it is possible to shorten the previously indicated procedure. Consider two *s* electrons belonging to different shells; namely forming an *nsn′s* configuration. Because $\ell_1 = 0$ and $\ell_2 = 0$, the resultant orbital angular momentum is zero, namely $L = 0$. The separate electron spins are either parallel or antiparallel, namely $S = 1$ or $S = 0$, respectively. The resultant angular momentum is then equal to the resultant spin, namely $J = S$. The possible Terms are a triplet $^3S_1$ or singlet $^1S_0$. On the other hand, if the two electrons are equivalent, belonging to the same shell (ns^2 configuration), the spins must be antiparallel, and the only possible Term is a singlet $^1S_0$.

A less simple example considers a system of two *p* electrons belonging to different shells, say a 2*p*3*p* configuration (meaning $n_1 = 2,\ n_2 = 3,\ \ell_1 = \ell_2 = 1$). The possible resultant orbital angular momentum numbers are $L = 0, 1, 2$, yielding *S*, *P*, and *D* states. The independent spins could be parallel or antiparallel, namely $S = 1$ or $S = 0$, respectively, per each value of *L*. The possible Terms are thus $^1S,\ ^1P,\ ^1D,\ ^3S,\ ^3P,\ ^3D$. Table 5.2 details the possible combinations for the independent electrons. Each combination lists the $(m_{\ell 1} m_{s1})(m_{\ell 2} m_{s2})$ projections, classified in rows by the resultant orbital angular momentum projection $M_L = m_{\ell 1} + m_{\ell 2}$, and in columns by the resultant spin angular momentum projection $M_S = m_{S1} + m_{S2}$. The gray-highlighted combinations are forbidden by Pauli principle for equivalent electrons, namely when $n_1 = n_2$. To calculate the different $|LM_L\rangle$ wave functions, use of Eq. (5.56) is required (also Eq. [4.117]). Table 5.3 details the Clebsch-Gordan coefficients needed for wave function calculations.

As an example, we consider the $|LM_L\rangle = |00\rangle$ case:

$$|00\rangle = \frac{1}{\sqrt{3}}\left[2p_{+1}(1)3p_{-1}(2) - 2p_0(1)3p_0(2) + 2p_{-1}(1)3p_{+1}(2)\right]. \tag{5.58}$$

The symmetric and antisymmetric combinations (S-superscript and A subscript, respectively) are

$$\begin{aligned}|00\rangle_A^S = \frac{1}{\sqrt{6}}&\left[2p_{+1}(1)3p_{-1}(2) \pm 2p_{+1}(2)3p_{-1}(1) - 2p_0(1)3p_0(2)\right.\\ &\left.\mp 2p_0(2)3p_0(1) + 2p_{-1}(1)3p_{+1}(2) \pm 2p_{-1}(2)3p_{+1}(1)\right]\end{aligned}. \tag{5.59}$$

Table 5.2 Possible m_ℓ and m_s combinations list of the $(np)(n'p)$ configuration. The gray-highlighted combinations are forbidden by Pauli principle for an np^2 configuration.

M_L \ M_S	−1	0	+1
+2	(+1, −1/2)(+1, −1/2)	(+1, +1/2)(+1, −1/2)	(+1, +1/2)(+1, +1/2)
+1	(+1, −1/2)(0, −1/2)	(+1, +1/2)(0, −1/2) (+1, −1/2)(0, +1/2)	(+1, +1/2)(0, +1/2)
0	(+1, −1/2)(−1, −1/2)	(+1, +1/2)(−1, −1/2) (0, +1/2)(0, −1/2) (+1, −1/2)(−1, +1/2)	(+1, +1/2)(−1, +1/2)
−1	(0, −1/2)(−1, −1/2)	(0, +1/2)(−1, −1/2) (0, −1/2)(−1, +1/2)	(+1, +1/2)(+1, +1/2)
−2	(−1, −1/2)(−1, −1/2)	(−1, +1/2)(−1, −1/2)	(−1, +1/2)(−1, +1/2)

Table 5.3 $\langle 1, m_1, 1, m_2 | Lm\rangle$ Clebsch–Gordan coefficients.

$\ell_1 = 1$	$\ell_2 = 1$	$L = 2$					$L = 1$			$L = 0$
m_1	m_2	$m = 2$	$m = 1$	$m = 0$	$m = -1$	$m = -2$	$m = 1$	$m = 0$	$m = -1$	$m = 0$
1	1	1								
1	0		$\sqrt{1/2}$				$\sqrt{1/2}$			
1	−1			$\sqrt{1/6}$				$\sqrt{1/2}$		$\sqrt{1/3}$
0	1		$\sqrt{1/2}$				$-\sqrt{1/2}$			
0	0			$\sqrt{2/3}$				0		$-\sqrt{1/3}$
0	−1				$\sqrt{1/2}$				$\sqrt{1/2}$	
−1	1			$\sqrt{1/6}$				$-\sqrt{1/2}$		$\sqrt{1/3}$
−1	0				$\sqrt{1/2}$				$-\sqrt{1/2}$	
−1	−1					1				

A plus sign provides the symmetric wave function and a minus sign provides the antisymmetric one. Here we used the $n\ell_m$ notation expressed by the letter convention instead of $|n\ell m\rangle$ for the single-electron wave function. It saves some writing space, but mostly prevents confusion among the different quantum numbers.

On multiplying the symmetric function of Eq. (5.59) with the antisymmetric spin function of Eq. (5.55), we obtain the wave function belonging to the ^1S Term:

$$\left|{}^1S00\right\rangle = \frac{1}{\sqrt{6}}\left[\left|2p_{+1}\uparrow;3p_{-1}\downarrow\right\rangle - \left|2p_{+1}\downarrow;3p_{-1},\uparrow\right\rangle - \left|2p_0\uparrow;3p_0\uparrow\right\rangle\right. \\ \left. + \left|2p_0\downarrow;3p_0\uparrow\right\rangle + \left|2p_{-1}\uparrow;3p_{+1}\downarrow\right\rangle - \left|2p_{-1}\downarrow;3p_{+1}\uparrow\right\rangle\right], \tag{5.60}$$

where the Terms noting method is in the form $|^{2S+1}LSM_S\rangle$, and each of the ket addends in the square parentheses is a Slater determinant; for example, the first addend is

$$\left|2p_{+1}\uparrow;3p_{-1}\downarrow\right\rangle = \frac{1}{\sqrt{2}}\begin{vmatrix} 2p_{+1}(1)\chi_+(1) & 3p_{-1}(1)\chi_-(1) \\ 2p_{+1}(2)\chi_+(2) & 3p_{-1}(2)\chi_-(2) \end{vmatrix}, \tag{5.61}$$

and the last addend is

$$\left|2p_{-1}\downarrow;3p_{+1}\uparrow\right\rangle = \frac{1}{\sqrt{2}}\begin{vmatrix} 2p_{-1}(1)\chi_-(1) & 3p_{+1}(1)\chi_+(1) \\ 2p_{-1}(2)\chi_-(2) & 3p_{+1}(2)\chi_+(2) \end{vmatrix}. \tag{5.62}$$

In the case of two equivalent electrons (an np^2 configuration), we set $n_1 = n_2$ in the Slater determinant such as in Eqs. (5.61) and (5.62). For Eqs. (5.61) and (5.62) specifically, we get

$$\left|np_{+1}\uparrow;np_{-1}\downarrow\right\rangle = -\left|np_{-1}\downarrow;np_{+1}\uparrow\right\rangle,$$

And for the other addends in Eq. (5.59) we get

$$|np_{+1}\downarrow; np_{-1}\uparrow\rangle = -|np_{-1}\uparrow; np_{+1}\downarrow\rangle;$$

$$|np_0\uparrow; np_0\downarrow\rangle = -|np_0\downarrow; np_0\uparrow\rangle.$$

Setting the latter results for the np^2 configuration in Eq. (5.59) yields

$$\left|{}^1S00\right\rangle = \frac{1}{\sqrt{3}}\left[|np_{+1}\uparrow; np_{-1}\downarrow\rangle - |np_{+1}\downarrow; np_{-1},\uparrow\rangle - |np_0\uparrow; np_0\downarrow\rangle\right]. \quad (5.63)$$

As to other Terms, some vanish for the np^2 configuration. Thus, of the six Terms possible for the $npn'p$ configuration: ${}^1S$, ${}^1P$, ${}^1D$, ${}^3S$, ${}^3P$, ${}^3D$, only three are left for the np^2 configuration: ${}^1S$, ${}^3P$, and ${}^1D$. This is obviously a result of Pauli exclusion principle.

Term and wave function calculation using the demonstrated procedure is not sufficient for determining the state energy. The single electron $n\ell_m$ orbital functions require correction by inclusion of an $\varsigma(r)$ Hartree screening function in the independent electron Hamiltonian (Eq. [5.13] and related text). Eventually, each state energy is calculated by adding the energies of all independent electrons. Alternatively, it is possible to calculate corrections to the unperturbed state functions using overlap and exchange integrals by a method described in relation to Eq. (5.16). The two methods may also be combined. The $\varsigma(r)$ screening function would eventually be determined by agreement between theory and experiment.

Calculation of Terms and state energies turn more complex for coupling between more than two electrons, especially when some of the electrons are not equivalent. Such calculations require mastering the angular momentum addition algebra. In the present frame, we satisfy by the above presented examples. In Table 5.4, Terms obtained for different numbers of equivalent electrons are listed without proof. The configurations s^2, p^6, and d^{10}, set in grey-shaded boxes, indicate filled subshells containing $2(2\ell + 1)$ independent electrons. The only possible Term for a filled subshell is ${}^1S$, namely $L = 0$, and $S = 0$, hence also $J = 0$. Therefore, in calculating the Terms of electrons of numbers exceeding a filled subshell, it is possible to ignore the contribution of all filled subshells. Notably, Terms generated by excess electrons occupying a subshell are identical with those generated if the same number of electrons is missing for filling up the subshell.

Table 5.4 Possible Terms for different configurations of equivalent electrons.

Configuration	Term	Configuration	Term
s	${}^2S$	d or d^9	${}^2D$
s^2	${}^1S$	d^2 or d^8	${}^1S$, ${}^1D$, ${}^1G$, ${}^3P$, ${}^3F$
p or p^5	${}^2P$	d^3 or d^7	${}^2P$, ${}^2D$, ${}^2F$, ${}^2G$, ${}^2H$, ${}^4P$, ${}^4F$
p^2 or p^4	${}^1S$, ${}^1D$, ${}^3P$	d^4 or d^6	${}^1S$, ${}^1D$, ${}^1F$, ${}^1G$, ${}^1I$, ${}^3P$, ${}^3D$, ${}^3F$, ${}^3G$, ${}^3H$, ${}^5D$
p^3	${}^4S$, ${}^2P$, ${}^2D$	d^5	${}^2S$, ${}^2P$, ${}^2D$, ${}^2F$, ${}^2G$, ${}^2H$, ${}^2I$, ${}^4P$, ${}^4F$, ${}^4D$, ${}^4G$
p^6	${}^1S$	d^{10}	${}^1S$

A continuation to the empiric **Hund's rules** allows for qualitative classification of Terms belonging to a certain configuration by an energy order (see also Eq. [5.48] and related text):

> **(i) A Term energy is a descending function of the $2S+1$ multiplicity. For identical multiplicities, the energy is a descending function of the resultant orbital angular momentum L.**
> **(ii) For a specific multiplet, the energy is an ascending function of the resultant angular momentum J for a less-than-half-filled subshell; it is a descending function of the resultant angular momentum J for a more-than-half-filled subshell** . (5.64)

As an example, Figure 5.8 shows schematically, and not necessarily to scale, an energy scheme of an np^2 configuration. The three Terms listed in Table 5.4 are split at an order conforming to Hund laws. As we have previously indicated, the reason for this splitting is the strength of the Coulomb repulsion between electron pairs, which increases when they become close. Accounting for the resultant angular momentum J, spin/orbit interaction causes further splitting among states. For the ^1S Term, only $J = 0$ is possible, namely the $^1S_0$ state. Thus, it exhibits no splitting by different J values. For the ^1D Term also, only a single J value is possible: it is $J = 2$. The possible state is thus $^1D_2$, with no splitting. On the other hand, the ^3P Term has three belonging values of the resultant angular momentum: $J = 0, 1, 2$ (Eq. [5.52]). Because a p^2 configuration involves two electrons, which is less than half the number of electrons forming a filled subshell (six electrons), the Term splits into three states arranged by an increasing value of J. Eventually, if an external magnetic field is applied (Section 4.9.1), a Zeeman splitting takes place, and each state splits into $2J + 1$ substates of separation depending on the magnetic field strength.

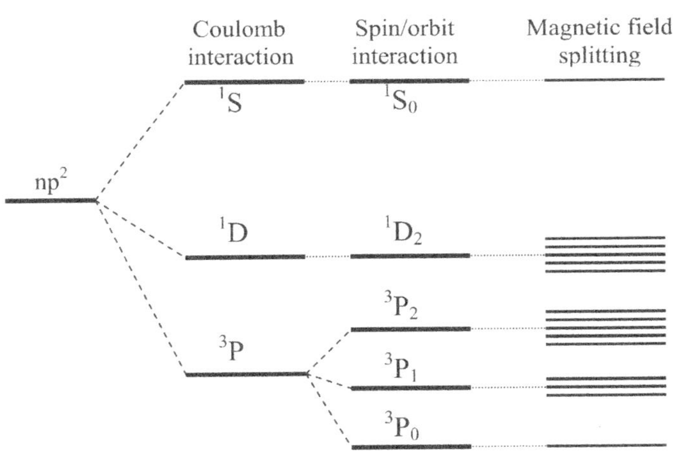

Figure 5.8 Energy scheme of an np^2 configuration.

Considerations presented so far enable finding the ground-state Term for each atom. Deviations from Hund rules occur only in a limited number of cases. Table 5.5 summarizes the electron configuration and ground-state Terms for all periodic table atoms and their first electron ionization energy (energy required to remove a single electron). High ionization energies are obtained for the noble gases, and low ones for the alkali metals. For emphasizing demonstration, Figure 5.9 shows a plot of the ionization energy dependence on the atomic number Z.

Determination of the wave functions and energies (not necessarily of the ground state) is a long and tedious task, often requiring use of sophisticated computation. Use of various approximation methods and introduction of empiric parameters allow considerable shortening of the required time.

Table 5.5 Configuration, Term, and ionization energy of the chemical elements.

Z	Chem. symbol	Ground-state configuration	g.s. Term	Ioniza. energy (eV)
1	H	1s	^2S	13.595
2	He	$1s^2$	^1S	24.581
3	Li	[He]2s	^2S	5.39
4	Be	$2s^2$	^1S	9.32
5	B	$2s^2\,2p$	$^2P_{1/2}$	8.296
6	C	$2s^2\,2p^2$	$^3P_0$	11.256
7	N	$2s^2\,2p^3$	^4S	14.545
8	O	$2s^2\,2p^4$	$^3P_2$	13.614
9	F	$2s^2\,2p^5$	$^2P_{3/2}$	17.418
10	Ne	$2s^2\,2p^6$	^1S	21.559
11	Na	[Ne]3s	^2S	5.138
12	Mg	$3s^2$	^1S	7.644
13	Al	$3s^2\,3p$	$^2P_{1/2}$	5.984
14	Si	$3s^2\,3p^2$	$^3P_0$	8.149
15	P	$3s^2\,3p^3$	^4S	10.484
16	S	$3s^2\,3p^4$	$^3P_2$	10.537
17	Cl	$3s^2\,3p^5$	$^2P_{3/2}$	13.01
18	Ar	$3s^2\,3p^6$	^1S	15.755
19	K	[Ar]4s	^2S	4.339
20	Ca	$4s^2$	^1S	6.111
21	Sc	$3d\,4s^2$	$^2D_{3/2}$	6.54
22	Ti	$3d^2\,4s^2$	$^3F_2$	6.83
23	V	$3d^3\,4s^2$	$^4F_{3/2}$	6.74
24	Cr	$3d^5\,4s$	^7S	6.764
25	Mn	$3d^5\,4s^2$	^6S	7.432
26	Fe	$3d^6\,4s^2$	$^5D_4$	7.87
27	Co	$3d^7\,4s^2$	$^4F_{9/2}$	7.86
28	Ni	$3d^8\,4s^2$	$^3F_4$	7.633
29	Cu	$3d^{10}\,4s$	^2S	7.724
30	Zn	$3d^{10}\,4s^2$	^1S	9.391
31	Ga	$3d^{10}\,4s^2\,4p$	$^2P_{1/2}$	6.00
32	Ge	$3d^{10}\,4s^2\,4p^2$	$^3P_0$	7.88
33	As	$3d^{10}\,4s^2\,4p^3$	^4S	9.81
34	Se	$3d^{10}\,4s^2\,4p^4$	$^3P_2$	9.75
35	Br	$3d^{10}\,4s^2\,4p^5$	$^2P_{3/2}$	11.84
36	Kr	$3d^{10}\,4s^2\,4p^6$	^1S	13.996
37	Rb	[Kr]5s	^2S	4.176
38	Sr	$5s^2$	^1S	5.692
39	Y	$4d\,5s^2$	$^2D_{3/2}$	6.377
40	Zr	$4d^2\,5s^2$	^3F	6.835
41	Nb	$4d^4\,5s$	$^6D_{1/2}$	6.811
42	Mo	$4d^5\,5s$	^7S	7.10
43	Tc	$4d^5\,5s^2$	^6S	7.228
44	Ru	$4d^7\,5s$	$^5F_5$	7.365
45	Rh	$4d^8\,5s$	$^4F_{9/2}$	7.461
46	Pd	$4d^{10}$	^1S	8.33
47	Ag	$4d^{10}\,5s$	^2S	7.574
48	Cd	$4d^{10}\,5s^2$	^1S	8.991
49	In	$4d^{10}\,5s^2\,5p$	$^2P_{1/2}$	5.785
50	Sn	$4d^{10}\,5s^2\,5p^2$	$^3P_0$	7.342
51	Sb	$4d^{10}\,5s^2\,5p^3$	^4S	8.639
52	Te	$4d^{10}\,5s^2\,5p^4$	$^3P_2$	9.01
53	I	$4d^{10}\,5s^2\,5p^5$	$^2P_{3/2}$	10.454
54	Xe	$4d^{10}\,5s^2\,5p^6$	^1S	12.127

Z	Chem. symbol	Ground-state configuration	g.s. Term	Ioniza. energy (eV)
55	Cs	[Xe]6s	^2S	3.893
56	Ba	$6s^2$	^1S	5.210
57	La	$5d\,6s^2$	$^2D_{3/2}$	5.61
58	Ce	$4f\,5d\,6s^2$	$^1G_4$	6.54
59	Pr	$4f^3\,6s^2$	$^4I_{9/2}$	5.48
60	Nd	$4f^4\,6s^2$	$^5I_4$	5.51
61	Pm	$4f^5\,6s^2$	$^6H_{5/2}$	5.55
62	Sm	$4f^6\,6s^2$	$^7F_0$	5.6
63	Eu	$4f^7\,6s^2$	^8S	5.67
64	Gd	$4f^7\,5d\,6s^2$	$^9D_2$	6.16
65	Tb	$4f^9\,6s^2$	$^6H_{5/2}$	6.74
66	Dy	$4f^{10}\,6s^2$	$^5I_8$	6.82
67	Ho	$4f^{11}\,6s^2$	$^4I_{15/2}$	6.02
68	Er	$4f^{12}\,6s^2$	$^3H_6$	6.10
69	Tm	$4f^{13}\,6s^2$	$^2F_{7/2}$	6.18
70	Yb	$4f^{14}\,6s^2$	^1S	6.22
71	Lu	$4f^{14}\,5d\,6s^2$	$^2D_{3/2}$	6.15
72	Hf	$4f^{14}\,5d^2\,6s^2$	$^3F_2$	7.0
73	Ta	$4f^{14}\,5d^3\,6s^2$	$^4F_{3/2}$	7.88
74	W	$4f^{14}\,5d^4\,6s^2$	$^5D_0$	7.98
75	Re	$4f^{14}\,5d^5\,6s^2$	^6S	7.87
76	Os	$4f^{14}5d^6 6s^2$	$^5D_4$	8.7
77	Ir	$4f^{14}\,5d^7\,6s^2$	$^4F_{9/2}$	9.2
78	Pt	$4f^{14}\,5d^8\,6s^2$	$^3D_3$	8.88
79	Au	$[Xe,4f^{14}\,5d^{10}]6s$	^2S	9.22
80	Hg	$6s^2$	^1S	10.434
81	Tl	$6s^2\,6p$	$^2P_{1/2}$	6.106
82	Pb	$6s^2\,6p^2$	$^3P_0$	7.415
83	Bi	$6s^2\,6p^3$	^4S	7.287
84	Po	$6s^2\,6p^4$	$^3P_2$	8.43
85	At	$6s^2\,6p^5$	$^2P_{3/2}$	10.19
86	Rn	$6s^2\,6p^6$	^1S	10.745
87	Fr	[Rn]7s	^2S	4.0
88	Ra	$7s^2$	^1S	5.277
89	Ac	$6d7s^2$	$^2D_{3/2}$	6.9
90	Th	$6d^27s^2$	$^3F_2$	6.0
91	Pa	$5f^26d7s^2$	$^4K_{11/2}$	5.89
92	U	$5f^3\,6d\,7s^2$	$^5L_6$	6.194
93	Np	$5f^4\,6d\,7s^2$	$^6L_{11/2}$	6.274
94	Pu	$5f^6\,7s^2$	$^7F_0$	5.8
95	Am	$5f^7\,7s^2$	^8S	6.0
96	Cm	$5f^7\,6d\,7s^2$	$^9D_2$	5.99
97	Bk	$5f^9\,7s^2$	$^6H_{15/2}$	6.23
98	Cf	$5f^{10}\,7s^2$	$^5I_8$	6.29
99	Es	$5f^{11}\,7s^2$	$^5L_{15/2}$	6.41
100	Fm	$5f^{12}\,7s^2$	$^3H_6$	6.5
101	Md	$5f^{13}\,7s^2$	$^2F_{7/2}$	6.58
102	No	$5f^{14}\,7s^2$	$^1S_0$	6.65
103	Lw	$5f^{14}\,7s^2\,7p^1$ (?)	$^2P_{1/2}$ (?)	4.6

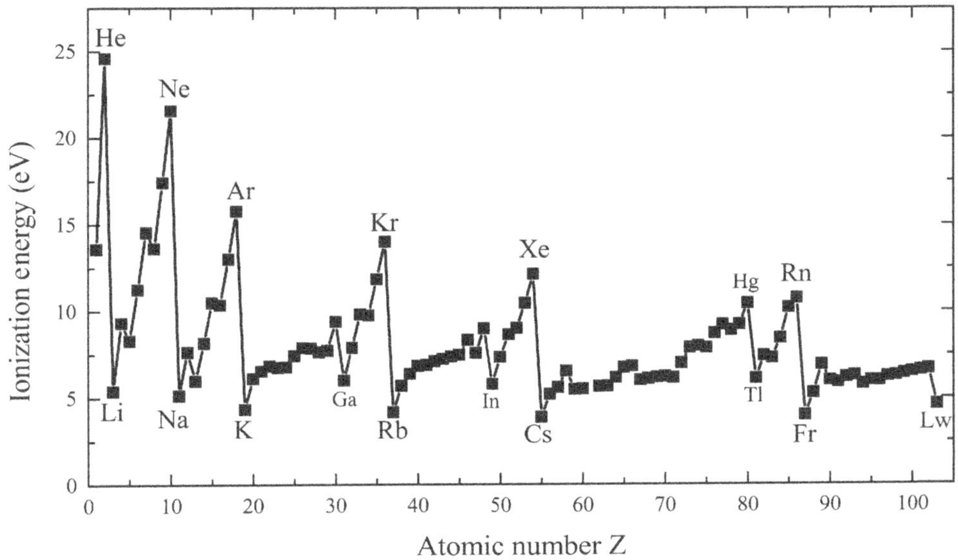

Figure 5.9 Ionization energy of the chemical elements as a function of their atomic number.

The theoretical calculations are confronted against experimental results, mostly obtained by light absorption and emission spectroscopy. Dipole allowed transitions in many-electron atoms are usually similar to those expressed for a single-electron atom (Eqs. [3.58] and [4.91]). The quantities playing the role are the resultant spin and orbital angular momentums, and their Z-axis projections:

$$\boxed{\Delta L = \pm 1; \quad \Delta S = 0; \quad \Delta J = 0, \pm 1; \quad \Delta M_J = 0, \pm 1}, \tag{5.65}$$

except for the $J = 0 \rightarrow J' = 0$ transition, which is forbidden, and the $\Delta M_J = 0$ transition, which is forbidden in cases where $\Delta J = 0$.

We won't prove these selection rules in the present frame. However, it should be stressed that they all are basically approximations, because they are based on the independent electron model, which in itself is an approximation. The deviations are toward allowing "forbidden" transitions, and their number increases with the increasing number of electrons forming the atom.

5.5 Excited-States Structure in Many-Electron Atoms

5.5.1 States Structure of Single Valence Atoms

To demonstrate the applicability of the previous considerations for many-electron states, we compare the states structure of two single electron valence atoms, lithium *Li* and sodium *Na*, with those of hydrogen *H*. Both lithium and sodium contain filled shells and an additional lone electron. The model used to describe such an atom considers the highest closed shell to form a spherical **core** close to the nucleus, whereas the last (lone) electron orbits mostly far from the nucleus. This electron is called **valence electron**.

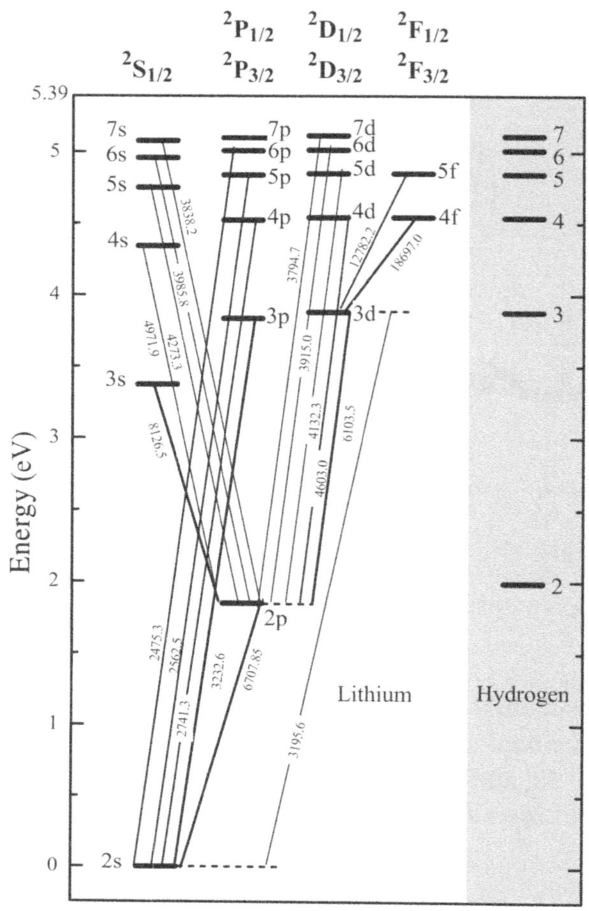

Figure 5.10 Energy scheme of the lithium *Li* atom compared to the hydrogen *H* atom scheme. The energy scale is set to match the free electron energy for both. The wavelength of each specified transition is noted using Ångström (Å) units.

Bonding energies of the core electrons are relatively high compared to that of the valence electron. Therefore, the core electrons are not involved in radiative excitations in the optical (near visible) vicinity, nor in forming a chemical bond. We have already indicated that a Term belonging to a closed shell or sub-shell is a ^1S one, namely $L = 0$ and $S = 0$. Therefore, the resultant orbital and spin angular momentums are determined solely by the valence electron, much like in the hydrogen atom of a single orbiting electron.

The lithium atom contains three protons, namely $Z = 3$. Its electronic configuration is $1s^2 2s$. The two core electrons form the $1s^2$ configuration, like in the helium *He* ground state. If the valence electron orbits far from the nucleus, its wave function rarely penetrates the core. The electrostatic screening effected by the core reduces the effective charge to only $Z_{eff} \cong 3 - 2 = 1$. Orbits belonging to a large ℓ satisfy this condition accurately. Energy states of the valence electron approach those of hydrogen as ℓ grows. The case $\ell = 0$, on the other hand, is characterized by a considerable penetration of the electron through the core space closer to the nucleus (Figure 3.6). The effective attractive charge experienced by such an electron increases, and its state energy becomes considerably lower than the parallel hydrogen state. In the sodium *Na* atom, the electronic configuration is $2s^2 2p^6 3s = [Ne]\ 3s$, and the qualitative considerations regarding the energy states are similar.

Figure 5.10 shows an energy scheme of the lithium atom, and Figure 5.11 shows the energy scheme of the sodium atom, each compared to the Balmer states of hydrogen (Eq. [3.50] and related text, and overlooking fine and hyperfine structure). The ground state of lithium belongs to the $n = 2$ principal quantum number, and in sodium – to the $n = 3$ principal quantum number. We don't present the energy states' schemes for potassium *K*, rubidium *Rb*, and cesium *Cs*, but note that their ground states belong to the $n = 4, 5, 6$ principal quantum numbers, respectively. In all alkali atoms, the degeneracy with respect to the angular momentum quantum number L is removed because of penetration differences through the core shells toward the nucleus. As noted, energy states for large L's are close to those of hydrogen for same n's. Each of the $P, D, F, \cdots$ states are split in two because of spin/orbit coupling. Such splitting is observed by light absorption or emission spectroscopy. For example, note the sodium $3^2S_{1/2} \rightarrow 3^2P_{1/2}$ and $3^2S_{1/2} \rightarrow 3^2P_{3/2}$ transitions; they are called **sodium D-lines**. They form a strong transition, and their splitting in two is almost 6.0 Å, as seen in the energy scheme (Figure 5.11). That was the most outstanding evidence suggesting the existence of a 1/2 angular momentum eigenvalue. This property was initially attributed to an *independent spinning* of the electron about an axis, which is still a convenient approximation for many purposes.

5.5.2 States Structure of Two-Valence Atoms

As a first example of atoms containing two valence electrons, we consider the beryllium *Be* atom. The beryllium nucleus contains four protons, $Z = 4$. The ground-state configuration is $1s^2 2s^2$. The two core electrons form a $1s^2$ configuration, like the helium in its ground state. The two valence electrons form a $2s^2$ configuration. If a valence electron orbits far from the nucleus, its wave function rarely penetrates the core. The electrostatic screening effected by the core reduces the effective charge to $Z_{eff} \cong 4 - 2 = 2$. That way, the two valence electrons form an energy scheme similar to the helium *He* atom (Figure 4.3 and related text). The energy scheme is provided in Figure 5.12. The ground state is a singlet of $S = 0$, yet belongs to the $n = 2$ principal quantum number (rather than $n = 1$ case in helium). Excited states form by excitation of one of the valence electrons to states of higher principal quantum numbers. Because various combinations of spin projections are possible, two separate manifolds are obtained: a singlet manifold of $S = 0$ and a triplet manifold of $S = 1$, where each state is triply degenerate according to the three resultant spin projections. These states are usually split into three because of spin/orbit interaction.

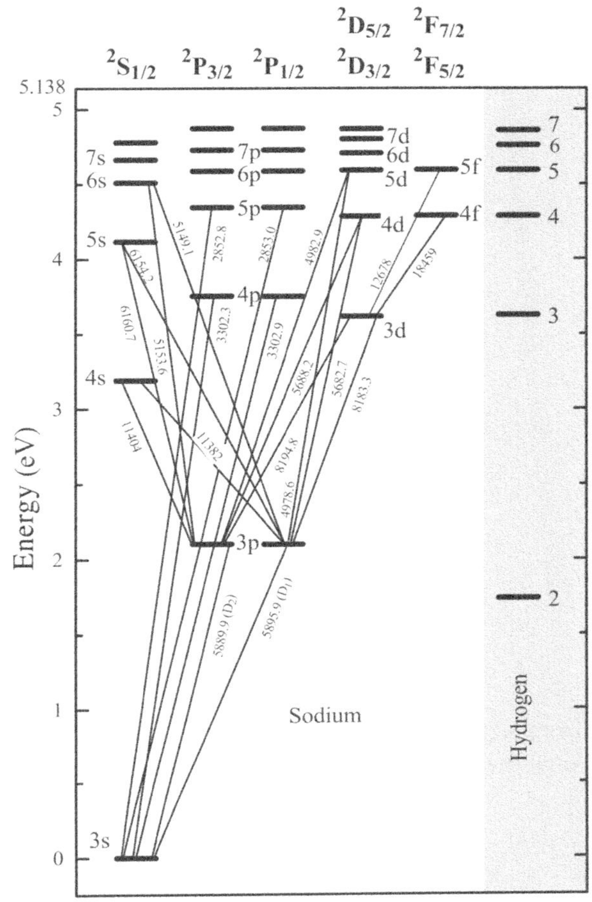

Figure 5.11 Energy scheme of the sodium *Na* atom compared to the hydrogen *H* atom. The energy scale is set to match the free electron energy for each. The wavelength of each specified transition is noted using ångström (Å) units.

Another phenomenon apparent in this atom (as well as in all many-electron atoms) is the occurrence of states generated by excitation of two or more simultaneously excited electrons. They are addressed as **complex Terms**, or sometimes **displaced Terms**. The lowest state belonging to those Terms is considerably higher than the ground state. The adjective "displaced" is used because some such states have positive energy; an atom thus excited will immediately ionize. They are therefore addressed as "auto-ionizing states". The beryllium atom has three bonded complex Terms. All belong to the $2p^2$ configuration. The energy order of the ^1D and ^3P Terms is reversed compared to that described in Figure 5.8, and the ^3P Term energy is rather higher than the ^1D Terms (it is one of many examples of departure from strict Hund rules).

The calcium *Ca* atom exhibits a more complicated energy scheme, shown in Figure 5.13. Its nucleus contains twenty protons ($Z = 20$). The ground-state configuration is $1s^2 2s^2 2p^6 3s^2 3p^6 4s^2 = [Ar]4s^2$. The core contains 18 electrons forming a configuration identical to the ground state of the noble argon *Ar* gas. The two valence electrons form a $4s^2$ configuration. When a valence electron orbits far from the nucleus, its wave function rarely penetrates the core. The electrostatic screening performed by the core electrons reduced the effective attractive nuclear charge to only $Z_{eff} \cong 20 - 18 = 2$. Thus, the two valence electrons form an energy scheme (Figure 5.13) similar to that of the helium *He* atom (Figure 5.3 and related text).

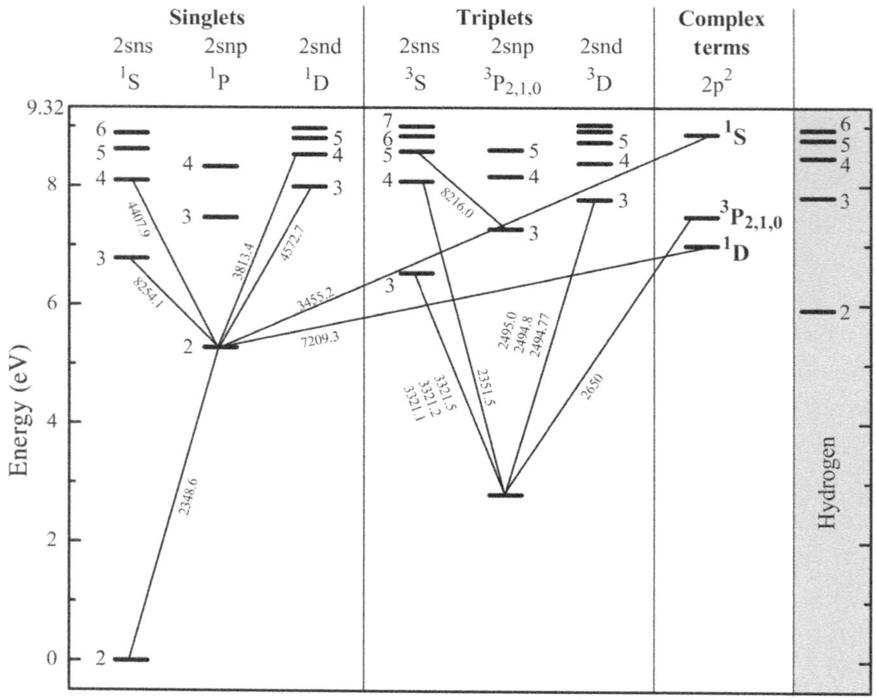

Figure 5.12 Energy scheme of the beryllium *Be* atom compared to the hydrogen *H* atom scheme. The energy scale is set to match the free electron energy for each. The wavelength of each specified transition is noted using ångström (Å) units.

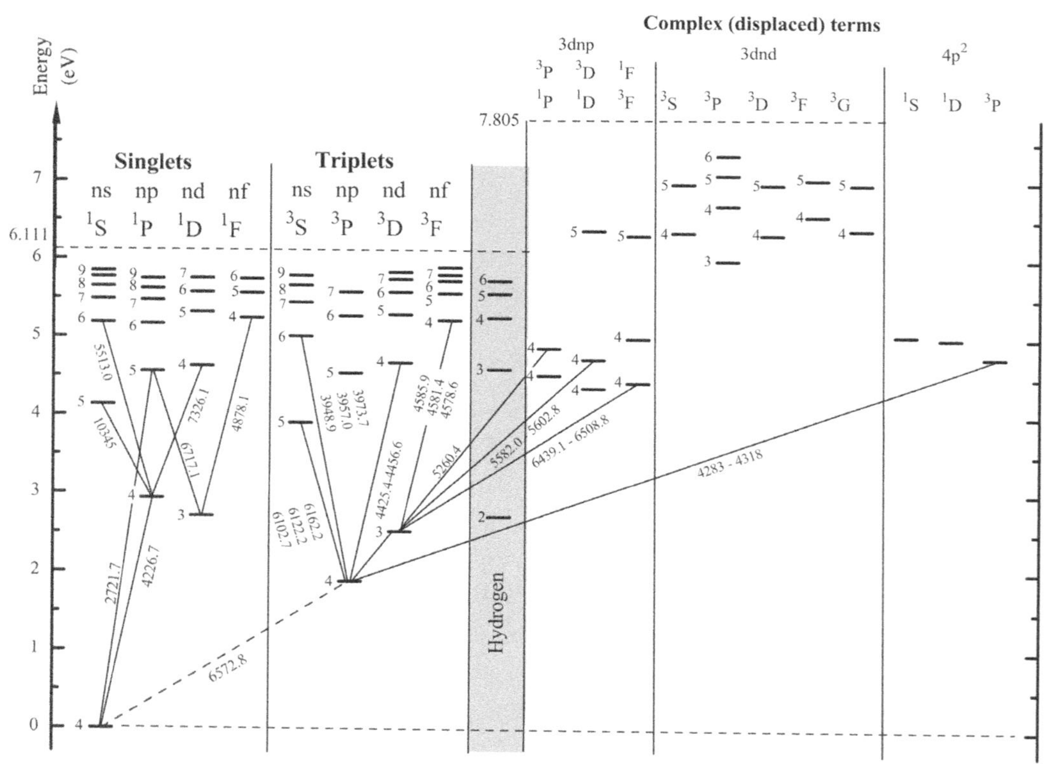

Figure 5.13 Energy scheme of the calcium *Ca* atom compared to the hydrogen *H* atom. The energy scale is set to match the free electron energy for each. Wavelengths of some specified transitions are noted using ångström (Å) units.

Like in helium, the ground sate is a singlet of $S = 0$, belonging however to the $n = 4$ principal quantum number (rather than $n = 1$ in helium). Excited states are obtained by excitation of a single valence electron to states of higher principal quantum number. Because various combinations of spin projections are possible, two separate manifolds are obtained: a singlet manifold of $S = 0$ and a triplet manifold of $S = 1$, where each state is triply degenerate according to the three resultant spin projections. These states are usually split into three because of spin/orbit interaction. Like in beryllium, complex Terms appear, related to simultaneous excitation of the two valence electrons, some exhibiting positive energy. As in beryllium, Hund laws are only approximately satisfied, and absorption and emission spectra also involve some dipole-forbidden transitions.

A most interesting energy scheme is that of the carbon *C* atom, especially for being the basis for the vast number of chemical compounds that are important in life creation. The scheme is presented in Figure 5.14. The atom ground-state configuration is $1s^2 2s^2 2p^2$, and the ground-state Term is ^3P. The same configuration has two excited states belonging to the ^1D and ^1S Terms (Figure 5.8). Excited states are obtained by exciting a single $2p$ electron to states of higher principal quantum number n. In Figure 5.14 it is the $2s^2 2pnp$ in the second column from the left.

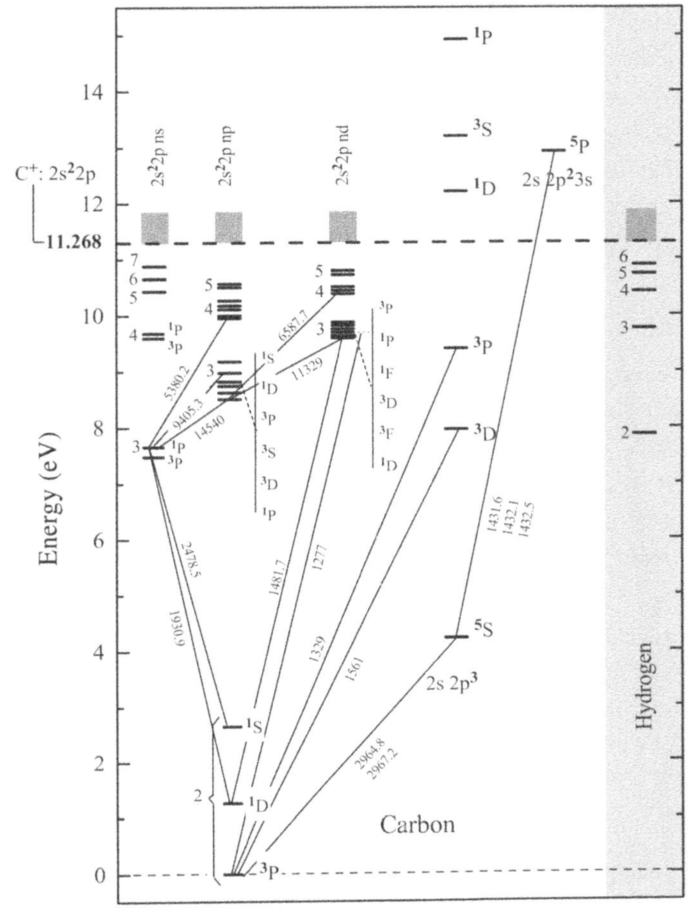

Figure 5.14 Energy scheme of the carbon *C* atom compared to the hydrogen *H* atom. The energy scale is set to match the free electron energy for each. Wavelengths of some specified allowed transitions are noted using ångström (Å) units.

The separate singlet and triplet manifolds emanating from coupling with the other $2p$ electron are joined in the figure to save space. Another series of excited states is obtained by excitation of a single $2p$ electron into *s* states of higher principal quantum number n; it is the $2s^2 2pns$ in the far left column. The lowest principal quantum number in that series is 3, because the $n = 1$ and $n = 2$ shells are incapable of hosting additional *s* electrons. This states series may also be classified into singlet and triplet manifolds generated by coupling the excited *s*-electron with the remaining $2p$ one; the two manifolds are joined in the figure. Another states series (third from the left in Figure 5.14) is obtained by excitation of a single $2p$ electron into *d* states of increasing principal quantum number n. This series starts at $n = 3$ because of the $n \geq (\ell + 1)$ condition (text following Eq. [3.37]). The singlet and triplet states generated because of coupling of the excited *d* electron with the remaining $2p$ one are joined in the figure.

In addition to the above states, there are two prominent series. One results from spin/orbit coupling of the $2s2p^3$ configuration created by excitation of one of the $2s$ electrons to a $2p$ state. The low-energy Term of that configuration is ^5S. This state is responsible for forming the vast number of chemical compounds created by carbon (see Section 6.6). Some of its excited states are autoionizing, that is, the carbon atom decomposes into a carbon ion C^+ and a free electron. The last presented series is created by excitation of one of the 2*s* electrons into a 3*s* state forming a $2s2p^23s$ configuration. The low-energy Term of that configuration is ^5P, and it is autoionizing.

This section won't provide further discussion of many-electron atoms. A considerable extension is provided in Sections 12.2 and 12.3. We just note that the ongoing development of powerful computers and efficient numerical algorithms has brought about many improvements in the understanding of experimental results, and in better matching between theory and experiment.

5.5.3 Classical Approximations

In conclusion, it is possible to analyze and understand a considerable part of atomic characteristics by using classical models. In fact, an atomic model based on the assumption that some of its electrons form a shell, is classical. We have seen that this model succeeds in accounting for the energy scheme structure of the valence electrons, both qualitatively and quantitatively. It is possible to proceed in that spirit, and assume that in the ground state, *all valence electrons* reside statically at uniform distribution over the surface of a sphere of a_{eff} radius. Thus, the ionization energy (binding energy) of a single electron would be $E_{B1} = e^2/a_{eff}$. The radius a_{eff} of that imaginary sphere will be addressed as the **effective ionization radius**. We have pointed out such possible interpretation in relation to the electronic binding energy in the hydrogen atom, given by Eq. (3.38). We further assume that the effective ionization radius is independent of the degree of ionization! For an atom containing several v > 1 valence electrons, the successive ionization energies of the second, third, fourth, and so on electrons will relate like the 1, 2, 3, · · ·, v numbers, namely,

$$\boxed{E_{Bi} = i\left(e^2/a_{eff}\right)}\ , \tag{5.66}$$

where i is the numerator of the ionization stage. The total ionization energy of $q \le \text{v}$ electrons will be given by

$$\boxed{E_B = \sum_{i=1}^{q} E_{Bi} = \frac{e^2}{2a_{eff}}\left(q^2 + q\right)}\ . \tag{5.67}$$

Recalling the model described in relation to Eq. (3.38), the electrostatic energy required to accumulate the missing charge of a q-fold ionized atom at a uniform distribution on a sphere of a_{eff} radius is $q^2e^2/2a_{eff}$, close to the expression given in Eq. (5.67). The binding energy of q electrons just compensates that energy.

The conclusions of that simplistic model, presented by **Rose**, are validated experimentally with a surprising accuracy. For demonstration, Figure 5.15 presents the ionization energy dependence of an additional single electron as a function of the ionization degree for several many-electron atoms: polonium *Po*, tellurium *Te*, selenium *Se*, and sulfur *S*. The dependence appears accurately linear (Eq. [5.66)]). Departures from such strict linearity occurs for some light atoms like oxygen *O*, fluorine *F*, and neon *Ne*, as well as for ionizations exceeding the valence electrons.

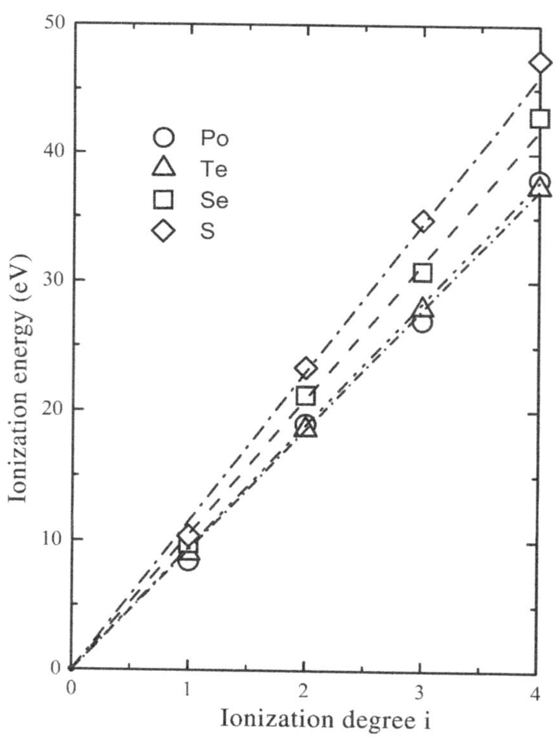

Figure 5.15 Ionization energy of a single electron as a function of the degree of ionization for atomic polonium *Po*, tellurium *Te*, selenium *Se*, and sulfur *S*.

Experimental slopes of curves like those in Figure 5.15 allow for an accurate assessment of the effective radius a_{eff} of Eq. (5.66). Table 5.6 summarizes results obtained for some of the periodic table elements, where all valence electrons belong to a p subshell. Apparently, the effective ionization radius varies in the $\sim 1 - 1.7$ Å range, $2 - 3.5$ times the a_0 Bohr radius (Eq. [3.38]). Usually, for identical valence in different atoms, a_{eff} increases with the atomic number Z.

The above provided example is not the only atomic quantity for which it is possible to use very simple classical considerations. It is a demonstration of the **Bohr correspondence principle**, stating that quantum phenomena converge toward the classical presentation when the system contains many particles. *Recall that the starting point for every quantum calculation is a classical model of the system, converted into a quantum one by replacing dynamic variables by differential operators. The Bohr correspondence principle expresses the insight that this procedure is self-consistent.*

Indeed, the classical limit considerations could not replace accurate quantum calculations in all cases. However, they are very useful as approximations, and for assessing the validity of accurate quantum calculations.

Table 5.6 Effective atomic radiuses for some chemical elements.

Z	Chemical symbol	Valence v	a_{eff} (Å)
7	N	3	0.95
15	P	3	1.46
33	As	3	1.54
51	Sb	3	1.73
83	Bi	3	1.73
16	S	4	1.25
34	Se	4	1.38
52	Te	4	1.58
84	Po	4	1.55
17	Cl	5	1.07
35	Br	5	1.21
53	I	5	1.39
85	At	5	1.41
18	Ar	6	0.96
36	Kr	6	0.98
54	Xe	6	1.27
86	Rn	6	1.31

Exercises for Chapter 5

5.1 Calculate the electrostatic energy required for accumulating a charge e at uniform spatial distribution inside a sphere of radius a_0. Follow the procedure demonstrated by Eqs. (5.7) through (5.9).

5.2 As a simplistic continuation of the effective screening idea (Section 5.2.2), it is possible to assume in first-order approximation that excited single-electron energies of a two-electron atom follow the relation

$$E_n = -\frac{Z^2e^2}{2a_0} - \frac{1}{n^2}\frac{(Z-1)^2e^2}{2a_0}. \tag{5.68}$$

The first addend describes the energy of a single, non-screened independent electron at the ground state. The second addened describes the energy of the second independent electron of a principal quantum number n, where the first electron has screened a part (one elementary charge) of the nucleus charge. Plot the corresponding energy scheme, and compare it with the experimental scheme of helium *He* provided in Figure 5.3.

5.3 (a) Calculate the $\psi_6(Q)$ wave function of the harmonic oscillator according to Eq. (5.31), and indicate the $\boldsymbol{H}_6(Q)$ Hermite polynomial (Eq. [5.32]).

(b) Plot graphs of the $\psi_0 \cdots \psi_6$ harmonic oscillator wave functions and their corresponding probability densities. Use proper graphic computing procedures.

(c) Calculate the expectation value of the squared harmonic oscillator amplitude $\langle Q^2 \rangle \equiv \langle n | Q^2 | n \rangle$ for $n = 0$ and $n = 1$.

5.4 Use the method demonstrated in Figure 5.5 to plot a diagram of the occupied electronic states of silicon *Si* and chlorine *Cl* atoms. Indicate the electronic configuration and ground-state Term for each.

5.5 Calculate the ionization energy of each of the five valence electrons of the chlorine atom. Use the effective ionization radius provided in Table 5.6.

6 Electron Orbits in Molecules

6.1 Preamble

Molecules are usually described as a collection of bonded atoms forming a constant structure. Fundamentally, however, it should be described as a collection of nuclei and electrons interacting by electrostatic forces, forming a stable structure, in other words a bonded solution of the time-independent Schrödinger equation. In the electrostatic approximation, the molecular Hamiltonian is given by

$$\mathbf{H} = \sum_{\substack{all\\nuclei}} \frac{\mathbf{p}_u{}^2}{2M_u} + \sum_{\substack{all\\electrons}} \frac{\mathbf{p}_i{}^2}{2m} + \sum_{\substack{all\\nucleus\\pairs}} \frac{Z_u Z_v e^2}{|\underline{R}_u - \underline{R}_v|} + \sum_{\substack{all\\electron\\pairs}} \frac{e^2}{|\underline{r}_i - \underline{r}_j|} + \sum_{\substack{all\\electron/nucleus\\pairs}} \frac{-Z_u e^2}{|\underline{R}_u - \underline{r}_i|}, \tag{6.1}$$

where $\mathbf{p}_u$, M_u, $\underline{R}_u$, and Z_u are the linear momentum, mass, position vector, and atomic number of the u nucleus, respectively, $\mathbf{p}_i$ and $\underline{r}_i$ are the linear momentum and position vectors of the i electron, respectively, and m is the electronic mass. The first two sums provide the kinetic energies of the nuclei and electrons, respectively; the third sum provides the electrostatic repulsion energy among electron pairs; and the last sum provides the electrostatic attraction energy between nuclei and electrons. Obviously, an accurate solution of such Hamiltonian is far more complex than any atomic many-electron one, which by itself is very complex and requires many approximations as described in Chapter 5.

Two assumptions are made as a zero-order approximation: one is that all atomic nuclei are resting at permanent positions. Under such assumption, the first sum vanishes. The third sum is therefore constant, and may be ignored in considering dynamic characteristics. The second assumption is that it is possible to classify the electrons into separate sets, such that electrons belonging to the same set are always closer to a specific nucleus. In other words, they may be permanently labeled as belonging to the same nucleus, and their interaction with electrons of other sets and other nuclei may be ignored. Under this assumption, the number of addends in the fourth and fifth sums is greatly reduced. Experiments prove that this assumption is reasonably justified for all atomic core electrons. The complications related to the interactions among all electrons and all nuclei are thus limited to the valence electrons. The separate identity of the atoms is fully preserved regarding the nuclei and core electrons. The situation is different for at least some of the valence electrons.

It is not at all simple to guess in advance what the assumed permanent locations of the nuclei would be. For example, experiments prove that the carbon dioxide CO_2 molecule is linear (that is, all nuclei are

Physics, Optics, and Spectroscopy of Materials, First Edition. Zeev Burshtein.

situated along a straight line). The water H_2O molecule, on the other hand, is bent, forming a V shape, with the oxygen atom O at the vertex. The ammonia NH_3 molecule forms a triangular pyramid, with the nitrogen N atom at the vertex. The benzene C_6H_6 molecule forms two concentric regular hexagons; the inner is formed by six carbon C atoms, the outer - by six hydrogen H atoms, each bonded to a corresponding carbon. In retrospect, these forms apparently provide the lowest ground-state energy. A similar interpretation relates to the internuclear distances. An additional issue that is difficult to predict by the fundamental presentation is whether a certain combination of atoms would be stable. For example, the hydrogen H_2 molecule is quite stable, yet no stable H_3 molecule exists. In view of all said difficulties, the theoretical calculations always depend on experimental results. Furthermore, description of molecules always includes classical considerations. We provide a survey of basic approaches for theoretical treatment of molecules, simple ones in the beginning, and more complex ones afterwards.

6.2 The Hydrogen Molecule Ion

6.2.1 The Hamiltonian of the Hydrogen Molecule Ion

The hydrogen molecule ion $H_2{}^+$ is the simplest molecule. It comprises two protons and a single electron. It is also the simplest in a family of diatomic molecules described as **homonuclear**, in contrast to molecules comprising different-type atoms described as **heteronuclear**. This ion is generated when a hydrogen atom captures a proton (the H^+ atom) by an $H + H^+ \rightarrow H_2{}^+$ chemical reaction. After being formed, it is not possible to determine which proton has been the original, and the wave function should thus be symmetric in relation to exchange between the protons.

Seemingly, we could consider similarity between the hydrogen molecule ion and the hydrogen negative ion H^-, the latter comprising a single proton and two electrons, in the sense that the two systems comprise two particles of same mass and electric charge, and an additional particle of a different mass and an opposite charge. However, in the system of our present consideration, the mass of the two identical particles is very large, and by a zero-order approximation are considered at rest, with a constant interdistance $\mathscr{r}$. Figure 6.1 displays the fundamental construction assumed for the hydrogen molecule ion.

Assuming that the protons are at rest, the system Hamiltonian is given by

$$\boxed{\mathsf{H} = \frac{\mathbf{p}^2}{2m} - \frac{e^2}{r_1} - \frac{e^2}{r_2} + \frac{e^2}{\mathscr{r}}}, \tag{6.2}$$

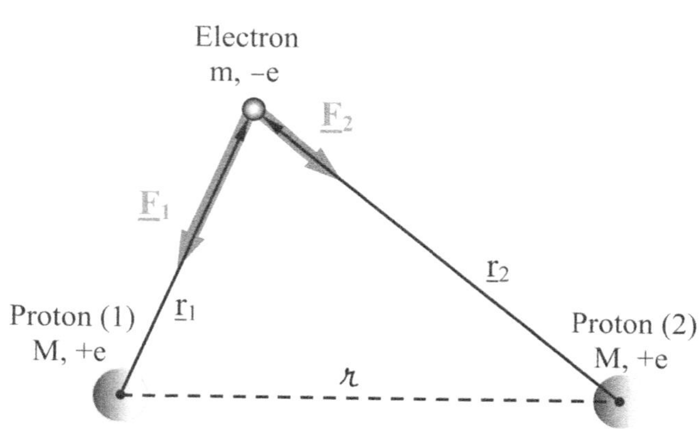

Figure 6.1 Schematic description of the hydrogen molecule ion, comprising two protons and a single electron, and orientation of forces acting on the electron.

where r_1 and r_2 are the dynamically varying distances of the electron from the first and second proton, respectively, and $\mathscr{r}$ is a stable parameter, describing the distance between the protons (not necessarily the equilibrium interdistance).

Wave functions solving a molecular Hamiltonian are called **molecular orbitals** (MO), as a logical extension of the **atomic orbitals** (AO) name often used for atomic wave functions.

6.2.2 A Qualitative Approach to Solution Using a Linear Combination of Atomic Orbitals

To understand the solving method to be presented, we first consider a case where the $\imath$ interprotons distance is very large. That way we may ignore electrostatic repulsion between the protons. The construction may be durable, though obviously not a Hamiltonian eigenstate, where the electron orbits only one of the protons and never comes close to the second one. Still, during such a case, it is possible to ignore the electrostatic attraction between the electron and the second proton. The (quasi) ground state of such combination is the $\psi \equiv 1s$ function belonging separately to the coordinates where the said proton resides at the origin. Generalizing, we denote by ψ_1 and ψ_2 the electron wave functions for the first and second proton, respectively; we also assume that each is properly normalized.

We now seek a means to calculate the *molecular* orbitals. Considering the entire system, we realize that the electron must have an equal probability to orbit the first or second proton (exchange symmetry between the protons). A molecular orbital must then be either symmetric or antisymmetric with respect to exchange between the protons (whereby its absolute square is independent of protons exchange). Up to multiplication by a normalizing constant, there are two linearly independent combinations of ψ_1 and ψ_2 satisfying the exchange symmetry requirement

$$\psi_{even} \equiv \psi_1 + \psi_2; \quad \psi_{odd} \equiv \psi_1 - \psi_2 \,. \tag{6.3}$$

The ψ_{even} function is symmetric with respect to protons exchange, and ψ_{odd} is antisymmetric to such exchange (namely, it reverses sign upon exchange). It is *assumed* that the linear combinations of atomic orbitals provided in Eq. (6.3) provide a reasonably close approximation to the correct molecular orbitals. The method that adopts that approach is called **linear combination of atomic orbitals** (LCAO), or sometimes **Roothaan method**. For demonstration, we recall that the hydrogen ground-state configuration is $1s$ (Table 5.5). The states related to the symmetric wave function and those related to the antisymmetric one are symbolized by $\sigma_g 1s$ and $\sigma_u 1s$, respectively. We will encounter several such symbols to be systematically explained in Chapter 8. Here we just indicate that the σ symbol describes a molecular orbital that is symmetric with respect to rotation about an axis formed by the line connecting two protons; the g and u subscripts are acronyms of the German words "gerade" and "ungerade", meaning "even" and "odd", respectively. In **group theory**, these labels are used to indicate symmetry or antisymmetry with respect to an inversion operation. For the present case, the inversion is about the midpoint on the line between the two protons.

For normalization, we calculate the absolute values of the ψ_{even} and ψ_{odd} functions (Eq. [6.3]):

$$\langle \psi_1 \pm \psi_2 | \psi_1 \pm \psi_2 \rangle = 2 \pm 2S = 2(1 \pm S) \,,$$

where $S = S(\imath) \equiv \langle \psi_1 | \psi_2 \rangle$ is the overlap integral between the ψ_1 and ψ_2 functions; the plus "+" and minus "-" signs belong to the even "g" and odd "u" cases, respectively. The above calculation assumes that the ψ_1 and ψ_2 atomic orbitals are normalized to a unity. In special cases where the atomic orbitals are real functions, the overlap integral is a positive number between zero and a unity. The normalized molecular wave functions (orbitals) would be given by

$$\boxed{\psi_{even} \equiv \frac{\psi_1 + \psi_2}{\sqrt{2(1+S)}}; \quad \psi_{odd} \equiv \frac{\psi_1 - \psi_2}{\sqrt{2(1-S)}}} \,. \tag{6.4}$$

The construction procedure of the molecular orbitals is demonstrated in Figure 6.2. Figures 6.2a and 6.2b portray the unperturbed atomic wave functions along the line connecting the two protons. To

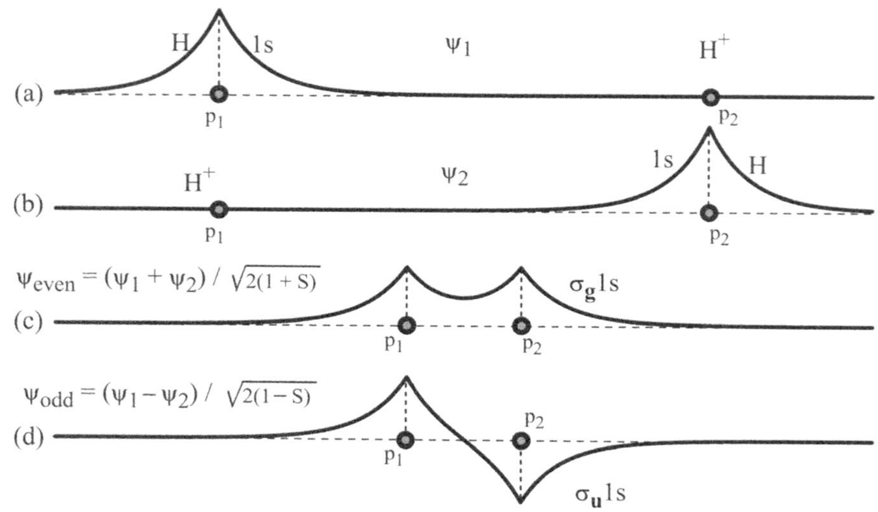

Figure 6.2 (a) and (b): Non-disturbed atomic orbitals along the distance axis between the protons. (c) and (d): Molecular orbitals construction for the H_2^+ hydrogen molecule ion for an $S = 0.2$ overlap integral.

demonstrate the absence of perturbation, the protons are placed at a large distance. Perturbation "ignition" is obtained by bringing the protons closer, and the resulting ψ_{even} and ψ_{odd} eigenfunctions are shown in Figures 6.2c and 6.2d, respectively.

Figure 6.3 portrays the orbitals' probability density variation along the line connecting the protons, and density lines (equiprobability contours) in a plane containing the two protons.

Qualitative considerations concerning the physical meaning of the difference between the ψ_{even} and ψ_{odd} molecular orbitals suggest that each belongs to a different energy. In both, the electron probability density distribution peaks at the nuclei (protons). However, the probability density of the odd ψ_{odd} state vanishes exactly at the midpoint between the nuclei, and is relatively small in the space between. The probability

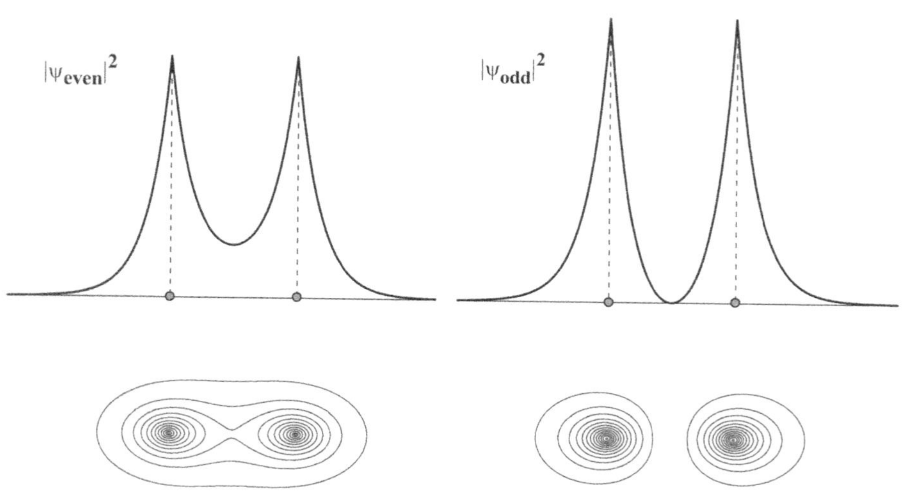

Figure 6.3 Probability density distribution in the different molecular orbitals of the hydrogen molecule ion H_2^+: top: the probability density along a line connecting the two protons; bottom: density lines (equiprobability contours) in a plane containing the two protons.

density of the even state ψ_{even} never vanishes, and exhibits considerable positive values in the space between the nuclei. When the electron happens to be between the nuclei, it reduces their electrostatic rejection, thus lowering the $\sigma_g 1s$ state energy (causing **state stabilization**). This effect is sometimes addressed as "gluing" of molecular parts. In conclusion, the $\sigma_g 1s$ state energy is expected to be lower than the $\sigma_u 1s$ state energy.

The energy difference between the $\sigma_g 1s$ and $\sigma_u 1s$ states depends on the distance between the protons. Without the electron, the electrostatic rejection between the protons grows as the inverse squared distance. However, the screening effect provided by the electron in ψ_{even} increases exponentially. Therefore, at least past a certain distance, the $\sigma_g 1s$ state energy and the screening, are expected to stabilize as the distance shortens. Past that region, shortening the distance between the protons is expected to rather increase the state energy due to the protons' repulsion. Thus, for the $\sigma_g 1s$ state there should be a certain distance $\mathscr{r}_0$ between the protons that exhibits a lowest energy. By similar considerations, rejection between the protons in a $\sigma_u 1s$ state is expected to grow continuously upon protons distance shortening. In view of these characteristics, the $\sigma_g 1s$ state is named **bonding**, whereas the $\sigma_u 1s$ state is named **antibonding**. Antibonding wave-function symbols are often labeled with a superscript asterisk (∗). The $\sigma_u 1s$ state is thus written as $\sigma_u^* 1s$.

6.2.3 Energy States Calculation by LCAO Method

For quantitative calculation of the states energy by the method described in Sub-Section 6.2.2, we recall the Hamiltonian expressed in Eq. (6.2). We assume that the separate atom ψ_1 and ψ_2 wave functions (Figure 6.2) belong to identical atomic states (could be $1s$ or other) of E_a energy. The molecular energy of any ψ state (even or odd) is given by

$$E = \int \psi^* \mathbf{H} \psi \, d^3\underline{r}.$$

Setting the specific forms of the even and odd wave-functions, we get

$$E = \frac{\langle \psi_1 \pm \psi_2 | \mathbf{H} | \psi_1 \pm \psi_2 \rangle}{2(1 \pm S)} = \frac{E_{11} + E_{22} \pm 2E_{12}}{2(1 \pm S)}, \tag{6.5}$$

where

$$E_{11} \equiv \langle \psi_1 | \mathbf{H} | \psi_1 \rangle; \;\; E_{22} \equiv \langle \psi_2 | \mathbf{H} | \psi_2 \rangle; \;\; E_{12} = E_{21} \equiv \langle \psi_1 | \mathbf{H} | \psi_2 \rangle = \langle \psi_2 | \mathbf{H} | \psi_1 \rangle.$$

On setting the specific Hamiltonian form (Eq. [6.2]), we get, for example, for E_{11}

$$E_{11} = \langle \psi_1 | \frac{-\hbar^2}{2m} \nabla^2 - \frac{e^2}{r_1} | \psi_1 \rangle - \langle \psi_1 | \frac{e^2}{r_2} | \psi_1 \rangle + \frac{e^2}{\mathscr{r}} \langle \psi_1 | \psi_1 \rangle.$$

Assume that the first proton ("1") resides at the origin, and the second at distance $+\mathscr{r}$ on the X-axis. We denote

$$A(\mathscr{r}) \equiv \langle \psi_1 | \frac{e^2}{r_2} | \psi_1 \rangle = \langle \psi_1(\underline{r}_1) | \frac{e^2}{\sqrt{(\mathscr{r} - x_1)^2 + {y_1}^2 + {z_1}^2}} | \psi_1(\underline{r}_1) \rangle, \tag{6.6}$$

and get $E_{11} = E_{22} = E_a - A(\mathcal{R}) + \frac{e^2}{\mathcal{R}}$. We also denote

$$B(\mathcal{R}) \equiv \langle \psi_1 | \frac{e^2}{r_1} | \psi_2 \rangle = \langle \psi_1(\underline{r}_1) | \frac{e^2}{r_1} | \psi_2(\underline{r}_1 - \underline{\mathcal{R}}) \rangle , \tag{6.7}$$

and get $E_{12} = E_{21} = E_a S(\mathcal{R}) - B(\mathcal{R}) + \frac{e^2}{\mathcal{R}} S(\mathcal{R})$. Setting in Eq. (6.5), we get

$$\boxed{E(\mathcal{R}) = E_a + \frac{e^2}{\mathcal{R}} - \frac{A(\mathcal{R}) \pm B(\mathcal{R})}{1 \pm S(\mathcal{R})}} , \tag{6.8}$$

where the plus (+) and minus (−) signs relate to the even and odd molecular orbitals, respectively.

The first addend in the states' energy expressed by Eq. (6.8) is the (obviously negative) binding energy of the electron orbiting a single proton. The second addend expresses the electrostatic repulsion between the protons. The $-A(\mathcal{R})$ quantity expresses the electrostatic attraction energy between an electron that orbits unperturbed around one proton, to the other proton. The $B(\mathcal{R})$ quantity is called **exchange energy**, or **exchange interaction**. It is important to realize that the quantities $B(\mathcal{R})$ and $S(\mathcal{R})$ exhibit no classical analogs. Such quantities appear in energy expressions when we *fail* in presenting wave functions and their eigenenergies that are direct solutions of the Hamiltonian system. The quantities $A(\mathcal{R})$, $B(\mathcal{R})$, and $S(\mathcal{R})$ depend on the overlap between the unperturbed electron wave functions belonging to the separate nuclei.♣

We limit ourselves now to the specific case of 1*s* states, and perform detailed calculations of $A(\mathcal{R})$, $B(\mathcal{R})$, and $S(\mathcal{R})$. We set the explicit 1*s* functions per Tables 3.2 and 3.3. We get

$$A(\mathcal{R}) = \frac{1}{\pi a_0{}^3} \int_{\mathcal{R}}^{\infty} \frac{e^2}{r_2} \cdot e^{-2r_1/a_0} d^3\underline{r}_1; \quad \underline{r}_2 = \underline{r}_1 - \underline{\mathcal{R}} . \tag{6.9}$$

This integral expresses an electrostatic attractive energy between the "second" proton and an electrostatic charge density $\rho(r_1) = (-e/\pi a_0{}^3) \cdot e^{-2r_1/a_0}$ distributed around the point occupying the "first" proton. It is the energy required to move the second proton from its $\mathcal{R}$ distance to infinity, given by $\int_{\mathcal{R}}^{\infty} e\mathcal{E}(r_1) dr_1$, where $\mathcal{E}(r_1)$ is the electric field strength at distance r_1 from the origin. Eqs. (5.7) and (5.8) describe a calculation of such field strength. For the present case (setting $Z = 1$), we get

$$\mathcal{E}(r_1) = \frac{(-e)}{r_1{}^2} \cdot \left[1 - e^{-(2r_1/a_0)} \left\{ 1 + \left(\frac{2r_1}{a_0} \right) + \frac{1}{2} \left(\frac{2r_1}{a_0} \right)^2 \right\} \right] . \tag{6.10}$$

Therefore,

$$A(\mathcal{R}) = e^2 \int_{\mathcal{R}}^{\infty} \frac{1}{r_1{}^2} \left[1 - e^{-(2r_1/a_0)} \left\{ 1 + \left(\frac{2r_1}{a_0} \right) + \frac{1}{2} \left(\frac{2r_1}{a_0} \right)^2 \right\} \right] dr_1 .$$

♣ The specific case of the hydrogen molecule ion is unique in the sense that it exhibits a direct analytic solution to the Hamiltonian when transformed into elliptic coordinates: $\lambda \equiv (r_1 - r_2)/\mathcal{R}$ and $\mu \equiv (r_1 + r_2)/\mathcal{R}$. However, we present that case solution by the LCAO method to demonstrate the procedure for general cases, where its use is unavoidable. Also demonstrated are its advantages and limitations.

We perform a variable conversion according to $\xi \equiv 2r_1/a_0$ and get

$$A(\mathcal{r}) = \left(\frac{2}{a_0}\right)e^2 \int_{2\mathcal{r}/a_0}^{\infty} \left[\frac{1 - e^{-\xi}(1+\xi)}{\xi^2} - \frac{1}{2}e^{-\xi}\right] d\xi = \left(\frac{2}{a_0}\right)e^2 \left[\frac{e^{-\xi} - 1}{\xi} + \frac{1}{2}e^{-\xi}\right]_{(2\mathcal{r}/a_0)}^{\infty};$$

$$A(\mathcal{r}) = \frac{e^2}{\mathcal{r}}\left\{1 - e^{-2(\mathcal{r}/a_0)}\left[1 + \left(\frac{\mathcal{r}}{a_0}\right)\right]\right\}. \tag{6.11}$$

To find $B(\mathcal{r})$ and $S(\mathcal{r})$, we convert to (μ, λ, φ) **elliptic coordinates**, defined by $\mu \equiv (r_1 + r_2)/\mathcal{r}$, $\lambda \equiv (r_1 - r_2)/\mathcal{r}$, and φ (the latter being the azimuth angle about the axis joining the two nuclei). Section 6.7 shows that an infinitesimal element in these coordinates is given by

$$d\tau = \frac{\mathcal{r}^3}{8}\left(\mu^2 - \lambda^2\right) d\mu\, d\lambda\, d\varphi, \tag{6.12}$$

where μ varies in the $(1, \infty)$ range, λ varies in the $(-1, +1)$ range, and φ varies in the $(0, 2\pi)$ range. Thus, we get (Eq. [6.7])

$$B(\mathcal{r}) = \frac{\mathcal{r}^2}{4\pi a_0^3} \int_1^{\infty} \int_{-1}^{+1} \int_0^{2\pi} (\mu - \lambda) e^{-\mu\mathcal{r}/a_0}\, d\mu\, d\lambda\, d\varphi;$$

$$B(\mathcal{r}) = \frac{e^2}{\mathcal{r}}\left(\frac{\mathcal{r}}{a_0}\right)\left(1 + \frac{\mathcal{r}}{a_0}\right)\cdot e^{-\mathcal{r}/a_0}. \tag{6.13}$$

Similarly (see Eq. [6.4]) and related text),

$$S(\mathcal{r}) = \frac{\mathcal{r}^3}{8\pi a_0^3} \int_1^{\infty} \int_{-1}^{+1} \int_0^{2\pi} \left(\mu^2 - \lambda^2\right) e^{-\mu\mathcal{r}/a_0}\, d\mu\, d\lambda\, d\varphi;$$

$$S(\mathcal{r}) = \left[1 + \left(\frac{\mathcal{r}}{a_0}\right) + \frac{1}{12}\left(\frac{\mathcal{r}}{a_0}\right)^2\right] e^{-\mathcal{r}/a_0}. \tag{6.14}$$

For brevity we define $\rho = \mathcal{r}/a_0$ (physically meaning the reduced interproton distance; it should not be confused with the electric charge density employing the same letter). Recall also that the energy of the 1s state is $E_a = -e^2/2a_0$ (Eq. [3.38]). Setting all the above in Eq. (6.8), we get

$$\boxed{\frac{E(\rho)}{|E_a|} = -1 + \frac{2}{\rho}\left[1 - \frac{1 - (1+\rho)\,e^{-2\rho} \pm \rho(1+\rho)\,e^{-\rho}}{1 \pm (1 + \rho + \rho^2/12)\,e^{-\rho}}\right]}. \tag{6.15}$$

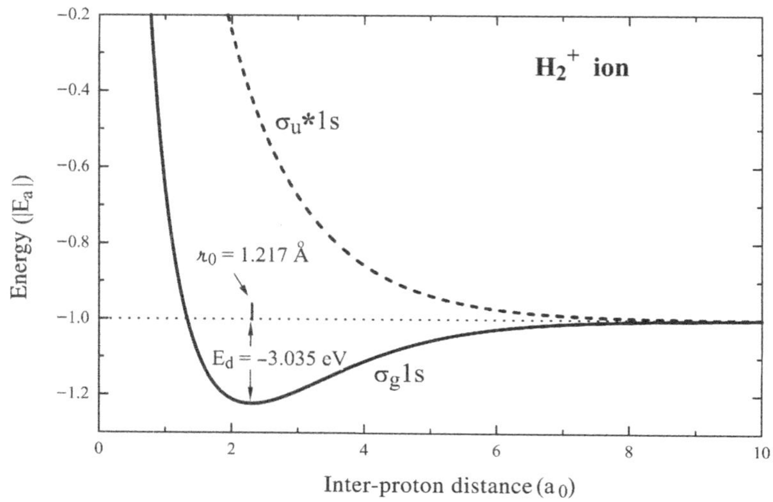

Figure 6.4 Reduced potential energy of the hydrogen molecule H_2^+ ion for the molecular $\sigma_g 1s$ and $\sigma_u^* 1s$ orbitals as functions of the reduced interprotons distance given by Eq. (6.15).

Figure 6.4 describes the energies of the $\sigma_g 1s$ and $\sigma_u^* 1s$ states of the hydrogen molecule ion $H_2{}^+$ as functions of the interprotons distance according to Eq. (6.15). Compared to the energy of infinitely distant protons, this energy is called **potential energy**. The $\sigma_g 1s$ potential energy displays a −3.035 eV minimum at an interproton distance $r_0 = 2.3a_0 = 1.217$ Å: namely, it is lower than in the case of infinitely distant protons. On further distance shortening, the potential energy grows steeply to infinity. The interproton distance r_0 at minimum potential energy obtained by the accurate solution of the Hamiltonian (Eq. [6.2]) is $r_0 = 2.0a_0 = 1.06$ Å. The minimum potential energy thus obtained is only $E_d = -2.468$ eV. This value has been accurately verified by experiment. The r_0 distance is called **bond length**, and the minimum depth $-E_d$ (inverse of the minimum potential) is called **dissociation energy**. We won't present the accurate solution here.

In these qualitative considerations, we anticipated that the $\sigma_u^* 1s$ state potential energy will be higher than the $\sigma_g 1s$ state potential energy. It is indeed verified in the calculations (Eq. [6.15] and Figure 6.4). Also as expected, the states' potential energies increase steeply upon shortening of the interproton distance. The same behavior is obtained by the accurate solution of the $H_2{}^+$ system Hamiltonian (Eq. [6.4]), that is also verified by experiment.

In relation to comparison between the accurate and the approximate LCAO solutions, we note that by simple physical considerations, and irrespective of the previous calculations (Eqs. [6.8] and [6.15]), it is possible to assess the potential energy dependence on the interproton distance in the limit of very short distances. Generally, the molecular energy has two components: electronic, and electrostatic rejection between the protons. Close to the limit where the interproton distance approaches zero, the electronic energy converges to that of a single-electron helium ion, which in fact is a hydrogen-like atom of $Z = 2$ (Section 3.3; Eq. [3.38]) – namely, $-4(e^2/2a_0) = -4E_a$. The electrostatic repulsive energy between the protons is obviously given by e^2/r. In terms of the reduced distance variable Eq. (6.15), the total potential energy for $\rho << 1$ thus converges to the $E(\rho)/|E_a| = -4 + 2/\rho$ form. On the other hand, Eq. (6.15) in this same limit converges to the $E(\rho)/|E_a| = -1 + 2/\rho$ form. This is a clear demonstration of the LCAO method limitation under conditions of large electronic overlap. The LCAO method agrees with experiment only under relatively small electronic overlap among the unperturbed elements of the wave functions.

Turning to discuss the molecular $\sigma_g 1s$ state, it is obvious that in the ground state, the molecule will vibrate about the interproton distance that provides the minimum potential energy ($r_0 = 1.06$ Å for the hydrogen molecule ion). Detailed accounts of molecular vibrations and motions are provided in Chapters 7 and 8.

It is possible to construct molecular orbitals of the hydrogen molecule ion based on other, excited atomic orbitals. For example, the $\sigma_g 2s$ and $\sigma_u^* 2s$ molecular orbitals would be constructed based on a $2s$ atomic orbital. The $\sigma_g 2p$ and $\sigma_u^* 2p$ molecular orbitals would be constructed based on a $2p_0$ atomic orbital (the zero "0" subscript in the present case indicates the projection of the atomic orbital angular momentum along the axis connecting the two protons). The $\pi_g^* 2p$ and $\pi_u 2p$ molecular orbitals would be constructed based on a $2p_1$ atomic orbital. The σ and π symbols express symmetry properties of the molecular orbitals relative to the molecular Z axis, and are adopted from **group theory**. Details and demonstrations concerning this issue are provided in Chapter 8.

6.2.4 Improvements in the LCAO Method

There are several means to improve the accuracy of the LCAO method predictions. First, the demand for accurate atomic orbitals is relaxed, and calculations use functions that are "quasi-atomic orbitals"; for example, by assuming that the charge Z of each atom is an adjustable parameter. Also, the demand for "pure" atomic orbitals, those belonging to steady (eigen) atomic states, is relaxed by considering linear combinations of several orbitals of close energies. The combination coefficients are considered adjustable parameters (obviously still demanding normalization of the recombined wave function). All changes are made for receiving the lowest energy for a given spatial structure of the molecular skeleton. Eventually, the molecular structure itself is an adjustable parameter: the one providing a minimum of the potential energy is assumed to exhibit best fit to the real structure. Complex computer software and multistep computing are needed to achieve the desired fit, often requiring long computing durations. Since the early 2000's, there has been a tremendous progress in computing technology – software sophistication and computing speed, allowing the achievement of fits with experimental results for larger and more complex molecules. This exciting issue is beyond the scope of this book.

6.2.5 Optical Transition Probabilities

The momentary electric dipole in the H_2^+ molecule is a sum of the two different electron dipole moments in relation to each proton: $e\underline{r}_1 + e\underline{r}_2 = e(\underline{r}_1 + \underline{r}_2)$. To calculate the optical transition probability, one needs to consider the $\left\langle \psi_M^{(f)} \middle| \underline{r}_1 + \underline{r}_2 \middle| \psi_M^{(i)} \right\rangle$ vector, where $\psi_M^{(i)}$ and $\psi_M^{(f)}$ are the molecular initial and final orbital states, respectively (see Section 3.6, especially Eq. [3.52] and related text). For example, to assess the probability for light absorption between the $\sigma_g 1s$ and $\sigma_u^* 1s$ states, one must consider the $\langle \sigma_u^* 1s | \underline{r}_1 + \underline{r}_2 | \sigma_g 1s \rangle$ vector. The $\underline{r}_1 + \underline{r}_2$ operator and the $\sigma_g 1s$ orbital are both even, so each maintains its sign on replacing the roles of the two protons. However, the $\sigma_u^* 1s$ orbital is odd, so it reverses sign upon exchanging the protons role, and so does the $\langle \sigma_u^* 1s | \underline{r}_1 + \underline{r}_2 | \sigma_g 1s \rangle$ vector. Because this vector may not depend on the numeration order of the protons, it vanishes, hence also the optical absorption probability. On the other hand, a transition dipole between "g" and "g" states or between "u" and "u" states does not necessarily vanish.

For specific calculation of transition dipoles, the molecular orbitals must be expressed specifically. Using LCAO approximation yields a sum of vectors emanating from one-atomic dipoles and from exchange dipoles. Correspondingly, each addend exhibits separate selection rules and effective selection rules. The calculation is complex, and won't be elaborated upon here.

It is important to note that generally, when a molecule exhibits bonding based on atomic excited states, its interatomic distances belonging to a minimum potential energy are considerably different. It is particularly different than the ground-state interatomic distance. This effect strongly influences the optical transitions probability, because the molecular orbitals change as functions of distance. Also important to note is that bonded states exhibit a dynamic distribution of the bond lengths due to thermal vibrations, hence a distribution of transition probabilities. These issues will be discussed in the following chapters.

6.3 Molecular Electronic Angular Momentum

6.3.1 Eigenfunctions of L^2 and L_z^2 in a Lone Atom

Analysis of molecular structure prefers an alternative description of the angular part of the wave functions belonging to its lone constituent atoms. Here we address this alternative possibility.

Schrödinger's solution of the hydrogen atom provided in Chapter 3 presented only a single choice for the basis functions of the azimuthal $\Phi(\varphi)$ component (Eq. [3.33]). However, instead of the $\Phi_m = \frac{1}{\sqrt{2\pi}} e^{im\varphi}$ functions set, with $m = -\ell, -\ell+1, -\ell+2, \cdots 0, 1, 2, \cdots +\ell$, it is equally possible to choose the $\left\{ \Phi_m^{(1)} = \frac{1}{\sqrt{\pi}} \cos|m|\varphi \ ; \ \Phi_m^{(2)} = \frac{1}{\sqrt{\pi}} \sin|m|\varphi \right\}$ set (in the special case where $|m| = 0$, the $\Phi_0^{(2)}$ function vanishes, and $\Phi_0^{(1)} = 1/\sqrt{2\pi}$). These functions may obviously be developed as linear combinations of the former set. In Schrödinger's solution to the hydrogen atom, the angular functions replacing the $Y_{\ell m}(\theta, \varphi)$ of Table 3.2 will correspondingly change, and they are summarized in Table 6.1 for $\ell = 0, 1, 2, 3$;

Table 6.1 $\mathcal{Y}_{\ell|m|}(\theta,\varphi)$ angular functions of the hydrogen atom for $\ell = 0, 1, 2, 3$, and the $|m|$ values possible for each ℓ. The $\ell = 0, 1, 2, 3, 4, 5, 6, 7$ states are noted by the s, p, d, f, g, h, i, j, letters, respectively. The function in Cartesian coordinates appearing as subscript label for each wave function transforms according to the same symmetry as the $\mathcal{Y}_{\ell|m|}(\theta,\varphi)$ function.

ℓ	$\lvert m\rvert$	Real angular functions $\mathcal{Y}_{\ell\lvert m\rvert}(\theta,\varphi)$
0 (s)	0	$s = 1/\sqrt{4\pi}$
1 (p)	0	$p_z = \sqrt{3/4\pi}\ \cos\theta$
	1	$p_x = \sqrt{3/4\pi}\ \sin\theta\cdot\cos\varphi$ $p_y = \sqrt{3/4\pi}\ \sin\theta\cdot\sin\varphi$
2 (d)	0	$d_{3z^2-r^2} = \frac{1}{2}\sqrt{5/4}(3\cos^2\theta - 1)$
	1	$d_{xz} = \sqrt{15/4\pi}\ \sin\theta\ \cos\theta\cdot\cos\varphi$ $d_{yz} = \sqrt{15/4\pi}\ \sin\theta\ \cos\theta\cdot\sin\varphi$
	2	$d_{x^2-y^2} = \frac{1}{4}\sqrt{15/\pi}\ \sin^2\theta\cdot\cos 2\varphi$ $d_{xy} = \frac{1}{4}\sqrt{15/\pi}\ \sin^2\theta\cdot\sin 2\varphi$
3 (f)	0	$f_{z(5z^2-3r^2)} = \frac{1}{4}\sqrt{7/\pi}(5\cos^3\theta - 3\cos\theta)$
	1	$f_{x(5z^2-r^2)} = \frac{1}{4}\sqrt{21/2\pi}\ \sin\theta\cdot(5\cos^2\theta - 1)\cdot\cos\varphi$ $f_{y(5z^2-r^2)} = \frac{1}{4}\sqrt{21/2\pi}\ \sin\theta\cdot(5\cos^2\theta - 1)\cdot\sin\varphi$
	2	$f_{z(x^2-y^2)} = \frac{1}{4}\sqrt{105/\pi}\cdot\cos\theta\ \sin^2\theta\cdot\cos 2\varphi$ $f_{zxy} = \frac{1}{4}\sqrt{105/\pi}\cdot\cos\theta\ \sin^2\theta\cdot\sin 2\varphi$
	3	$f_{x(x^2-3y^2)} = \frac{1}{4}\sqrt{35/2\pi}\cdot\sin^3\theta\cdot\cos 3\varphi$ $f_{y(y^2-3x^2)} = \frac{1}{4}\sqrt{35/2\pi}\cdot\sin^3\theta\cdot\sin 3\varphi$

they are written as $\mathcal{Y}_{\ell|m|}(\theta,\varphi)$. Usually, these angular functions are designated by the lettering convention for ℓ, however with a subscript label presenting a Cartesian function that transforms according to the same wave function. The use of the "transforms according to" phrase stems from **group theory**. It means that a point-symmetry operation, like rotation or reflection of the coordinate axes, preserves both values and functional forms of $\mathcal{Y}_{\ell|m|}(\theta,\varphi)$ and its subscript function.

While the $\boldsymbol{Y}_{\ell m}(\theta,\varphi)$ spherical harmonics are mathematically complex, the $\mathcal{Y}_{\ell|m|}(\theta,\varphi)$ angular functions are always real. Recall that the $\boldsymbol{Y}_{\ell m}(\theta,\varphi)$ functions are eigenfunctions of both $\mathbf{L}^2$ and $\mathbf{L}_Z$ (Eqs. [4.6] and [4.7]). The $\mathcal{Y}_{\ell|m|}(\theta,\varphi)$ angular function are also eigenfunctions of $\mathbf{L}^2$, but not of $\mathbf{L}_Z$. They are, however, eigenfunctions of $\mathbf{L}_Z{}^2$, with $m^2\hbar^2$ eigenvalues. Notably, the atomic orbitals belonging to the $\boldsymbol{Y}_{\ell m}(\theta,\varphi)$ spherical harmonics exhibit the same energies as the $\mathcal{Y}_{\ell|m|}(\theta,\varphi)$ angular functions. Like the $\boldsymbol{Y}_{\ell m}(\theta,\varphi)$ functions, their absolute squared value $\left|\mathcal{Y}_{\ell|m|}(\theta,\varphi)\right|^2$ provides the probability density per unit solid angle (Section 3.4). In the former case, the probability density is independent of the φ azimuth angle. The probability density in the second case, however, is φ dependent. In polar figures, each function exhibits several lobes, as portrayed in Figure 6.5. It is worthwhile to compare that figure with Figures 3.5(1) and 3.5(2).

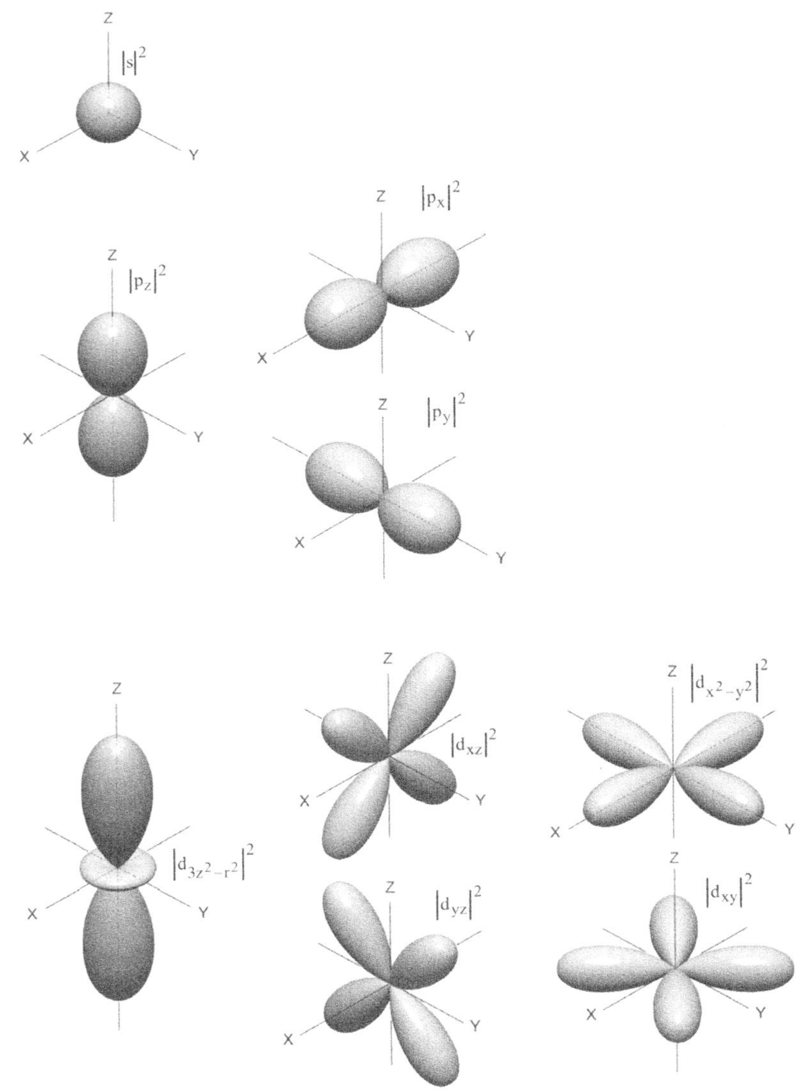

Figure 6.5 Spatial probability density per solid angle unit of the s ($\ell = 0$), p ($\ell = 1$), d ($\ell = 2$) angular wave functions of Table 6.1.

The $\mathcal{Y}_{\ell|m|}(\theta, \varphi)$ angular functions can replace the $\boldsymbol{Y}_{\ell m}(\theta, \varphi)$ spherical harmonics not only in hydrogen-like atoms, but also in all models of many-electron atoms once using the independent electron approach. Furthermore, the presence of oriented lobes in the $\left|\mathcal{Y}_{\ell|m|}(\theta, \varphi)\right|^2$ functions has an advantage in constructing molecular orbitals based on their constituting atoms. Recall that an electron belonging to a certain molecule, participates in motion through all constituting atoms. It is then just natural to assume that its independent orbit reflects the skeletal structure; that is, it exhibits increased or reduced probabilities along bond lines (Figure 6.3). Most probably, atomic orbitals having lobes oriented toward neighboring atoms will provide improved approximation in construction of molecular orbitals by using the LCAO method, compared to using atomic orbitals lacking that trait. This issue will be further discussed in following sections.

6.3.2 Orbital Angular Momentum of an Independent Electron in a Molecule

An independent electron in a diatomic molecule, like the lone electron in the hydrogen molecule ion (Section 6.2), is subject to a noncentral field force. Therefore, its orbital angular momentum $\underline{\boldsymbol{L}}$ is not a constant of motion. However, it is possible, even by just qualitative considerations, to point out a related constant of motion. We select an orientation along the line connecting the two protons as the Z-axis. As shown in Figure 6.6, the electron resultant force orientation due to attraction by the two protons is always oriented toward that axis. In other words, it is an **axial force**. The rotation force moment (torque) acting on the electron in relation to the "**O**" central point between the protons, is always perpendicular to Z. Therefore, $\underline{\boldsymbol{L}}_Z$, the angular momentum projection on the Z-axis, is a constant of motion. Figure 6.6 provides a geometric demonstration to the constancy of $\underline{\boldsymbol{L}}_Z$.

Mathematically, $\mathbf{L}_Z$ being a constant of motion means that the $\underline{\mathbf{L}}_Z$ operator commutes with the Hamiltonian operator, and its eigenvalue is identical to its expectation value. The eigenvalues would be $L_Z = m_\ell \hbar$, where $m_\ell = 0, \pm 1, \pm 2, \cdots$. The sign of m_ℓ reflects the electron sense of rotation about the Z axis. The electron energy, however, is independent of its sense of rotation. For indication of the state energy,

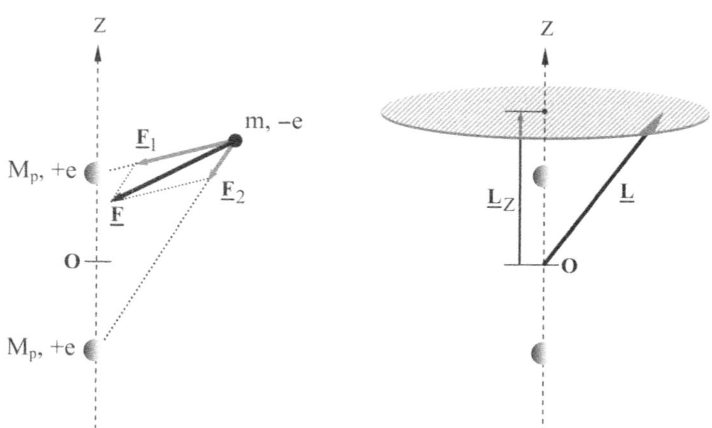

Figure 6.6 Geometric demonstration that the resultant force acting on a lone independent electron in a diatomic molecule is always oriented toward the axis. Its axial orbital angular momentum component $\underline{\boldsymbol{L}}_Z$ is a constant of motion. The resultant angular momentum $\underline{\boldsymbol{L}}$ is not constant, but its edge must always reside on the constant (gray-lined) plane perpendicular to the axis.

the absolute value of m_ℓ, namely $\lambda \equiv |m_\ell|$, is sufficient. The angular momentum projection states, noted in relation to λ, are as follows:

$$\boxed{\begin{array}{rccccc} m_\ell: & 0, & \pm 1, & \pm 2, & \pm 3, & \cdots \\ \lambda: & 0, & 1, & 2, & 3, & \cdots \\ State-symbol: & \sigma, & \pi, & \delta, & \phi, & \cdots \end{array}} \quad . \tag{6.16}$$

The selection of the $\sigma,\ \pi,\ \delta,\ \phi,\ \cdots$ symbols originates in group theory, reflecting symmetry properties of the molecular orbitals. This is the link to the consideration of using atomic orbitals represented by $\mathcal{Y}_{\ell|m|}(\theta, \varphi)$ functions. If the molecular orbitals are expected to transform according to a certain symmetry, it is preferable to construct it by an LCAO method that uses atomic orbitals transforming according to the same symmetry.

6.3.3 Electronic Spin in a Diatomic Molecule

A diatomic molecule exhibits a well-defined Z axis oriented along the bond connecting the two nuclei. The spin projection of an independent electron may exhibit two values, $\pm 1/2$, along this axis. If there is only one orbiting electron, all states are doublets, namely, the energy states are all split in two, except for the σ state for which the spin/orbit interaction vanishes. It is still possible to split the σ state by applying an external magnetic field (Zeeman effect – see Section 4.9). The $\pi,\ \delta,\ \phi,\ \cdots$ states obviously exhibit spin/orbit splitting in the absence of any external magnetic field. In the present frame we won't engage in the quantitative aspects of spin/orbit and Zeeman splitting in molecules.

6.4 Many-Electron Homonuclear Diatomic Molecules

The hydrogen molecule ion H_2 is the simplest many-electron homonuclear diatomic molecule. It comprises two protons and two electrons. As a starting point we consider a general homonuclear diatomic molecule, each of Z atomic number, to be noted separately by a and b subscripts. The electrons will be numerated separately by the "1" and "2" numeral subscripts. The system Hamiltonian will be

$$\mathsf{H} = \frac{\mathbf{p}_1{}^2}{2m} + \frac{\mathbf{p}_2{}^2}{2m} - \frac{Ze^2}{r_{1a}} - \frac{Ze^2}{r_{1b}} - \frac{Ze^2}{r_{2a}} - \frac{Ze^2}{r_{2b}} + \frac{e^2}{r_{12}} + \frac{Z^2e^2}{r_{ab}};$$

$$\boxed{\mathsf{H} = \left[\frac{\mathbf{p}_1{}^2}{2m} - \frac{Ze^2}{r_{1a}} - \frac{Ze^2}{r_{1b}}\right] + \left[\frac{\mathbf{p}_2{}^2}{2m} - \frac{Ze^2}{r_{2a}} - \frac{Ze^2}{r_{2b}}\right] + \frac{e^2}{r_{12}} + \frac{Z^2e^2}{r_{ab}}}\ , \tag{6.17}$$

where $\mathbf{p}_1$ and $\mathbf{p}_2$ are the linear momentum operators of the first and second electrons, respectively, and r_{ij} is the distance between two particles (electrons noted by numerals, nuclei noted by Latin letters; the internuclei distance r_{ab} has been denoted $\mathcal{r}$ in previous sections).

For construction of the wave functions and their eigenenergies, we adopt an independent electron approximation model. It is assumed that each such electron motion follows an individual molecular orbital, whose construction has been described in the previous section. Correspondingly, a molecular state configuration may be presented in a format like that used for electrons. For example, if two independent electrons belong to a $\sigma_g 1s$ orbital, the electronic configuration will be noted as $(\sigma_g 1s)^2$. A reasonably satisfying procedure is like that described for the Helium atom (Section 5.2). Neglecting the electrostatic interactions between the electrons and between the nuclei, each expression enclosed within square parentheses in Eq. (6.17) describes a Hamiltonian of an independent electron orbiting in the molecule. Initially, such Hamiltonian is solved using the LCAO method. We denote the orbitals of those independent electrons as $\psi_\alpha(1)$ and $\psi_\beta(2)$ belonging to $E_\alpha(r_{ab})$ and $E_\beta(r_{ab})$ energies, respectively, where the α and β subscripts indicate the spatial quantum state of each electron, and the "1" and "2" numerals are the tagging numbers of the electrons, also collectively symbolizing the spatial coordinates of each. For independence upon exchange between electrons, the molecular orbitals are expressed by symmetric and antisymmetric linear combinations of the ψ_α and ψ_β products, as follows (see Eq. [5.14]):

$$\left.\begin{aligned} \psi^{(S)}_{molecule}(1,2) = \psi_S(1,2) \equiv \frac{\psi_\alpha(1)\psi_\beta(2) + \psi_\alpha(2)\psi_\beta(1)}{\sqrt{2}} \\ \psi^{(A)}_{molecule}(1,2) = \psi_A(1,2) \equiv \frac{\psi_\alpha(1)\psi_\beta(2) - \psi_\alpha(2)\psi_\beta(1)}{\sqrt{2}} \end{aligned}\right\} . \tag{6.18}$$

Additionally, overlooking the repulsion between the electrons (still not between the nuclei), the ψ_S and ψ_A functions are energy-degenerate, having $E_{\alpha\beta}(r_{ab}) = E_\alpha(r_{ab}) + E_\beta(r_{ab}) + Z^2e^2/r_{ab}$ as the molecular energy. If the interelectrons repulsion is included as a perturbation, and its expectation value added to the molecular energy, we get (see Eq. [5.16])

$$\boxed{E_{\alpha\beta}(r_{ab}) = E_\alpha(r_{ab}) + E_\beta(r_{ab}) + C_{\alpha\beta}(r_{ab}) \pm K_{\alpha\beta}(r_{ab}) + \frac{Z^2e^2}{r_{ab}}} , \tag{6.19}$$

where the Coulomb integral $C_{\alpha\beta}(r_{ab})$ and the exchange integral $K_{\alpha\beta}(r_{ab})$ are given by

$$C_{\alpha\beta}(r_{ab}) = \int d^3\underline{r}_1 d^3\underline{r}_2 \frac{e^2}{|\underline{r}_1 - \underline{r}_2|} |\psi_\alpha(\underline{r}_1)|^2 |\psi_\beta(\underline{r}_2)|^2 ;$$

$$K_{\alpha\beta}(r_{ab}) = \int d^3\underline{r}_1 d^3\underline{r}_2 \frac{e^2}{|\underline{r}_1 - \underline{r}_2|} \psi^*_\alpha(\underline{r}_1)\psi_\beta(\underline{r}_1)\cdot\psi^*_\beta(\underline{r}_2)\psi_\alpha(\underline{r}_2) .$$

The (+) and (-) signs preceding the exchange integral in Eq. (6.19) refer to the symmetric $\psi_S(1,2)$ and antisymmetric $\psi_A(1,2)$ molecular orbitals, respectively. Because the overlap integral is usually positive, the symmetric state energy is higher than the antisymmetric one by twice the exchange integral. In other words, the spatially antisymmetric state is more stable!

Now we include the electronic spin states in the wave functions description. Each independent electron spin projection along the Z axis may be co-oriented (spin-up) or inversely oriented (spin-down) relative to the axis. By considerations of exchange symmetry between the electrons, a resultant spin wave function could be either symmetric or antisymmetric with respect to exchange. Linear combinations of independent spin functions providing such a feature would be (see Eq. [5.17])

$$\left.\begin{array}{ll}\chi_A = \dfrac{\chi_+(1)\chi_-(2) - \chi_+(2)\chi_-(1)}{\sqrt{2}}; & M_S = 0 \\ \chi_S = \begin{cases} \chi_+(1)\chi_+(2); \\ \dfrac{\chi_+(1)\chi_-(2) + \chi_+(2)\chi_-(1)}{\sqrt{2}}; \\ \chi_-(1)\chi_-(2); \end{cases} & \begin{array}{l} M_S = +1 \\ M_S = 0 \\ M_S = -1 \end{array} \end{array}\right\}. \tag{6.20}$$

The M_S quantity indicated to the right of each function is the resultant spin $\underline{S} = \underline{s}_1 + \underline{s}_2$ projection along the Z axis. There is only a single function that is antisymmetric to exchange between the electrons, the one whose resultant spin (and its Z projection) vanish. On the other hand, there are three different symmetric functions belonging to the resultant $S = 1$ spin, which differ by their Z-axis projection. A molecular wave function is a product between an orbital molecular function and a spin function. Consistency with Pauli Exclusion principle requires the resultant wave function to be antisymmetric with respect to electron exchange. Therefore, the only permitted combinations are $\psi_A\chi_S$, providing a triplet, and $\psi_S\chi_A$, providing a singlet. Like in the helium atom, the energy states are classified by two manifolds: a singlet manifold and a triplet manifold. Following Eq. (6.19) we may conclude that triplet states energy are lower than singlet state energy when both belong to the same principal quantum number n of the atomic independent electron. However, the ground state is a singlet, because it belongs to parent molecular orbitals belonging to a principal quantum number $n = 1$, while Pauli exclusion principle requires the two independent electron spins to be inversely oriented (note that atomic $1s$ orbitals permit only molecular σ states). Thus, in the ground state, each contributing molecular state will be $\sigma_g 1s$.

The orbital angular momentum of the molecule must still be characterized. The Z-axis component of the resultant angular momentum is a sum of angular momentum components of the individual independent electrons with respect to that axis. It is readily seen that the possible components of the resultant Z-axis component are $\lambda = |\lambda_1 \pm \lambda_2|$. Symbols now employed to indicate the resultant component states are capital letters (see Eq. [6.16]), namely

$$\boxed{\begin{array}{rccccc} \lambda: & 0, & 1, & 2, & 3, & \cdots \\ State-symbol: & \Sigma, & \Pi, & \Delta, & \Phi, & \cdots \end{array}}. \tag{6.21}$$

The ground state of the hydrogen H_2 molecule satisfies $\lambda_1 = \lambda_2 = 0$; therefore, the only option is $\lambda = 0$, namely a Σ state. The ground-state Term is written as ${}^1\Sigma_g^+$. The Σ symbol is adopted from group theory, indicating that the molecular orbital is symmetric with respect to any rotation about the inter-nuclei axis. The left superscript label presents the spin-state multiplicity $2S + 1$. In the present case $S = 0$. The right subscript label "g" indicates an even parity of the molecular orbital with respect to exchange between electrons. This must be since each of the forming orbitals is of an even parity, and the relevant linear combination is the symmetric one. The right "+" superscript indicates that the molecular orbital is symmetric with respect to reflection about any mirror containing the molecular axis.

The experimental energy scheme of the hydrogen H_2 molecule is presented in Figure 6.7. All state energies relate to the interatomic distance of the molecular ground state. For comparison see Figure 5.3, which shows the helium atom energy scheme. Notable is the triplet $2p\sigma^3\Sigma_u^+$ state, an unstable state that virtually describes two separate hydrogen atoms, each in its ground state in an antibonding molecular orbital. This fact is emphasized by the continuous gray bar up to ionization.

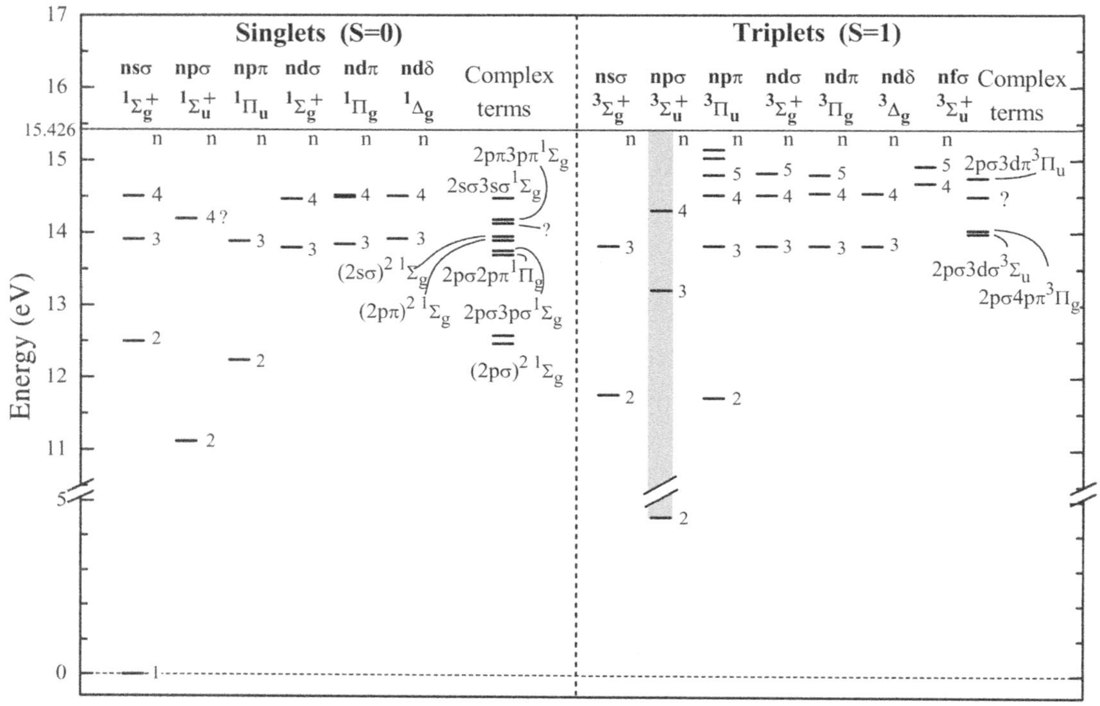

Figure 6.7 Electronic energy scheme in the hydrogen H_2 molecule. The state energies of the complex states relate to simultaneous excitation of both independent electrons. See text for details. Not shown are vibrational and rotational states of the molecule. Detailed discussion is provided in Chapters 7 and 8.

The *dissociation energy* of that state, as described by its location on the electronic energy scale, is 4.476 eV. The ionization energy of a first single ground-state electron is rather 15.426 eV.

Discussion of electronic configuration of several valence electrons in a diatomic molecule requires attention to states' occupation due to the need to conform to the Pauli exclusion principle. A σ-state could occupy up to two equivalent independent electrons, of opposite spin orientation relative to the molecular axis ($m_s = +1/2$ or $m_s = -1/2$). The other, π, δ, ϕ, $\cdots$ states could accommodate up to four such electrons, because each allows only two m_ℓ states – positive and negative (Eq. [6.16]), and each may accommodate two electrons of opposite spins. Correspondingly, each of the π, δ, ϕ, $\cdots$ states may exhibit two projection states of the resultant angular momentum: $J_Z = \lambda \pm \frac{1}{2}$.

Hund rules are applicable in most cases. To be thorough, we present the form of Hund rules relevant to diatomic molecules:

$$\boxed{\begin{array}{l}\textbf{(i) A Term energy is a decreasing function of the multiplicity } 2S+1\textbf{. For equal multiplicities,}\\ \textbf{the energy is a decreasing function of } L_Z.\\ \textbf{(ii) For a specific multiplet, the energy is an increasing function of } J_Z \textbf{ for a less than half-filled}\\ \textbf{sub-shell, and a decreasing function of } J_Z \textbf{ for a more than half-filled subshell.}\end{array}} \quad . \tag{6.22}$$

Like electrons in an atom, the σ, π, δ, ϕ, $\cdots$ states form subshells of close-energy Terms (Sub-Section 5.4.1.2). Finding of possible Terms per each electronic configuration is made by a procedure like that used for many-electron atoms.

Table 6.2 summarizes data on electronic configuration and ground-state Terms of several homonuclear diatomic molecules, their bond length, and dissociation energy. Following previous discussions, the hydrogen molecule occupies two electrons, forming a $(\sigma_g 1s)^2$ configuration. Because of the Pauli exclusion principle, the two independent electrons have opposite spins, a property demonstrated in Table 6.2 by the two oppositely oriented arrows. Because both electrons may occupy bonding $\sigma_g 1s$ orbitals, a stable molecule is formed, with a 0.74 Å bond length. It is still possible to form a state of two co-oriented spins belonging to a $(\sigma_g 1s)(\sigma_u^* 1s)$ configuration, namely, one of the electrons occupies an antibonding orbital. That configuration is unstable.

We consider now the He_2^+ molecule having three orbiting electrons. Two electrons occupy $\sigma_g 1s$ bonding orbitals, and a third one in an antibonding $\sigma_u^* 1s$ orbital. The total configuration is thus $(\sigma_g 1s)^2 \sigma_u^* 1s$. The result is a stable molecule of 3.1 eV dissociation energy. In comparison, the H_2^- negative hydrogen ion (not listed in Table 6.2) also contains three orbiting electrons, and also forms a stable molecule; however, because the summed charge on the participating nuclei is smaller, the Coulomb attraction potential is smaller. Therefore, the dissociation energy is quite tiny.

There are four electrons orbiting the He_2 helium molecule. In the ground state, two electrons occupy $\sigma_g 1s$ bonding orbitals, and two others occupy $\sigma_u^* 1s$ antibonding orbitals; the summed configuration is $(\sigma_g 1s)^2(\sigma_u^* 1s)^2$. Because half of the electrons occupy antibonding orbitals, the result is an unstable

Table 6.2 Electronic configuration, ground-state (g.s.) Term, bond length, and dissociation energy of various homonuclear diatomic molecules.

Z	Molecule	Electronic configuration								g.s. Term	Bond length (Å)	Dissoc. energy (eV)
		$\sigma_g 1s$	$\sigma_u^* 1s$	$\sigma_g 2s$	$\sigma_u^* 2s$	$\pi_u 2p$	$\sigma_g 2p$	$\pi_g^* 2p$	$\pi_u^* 2p$			
1	H_2^+	↑								$^2\Sigma_g$	1.06	2.786
1	H_2	↑↓								$^1\Sigma_g$	0.742	4.476
2	He_2^+	↑↓	↑							$^2\Sigma_u$	1.08	3.1
2	He_2	↑↓	↑↓							$^1\Sigma_g$	Unstable	
3	Li_2	↑↓	↑↓	↑↓						$^1\Sigma_g$	2.672	1.03
4	Be_2	↑↓	↑↓	↑↓	↑↓					$^1\Sigma_g$	Unstable	
5	B_2	↑↓	↑↓	↑↓	↑↓	↑↑				$^3\Sigma_g$	1.589	3.0
6	C_2	↑↓	↑↓	↑↓	↑↓	↑↓↑	↑			$^3\Pi_u$	1.31	5.9
7	N_2	↑↓	↑↓	↑↓	↑↓	↑↑↓↓	↑↓			$^1\Sigma_g$	1.09	9.756
8	O_2	↑↓	↑↓	↑↓	↑↓	↑↑↓↓	↑↓	↑↑		$^3\Sigma_g$	1.207	5.08
9	F_2	↑↓	↑↓	↑↓	↑↓	↑↑↓↓	↑↓	↑↑↓↓		$^1\Sigma_g$	1.435	1.60
10	Ne_2	↑↓	↑↓	↑↓	↑↓	↑↑↓↓	↑↓	↑↑↓↓	↑↓	$^1\Sigma_g$	Unstable	

molecule. This is why the helium gas is monatomic. However, it is possible to form a stable molecule by exciting one antibonding electron off its $\sigma_u^* 1s$ orbital into a $\sigma_g 2s$ bonding orbital. The summed configuration is then $(\sigma_g 1s))^2(\sigma_u^* 1s)(\sigma_g 2s)$.

On further consideration of Table 6.2, some systematic properties of the listed molecules become apparent. Generally, stabilization of molecular bonding is obtained when two electrons of opposite spins occupy bonding orbitals; having the opposite spins allows the two bonding electrons to spend more time between the nuclei. This is only an approximate "rule", because for example, in the Boron B_2 and the oxygen O_2 molecules, the "last" two electrons occupy π orbitals, and their spins are rather parallel. This is because π orbitals can accommodate up to four electrons. As with atomic orbitals (Section 5.3), parallel spin states exhibit a stabilizing effect, because the electrons may not come close, and their electrostatic repulsion reduces. That effect is also expressed by Hund rules. The latter contribution is dominant in forming the ground states. The resultant spin functions of the boron B_2 and oxygen O_2 molecules is thus $S = 1$. Correspondingly, the oxygen molecule exhibits a permanent magnetic dipole. Therefore, the oxygen gas is paramagnetic, whereas most other diatomic molecules are diamagnetic♣. It is also reasonably expected that molecular structure stability, as expressed by the dissociation energy, depends on the difference between the numbers of bonding and antibonding orbiting electrons. For example, the nitrogen N_2 molecule exhibits five pairs of bonding electrons, but only two pairs of antibonding ones, and its dissociation energy is high: 9.756 eV. The oxygen O_2 molecule exhibits five pairs of bonding electrons and three pairs of antibonding ones; its dissociation energy is smaller: 5.08 eV. The fluorine F_2 molecule exhibits five pairs of bonding electrons against four pairs of antibonding ones; its dissociation energy is only 1.6 eV. The neon Ne_2 molecule exhibits five pairs of bonding electrons against five pairs of antibonding ones, and it is unstable!

A bond generated by an electronic orbital that is symmetric with respect to neighboring nuclei is described as **covalent**. For molecular orbitals written as linear combinations of atomic orbitals (LCAO method), it means that the coefficients of all atomic orbitals forming the molecular orbitals are equal. Particularly, all bonds of the homonuclear diatomic molecules are covalent.

The accuracy of numerical dissociation data in Table 6.2 should be addressed with some caution. If measured by different methods, or by employing different interpretations to same measurements, slightly different values are deduced. Term assignment to measured energy states (as in the hydrogen molecule electronic states presented in Figure 6.7) is also sometimes questionable. The uncertainties increase with the number of participating electrons. Fundamentally, the Term concept is only a simplifying approximation! Measured interatomic distances, however, are considered very accurate, and deviations among independent measurements are extremely small.

6.5 Many-Electron Heteronuclear Diatomic Molecules

In this section, we consider molecules formed by a pair of different atoms, like hydrogen chloride HCl, carbon monoxide CO, and sodium chloride $NaCl$. Like in the homonuclear diatomic molecules, the Z-axis is defined by the line connecting the two nuclei. Because the electrons Coulomb attractions towards the nuclei are different, the molecule is nonsymmetric, and it is impossible to classify molecular orbitals as even "g" or odd "u". It is still possible to point out symmetries related to orbital rotations about the Z-axis, or to mirror reflections in a plane that contains the Z-axis. Thus, the molecular orbitals are classified and labeled as $\sigma, \pi, \delta, \cdots$.

♣ Except for similar reminders, this book doesn't deal with magnetic properties of matter.

The energy states and assigned Terms of molecules comprising different atoms of close atomic numbers are similar when the summed atomic numbers are identical. For example, the carbon monoxide CO molecule of $Z_C = 6$ and $Z_O = 8$ exhibits similarity to the nitrogen N_2 molecule of $Z_N = 7$ (Table 6.2). The carbon monoxide CO molecule has five pairs of bonding electrons and two pairs of antibonding ones, exactly like the nitrogen N_2 molecule. The dissociation energies are similar: 9.141 eV compared to 9.756 eV, respectively. In retrospect, it is quite justified to expect such similarity, because both systems contain identical numbers of positive and negative charges (although differently distributed in space).

An extreme case of heteronuclear diatomic molecules is the **ionic bonding**. Experiment and qualitative chemical considerations show that such molecules exhibit a permanent edge of a positive charge, against an opposite edge of a negative charge. An electron may leave an atom and turn free by a small investment of energy if it leaves behind a filled shell or a subshell. A free electron may be captured by an atom and gain some energy♣ if by joining it completes a shell or a subshell. For sufficiently close atoms, a transition of an electron from a former-type atom to a latter-type one may occur by investing the difference between the energies. The two ions thus formed will experience a simple mutual Coulomb attraction and move closer, initially compensating for the invested energy. Additional energy is emitted while nearing, until the electronic orbital exchange introduces sufficient repulsion (see Eq. [6.8] and related text).

For demonstration, we consider the sodium chloride $NaCl$ molecule. The sodium Na^+ ion electronic configuration is identical to that of the noble neon Ne gas. The energy invested for its creation (the ionization energy of the valence electron in the neutral sodium) is 5.138 eV (Table 5.5). The chlorine Cl^- negative ion electronic configuration is identical with that of the noble argon Ar gas. Its affinity is 3.61 eV. The energy invested upon forming a Na^+ and Cl^- pair at the large distance limit is $(5.138 - 3.61)$ eV $= 1.528$ eV. On nearing, the invested energy is compensated by the electrostatic $-e^2/r$ attraction, r being the interionic distance. In sodium chloride, compensation is achieved already at $r = 10.15$ Å. Further adiabatic nearing (kinetic energy given up) brings about the formation of molecular orbitals. The bond contains equal numbers of bonding and antibonding orbitals. At a large distance, their contribution to bonding is negligible (like between neon and argon atoms). However, increasing orbital overlap generates repulsive exchange energy. Eventually, minimum potential is obtained at a certain distance, and the molecule is stabilized at that bond length: $r_0 = 2.51$ Å for sodium chloride $NaCl$.

An ionic bonding of a molecule expresses itself by a steady dipole moment. Table 6.3 provides comparative data for the dipole moments of various diatomic molecules together with other relevant characteristics. The table is arranged by increasing order of the electric dipole moment. The latter vanishes in homonuclear diatomic molecules, where the bonds are entirely covalent. It is small in molecules of similar atoms and large in molecules whose constituent atoms belong to diametrically opposite left and right ends of the periodic table of the elements (alkali metals versus nonmetallic halides). As to the sodium chloride $NaCl$ molecule (Table salt): if the additional electron on the chlorine ion resided there permanently, the generated dipole $e \cdot r_0$ would be 12×10^{-18}esu · cm. The experimental value (Table 6.3) is only 71% of the former. Thus, the **bond ionicity** is determined as 71%; correspondingly, the **bond covalency** is determined as 29%.

A notable phenomenon is the low ionicity of hydrogen halide bonds like in hydrogen iodide HI and hydrogen chloride HCl molecules. It may have been expected that the hydrogen atom is chemically similar to alkali metals by having a single valence electron. However, the hydrogen H ionization energy is very high – 13.6 eV, compared, for example, to that of sodium, which is only 5.138 eV (see above text). The energy required to invest in forming a couple of ions comprising a hydrogen H^+ ion and a chlorine Cl^- ion

♣ In fact, equal to the ionization energy of the *negative ion*. It is called **chemical affinity**.

Table 6.3 Comparison among dissociation energies, bond lengths, and permanent electric dipole of various diatomic molecules that are bonded at different degrees of ionicity and covalency.

Mostly covalent bonds				Mostly ionic bonds			
Molecule	dissociation energy (eV)	Bond length (Å)	Electric dipole (10^{-18} esu·cm)	Molecule	dissociation energy (eV)	Bond length (Å)	Electric dipole (10^{-18} esu·cm)
H_2	4.48	0.74	0	HCl	4.43	1.27	1.07
Li_2	1.03	2.67	0	KBr	3.96	2.94	1.29
O_2	5.08	1.21	0	LiH	2.5	1.6	5.88
N_2	7.37	1.09	0	KCl	4.92	2.79	8.0
Cl_2	2.47	1.99	0	$NaCl$	3.58	2.51	8.5
CO	11.11	1.13	0.12	KF	5.9	2.55	8.60
NO	5.3	1.15	0.15	KI	3.0	3.23	9.24
HI	3.06	1.61	0.38	$CsCl$	3.76	3.06	9.97

at large distance is (13.6 − 3.72) eV = 9.88 eV. The ions will have to come very close to compensate for that invested energy, thus forming quite symmetric, largely covalent molecular orbitals (small electric dipole). An interesting contrast to that case is demonstrated by the lithium hydride LiH molecule. This molecule is highly ionic, of 76% ionicity, yet the hydrogen edge is the *negative* ion! This is because lithium ionization energy is only 5.39 eV compared to 13.6 eV of hydrogen (Table 5.5), whereas the hydrogen affinity is 0.749 eV.

On a theoretical aspect it should be noted, that usually, molecular orbitals calculated based on separate ionic orbitals, yield potential and state energies that are different than those calculated based on the originally neutral atomic orbitals. The latter approach would provide satisfactory results if the employed LCAO method is modified by selecting coefficients that favor a long time-spending probability for the valence electron forming the negative ion. We won't elaborate further on that issue.

6.6 Multiatomic Molecules

Experiments indicate that multiatomic molecules exhibit various angles between their bonds. Simplifying approximations are needed to theoretically describe such structures. In certain cases, an independent atomic electron is assumed to participate in bonding with only the closest nuclei. Molecules that are well described that way are called **nonconjugated**. A diametrically opposite assumption is used for a family of molecules called **conjugated**, for which all valence electrons are assumed to participate equally in orbiting about all atoms. (The same assumption is also used to describe many macroscopic crystals). It is often further assumed that an atomic independent electron participating in forming a molecular bond is initially a **hybrid** of several atomic orbitals of close energy. (Orbitals hybridization is their linear combination; this issue is discussed later in more detail).

6.6.1 Nonconjugated Molecules

The former discussions indicate that presentation of bonding among molecular atoms is more accurate when overlap exists among the separate atomic orbitals. A symmetric molecular orbital increases the probability of the independent electron to spend time between the nuclei, and thus contributes to the assessment of the real bond strength. Molecular bonds will thus form in orientations of higher electronic overlap between the electronic orbitals of adjacent atoms. To demonstrate the above, we consider two simple examples: the water H_2O molecule and the ammonia NH_3 molecule. For bonding, the water H_2O molecule has two nonconjugated $2p$ electrons of the oxygen L shell (Figures 5.4 and 5.5), each bonding the oxygen to a different hydrogen atom, and $1s$ electron from each hydrogen atom bonding it to the oxygen. Because the bonds are oriented, use is made of the angular part of the atomic orbitals described by the angular $\mathcal{Y}_{\ell|m|}(\theta, \varphi)$ functions of Table 6.1, and not by the $\boldsymbol{Y}_{\ell m}(\theta, \varphi)$ spherical harmonics. The molecular orbital of an electron connecting between the oxygen and a "first" hydrogen will be a linear combination of an oxygenic $2p_x$ orbital and a $1s$ orbital of the "first" hydrogen, oriented in the X-direction, thereby increasing the overlap. There will be two such electrons: one contributed by the oxygen, the other by the "first" hydrogen, with opposite spins. A molecular orbital connecting the oxygen to a "second" hydrogen will be a linear combination of an oxygenic $2p_y$ orbital and a $1s$ orbital of the "second" hydrogen, oriented in the Y-direction to increase the overlap. There will also be two such electrons: one contributed by the oxygen, the other by the "second" hydrogen with an opposite spin. The above suggests that the two oxygen-hydrogen bonds will be equally long, and separately oriented along the X and Y axes. The molecular shape is thus expected to display a right (90°) angle, with oxygen O at its vertex, as shown in Figure 6.8. Notably, we have disregarded possible bonding between the hydrogen atoms, making the present approach of only a zero-order nature approximation. The angle between the $O - H$ bonds is actually 104.5°, not too much larger than 90°. It is qualitatively interpreted as due to enhanced electron probability between the two $O - H$ bonds. It causes some exposure of the hydrogen nuclei, entailing electrostatic repulsion, hence increase of the interbond angle.

In relation to Figure 6.8 and similar ones to be shown, the meticulous reader is reminded that the angular electronic probabilities are described by the distance of the presented body surface from the corresponding nucleus. The axis units of that quantity are steradian^{-1}. The hydrogen nuclei in Figure 6.8 have been placed such that the angular probability density surface for a hydrogenic electron is tangent to the lobe surface describing the angular probability density of an oxygenic electron. A Cartesian system describing internuclear distances has obviously length units, and there is no straightforward conversion relation between the described lobe lengths of Figure 6.8 and the internuclear distances. The latter depend on the bonding type, as described for example in Section 6.2 (specifically Figures 6.3 and 6.4, and the related text).

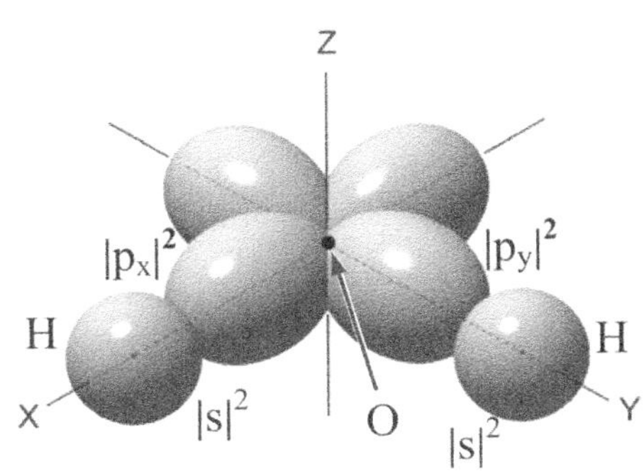

Figure 6.8 Angular part of the electronic probability distribution in the oxygen and hydrogen atoms that form the bonds in the water H_2O molecule.

Another result of the increased electronic probability between the oxygen-hydrogen bonds is that the electronic mass center deviates from the nuclei mass center. Detailed experiment-supported calculations show that it tends towards the oxygen. A permanent

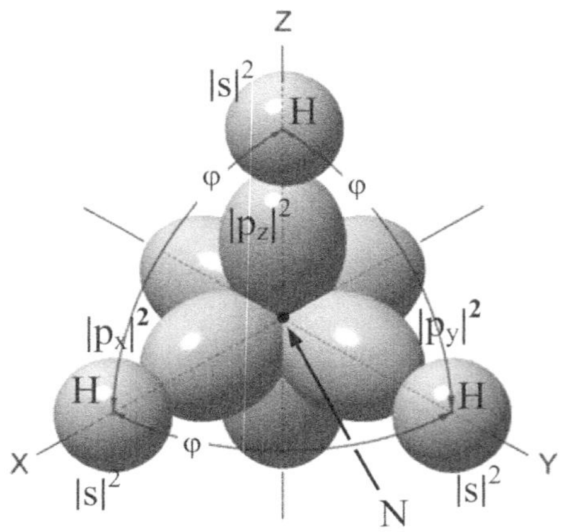

Figure 6.9 Angular part of the electronic probability distribution in the nitrogen N and hydrogen H atoms that form the bonds in the ammonia NH_3 molecule. The φ angle described as a right angle in the figure changes into 107° because of the molecular bonding nature.

dipole is created, oriented parallel to the bisector of the two $O-H$ bonds angle. Its experimental value is 1.86×10^{-18} esu·cm. Its presence supports the conclusion that the water molecule is not linear (atoms not arranged along a straight line); by simple symmetry considerations, a linear molecule could not exhibit a permanent electric dipole.

We turn to describe the ammonia molecule. The nitrogen N has three uncoupled $2p$ electrons (Figure 5.5) belonging to three separate orbitals $2p_x$, $2p_y$, and $2p_z$. A chemical bonding would be formed with hydrogen atoms situated along the X, Y, and Z directions, respectively. The ammonia molecule would thus have a pyramidal shape with the nitrogen at one of its vertices and the hydrogen atoms at the others, as shown in Figure 6.9. The angles among the nitrogen/hydrogen bonds are 107.3°, instead of the 90° suggested by the previous simplistic discussion; it is a result of the electrostatic repulsion among the hydrogen atoms. Much like for the water molecule, this molecule carries a permanent electric dipole, whose value is 1.5×10^{-18} esu·cm, oriented along the pyramid axis.

Understanding the spatial structure of carbon compounds requires a different approach, because the ground state of this atom has only two uncoupled $2p$ electrons, whereas it tends to form three or four bonds with neighboring atoms, like in the methane CH_4 molecule. It is interpreted as due to close carbon excited states that include one $2s$ electron and three $2p$ electrons (Figures 5.6 and 5.14). These are the electrons that form the chemical bonds. Proximity of neighboring atoms induces perturbation, which mixes the orbitals. A first-order approximation to the mixed orbitals is a linear combination of the original orbitals, called **hybridization**; an atomic orbital thus formed is called **hybrid wave function**. Hybridization is in fact an intermediate step in forming a linear combination of atomic orbitals involving different atoms (LCAO), aimed at eventually forming a molecular orbital. The different hybrid wave functions exhibit prominent lobes along relative spatial orientations that are not necessarily perpendicular. Eq. (6.23-i) provides the hybridization of a $2s$ orbital with $2p_x$, $2p_y$, and $2p_z$ orbitals, forming prominent lobes along tetrahedral orientations exhibiting 109°28' angle between each pair. This is called sp^3 hybridization. The mathematical expressions for the orbitals belonging to the sp^3 hybridization are

$$\left.\begin{aligned}
[sp^3]_1 &= \frac{1}{2}\left(s + p_x + p_y + p_z\right)\\
[sp^3]_2 &= \frac{1}{2}\left(s + p_x - p_y - p_z\right)\\
[sp^3]_3 &= \frac{1}{2}\left(s - p_x + p_y - p_z\right)\\
[sp^3]_4 &= \frac{1}{2}\left(s - p_x - p_y + p_z\right)
\end{aligned}\right\} . \tag{6.23-i}$$

A polar spatial image comparing among the original orbitals and their hybrid ones is provided in Figure 6.10. Each prominent lobe of a hybrid wave function is accompanied by a secondary lobe, almost four times shorter, pointing in the opposite direction. The four sp^3 wave functions are orthonormal, and

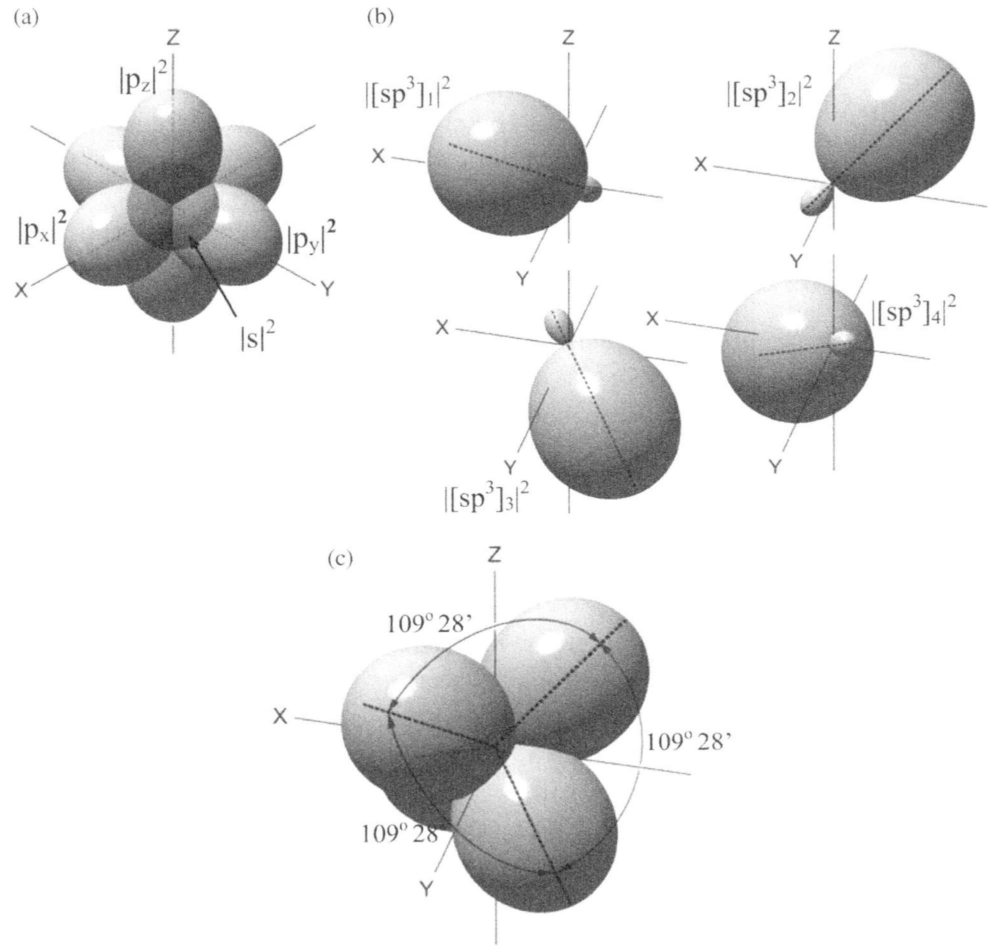

Figure 6.10 (a) Polar description of the angular functions forming the basis for sp^3 hybridizations, plotted on a common axes system. (b) The four obtained hybrid orbitals; note the axes rotations compared to (a); it is done to allow noticing the secondary lobes in each hybrid orbital. (c) The four hybrid orbitals of (b) plotted on a common Cartesian axis system; the tetrahedral structure formation is obvious. To prevent data overcrowding, the secondary lobes were omitted.

span the same functions space spanned by $2s$, $2p_x$, $2p_y$, and $2p_z$. However, it is important to note that because the s and p functions belong to different orbital angular momentums, an electron orbiting in an sp^3 hybrid wave function has no specific orbital angular momentum.

The methane CH_4 molecule is formed when hydrogen H atoms bond to the carbon C in the orientations of the sp^3 hybrid wave functions: a $1s$ electron of each hydrogen bonds with an electron belonging to one of the hybrid orbitals (prominent lobes). The result is the tetrahedral structure of the methane molecule, whose schematic image is provided in Figure 6.11.

One of the carbon electrons in a hybrid sp^3 orbital can bond to a corresponding electron of another carbon atom in a form similar to a σ bond between two hydrogen atoms (Figure 6.3 and corresponding text). If the other orbitals bond with oxygen atoms, they form an ethane $H_3C - CH_3$ molecule. Its forming electronic orbitals image is presented in Figure 6.12. Here as well, the secondary lobes of the sp^3 hybrid

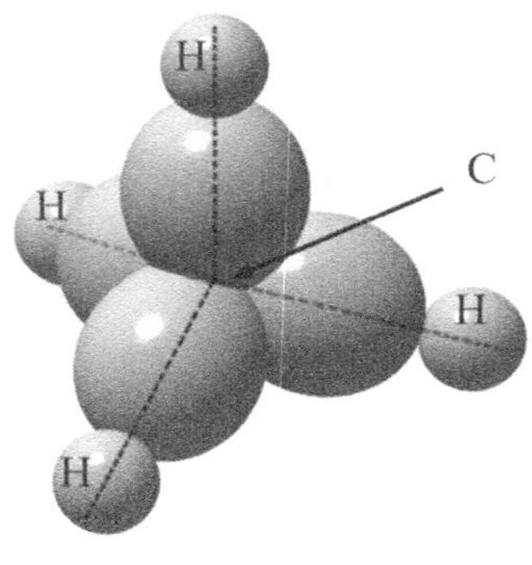

Figure 6.11 Angular part of the electronic probability distributions in the carbon *C* and hydrogen *H* atoms that form the bonds in the methane CH_4 molecule. The secondary lobes in the carbon sp^3 orbitals were omitted to prevent data overcrowding. The angles among the bonds is 109°28', as expected for a regular tetrahedral shape.

wave functions have been omitted to avoid data overcrowding. It is worth noting that the molecule lacks simple mirror symmetry relative to a plane perpendicular to the axis connecting the two carbon atoms, bisecting the axis in the middle. In other words, the sets of hydrogen atoms bonded to the different carbons are not forming mirror-like images. The cause is electrostatic repulsion among the hydrogen atoms, whose positive charges are exposed after being bonded to the carbon. On the other hand, the molecule exhibits symmetry in relation to two successive operations: reflection by the plane, then rotation of the reflected image by 60° about the intercarbon axis. Chapter 8 is devoted to a systematic classification of symmetry operations on molecules.

Hybrid sp^3 orbitals are not unique to carbon, and they appear in chemical bonds forming for other atoms exhibiting electronic configurations that are like that of carbon. For example, the nitrogen cation N^+ has a configuration identical to the carbon's; the occurrence of the chemical N^+H_4 complex appearing as a radical in many chemical compounds is interpreted on that grounds.

It is possible to devise other hybridizations different than sp^3, that are linear combinations of several atomic orbitals belonging to a larger than unity orbital quantum number. In the present frame we address only the sp^2 and sp ones. The sp^2 hybridization is a linear combination of s, p_x, and p_y wave functions that generates three new wave functions that exhibit large lobes in the XY plane, having 120° between their orientations, as shown in Figure 6.13. The mathematical expressions for these wave functions are

$$\left.\begin{aligned}
[sp^2]_1 &= \frac{1}{\sqrt{3}}\left(s+\sqrt{2}\,p_x\right)\\
[sp^2]_2 &= \frac{1}{\sqrt{3}}\left(s-\frac{1}{\sqrt{2}}\,p_x+\sqrt{\frac{3}{2}}\,p_y\right)\\
[sp^2]_3 &= \frac{1}{\sqrt{3}}\left(s-\frac{1}{\sqrt{2}}\,p_x-\sqrt{\frac{3}{2}}\,p_y\right)
\end{aligned}\right\}. \tag{6.23-ii}$$

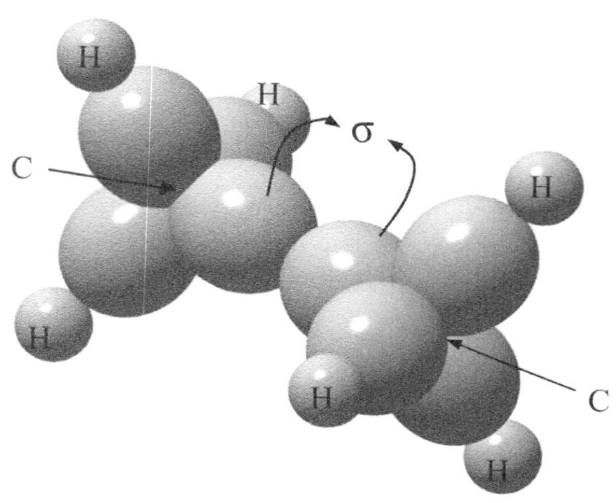

Figure 6.12 Angular part of the electronic probability distributions in the carbon *C* and hydrogen *H* atoms that form the bonds in the ethane C_2H_6. molecule.

The fourth electron occupies an undisturbed p_z orbital. Each prominent lobe exhibits a secondary lobe, about 5.6 times smaller, pointing in the opposite direction. The three sp^2 wave functions are orthonormal, and together with the p_z wave function they span the same functions space spanned by $2s$, $2p_x$, $2p_y$, and $2p_z$.

The sp^2 described hybridization accounts for the formation of molecules like the ethylene $H_2C=CH_2$; the angular spatial probability distribution of the orbitals forming the chemical bonding is provided in

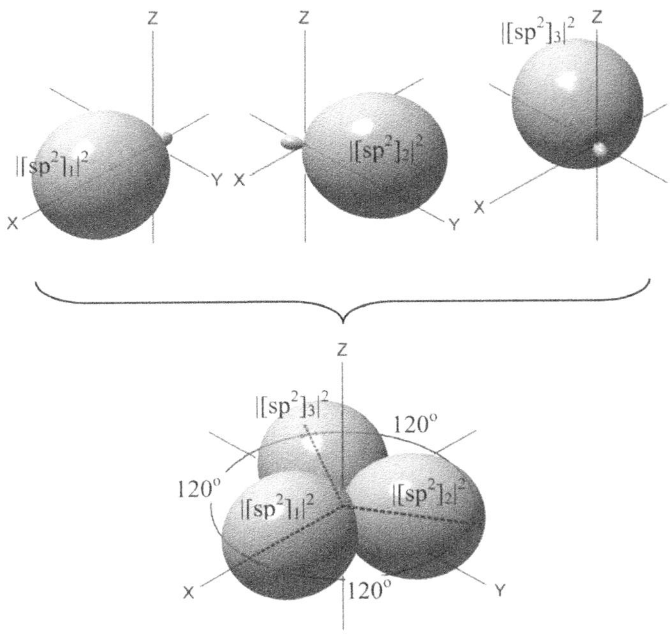

Figure 6.13 Upper part: the three sp^2 hybrid orbitals. Lower part: the prominent sp^2 hybrid orbitals plotted on a common Cartesian axis system, exhibiting a threefold rotational symmetry about the Z axis. The secondary lobes were omitted to prevent data overcrowding.

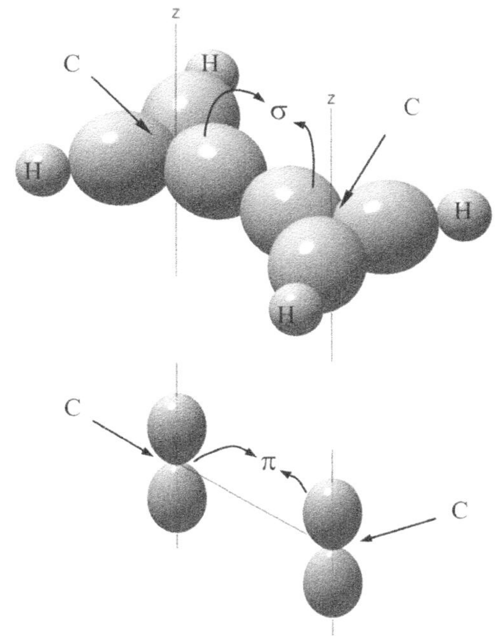

Figure 6.14 Angular part of the electronic probability distributions in the carbon C and hydrogen H atoms that form the bonds in the ethylene C_2H_4 molecule. The distributions, whose origins are the p_z carbon atoms, were shifted down to prevent confusion due to data overcrowding.

Figure 6.14. One bond between the two carbon atoms is caused by overlap of two sp^2 hybrid orbitals, each originating from an opposite carbon along the line connecting the two. The resultant linear combination of these two orbitals is like the symmetric σ bond in the hydrogen H_2 molecule (Section 6.2), and it is indeed addressed as such. A second bond is formed by a linear combination of p_z orbitals of the two carbons, forming a π bond, addressed so because of the similarity with the π orbitals in diatomic molecules. Two electrons, one from each carbon, occupy the σ bond; and two electrons, one from each carbon, occupy the π bond. The hydrogen atoms form bonds through the remaining sp^2 hybrid wave functions of the two carbon atoms. Though the σ bond exhibits a full rotational symmetry about the axis connecting the two carbons, it is not so for the π bond. However, this bond is symmetric with respect to reflection in the XY plane. This property stabilizes the entire molecule in the XY plane, because any twist of the molecule about the C–C axis increases the length of the π orbital that bonds the two carbon atoms, requiring investment of energy. On the other hand, the contribution of π bonds to chemical bonding strength is considerably smaller than the contribution of σ bonds. For example, the dissociation energy of a σ bond between to carbon atoms is about 6.33 eV, compared to only 3.98 eV required for dissociating a $\sigma\pi$ double bond, while leaving a single σ bond.

In a different, sp hybridization, the s and p_z orbitals are used to form two new wave functions by a linear combination. They form prominent lobes in the $\pm Z$ orientations. Mathematically,

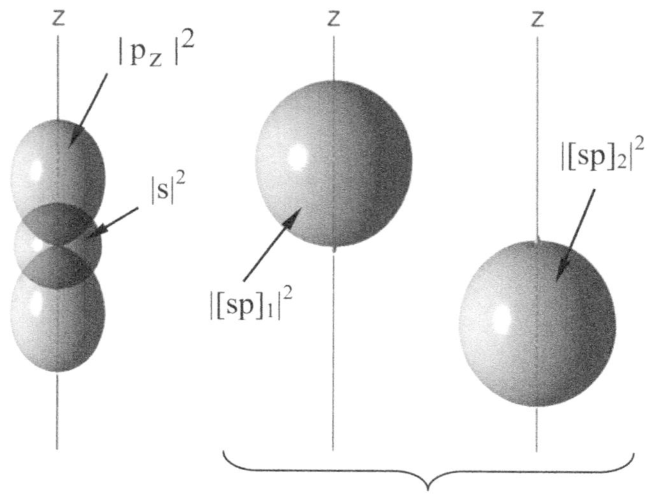

Figure 6.15 Left: angular probability distributions of the two atomic orbitals that form the basis for *sp* hybridization; right: angular probability distributions of the two separate *sp* hybrid orbitals.

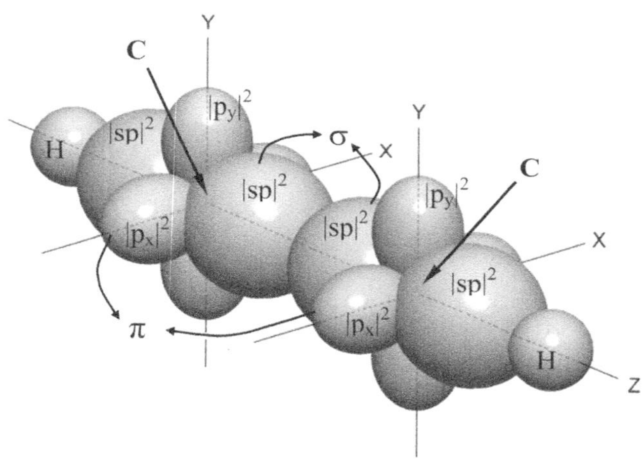

Figure 6.16 Angular part of the electronic probability distribution in the carbon *C* and hydrogen *H* atoms that form the bonds in the acetylene C_2H_2 molecule. Only a part of the bonding types is indicated. All bond types are detailed in the text.

$$[sp]_1 = \frac{1}{\sqrt{2}}(s + p_z); \qquad [sp]_2 = \frac{1}{\sqrt{2}}(s - p_z). \tag{6.24}$$

An angular electron probability distribution image of the *sp* hybrid wave functions is provided in Figure 6.15. Together with each prominent lobe, a tiny, 15 times smaller secondary lobe exists, oriented oppositely. The two *sp* wave functions are orthonormal, and together with p_x and p_y orbitals they span the same functions space spanned by the $2s$, $2p_x$, $2p_y$, and $2p_z$ wave functions. The *sp* hybridization accounts for the appearance of molecular structures like the acetylene $HC \equiv CH$ described in Figure 6.16. A triple bonding between the two carbon atoms involves one overlapping *sp* hybrid orbital of each atom, which forms a σ-bond, and overlapping p_x and p_y orbitals of each atom, forming two π-bonds. Each of the two hydrogen atoms that bond to a different carbon by a σ-bond is generated between its *s*-orbital and the remaining *sp*-orbital of its adjacent carbon. The acetylene $HC \equiv CH$ molecule is thus linear.

6.6.2 Conjugated Molecules

Conjugated molecules are characterized by a spatial structure, whose electronic configuration is better approximated by assuming that some of their valence electrons are equally shared by several atoms. For demonstration, we consider the structure of the butadiene C_4H_6 molecule described in Figure 6.17. Four carbon atoms form a zigzagged linear chain of approximately 120° between adjacent inter-carbon bonds. One component forming the molecular bonding is a σ bond, formed by sp^2 hybrid orbitals of the carbon atoms. The hydrogen atoms also bond to the carbon atoms by σ bonds formed by an *s* hydrogen orbital and an sp^2 hybrid carbon orbital. Notably, each chain-edge carbon atom is left with two such orbitals, and each of the center carbons is left by one; see Figure 6.17a. In contrast to the electrons involved in forming the σ bonds, the π electrons, one contributed by each carbon, as shown in Figure 6.17b, are free to move along the carbon chain. The presence of the π orbitals stabilizes the molecular planar structure, as noted in relation to the ethylene C_2H_4 molecule. As a result of the π electron's free movement along the carbon skeleton chain, the molecular polarizability along the chain is considerably larger than perpendicularly to that chain. The definition of polarizability is provided in Chapter 4; see particularly Eqs. (4.110) and (4.111) and their related text.

We consider now the molecular orbitals generated by the π electrons. The approach is quite like that presented in relation to the construction of the molecular orbitals of the hydrogen molecule ion H_2^+ and the hydrogen molecule H_2 (Section 6.2). We label the p_z orbitals of the chain carbon atoms by C_j, where j is the ordinal number of the carbon C atom along the chain. When the interatomic distances are sufficiently short, their interaction induces molecular orbitals through linear combinations of the atomic orbitals. In the present case it means combinations among four wave functions, instead of just two in the cases of H_2^+ and H_2. Because of the molecular symmetry, the wave functions' combination must be either symmetric or antisymmetric with respect to inversion about the molecule's mass center. There would be four possible combinations:

$$
\begin{aligned}
\varphi_1^{(S)} &= \frac{1}{\sqrt{4+6S}}(\psi_{C_1} + \psi_{C_2} + \psi_{C_3} + \psi_{C_4}) \\
\varphi_2^{(A)} &= \frac{1}{\sqrt{4+2S}}(\psi_{C_1} + \psi_{C_2} - \psi_{C_3} - \psi_{C_4}) \\
\varphi_3^{(S)} &= \frac{1}{\sqrt{4-2S}}(\psi_{C_1} - \psi_{C_2} - \psi_{C_3} + \psi_{C_4}) \\
\varphi_4^{(A)} &= \frac{1}{\sqrt{4-6S}}(\psi_{C_1} - \psi_{C_2} + \psi_{C_3} - \psi_{C_4}),
\end{aligned}
\tag{6.25}
$$

where S is the electronic overlap integral between adjacent atomic orbitals ψ originating in carbon atoms (Section 6.2.2). The (S) and (A) superscripts indicate the symmetry property of the resulted molecular orbital: symmetric or antisymmetric, respectively, to exchange between nuclei relative to the molecular center. The wave functions' normalization is performed under the assumption that overlap between atomic orbitals of distant nuclei may be ignored, and that the chain interatomic distances are equal.

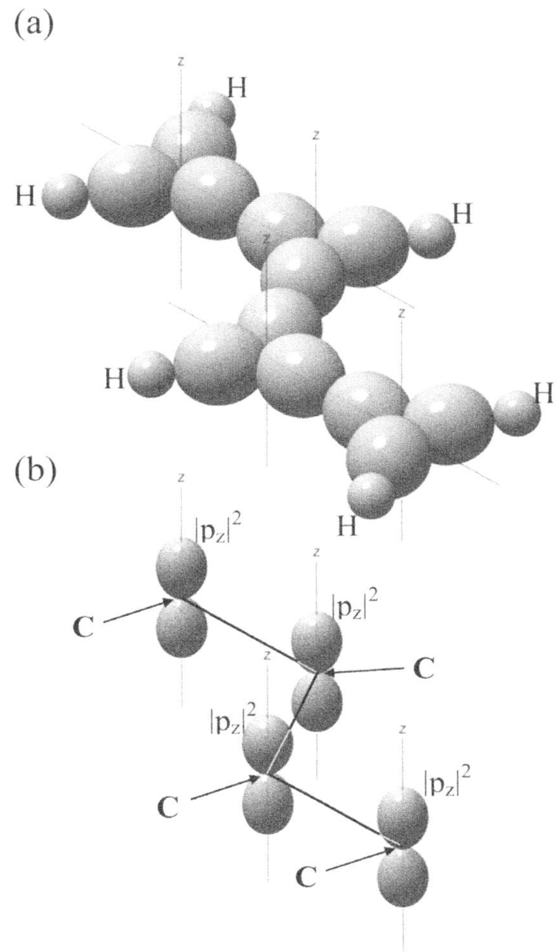

Figure 6.17 Angular part of the electronic probability distribution of the carbon C and hydrogen H atoms that form the bonds in the butadiene C_4H_6 molecule. (a) provides the distributions emanating from sp^2 orbitals of the carbon atoms and the s orbitals of the hydrogen atoms. To prevent data overcrowding, and to emphasize that these orbitals form the π bonds, the probability distributions emanating from the p_z orbitals of the carbon atoms are displaced downwards (b). See text for a more detailed description.

Each of the $\varphi_1^{(S)}$, $\varphi_2^{(A)}$, $\varphi_3^{(S)}$, $\varphi_4^{(A)}$ functions belongs to a different energy of the molecule. To justify that statement, and to qualitatively determine the energies order, we discuss some fundamental properties of these functions. For simplicity, we assume that all carbon atom positions are equally spaced along a straight line. The top line of Figure 6.18 displays the carbon separate wave functions ψ_{C_j} as if they exhibit a 1s form. In carbon, the relevant functions are obviously 2p ones; however, the fundamental consideration is independent of the wave functions' shape details. The presentation is made that way just to simplify the figure plotting. The following lines from top to bottom

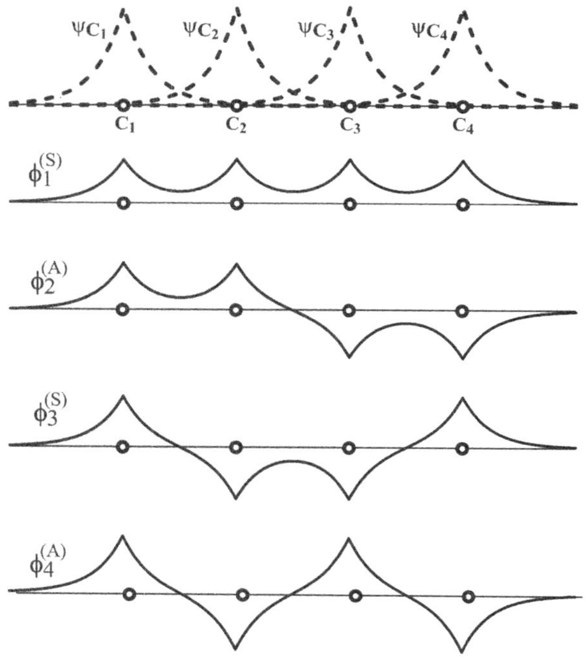

Figure 6.18 Molecular electronic orbitals constructed using atomic electronic 1*s* orbitals on a linear chain of four close, equally spaced atoms. See text for details.

display the molecular wave functions by the order presented in Eq. (6.25), each along the line connecting the carbon nuclei.

It should be noted that the $\varphi_1^{(S)}$ function never attains a zero value, except in the $\pm\infty$ limits. The $\varphi_2^{(A)}$ function exhibits a single zero-point between those limits; the $\varphi_3^{(S)}$ function exhibits two zero-points, and $\varphi_4^{(A)}$ exhibits three zero-points between those limits. We assume that these functions solve quite reasonably the molecular Hamiltonian, which is an inhomogeneous differential equation of the second order. A mathematical theorem states that the eigenvalues of functions that solve such differential equation increase as a function of the number of zero-points. We have already encountered such a characteristic for two important cases. One case concerns the energy of hydrogen electronic levels, which increases with the principal quantum number n (Eq. [3.38]); correspondingly, the number of zeros in the radial wave function $R_{n\ell}(r)$ (Eq. [3.42] and Table 3.3) is $n-1$. The reader may corroborate that characteristic by simply observing the shape of the squared wave functions for small n values provided in Figure 3.6. As a second example, the electronic orbital angular momentum increases with the orbital angular quantum number ℓ (Eq. [4.7]). Correspondingly, the number of zeros in the $\boldsymbol{Y}_{\ell 0}(\theta, \varphi)$ spherical harmonics also equals ℓ (see Eq. [3.35] and Table 3.2). The reader may corroborate that characteristic simply by observing the shapes of the squared spherical harmonics provided in Figure 3.4. It is often qualitatively said that the number of quanta occupied in a physical system is given by the number of zeros appearing in its descriptive wave function. The reader is referred to observe this behavior for the case of the harmonic oscillator, and consider it as a complementing part of exercise 5.3 (b).

In view of the above considerations, we may conclude that the $\varphi_1^{(S)}$ orbital provides the lowest possible energy, and the $\varphi_2^{(A)}$, $\varphi_3^{(S)}$, and $\varphi_4^{(A)}$ orbitals provide increasing state energies, respectively. Thus, energy is split into four states. We could have reached the same conclusion by using qualitative considerations that are similar to those by which we concluded the states' energy order between the $\sigma_g 1s$ and $\sigma_u 1s$ states of the hydrogen molecule ion H_2^+ (namely, consideration involving the presence or absence of electronic screening to the internuclear attraction – see Section 6.2). For comparison, in the latter case, the states in the hydrogen molecule ion H_2^+ split into two, as displayed in Figure 6.4.

Each molecular orbital provides a different curve for the potential energy as a function of the intercarbon atom's distance, as displayed in Figure 6.19. Such curves are obtained by calculating the expectation value of the molecular Hamiltonian, in a procedure like that described in Section 6.2.3. It should be recognized that the internuclear distance, which yields an energy minimum, is different among different molecular orbitals (states); furthermore, the $\varphi_4^{(A)}$ is in fact antibonding. Each state can occupy two electrons of opposite spins. Thus, the ground state of butadiene C_4H_6 contains two electrons occupying the $\varphi_1^{(S)}$ orbital, and two more occupying the $\varphi_2^{(A)}$ orbital. That occupation is indicated schematically on the curves in

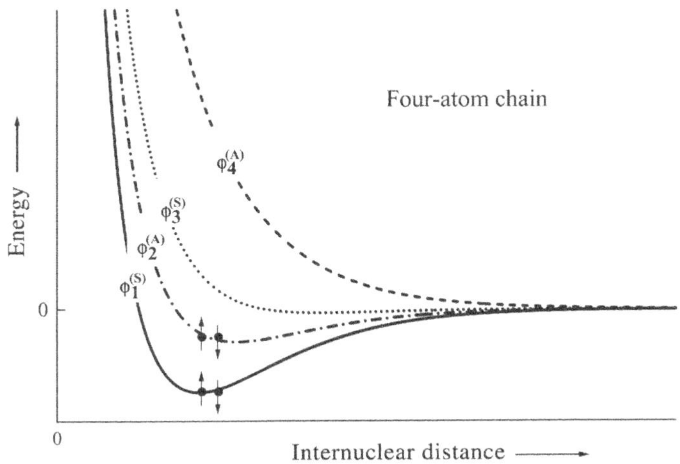

Figure 6.19 Schematic illustration of the potential energy curves as functions of the interatomic distance in molecular orbitals generated by a linear equidistant chain of identical atoms, such as for $2p_z$ carbon C electrons in the butadiene C_4H_6 molecule, which form π molecular orbitals. See text for discussion.

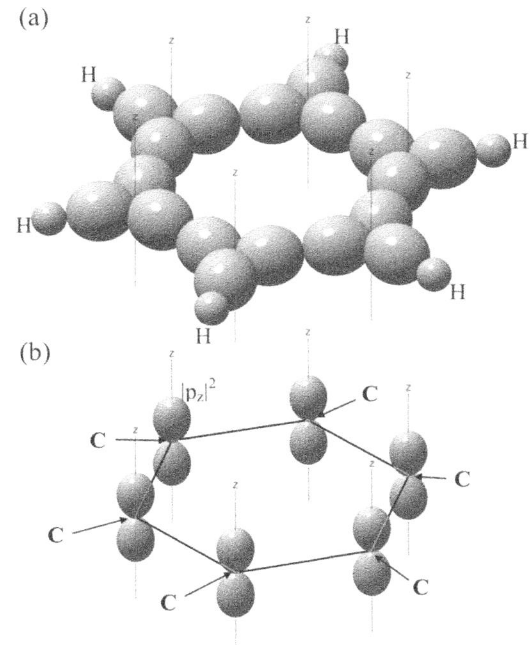

Figure 6.20 Angular part of the electronic probability distribution of the carbon C and hydrogen H atoms that form the bonds in the benzene C_6H_6 molecule. The top part (a) provides the distributions emanating from sp^2 orbitals of the carbon atoms and the s orbitals of the hydrogen atoms. The probability distributions emanating from the p_z orbitals of the carbon atoms were displaced downward (b) to avoid data overcrowding, and emphasize that these orbitals form the π bonds. See text for more-detailed description

Figure 6.19 by the arrows related to the spin-up or spin-down states. Because each orbital exhibits a different internuclear (interatomic) distance that yields a potential energy minimum, the actual distance is the one that provides a minimum of the summed energy of all molecular electrons, not least of which are the σ bonding electrons. Further note that the $\varphi_1^{(S)}$ orbital is bonding between all adjacent carbon pairs. The $\varphi_2^{(A)}$ orbital, on the other hand, is bonding only between adjacent end pairs, and it is antibonding between the center pair. As a result, it is more difficult to separate carbon end pairs than separate between the center pair. Correspondingly, it has been found that the distance between an end carbon and its adjacent one in butadiene is short - 1.35 Å. It is longer, 1.46 Å, between the center chain carbons.

It is possible to extend the previous discussion to include a large family of conjugated carbon molecules, called polyenes, whose skeletons form a chain of $2n$ carbon atoms (n=integer). Schematically, the chain may be described by a $C - C = C - ... - C = C - C$ form, where each horizontal line represents a σ bond between adjacent carbon atoms. Like the butadiene C_4H_6 molecule described in the previous sections, there are $2n$ additional π electrons moving "freely" along the chain (sometimes described graphically as being "smeared" along the chain). Thus, $2n$ energetically close energy states are generated, belonging to different linear combinations of separate wave functions from the separate carbon atoms, being either symmetric or antisymmetric relative to the chain center. Each such state may occupy two electrons of opposite spin states. The entire states system may thus occupy $4n$ electrons. However, in the ground state only half of the states, those of lower energy, are occupied. On the other hand, if the number of π electrons in the molecule would be $4n$, all said states would be occupied. Excitation of the molecule to a higher state requires transition to a different electronic configuration, usually requiring quite a large investment of energy compared to the previous case.

As a further example of a conjugated molecule, we consider the benzene C_6H_6 molecule. The six carbon atoms in this molecule are positioned in the vertices of a regular hexagon, and are bonded by molecular σ

orbitals formed by atomic sp^2 hybrid orbitals, as displayed in Figure 6.20. A hydrogen atom bonds to each carbon through a σ bond with its remaining sp^2 orbital. There are six additional π electrons, each contributed by a different carbon atom, whose origin is the atomic p_z electron (the Z orientation is perpendicular to the molecular plane). These π electrons are shared among all carbon atoms, forming closed current loops. That structure accounts for the typical diamagnetic properties of benzene, as well as of other conjugated circular molecules.

In the present frame we won't elaborate on the issue of molecular electronic configurations. The presented examples demonstrate the major problems and means administered in analysis of the chemical bonding and states energy spectra, and their understanding.

6.7 Appendix: Calculation of an Infinitesimal Volume Element in Elliptic Coordinates

Elliptic coordinates (μ, λ, φ) for a point whose Cartesian coordinates are (x, y, z) are defined by the relations $\mu \equiv (r_1 + r_2)/\mathcal{R}$; $\lambda \equiv (r_1 - r_2)/\mathcal{R}$, and φ is the azimuth angle of a plane containing the Z -axis about that axis in comparison to the XY plane, where r_1 and r_2 are the distances of the (x, y, z) point from the first and second focal points located at $(0, 0, -\mathcal{R}/2)$ and $(0, 0, +\mathcal{R}/2)$, respectively. The μ variable range is $(1,\infty)$, the λ variable range is $(-1, +1)$, and the φ variable range is $(0, 2\pi)$. The definitions of r_1 and r_2 are described schematically in Figure 6.21.

The inverse conversions read

$$
\begin{aligned}
x &= \frac{\mathcal{R}}{2}\sqrt{(\mu^2-1)(1-\lambda^2)}\cdot \boldsymbol{\cos}\ \varphi \\
y &= \frac{\mathcal{R}}{2}\sqrt{(\mu^2-1)(1-\lambda^2)}\cdot \boldsymbol{\sin}\ \varphi \\
z &= \frac{\mathcal{R}}{2}\mu\lambda.
\end{aligned}
\qquad (6.26)
$$

For static focuses, the condition $\mu = \boldsymbol{const}.$ defines an ellipse on any plane containing the Z- axis; the condition $\lambda = \boldsymbol{const}.$ defines a hyperbola belonging to the same focal points. For example, the ellipse and hyperbola equations in the XY plane would be given by

$$
\begin{aligned}
&\frac{z^2}{\mu^2} + \frac{y^2}{\mu^2-1} = \left(\frac{\mathcal{R}}{2}\right)^2 \quad ; \quad \textit{Ellipse for} \quad 1 < \mu < \infty \\
&\frac{z^2}{\lambda^2-1} + \frac{y^2}{\lambda^2} = -\left(\frac{\mathcal{R}}{2}\right)^2 ; \quad \textit{Hyperbola for} \quad -1 < \lambda < +1 .
\end{aligned}
\qquad (6.27)
$$

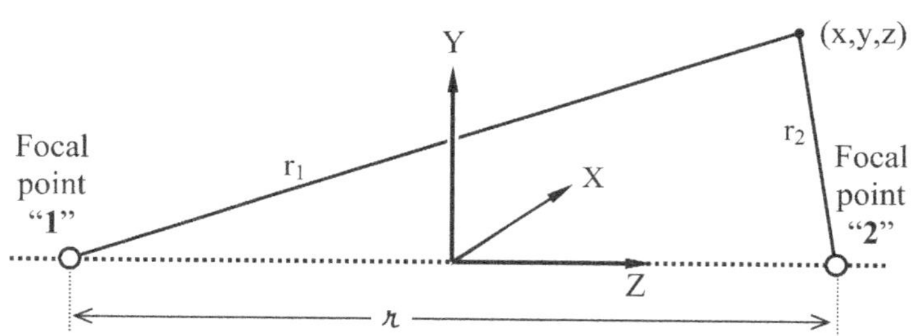

Figure 6.21 Schematic illustration aimed at clarifying the conversion of Cartesian (x, y, z) coordinates into elliptic (μ, λ, φ) coordinates.

To avoid double definition, it is accepted that the hyperbolic branch belonging to the right half space $(z > 0)$ will be attributed to positive λ values (Figure 6.22), and its symmetric branch belonging to the left half space $(z < 0)$ will be attributed to negative λ values. The tangents to the ellipse and hyperbola defined by Eq. (6.27) are mutually perpendicular at the intersection point between the two curves. As a result, an infinitesimal area dA enclosed between two adjacent ellipses defined by μ and

$\mu + d\mu$, and two adjacent hyperbolas defined by λ and $\lambda + d\lambda$, is given by a product between the two arc lengths ds_μ and ds_λ defining that area, as described in Figure 6.22. The said arc lengths are given by $ds_\mu = h_\mu d\mu$ and $ds_\lambda = h_\lambda d\lambda$, where generally, the quantities h_μ and h_λ are functions of the intersection point coordinates. As a result of the tangents' perpendicularity, the diagonal length ds of the grey shaded area (dA hereafter), satisfies

$$(ds)^2 = ds_\mu{}^2 + ds_\lambda{}^2 = h_\mu{}^2 d\mu^2 + h_\lambda{}^2 d\lambda^2 . \quad (6.28)$$

We denote by (z, y) the intersection coordinates of the μ ellipse with the λ hyperbola, and by $(z + dz, y + dy)$ the intersection coordinates of the $\mu + d\mu$ ellipse with the $\lambda + d\lambda$ hyperbola. Then

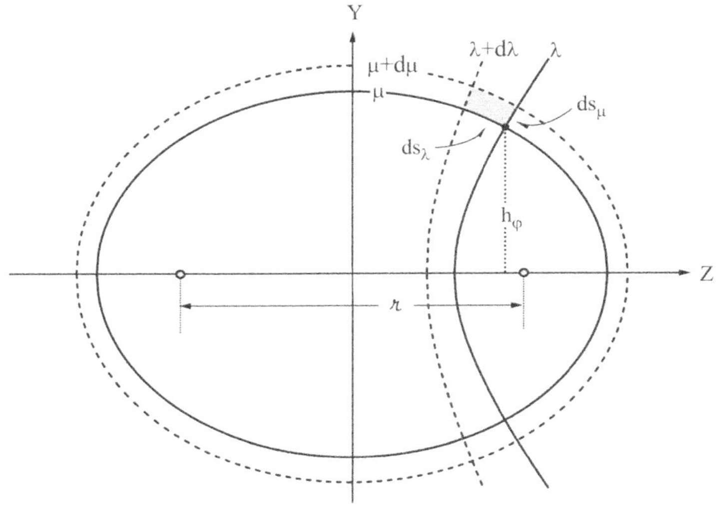

Figure 6.22 Schematic illustration aimed at clarifying the calculation of an infinitesimal volume in elliptic coordinates. The gray-colored zone indicates an infinitesimal area generated between two infinitesimally close parabolas and two infinitesimally close hyperbolas of identical focal points.

$$(ds)^2 = (dz)^2 + (dy)^2 = \left(\frac{\partial z}{\partial \mu} d\mu + \frac{\partial z}{\partial \lambda} d\lambda\right)^2 + \left(\frac{\partial y}{\partial \mu} d\mu + \frac{\partial y}{\partial \lambda} d\lambda\right)^2 . \quad (6.29)$$

By expanding, compiling terms, and comparing the multipliers of $d\mu^2$ and $d\lambda^2$ with Eq. (6.28), we get

$$h_\mu{}^2 = \left(\frac{\partial z}{\partial \mu}\right)^2 + \left(\frac{\partial y}{\partial \mu}\right)^2 = \left(\frac{\varkappa}{2}\right)^2 \frac{\mu^2 - \lambda^2}{\mu^2 - 1};$$

$$h_\lambda{}^2 = \left(\frac{\partial z}{\partial \lambda}\right)^2 + \left(\frac{\partial y}{\partial \lambda}\right)^2 = \left(\frac{\varkappa}{2}\right)^2 \frac{\mu^2 - \lambda^2}{1 - \lambda^2} .$$

We thus obtain

$$dA = h_\mu h_\lambda d\mu \, d\lambda = \left(\frac{\varkappa}{2}\right)^2 \frac{\mu^2 - \lambda^2}{\sqrt{(\mu^2 - 1)(1 - \lambda^2)}} d\mu \, d\lambda . \quad (6.30)$$

An infinitesimal $ds_\varphi = h_\varphi \cdot d\varphi = (\varkappa/2)\sqrt{(\mu^2 - 1)(1 - \lambda^2)}$ length element is obtained on performing an infinitesimal rotation $d\varphi$, which is perpendicular to the dA plane area. The infinitesimal volume element obtained by multiplication would be

$$\boxed{d\tau = \frac{\varkappa^3}{8}\left(\mu^2 - \lambda^2\right) d\mu \cdot d\lambda \cdot d\varphi} \quad . \quad (6.31)$$

This expression is the one quoted at the above Eq. (6.12).

Exercises for Chapter 6

6.1 The following process equation expresses the dissociation of the hydrogen molecule cation H_2^+ into its components: two hydrogen cations H^+ and one free electron e, all at rest separated by a long distance:

$$H_2^+ \xrightarrow[+\,2.8\ \text{eV}]{} H^+ + H \xrightarrow[+\,13.6\ \text{eV}]{} H^+ + H^+ + e\,. \tag{6.32}$$

Energy data for the consecutively involved stages are provided in Tables 6.2 and 5.5, respectively. The summed process thus reads

$$H_2^+ \xrightarrow[16.4\ \text{eV}]{} H^+ + H^+ + e\,. \tag{6.33}$$

The H_2^+ cation dissociation energy is therefore calculated as 16.4 eV.

(a) Using a calculation method like the above, determine the energy required to transform two free hydrogen atoms into a hydrogen molecule cation and a free electron, resting at long distances.
(b) Calculate the ionization energy of the hydrogen H_2 molecule.
(c) Given that the doubly ionized helium He_2^{2+} molecule dissociates immediately upon formation, calculate the ionization energy of the singly ionized helium He_2^+ cation.

6.2 Write down the electronic configuration and ground-state Terms of the diatomic Al_2, S_2, and Cl_2 molecules. Make use of the writing method presented in Table 6.2 and the following text. Calculate the difference between the number of bonding and antibonding molecular orbitals, respectively, and estimate an order of magnitude of their dissociation energy.

6.3 Consider the nitrogen monoxide *NO* molecule, and discuss its ground-state electronic configuration. Compare it to the N_2 nitrogen molecule, and determine which is more stable.

6.4 Write down the ground-state electronic configurations of the heterogeneous diatomic molecules lithium hydride *LiH*, carbon nitride *CN*, sulfur monoxide *SO*, chlorine fluoride *ClF*, and hydrogen iodide *HI*. Use Table 6.4, listing the chemical affinity of all mentioned atoms. Determine which molecule is polar, and which end of a polar molecule is positively charged.

Table 6.4 Chemical affinity of several free atoms.

Atom	*H*	*Li*	*C*	*N*	*O*	*F*	*S*	*Cl*	*I*
Z	1	3	6	7	8	9	16	17	53
Affinity (eV)	0.754	0.62	1.26	< 0	1.46	3.40	2.08	3.61	3.06

7 Molecular (Especially Diatomic) Internal Oscillations

7.1 Preamble

Chapter 6 indicates that the molecular constituting atoms don't form a rigid structure; they rather move and oscillate about positions that form a minimum of the molecular energy. This chapter presents the classification of the molecular modes, and their implications on their light absorption and emission spectra. The initial classification is to **translational** (linear), **rotational**, and **vibrational** modes. In translational modes, all the atoms move linearly with a common constant velocity. Such motions cause Doppler shifting of absorption and emission spectral lines, discussed in detail in Section 1.2.2. Rotational modes are characterized by constant interatomic bond lengths and interbond angles, while all atoms form a circular motion about an axis crossing the molecular mass center, with a common phase and angular velocity; vibrational modes are oscillations of all atoms about their equilibrium positions, with a common frequency and phase, both with a resting mass center. Any general motion may be described as a sum of independent contributions according to this classification. Rotational and vibrational modes are addressed as **internal modes** because they preserve the position of the molecular mass center.

7.2 The Born-Oppenheimer Approximation

We return to discuss some approximations performed in Chapter 6, and approximations to be performed later, in a form that exposes the underlying assumptions. Thus, we rewrite the molecular Hamiltonian, which, in the electrostatic approximation, is written by Eq. (6.1):

$$\mathbf{H} = \sum_{\substack{all\\nuclei}} \frac{\mathbf{p}_u{}^2}{2M_u} + \sum_{\substack{all\\electrons}} \frac{\mathbf{p}_i{}^2}{2m} + V(\underline{R}_u, \underline{r}_i)\,, \tag{7.1}$$

where the system potential energy is

$$V(\underline{R}_u, \underline{r}_i) = \sum_{\substack{all\\nucleon\\pairs}} \frac{Z_u Z_v e^2}{|\underline{R}_u - \underline{R}_v|} + \sum_{\substack{all\\electron\\pairs}} \frac{e^2}{|\underline{r}_i - \underline{r}_j|} + \sum_{\substack{all\\electron/nucleon\\pairs}} \frac{-Z_u e^2}{|\underline{R}_u - \underline{r}_i|}\,. \tag{7.2}$$

In Eqs. (7.1) and (7.2), $\mathbf{p}_u$, M_u, $\underline{R}_u$, and $\underline{R}_v$, Z_u, and Z_v are the linear momentum, mass, position vector, and atomic number of the u-, or v-tagged nucleus, respectively; $\mathbf{p}_i$ and $\underline{r}_i$ are the linear momentums and position vectors of the i-th electron, respectively; and m is the electronic mass. The two summations in

Eq. (7.1) describe the kinetic energy of the nuclei and electrons, respectively. The first summation in Eq. (7.2) describes the electrostatic repulsion energy between nuclei pairs; the following one describes the electrostatic repulsion energy between electron pairs; and the last summation describes the electrostatic attractive energy between electrons and nuclei.

Without specifically indicating, the approximation used in Chapter 6 is called **Born-Oppenheimer** approximation, or the **adiabatic approximation**. The first step is to assume that all nuclei are at rest. That way, the first summation in Eq. (7.1) is zero. The other Hamiltonian elements are called **electronic Hamiltonian** $\mathbf{H}_e$. The second step is to solve a *set of Schrödinger equations* with the following electronic Hamiltonians:

$$\mathbf{H}_e\psi(\underline{R}_u,\underline{r}_i) \equiv \left[\sum_i \frac{\mathbf{p}_i^{\,2}}{2m} + V(\underline{R}_u,\underline{r}_i)\right]\psi(\underline{R}_u,\underline{r}_i) = E_{el}(\underline{R}_u)\psi(\underline{R}_u,\underline{r}_i)\,, \tag{7.3}$$

where the electronic position vectors $\underline{r}_i$ are the *dynamic variables*, and the nuclei position vectors $\underline{R}_u$ are *parameters*. Each combination of $\underline{R}_u$ vectors provides a different eigenenergy $E_{el}(\underline{R}_u)$. The subsequent step is to assume that the $\boldsymbol{\Phi}(\underline{R}_u,\underline{r}_i)$ wave function, which solves Schrödinger's equation of the full Hamiltonian (Eq. [7.1]), may be expressed by *variables separation* of the form

$$\boldsymbol{\Phi}(\underline{R}_u,\underline{r}_i) = \psi(\underline{R}_u,\underline{r}_i)\cdot\boldsymbol{\mathcal{S}}(\underline{R}_u)\,, \tag{7.4}$$

where the $\boldsymbol{\mathcal{S}}(\underline{R}_u)$ function depends only on the nuclei positions $\underline{R}_u$ which form the molecular skeleton. To calculate $\boldsymbol{\mathcal{S}}(\underline{R}_u)$, the full Schrödinger Eq. (7.1) needs to be solved:

$$\left[\sum_u \frac{\mathbf{p}_u^{\,2}}{2M_u} + \sum_i \frac{\mathbf{p}_i^{\,2}}{2m} + V(\underline{R}_u,\underline{r}_i)\right]\psi(\underline{R}_u,\underline{r}_i)\boldsymbol{\mathcal{S}}(\underline{R}_u) = E\psi(\underline{R}_u,\underline{r}_i)\boldsymbol{\mathcal{S}}(\underline{R}_u)\,. \tag{7.5}$$

Now we consider one of the above addends in the following equation:

$$\frac{\mathbf{p}_u^{\,2}}{2M_u}\psi(\underline{R}_u,\underline{r}_i)\boldsymbol{\mathcal{S}}(\underline{R}_u) = -\frac{\hbar^2}{2M_u}\underline{\nabla}_u^{\,2}\psi(\underline{R}_u,\underline{r}_i)\boldsymbol{\mathcal{S}}(\underline{R}_u)$$

$$= -\frac{\hbar^2}{2M_u}\left[\psi(\underline{R}_u,\underline{r}_i)\cdot\underline{\nabla}_u^{\,2}\boldsymbol{\mathcal{S}}(\underline{R}_u) + \underline{\nabla}_u\psi(\underline{R}_u,\underline{r}_i)\cdot\underline{\nabla}_u\boldsymbol{\mathcal{S}}(\underline{R}_u) + \boldsymbol{\mathcal{S}}(\underline{R}_u)\underline{\nabla}_u^{\,2}\psi(\underline{R}_u,\underline{r}_i)\right].$$

We *further assume* that it is possible to neglect the two last addends of the expression enclosed in the square brackets of the above last expression, namely that it is possible to write

$$\frac{\mathbf{p}_u^{\,2}}{2M_u}\psi(\underline{R}_u,\underline{r}_i)\boldsymbol{\mathcal{S}}(\underline{R}_u) \cong -\frac{\hbar^2}{2M_u}\psi(\underline{R}_u,\underline{r}_i)\cdot\underline{\nabla}_u^{\,2}\boldsymbol{\mathcal{S}}(\underline{R}_u) = \psi(\underline{R}_u,\underline{r}_i)\frac{\mathbf{p}_u^{\,2}}{2M_u}\boldsymbol{\mathcal{S}}(\underline{R}_u)\,. \tag{7.6}$$

The last assumption is the core of the Born-Oppenheimer approximation. Its justification stems from the fact that the mass of each nucleus is very large compared to an electronic mass; in classical terms, this means that the nuclei velocities are tiny compared to those of the electrons. Thus, the **electronic wave function** $\psi(\underline{R}_u,\underline{r}_i)$ adjusts itself quasistatically to the positions of the nuclei $\underline{R}_u$, independently of the nuclei velocities or accelerations. Physicists refer to such approximations as "**adiabatic approximations**". On setting Eq. (7.6) approximation into Eq. (7.5), we get

$$\psi(\underline{R}_u,\underline{r}_i)\left[\sum_u \frac{\mathbf{p}_u^{\,2}}{2M_u}\right]\boldsymbol{\mathcal{S}}(\underline{R}_u) + \left[\sum_i \frac{\mathbf{p}_i^{\,2}}{2m} + V(\underline{R}_u,\underline{r}_i)\right]\psi(\underline{R}_u,\underline{r}_i)\boldsymbol{\mathcal{S}}(\underline{R}_u) = E\psi(\underline{R}_u,\underline{r}_i)\boldsymbol{\mathcal{S}}(\underline{R}_u)\,.$$

When we further set Eq. (7.3) instead of the second addend on the left side, and eliminate $\psi(\underline{R}_u, \underline{r}_i)$ on both sides, we get the equation

$$\left[\sum_u \frac{\mathbf{p}_u^{\,2}}{2M_u} + E_{el}(\underline{R}_u)\right]\mathcal{S}(\underline{R}_u) = E\mathcal{S}(\underline{R}_u)\,. \tag{7.7}$$

The physical meaning of Eq. (7.7) is that the $\mathcal{S}(\underline{R}_u)$ wave equation for the motions of the molecular nuclei may be calculated as if they are subject to an effective potential $E_{el}(\underline{R}_u)$; recall that these are the eigenfunctions of the electronic states, obtained separately per each arrangement of the nuclei positions (skeleton structure).

An interesting result is obtained when calculating the *effective force field* acting on the nuclei. For its calculation we return to consider the electronic Hamiltonian $\mathbf{H}_e$. The effective potential is obviously its expectation value

$$E_{el}(\underline{R}_u) = \int \psi^*(\underline{R}_u, \underline{r}_i)\, \mathbf{H}_e \psi(\underline{R}_u, \underline{r}_i)\, d\tau_r\,, \tag{7.8}$$

where $d\tau_r$ is the infinitesimal volume in the generalized all electrons $\underline{r}_i$ positions space (the configuration space). The force acting on a particular u-tagged nucleus under the effective potential $E_{el}(\underline{R}_u)$ is the potential gradient $\underline{\mathcal{F}}_u(\underline{R}_u) = -\underline{\nabla}_u E_{el}(\underline{R}_u)$. Thus, we operate with $\underline{\nabla}_u$ on both sides of Eq. (7.8) and get

$$\begin{aligned}\underline{\nabla}_u E_{el}(\underline{R}_u) &= \underline{\nabla}_u \int \psi^*(\underline{R}_u, \underline{r}_i)\, \mathbf{H}_e \psi(\underline{R}_u, \underline{r}_i) d\tau_r \\ &= \int \underline{\nabla}_u \psi^*(\underline{R}_u, \underline{r}_i) \mathbf{H}_e \psi(\underline{R}_u, \underline{r}_i) d\tau_r + \int \psi^*(\underline{R}_u, \underline{r}_i) \underline{\nabla}_u \mathbf{H}_e \psi(\underline{R}_u, \underline{r}_i) d\tau_r \\ &\quad + \int \psi^*(\underline{R}_u, \underline{r}_i) \mathbf{H}_e \underline{\nabla}_u \psi(\underline{R}_u, \underline{r}_i) d\tau_r\,.\end{aligned} \tag{7.9}$$

We shall see immediately, that the sum of the first and third addends vanishes. First, we treat the third addend. Because the Hamiltonian operator is Hermitian, it may be written in the form

$$\begin{aligned}\int \psi^*(\underline{R}_u, \underline{r}_i) \mathbf{H}_e \underline{\nabla}_u \psi(\underline{R}_u, \underline{r}_i) d\tau_r &= \int \underline{\nabla}_u \psi(\underline{R}_u, \underline{r}_i) \mathbf{H}_e \psi^*(\underline{R}_u, \underline{r}_i) d\tau_r \\ &= \int \underline{\nabla}_u \psi(\underline{R}_u, \underline{r}_i) E_{el}(\underline{R}_u) \psi^*(\underline{R}_u, \underline{r}_i) d\tau_r \\ &= E_{el}(\underline{R}_u) \int \underline{\nabla}_u \psi(\underline{R}_u, \underline{r}_i) \psi^*(\underline{R}_u, \underline{r}_i) d\tau_r\,.\end{aligned}$$

We now add the latter result and the first addend in Eq. (7.9) and get

$$\begin{aligned}\int \underline{\nabla}_u \psi^* \mathbf{H}_e \psi d\tau_r + \int \psi^* \mathbf{H}_e \underline{\nabla}_u \psi d\tau_r &= E_{el} \int \left(\underline{\nabla}_u \psi^* \psi + \psi^* \underline{\nabla}_u \psi\right) d\tau_r \\ &= E_{el} \int \underline{\nabla}_u \left(\psi^* \psi\right) d\tau_r = E_{el} \underline{\nabla}_u \int \left(\psi^* \psi\right) d\tau_r = 0\,.\end{aligned}$$

Therefore, $\underline{\mathcal{F}}_u(\underline{R}_u) = -\int \psi^*(\underline{R}_u, \underline{r}_i)[\underline{\nabla}_u \mathbf{H}_e]\psi(\underline{R}_u, \underline{r}_i) d\tau_r$. The kinetic energy operator of the electrons is independent of the nuclei coordinates. Thus, the latter expression reduces to

$$\underline{\mathcal{F}}_u(\underline{R}_u) = -\int \psi^*(\underline{R}_u, \underline{r}_i)\,[\underline{\nabla}_u V(\underline{R}_u, \underline{r}_i)]\,\psi(\underline{R}_u, \underline{r}_i)\,d\tau_r\,. \tag{7.10}$$

The last result is called the **Hellmann-Feynman** theorem. Its meaning is that under the Born-Oppenheimer approximation, the force field operating on the nuclei at every electronic state is the expectation value of their position gradient of the classical electronic potential at the particular electronic state.

Before addressing general aspects of molecular internal vibrations, we demonstrate some of the problems and physical interpretations by studying the vibrational and rotational motions of diatomic molecules.

7.3 Vibrational and Rotational Modes of Diatomic Molecules

7.3.1 Empiric Analytic Potential

As seen in the previous section, it is possible to describe the molecular skeleton of nuclei as if each nucleus is subject to an effective potential, whose value at each construction is the electronic energy of that construction. In Section 6.2.3 we have calculated such potential for the hydrogen molecule cation H_2^+ by the LCAO approximation (Eq. [6.15] and Figure 6.4). For the electronic ground state, the potential approaches asymptotically a constant value when the internuclei distance is infinite. On reducing the distance, it attains a minimum for a calculated internuclei distance $\imath_0 = 1.217$ Å. It then increases steeply toward infinity in the limit where the internuclei distance vanishes. We have noted that the minimum distance and value are different, but the general curve behavior is identical. In view of the difficulty in calculating the potential based on fundamental principles, and even then never achieving an absolute match to experimental evidence, a quasi-classical approach is adopted. One establishes an analytic empiric function that provides a reasonable approximation to the actual potential, consequently allowing analytic calculations of the wave functions and energy states. **Morse** suggested the use of the following, relatively simple function, whose general form fits many diatomic molecules:

$$U(\imath) \equiv E_{el}(\imath) = D\left[1 - e^{-\beta(\imath - \imath_0)}\right]^2, \tag{7.11}$$

where $\imath$ is the internuclei distance. D, β, and $\imath_0$ are positive parameters, adjusted for each molecule to provide the best fit to its vibrational and rotational states. Morse potential assumes the D value for an infinitely large distance between the nuclei, $\imath = \infty$, and a zero minimum at an internuclei distance $\imath = \imath_0$. Obviously, D then expresses the dissociation energy of the molecule, and $\imath_0$ expresses the equilibrium bond length (Table 6.3). Still, the potential assumes a finite value $D\left[1 - e^{+\beta\,\imath_0}\right]^2$ for a zero internuclei distance $\imath = 0$, a behavior inconsistent with strict reality, because the potential must grow steeply to infinity as $\imath$ approaches zero (Figure 6.4 and related text). However, because usually $\beta\imath_0 > 1$, it turns out that $U(0) >> D$. Thus, as long as the discussed cases don't involve the nuclei coming very close, the Morse potential may predict many molecular properties with great accuracy. The β parameter relates to the vibrational frequencies of the molecule, as will be pointed out specifically in the following analyses.

Expanding the Morse potential to a Tailor series about $\imath_0$, the potential assumes the form

$$U(\imath) \cong \beta^2 D(\imath - \imath_0)^2, \tag{7.12}$$

establishing a useful approximation for small deviation of $\imath$ from the equilibrium position $\imath_0$. We will make use of this approximation shortly.

Hulburt and **Hirschfelder** have proposed a corrected potential function based on the Morse potential, that brings about improved agreement with experimental results:

$$U(x) = D\left(1 - e^{-\beta x}\right)^2 + c\beta^2 x^3 e^{-2\beta x}(1 + b\beta\, x)\,, \tag{7.13}$$

where $x \equiv r - r_0$ is the deviation from the equilibrium position, and b and c are additional parameters used for fitting experimental results. **Linett** suggested a different practical potential function of the form:

$$U(r) = \frac{a}{r^m} - be^{-nr}\,, \tag{7.14}$$

where a, b, m, and n are positive parameters used for fitting experimental results. The Linett potential differs from the Morse and Hulburt-Hirschfelder ones by growing to infinity in the limit of zero internuclei distance. That way it provides agreement with the fundamental reality. In other properties it behaves like those of Morse and Hulburt-Hirschfelder.

7.3.2 Molecular Vibrational Modes

Recall that we limit our consideration to a diatomic molecule, and we now further assume that the motions of the nuclei are limited to their bonding axis orientation that is fixed in space. Because the effective force acting between the nuclei is central (for the present consideration, any point on their connecting axis may be considered as center), it is possible to describe their motion as an effective motion of a *single particle*, acted on by a force always oriented toward or away from their mass center. The effective mass of that motion is the reduced mass $\mu = M_1M_2/(M_1 + M_2)$, where M_1 and M_2 are the masses of the first and second nucleus, respectively. The potential energy is $U(r)$, where r is the internuclei distance. Schrödinger's equation is (see Eq. [7.7])

$$\left[\frac{\mathbf{P}_\mu{}^2}{2\mu} + U(r)\right]\mathcal{V}(r) = E\mathcal{V}(r)\,, \tag{7.15}$$

where $\mathbf{P}_\mu$ and $\mathcal{V}(r)$ are the linear momentum operator and wave function of the reduced mass, respectively. The physical meaning of $\mathcal{V}(r)$ is that $|\mathcal{V}(r)|^2$ is the probability density for the two nuclei to assume an r distance. Changing to the $x \equiv r - r_0$ variable, and using the small deviation approximation (harmonic approximation – Eq. [7.12]), we get

$$\left[\frac{\mathbf{P}_\mu{}^2}{2\mu} + \frac{1}{2}\mu\left(\frac{2\beta^2 D}{\mu}\right)x^2\right]\mathcal{V}(x) = E\mathcal{V}(x)\,. \tag{7.16}$$

Eq. (7.16) provides a Hamiltonian of a harmonic oscillator belonging to the characteristic $\omega_0 = (2\beta^2 D/\mu)^{1/2}$ frequency (Section 5.3, especially Eq. [5.18]). The eigenenergies are given by $E_\mathrm{v} = \left(\mathrm{v} + \frac{1}{2}\right)\cdot\hbar\omega_0$, where $\mathrm{v} = 0,\ 1,\ 2,\ \cdots\ \infty$. The v - quantity is called **vibrational quantum number**, or **vibrational order**. The molecule vibrates with and energy $(1/2)\,\hbar\omega_0$ even at its ground state, and the higher vibrational orders are equidistantly separated by $\hbar\omega_0$. Deviations from the latter property are seen experimentally, especially for the higher vibrational orders, resulting from moderation of the potential rise at large internuclei distances. The states' energy separations become smaller, until eventually the molecule dissociates. According to the Morse potential, above the dissociation energy D (physically meaning a dissociation-causing vibration), a continuity of states is obtained. The actual energy needed for inducing dissociation, D_{ex}, is smaller than D by $(1/2)\hbar\omega_0$, because the molecule has such energy as its ground state. A schematic demonstration of the Morse potential and the states structure is shown in Figure 7.1. The states are marked by straight horizontal lines, whose edges are located at the two sides of a classical oscillator amplitude of same energy.

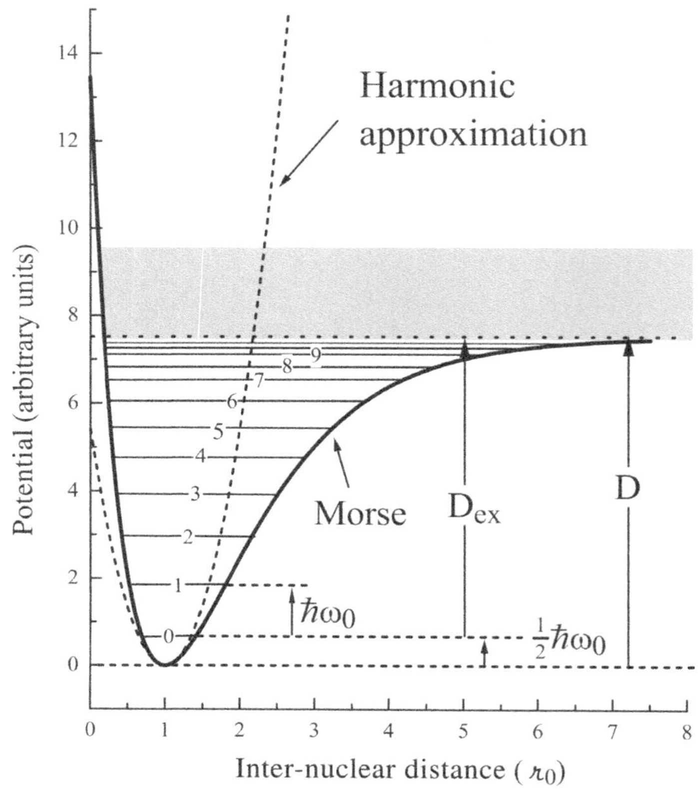

Figure 7.1 Schematic illustration of the diatomic Morse potential compared to the harmonic approximation, and energy states in the Morse potential. The gray area illustrates the continuity of vibrational states for energies exceeding dissociation. For more discussion and details see text.

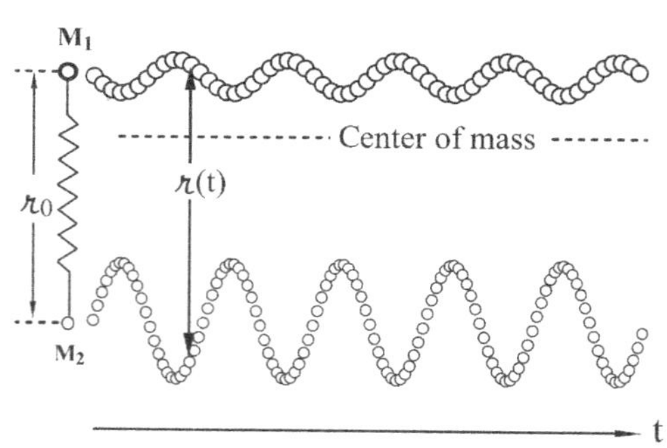

Figure 7.2 Demonstration of the time dependence of the separate motions of each nucleus in a diatomic molecule, by a mechanical model whereby the nuclei, of different masses, are connected by a spring of r_0 equilibrium length.

It is interesting to note, that the wave equation for the Morse potential is analytically solvable. There is a finite number of bound states, their energies given by

$$E_v = \hbar\left(2\beta^2 D/\mu\right)^{1/2}\left(v + \frac{1}{2}\right) - \frac{2\hbar^2\beta^2}{4\mu}\left(v + \frac{1}{2}\right)^2, \tag{7.17}$$

where v is an integer starting at zero and ending at the integral number of $k \equiv (2\mu D)^{1/2}/\beta\hbar - 1/2$. Apparently, the reduction in states' energy separation doesn't end up with an infinite number of states prior to dissociation, as is the case, for example, for the electronic states in the hydrogen H atom (Eq. [3.38]). We won't describe here the development of Eq. (7.17).

By the way, Morse also pointed out a further empiric property of diatomic molecules. All bond lengths abide by the relation $r_0{}^3\omega_0 = 5.65 \times 10^{-10}\ \mathrm{cm^3\ rad/s}$ with less than 1% deviation. No apparent classical model is known to predict such behavior, yet it is useful, and an example for many simple empiric relations existing among compounds belonging to well-defined chemical families.

It is important to note that the oscillations described by the $\mathcal{V}(r)$ wave function relates to the distance *between* the nuclei, and not the separate positions of each nucleus. To emphasize that issue, Figure 7.2 describes the separate motion of each nucleus in a diatomic molecule. The two nuclei are connected by a spring of r_0 equilibrium length. They oscillate about their equilibrium positions with opposite phases. The amplitude of each depends on its relative mass: the larger amplitude belongs to the lighter nucleus, such that their mass center remains at rest.

7.3.3 Molecular Rotational Modes

In contrast to the model of the previous subsection, here we assume that the distance between nuclei (namely the bond length) remains constant, but the axis connecting the two nuclei is free to point anywhere in space. The molecule then establishes a "rigid body", and the effective potential $E_{el}(\underline{R}_u)$ of Eq. (7.7) assumes a constant value. When the mass center position stays constant, the remaining dynamic term in the Hamiltonian is the energy of a rigid molecular rotation. The classical kinetic energy of a rotating rigid body is given by

$$E_{kin} = \frac{L_A{}^2}{2I_A} + \frac{L_B{}^2}{2I_B} + \frac{L_C{}^2}{2I_C}, \tag{7.18}$$

where L_A, L_B, and L_C are the components of the angular momentum vector $\underline{L}$ along the principal axes A, B, and C of the moment of inertia tensor, respectively. Correspondingly, I_A, I_B, and I_C are the moments of inertia of the body about these A, B, and C axes. The axis connecting the two nuclei in a diatomic molecule is a principal axis of the moment of inertia tensor; let it be selected as the C axis. For point-size dimensions of nuclei, its moment of inertia I_C vanishes (in fact it is just tiny compared to I_A and I_B, practically involving solely the electronic rotations). The two other principal axes are perpendicular to the C axis, located on a plane crossing the mass center. Their specific orientations may be chosen arbitrarily, provided being mutually perpendicular. Their moments of inertia I_A and I_B are equal, given by

$$I_A = I_B = \frac{M_1 M_2}{M_1 + M_2} r_0{}^2 = \mu\, r_0{}^2 . \tag{7.19}$$

Because I_C would be tiny compared to I_A and I_B, the contribution of its angular-momentum related energy addend would be relatively huge. We thus overlook its contribution to the dynamics of the "rigid" nuclei motions about the A and B axes. Their contributing kinetic energy would be

$$E_{kin} = \frac{L_A{}^2 + L_B{}^2}{2I_A} = \frac{L^2}{2I_A} . \tag{7.20}$$

Correspondingly, their wave equation obtained by representing the angular momentum $\underline{L}$ by its quantum operator $\mathbf{L}$ is given by

$$\frac{\mathbf{L}^2}{2I_A} \mathcal{Y}(\theta, \varphi) = E\mathcal{Y}(\theta, \varphi), \tag{7.21}$$

where $\mathcal{Y}(\theta, \varphi)$ is the wave function, whose absolute square represents the probability distribution per solid angle unit that the angular momentum vector would point at a (θ, φ) angle relative to the C axis (Z axis). The Hamiltonian eigenvalues would be

$$E_\ell = \frac{\hbar^2}{2I_A} \cdot \ell(\ell + 1); \quad \ell = \quad 0, \quad 1, \quad 2, \quad \cdots, \tag{7.22}$$

where ℓ is the **orbital quantum number**. The physical meaning of this quantity is like that of the electronic orbital quantum number, which in fact relates to a joint rotation of an electron and nucleus about their mass center. The eigenfunctions are $\mathcal{Y}_{\ell m}(\theta, \varphi) = \mathbf{Y}_{\ell m}(\theta, \varphi)$, where $\mathbf{Y}_{\ell m}(\theta, \varphi)$ are the spherical harmonics of Table 3.2, and $m = -\ell, -\ell + 1, \cdots, 0, 1, 2, \cdots, \ell$ is the **magnetic quantum number**. By Eq. (7.22), the rotational energy is independent of m.

Eq. (7.22) describes a series of eigenenergies whose separations increase with ℓ. Figure 7.3 shows the rotational energies structure in a diatomic molecule.

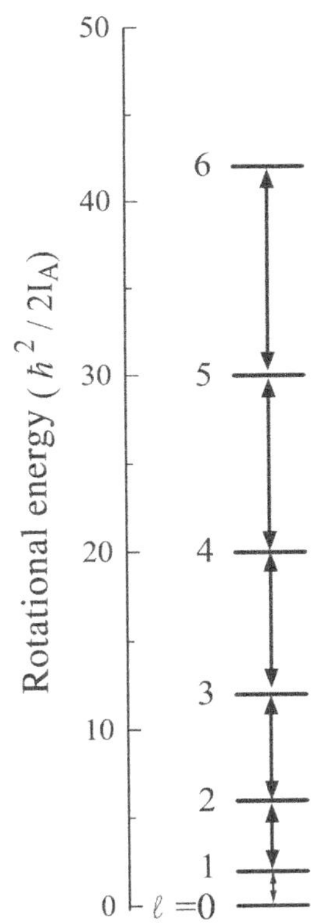

Figure 7.3 Rotational energy scheme in a diatomic molecule. Two-sided arrows indicate dipole-allowed transitions between states.

7.3.4 Molecular Vibrational/Rotational Modes

If we indeed only require that the molecular mass center remains at rest, the molecule will perform simultaneous vibrational and rotational motions. We refrain from the simplifying assumptions of absence of rotation or alternatively absence of vibration. Section 4.1 showed that the Schrödinger equation for two particles bonded by a central force may be transformed to that of a single particle whose mass is the reduced two-particles mass, and perform the variables separation $\mathcal{S}(r,\theta,\varphi) = r\cdot\mathcal{V}(r)\mathbf{Y}_{\ell m}(\theta,\varphi)$ (see Eq. [7.7]). Schrödinger's equation then reduces to a form called radial Schrödinger equation (Eq. 4.16):

$$-\frac{\hbar^2}{2\mu}\frac{d^2\mathcal{V}(r)}{dr^2} + \left[U(r) + \frac{\hbar^2\ell(\ell+1)}{2\mu r^2}\right]\mathcal{V}(r) = E\mathcal{V}(r)\,. \tag{7.23}$$

The normalization condition on the $\mathcal{V}(r)$ wave function would be (see Eq. [3.43])

$$\int_0^\infty \mathcal{V}_{n\ell}^*(r)\mathcal{V}_{n\ell}(r)r^4 dr = 1\,. \tag{7.24}$$

Eq. (7.23) differs from Eq. (7.15) by the inclusion of an additional component to the potential: the $\hbar^2\ell(\ell+1)/2\mu r^2$ quantity. It describes the kinetic energy of the molecular rotation having a moment of inertia that *changes dynamically* with the radius r (Eqs. [7.19] and [7.20]). We now transform Eq. (7.23) by transforming it to the $x \equiv r - r_0$ variable, while assuming that always $x << r_0$. The latter assumption establishes a good approximation, especially for low vibrational orders (Eq. [7.17], Section 7.3.2). By expanding the rotational energy component to a Taylor series about r_0, and neglecting all powers of x from the second order and above, we get

$$-\frac{\hbar^2}{2\mu}\frac{d^2\mathcal{V}(x)}{dx^2} + \left[U(x) - \frac{\hbar^2\ell(\ell+1)}{\mu r_0^2}\left(\frac{x}{r_0}\right)\right]\mathcal{V}(x) = \left[E - \frac{\hbar^2\ell(\ell+1)}{2\mu r_0^2}\right]\mathcal{V}(x)\,. \tag{7.25}$$

For obtaining a zero-order solution to this equation, we address the $\frac{\hbar^2\ell(\ell+1)}{\mu r_0^2}\left(\frac{x}{r_0}\right)$ addend as a potential perturbation, which expresses a *coupling* between vibrations and rotations. The unperturbed Schrödinger equation would be

$$\left[-\frac{\hbar^2}{2\mu}\frac{d^2}{dx^2} + U(x)\right]\mathcal{V}(x) = \left[E - \frac{\hbar^2\ell(\ell+1)}{2\mu r_0^2}\right]\mathcal{V}(x)\,. \tag{7.26}$$

This equation exhibits the same structure as Eq. (7.15) for molecular vibrations. The eigenfunctions are independent of ℓ, and will then be written as $\mathcal{V}_\mathrm{v}(x)$ without an ℓ tagging. The eigenenergy will correspondingly be written as E_v. The eigenfunctions and eigenenergies obviously depend on the specific form of $U(x)$, whether it is either the Morse potential, its harmonic approximation, or another (see Figure 7.1 and related text). In any case, for Eq. (7.26) we get

$$E = E_\mathrm{v} + \frac{\hbar^2\ell(\ell+1)}{2\mu r_0^2} = E_\mathrm{v} + E_\ell\,. \tag{7.27}$$

The physical meaning of the last result is that in the *zero-order approximation*, an eigenenergy in a diatomic molecule is a sum of a vibrational energy calculated as if the molecule doesn't rotate, and a rotation energy calculated as if the molecule doesn't vibrate. For higher vibrational orders, where the probability for the molecule to deviate from the equilibrium bond length r_0 is large, there is a need to consider the contribution of the perturbation potential appearing in Eq. (7.25). We won't elaborate further on that issue. We only note that when the potential $U(x)$ is symmetric with respect to r_0, such as in the harmonic approximation of the Morse potential (Eq. [7.12]), the linear correction vanishes in the first order perturbation theory (Eq. [2.6]). The occurrence of nonzero corrections to Eq. (7.27) by first-order perturbation, reflects the asymmetry of the molecular potential with respect to the equilibrium bond length.

Eq. (7.27) provides the energy spectrum of the integrated vibrational/rotational molecular motions in the zero-order approximation. Calculations and experiments show that subject to a Morse potential, a vibrational energy ration of a diatomic molecule in the lower vibrational modes $\hbar\omega_0 = \hbar(2\beta^2 D/\mu)^{1/2}$ is very large compared to a rotational energy ration $\left(\hbar^2/2\mu r_0^2\right)$. For demonstration, Table 7.1 lists the values of vibrational and rotational rations in several diatomic molecules.

Energy schemes of vibrational/rotational levels (Eq. [7.27]) are usually sketched such that the vibrational levels are plotted as relatively long horizontal lines, and the rotational levels belonging to the same vibrational one are plotted by shorter lines (Figure 7.4). Notably, for each vibrational level, the rotational higher order levels may exhibit higher energy than an adjacent higher order vibrational level. Thus, energy proximity may be realized among different vibration/rotation combinations. Because of coupling between states, transitions between such combinations may occur. We will not elaborate on this issue.

Table 7.1 Vibrational and rotational rations in various diatomic molecules.

Molecule	Vibrational ration $\hbar\omega_0 = \hbar(2\beta^2 D/\mu)^{1/2}$ (eV)	Rotational ration $\hbar^2/2\mu r_0^2$ (eV)	Molecule	Vibrational ration $\hbar\omega_0 = \hbar(2\beta^2 D/\mu)^{1/2}$ (eV)	Rotational ration $\hbar^2/2\mu r_0^2$ (eV)
H_2	0.543	8.0×10^{-3}	HBr	0.328	1.05×10^{-3}
LiH	0.174	9.32×10^{-4}	BeF	0.151	1.84×10^{-4}
Li_2	0.0434	8.3×10^{-4}	KF	0.0483	2.51×10^{-5}
O_2	0.0194	1.78×10^{-4}	KBr	0.0286	9.2×10^{-7}
N_2	0.292	2.48×10^{-4}	$CsBr$	0.024	4.25×10^{-6}
Cl_2	0.070	3.02×10^{-5}	$NaCl$	0.0471	2.378×10^{-5}
CO	0.268	2.38×10^{-4}	KCl	0.0347	1.443×10^{-5}
HF	0.510	2.48×10^{-3}	KI	0.0263	6.7×10^{-6}
HCl	0.369	1.31×10^{-3}			

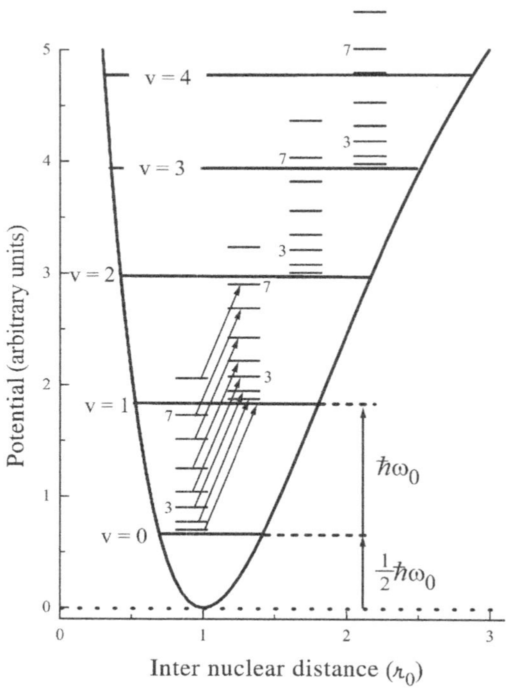

Figure 7.4 Illustration of fundamental vibration/rotation energy scheme structure in a diatomic molecule. Dipole-allowed transitions are shown between states belonging to a spectroscopic group called "P-branch". Detailed discussion will follow.

7.3.5 Transition Probabilities and Selection Rules

Calculation of selection rules and transition probabilities between vibrational/rotational states is analogous to calculations already performed for the hydrogen H electronic states. As in Eq. (3.51), we could write that the probability per unit time for a stimulated transition $W_{v'\ell'm',v\ell m}$ from the initial $|v\ell m\rangle$ state to the final $|v'\ell'm'\rangle$ state is given by

$$W_{v'\ell'm',v\ell m} \cong \frac{2\pi^2}{2\hbar^2}\left|\left\langle v'\ell'm'\middle|\underline{M}_{Sk}|v\ell m\rangle\cdot\underline{u}\right|^2\cdot\rho_\omega(\omega_0);$$

$$\omega_0 \equiv \frac{E_{v'\ell'm'} - E_{v\ell m}}{\hbar}, \tag{7.28}$$

where $\underline{u}$ is a unit vector co-oriented with the electric field of the electromagnetic radiation, ρ_ω is the volume concentration of the energy in the electromagnetic radiation field per unit angular frequency, and $\underline{M}_{Sk}$ is the electric dipole moment (recall that the latter quantity has been symbolized by Q in Chapter 1 and by μ in Chapter 2). This symbol is commonly used in the current discussed discipline, most likely because it avoids confusion with the molecular reduced mass, which is usually symbolized by the letter μ.

Eq. (7.28) is fundamentally different than Eq. (3.51) in that the electric dipole moment of the molecular skeleton $\underline{M}_{Sk}$ is written in a general form (see also Eq. [7.44]), while in connection to the hydrogen atom, it is specified as $\underline{M} = e\underline{r}$. In a general molecule, the electric dipole may vanish permanently (like in a homonuclear molecule) or exhibit a complex dependence on the internuclei distances. The transition oscillator strength or its mere existence thus depends on the size of the $\langle v'\ell'm'|\underline{M}_{Sk}|v\ell m\rangle$ vector.

The following treatment is analogous to the calculations appearing in Section 3.6, which are based on the wave function r, θ, φ variables separation. It may thus be summarized, that concerning transitions between different ℓ and m states, the following selections rules are in effect:

$$\boxed{\Delta\ell = \pm 1 \text{ and } \Delta m = 0, \pm 1}\ . \tag{7.29}$$

The $\Delta m = 0$ case relates to absorption or emission of a linearly polarized light. The $\Delta m = \pm 1$ cases relate to absorption or emission of right or left circularly polarized light. The $\Delta\ell = \pm 1$ cases relate to electric dipole transition between different rotational states belonging to the same vibrational state (to be dealt with just below); Figure 7.3 emphasizes such transitions.

In correspondence to the development described in Section 3.6, for the present case we may write

$$\left.\begin{aligned}\langle \mathrm{v}', \ell+1, m|\underline{M}_{Sk}|\mathrm{v}\ell m\rangle_Z &= \sqrt{\frac{(\ell+1)^2-m^2}{(2\ell+3)(2\ell+1)}}\left\langle M_{\mathrm{v}}^{\mathrm{v}'}\right\rangle \\ \langle \mathrm{v}', \ell-1, m|\underline{M}_{Sk}|\mathrm{v}\ell m\rangle_Z &= \sqrt{\frac{\ell^2-m^2}{(2\ell+1)(2\ell-1)}}\left\langle M_{\mathrm{v}}^{\mathrm{v}'}\right\rangle\end{aligned}\right\}, \tag{7.30}$$

where $\left\langle M_{\mathrm{v}}^{\mathrm{v}'}\right\rangle \equiv \int_0^\infty M_{Sk}(r)\mathcal{V}_{\mathrm{v}'}(r)\mathcal{V}_{\mathrm{v}}(r)\cdot r^4 dr$ is the electric dipole transition of the molecule. Also,

$$\left.\begin{aligned}\langle \mathrm{v}', \ell+1, m+1|\underline{M}_{Sk}|\mathrm{v}\ell m\rangle_{Q^+} &= \sqrt{\frac{(\ell+m+2)(\ell+m+1)}{(2\ell+3)(2\ell+1)}}\langle M_{\mathrm{v}}^{\mathrm{v}'}\rangle \\ \langle \mathrm{v}', \ell-1, m+1|\underline{M}_{Sk}|\mathrm{v}\ell m\rangle_{Q^+} &= -\sqrt{\frac{(\ell-m)(\ell-m-1)}{(2\ell+1)(2\ell-1)}}\langle M_{\mathrm{v}}^{\mathrm{v}'}\rangle \\ \langle \mathrm{v}', \ell+1, m-1|\underline{M}_{Sk}|\mathrm{v}\ell m\rangle_{Q^-} &= -\sqrt{\frac{(\ell-m+2)(\ell-m+1)}{(2\ell+3)(2\ell+1)}}\langle M_{\mathrm{v}}^{\mathrm{v}'}\rangle \\ \langle \mathrm{v}', \ell-1, m-1|\underline{M}_{Sk}|\mathrm{v}\ell m\rangle_{Q^-} &= \sqrt{\frac{(\ell+m)(\ell+m-1)}{(2\ell+1)(2\ell-1)}}\langle M_{\mathrm{v}}^{\mathrm{v}'}\rangle\end{aligned}\right\}, \tag{7.31}$$

where $Q^+ \equiv X + iY$ and $Q^- \equiv X - iY$ are coordinates in the complex coordinate system, whose quantities describe the electric transition dipole components for right or left circularly polarized light, respectively.

Now we turn to calculate the electric transition dipole. To simplify the discussion, we assume the harmonic approximation of the Morse potential. Then, the $\mathcal{V}_{\mathrm{v}}(r)$ wave function would be proportional to the corresponding $\psi_{\mathrm{v}}(r - r_0)$ harmonic oscillator wave function (Table 5.1). To connect between variable symbols, we define ♣ $Q = (\mu\omega_0/\hbar)^{1/2}(r - r_0)$ (see Table 5.1). We then get that $r = r_0 + Q/(\mu\omega_0/\hbar)^{1/2}$, namely that $\mathcal{V}_{\mathrm{v}}\left(r_0 + \dfrac{Q}{(\mu\omega_0/\hbar)^{1/2}}\right) = N_{\mathrm{v}}\psi_{\mathrm{v}}(Q)$, where N_{v} is a normalization factor that adjusts the harmonic oscillator functions ψ_{v} to serve as the radial wave function. For calculating N_{v} we recall the normalization condition given in Eq. (7.24): $\int_0^\infty |\mathcal{V}_{\mathrm{v}}(r)|^2 r^4 dr = 1$. Performing the required variable change, the normalization condition becomes

$$\frac{N_{\mathrm{v}}{}^2\cdot r_0{}^4}{(\mu\omega_0/\hbar)^{1/2}} \int_{-r_0(\mu\omega_0/\hbar)^{1/2}}^{\infty} |\psi_{\mathrm{v}}(Q)|^2 \left[1 + \frac{Q}{r_0(\mu\omega_0/\hbar)^{1/2}}\right]^4 dQ = 1\,. \tag{7.32}$$

We now perform two additional approximations. One is setting the lower limit of the integral as $-\infty$. This approximation is justified because for all known cases, the precise lower limit is very large compared to a unity; for example, $r_0(\mu\omega_0/\hbar)^{1/2} = 6$ for the hydrogen H_2 molecule; it is $r_0(\mu\omega_0/\hbar)^{1/2} = 31.6$ for the table-salt *NaCl* molecule. Simultaneously, the $|\psi_{\mathrm{v}}(Q)|^2$ functions attain non-zero values over practically a $\sqrt{2}$ range (twice the mean deviation) in the vicinity of $Q = 0$. Beyond that range, the $|\psi_{\mathrm{v}}(Q)|^2$ values practically diminish.

♣ Notably, the $(\hbar/\mu\omega_0)^{1/2}$ quantity is the classical amplitude of the zero-order vibration; the Q variable is thus the ratio between the temporal deviation from the equilibrium position and the classical amplitude.

It is not fundamentally complicated to perform an accurate calculation of the integral appearing in Eq. (7.32) with a $-\infty$ lower limit; one has to perform the indicated fourth power calculation of the expression inside the square brackets, getting a fourth-order polynomial; then utilize the recurrence and orthonormality relations among $\psi_v(Q)$ functions appearing in Eqs. (5.32) and (5.29), respectively. We won't elaborate on the detailed calculation, and we advise the reader to perform it as an exercise. For simplicity, we present a much simpler calculation by neglecting all Q powers in the polynomial (justified by the same arguments used for changing the lower limit of the integral in Eq. [7.32]). The difference between the proposed calculation method and the more accurate one amounts to only several percentage points. Thus, by leaving only the unity addend in the square brackets, the integral is a unity, and the approximate normalization factor, identically for all v-states, becomes

$$N_v \cong \frac{(\mu\omega_0/\hbar)^{1/4}}{r_0^{\,2}} . \tag{7.33}$$

To calculate the transition dipole, we assume that the molecular skeleton electric dipole (constant on the average) exhibits two components: one a constant, the other oscillating in-phase with the vibration; namely

$$M_{Sk} = M_0 + M_1(r - r_0) , \tag{7.34}$$

where M_0 and M_1 are constants. The M_1 quantity has electric charge dimensions, whose typical value is of the order of a single elementary charge e. When $M_1 \neq 0$, the vibration is addressed as "polar". The transition dipole would be

$$\left\langle M_v^{v'} \right\rangle = \int_0^{\infty} [M_0 + M_1(r - r_0)] \mathcal{V}_{v'}(r)\mathcal{V}_v(r) \cdot r^4 dr$$

$$\cong M_0\delta_{v'v} + M_1 N_v N_{v'} r_0^{\,4} \int_{-\infty}^{\infty} \frac{Q}{(\mu\omega_0/\hbar)} \psi_{v'}(Q)\psi_v(Q) \left[1 + \frac{Q}{r_0(\mu\omega_0/\hbar)^{1/2}}\right]^4 dQ$$

$$\cong M_0\delta_{v'v} + \frac{M_1}{(\mu\omega_0/\hbar)^{1/2}} \int_{-\infty}^{\infty} \psi_{v'}(Q) \cdot Q\psi_v(Q) dQ .$$

The latter approximate result is obtained by neglecting all higher-than-zero Q powers, as performed in relation to Eq. (7.33). Recall now Eq. (5.32): $Q\psi_v(Q) = \sqrt{\frac{v+1}{2}}\psi_{v+1}(Q) + \sqrt{\frac{v}{2}}\psi_{v-1}(Q)$. We set it in the last result, and use the orthonormality relations among the ψ_v functions. Then

$$\left\langle M_v^{v'} \right\rangle \cong M_0\delta_{v'v} + \frac{M_1}{(\mu\omega_0/\hbar)^{1/2}} \left\{ \sqrt{\frac{v+1}{2}} \cdot \delta_{v',v+1} + \sqrt{\frac{v}{2}} \cdot \delta_{v',v-1} \right\} . \tag{7.35}$$

The molecular electric dipole transitions would be non-zero when $v' = v$, or $\Delta v \equiv v' - v = \pm 1$ (vibration/rotation transition being allowed); obviously, provided that the selection rules of Eq. (7.29) are fulfilled. In summary, the selection rules required for vibration/rotation transitions are

$$\boxed{\Delta v = 0, \pm 1; \ \Delta \ell = \pm 1; \ \Delta m = 0, \pm 1} . \tag{7.36}$$

These selection rules were calculated for the harmonic shape of the vibration potential; however, they provide a valid description of most practical cases, especially for the low vibrational order transitions. Transitions belonging to $|\Delta v| \geq 2$, called **vibrational overtones**, may sometimes also be observed, but they are very faint, due to being forbidden by the selection rules.

7.4 Vibrational/Rotational Absorption Spectra

7.4.1 Pure Rotational Transitions

Transitions belonging to $\Delta v = 0$, during which no change occurs in the molecular vibrational state, are addressed as **pure rotational**. The transitions displayed in Figure 7.3 are plotted accordingly. However, if the permanent molecular dipole M_0 vanishes, the oscillator strength of these transitions also vanishes. Thus, in homonuclear diatomic molecules, pure rotational transitions are forbidden. Values of the permanent dipole moment in several diatomic molecules are listed in Table 6.3. If ℓ is the rotational quantum number of the lower rotational state, the transition energy to the strongest transition-allowed state belonging to $\ell' = \ell + 1$ ($\Delta\ell = +1$) would be given by

$$\Delta E_{\ell,\ell+1} = \frac{\hbar^2}{2I_A}[(\ell+1)(\ell+2) - \ell(\ell+1)] = \frac{2\hbar^2}{2I_A}(\ell+1), \tag{7.37}$$

for $\ell = 0, 1, 2, \cdots$.

The absorption spectrum belonging to pure rotational transitions in molecules would thus exhibit a series of equidistant absorption lines. The separation between lines is $2\hbar/2I_A$, which is twice the basic ration provided in Table 7.1. Such a spectrum is shown schematically in Figure 7.5. The spectral region of measured rotational spectra spans the far infrared down to the millimeter wavelength range of the electromagnetic radiation. For example, the first transition ($0 \rightarrow 1$) in the hydrogen fluoride molecule takes place at the 0.025 cm wavelength; in the table-salt *NaCl* molecule, it would be at 2.6 cm (Table 7.1). The other transitions, often addressed as **rotational overtones**, will obviously take place at shorter wavelengths.

7.4.2 Temperature Dependence of Pure Rotational Transitions

Figure 7.5 describes all transition lines as if they are of same strength. In fact, they are of different strengths. Each transition strength results by multiplication of three factors: degeneracy of the initial state, its thermal occupation probability, and the squared transition dipole. In continuation of the requested proof in Exercise 3.7, we adjust the stated equation to the diatomic molecule case (Eqs. [7.30] and [7.31]). We get that summing of squared transition dipoles from a certain $|v\ell m\rangle$ state to all m' substates of $|v, \ell+1\rangle$, satisfying the selection condition $\Delta m = 0, \pm 1$ (Eq. [7.29]) irrespective of the electromagnetic radiation polarization state, would be given by (see also Eq. [7.34])

$$M^2(\ell m) = \frac{\ell+1}{2\ell+1} M_0{}^2. \tag{7.38}$$

Notably, $M^2(\ell m)$ is independent of m. The $|v\ell m\rangle$ state is $2\ell + 1$ fold degenerate, according to the possible m numbers. Thus, the thermal occupation probability of a $|v\ell\rangle$ state for any specific ℓ would be given by

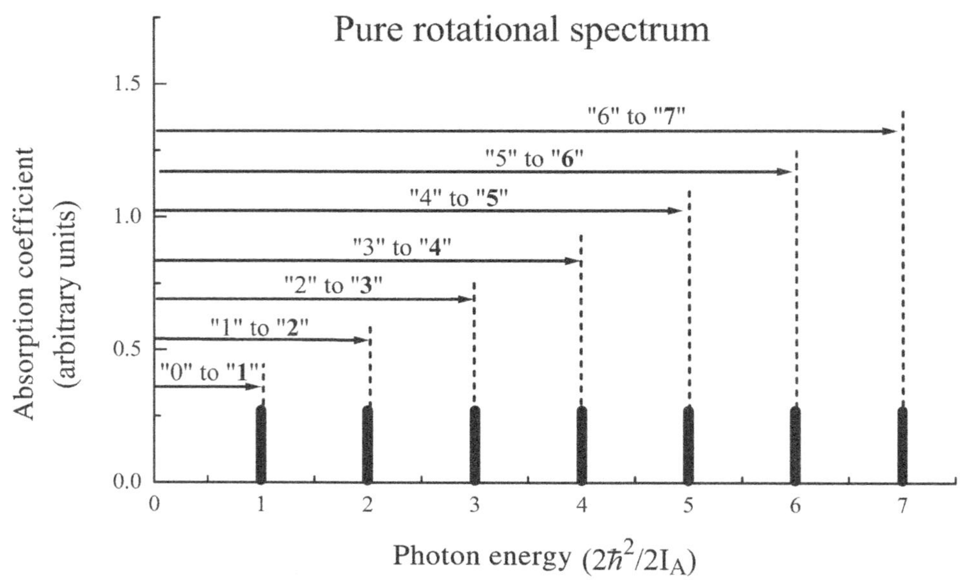

Figure 7.5 Schematic Illustration of a typical absorption spectrum belonging to pure rotational transitions in a molecule. It relates to the lines energy positions, not to their strengths.

$$P(\ell, T) = \frac{(2\ell + 1)\,\mathbf{exp}\left(-\dfrac{\ell(\ell+1)\hbar^2/2I_A}{k_B T}\right)}{\sum\limits_{s=0}^{\infty}(2s+1)\,\mathbf{exp}\left(-\dfrac{s(s+1)\hbar^2/2I_A}{k_B T}\right)}. \tag{7.39}$$

Eq. (7.39) is fundamentally like Eq. (2.25), describing the probability for n-photons occupation, except for inclusion of each state degeneracy in the present equation. Eq. (7.39) assumes simple forms at two different limits: high temperatures satisfying $k_B T >> \hbar^2/2I_A$, and low temperatures satisfying $k_B T << \hbar^2/2I_A$. At *high temperatures* the denominator term becomes

$$\mathcal{Z}(T) \equiv \sum_{s=0}^{\infty}(2s+1)\,\mathbf{exp}\left(-\frac{s(s+1)\hbar^2/2I_A}{k_B T}\right) \rightarrow \int_0^{\infty} dx(2x+1)\,\mathbf{exp}\left(-\frac{x(x+1)\hbar^2/2I_A}{k_B T}\right).$$

Note that $\mathcal{Z}(T)$ is in fact the thermodynamic distribution function of the molecule rotational states. We now make a variable change: we denote $\xi = x(x+1)$ and get

$$\mathcal{Z}(T) = \int_0^{\infty} d\xi\,\mathbf{exp}\left(-\frac{\hbar^2/2I_A}{k_B T}\,\xi\right) = \frac{k_B T}{\hbar^2/2I_A}. \tag{7.40}$$

Joining all calculated factors, the average effective squared transition dipole is

$$\boxed{\left\langle M^2(\ell)\right\rangle_T = M_0{}^2\,\frac{\hbar^2/2I_A}{k_B T}(\ell+1)\,\mathbf{exp}\left(-\frac{\ell(\ell+1)\hbar^2/2I_A}{k_B T}\right)}, \tag{7.41}$$

for $k_B T >> \hbar^2/2I_A$.

For low ℓ transition orders, namely as long as $\ell(\ell+1)\hbar^2/2I_A \lesssim k_BT$, the exponential in Eq. (7.41) may be approximated as a unity, and the absorption lines' intensities will grow with ℓ. Beyond that limit, the absorption lines' intensities will reduce exponentially. Lowering of the temperature T increases the pre-exponential factor in Eq. (7.41), acting to increase the absorption lines' intensity. However, the exponential reduction in intensity starts at lower orders. Eq. (7.41) is no longer valid at the very low temperatures limit. To calculate $\mathcal{Z}(T)$ at intermediate temperatures it is possible to obtain a fair approximation by using only a small number of addends. At extremely low temperatures, $\mathcal{Z}(T)$ may be approximated as a unity. Then we get (see Eq. [7.39])

$$P(\ell,T) \cong (2\ell+1)\,\mathbf{exp}\left(-\frac{\ell(\ell+1)\hbar^2/2I_A}{k_BT}\right).$$

Then, for $k_BT << \hbar^2/2I_A$

$$\boxed{\left\langle M^2(\ell)\right\rangle_T \cong {M_0}^2(\ell+1)\,\mathbf{exp}\left(-\frac{\ell(\ell+1)\hbar^2/2I_A}{k_BT}\right)}. \tag{7.42}$$

The calculations of Equations (7.41) and (7.42) did not include the thermal occupation probabilities of the involved *vibrational* level. To complement calculation, each of the expressions there must be multiplied by the proper occupation probability; the reader is advised to carry out the calculation as an exercise.

Figure 7.6 shows the thermal average of the squared effective transition dipoles for pure rotational levels according to Eqs. (7.41) and (7.42) as a function of transition energy for several temperatures. Recall that the transition energy of order ℓ is given by $(\ell+1)\cdot\hbar^2/2I_A$ per Eq. (7.37). As expected, at high temperatures $k_BT = 10\hbar^2/2I_A$ and $k_BT = 30\hbar^2/2I_A$, the squared dipole moment (which is proportional to the absorption

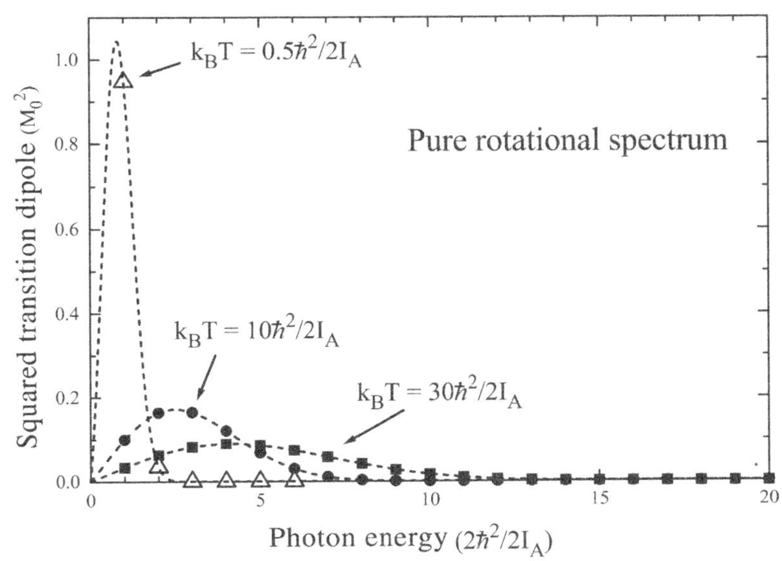

Figure 7.6 Spectral distribution of the effective squared transition moment as function of absorbed photon energy in pure rotational transition of a molecule for several high and low temperature limits. For details see text.

strength) exhibits a rise with growing order, reaching a peak and then falling. Particularly at low temperatures satisfying $k_BT = 0.5\ \hbar^2/2I_A$, only two transition lines may be observed, the $0 \to 1$ and $1 \to 2$.

A typical rotational energy ration in molecules is quite small. For example, in the hydrogen fluoride *HF* molecule it is 0.00248 eV (Table 7.1). For comparison, the room temperature characteristic energy is 0.026 eV, namely more than ten times greater. Therefore, the room temperature exhibits high-temperature conditions for hydrogen fluoride, but also for most other molecules, for which the rotational ration is much smaller (Table 7.1). Low-temperature conditions are the ones difficult to achieve, because most gases are already liquid or solid at the required low temperatures.

7.4.3 Mixed Vibration/Rotation Transitions

If the Δv value is non-zero, for example if $\Delta v = +1$ (a vibrational absorption), the spectrum will exhibit *two sets* of mixed vibration rotation transitions: one belonging to $\Delta \ell = -1$, called **P-branch**, and second belonging to $\Delta \ell = +1$, called **R-branch**. The P-branch transitions belonging to the vibrational $v = 0 \to v' = 1$ transition have been plotted on the energy levels scheme of Figure 7.4. The first transition energy in the P-branch is $\hbar\omega_0 - 2\hbar^2/2I_A$, and the transition energies become smaller, separated by a constant energy step $2\hbar^2/2I_A$. The first transition energy in the R-branch is $\hbar\omega_0 + 2\hbar^2/2I_A$, and the transition energies become larger, separated by a constant energy step $2\hbar^2/2I_A$. The entire spectrum is often addressed as a "single" absorption band that exhibits a fine structure of many, equally spaced lines. However, the absorption line of $\hbar\omega_0$, belonging to $\Delta\ell = 0$, is missing, being forbidden by the selection rules. The missing line position is called **band origin**, as it experimentally defines the vibrational energy separation. Figure 7.7 provides a schematic illustration of a mixed vibration/rotation absorption spectrum.

The spectral region belonging to the vibration/rotation transitions belongs to the infrared (IR) and the far infrared (FIR). The band center is determined by the vibrational energy $\hbar\omega_0$. For example, the $v = 0 \to v' = 1$ absorption transition in the hydrogen fluoride *HF* molecule belongs to the $\lambda = 2.43 \times 10^{-4}$ cm wavelength; in the sodium chloride *NaCl* molecule it would take place at $\lambda = 16.7 \times 10^{-4}$ cm (Table 7.1). Absorption transitions where the initial vibrational state is of a larger order will usually occur at longer wavelengths, because the energy separations between higher order states become smaller.

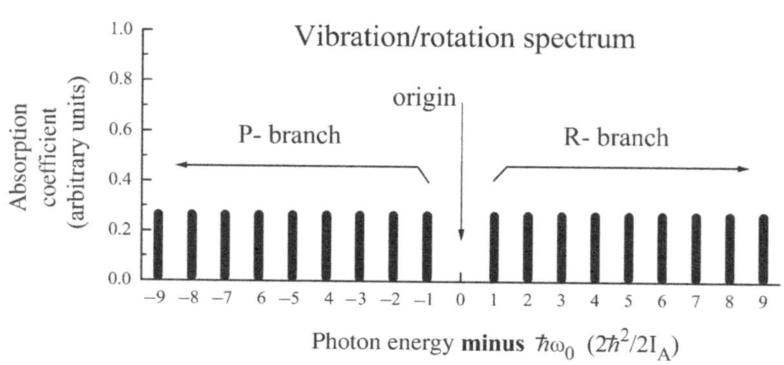

Figure 7.7 Schematic illustration of a typical absorption spectrum belonging to mixed vibration/rotation transitions in a diatomic molecule. The lines refer to the spectral energy positions, not to their relative strengths.

7.5 Electronic Transitions and the Franck-Condon Principle

7.5.1 General Considerations

Recall now that each molecular orbital exhibits its "private" potential function. In Section 6.2.3 we have calculated such potential curve using the LCAO approximation for the hydrogen molecule ion H_2^+ (see also Figure 6.4). We now consider optical transitions between electronic states belonging to different molecular orbitals. The molecular energy per orbital has an electronic component – which is the potential energy at the equilibrium position $E_{el}(\underline{R}_{u0})$; the vibrational energy, given as $E_{\rm v} = \left({\rm v} + \frac{1}{2}\right)\hbar\omega_0$; and the rotational energy, given as $E_\ell = \ell(\ell+1)(\hbar^2/2I_A)$ for a diatomic molecule. Still, the equilibrium positions $\underline{R}_{u0}$ at different molecular orbitals are also different, the vibrational frequencies ω_0 are different, and the moments of inertia I_A are different. It is also important to recall that most higher-electronic orbitals don't have a bonding potential (for example, the hydrogen molecule ion – see Figure 6.4). If, on the other hand, both involved electronic states do have a bonding potential, the absorption spectrum satisfies the following relation, required by the law of energy conservation:

$$\hbar\omega_T = E_{el}' - E_{el} + \left({\rm v}' + \frac{1}{2}\right)\hbar\omega_0' - \left({\rm v} + \frac{1}{2}\right)\hbar\omega_0 + \frac{\hbar^2}{2I_A'}\ell'\left(\ell'+1\right) - \frac{\hbar^2}{2I_A}\ell(\ell+1),$$

where ω_T is the angular frequency of the absorbed light, and the prime superscript refers to the quantum numbers and parameters of the upper electronic state. It is important to remember that electronic transition energies are large compared to both vibrational and rotational transition energies; they usually occur in the visible and ultraviolet. Vibrational transition energies are large compared to rotational transitions energies, and usually occur in the near infrared. Rotational transition energies usually occur in the far infrared and in the millimeter waves region.

Figure 7.8 displays an energy scheme belonging to two bonded molecular electronic states. It demonstrates a common case, where the vibrational energies belonging to the higher electronic state are smaller than those belonging to the lower electronic state. However, there are cases in which the higher electronic state doesn't have a bonding state. Section 6.2 discusses such a condition for the hydrogen molecule ion (specifically see Figure 6.4). An opposite case occurs, for example in diatomic molecules of the noble gases: these molecules have no bonded state based on ground-state orbitals. However, if one atom is electronically excited, it sometimes bonds to another atom, forming together a bonded molecular state that exhibits vibrational states. A molecule formed that way is called **excimer** (and sometimes **dimer**, in particular cases of only two bonded atoms). The potential function of the excimeric state in the $\imath \to \infty$ limit would be larger

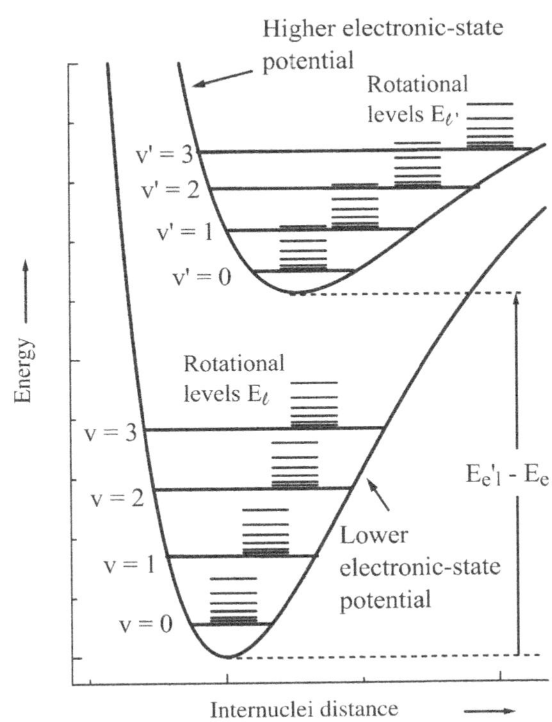

Figure 7.8 Fundamental scheme of potentials and vibrational and rotational states belonging to two different electronic states of a diatomic molecule.

than that of the two independent atoms in their ground state by the excitation energy of a single atom. There are also intermediate states, whereby bonded molecular states are formed by different electronic states of the two separate atoms; some, where the vibrational energies of the higher bonded states are larger than those belonging to the lower electronic state.

7.5.2 Selection Rules for Electronic Transitions

We turn now to reveal the selection rules for radiative transitions between two electronic states. The electric dipole formed when a specific independent electron orbits in a molecule has primarily two contributions: one from the distribution of the positive nuclei, the other from the distribution of the negative electrons. The contribution of the positive nuclei would be

$$\underline{M}^{+} = \sum\nolimits_u eZ_u(\underline{r} - \underline{R}_u) = e\underline{r}Z - eZ\underline{R}_{av} = eZ(\underline{r} - \underline{R}_{av}),$$

where $Z \equiv \sum_u Z_u$ is the sum of all atomic numbers of the nuclei Z_u; $\underline{R}_{av} \equiv \frac{1}{Z}\sum_u Z_u\underline{R}_u$ is the "electrostatic nuclei charge center"; $\underline{r}$ is the dynamic position of the electron $\underline{r} \equiv \underline{r}_{i_0}$, where i indicates the ordinal number of each electron and i_0 is the ordinal number of the specific electron under consideration. As to the contribution of the negative electrons, it would be

$$\underline{M}^{-} = -\sum_{i \neq i_0} e(\underline{r} - \underline{r}_i) = -e(Z-1)(\underline{r} - \underline{r}_{av}) = -eZ(\underline{r} - \underline{r}_{av}) + e(\underline{r} - \underline{r}_{av}),$$

where $\underline{r}_{av} = \frac{1}{Z-1}\sum_{i \neq i_0} \underline{r}_i$ is the "electrostatic electrons charge center", *except* for the i_0 electron. Summing the two dipole moment contributions, the total dipole moment is given by

$$\underline{M} = \underline{M}^{+} + \underline{M}^{-} = eZ(\underline{r}_{av} - \underline{R}_{av}) + e(\underline{r} - \underline{r}_{av}). \tag{7.43}$$

Statistical considerations are sufficient to conclude that dynamic variations in location of the average electrostatic charge centers $\underline{r}_{av}$ and $\underline{R}_{av}$ for time resolutions short compared to even a single period of the specific i_0 electron are extremely small compared to this particular electron amplitude. For that reason, we consider the said charge centers as averages over a small, yet integer number of its oscillations. Because all electrons exchange roles through *all* orbitals, $\underline{r}_{av}$ is in fact the electrostatic charge center of *all* electrons. The first addend on the right side of Eq. (7.43), $\underline{M}_{Sk} \equiv eZ(\underline{r}_{av} - \underline{R}_{av})$, is the dipole moment of the molecular skeleton (which is the one related to the vibration/rotation transitions; see Section 7.3.5); we presented its empiric analytical dependence on the internuclei distance in a diatomic molecule in Eq. (7.34). The second addend on the right side of Eq. (7.43) is the electronic dipole moment $\underline{M}_e \equiv e(\underline{r} - \underline{r}_{av})$; in summary, we may write

$$\underline{M} = \underline{M}_{Sk} + \underline{M}_e. \tag{7.44}$$

The electronic transition moment between the initial "i" and final "f" states is given by $\langle \boldsymbol{\Phi}^{\mathbf{f}} | \underline{M} | \boldsymbol{\Phi}^{\mathbf{i}} \rangle$ (see for example Eq. [2.35]). In Eq. (7.4) we performed variables separation in the molecular wave function into an "electronic" part and a "nuclei" part, in the form $\boldsymbol{\Phi}(\underline{R}_u, \underline{r}_i) = \psi(\underline{R}_u, \underline{r}_i) \cdot \boldsymbol{S}(\underline{R}_u)$. We have further used the Born-Oppenheimer approximation to ignore all derivatives of the **electronic wave function** $\psi(\underline{R}_u, \underline{r}_i)$ by the nuclei $\underline{R}_u$ coordinates; namely assuming that the $\psi(\underline{R}_u, \underline{r}_i)$ dependence on the nuclei coordinates is negligibly weak. Accordingly, the equilibrium position coordinates $\underline{R}_{u0}$ are selected as representative of $\underline{R}_u$ in $\psi(\underline{R}_u, \underline{r}_i)$, and for brevity, their inclusion as parameters of $\psi(\underline{R}_u, \underline{r}_i)$ is removed. The reason for selecting the $\underline{R}_{u0}$ set rather than any other set of $\underline{R}_u$ coordinates is that the variation of the molecular potential exhibits a zero tangent at that very set. The electronic functions are written simply as $\psi(\underline{r}_i)$. In continuation to the assumption by Eq. (7.4), we write $\langle \Phi^f | \underline{M} | \Phi^i \rangle = \left\langle \psi^f \mathcal{S}^f \middle| \underline{M}_{Sk} + \underline{M}_e \middle| \psi^i \mathcal{S}^i \right\rangle$

$$= \left\langle \psi^f \mathcal{S}^f \middle| \underline{M}_{Sk} \middle| \psi^i \mathcal{S}^i \right\rangle + \left\langle \psi^f \mathcal{S}^f \middle| \underline{M}_e \middle| \psi^i \mathcal{S}^i \right\rangle = \left\langle \psi^f | \psi^i \right\rangle \left\langle \mathcal{S}^f \middle| \underline{M}_{Sk} \middle| \mathcal{S}^i \right\rangle + \left\langle \mathcal{S}^f | \mathcal{S}^i \right\rangle \left\langle \psi^f \middle| \underline{M}_e \middle| \psi^i \right\rangle.$$

In the last expression, the first addend vanishes because of orthogonality among the electronic ψ functions. The physical meaning is that the selection rules for vibrational/rotational transitions discussed in the previous section (Eq. [7.35]) are not valid when the vibration/rotation states belong to different electronic states. The transition dipole will eventually be given as

$$\boxed{\langle \underline{M} \rangle = \left\langle \mathcal{S}^f | \mathcal{S}^i \right\rangle \left\langle \psi^f \middle| \underline{M}_e \middle| \psi^i \right\rangle} \quad . \tag{7.45}$$

Thus, transitions selection rules result from those related to the mutually independent electronic dipole transition $\langle \psi^f | \underline{M}_e | \psi^i \rangle$, and the $\left\langle \mathcal{S}^f | \mathcal{S}^i \right\rangle$ integral. The latter integral is expected to express the influence of the vibrational/rotational states on the probability for electronic transitions. It appears valid concerning the vibrational states. However, for rotational states, the considerations require modification because the angular momentum of the molecular skeletal rotation should be coupled with the electronic angular momentum. Section 7.5.6 elaborates on that issue.

In Section 7.3.4, we performed variable separation in the $\mathcal{S}$ wave function for a diatomic molecule in the form

$$\mathcal{S}(\mathcal{r}, \theta, \varphi) = \mathcal{r} \cdot \mathcal{V}(\mathcal{r}) \boldsymbol{Y}_{\ell m}(\theta, \varphi). \tag{7.46}$$

We should thus write

$$\left\langle \mathcal{S}^f | \mathcal{S}^i \right\rangle = \left\langle \mathcal{r} \mathcal{V}^f_{v'}(\mathcal{r}) | \mathcal{r} \mathcal{V}^i_{v}(\mathcal{r}) \right\rangle \langle \boldsymbol{Y}_{\ell' m'}(\theta, \varphi) | \boldsymbol{Y}_{\ell m}(\theta, \varphi) \rangle . \tag{7.47}$$

The first factor on the right side of Eq. (7.47) is the **overlap integral** or **Franck-Condon integral**. We now elaborate on the rules emerging from that integral concerning *vibrational transition*. We momentarily overlook conclusions related to the second factor; we have already indicated that they are not valid unless the molecular rotation angular momentum is coupled to the electronic angular momentum. Importantly, each vibrational wave function $\mathcal{V}_v(\mathcal{r})$ is labeled by its relation to the final *f* state or the initial *i* state. It emphasizes that the $\mathcal{r}\mathcal{V}_v(\mathcal{r})$ functions belonging to different electronic states are not at all orthonormal! It is sufficient for the equilibrium distances $\mathcal{r}_0$ to be different, to "spoil" any orthonormality even if the potential shape would be otherwise identical, or even harmonic; especially because the potential shapes are never identical. The $\left\langle \mathcal{r} \mathcal{V}^f_{v'}(\mathcal{r}) | \mathcal{r} \mathcal{V}^i_{v}(\mathcal{r}) \right\rangle$ integral never vanishes for any combination between v and v′. Therefore, there are no definite selection rules for transitions between different electronic orbitals that involve vibrational states changes, and *all transitions are allowed to a certain degree*. Differences among transition strengths depend on the $\left\langle \mathcal{r} \mathcal{V}^f_{v'}(\mathcal{r}) | \mathcal{r} \mathcal{V}^i_{v}(\mathcal{r}) \right\rangle$ integral value.

Figure 7.9 presented vibrational energy schemes for different electronic orbitals. The separations among the vibrational states have been grossly exaggerated compared to the electronic excitation energy. The rotational levels are not presented; they form a fine structure of the vibrational ones. If the transitions are measured by a low spectroscopic resolution, the vibrational absorption (or emission) transitions would look like continuous **bands**. The absorption bands are classified into two sets. One set includes transitions whereby the final vibrational state remains the same, but the initial vibrational state grows. This set is called **v-progression**. The transition energies in the v-progression turn smaller, with separations

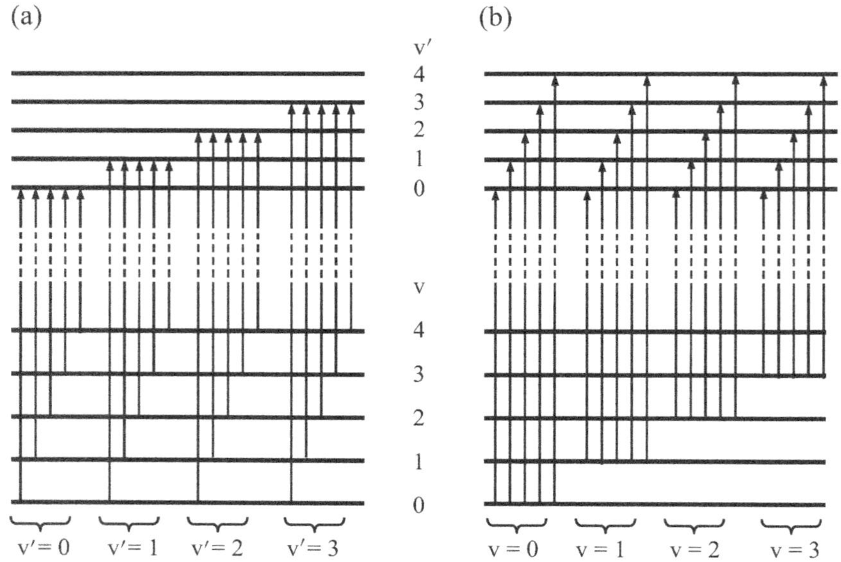

Figure 7.9 Vibrational energy level schemes belonging to different electronic states. The scheme demonstrates absorption band progressions belonging to different vibrational states, (a) v-bands progression; (b) v'-bands progression. Only five states are shown per electronic orbital.

representing the energetic differences between adjacent vibrational states of the lower electronic orbital. The second set includes transitions whereby the initial vibrational state remains the same, but the final vibrational state grows. This set is called **v′-progression**. The transition energies in the v′-progression grow larger, with separations representing the energetic differences between adjacent vibrational states of the higher electronic orbital. Experimentally, it is possible to discriminate between the two progressions by the fact that the energy separations among states in the v′ progression are usually smaller than in the v-progression. The $v = 0 \rightarrow v' = 0$ transition is called **zero-zero transition**, and it appears common in both v-, and v′-progressions.

Figure 7.10 demonstrates a special, yet common, case, in which the vibrational energy of the upper electronic state is indeed smaller than of the lower electronic state, but not by much. The absorption spectrum will then exhibit lines gatherings, each gathering belonging to a constant $v' - v$ difference. These gatherings are called **sequence**. The figure shows the $v' - v = 1$ sequence; the sequence of its left belongs to $v' - v = 0$, and the one on its right obviously belongs to $v' - v = 2$.

7.5.3 Temperature Dependence of the Electronic Transitions Spectrum

Experimental discrimination between the different spectral progressions is obtained by varying the molecular gas temperature. Recall the Boltzmann population ratio (Eq. [2.38]): $N_{v2}/N_{v1} = \exp(-(E_{v2} - E_{v1})/k_B T)$, where N_{v1} and N_{v2} are the molecules' populations in the vibrational E_{v1} and E_{v2}, respectively, and k_B is the Boltzmann constant. Lowering of the gas absolute temperature T reduces the population of the higher vibrational states of the lower electronic state. Absorption intensities of transitions starting in those levels are then reduced. The mere fact that the temperature affects absorption intensities allows discrimination among the temperature-dependent transitions. It is often possible to lower the temperature to such an extent, that the thermal energy $k_B T$ becomes smaller than the vibrational ration $\hbar\omega_0$. All v′ progressions, except for that belonging to $v = 0$ one, will practically disappear, and the v-progression will exhibit a single

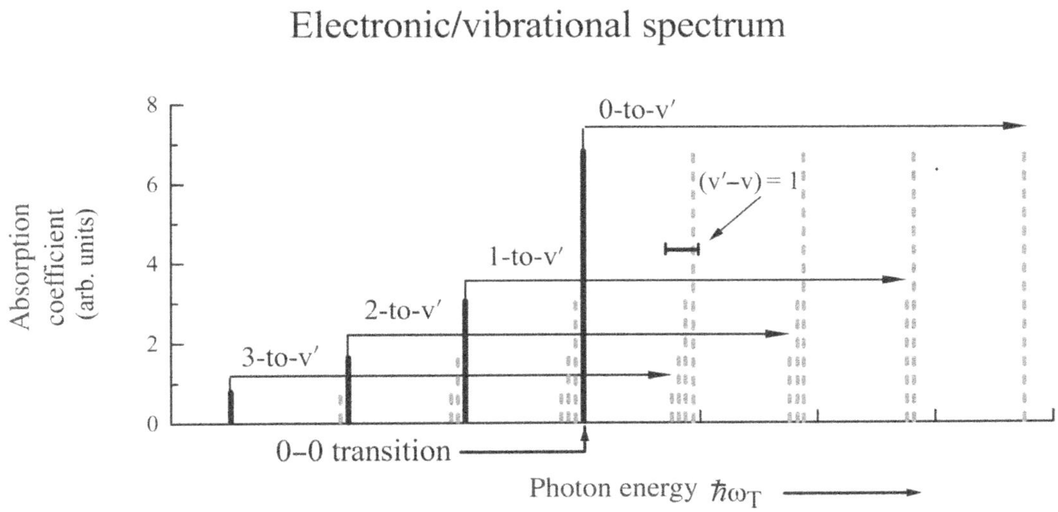

Figure 7.10 Schematic of a typical absorption spectrum belonging to electronic transitions that involve different vibrational levels: v-progression bands shown by solid lines; corresponding v'-progression bands – by same height bright dashed lines. A case of relatively low temperature is demonstrated, causing the v-progression bands to become smaller.

line. For example, any temperature lower than 5900°K suffices for that purpose for *HF* molecules. For sodium chloride (table salt) *NaCl* molecules, only a temperature lower than 550°K is needed (Table 7.1). The latter is quite impractical, because the boiling temperature of *NaCl* is 1686°K. We won't elaborate further on developing the analytical expressions describing the temperature dependence of the relevant absorption spectra; this is because differently from the rotational spectra (Section 7.4.2), the dipole transition calculations must involve the overlap integral that lacks a simple analytical structure. We recommend addressing that issue as an exercise.

7.5.4 The Franck-Condon Principle

Based on the understanding that the ground-state population is highest at all temperatures, the zero-zero transition of the v-progression was drawn highest in Figure 7.10. However, this transition strength is usually smaller than others because of the relative size of the $\left\langle \mathscr{r}\mathcal{V}^{f}_{v'}(\mathscr{r}) | \mathscr{r}\mathcal{V}^{i}_{v}(\mathscr{r}) \right\rangle$ overlap integrals (the transition strength is proportional to the squared size of that integral). We thus concentrate on a fundamental understanding of the radial part of the overlap integral.

The $\mathcal{V}_v(\mathscr{r})$ functions are non-zero, changing strongly only across a narrow region of the internuclei distance. Therefore, multiplication by $\mathscr{r}$ or its powers provides no considerable change among the relative sizes of the overlap integrals. Consideration of the $\left\langle \psi^{f}_{v'}\left(\mathscr{r} - \mathscr{r}^{f}_{0}\right) \mid \psi^{i}_{v}\left(\mathscr{r} - \mathscr{r}^{i}_{0}\right) \right\rangle$ integral is thus sufficient for a qualitative understanding (similar considerations were employed in the N_v normalization constant calculation; see Eqs. (7.32) and (7.33), and relevant text). Figure 7.11 provides a fundamental energy scheme of two electronic states and four vibrational states belonging to each. The function $|\psi_v(\mathscr{r} - \mathscr{r}_0)|^2$ for each vibrational level is sketched according to the harmonic approximation, where $\mathscr{r}_0$ is chosen as the center position between the edges of the vibrational classical amplitude. (In principle, $\psi_v(\mathscr{r} - \mathscr{r}_0)$ should have been plotted, rather than its square; however, for the following simplistic qualitative consideration, this presentation is quite useful). The $|\psi_v(\mathscr{r} - \mathscr{r}_0)|^2$ is approximately proportional to the nuclei probability to be

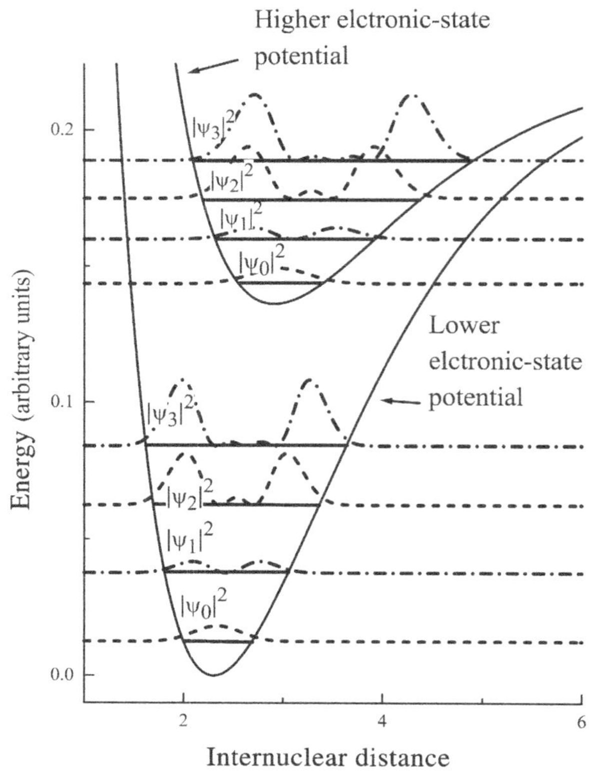

Figure 7.11 Squared wave functions (probability density of the internuclei distance) related to vibrational energy states of two different electronic states in a diatomic molecule. Cases belonging to symmetric wave functions (relative to the vibration center) are plotted by dashed lines; the antisymmetric cases are plotted by dash-dotted lines.

at distance r. In the zero-order vibration, most probable is the vibration center. As the vibrational order increases, the probability for the center position is reduced, and that of the classical edges increases. The vibrations then turn like a classical vibration, where the vibrating body crosses fast through the equilibrium position, but stays longer near the edges, where the moving speeds diminish.

For example, we assess the expected size of the $\left\langle \psi_0^f\left(r - r_0^f\right) | \psi_0^i\left(r - r_0^i\right)\right\rangle$ integral (the zero-zero transition) for the case described in Figure 7.11. Because each probability is distributed about a different point, the wave functions product is small, and its end size is very small compared to an assumed case of full overlap. Larger overlaps are obtained with the wave functions of higher order belonging to the upper electronic state. It will be expressed in the transition strengths of the v′ progression belonging to v = 0: the zero-zero transition would be weak, and the higher-order transitions stronger. They would reach a maximum at some order, and then turn smaller.

As another example, we assess the expected size of the $\left\langle \psi_{v'}^f\left(r - r_0^f\right) \mid \psi_1^i\left(r - r_0^i\right)\right\rangle$ integral (the $1 \rightarrow v'$ transition). The right-side lobe of $|\psi_1^i|^2$ exhibits growing overlaps with the left-side lobes of $\left|\psi_{v'}^f\right|^2$, and they all are larger than the $|\psi_0^i|^2$ overlaps with the various $\left|\psi_{v'}^f\right|^2$ ones. In certain cases of absorption spectroscopy measurements, this might compensate against the lower thermal population of the higher-order vibrational states belonging to the lower electronic state.

The presented qualitative consideration is suitable to lower-order vibrational transitions between the two electronic states. If the difference between the orders is large, the higher-order wave function will oscillate (that is, inverse sign) several times more than lower order one. The overlap integral becomes smaller, correspondingly altering the transition bands intensities. A complete accurate image of the absorption bands' intensities requires direct theoretical calculations of the overlap integrals, with additional calculation of the Boltzmann thermal population relations.

Franck and Condon proposed a simplistic method for qualitative assessment of the overlap integrals. It is assumed that in a zero-order vibration, the system is *always resting* at the equilibrium internuclear distance. For higher-order vibrations, the system is half-time resting at one side edge of the classical amplitude, and half-time resting at the other side of the classical amplitude. These positions are addressed as **exit points**. Using the energy scheme of Figure 7.11, vertical lines are plotted from an exit point of the lower electronic vibrational state till crossing the upper electronic potential function. A vibrational upper electronic state whose exit point is closest to the potential crossing point will provide the largest overlap integral. Because the higher-order vibrational states have two exit points, and because the potential functions are nonsymmetric with respect to r_0, there will always be two upper-state levels providing maximal overlap integral.

The Franck-Condon simplistic approach exhibits a reasonable agreement with experimental results, as well as with more-accurate theoretical calculations, especially regarding general relations among optical transitions strengths. The agreement may be classically accounted for by assuming that an optical transition between states is extremely fast, requiring a single electronic oscillation time period, namely only several femtoseconds ($1\text{fs} \equiv 10^{-15}\text{s}$); a vibrational oscillation time period is orders of magnitude longer: several picoseconds ($1\text{ps} \equiv 10^{-12}$ s). By that consideration, the start of absorption would be most probable from an exit point of the lower vibrational states. The internuclei distance won't exhibit any significant change during transition (which is the reason for the plotting of the vertical lines). The transition would stimulate an upper-state vibration, whose amplitude edge is closest to the internuclei distance from which the absorption started. This interpretation is called **Franck-Condon principle**.

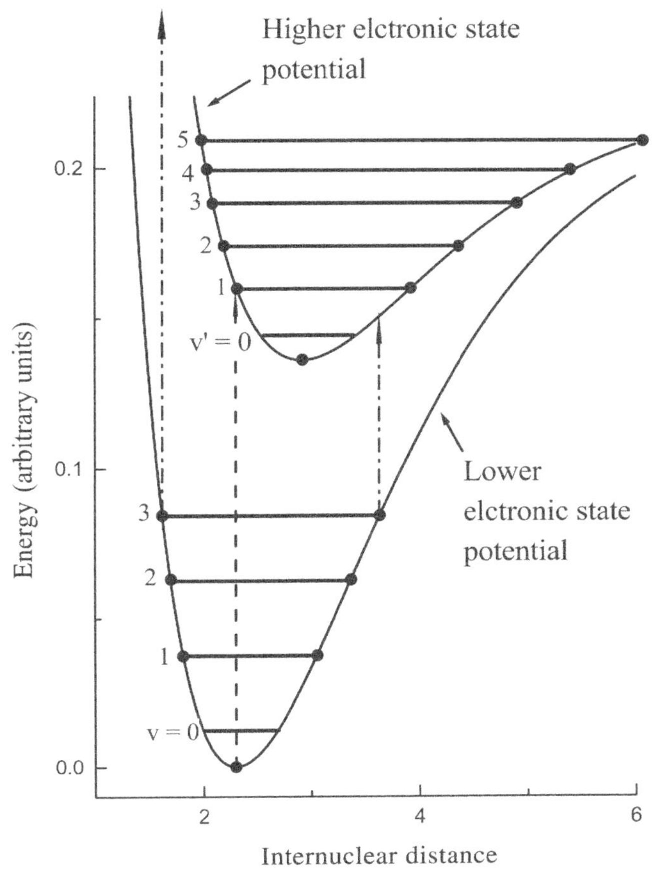

Figure 7.12 Demonstration of the Franck-Condon principle use for finding the dominant vibration-vibration absorptions between two different electronic levels. The transitional exit points are marked by full circles. See text for detailed discussion.

To demonstrate the use of the Franck-Condon principle, we consider Figure 7.12. The figure describes the potential functions of two electronic states and several belonging vibrational levels. The exit points of each vibrational level is marked by full circles. Two absorption transitions are indicated: one from v = 0 state, the second from the v = 3 state. The vertical arrow plotted from the exit point of the ground vibrational v = 0 state crosses the upper electronic state potential close to the left exit point of the v′ = 1. The most probable (strongest) transition would then be 0 → 1. Two arrows are plotted for the v = 3 vibrational state. The arrow emerging from the left exit point crosses the upper electronic state potential at a very high vibrational state, a one belonging to the vibrational continuum causing molecular dissociation (Figure 7.1). The probability (strength) for such transition is small, because of their continuum nature. In most diatomic molecules, these transitions belong to the ultraviolet region. The arrow emerging from the right exit point crosses the upper electronic state potential close to the v′ = 1 vibrational state. By the Franck-Condon principle, the most probable absorption transition would be 3 → 1.

7.5.5 Fluorescence and Stokes-Shift

Considerations concerning emissions transitions strengths are identical to those of the absorption ones. In many practical cases, it is possible to assume that in a molecular gas, the populations of the upper electronic state vibrations (as well as those of the lower electronic state) closely simulate a thermal equilibrium; it is thus addressed as "quasithermal equilibrium". Because of the need to involve vibrational transitions, radiative lifetimes of electronically excited molecules are usually longer than several hundred nanoseconds (long compared to the 10^{-8} *s* of an excited electric dipole lifetime). Many spectroscopic studies are performed at room temperature under atmospheric pressure. Under such conditions, the mean time between collisions among molecules is much shorter – of the order of only 10 ns. Overlooking the need to specify the interactions that define the "collision" concept, the vibrational states of each molecule changes on collision, and the vibrational states quickly reach a quasithermal equilibrium distribution (Boltzmann distribution – Eq. [2.38]). The mean time between collisions lengthens when pressure is lowered; however, the gas molecules still collide with the gas-container walls, leading to the same distribution. It is possible to achieve meaningful deviation from quasi-thermal equilibrium distributions only if the gas is kept under very low pressure, and/or under very low temperatures, in very large containers.

The above considerations express themselves in fluorescence processes of molecular gases. The term **fluorescence** describes a radiative decay of excited electronic states of an atom or a molecule. Excited states could be formed in several ways, like a chemical reaction whereby the molecule is the final product, by causing an electrical discharge through the gas, or by light absorption. The latter is usually called **photo-fluorescence/luminescence**, or **resonance fluorescence/luminescence**; a monochromatic light at proper wavelength may be absorbed by a molecule, transforming it from the zero-order vibrational state of the lower electronic level to a higher vibrational state of the upper electronic level. If the molecular gas has a very low pressure, the excited state will remain unaltered, except by light emission returning the molecule into a vibrational level of the lower electronic state. The fluorescence spectrum shows v-progression bands, whose wavelengths are longer than or equal to the exciting light wavelength. The shift of emission bands toward longer wavelengths compared to the exciting light wavelength is called **Stokes-shift**. Figure 7.13b demonstrates such case on the same energy scheme as in Figure 7.13a. If the gas pressure increases such that the vibrational levels of the upper electronic states assume a quasithermal

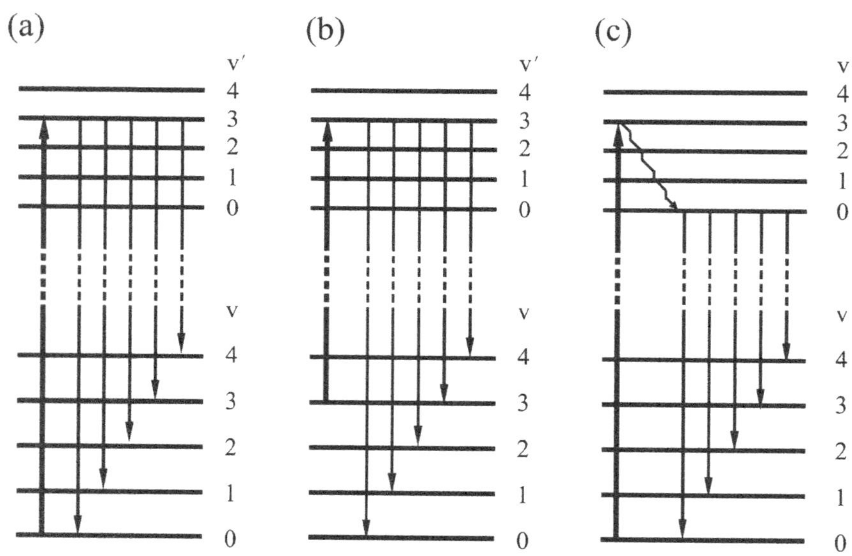

Figure 7.13 Fundamental energy scheme of vibrational levels belonging to different electronic orbitals and different optical transitions. Absorption transitions – thick lines; emission transitions – thin lines. (a) v-Progression of photo-luminescence bands exhibiting a Stokes shift; (b) v-progression of photo-luminescence bands partially exhibiting an anti-Stokes shift; (c) v-progression of photo-luminescence bands always exhibiting a Stokes shift, because all belong to an initial zero vibrational level of the upper electronic state, reached by nonradiative decay from higher vibrational levels toward a quasithermal equilibrium distribution.

equilibrium distribution, the larger portion of the fluorescence band will start from its lowest vibrational level. Figure 7.13c demonstrates such a case on the same energy scheme as in Figure 7.13a. The Stokes shift is larger than in Figure 7.13a. Photoluminescence measurements of molecular gases as a function of temperature and pressure establishes a means to identify the states' scheme structure in molecules.

The energy scheme in Figure 7.13 refers only to the states' energies and the optical transitions. To assess the transitions' relative strengths, the overlap integrals must be calculated independently.

7.5.6 Selection Rules for Electronic Transitions Including Vibrations and Rotations

To obtain the selection rules concerning electronic transitions that involve rotations of the molecular skeleton, it is imperative to couple the molecular angular momentum with the electronic one. This coupling has only a tiny effect on the energy states, but it is important for determining the states' wave functions symmetry. The vibrational states play no role in this coupling, as described by the Franck-Condon integral (Eq. [7.47] and following text). Without the said coupling, the selection rules would be derived from the $\langle \boldsymbol{Y}_{\ell'm'}(\theta,\varphi)|\boldsymbol{Y}_{\ell m}(\theta,\varphi)\rangle$ product of Eq. (7.47). The $\boldsymbol{Y}_{\ell m}$ spherical harmonics are orthonormal, namely $\langle \boldsymbol{Y}_{\ell'm'}(\theta,\varphi)|\boldsymbol{Y}_{\ell m}(\theta,\varphi)\rangle = \delta_{\ell'\ell}\delta_{m'm}$. The resulting selection rules would be $\Delta\ell = 0$; $\Delta m = 0$, with additional separate and independent selection rules for the electronic transition. As we will see, in special cases, transitions fulfilling these rules are indeed allowed.

Recall Section 4.8, where we made a detailed calculation of the coupling between the electronic spin and the orbital angular momentum of the hydrogen atom. It is possible to approach the present issue in an analogous way. However, in the present case, instead of using the zero-order assumption that the rotations are independent, and perform the coupling as a correction, we assume the presence of the coupled state as a starting point!

We consider a free molecule, namely a species not being subjected to any external rotational moment. The total angular momentum is then an invariant of motion, and not each of the separate contributions of neither the molecular skeleton rotation nor of the electronic orbitals. We return to the classical view, and describe the diatomic molecule as a rigid symmetric top, while overlooking the nuclei and electrons substructures contributions. We rewrite the expression for the system kinetic energy (Eq. [7.18]):

$$E_{kin} = \frac{{L_A}^2 + {L_B}^2}{2I_A} + \frac{{L_C}^2}{2I_C}, \tag{7.48}$$

where L_A, L_B, and L_C are the angular momentum vector $\underline{L}$ components along the A, B, and C principal axes of the moments of inertia tensor, respectively. In the specific case of the diatomic molecule, the C axis is selected as the one connecting the two nuclei; it is obviously a principal axis of the moments of inertia tensor, and I_C is the moment of inertia about that axis. No more do we assume that I_C diminishes, as we did in relation to Eq. (7.19); we accept it to have a positive final value determined by the electrons' mass distribution about the C axis. The two other axes, A and B, lie on a plane perpendicular to the C axis, crossing the molecular mass center. Their actual orientation in that plane can be chosen arbitrarily, as long as they are mutually perpendicular. The moments of inertia about those axes are identical, marked I_A. The electrons distribution to I_A is tiny compared to the contribution of the nuclei, but they provide the dominant contribution to I_C. Eq. (7.48) may be rewritten in the form

$$E_{kin} = \frac{L^2}{2I_A} + {L_C}^2\left(\frac{1}{2I_C} - \frac{1}{2I_A}\right), \tag{7.49}$$

where $L^2 = L_A{}^2 + L_B{}^2 + L_C{}^2$. The eigenvalues of the molecular kinetic energy would then be given as

$$E_{kin} = \frac{\hbar^2}{2I_A}\ell(\ell+1) + \hbar^2\left(\frac{1}{2I_C} - \frac{1}{2I_A}\right)\lambda(\lambda+1). \tag{7.50}$$

In the specific case of a diatomic molecule, λ expresses the quantum number of the electron's effective *orbital* angular momentum component along the *C*-axis (see Sections 6.3.2 and 6.4, especially Eq. [6.21]). Overlooking possible electronic spins contribution, $\lambda = 0, 1, 2, 3, \cdots$. Correspondingly, $\ell = \lambda, \lambda+1, \lambda+2, \cdots$ is the total orbital angular momentum of the *molecule*. In a specific case where $\lambda = 0$, the expression for the kinetic energy is identical to Eq. (7.22). Note that the electronic angular momentum described by the λ quantum number is actually a component of the total molecular angular momentum. Because any vector size could not be smaller than it components, a situation where $\lambda < \ell$ is simply impossible. Thus, when $\lambda > 0$, the lower molecular rotational states will be absent from the molecular energy scheme.

Performing detailed wave functions calculations of a symmetric top (in either diatomic or multiatomic form), their resulting transitional strengths, and selection rules, requires the solving of a differential hypergeometric equation. The calculations are long and complicated, and won't be presented in this book. We just quote their resulting selection rules,

$$\boxed{\begin{array}{l} (i)\ \Delta\ell = \pm 1 \text{ if the two electronic states belong to } \lambda = 0; \\ (ii)\ \Delta\ell = 0, \pm 1 \text{ if at least one electronic state belongs to } \lambda \neq 0. \end{array}} \tag{7.51}$$

As to the spatial part of the electronic transition, it is calculated independently of the rotational transitions according to the principles of the Born-Oppenheimer approximation.

The selection rules obviously cast on the radiative absorption and emission oscillator strengths. In the $\lambda = 0$ case, the $\Delta\ell = \pm 1$ selection rule for the initial and final electronic states (that may be the same) is identical to Eqs. (7.29) and (7.36). Pure rotational transitions will look the same as described in Sections 7.4.1 and 7.4.2. Mixed vibration/rotation transitions would exhibit the **P** and **R** branches described in Section 7.4.3, which belong to the $\Delta\ell = -1$, and $\Delta\ell = +1$ cases, respectively. However, if $\lambda = 1$ in at least one of the electronic states (initial or final), an additional branch will appear, belonging to the $\Delta\ell = 0$ case, called **Q-branch**. The Q-branch lines will be given by

$$\Delta E_\ell = \frac{\ell(\ell+1)\hbar^2}{2I_A^f} - \frac{\ell(\ell+1)\hbar^2}{2I_A^i} = \ell(\ell+1)\hbar^2\left[\frac{1}{2I_A^f} - \frac{1}{2I_A^i}\right], \tag{7.52}$$

for $\ell = \lambda, \lambda+1, \lambda+2, \cdots$. The f and i superscripts relate, respectively, to the final and initial states' moments of inertia. The differences between the moments of inertia are tiny, especially if the transition does not include an electronic transition, but only a vibrational level transition within the same electronic state. Therefore, the line positions are very dense compared to those of the **P** or **R** branches. Measuring their increment allows accurate estimation of the difference between the molecular moments of inertia in the different states. Because $I_A = \mu r_0{}^2$, it allows estimation of the difference between the internuclei average distances. Usually, (but not always!), the moments of inertia of higher states are the larger. Then, in absorption, the separation between the inverse moments appearing in Eq. (7.52) is negative. An increasing order of ℓ in the Q branch will thus be expressed by a decrease in the transition energy. The zero-order transition will always be absent.

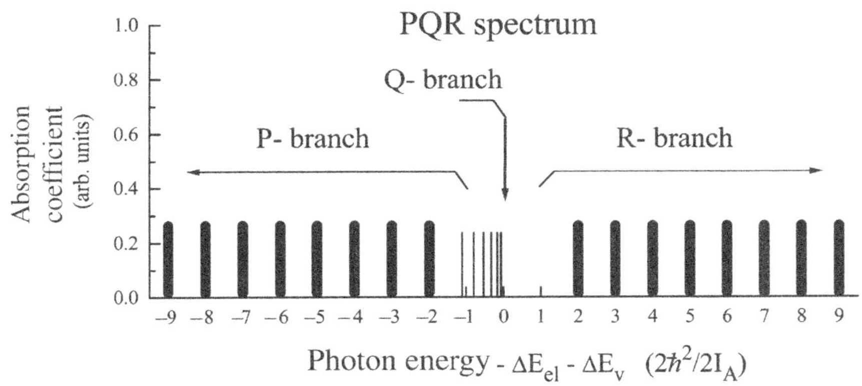

Figure 7.14 Schematic illustration of a **PQR** spectrum in mixed vibration/rotation transitions where $\lambda = 1$ in both initial and final states. The separation between adjacent lines of the **Q**-branch is exaggerated compared to separations between lines in both **P** and **Q** branches. No strength differences among lines have been expressed.

An absorption band containing the three branches, P, Q, and R, is called **PQR-band**. Figure 7.14 illustrates a schematic structure of such PQR band structure, for a case where the electronic orbital angular momentum quantum number is $\lambda = 1$ for both initial and final states. Such a case occurs if the electronic configuration remains the same after the electronic transition, for example, in a $\Pi \rightarrow \Pi$ transition; obviously also in vibration/rotation transitions belonging to the same electronic Π state. In such a case, the first absorption transitions in the P and R branches are absent. If, on the other hand, the electronic transitions belong to a $\Pi \rightarrow \Sigma$ transition, only the first transition in the R branch will be missing. Thus, the absence or presence of certain spectral transitions reveals the orbital angular momentums of the involved electronic states.

As in the case of the pure rotational transitions, the temperature dependence of the Q-transitions comprising the Q-branch depend on the relative occupation of the initial rotational states (Section 7.4.2). Eqs. (7.41) and (7.42) then provide a faithful description of the Q-branch spectral shape in the high and low temperature ranges, respectively.

It should be stressed that there are two additional, very important factors, expressed in the spectrum: the electronic spin states and the internal nuclear spin states. They may be expressed by bands splitting, and in absence or additions of spectral absorption lines. Such experimentally measured effects assisted in revealing the occurrence of elemental atomic isotopes, and assessment of their relative abundance. We won't elaborate on this issue.

Finally, note that spectroscopic resolution of different absorption lines in the Q-branch is quite difficult. Without specially designed spectrometers, the Q-branch appears as a single band, broader and stronger than the separate lines of the P and R branches.

Exercises for Chapter 7

7.1 Calculate a general expression for the equilibrium internuclei distance of a diatomic molecule whose nuclei are bonded by a Linett potential (Eq. [7.14]), its dissociation energy, and the zero-order frequency.

7.2 Calculate the first-order perturbation theory correction to the state energy of the Hamiltonian of Eq. (7.26), resulting from the *square* component in the series expansion of the $\hbar^2\ell(\ell+1)/2\mu r^2$ element appearing in the exact Hamiltonian of Eq. (7.23); assume the approximated harmonic potential $U(r)$ about $r = r_0$.

7.3 Make an accurate calculation of the integral appearing in the normalization relation of the vibrational wave function in Eq. (7.32) for the $\mathrm{v} = 1$ vibrational order, and determine the N_1 normalization factor.

7.4 Calculate numerically the wavelength and angular frequency of the pure rotational $5 \longrightarrow 6$ transition in the *HCl* molecular gas. Calculate the average oscillator strength when the gas temperature is 200°C. Use Eq. (7.41) for the calculation of the transition dipole moment; see for example the text preceding Eq. (2.58) for calculation of the oscillator strength.

7.5 Correct the calculation of the transition dipole moment for pure rotational transitions of Eqs. (7.41) and (7.42) by inclusion of the thermal occupation probability of the v-order vibrational level. Use Eq. (2.25).

7.6 Consider a diatomic molecule whose lowest vibrational state energy is 0.05 eV. The overlap integral for the $3 \longrightarrow 0$ transition is five times larger than the overlap integral of the $0 \longrightarrow 0$ transition of the same progression. What should be the temperature of that diatomic gas to achieve identical strength for these two transitions?

7.7 The characteristic vibrational ground-state energy in the carbon monoxide *CO* molecule, which is a Σ state, is 2100 cm^{-1}. It is 1530 cm^{-1} in one of the electronically excited states, which is a Π state.
(a) Use the empiric Morse relation (First paragraph following Eq. [7.17]) to calculate the molecular bond length in each of the states.
(b) Calculate the rotational energy portion of the molecule existing in each of the electronic states.
(c) Describe graphically the Q-branch electronic absorption spectrum at a 500°C gas temperature. Make sure to characterize the lines by their energetic separations and relative strengths.

8 Internal Oscillations of Polyatomic Molecules

8.1 Preamble

This chapter details the vibrational and rotational motions modes of polyatomic molecules, and relates their symmetry to that of their ideal-rigid structure. The mathematical instrument for forming that relation is group theory. The use of group theory extends to many other physical areas; because the electronic part of the molecular Hamiltonian carries the same symmetry properties, a notable use is classification of electronic energy levels. Classification of the molecular oscillations further allows the presentation of selection rules for transitions between vibration-rotation states: electric dipole transitions, as well as higher order ones like Raman scattering. We won't present proofs of the mathematical theorems we use, and the interested reader should refer to the vast literature and books dealing with the subject. However, we attempt to demonstrate physical consequences of the used abstract mathematical expressions. Organized methods will be presented to allow even non-experts to calculate the symmetry species of a molecule and of its oscillations.

8.2 Zero-Order Mechanical Energy Approximation of a Polyatomic Molecule

The starting point to the following calculations is the Born-Oppenheimer approximation of Section 7.2. The approximation obviates the need to provide a simultaneous description of all nuclei and electrons, and allows a simplified presentation of the molecular motions. The state of a molecule comprising N nuclei is determined by $3N$ position coordinates of the nuclei. The linear translational motion of a rigid molecule consumes three degrees of freedom, and the rigid molecule rotation consumes three additional degrees of freedom. We are thus left with $3N - 6$ degrees of freedom to describe the vibrational motions. In a linear molecule, where all atoms are positioned along a straight line, rotation about the molecular axis is overlooked concerning the molecular skeletal dynamics (see also Section 7.3.3). There are thus $3N - 5$ left degrees of freedom.

As in the diatomic molecule, the rotational and vibrational motions are addressed separately in the zero-order approximation (Sections 7.3.2 to 7.3.4, specifically Eq. [7.27]). We start by performing a complete classical calculation that formally and explicitly presents all underlying assumptions and approximations involved in reaching the zero-order approximation. We don't detail all developments and proofs, because it belongs to the analytical mechanics theory.

Let $\underline{G}$ be the position vector of the molecular mass center "**o**" in an arbitrary resting XYZ coordinate system, to be addressed as the **lab frame**. We indicate by $\underline{R}_u$ the equilibrium position vector of the u-numbered nucleus *relative to the molecular mass center*. The Cartesian components of $\underline{R}_u$ in the lab frame are (R_{ux}, R_{uy}, R_{uz}). Note that the **nuclei equilibrium positions system is a rigid body**, because the positions difference vectors $\underline{R}_u - \underline{R}_{u'}$ are constants of time (actually in all coordinate systems). We denote by

Physics, Optics, and Spectroscopy of Materials, First Edition. Zeev Burshtein.

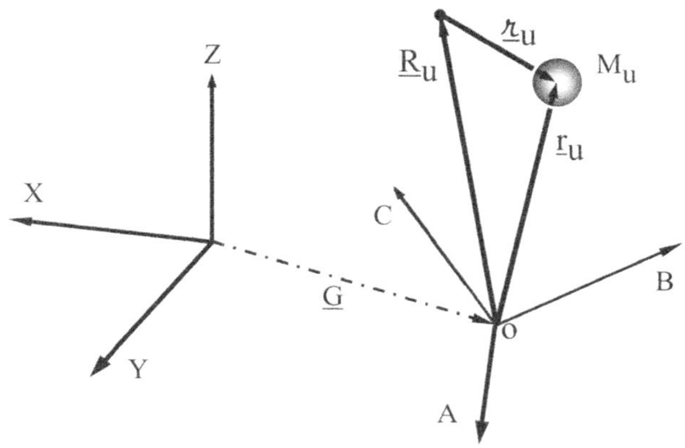

Figure 8.1 coordinate frames and vector noting used to analyze the nuclei motions in a molecule. XYZ is the lab frame; ABC is the molecular frame. See text for more details.

$\underline{r}_u$ the *actual* position vector of the u-numbered nucleus, also relative to the molecular mass center. The Cartesian components of $\underline{r}_u$ in the lab frame are (r_{ux}, r_{uy}, r_{uz}). Now we select an orthogonal coordinate system ABC, rigidly attached to the nuclei equilibrium positions system, whose origin is located at the molecular mass center. This is called **the molecular coordinate system**. The angular motion velocity of the molecular system is denoted by $\underline{\omega}$. A vector whose components relate to the molecular frame will be marked by a prime superscript ($'$). Correspondingly, the Cartesian components of $\underline{R}_u$ in the molecular system ($\underline{R}'_u$ components) will be denoted as A_u, B_u, and C_u; the Cartesian components of $\underline{r}_u$ in the molecular system (r'_u components) will be denoted as a_u, b_u, and c_u.

Figure 8.1 displays the coordinate systems and vectors. The velocity vector $\underline{v}_u$ of the u nucleus in the lab frame is given by

$$\underline{v}_u = \dot{\underline{G}} + \underline{\omega} \times \underline{r}_u + \underline{v}'_u\,, \tag{8.1}$$

where $\underline{v}'_u \equiv (\dot{a}_u, \dot{b}_u, \dot{c}_u)$ is the velocity vector of the u nucleus in the *molecular frame.* Recall that a dot symbol above a variable indicates its full time derivative. The molecular kinetic energy $\boldsymbol{T} = \frac{1}{2}\sum_u M_u \underline{v}_u{}^2$, where M_u is the mass of the u nucleus, will thus be given by

$$\begin{aligned}\boldsymbol{T} = {} & \frac{1}{2}\dot{G}^2\sum\nolimits_u M_u + \frac{1}{2}\sum\nolimits_u M_u(\underline{\omega} \times \underline{r}_u)\cdot(\underline{\omega} \times \underline{r}_u) + \frac{1}{2}\sum\nolimits_u M_u v'^2_u \\ & + \dot{\underline{G}}\cdot\underline{\omega} \times \sum\nolimits_u M_u\underline{r}_u + \dot{\underline{G}}\cdot\sum\nolimits_u M_u\underline{v}'_u + \underline{\omega}\cdot\sum\nolimits_u\left(M_u\underline{r}_u \times \underline{v}'_u\right).\end{aligned} \tag{8.2}$$

Because the origin of the molecular coordinate system coincides with the mass center, the relation $\sum_u M_u\underline{r}_u = 0$ is always valid; in fact, **this condition defines of the mass center position!** Thus, the fourth addend in Eq. (8.2) vanishes identically. We will see that the relation $\sum_u M_u\underline{v}'_u = 0$ also exists. For proof, we perform a full time differentiation of the $\sum_u M_u\underline{r}_u = 0$ expression:

$\sum_u M_u\dot{\underline{r}}_u = \sum_u M_u[\underline{\omega} \times \underline{r}_u + \underline{v}'_u] = \underline{\omega} \times \sum_u M_u\underline{r}_u + \sum_u M_u\underline{v}'_u = \sum_u M_u\underline{v}'_u = 0$; **Q.E.D.**

Thus, the fifth addend in the sum of Eq. (8.2) also vanishes. We overlook the first addend because it describes the kinetic energy of the mass center; in a closed system, namely in the absence of external forces, it is a constant component (because $\dot{\underline{G}}$ is constant), and it may be separated from the others in relation to a study of dynamic conditions of the molecular nuclei.

We set the relation $\underline{\rho}_u \equiv \underline{r}_u - \underline{R}_u$ in Eq. (8.2), where $\underline{\rho}_u$ is the displacement vector of the u nucleus from the equilibrium position. We get

$$\boldsymbol{T} = \frac{1}{2}\sum\nolimits_u M_u(\underline{\omega} \times \underline{r}_u)\cdot(\underline{\omega} \times \underline{r}_u) + \frac{1}{2}\sum\nolimits_u M_u v'^{\,2}_u + \underline{\omega}\cdot\sum\nolimits_u M_u\left(\underline{\rho}_u \times \underline{v}'_u\right) + \underline{\omega}\cdot\sum\nolimits_u M_u\left(\underline{R}_u \times \underline{v}'_u\right).$$

In the following, $\underline{\rho}_u$ Cartesian components in the molecular system will be denoted by $(\alpha_u, \beta_u, \gamma_u)$. Because the $\underline{R}_u$ vectors describe the equilibrium positions of the various nuclei in an axes system that rotates rigidly with them, they will always satisfy the relation $\sum_u M_u(\underline{R}_u \times \underline{v}'_u) = 0$; in fact, **this condition defines the nuclei equilibrium positions!** To demonstrate its physical significance, we note that the sum expresses a total angular momentum of the molecule in the molecule's axes frame at the very moment where all its nuclei simultaneously cross their corresponding equilibrium points. The definition of the equilibrium positions require that this sum remains unchanged if at some point in time the nuclei actual positions are different. Thus, we get

$$\boldsymbol{T} = \frac{1}{2}\sum_u M_u(\underline{\omega} \times \underline{r}_u)\cdot(\underline{\omega} \times \underline{r}_u) + \underline{\omega}\cdot\sum_u M_u\left(\underline{\lambda}_u \times \underline{v}'_u\right) + \frac{1}{2}\sum_u M_u \underline{v}'^{\,2}_u . \tag{8.3}$$

The first addend in Eq. (8.3) describes a pure rotational energy of the molecule $\boldsymbol{T}_{rot}$. The second addend describes a coupling between vibrational and rotational motions, and is called **Coriolis energy $\boldsymbol{T}_{cor}$**. The third and last addend describes a pure vibrational energy $\boldsymbol{T}_{vib}$. On developing the expression for the pure rotational energy, one gets

$$\boldsymbol{T}_{rot} = \frac{1}{2}I_{aa}\omega_x^2 + \frac{1}{2}I_{bb}\omega_y^2 + \frac{1}{2}I_{cc}\omega_z^2 - I_{ab}\omega_x\omega_y - I_{bc}\omega_y\omega_z - I_{ca}\omega_z\omega_x ,$$

where $I_{aa}, I_{bb}, I_{cc}, -I_{ab}, -I_{bc}$, and $-I_{ca}$ are the elements of the molecular tensor of inertia. These quantities vary dynamically during the nuclei motion; we have encountered such a case while studying the special case of a diatomic molecule (Eq. [7.23] and following text). On setting $a_u = A_u + \alpha_u$, $b_u = B_u + \beta_u$, and $c_u = C_u + \gamma_u$, each addend in the tensor of inertia elements will comprise products between coordinates belonging to the $\underline{R}'_u$ vectors like A_u^2 and A_uB_u, and cross products between $\underline{R}'_u$ and $\underline{\lambda}'_u$ vectors coordinates like $A_u\beta_u$. Assuming that the deviation sizes of the nuclei from their equilibrium positions are tiny compared to the distances between any two nuclei, the products containing a coordinate of an $\underline{\lambda}'_u$ vector are very small compared to those containing only components of $\underline{R}'_u$ (perhaps except for a single atom residing at the molecular mass center itself, or very close to it). However, the contribution of such nucleus to the summations defining the tensor of inertia elements is negligibly small. Thus, disregarding the crossed products, we may write the following (quite accurate) approximations: $I_{aa} \cong \sum_u M_u(B_u^2 + C_u^2)$; $I_{bb} \cong \sum_u M_u(C_u^2 + A_u^2)$; $I_{cc} \cong \sum_u M_u(A_u^2 + B_u^2)$; $I_{ab} \cong \sum_u M_u A_u B_u$; $I_{bc} \cong \sum_u M_u B_u C_u$; and $I_{ca} \cong \sum_u M_u C_u A_u$. That approximate tensor of inertia elements is calculated as if all nuclei are fixed at their equilibrium positions, as if the molecule is a rigid body. Therefore, its value is a time-independent constant. Recall that we have already performed a calculation based on the same considerations for the diatomic molecule (Eqs. [7.23] and [7.26]). By statistical considerations, the approximation grows more accurate for a larger number of atoms comprising the molecule, because the relative mean deviation of each dynamic element of the tensor of inertia from its mean constant value is inversely proportional to $\sqrt{N}$, where N is the number of nuclei.

A simplified form of the expressions for the tensor of inertia elements is obtained when the *ABC* coordinates system is chosen to match the principal axes of the approximate tensor; that is, the coordinates system that diagonalizes the tensor of inertia. The diagonal elements♣ will correspondingly be denoted as I_A, I_B, and I_C. The off-diagonal elements vanish. Thus, we have

$$\boldsymbol{T}_{rot} \cong \frac{1}{2}I_A\omega_A^2 + \frac{1}{2}I_B\omega_B^2 + \frac{1}{2}I_C\,\omega_C^2 ,$$

♣ As a reminder, the I_A, I_B, and I_C moments of inertia are solutions of the unknown parameter λ in the secular equation.

$$\begin{vmatrix} I_{aa}-\lambda & -I_{ab} & -I_{ac} \\ -I_{ab} & I_{bb}-\lambda & -I_{bc} \\ -I_{ac} & -I_{bc} & I_{cc}-\lambda \end{vmatrix} = 0,$$

where the original tensor-of-inertia elements were calculated for an arbitrary orthogonal axes system whose origin coincides with the molecular mass center.

where ω_A, ω_B, and ω_C are the angular rotation frequencies about constant axes that momentarily coincide with the molecular *ABC* axes. Using the $\underline{L} \equiv (I_A\omega_A, I_B\omega_B, I_C\omega_C)$ definition, where $\underline{L}$ is the molecular angular momentum vector, we get (see also Eq. [7.18])

$$\boldsymbol{T}_{rot} \cong \frac{L_A{}^2}{2I_A} + \frac{L_B{}^2}{2I_B} + \frac{L_C{}^2}{2I_C}. \tag{8.4}$$

The $\sum_u M_u(\underline{\imath}_u \times \underline{v}'_u)$ sum in Eq. (8.3) expresses a kind of angular momentum of the nuclei. Based on the assumption $\imath_u \ll r_u$ for all u, common angular velocities indeed satisfy the relation $v'_u \sim |\underline{\omega} \times \underline{r}_u|$ (see Eq. [8.1]). Therefore, $\boldsymbol{T}_{cor} \ll \boldsymbol{T}_{rot}$. Thus, in a zero-order approximation, we neglect the Coriolis energy. The kinetic energy in the zero-order approximation is then

$$\boldsymbol{T} \cong \frac{L_A{}^2}{2I_A} + \frac{L_B{}^2}{2I_B} + \frac{L_C{}^2}{2I_C} + \frac{1}{2}\sum_u M_u v'^2_u .$$

We denote the total potential energy of the N molecular nuclei as $U(\underline{r}_1, \underline{r}_2, \cdots, \underline{r}_N)$. The molecular energy in the zero-order approximation would then be

$$\boxed{E \cong \left[\frac{L_A{}^2}{2I_A} + \frac{L_B{}^2}{2I_B} + \frac{L_C{}^2}{2I_C}\right] + \frac{1}{2}\sum_{u=1}^{N} M_u v'^2_u + U(\underline{r}_1, \underline{r}_2, \cdots, \underline{r}_N)} . \tag{8.5}$$

8.3 Molecular Vibrational Modes

We consider a specific case where all *equilibrium positions* of the nuclei are resting, namely that the molecular mass center is at rest, and that the molecule is not rotating. A similar approximation is described in Section 7.3.2 for the diatomic molecule. Thus, the rotational energy given in the square brackets of Eq. (8.5) vanishes. We assign to each atom a separate resting Cartesian coordinate system, whose origin coincides with its equilibrium position. The molecule thus has $3N$ such coordinates. We denote by ξ_s the displacement component of the s–numbered nucleus from its equilibrium position in an arbitrary numbering method of all coordinates. We assign a zero potential energy to jointly all equilibrium positions, namely when all ξ_s simultaneously vanish. The molecular potential energy may be then Taylor-expanded about the equilibrium $\xi_s = 0$ positions in the form

$$U(\xi_1,\ \xi_2\ \cdots\ \xi_{3N}) \cong \frac{1}{2}\sum_{s,t}^{3N} \left.\frac{\partial^2 U}{\partial\xi_s\partial\xi_t}\right]_0 \xi_s\xi_t ,$$

where all terms containing higher-than-two powers of ξ_s have been neglected (harmonic approximation). The total molecular energy would thus be given by

$$E = \boldsymbol{T} + U \cong \frac{1}{2}\sum_{s=1}^{3N} M_s\dot{\xi}_s{}^2 + \frac{1}{2}\sum_{s,t}^{3N} \left.\frac{\partial^2 U}{\partial\xi_s\partial\xi_t}\right]_0 \xi_s\xi_t , \tag{8.6}$$

where M_s is the mass of the nucleus belonging to the s- numbered axis. It is important to remember that there is always an option to classify the axes in groups of three elements belonging to the same M_s mass. We now denote $g_s \equiv \sqrt{M_s} \cdot \xi_s$. Eq. (8.6) will now be written as

$$E = \frac{1}{2}\sum_{s=1}^{3N} \dot{g}_s^{\ 2} + \frac{1}{2}\sum_{s,t}^{3N} U_{st} g_s g_t \ , \tag{8.7}$$

where $U_{st} \equiv \left(\partial^2 U/\partial g_s \partial g_t\right)_0$.

The last result may be represented in a matrix language; to that end, we denote

$$g \equiv \begin{pmatrix} g_1 \\ g_2 \\ \vdots \\ g_{3N} \end{pmatrix}; \ \tilde{g} \equiv (g_1, \ g_2, \ \cdots, \ g_{3N}); \ \mathbf{U} \equiv \begin{pmatrix} U_{11} & U_{12} & \cdots & U_{1,3N} \\ U_{21} & \cdots & \cdots & \cdots \\ \vdots & \cdots & \cdots & \cdots \\ U_{3N,1} & \cdots & \cdots & U_{3N,3N} \end{pmatrix}, \tag{8.8}$$

and then may write

$$E = \frac{1}{2}\tilde{\dot{g}}\dot{g} + \frac{1}{2}\tilde{g}\mathbf{U}g \ . \tag{8.9}$$

By its definition, the $\mathbf{U}$ matrix is symmetric because $U_{st} = U_{ts}$; then there exists an orthogonal matrix $\mathbf{A}$ which diagonalizes it, satisfying

$$\mathbf{A}^{-1}\mathbf{U}\mathbf{A} = \mathbf{\Omega} \ , \tag{8.10}$$

where $\mathbf{\Omega}$ is a diagonal matrix whose elements $\omega_k^{\ 2}$ for $k = 1, \ 2, \ \cdots \ 3N$ are the eigenvalues of the $\mathbf{U}$ matrix. The replacement of s or t letter that were used as subscripts for the matrix elements by the k letter is used to emphasize that the matrix eigenvalues belong to new q-coordinates, where each is a specific different linear combination of the original coordinates. We will elaborate on that issue. The new q-coordinates are defined by

$$q \equiv \mathbf{A}^{-1}g \leftrightarrow g = \mathbf{A}q \ . \tag{8.11}$$

We set this definition in Eq. (8.9) and get

$$E = \frac{1}{2}\left(\widetilde{\mathbf{A}\dot{q}}\right)(\mathbf{A}\dot{q}) + \frac{1}{2}\left(\{\widetilde{\mathbf{A}q}\}\right)\mathbf{U}(\mathbf{A}q) = \frac{1}{2}\tilde{\dot{q}}\tilde{\mathbf{A}}\mathbf{A}\dot{q} + \frac{1}{2}\tilde{q}\tilde{\mathbf{A}}\mathbf{U}\mathbf{A}q$$

$$\boxed{E = \frac{1}{2}\tilde{\dot{q}}\dot{q} + \frac{1}{2}\tilde{q}\mathbf{\Omega}q} \ . \tag{8.12}$$

The last transition is based on Eq. (8.10), and on the fact that being orthogonal, the $\mathbf{A}$ matrix satisfies the relation $\tilde{\mathbf{A}} = \mathbf{A}^{-1}$. By specifying the matrix elements, we may write

$$E = \frac{1}{2}\sum_{k=1}^{3N}\left(\dot{q}_k^{\ 2} + \omega_k^{\ 2}q_k^{\ 2}\right) = \frac{1}{2}\sum_{k=1}^{3N}\dot{q}_k^{\ 2} \ + \ \frac{1}{2}\sum_{k=1}^{3N}\omega_k^{\ 2}q_k^{\ 2} \ . \tag{8.13}$$

The first sum is the molecular kinetic energy, and the second sum is its potential energy. It may be readily verified that the $\dot{q}_k$ variables are the generalized linear momentums p_k of the Hamiltonian, which are canonical conjugates of the corresponding q_k generalized coordinates; note that $p_k \equiv \partial \boldsymbol{L}/\partial \dot{q}_k = \partial \boldsymbol{T}/\partial \dot{q}_k = \dot{q}_k$, where $\boldsymbol{L} = \boldsymbol{T} - U$ is the Lagrangian. The classical Hamiltonian will thus be written in the form

$$\boxed{\mathrm{H} = \frac{1}{2}\sum_{k=1}^{3N}\left(p_k^{\ 2} + \omega_k^{\ 2} q_k^{\ 2}\right)} \quad . \tag{8.14}$$

For acquisition of a complete presentation of the subject classical aspect, recall that the dynamic equations of the coordinates are given by Hamilton's equations $\dot{p}_k = -\partial\mathrm{H}/\partial q_k$. Therefore,

$$\ddot{q}_k = -\omega_k^{\ 2} q_k \ for\ k = 1, 2, \cdots 3N \ . \tag{8.15}$$

Solution of each equation in Eq. (8.15) is a harmonic oscillation of the generalized coordinate q_k by an ω_k angular frequency. A solution of that kind for a single specific coordinate while the amplitudes of all other coordinates vanish, means that all nuclei oscillate at the same frequency and phase. Thus, the q_k coordinates for $k = 1, 2, \cdots 3N$ are actually the normal oscillation coordinates of the molecular vibrations. A general motion can be described as a linear combination of independent normal vibrations. We thus see that the Eq. (8.14) Hamiltonian is a sum of Hamiltonians of independent harmonic oscillators.

The transition to a quantum representation of a harmonic oscillator has already been performed in Section 5.3.1. Recall that for convenience, new operators are defined per each eigenfrequency as $\mathbf{Q}_k \equiv Q_k = \sqrt{\frac{\omega_k}{\hbar}} q_k$ and $\mathbf{P}_k \equiv \frac{\mathbf{p}_k}{\sqrt{\hbar\omega_k}}$, where $\mathbf{p}_k \equiv -i\hbar\,\partial/\partial q_k$ is the quantum linear momentum operator that replaces the p_k variable (Table 3.1). The Hamiltonian is represented in Eq. (5.19), and the solved wave functions are represented in Table 5.1. The eigenenergies of v-order per vibrational mode frequency are given by

$$\boxed{E_{\mathrm{v}k} = \left(\frac{1}{2} + \mathrm{v}\right)\hbar\omega_k} \quad . \tag{8.16}$$

It should be emphasized, that six of the **U** matrix eigensolutions are zero, and they relate to the three translation and three rotation degrees of freedom mentioned in the beginning of Section 8.2 (only five for a linear molecule). Accordingly, the Hamiltonian expression Eq. (8.14) may be reduced to only $3N - 6$ addends (or to only $3\mathrm{N} - 5$ ones for a linear molecule). As to the rotational levels, they abide by similar considerations and conclusions concerning the diatomic molecules, addressed in detail in Chapter 7. We will address this subject shortly.

The molecule's electric dipole moment also depends on its vibrational state. Like the underlying assumptions used for Eqs. (7.43) and (7.44), we assume that the temporal electric dipole (whose average is constant over time) contains two contributions: one is constant, and the other is a sum of vibrating dipoles, each being in phase with one of the different vibrational modes; namely,

$$\underline{M} = \underline{M}_0 + \sum_{k=1}^{3N} \underline{M}_k Q_k \ , \tag{8.17}$$

where $\underline{M}_0$ and $\underline{M}_k$ are constants (see also Eq. [7.34] and related text). Because the normal vibrational coordinates Q_k are dimensionless, the dimensions of the $\underline{M}_k$ components are of an electric dipole, namely $\mathrm{esu}\cdot\mathrm{cm}$ or $(\mathrm{erg}\cdot\mathrm{cm}^3)^{1/2}$ in cgs units. Their typical size is of the order of an elementary charge "e" times the Bohr radius "a_0". As with the diatomic molecules case, the selection rules and strength between pure rotational states depend on the existence of $\underline{M}_k \neq 0$ (Eq. [7.35]), as will be detailed in Section 8.9.

8.4 Vibrational Energy Scheme

In contrast to the diatomic case, a polyatomic molecule has many vibrational modes. To present an energy scheme, the modes are sorted at an ascending order of their frequency. The normal coordinate belonging to the highest mode frequency is called **configuration coordinate**. Accordingly, the potential energy dependence on the configuration coordinate is called **configuration potential**. The energy states belonging to the configuration potential are usually plotted in the same way as done for the vibrational states of a diatomic molecule (Figures 7.1 and 7.5), namely by horizontal lines whose edges coincide with the edge positions of a classical oscillator of the same mechanical energy. The vibrational states of lower frequencies are described by shorter lines according to a descending frequency order, related to the closest higher energy state as their potential minimum. Because of lines overcrowding, the number of modes presented that way is limited to only two or three. An example of such an arranged energy scheme is presented in Figure 8.2. Frequency data on additional modes (if they exist) are listed in tables (see Table 18.1 as a simple example).

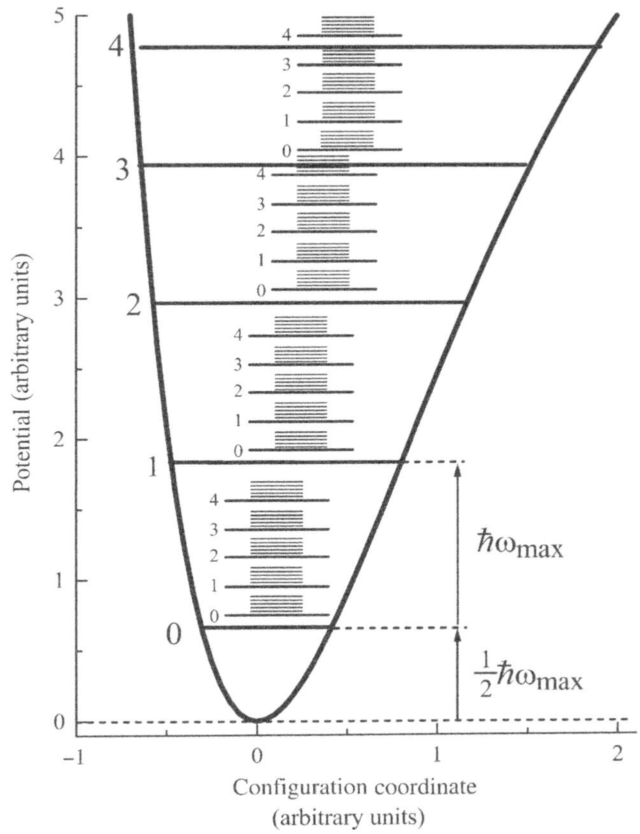

Figure 8.2 Demonstration of vibrational energy scheme in a polyatomic molecule in relation to its configuration potential. To prevent data overcrowding, only states related to the three highest frequencies are demonstrated.

8.5 Rayleigh and Raman Scattering

8.5.1 General Rayleigh Scattering by Molecules

Before addressing the mechanisms underlying the process called **Raman scattering**, we need to refine the general Rayleigh scattering presentation described in Section 1.4. An underlying assumption presented there was that the oscillating electric field of the electromagnetic radiation acting on a molecule induces an oscillating electric dipole, co-oriented with the electric field polarization (see Eqs. [1.46] and [1.47]). That behavior is valid for a single atom, which always exhibits a full rotational symmetry. However, molecules rarely exhibit a full rotational symmetry. A molecular response to the action of an electric field thus depends on the relation between the field polarization and the molecular spatial posture.

The **polarizability** α of an atom or a molecule is defined by the relation $\underline{\mu} = \alpha\underline{\mathcal{E}}$, where $\underline{\mathcal{E}}$ is the electric field vector and $\underline{\mu}$ is the induced molecular dipole vector. If $\underline{\mathcal{E}}$ and $\underline{\mu}$ are always co-oriented, α is a scalar quantity of volume units (cm^3 in cgs units system; see also Eq. [4.111] and the relevant text). Using Eq. (1.47), the polarizability of such case is given by

$$\alpha = \frac{e^2/m}{{\omega_0}^2 - \omega^2 - i\gamma\,\omega} = 2r_0c^2\,\frac{{\omega_0}^2 - \omega^2 + i\gamma\,\omega}{({\omega_0}^2 - \omega^2)^2 + \gamma^2\,\omega^2}, \quad (8.18)$$

where ω is the angular frequency of the external (light) electric field, ω_0 is the resonance frequency of the atom or molecule, r_0 is the electron electrostatic radius (see Section 1.1), and γ is a damping parameter. We are now interested in the Rayleigh spectral range, where the light frequency is sufficiently far from resonance, satisfying $(\omega_0 - \omega) >> \gamma$. It is then possible to assume that α is not complex (see also Eqs. [1.51] and [1.55]). In a general molecule, the $\underline{\mathcal{E}}$ and $\underline{\mu}$ vectors may be differently oriented. The α quantity is then a tensor. In a full matrix language, it reads

$$\begin{pmatrix} \mu_x \\ \mu_y \\ \mu_z \end{pmatrix} = \begin{pmatrix} \alpha_{xx} & \alpha_{xy} & \alpha_{xz} \\ \alpha_{yx} & \alpha_{yy} & \alpha_{yz} \\ \alpha_{zx} & \alpha_{zy} & \alpha_{zz} \end{pmatrix} \begin{pmatrix} \mathcal{E}_x \\ \mathcal{E}_y \\ \mathcal{E}_z \end{pmatrix}. \tag{8.19}$$

The polarizability tensor α is symmetric, so its elements satisfy the relation $\alpha_{ij} = \alpha_{ji}$ for $i, j = x, y, z$. There would be three combinations where the light polarization and induced dipole orientation coincide; that is when the electric field orientation coincides with one of the principal axes of the polarizability tensor.

Now we consider a linearly polarized radiation incident on a molecule. The induced oscillating electric dipole radiates in all directions, as described in Section 1.1, especially Eq. (1.1). We limit ourselves to consider only a special case of radiation, emitted (scattered) perpendicularly to the incident radiation, and two perpendicular polarization orientations. Two of the possible combinations yielding such formation are demonstrated in Figure 8.3. They are marked by the **Porto notation** method. In that method, the propagation directions of the incident and scattered light are correspondingly indicated to the left and right of a parentheses symbol. Inside the parentheses, the polarizations of the incident and scattered light are correspondingly indicated from left to right. For example, a formation where the incident light propagates in the Y direction and is polarized in the X direction, and the scattered light is observed propagating in the X direction, being polarized in the Z direction, is noted $Y(XZ)X$. Figure 8.3 demonstrates the $Y(XY)X$ and $Y(ZZ)X$ formations. In the Porto notation, scattering orientations opposite an axis are indicated by a bar above the axis symbol. For example, a formation in which the incident radiation propagates in the Y direction being X polarized, and the scattered radiation propagates in the $-Y$ orientation (opposite the incident light) being Z polarized, is noted by $Y(XZ)\overline{Y}$. Scattering formation where the scattered light

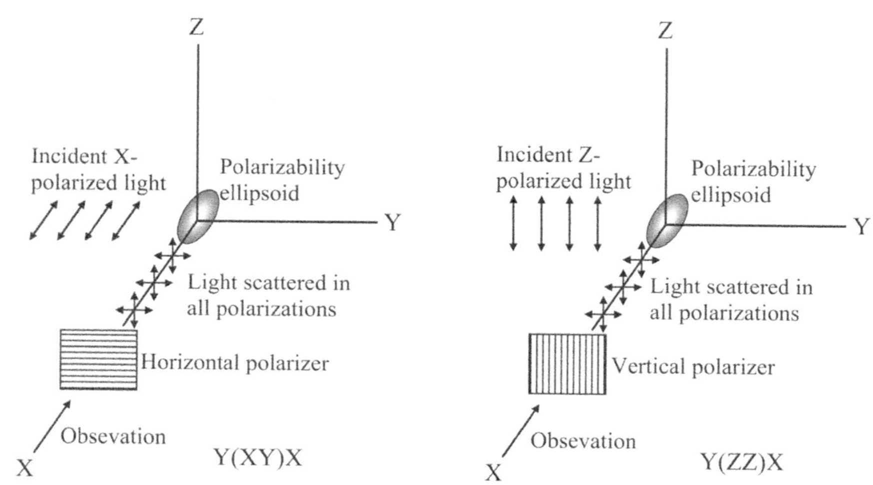

Figure 8.3 Schematic illustration of light-scattering by a molecule, and viewing (observing) formations of the scattered light. The relations among the polarization-states and the incident and scattered light propagation directions are indicated by the Porto notation.

propagating perpendicularly to the incident light propagation has a polarization component perpendicular to that of the incident light is addressed as **cross-polarized**.

We turn to calculating the scattered light intensity per each Porto formation. For demonstration, consider the *Y(XY)X* formation (left side of Figure 8.3). Let the maximal electric field of the incident light be $\underline{\mathcal{E}} = (\mathcal{E}_0, 0, 0)$. Correspondingly, the maximal induced dipole moment would be $\underline{\mu} = \mathcal{E}_0 \left(\alpha_{xx}, \alpha_{yx}, \alpha_{zx}\right)$ (see Eq. [8.19]). The light crossing the Y oriented polarizer is generated by the Y component of the induced dipole moment (Figure 1.1). The power density of the crossing scattered light is given by the average over a complete period of the Poynting vector (see text related to Eqs. [1.2] and [1.3]). Thus, we get

$$I_{Y(XY)X} = \left(\frac{1}{2}\right) \frac{c}{4\pi} \omega^4 \mathcal{E}_0{}^2 \cdot \frac{\alpha_{yx}{}^2}{c^4 r^2} = \frac{1}{r^2} \cdot \frac{\omega^4 \mathcal{E}_0{}^2}{8\pi c^3} \alpha_{yx}{}^2 = \frac{I_{inc}}{r^2} \left(\frac{\omega}{c}\right)^4 \alpha_{yx}{}^2 ,$$

where $I_{inc} = \dfrac{c}{8\pi} \mathcal{E}_0{}^2$ is the power density (energy per unit area per unit time) of the incident radiation, and r is the observer (viewer) distance from the scattering molecule. Note that near a resonance frequency, the polarizability tensor elements are complex (Eq. [8.18]). For light scattering intensity assessment, however, all elements may be replaced by their absolute value, because the phases of all elements are identical. Recalling that the polarizability tensor is symmetric, we get the following expression for all formations of incident light propagating in the Y direction:

$$I_{Y(XY)X} = \frac{I_{inc}}{r^2} \left(\frac{\omega}{c}\right)^4 \alpha_{xy}{}^2 \ ; \tag{8.20[i]}$$

$$I_{Y(XZ)X} = \frac{I_{inc}}{r^2} \left(\frac{\omega}{c}\right)^4 \alpha_{xz}{}^2 \ ; \tag{8.20[ii]}$$

$$I_{Y(ZY)X} = \frac{I_{inc}}{r^2} \left(\frac{\omega}{c}\right)^4 \alpha_{zy}{}^2 \ ; \tag{8.20[iii]}$$

$$I_{Y(ZZ)X} = \frac{I_{inc}}{r^2} \left(\frac{\omega}{c}\right)^4 \alpha_{zz}{}^2 \ . \tag{8.20[iv]}$$

The last result is valid for a molecule of a steady posture, as is often obtained for molecules hosted in a specific crystalline site. If the volume from which the scattered light is collected includes N molecules of identical posture, and the said volume dimensions are small compared to r, then all expressions on the right side of Eqs. (8.20) should be multiplied by N, provided that scattering by different molecules is independent. Eq. (8.20-[i] to [iv]) provides a method for calculating the polarization tensor elements α_{ij} from measured intensity ratios between the incident and scattered light. The situation is more complex for a gas, where the molecules exhibit a random distribution of postures relative to the *XYZ* axes system (the lab frame; see Figure 8.1). The same holds for liquid-dissolved or glass-hosted molecules. Calculation of the scattered light at each formation requires averaging over all possible postures of the molecule relative to the lab frame. We will present such a calculation.

For convenience, we address as "initial" posture the one for which the principal axes of the molecular polarization tensor $X'\ Y'\ Z'$ (the **molecular frame**) coincide with the lab frame *XYZ*. In that initial posture, the tensor has only three non-zero elements in the lab frame; these are the diagonal α_1, α_2, and α_3 elements belonging to the X', Y', and Z' axes, respectively. We denote that diagonal tensor by α_d. Now we describe a situation in which the molecule changes it posture while preserving its mass center position. We denote by T the matrix that transforms any vector in the XYZ lab frame to the molecular coordinate system, the $X'\,Y'\,Z'$ system:

The electric field in the molecular coordinates frame is then $\underline{\mathcal{E}}' = T\underline{\mathcal{E}}$;

The induced dipole moment in the molecular coordinates frame is $\underline{\mu}' = \alpha_d T\underline{\mathcal{E}}$;

The induced dipole moment in the lab frame is $\underline{\mu} = T^{-1}\underline{\mu}' = \left(T^{-1}\alpha_d T\right)\underline{\mathcal{E}}$;

Thus, the general polarization tensor in the lab frame is given by $\alpha = T^{-1}\alpha_d T$. The T matrix is orthogonal, satisfying $T^{-1} = \widetilde{T}$. We then get

$$\alpha_{i\,j} = \sum_{u=1}^{3} \alpha_u T_{u\,i} T_{u\,j} \Rightarrow \alpha_{i\,j}^{\;2} = \sum_{u,\,u'=1}^{3} \alpha_u \alpha_{u'} T_{u\,i} T_{u'i} T_{u\,j} T_{u'j}\,, \tag{8.21}$$

where T_{uv} for u, $v = 1, 2, 3$ are the T matrix elements. To average over all possible molecular postures, we write

$$\overline{\alpha_{i\,j}^{2}} = \sum_{u,\,u'=1}^{3} \alpha_u \alpha_{u'} \overline{T_{u\,i} T_{u'i} T_{u\,j} T_{u'j}}\,, \tag{8.22}$$

where the bar symbol above variables indicates the averaging operation. First, we consider the special case of the squared diagonal elements $i = j$. We get

$$\overline{\alpha_{i\,i}^{2}} = \sum_{u,\,u'=1}^{3} \alpha_u \alpha_{u'} \overline{T_{u\,i}^{2} T_{u'i}^{2}}\,. \tag{8.23}$$

The elements occupying each line in the T matrix are the orientation cosines (projections) in the lab frame of a unity vector on one of the axes forming the molecular coordinates frame. We concentrate on an arbitrary such unity vector; averaging over all cosines is performed for the said unity vector points through all directions on a unity sphere. We temporarily denote $T_{ui} \equiv \cos\theta$, and average over $T_{ui}^{\;4}$ in spherical coordinates (see Figure 3.3):

$$\overline{T_{ui}^{\;4}} = \overline{\cos^4\theta} = \frac{1}{4\pi}\int_0^{2\pi} d\varphi \int_0^{\pi} \cos^4\theta \,\sin\theta \, d\theta = \frac{1}{2}\frac{\cos^5\theta}{5}\bigg]_0^{\pi} = \frac{1}{5}\,. \tag{8.24}$$

Being orthogonal, the T matrix elements satisfy $\sum_{u=1}^{3} T_{ui}^{\;2} = 1$. We square that relation and average. We recall further that by symmetry considerations, averages of all elements' powers, or similar products of their powers, are equal. Thus, we get

$$3\,\overline{T_{ui}^{\;4}} + 6\,\overline{T_{ui}^{\;2} T_{u\prime i}^{\;2}} \underset{u'\neq u}{=} 1;\; \overline{T_{ui}^{\;2} T_{u\prime i}^{\;2}} \underset{u'\neq u}{=} \frac{1}{15}\,. \tag{8.25}$$

We set the previous result and get

$$\overline{T_{ui}^{\;2} T_{u\prime i}^{\;2}} \underset{u'\neq u}{=} \frac{1}{15}, \text{ and similarly } \overline{T_{ui}^{\;2} T_{uj}^{\;2}} \underset{i\neq j}{=} \frac{1}{15}\,. \tag{8.26}$$

Setting in Eq. (8.23) we immediately get

$$\boxed{\overline{\alpha_{i\,i}^{2}} = \frac{1}{5}\left(\alpha_1^{\;2} + \alpha_2^{\;2} + \alpha_3^{\;2}\right) + \frac{2}{15}\left(\alpha_1\alpha_2 + \alpha_1\alpha_3 + \alpha_2\alpha_3\right)}\,. \tag{8.27}$$

To calculate the other averages, we adopt the same method; being orthogonal, the T matrix elements satisfy $\sum_{u=1}^{3} T_{ui}T_{uj} \underset{i \neq j}{=} 0$. We square that relation and average, activate the symmetry considerations, and get

$$3\,\overline{T_{ui}{}^2 T_{uj}{}^2} + 6\,\overline{T_{ui}T_{u'i}T_{uj}T_{u'j}} \underset{u' \neq u}{=} 0\,. \tag{8.28}$$

We set the right-side relation of Eq. (8.26) in the above results, and get

$$\overline{T_{ui}T_{u'i}T_{uj}T_{u'j}} \underset{i \neq j}{=} -\frac{1}{30}\,;$$

$$\overline{T_{ui}T_{u'i}T_{uj}T_{u'j}} \underset{i = j}{=} +\frac{1}{15}\,.$$

Setting the latter in Eq. (8.22), we immediately get

$$\boxed{\overline{\alpha_{ij}{}^2} \underset{i \neq j}{=} \frac{1}{15}\left(\alpha_1{}^2 + \alpha_2{}^2 + \alpha_3{}^2\right) - \frac{1}{15}(\alpha_1\alpha_2 + \alpha_1\alpha_3 + \alpha_2\alpha_3)}\,. \tag{8.29}$$

We set Eqs. (8.27) and (8.29) in Eq. (8.20) and get

$$\left.\begin{aligned} &I_{Y(XY)X} = I_{Y(XZ)X} = I_{Y(ZY)X} = \\ &= \frac{I_{inc}}{r^2}\left(\frac{\omega}{c}\right)^4\left[\frac{1}{15}\left(\alpha_1{}^2 + \alpha_2{}^2 + \alpha_3{}^2\right) - \frac{1}{15}(\alpha_1\alpha_2 + \alpha_1\alpha_3 + \alpha_2\alpha_3)\right] \end{aligned}\right\}; \tag{8.30[i] – [iii]}$$

$$\left.\begin{aligned} &I_{Y(ZZ)X} = \\ &= \frac{I_{inc}}{r^2}\left(\frac{\omega}{c}\right)^4\left[\frac{1}{5}\left(\alpha_1{}^2 + \alpha_2{}^2 + \alpha_3{}^2\right) + \frac{2}{15}(\alpha_1\alpha_2 + \alpha_1\alpha_3 + \alpha_2\alpha_3)\right] \end{aligned}\right\}. \tag{8.30[iv]}$$

As expected, in an isotropic molecule, where $\alpha_1 = \alpha_2 = \alpha_3 = \alpha$, one gets $\alpha_{i\,i} = \alpha$, and $\alpha_{ij} \underset{i \neq j}{=} 0$. Thus, we get $I_{Y(XY)X} = I_{Y(XZ)X} = I_{Y(ZY)X} = 0$, however $I_{Y(ZZ)X} = (I_{inc}/r^2)\,(\omega/c)^4\alpha^2$.

The occurrence of non-zero scattered intensities in crossed polarizations is called **depolarization**; it stems from anisotropy in the molecule polarizability tensor. A case of zero scattering in all crossed polarizations is called **totally polarized scattering**; if crossed polarization scatterings appear, but are small compared to the polarized scattering, the case is called **partially polarized scattering**; if crossed polarization intensities are comparable to polarized scattering, the case is addressed as **non-polarized scattering**.

Now we calculate the *total scattered light intensity* for a case where a collimated polarized light beam is incident on a medium containing the scattering molecules. Let the light be X-polarized as described on the left side of Figure 8.3. The electric field of the incident light is then $(\mathcal{E}_0, 0, 0)$. The induced dipole moment would be $\underline{\mu} = \mathcal{E}_0\left(\alpha_{xx}, \alpha_{yx}, \alpha_{zx}\right)$ (see Eq. [8.19]); its square is $\mu^2 = \mathcal{E}_0{}^2\left(\alpha_{xx}{}^2 + \alpha_{yx}{}^2 + \alpha_{zx}{}^2\right)$. By Eq. (1.5), the total power radiated by the dipole is given by

$$P_{total} = \frac{1}{3}\frac{\omega^4}{c^3}\mathcal{E}_0{}^2\left(\alpha_{xx}{}^2 + \alpha_{yx}{}^2 + \alpha_{zx}{}^2\right) = \frac{8\pi}{3}I_{inc}\left(\frac{\omega}{c}\right)^4\left(\alpha_{xx}{}^2 + \alpha_{yx}{}^2 + \alpha_{zx}{}^2\right)\,.$$

Averaging over all molecular postures, we get (Eqs. [8.27] and [8.29]):

$$\boxed{\overline{P}_{total} = \frac{8\pi}{9} I_{inc} \left(\frac{\omega}{c}\right)^4 \left({\alpha_1}^2 + {\alpha_2}^2 + {\alpha_3}^2\right)} \quad . \tag{8.31}$$

We saw that when all molecules exhibit the same posture, experiment allows calculating *all* elements of the polarizability tensor (Eqs. [8.20]). Consideration of Eq. (8.30) shows that *for randomly postured molecules*, measurements of scattered light intensities in parallel or crossed polarizations allows obtaining only the sums $({\alpha_1}^2 + {\alpha_2}^2 + {\alpha_3}^2)$ and $(\alpha_1\alpha_2 + \alpha_1\alpha_3 + \alpha_2\alpha_3)$. An additional independent relation is needed to allow calculation of *all* tensor elements, but it can't be obtained by scattering measurements. Total scattered power measurement, however, allows an independent assessment of the $({\alpha_1}^2 + {\alpha_2}^2 + {\alpha_3}^2)$ sum.

Finally, we present the general expression for the scattered light intensity under any specific combination of incident and scattered light orientations, and their polarizations, for a molecule of a steady posture. As in the case of Eqs. (8.20), the scattered light intensity depends only on polarization relations, and not on light propagation directions, and is given by

$$\boxed{I_{\underline{A}_i(\underline{u}_i\underline{u}_s)\underline{A}_s} = \frac{I_{inc}}{r^2}\left(\frac{\omega}{c}\right)^4 S^2} \quad , \tag{8.32}$$

where $\underline{A}_i$ and $\underline{A}_s$ are unity vectors in the incident and scattered light orientations, respectively, $\underline{u}_i$ and $\underline{u}_s$ are unity vectors in the incident and scattered light polarization orientations, respectively, and

$$S \equiv \widetilde{\underline{u}}_i(\alpha)\underline{u}_s = \begin{pmatrix} u_i^x, & u_i^y, & u_i^z \end{pmatrix} \begin{pmatrix} \alpha_{xx} & \alpha_{xy} & \alpha_{xz} \\ \alpha_{yx} & \alpha_{yy} & \alpha_{yz} \\ \alpha_{zx} & \alpha_{zy} & \alpha_{zz} \end{pmatrix} \begin{pmatrix} u_s^x \\ u_s^y \\ u_s^z \end{pmatrix} . \tag{8.33}$$

8.5.2 Raman Scattering

The former calculation of light scattering by molecules was performed under an assumption that the molecular polarizability is a time-independent quantity. It would be a proper approximation for a rigid molecular structure, but requires corrections for accounting for the molecular vibrational motions. The molecular polarizability values vary during vibration. We assume that these variations are in phase with the normal vibration coordinates Q_k (see text related to the paragraph preceding Eq. [8.17]); namely

$$\alpha_{ij} = \alpha_{ij}^0 + \sum_{k=1}^{3N-6} \left(\frac{\partial \alpha_{ij}}{\partial Q_k}\right)_0 Q_k \, . \tag{8.34}$$

Because the normal coordinates are dimensionless, a polarizability derivative by a normal coordinate has the same dimensions as the polarizability, namely cm^3 in the cgs units system.

For demonstration and simplicity, we describe the incident light as monochromatic, X-polarized, propagating in the Y direction, as displayed in the formation on the left side of Figure 8.3. Its electric field is thus given as $\underline{\mathcal{E}}(t) = (\mathcal{E}_0 \cos \omega t, 0, 0)$. The molecular vibration normal coordinates read $Q_k(t) = Q_k^0 \cos(\omega_k t + \varphi_k)$, where Q_k^0 is the vibration amplitude of the k-mode, ω_k is its angular vibration frequency (see Eq. [8.15]) and φ_k is its phase. As described in Section 8.5.1, the induced electric dipole would be $\underline{\mu} = \mathcal{E}_0(\alpha_{xx}, \alpha_{yx}, \alpha_{zx}) \cos \omega t$. The scattered electric field in the $Y(XY)X$ formation is generated by the Y component of the induced dipole, namely by $\mu_y(t) = \mathcal{E}_0\alpha_{yx} \cos \omega t$. Setting the latter in Eq. (8.34) yields

$$\mu_y(t) = \mathcal{E}_0\left\{ \overset{0}{\alpha}_{yx} \cos \omega t + \sum_{k=1}^{3N-6} \left(\frac{\partial \alpha_{yx}}{\partial Q_k}\right)_0 Q_k^0 \cos \omega t \cos(\omega_k t + \varphi_k) \right\} \tag{8.35}$$

$$= \mathcal{E}_0 \overset{0}{\alpha}_{yx} \cos \omega t + \frac{1}{2}\mathcal{E}_0 \sum_{k=1}^{3N-6} \left(\frac{\partial \alpha_{yx}}{\partial Q_k}\right)_0 Q_k^0 \cos[(\omega + \omega_k)t + \varphi_k]$$

$$+ \frac{1}{2}\mathcal{E}_0 \sum_{k=1}^{3N-6} \left(\frac{\partial \alpha_{yx}}{\partial Q_k}\right)_0 Q_k^0 \cos[(\omega - \omega_k)t - \varphi_k] \ .$$

It is already clear, that in addition to vibration at the incident light frequency ω, the electric dipole includes two other vibrational frequency components: $\omega + \omega_k$ and $\omega - \omega_k$. Thus, in addition to a component vibrating at the same frequency as the incident light, the electric field of the scattered light radiated by the vibrating induced dipole comprises frequency components shifted up and down by the normal vibrational frequencies of the molecule. The amplitudes of these components are very small, as experiment proves that $\left(\frac{\partial \alpha_{ij}(\omega \pm \omega_k)}{\partial Q_k}\right) Q_k^0 << \alpha_{ij}^0(\omega)$ for all k. The component of ω frequency is the usual Rayleigh scattering, addressed in Section 8.5.1. The components of the shifted frequencies are called **Raman scattering**. The up-shifted frequency components are called **Raman anti-Stokes shifts**; the down-shifted components are called **Raman Stokes shifts** (see also Section 7.5.5). Importantly, measurements of the shifted frequencies reflect the vibrational frequencies ω_k of the molecular normal modes. The intensities of the scattered light in the different formations relate to the $\left(\partial\alpha_{ij}/\partial Q_k\right)_0 Q_k^0$ quantities. For a specific normal vibration k, the $\left(\partial\alpha_{ij}/\partial Q_k\right)_0 Q_k^0$ quantity for $i, j = x, y, z$ is a tensor; selection rules relate to its components being either non-zero or vanishing. These selection rules will be detailed in Section 8.9. For brevity, when dealing solely with Raman scattering, the $\left(\partial\alpha_{ij}/\partial Q_k\right)_0 Q_k^0$ tensor is simply denoted as (α_{ij}).

Despite the success of the classical model in describing the fundamentals of Raman scattering, it is insufficient, especially because of its failure in predicting major differences between Stokes and anti-Stokes shifts achieved by experiment. The presented classical model predicts identical intensities for both, and parallel intensity increases per increase in vibrational amplitudes Q_k^0 of the molecular oscillations at each mode (brought about by increase in temperature). In reality, Raman anti-Stokes scattering is *very small* compared to the Stokes scattering; increase of the Raman Stokes intensity changes only slightly with temperature, whereas the Raman anti-Stokes intensity increases steeply with temperature. In addition, the Raman anti-Stokes intensities decrease as a function of the shift frequency, and the decrease is temperature dependent. These contradictions are fully settled by treating Raman scattering through quantum mechanics. Figure 8.4 schematically displays the fundamental assumptions underlying the quantum mechanical description of light scattering.

In Rayleigh scattering, light incident on a molecule at a ground state, whose photons lack sufficient energy to induce transition to the excited resonance state, does however excite it to a lower, *non-resonance state*, called **virtual state**. The molecule returns "immediately" to the ground state by emitting a photon of the same energy (frequency), but of propagation orientation and polarization that could differ from the original inducing photon. In Raman-Stokes scattering, the molecule does not return to the ground electronic state, but rather to an excited vibrational state. The emitted photon energy is then smaller than that of the incident light by the vibrational phonon energy E_{vk}. In Raman anti-Stokes scattering, the incident photon interacts with a molecule that is *already excited* by a single phonon E_{vk}. It then excites the molecule to a virtual electronic state. The molecule does not return to the former excited vibrational state, but rather

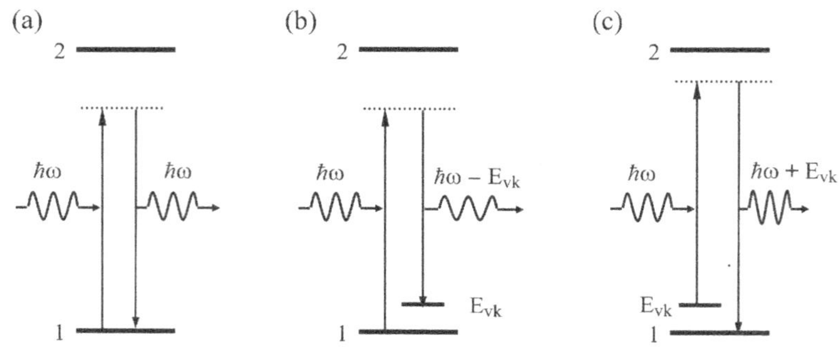

Figure 8.4 Schematic illustration of energy states and optical transitions involved in molecular Rayleigh and Raman scattering. (a) Rayleigh scattering; (b) Raman Stokes scattering; (c) Raman anti-Stokes scattering. Molecular ground and excited electronic states are marked as "1" and "2," respectively, and the relevant "virtual states" by dotted lines. See text for a detailed discussion.

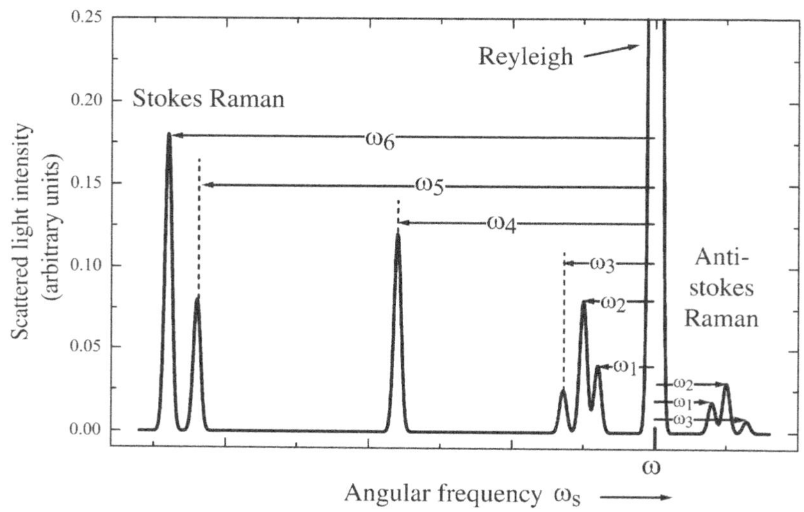

Figure 8.5 Schematic illustration of scattered light spectrum from a cluster of identical molecules at thermal equilibrium. The very high peak at the monochromatic incident light frequency ω belongs to Rayleigh scattering. Its left side peaks belong to Raman Stokes scattering; those on the right belong to Raman anti-Stokes scattering. See details in text.

to the ground electronic state. The emitted photon energy is then larger than that of the incident light by the vibrational phonon energy E_{vk}. The temperature dependence of scattering intensity is interpreted as resulting from vibrational states' occupation dependence on the temperature.

Figure 8.5 schematically shows an intensity spectrum of scattered light obtained off a cluster of identical molecules at thermal equilibrium, resulting from the incidence of a monochromatic light beam at ω frequency. At scattered frequency $\omega_s = \omega$, the Rayleigh scattered intensity is very large relative to other frequencies. Various, smaller peak heights are obtained at frequencies ω_s smaller than ω (Stokes region) due to Raman scattering. The frequency shifts relative to ω reflect the vibrational frequencies of the molecules ω_k. At frequencies larger than ω (the anti-Stokes region), various peaks are also obtained, forming a kind of mirror reflection to the Stokes region ones; however, they are much smaller, and their height relations express the intensity lowering for larger shifts. Particularly, the ω_4, ω_5, and ω_6 are already too small to be recorded. Use of different Porto formations could entirely change the intensity ratios, and reveal other different vibrational frequencies. Note that because of the relatively huge Rayleigh peak intensity compared to the Raman ones, the measuring of Raman spectra requires use of monochromators of very high resolution, at least $0.1 - 1.0\ \text{cm}^{-1}$.

At this point it is worthwhile to discuss the meaning of the **virtual-state** term used previously. A virtual state is simply a state of a bound electron (to an atom or to a molecule) that is *not a resonance state*. We have already treated the issue classically without naming, when we wrote the dynamic equation of an electron bound by a harmonic potential under the influence of a driving force (Eq. [1.28]). The equation solution under a harmonic driving force is always a harmonic oscillation of the electron; however, for far-from-resonance frequencies, the driven oscillation amplitude is tiny compared to that obtained at resonance. In quantum terms, the tiny amplitude means a tiny probability for the absorption transition to take place. However, if the transition occurs, the oscillating electron radiates, and returns the energy to the electromagnetic field. That emission probability is independent of the resonance frequency, which is thus very large compared to the former "absorption" probability. That way, the electron does not stay long in a virtual state, and the emission phase is practically the same as the absorbed light phase. That contrasts with a case of a resonant state (namely a "usual" state), where the excited electron oscillates with a large amplitude, and in most cases long enough to disconnect between the absorption and emission phases. In the latter, the light emission process is called fluorescence (see Section 7.5).

The classical description of Raman scattering is sufficient to obtain the selection rules for its occurrence. This book does not elaborate on the quantum approach to Raman scattering.

8.6 Point Symmetry

The molecular Hamiltonian must be preserved under all symmetry operations that are valid for the molecule at its equilibrium positions. This property is also valid for the normal coordinates describing the molecular internal oscillations. The quantity that determines the oscillations Hamiltonian is the potential function U discussed in previous sections. This section provides a detailed study of molecular symmetry properties.

A rigid displacement of a molecule preserves the distance and orientation relations among the nuclei. A **symmetry operation** is a rigid displacement that provides an identically shaped structure, indistinguishable from the original one. A symmetry operation that leaves some single point at its original place is called a **point-symmetry** operation. The molecular mass center is a singular point, and it is thus the one left undisplaced for all point-symmetry operations. The number of point symmetry operations is limited. Group theoretical notation of point symmetry operations uses **Schoenflies notation** as summarized in Table 8.1.

The E operation and all C_p's are called **proper rotations**; i, all S_p's, and all σ ones, are called **improper rotations**. Notably, all rotational axes and reflection mirrors contain the mass center point. All above symmetry operations relate to a point, axis, or plane. If a molecule exhibits a symmetry relative to a certain point, axes, or planes, these objects are called **molecular symmetry elements**. A point being a symmetry element is called **symmetry center**. A line being a symmetry element is called **symmetry axis**. A plane being a symmetry element is called **symmetry plane**. The highest-order C_p axis is called **principal axis**; it is usually imagined to be vertical, assumed to coincide with the Z -axis of the molecular coordinate system. If a molecule exhibits more than a single **symmetry element**, it often necessarily exhibits certain other ones.

Table 8.1 Point symmetry operations on a molecule.

Symbol	Operation name	Description
E	Identity	No operation, or a $(2\pi)\cdot n$ rotation about any axis crossing the mass center, where $n = 1, 2, 3, \cdots$
C_p	Rotation of order p (p-fold rotation)	$2\pi/p$ rotation about an axis crossing the mass center, where p is an integer.
S_p	p-fold rotation/reflection	$2\pi/p$ rotation about an axis crossing the mass center followed by reflection in a perpendicular plane crossing the mass center, where p is an integer.
i	Inversion	Transformation of any space point $\underline{r}$ to $-\underline{r}$, the mass center as the origin.
σ_v	Vertical mirror reflection	Mirror reflection in a plane that contains the highest-order C_p axis.
σ_h	Horizontal mirror reflection	Mirror reflection in a plane perpendicular to the highest-order C_p axis, crossing the mass center.
σ_d	Dihedral mirror reflection	A σ_v plane bisecting an angle between two C_2 axes.

Multiplication between two symmetry operations is defined as a successive operation by the two; if a and b are symmetry operations, the "product" ab symbolizes a successive operation, where b operates first and a operates next. The mathematical set of all symmetry operations belonging to a molecule establishes a mathematical **group**, because it satisfies the following requirements: (i) multiplication between two symmetry operations of a molecule is also a symmetry operation of the same molecule (namely, if a and b are symmetry operations, so are ab and ba); (ii) any such a, b, and c, satisfy the relation $(ab)c = a(bc)$; (iii) there exists a unity element – it is the identity operation E; (iv) per each a operation, there exists an inverse operation a^{-1} such that $a^{-1}a = aa^{-1} = E$. Because all molecular symmetry operations are point symmetries, the symmetry groups describing molecules are called **point groups**.

Only very few groups are known to exhibit symmetry axes of order larger than six (except for a $p = \infty$ order, existing for many molecules; for example, the axis connecting the two nuclei in all diatomic molecules is a C_∞ axis). Thus, there exist only a limited number of point groups of practical interest. Such symmetry groups are listed in Table 8.2. Most are the **crystalline point groups** – those belonging to crystalline sites consistent with lattice translational symmetry. The table presents the groups by their **Schoenflies notation** and lists the symmetry elements belonging to each group, along with examples of belonging molecules.

There is no practical need to memorize all the group names and list of elements; however, it is worthwhile to observe several features:

1) The operations in each C_p and S_p group relate to a singular axis; special cases are the C_1 group describing a case of no symmetry except for the identity, and the C_i group whose sole element is the inversion i, which is actually identical to the S_2 operation ♣.

♣ Meticulously, the identity between the S_2 and i operations makes the latter redundant. However, it is still used by its specific symbol for clarity, but also for its importance in determining many spectroscopic selection rules.

Table 8.2 Important symmetry groups for molecular spectroscopy, and their elements.

No	Group name	Symmetry operations		Examples of molecules
		Fundamental	Derived	
1	C_1	E		$CHFClBr$; $CH_3 \cdot CHO$
2	C_2	C_2		H_2O_2
3	C_3	C_3		
4	C_4	C_4	$C''_2(\equiv C_4)$	
5	C_6	C_6	$C''_2(\equiv C_6)$; $C_3(\equiv C_6)$	
6	$C_i\ (=S_2)$	i		$trans-ClBrHC \cdot CHBrCl$
7	S_4	S_4	$C''_2(\equiv S_4)$	
8	S_6	S_6	$C_3(\equiv S_6)$; i	
9	$C_S\ (\equiv C_{1v})\ (\equiv S_2)$	σ		$NOCl$; NH_2D; C_2H_5Cl
10	C_{2v}	C_2; σ_v	$\sigma'_v(\perp\sigma_v)$	H_2O; H_2CCl_2; SO_2 chlorobenzene
11	C_{3v}	C_3; $3\sigma_v$		NH_3; $HCCl_3$; $OPCl_3$
12	C_{4v}	C_4; $2\sigma_v$	$2\sigma'_v(\perp\sigma_v)$; $C''_2(\equiv C_4)$	IF_5; $Fe(CO)_5$
13	C_{6v}	C_6; $3\sigma_v$	$3\sigma'_v(-\!\!\!\times\sigma_v)$; $C''_2(\equiv C_6)$; $C_3(\equiv C_6)$	
14	$C_{\infty v}$	C_∞; $\infty\sigma_v$	$\forall\ C_p(\equiv C_\infty)$	HCl; HCN; $HCCCl$
15	C_{2h}	C_2; σ_h	i	$trans-ClHC=CHCl$; C_3O_2
16	$C_{3h}(\equiv S_3)$	C_3; σ_h	$S_3(\equiv C_3)$	H_3BO_3; $C^+(NH_2)_3$
17	C_{4h}	C_4; σ_h	$C''_2(\equiv C_4)$; $S_4(\equiv C_4)$; i	
18	C_{6h}	C_6; σ_h	$C''_2(\equiv C_6)$; $C_3(\equiv C_6)$; $S_6(\equiv C_6)$; $S_3(\equiv C_6)$; i	
19	$D_2\ (\equiv V)$	C_2; $C'_2(\perp C_2)$	$C''_2(\perp C_2\ \&\ C'_2)$	
20	D_3	C_3; $3C_2(\perp C_3)$		
21	D_4	C_4; $2C_2(\perp C_4;+)$	$2C'_2(-\!\!\!\times C_2)$; $C''_2(\equiv C_4)$	
22	D_6	C_6; $3C_2(\perp C_6)$	$3C'_2(-\!\!\!\times C_2)$; $C''_2(\equiv C_6)$ $C_3(\equiv C_6)$	
23	$D_{2d}\ (\equiv V_d)$	C_2; $2C'_2(\perp C_2;+)$; $2\sigma_d(\supset C_2)$	$S_4(\equiv C_2)$	B_2Cl_4; $H_2C_3H_2$; N_4S_4

(Continued)

Table 8.2 (Continued)

No	Group name	Symmetry operations		Examples of molecules
		Fundamental	Derived	
24	D_{3d}	C_3; $3C_2(\perp C_3)$; $3\sigma_d$	$S_6(\equiv C_3)$; i	C_2H_6; C_6H_{12} (cyclohexane); $(SiH_3)_2O$
25	D_{4d}	C_4; $4C_2(\perp C_4)$; $4\sigma_d$	$C''_2(\equiv C_4)$; $S_8(\equiv C_4)$	S_8
26	D_{2h} $(\equiv V_h)$	C_2; $C'_2(\perp C_2)$; $\sigma(\supset C'_2 \,\&\, C''_2)$	$C''_2(\perp C_2 \,\&\, C'_2)$; $\sigma'(\supset C_2 \,\&\, C'_2)$; $\sigma''(\supset C_2 \,\&\, C''_2)$; i	C_2H_4; p-dichlorobenzene; naphthalene
27	D_{3h}	C_3; $3C_2(\perp C_3)$; σ_h	$S_3(\equiv C_3)$; $3\sigma_v$	BF_3; 1, 3, 5 – $C_6H_3Cl_3$
28	D_{4h}	C_4; $2C_2(\perp C_4)$; $2C'_2(\perp C_4)$; σ_h	$C''_2(\equiv C_4)$; $S_4(\equiv C_4)$; $2\sigma_v$; $2\sigma_d$; i	C_4H_8; $(PtCl_4)^{2-}$
29	D_{5h}	C_5; $5C_2(\perp C_5)$; σ_h	$S_5(\equiv C_5)$; $5\sigma_v$	Cyclopentane; ferrocene
30	D_{6h}	C_6; $3C_2(\perp C_6)$; $3C'_2(\perp C_6)$; σ_h	$C''_2(\equiv C_6)$; $C_3(\equiv C_6)$; $S_3(\equiv C_6)$; $S_6(\equiv C_6)$; $3\sigma_v$; $3\sigma_d$; i	C_6H_6 (benzene)
31	$D_{\infty h}$	C_∞; $\infty C_2(\perp C_\infty)$; σ_h	$\forall C_p(\equiv C_\infty)$; $\forall S_p(\equiv C_\infty)$; $\infty\sigma_v$; i	O_2; CO_2; C_2H_2
32	T	$4C_3$; $3C_2(-\!\!\times\!\!- C_3;\ +)$		
33	T_d	$4C_3$; $3C_2(-\!\!\times\!\!- C_3;\ +)$; $3S_4(\equiv C_2)$; $6\sigma_d(\supset 2C_3)$		CH_4; $BF_4{}^-$; $NH_4{}^+$; $SO_4{}^{2-}$
34	T_h	$4C_3$; $3C_2(+)$; i	$3\sigma(\supset 2C_2)$; $4S_6(\equiv C_3)$	
35	O	$4C_3$; $3C_4(+)$	$6C_2$; $3C'_2(\equiv C_4)$	
36	$\mathrm{O_h}$	$3C_4(+)$; $4C_3(p.\ /)$; i	$3C_2(\equiv C_4)$; $6C_2(\supset\circ\vert\leftrightarrow\vert)$; $3S_4(\equiv C_4)$; $4S_6(\equiv C_3)$; $3\sigma_h(\perp C_4)$; $6\sigma_d(\supset\vert\leftrightarrow\vert)$;	SF_6; $PtCl_6{}^{2-}$; UF_6

2) The C_{pv} and C_{ph} groups are formed by adding σ_v and σ_h operations to the C_p group, respectively; a special case is the C_S group, where its symmetry plane may be equally considered either horizontal or vertical. This property justifies its alternative naming as C_{1v} or as C_{1h}.
3) The D_p groups (called "dihedral" groups) are formed by adding a C_2 symmetry axis perpendicularly to the principal axis in the C_p group. Operations using these axes generate additional C_2 axes.
4) The D_{pd} groups are formed by adding σ_d mirror planes bisecting adjacent C_2 axes in D_p groups, respectively.
5) The D_{ph} groups are formed by adding a σ_h mirror plane to the D_p groups, respectively.
6) All above groups are called **axial groups** because they include at most a single axis of order of three or more. Groups including more than one axis of order three or four are called **cubic groups**. Many free molecules belong to the T_d group, which exhibits the symmetry of the regular tetrahedron; and to the O_h group, which exhibits the symmetry of the cube and the regular octahedron. The T, T_h, and O groups are important to describe the symmetry of molecules residing in crystalline sites.

The symmetry operations belonging to each group are listed in the second column of Table 8.2. It is divided into two sub-columns. The first, lists the set of **fundamental operations**, namely the smallest set of operations defining the group. The second one lists operations derived by the existence of the first set.

Table 8.2 lists 36 point groups, of which only 32 are consistent with the crystalline translational symmetry; they are called **crystalline point groups** (in fact, they constitute the basis for obtaining the 230 crystalline **space groups** obtained by adding "spatial" symmetry operations: translations by lattice vectors, mirror plane reflections joined by translations by other non-lattice vectors [**glide plane**], and rotations joined by translations by other non-lattice vectors [**screw axis**]). The non-crystalline groups are indicated by thick frames. A parenthetical comment that follows a symmetry symbol, details its geometric relation to other symmetry elements, derived, independent, or equivalent. Table 8.3 provides the meaning of shorthand symbols used for writing the comments.

It is often possible, at first order approximation, to describe molecules residing in solids as free ones, but of reduced symmetry according to the crystalline site symmetry they occupy.

Note that:

1) The identity operation E is a member *of all groups*. For brevity however, it is not specified in the elements lists except in the C_1 group, where it is the sole element.
2) Whenever a numeral or another symbol appears in front of a symmetry axis or a symmetry plane, it indicates the number of different axes (or respectively the number of different planes) in space, which are equivalent in the sense that each may be obtained by performing symmetry operations belonging to the group on each one.

Table 8.3 Translation of the shorthand symbols used in specifying the position relations among symmetry elements in Table 8.2.

Symbol	Meaning
$\equiv$	Coincides with ....
$\perp$	Perpendicular to
$\nrightarrow\!\!\!\!\times$	Bisects two adjacent planes or Bisects two adjacent axes
+	Axes mutually perpendicular
$\forall$	All
&	As well as
$\supset$	Includes the (axis/axes; points)
p. /	A principal cube diagonal
∘	The centers of
$\vert\leftrightarrow\vert$	Two opposite cube corners

3) In the C_{2v}, C_{4v}, and C_{6v} groups, the σ_v plane will be selected to contain more atoms than σ'_v (which in the C_{4v}, and C_{6v} is sometimes noted as σ_d). If for a specific molecule it is impossible to apply that rule, the σ_v plane will be selected (if possible) to cross more interatomic bonds than σ'_v. Similar rules apply to relations among C_2, C'_2, and C''_2 in the D_2 and D_{2h} groups, and between C_2 and C'_2 in the D_4, D_6, D_{2d}, D_{4h}, and D_{6h} groups.
4) The symmetry operations and group name symbols in Table 8.2 are the ones commonly used in molecular spectroscopy. There are other recognized noting systems, particularly the **Hermann-Mauguin** system, used especially in crystallography. The groups' presentation order in crystallography is also different, relating to Bravais sorting of crystal lattices. It will be presented in detail along with the Hermann-Mauguin one when we consider the influence of crystal lattice vibrations on light absorption and reflection, and on Raman scattering (Chapters 9 through 11, especially Section 11.5).

8.7 Group Representations, Characters, and Reduction Equation

When dealing with spatial constructions, each symmetry operation can be represented by a matrix operating on all the construction points' coordinates. A set of numerical matrices whose multiplications exhibit the same relations between the elements in the group belonging to a particular construction, is also a group, and is addressed as **representation**. The **representation dimension** is the order of the forming matrices. It is possible to convert the axes orthogonally, such that all matrix operations appear simultaneously as square matrices of lower order, located along the larger matrix diagonal, and all other matrix elements are zero.

$$\begin{pmatrix} (A^1_{11}) & & & & \\ & \begin{pmatrix} F^1_{11} & F^1_{12} & F^1_{13} \\ F^1_{21} & F^1_{22} & F^1_{23} \\ F^1_{31} & F^1_{32} & F^1_{33} \end{pmatrix} & & & \\ & & (A^2_{11}) & & \\ & & & \begin{pmatrix} E_{11} & E_{12} \\ E_{21} & E_{22} \end{pmatrix} & \\ & & & & \begin{pmatrix} F^2_{11} & F^2_{12} & F^2_{13} \\ F^2_{21} & F^2_{22} & F^2_{23} \\ F^2_{31} & F^2_{32} & F^2_{33} \end{pmatrix} \end{pmatrix} \tag{8.36}$$

Eq. (8.36) demonstrates schematically the decomposition of a matrix belonging to a 10×10 order representation, to submatrices of 1×1, 3×3, 1×1, 2×2, and 3×3 orders. All submatrices obtained from original matrices belonging to other operations of the same representation will look similar; namely, exhibit submatrices of same order along the diagonal – the first 1×1, second 2×2, and so on. The numerical quantities inside respective matrices along the diagonal may be different, however. A set of submatrices belonging to the same position along the diagonal also establish a regular representation of the group, though of a lower dimension. Manipulation of any matrix representation using orthogonal matrices into a set of lower-order representations is called **reduction**. The reduction process of an original representation ends when further reduction is mathematically impossible. Representations thus obtained are called **irreducible representations**.

In matrix algebra, the sum of elements along a diagonal is usually called **trace**. In connection to group theory, however, where the matrix serves as an element in a group representation, it is called **character**. The character is invariant under orthogonal conversions of the matrix. That trait forms the importance of the character parameter. Let g denote the number of different operations in the group. The character of the group element whose number in a *reducible* representation is i, is denoted χ_i for $i = 1, 2, 3, \cdots, g$. The character of the group element whose number in a certain *irreducible representation* γ of the same group is i, is denoted as χ_i^γ for $i = 1, 2, 3, \cdots, g$. On conclusion of a reduction process, the same irreducible representation γ could appear several (N^γ) times. There exists a simple relation among χ_i, χ_i^γ and N^γ, called **reduction equation**:

$$N^\gamma = \frac{1}{g}\sum_{i=1}^{g}\chi_i^\gamma \cdot \chi_i \quad . \tag{8.37}$$

There is a relation between molecular vibrations and the irreducible representations of the group that describe its stiff construction. It stems from the following property:

The nuclei *displacement vectors* from their equilibrium positions in a molecular *normal vibration* satisfy symmetry requirements of *one* of the irreducible representations of the stiff molecule, where all nuclei reside in their equilibrium positions. (8.38)

We will not prove this important theorem in this book.

8.8 Similarity Classes, Irreducible Representations, and Character Tables

Two group elements a and b are considered **similar**, if there exists a group element x forming the relation $b = xax^{-1}$. Nonsimilar elements are called **dissimilar elements**. It is thus possible to split and classify the group elements into sets of similar elements, called **similarity classes**, or just **classes**. Elements belonging to a certain class are similar, and are dissimilar to all elements belonging to other classes. A group-theoretical theorem states that **the number of similarity classes in a group is equal to the number of possible irreducible representations**. Furthermore, the characters of all members of a similarity class are equal. Accordingly, each group may be described by a table detailing the characters classified according to their similarity classes. Such a table is called a **character table**. Section 8.12 presents the character tables of all groups listed in Table 8.2.

For demonstration, in Table 8.4 we provide separately, and describe in detail, the D_{3h} group character table. The upper-left cell is used to write the group name according to Schoenflies notation. All group symmetry operations, classified by similarity, are listed from left to right in the first row. The identity operation E is written first; it is a lone element belonging to its similarity class. Then written is a similarity class containing two elements, C_3^1 and C_3^2, that perform 120° and 240° rotations, respectively, about the C_3

Table 8.4 Character table of the D_{3h} group.

D_{3h}	E	$2C_3$	$3C_2$	σ_h	$2S_3$	$3\sigma_v$		
A'_1	1	1	1	1	1	1		$\alpha_{xx} + \alpha_{yy}, \alpha_{zz}$
A'_2	1	1	−1	1	1	−1	R_z	
E'	2	−1	0	2	−1	0	(T_x, T_y)	$\left(\alpha^y_{xx} - \alpha^y_{yy}; \alpha^x_{xy}\right)$
A''_1	1	1	1	−1	−1	−1		
A''_2	1	1	−1	−1	−1	1	T_z	
E''	2	−1	0	−2	1	0	(R_x, R_y)	$\left(\alpha_{yz}; \bar{\alpha}_{xz}\right)$

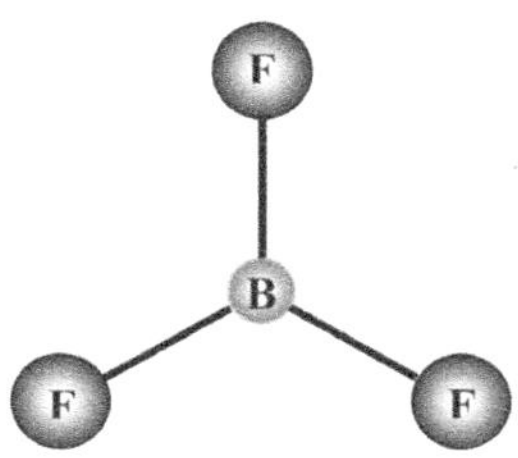

Figure 8.6 The spatial construction of the boron trifluoride BF_3 molecule.

axis. The number 2 appearing to the left of the class notation indicates the number of different elements (operations) it contains. Then to the right another class is noted, containing three different C_2-type rotations, each about a different C_2 axis. The list also includes a similarity class containing a lone element of reflection in a "horizontal" mirror plane; a class containing two improper rotations S_3^1 and S_3^2 about an S_3 axis that coincides with the C_3 axis; and a last similarity class containing three reflections in horizontal planes, all containing the C_3 axis, each separately containing a different C_2 axis. The group thus contains a total of 12 different operations, namely $g = 12$.

To assist in demonstrating these operations, Figure 8.6 shows the spatial construction of the boron trifluoride (BF_3) molecule, belonging to the D_{3h} symmetry (see also Table 8.2). All belonging atoms lie on a common plane – the "horizontal" σ_h plane. All $B - F$ bonds are of equal length, forming 120° interbond angles. The C_3 and S_3 axes coincide; both are perpendicular to the molecular plane, crossing through the boron B atom (which is also at the mass center). The three C_2 axes coincide each with one of the three $B - F$ bonds.

The following rows in the character table describe the irreducible group representations; the first cell in each row (from left to right) hosts the representation name (symbol), and then lists all characters belonging to each similarity class. An irreducible representation is also called **symmetry species**. The species symbol provides some information, though not always complete, about its properties. The symbol is constructed of a capital letter, Latin or Greek, attached with various upper or lower subscripts according to the following principles:

1) One-dimensional symmetry species (namely of order 1) will be called *symmetric* or *antisymmetric* in relation to a certain symmetry operation of the group if the related character is +1 or −1, respectively. The two-dimensional species are called *symmetric* or *antisymmetric* in relation to a certain symmetry operation of the group if the related character is +2 or −2, respectively. The same principle applies as well for higher-dimensional species.
2) The letters A or B are used to symbolize one-dimensional species that are *symmetric* or *antisymmetric*, respectively, in relation to rotations about the principal symmetry axis C_p.

3) The letter E is used to symbolize two-dimensional species, as well as pairs of one-dimensional species whose corresponding character sets are complex conjugates of each other.
4) The letter F is used to symbolize three-dimensional species; sometimes the letter T is used instead.
5) "1", "2", or "3" subscripts are *added* to all species symbols for numbering. For the A and B species, however, the "1" and "2" numbers are also used to indicate their symmetry or antisymmetry, respectively, in relation to rotations about C_2 axes that are perpendicular to the principal C_p axis. If no such C_2 axes are present, the same subscript numbers indicate symmetry or antisymmetry, respectively, in relation to mirror reflections in a σ_v plane.
6) "g" or "u" subscripts are *added* to all species to indicate whether they are *symmetric* or *antisymmetric*, respectively, in relation to inversion about the molecular mass center (if such symmetry operation is valid).
7) Single or double superscript primes are added to all species symbols to indicate their symmetry or antisymmetry, respectively, in relation to reflection in a σ_h plane.
8) Σ^+ and Σ^- symbolize one-dimensional species of groups belonging to linear molecules that are *symmetric* or *antisymmetric*, respectively, in relation to reflection in a σ_v plane.
9) Π, Δ, Φ, $\cdots$ symbolize different two-dimensional species of groups belonging to linear molecules.

Eq. (8.38) presents only a part of the physical significance expressed by the symmetry species. Other meanings become clearer when we detail their relations to light absorption and Raman scattering. Two added last right columns in the character table relate the species to various vibrational modes of the molecule. Species whose symmetry characters are identical to vibrational modes that generate electric dipoles along the X, Y, and Z orientations are denoted by T_x, T_y, and T_z, respectively (it is customary to say that they "belong" to T_x, T_y, or T_z). Degenerate components are enclosed by parentheses. For brevity, the T letter indicates the set of all three. Species "belonging" to rotations are denoted by R_x, R_y, and R_z, namely rotations about the X, Y, and Z axes, respectively. Degenerate components are enclosed in parentheses. The belonging of T or R components to a certain species is indicated in the second-to-last column. The last column indicates non-zero polarization tensor elements α_{xx}, α_{yy}, α_{zz}, α_{xy}, α_{yz}, or α_{zx} that are active in the molecule Raman scattering (recall that they are in fact the $(\partial\alpha_{ij}/\partial Q_k)_0 Q_k^0$ quantities which were denoted α_{ij} for brevity; see Section 8.5.2). Degenerate (equal sized) components are enclosed by parentheses. For brevity, the α letter indicates the set of all tensor elements. In cases where the symmetry species belong to both α and T elements, the proper x, y, or z superscript is added to the corresponding specific α element. In other words, the superscript label relates a Raman scattering formation for which a certain α_{ij} element appears in the scattering equation, with the electric dipole polarization belonging to the same vibrational mode (Eqs. [8.20]). The listed α elements determine the Raman scattering selection rules. They are further discussed in detail in Section 8.9.

In the present case, the two top-right cells of the character table are empty; they are sometimes used for adding comments. Notably, the core table is constituted by the 6×6 inner table of numbers. Section 8.12 provides character tables for all the groups listed in Table 8.2.

8.9 Selection Rules for Electric Dipole Absorption and Raman Scattering

We have indicated that selection rules related to electric dipole transitions between molecular vibrational states, and Raman scattering, which require change in the molecular vibrational state (Section 8.5.2), are linked to the molecular symmetry. This section presents some details, yet without formal proofs. Keep in mind that vibrational transitions belong to the infrared region, whereas the incident and scattered light in Raman scattering usually belong to the visible range. Therefore, if a normal vibration transition is electric-dipole allowed, it is addressed as "infrared-active"; if a normal vibration is allowed for involvement in Raman scattering, it is addressed as "Raman-active".

Eq. (8.38) states that the nuclei displacement vectors from their equilibrium positions in a molecular normal vibration satisfy symmetry requirements of *one* of the irreducible representations of the stiff molecule. On talking practice in that research field, each normal vibration "belongs" to one of the symmetry species. Classification of vibrational states accordingly exhibits further significant aspects as listed below.

1) Vibrational classification by symmetry species assists, at least partially, to design a "figurative" demonstration of the displacement vectors of every atom from its equilibrium position in a state of a normal mode. If the character of a symmetry operation belonging to the species is "symmetric" (namely +1, +2, and so on), the symmetry operation will result in full congruency of the operated-upon vectors with the original ones. If the character is antisymmetric (namely −1, −2, and so on), the symmetry operation will bring each operated-upon vector into an orientation opposite to the original displacement vector of the original point.
2) The degeneracy order of a normal vibrational species (belonging to a certain representation) is given by its order; degeneracy means that the different vibrational modes exhibit the same frequency.
3) A vibrational species that transforms like any translation T component, generates an electric dipole vibrating along the same orientation as the T component. Such a species is often addressed as **polarized vibration**, and it is infrared active.
4) Gas molecules having a spatial structure of a symmetric top, will exhibit a PQR absorption spectrum for vibrational species transforming like T_z (Section 7.5.6, especially Figure 7.14).
5) Vibrational or rotational species that transform like any component of the polarizability tensor will be Raman-active. The tensor elements are presented using a compact form, based on conventions that allow reconstruction of the full tensor structure. Elements written continually from left to right are *equal*. If separated by a *comma*, they are *different*. Thus, for example, the polarizability tensor of the A_1' species of the D_{3h} group (Table 8.4) looks like $\alpha_1 = \begin{pmatrix} a & 0 & 0 \\ 0 & a & 0 \\ 0 & 0 & b \end{pmatrix}$, where a and b are different positive quantities. The E' species is doubly degenerate, that is, it exhibits two separate polarizability tensors: components belonging to *different tensors* are separated by a semicolon (;). They would be $\alpha_2 = \begin{pmatrix} 0 & c & 0 \\ c & 0 & 0 \\ 0 & 0 & 0 \end{pmatrix}$ for a vibration forming an X-oriented electric dipole, and $\alpha_3 = \begin{pmatrix} c & 0 & 0 \\ 0 & -c & 0 \\ 0 & 0 & 0 \end{pmatrix}$ for a vibration forming a Y-oriented electric dipole, where c is a positive quantity, not necessarily equal to a or b. Equality of absolute values of the quantities written for the two tensors manifests the double "degeneracy" of the E' species; this is indicated in the character table by enclosing the two written tensor components inside parentheses. The E'' species is also doubly degenerate. The two belonging polarizability tensors are $\alpha_4 = \begin{pmatrix} 0 & 0 & 0 \\ 0 & 0 & d \\ 0 & d & 0 \end{pmatrix}$ where a corresponding motion transforms like a rotation about the X-axis, and $\alpha_5 = \begin{pmatrix} 0 & 0 & -d \\ 0 & 0 & 0 \\ -d & 0 & 0 \end{pmatrix}$, where a corresponding motion transforms like a rotation about the Y-axis; d is a positive quantity, not necessarily equal to a, b, or c. Enclosing parentheses indicate degeneracy; the bar symbol above the α_{xz} component belonging to that species indicates that its size (absolute value) is indeed identical to that of α_{yz}, but is of an opposite sign.

8.10 Method for Calculation and Description of Molecular Vibrational Species

This section provides methods for calculating the number, polarization states, Raman scattering, and infrared light absorption of species describing normal vibrations of a known-structured molecule.

Step 1: Assignment of a symmetry group to a known-structured molecule.

A clear spatial figure of the molecular skeleton must be portrayed, preferably as a mechanical model; this will be handy in identifying the molecular symmetry operations. This step requires checking consecutively the symmetry operations listed in Table 8.1. Initially it deserves finding the rotational axis of the highest order (the principal axis). Then the other C_p operations valid for the same axis must be written up. If no such axis exists, the molecule may belong to either C_S, C_i, or C_1. Finding of a mirror plane, an inversion center, or neither, will determine which of the above it is. If there are several C_p axes, all groups exhibiting such a feature should be marked in Table 8.2 from higher to lower order. The list of fundamental operations may guide in seeking for additional operations. Absence of operations in that list will help reduce the list of options. That procedure should be continued until only a single possibility is left!

Step 2: Calculating the number of atoms **not changing position** under symmetry operations.

a) For E and C_p operations: the number of atoms *not changing position* should be counted, **then deduct 2**. The result should be denoted u_k^{vib}, where $k = 1, \cdots, s$ is the position numerator of the similarity class by the order appearing in the character table. "s" is the number of classes, namely the number of columns in the core character table (recall that it is also the number of rows in the core table). The "vib" superscript is a reminder of the "2" number deduction from the count (if "2" has not been deducted, the end result will include the translation and rotational modes).

b) For i, S or σ operations: the number of atoms *not changing positions* should be counted, and the result should as before be denoted u_k^{vib}.

Notably, the u_k^{vib} quantity is identical for all operations belonging to the same k class!

Step 3: Calculating the number of normal oscillations belonging to each species.

The number of normal oscillations N^γ belonging to each species is given by the weighted reduction equation

$$\boxed{N^\gamma = \frac{1}{g}\sum_{k=1}^{s} n_k u_k^{vib} \chi_k^{one} \chi_k^{\gamma}}\ , \tag{8.39}$$

where n_k is the number of symmetry operations included in the k-th class; recall that the classes and the number of operations they contain are written in the top line of the character table. $g = \sum_{k=1}^{s} n_k$ is the total number of different group operations (the group order); χ_k^γ is the k-class character of the γ irreducible representation; and χ_k^{one} is the contribution to the class character by a reducible representation, where only a single atom doesn't change position under a class-member operation. The χ_k^{one} values are summarized in Table 8.5.

In certain cases, the character table presents complex numbers for the two-dimensional E species; however, they always appear as a pair of complex conjugate numbers per similarity class. The quantity to be set in Eq. (8.39) for χ_k^γ is their sum, which is always real. Specifically,

$$i + (-i) = 0\ ;\quad e^{2\pi i/p} + e^{-2\pi i/p} = 2\,\mathbf{cos}\,(2\pi/p)\,.$$

It is possible to outline a convenient method for calculating the number of normal oscillations N^γ by Eq. (8.39). Prepare five assisting tables: the original character table, and four additional "mute" character

Table 8.5 Contribution to a class character in a reducible representation when only one atom preserves its position under that class symmetry operation (χ_k^{one}).

Operation in *k*-th similarity class	χ_k^{one}	Operation in *k*-th similarity class	χ_k^{one}
E	3	σ	1
C_2	−1	i	−3
C_3^1, C_3^2	0	S_3^1, S_3^2	−2
C_4^1, C_4^3	1	S_4^1, S_4^2	−1
C_5^1, C_5^4	1.618	S_5^1, S_5^4	−0.382
C_5^2, C_5^3	−0.618	S_5^2, S_5^3	−2.618
C_6^1, C_6^2	2	S_6^1, S_6^2	0

tables, namely disposed of all characters, yet preserving the top-row listing of the similarity classes, and the left column indicating the group name followed by the list of irreducible representations (species). In the *first* mute table, write down in all k-column cells the u_k^{vib} value calculated by Step 2 (identical for all column cells of each k). In the *second* mute table, write down in all k-column cells the corresponding χ_k^{one} copied from Table 8.5 (identical for all column cells of each k). In the *third* mute table, write down in all k-column cells the corresponding n_k values (just copied from the top row of the character table as the number of similar operations belonging to the k-th class; they are identical for all column cells of each k). In the *fourth* mute table, write down the result of *multiplication* among all numbers belonging to the same-positioned cells in first, second, and third mute tables; they are in fact the product elements $n_k u_k^{vib} \chi_k^{one} \chi_k^{\gamma}$ serving as addends in Eq. (8.39).

Finally, in the *fourth* mute table, add up all n_k values appearing in any row of the third mute table, thus getting the group order g; enter it in the top right cell. Then add up all numbers appearing in each row, divide the sum by g, and write down the result in the last right cell of the same row. This number is the N^{γ} sought for quantity by Eq. (8.39).

Section 8.11 provides several examples utilizing the above method; however, for brevity we present only the final step of using the fourth mute table.

Step 4: Identifying Raman and infrared-active vibrational modes.

a) Infrared-active species are those transforming like T_x, T_y, or T_z. Species transforming like (T_x, T_y) or like (T_x, T_y, T_z) are doubly or triply degenerate, respectively.
b) Raman-active species are those transforming like any component of the polarizability tensor α. In the gas phase, Raman scattering belonging to species that transform like α_{xx}, α_{yy}, or α_{zz} exhibit polarized scattering: totally polarized in the cubic groups and partially polarized in the noncubic groups. Species transforming like α_{xy}, α_{yz}, or α_{zx} exhibit nonpolarized scattering.

Notably, there exist species that are only infrared-active, and others that are only Raman active. There are still others that are neither infrared nor Raman active; they are called **silent species**. In groups having an inversion center, infrared active species are always antisymmetric under inversion, and marked by a "u" subscript (an acronym of the German **ungerade** term, meaning "odd"). These species are never *Raman-active*! Raman-active species are always symmetric under inversion, and marked by a "g" subscript (an acronym of the German **gerade** term, meaning "even"). These species are always *infrared inactive*! In that respect, a combined study of infrared absorption and Raman scattering assists in obtaining a complete description of molecular vibrational states.

Step 5: Plotting of displacement vectors for one-dimensional vibrational modes.

It should be emphasized that the molecular skeleton plus the displacement vectors is in fact a new rigid body. Knowing the symmetry features of each vibrational species indeed allows to propose a possible pattern for the displacement vectors, but this pattern in not absolute; an absolute solution to the pattern requires an advanced knowledge of the potential energy function – or for the present case, the potential matrix of Eq. (8.8). The approach requires modeling the forces among all atoms; getting solutions is a long and tedious process, and not always practical. Especially for degenerate states, any linear combination of linearly independent solutions is also a valid solution. Plotting the displacement vectors indeed provides quite a useful demonstration, but its benefits are marginal. In this section we limit ourselves to describing the method of plotting the displacement vectors for one-dimensional species. It is possible to extend the method to doubly and triply degenerate vibrations, but this requires the use of concepts and methods the reader in not yet acquainted with.

The molecular skeleton figure prepared during the *first step* is used separately for each species. If a nucleus resides on a symmetry plane, and the species is symmetric with respect to mirror reflection by that plane, the displacement vector is plotted in the plane (there is a continuum of orientations for each such nucleus). However, if the species is antisymmetric with respect to reflection by that plane, the displacement vector is plotted perpendicularly to the plane. If the nucleus belongs to several symmetry planes, it must be verified that the displacement vector satisfies the above requirements for all planes; it of course reduces the number of degrees of freedom for the displacement vector orientation. This procedure should be executed for all nuclei. Then, the species symmetry is examined against rotations (proper or improper) about the symmetry axes. A symmetric rotational operation should bring every transformed displacement vector to congruence with the original vector that belonged to the nucleus there resided. An antisymmetric rotational operation should bring every transformed displacement vector to an opposite orientation relative to the original vector that belonged to the nucleus there resided. After that examination, orientation corrections, if needed, further reduce the freedom in selecting the displacement vectors' orientations. Often, the displacement vectors' orientation in the eventually obtained pattern is unique for some, but allows some freedom in orientation selection for others.

8.11 Examples of Molecular Vibrational Symmetry Species

8.11.1 The Ammonia NH_3 Molecule

Step 1: The ammonia NH_3 molecule exhibits a pyramidal shape with the nitrogen atom N at its vertex, and the three hydrogen H atoms at the vertices of an equilateral triangular base (Figures 6.9 and 8.7). The molecule exhibits a single principal C_3 axis crossing the nitrogen perpendicularly to the pyramid base. In addition, the three vertical symmetry planes, $3\sigma_v$, all include the C_3 axis, and each separately includes an $N-H$ bond. Observation of Table 8.2 indicates that this structure belongs to the C_{3v} group (No. 11 in Table 8.2).

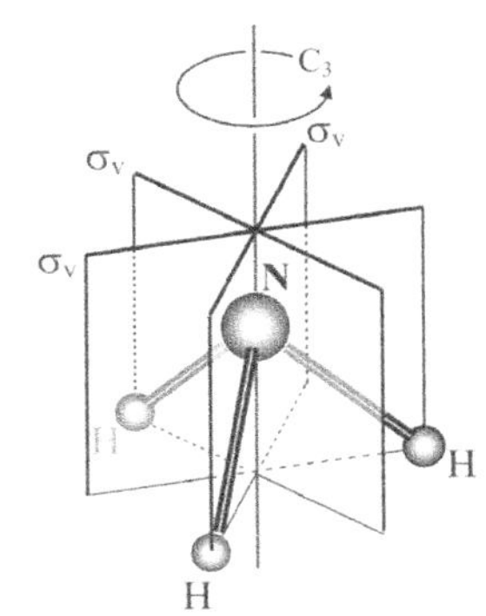

Figure 8.7 Construction and symmetry elements of the ammonia NH_3 molecule.

Step 2: We find the u_k^{vib} quantities by counting the number of atoms preserving their positions for every group operation according to the order the classes are written in the top row of the character table (Section 8.12.3): E; C_3; and σ_V. Thus, $u_E = 4 - 2 = 2$ (all four atoms); $u_{C_3} = 1 - 2 = -1$(The nitrogen atom); and $u_{\sigma_V} = 2$ (the nitrogen N atom and a single separate hydrogen H atom).

Step 3: For brevity, we show only the fourth of the assisting tables suggested in section 8.10, while writing down all multiplying numbers of the previous tables.

C_{3V}	E	$2C_3$	$3\sigma_V$	$g = 1 + 2 + 3 = 6$
A_1	$1 \times 2 \times 3 \times 1$	$2 \times (-1) \times 0 \times 1$	$3 \times 2 \times 1 \times 1$	$N^{A_1} = \frac{1}{6}[6 + 0 + 6] = 2$
A_2	$1 \times 2 \times 3 \times 1$	$2 \times (-1) \times 0 \times 1$	$3 \times 2 \times 1 \times (-1)$	$N^{A_2} = \frac{1}{6}[6 + 0 - 6] = 0$
E	$1 \times 2 \times 3 \times 2$	$2 \times (-1) \times 0 \times (-1)$	$3 \times 2 \times 1 \times 0$	$N^{E} = \frac{1}{6}[12 + 0 + 0] = 2$

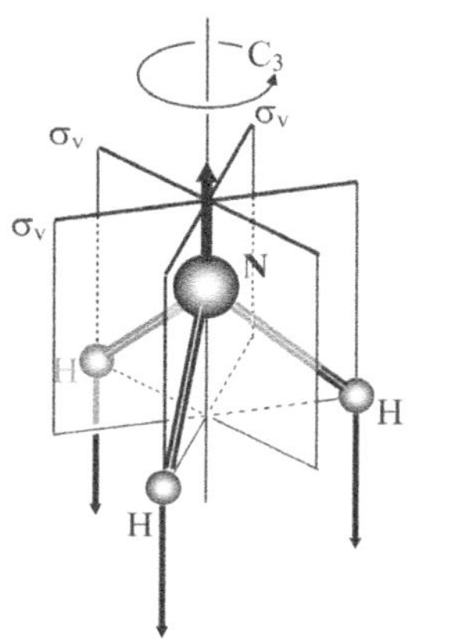

Figure 8.8 Displacement vectors of the atoms in the ammonia NH_3 molecule for the totally symmetric A_1 vibrations.

The vibrational species of the molecule are then written as

$$\Gamma = 2A_1 \oplus 2E \ . \tag{8.40}$$

Step 4: Observation of the character table reveals that all vibrational species are both Raman and infrared active. The four absorption frequency lines in the infrared, and the four Raman frequency shift lines, are identical. In the Raman scattering, the A_1 species are partially polarized, and the E lines are entirely nonpolarized. Each of the two A_1 absorption lines will exhibit a PQR structure.

Step 5: Figure 8.8 displays a possible pattern of the displacement vectors for one of the "totally symmetric" A_1 one-dimensional species of ammonia NH_3. This pattern is consistent with the symmetry property of the species: the nitrogen N atom displacement vector must point along the C_3 axis because it must belong equally to the three σ_V reflection planes crossing that axis. The three hydrogen H displacement vectors are plotted parallel to the C_3 axis as a special case. However, this is not generally necessary. It is sufficient that the displacement vector of each of the hydrogen atoms is placed on its containing reflection plane. The angle between the displacement vector and the nitrogen-hydrogen $N - H$ bond may assume an arbitrary value as long as the angle is identical for all hydrogen atoms. This stems from the C_3 symmetry requirement. The relations among the lengths and vector orientations should preserve the mass center position. The two nondegenerate A_1 species written in Eq. (8.40) differ by the angle between the hydrogen displacement vectors and their (common) $N - H$ bonds, and in the relation between lengths of the hydrogen atom's displacement vectors and that of the nitrogen N atom.

8.11.2 The Ethylene C_2H_4 Molecule

Step 1: The ethylene molecule structure appears as a combination of two separate isosceles triangles with a carbon C atom at the vertex, and hydrogen H atoms at the two base vertices. The two triangles connect by bonding their carbon vertices, and all atoms lie on a common plane; see Figures 6.14 and 8.9. The molecule lacks any C_3 axis or higher-order symmetry axes. However, it has three mutually perpendicular C_2 axes crossing a common point at the center between the $C = C$ double bond: one axis (Z) coincides with the bond; the second (X) is perpendicular to it, perpendicularly to the molecular plane; and the third (Y) is parallel to the molecular plane. Accordingly, the belonging groups to consider per Table 8.2 are D_2,

D_{2h} or D_{2d}. The existence of three symmetry planes formed by pairs of C_2 axes shows that it belongs to the D_{2h} group (No. 26 in Table 8.2).

Step 2: We find the u_k^{vib} quantities by counting the number of atoms preserving their positions for every group operation by the order that the classes are listed in the top row of the character table (Section 8.12.7): E, C_2, C_2', C''_2, i, σ, σ', and σ''. Thus, $u_E = 6 - 2 = 4$, $u_{C_2} = 2 - 2 = 0$, $u_{C_2'} = 0 - 2 = -2$, $u_{C''_2} = 0 - 2 = -2$, $u_i = 0$, $u_\sigma = 0$, $u_{\sigma''} = 2$, and $u_{\sigma'} = 6$.

Step 3: For brevity, we show only the fourth of the assisting tables suggested in Section 8.10, while writing down all multiplying numbers from the previous tables.

D_{2h}	E	$C_2(z)$	$C_2'(y)$	$C''_2(x)$	i	$\sigma(xy)$	$\sigma''(zx)$	$\sigma'(yz)$	$g = 8$
A_g	$1\times 4 \times 3\times 1$	$1\times 0\times (-1)\times 1$	$1\times(-2)\times (-1)\times 1$	$1\times(-2)\times (-1)\times 1$	$1\times 0\times (-3)\times 1$	$1\times 0\times 1\times 1$	$1\times 2\times 1\times 1$	$1\times 6\times 1\times 1$	$\frac{1}{8}\times 24 = 3$
B_{1g}	$1\times 4 \times 3\times 1$	$1\times 0\times (-1)\times 1$	$1\times(-2)\times (-1)\times(-1)$	$1\times(-2)\times (-1)\times(-1)$	$1\times 0\times (-3)\times 1$	$1\times 0\times 1\times 1$	$1\times 2\times 1\times(-1)$	$1\times 6\times 1\times(-1)$	$\frac{1}{8}\times 0 = 0$
B_{2g}	$1\times 4 \times 3\times 1$	$1\times 0\times (-1)\times(-1)$	$1\times(-2)\times (-1)\times 1$	$1\times(-2)\times (-1)\times(-1)$	$1\times 0\times (-3)\times 1$	$1\times 0\times 1\times(-1)$	$1\times 2\times 1\times 1$	$1\times 6\times 1\times(-1)$	$\frac{1}{8}\times 8 = 1$
B_{3g}	$1\times 4 \times 3\times 1$	$1\times 0\times (-1)\times(-1)$	$1\times(-2)\times (-1)\times(-1)$	$1\times(-2)\times (-1)\times 1$	$1\times 0\times (-3)\times 1$	$1\times 0\times 1\times(-1)$	$1\times 2\times 1\times(-1)$	$1\times 6\times 1\times 1$	$\frac{1}{8}\times 16 = 2$
A_u	$1\times 4 \times 3\times 1$	$1\times 0\times (-1)\times 1$	$1\times(-2)\times (-1)\times 1$	$1\times(-2)\times (-1)\times 1$	$1\times 0\times (-3)\times(-1)$	$1\times 0\times 1\times(-1)$	$1\times 2\times 1\times(-1)$	$1\times 6\times 1\times(-1)$	$\frac{1}{8}\times 8 = 1$
B_{1u}	$1\times 4 \times 3\times 1$	$1\times 0\times (-1)\times 1$	$1\times(-2)\times (-1)\times(-1)$	$1\times(-2)\times (-1)\times(-1)$	$1\times 0\times (-3)\times(-1)$	$1\times 0\times 1\times(-1)$	$1\times 2\times 1\times 1$	$1\times 6\times 1\times 1$	$\frac{1}{8}\times 16 = 2$
B_{2u}	$1\times 4 \times 3\times 1$	$1\times 0\times (-1)\times(-1)$	$1\times(-2)\times (-1)\times 1$	$1\times(-2)\times (-1)\times(-1)$	$1\times 0\times (-3)\times(-1)$	$1\times 0\times 1\times 1$	$1\times 2\times 1\times(-1)$	$1\times 6\times 1\times 1$	$\frac{1}{8}\times 16 = 2$
B_{3u}	$1\times 4 \times 3\times 1$	$1\times 0\times (-1)\times(-1)$	$1\times(-2)\times (-1)\times(-1)$	$1\times(-2)\times (-1)\times 1$	$1\times 0\times (-3)\times(-1)$	$1\times 0\times 1\times 1$	$1\times 2\times 1\times 1$	$1\times 6\times 1\times(-1)$	$\frac{1}{8}\times 8 = 1$

The vibrational species of the molecule are then written as

$$\Gamma = 3A_g \oplus B_{2g} \oplus 2B_{3g} \oplus A_u \oplus 2B_{1u} \oplus 2B_{2u} \oplus B_{3u}\ . \tag{8.41}$$

Step 4: Observation of the D_{2h} character table reveals that the ethylene C_2H_4 molecule exhibits five absorption lines in the infrared: $2B_{1u} \oplus 2B_{2u} \oplus B_{3u}$. In addition, it exhibits six Raman active lines: $3A_g \oplus B_{2g} \oplus 2B_{3g}$. All lines are separate, so there is no degeneracy among the line's frequencies (except for accidental ones). The A_u vibrational species is silent, that is, it is expressed by neither infrared absorption nor by Raman scattering. (It will express itself, however, in interelectronic states' transitions). The three A_g Raman scattering lines are polarized; the other three Raman lines belonging to other species are nonpolarized.

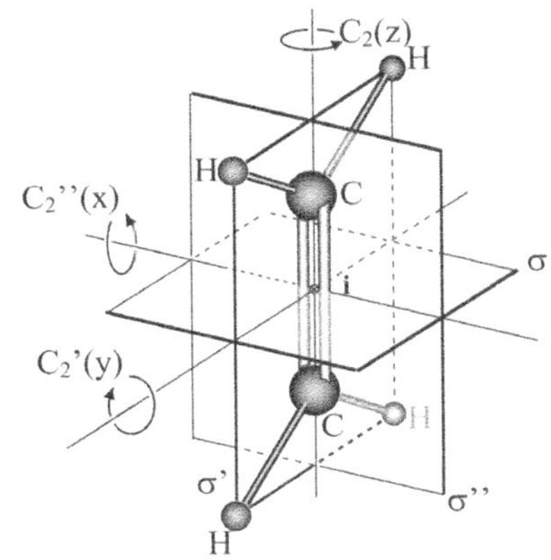

Figure 8.9 Construction and symmetry elements of the ethylene C_2H_4 molecule.

Step 5: In this example, we demonstrate plots of possible displacement vectors of two different species: the totally symmetric A_g species, and the B_{2g} species; they are presented in Figure 8.10.

For the particular A_g species, the two carbon C displacement vectors must be oriented along the axis connecting the two. This is because each vector must belong equally to two mirror planes – $\sigma''(zx)$ and $\sigma'(yz)$ – which intersect at this axis. They should be oppositely oriented to abide by the $\sigma(xy)$ mirror symmetry requirement. The displacement vector of

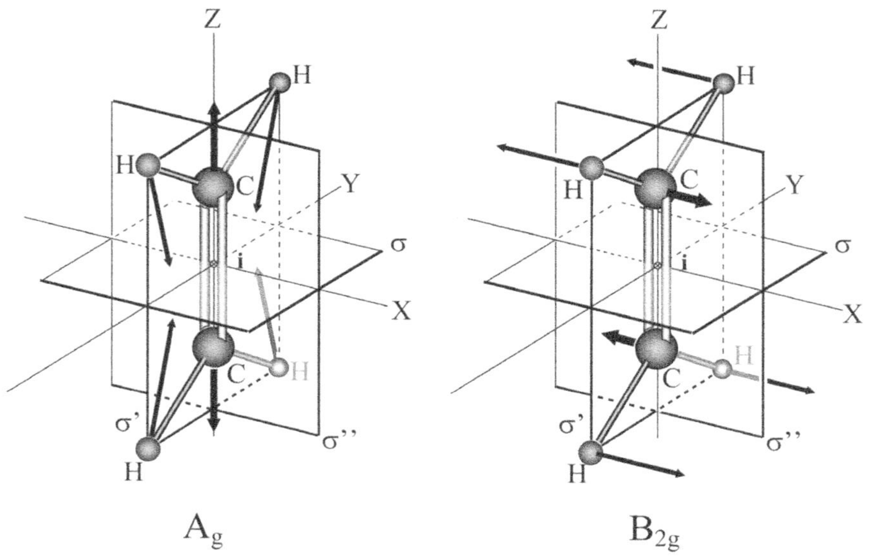

Figure 8.10 Displacement vectors of the ethylene C_2H_4 molecule atoms in the A_g and B_{2g} vibrational modes.

each hydrogen H atom must be placed on the $\sigma'(yz)$ mirror reflection plane. Symmetry considerations alone are insufficient for determining the angle between the vector and its $C-H$ bond orientation; still, to abide by the symmetry requirement for reflection in the $\sigma''(zx)$ plane, these angles, as well as the vectors' length, must be identical for all hydrogen atoms. The length relations between the carbon and hydrogen displacement vectors should preserve the mass-center position. The three different, nondegenerate A_g vibrational species, differ by the angle between the hydrogen vector and the $C-H$ bond, and by the hydrogen vector length compared to that of each carbon C displacement vector.

For the particular B_{2g} species, the two carbon C displacement vectors must be oriented perpendicularly to the $\sigma'(yz)$ plane, because that species is antisymmetric to reflection in that plane. They must point in opposite directions to be antisymmetric in relation to reflection in the horizontal $\sigma(xy)$ plane. The two hydrogen H displacement vectors must also be oriented perpendicularly to the $\sigma'(yz)$ plane to be antisymmetric with respect to reflection in that plane. Each pair of vectors belonging to hydrogen H atoms that are bonded to the same carbon C atom should point in the same direction; that is, in order to be symmetric to mirror reflection in the $\sigma''(zx)$ plane, which is perpendicular to the molecular plane and contains the two carbon atoms. However, displacement vectors of hydrogen H atoms bonded to different carbon C atoms should point in opposite directions, to be antisymmetric in respect to mirror reflection in the horizontal $\sigma(xy)$ plane. Furthermore, these vectors must be of the same length. If the vectors point in same direction as their bonded carbon, it would mean a molecular rotation; thus, a vibrational motion is obtained only when they point opposite to their bonded carbon. The length relations among all displacement vectors must preserve the mass center position.

8.11.3 The Carbon Tetrachloride CCl_4 Molecule

Step 1: The carbon tetrachloride CCl_4 has the form of a regular tetrahedron, with the chlorine Cl atoms at its vertices, and the carbon C atom at its center (Figure 8.11). This structure may be enclosed inside a cube, with the carbon C atom at its center, and the chlorine Cl atoms occupying four of the eight cube vertices. Each chlorine atom relates to the other three ones situated at other edges of the cube square side-facet diagonals. Each of the $C-Cl$ bonds coincides with a C_3 symmetry axis, and the molecule exhibits no C_4 axes. The groups exhibiting such features are T, T_h, and T_d (Table 8.2). The fact that each combination of two C_3 axes also defines a reflection symmetry plane (σ_d; only one shown in Figure 8.11) unequivocally determines that the representing group is T_d (No. 33 in Table 8.2). It is easy to verify the presence of three C_2 axes, each oriented along an angle bisector between each two C_3 axes (only two are portrayed in Figure 8.11).

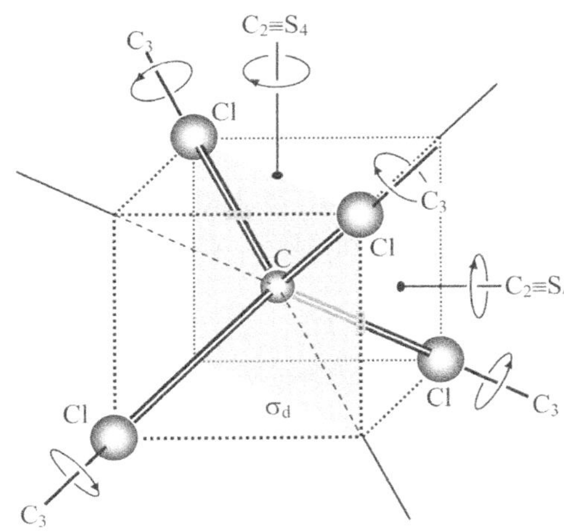

Figure 8.11 Construction and symmetry elements of the carbon tetrachloride CCl_4 molecule.

Step 2: We find the u_k^{vib} quantities by counting the number of atoms preserving their positions for every group operation by the order that the classes are written in the top row of the character table (Section 8.12.8): E, C_3, C_2, S_4, and σ_d. Thus, $u_E = 5 - 2 = 3$: (all five atoms); $u_{C_3} = 2 - 2 = 0$: (a chlorine atom and a carbon atom in a $C-Cl$ bond); $u_{C_2} = 1 - 2 = -1$: (the carbon atom); $u_{S_4} = 1$: (the carbon atom); $u_{\sigma_d} = 3$: (two chlorine atoms and the carbon atom).

Step 3: For brevity, we show only the fourth of the assisting tables suggested in Section 8.10, while writing down all multiplying numbers from the previous assisting tables.

T_d	E	$8C_3$	$3C_2$	$6S_4$	$6\sigma_d$	$g = 24$
A_1	$1\times3\times3\times1$	$8\times0\times0\times1$	$3\times(-1)\times(-1)\times1$	$6\times1\times(-1)\times1$	$6\times3\times1\times1$	1
A_2	$1\times3\times3\times1$	$8\times0\times0\times1$	$3\times(-1)\times(-1)\times1$	$6\times1\times(-1)\times(-1)$	$6\times3\times1\times(-1)$	0
E	$1\times3\times3\times2$	$8\times0\times0\times(-1)$	$3\times(-1)\times(-1)\times2$	$6\times1\times(-1)\times0$	$6\times3\times1\times0$	1
F_1	$1\times3\times3\times3$	$8\times0\times0\times0$	$3\times(-1)\times(-1)\times(-1)$	$6\times1\times(-1)\times1$	$6\times3\times1\times(-1)$	0
F_2	$1\times3\times3\times3$	$8\times0\times0\times0$	$3\times(-1)\times(-1)\times(-1)$	$6\times1\times(-1)\times(-1)$	$6\times3\times1\times1$	2

The vibrational species of the molecule are then written as

$$\Gamma = A_1 \oplus E \oplus 2F_2\,. \tag{8.42}$$

Step 4: Observation of the T_d character table reveals that the four $A_1 \oplus E \oplus 2F_2$ vibrational species of the carbon tetrachloride CCl_4 molecule are Raman active. However, only two of them, $2F_2$, are also infrared active. In the gas phase, the Raman scattering line of A_1 species vibration is fully polarized; the other three are unpolarized.

Step 5: For this example we will demonstrate plots of possible displacement vectors of the totally symmetric A_1 species, presented in Figure 8.12. The carbon atom must remain at rest (displacement vector vanishes) because the displacement vector must be placed equally in all σ_d reflection mirror planes, and these planes intersect at a single common point. The displacement vector of each chlorine Cl atom must be oriented along its $C-Cl$ bond, because each such bond coincides with a C_3 axis, and the A_1 species is symmetric with respect to the corresponding C_3 operation. All vectors must commonly point toward the carbon atom or away from it, and be of same size, to abide by the symmetry requirements concerning all C_2 operations.

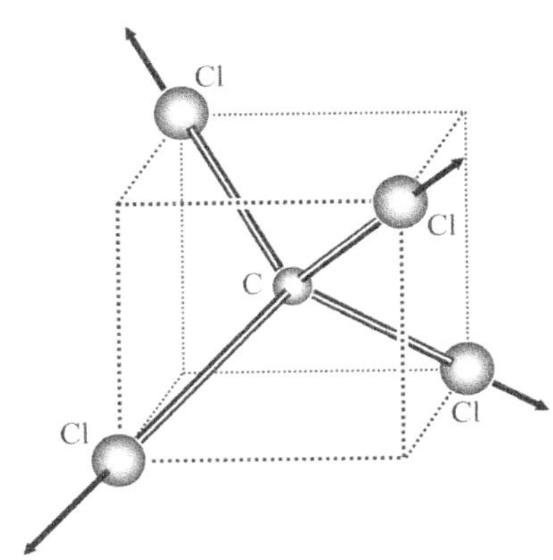

Figure 8.12 Displacement vectors of the carbon tetrachloride CCl_4 molecule atoms in the A_1 vibrational mode.

In molecular vibrations spectroscopy, the A_1 species is called "breathing species" or "breathing mode", because during vibration, the tetrahedron volume formed by the four chlorine atoms increases or decreases without change of shape, and this is like a volume change in the lungs when breathing. Structures like the carbon tetrachloride CCl_4 molecule are common. Therefore, special accepted names are assigned to their vibrational modes as follows: $\nu_1(A_1)$, $\nu_2(E)$, and $\nu_3(F_2)$ and $\nu_4(F_2)$ for the two nondegenerate vibrations of the F_2 species.

8.12 Point Groups, Character Tables, and Selection Rules

8.12.1 The C_p groups

(1)

C_1	E		
A	1	T; R	$\alpha^{x,\ y,\ z}$

(2)

C_2	E	C_2		
A	1	1	T_z; R_z	α^z_{xx}, α^z_{yy}, α^z_{zz}, α^z_{xy}
B	1	−1	T_x, T_y; R_x, R_y	$\alpha^{x,y}_{xz}$, $\alpha^{x,y}_{yz}$

(3)

C_3	E	C_3	C_3^2		$\varepsilon \equiv e^{2\pi i/3}$
A	1	1	1	T_z; R_z	$\alpha^z_{xx} + \alpha^z_{yy}$, α^z_{zz}
E	$\begin{Bmatrix} 1 \\ 1 \end{Bmatrix}$	$\begin{matrix} \varepsilon \\ \varepsilon^* \end{matrix}$	$\begin{matrix} \varepsilon^* \\ \varepsilon \end{matrix}$	(T_x, T_y); (R_x, R_y)	$\begin{Bmatrix} \left(\alpha^x_{xx} - \alpha^x_{yy};\ \bar{\alpha}^y_{xy}\right), \left(\alpha^x_{xz};\ \bar{\alpha}^y_{yz}\right) \\ \left(\alpha^y_{xx} - \alpha^y_{yy};\ \alpha^x_{xy}\right), \left(\bar{\alpha}^y_{xz};\ \alpha^x_{yz}\right) \end{Bmatrix}$

(4)

C_4	E	C_4	C_4^3	C''_2		
A	1	1	1	1	T_z; R_z	$\alpha^z_{xx} + \alpha^z_{yy}$, α^z_{zz}
B	1	−1	−1	1		$\alpha_{xx} - \alpha_{yy}$, α_{xy}
E	$\begin{Bmatrix} 1 \\ 1 \end{Bmatrix}$	$\begin{matrix} i \\ -i \end{matrix}$	$\begin{matrix} -i \\ i \end{matrix}$	$\begin{matrix} -1 \\ -1 \end{matrix}$	(T_x, T_y); (R_x, R_y)	$\begin{Bmatrix} \left(\alpha^x_{xz};\ \alpha^y_{yz}\right) \\ \left(\bar{\alpha}^y_{xz};\ \alpha^x_{yz}\right) \end{Bmatrix}$

(5)

C_6	E	C_6	C_3	C''_2	C_3^2	C_6^5		$\varepsilon \equiv e^{2\pi i/6}$
A	1	1	1	1	1	1	T_z; R_z	$\alpha_{xx}^z + \alpha_{yy}^z$, α_{zz}^z
B	1	−1	1	−1	1	−1		
E_1	1	ε	$-\varepsilon^*$	−1	$-\varepsilon$	ε^*	$\left\{(T_x, T_y);\right.$	$\left\{(\alpha_{xz}^x; \alpha_{yz}^y)\right.$
	1	ε^*	$-\varepsilon$	−1	$-\varepsilon^*$	ε	$\left.(R_x, R_y)\right\}$	$\left.(\overline{\alpha}_{xz}^y; \alpha_{yz}^x)\right\}$
E_2	1	$-\varepsilon^*$	$-\varepsilon$	1	$-\varepsilon^*$	$-\varepsilon$		$\left\{(\alpha_{xx}^I - \alpha_{yy}^I; \overline{\alpha}_{xy}^{II})\right.$
	1	$-\varepsilon$	$-\varepsilon^*$	1	$-\varepsilon$	$-\varepsilon^*$		$\left.(\alpha_{xx}^{II} - \alpha_{yy}^{II}; \alpha_{xy}^I)\right\}$

8.12.2 The S_p groups

(6)

Ci	E	i		
A_g	1	1	R	α
A_u	1	−1	T	

(7)

S_4	E	S_4	C''_2	S_4^3		
A	1	1	1	1	R_z	$\alpha_{xx} + \alpha_{yy}$, α_{zz}
B	1	−1	1	−1	T_z	$\alpha_{xx}^z - \alpha_{yy}^z, \alpha_{xy}^z$
E	1	i	−1	$-i$	(T_x, T_y); (R_x, R_y)	$\left\{(\alpha_{xz}^x;\ \overline{\alpha}_{yz}^y)\right.$
	1	$-i$	−1	i		$\left.(\alpha_{xz}^y;\ \alpha_{yz}^x)\right\}$

(8)

S_6	E	C_3	C_3^2	i	S_6^5	S_6		$S_6 = C_3 \times i$; $\varepsilon \equiv e^{2\pi i/3}$
A_g	1	1	1	1	1	1	R_z	$\alpha_{xx} + \alpha_{yy}$, α_{zz}
E_g	1	ε	ε^*	1	ε	ε^*	(R_x, R_y)	$\left\{(\alpha_{xx}^I - \alpha_{yy}^I;\ \overline{\alpha}_{xy}^{II}), (\alpha_{xz}^I;\ \overline{\alpha}_{yz}^{II})\right.$
	1	ε^*	ε	1	ε^*	ε		$\left.(\alpha_{xx}^{II} - \alpha_{yy}^{II};\ \alpha_{xy}^I), (\overline{\alpha}_{xz}^{II};\ \alpha_{yz}^I)\right\}$
A_u	1	1	1	−1	−1	−1	T_z	
E_u	1	ε	ε^*	−1	$-\varepsilon$	$-\varepsilon^*$	(T_x, T_y)	
	1	ε^*	ε	−1	$-\varepsilon^*$	$-\varepsilon$		

8.12.3 The C_{pv} groups

(9)

C_S	E	σ		
A'	1	1	$T_x, T_y; R_z$	$\alpha_{xx}^{x,y}, \alpha_{yy}^{x,y}, \alpha_{zz}^{x,y}, \alpha_{xy}^{x,y}$
A''	1	−1	$T_z; R_x, R_y$	$\alpha_{xz}^{z}, \alpha_{yz}^{z}$

(10)

C_{2v}	E	C_2	$\sigma_v(zx)$	$\sigma'_v(yz)$		
A_1	1	1	1	1	T_z	$\alpha_{xx}^{z}, \alpha_{yy}^{z}, \alpha_{zz}^{z}$
B_1	1	−1	1	−1	$T_x; R_y$	α_{xz}^{x}
A_2	1	1	−1	−1	R_z	α_{xy}
B_2	1	−1	−1	1	$Ty; Rx$	α_{yz}^{y}

(11)

C_{3v}	E	$2C_3$	$3\sigma_v$		
A_1	1	1	1	T_z	$\alpha_{xx}^{z} + \alpha_{yy}^{z}, \alpha_{zz}^{z}$
A_2	1	1	−1	R_z	
E	2	−1	0	$(T_x, T_y); (R_x, R_y)$	$\left(\alpha_{xx}^{y} - \alpha_{yy}^{y}; \alpha_{xy}^{x}\right), (\alpha_{yz}^{y};\ \alpha_{zx}^{x})$

(12)

C_{4v}	E	$2C_4$	C''_2	$2\sigma_v$	$2\sigma'_v$		
A_1	1	1	1	1	1	T_z	$\alpha_{xx}^{z} + \alpha_{yy}^{z}, \alpha_{zz}^{z}$
A_2	1	1	1	−1	−1	R_z	
B_1	1	−1	1	1	−1		$\alpha_{xx} - \alpha_{yy}$
B_2	1	−1	1	−1	1		α_{xy}
E	2	0	−2	0	0	$(Tx, Ty); (Rx, Ry)$	$(\alpha_{yz}^{y};\ \alpha_{zx}^{x})$

(13)

C_{6v}	E	$2C_6$	$2C_3$	C''_2	$3\sigma_v$	$3\sigma'_v$		
A_1	1	1	1	1	1	1	T_z	$\alpha_{xx}^{z} + \alpha_{yy}^{z}, \alpha_{zz}^{z}$
A_2	1	1	1	1	−1	−1	R_z	
B_1	1	−1	1	−1	1	−1		
B_2	1	−1	1	−1	−1	1		
E_1	2	1	−1	−2	0	0	$(Tx, Ty); (Rx, Ry)$	$(\alpha_{yz}^{y};\ \alpha_{zx}^{x})$
E_2	2	−1	−1	2	0	0		$(\alpha_{xx} - \alpha_{yy}; \alpha_{xy})$

(14)

$C_{\infty v}$	E	$2C_{\infty}^{\varphi}$	...	$\infty\sigma_v$		
$A_1 \equiv \Sigma^+$	1	1	...	1	T_z	$\alpha_{xx}^z + \alpha_{yy}^z, \alpha_{zz}^z$
$A_2 \equiv \Sigma^-$	1	1	...	−1	R_z	
$E_1 \equiv \Pi$	2	$2\cos\varphi$	...	0	$(T_x, T_y); (R_x, R_y)$	$(\alpha_{yz}^y;\ \alpha_{zx}^x)$
$E_2 \equiv \Delta$	2	$2\cos 2\varphi$	...	0		$(\alpha_{xx} - \alpha_{yy}; \alpha_{xy})$
$E_3 \equiv \Phi$	2	$2\cos 3\varphi$	...	0		
⋮	⋮	⋮	⋮	⋮		

8.12.4 The C_{ph} groups

(15)

C_{2h}	E	C_2	i	σ_h		
A_g	1	1	1	1	R_z	$\alpha_{xx}, \alpha_{yy}, \alpha_{zz}, \alpha_{xy}$
B_g	1	−1	1	−1	R_x, R_y	α_{xz}, α_{yz}
A_u	1	1	−1	−1	T_z	
B_u	1	−1	−1	1	$T_x,, T_y$	

(16)

C_{3h}	E	C_3	C_3^2	σ_h	S_3	S_3^5		$\varepsilon \equiv e^{2\pi i/3}$
A'	1	1	1	1	1	1	R_z	$\alpha_{xx} + \alpha_{yy}, \alpha_{zz}$
E'	1 1	ε ε^*	ε^* ε	1 1	ε ε^*	ε^* ε	(T_x, T_y)	$(\alpha_{xx}^x - \alpha_{yy}^x; \overline{\alpha}_{xy}^y)$ $(\alpha_{xx}^y - \alpha_{yy}^y; \alpha_{xy}^x)$
A''	1	1	1	−1	−1	−1	T_z	
E''	1 1	ε ε^*	ε^* ε	−1 −1	$-\varepsilon$ $-\varepsilon^*$	$-\varepsilon^*$ $-\varepsilon$	(R_x, R_y)	$(\alpha_{xz}^I; \alpha_{yz}^{II})$ $(\overline{\alpha}_{xz}^{II}; \alpha_{yz}^I)$

(17)

C_{4h}	E	C_4	C''_2	C_4^3	i	S_4^3	σ_h	S_4		
A_g	1	1	1	1	1	1	1	1	R_z	$\alpha_{xx} + \alpha_{yy}, \alpha_{zz}$
B_g	1	−1	1	−1	1	−1	1	−1		$\alpha_{xx} - \alpha_{yy}, \alpha_{xy}$
E_g	1 1	i $-i$	−1 −1	$-i$ i	1 1	i $-i$	−1 −1	$-i$ i	(R_x, R_y)	$(\alpha_{xz}^I; \alpha_{yz}^{II})$ $(\overline{\alpha}_{xz}^{II}; \alpha_{yz}^I)$
A_u	1	1	1	1	−1	−1	−1	−1	T_z	
B_u	1	−1	1	−1	−1	1	−1	1		
E_u	1 1	i $-i$	−1 −1	$-i$ i	$-i$ i	1 1	i $-i$	i $-i$	(T_x, T_y)	

(18)

C_{6h}	E	C_6	C_3	C''_2	C_3^2	C_6^5	i	S_3^5	S_6^5	σ_h	S_6	S_3		$\varepsilon \equiv e^{2\pi/3}$
A_g	1	1	1	1	1	1	1	1	1	1	1	1	R_z	$\alpha_{xx} + \alpha_{yy}$, α_{zz}
B_g	1	−1	1	−1	1	−1	1	−1	1	−1	1	−1		
E_{1g}	$\left\{ 1 \right.$	ε	$-\varepsilon^*$	−1	$-\varepsilon$	ε^*	1	ε	$-\varepsilon^*$	−1	$-\varepsilon$	$\left. \varepsilon^* \right\}$	$(R_x,$	$\left\{ \left(\alpha_{xz}^{I}; \alpha_{yz}^{II}\right) \right.$
	$\left. 1 \right\}$	ε^*	$-\varepsilon$	−1	$-\varepsilon^*$	ε	1	ε^*	$-\varepsilon$	−1	$-\varepsilon^*$	$\left. \varepsilon \right\}$	$R_y)$	$\left. \left(\overline{\alpha}_{xz}^{II}; \alpha_{yz}^{I}\right) \right\}$
E_{2g}	$\left\{ 1 \right.$	$-\varepsilon^*$	$-\varepsilon$	1	$-\varepsilon^*$	$-\varepsilon$	1	$-\varepsilon^*$	$-\varepsilon$	1	$-\varepsilon^*$	$\left. -\varepsilon \right\}$		$\left\{ \left(\alpha_{xx}^{I} - \alpha_{yy}^{I}; \overline{\alpha}_{xy}^{II}\right) \right.$
	$\left. 1 \right\}$	$-\varepsilon$	$-\varepsilon^*$	1	$-\varepsilon$	$-\varepsilon^*$	1	$-\varepsilon$	$-\varepsilon^*$	1	$-\varepsilon$	$\left. -\varepsilon^* \right\}$		$\left. \left(\alpha_{xx}^{II} - \alpha_{yy}^{II}; \alpha_{xy}^{I}\right) \right\}$
A_u	1	1	1	1	1	1	−1	−1	−1	−1	−1	−1	T_z	
B_u	1	−1	1	−1	1	−1	−1	1	−1	1	−1	1		
E_{1u}	$\left\{ 1 \right.$	ε	$-\varepsilon^*$	−1	$-\varepsilon$	ε^*	−1	$-\varepsilon$	ε^*	1	ε	$\left. -\varepsilon^* \right\}$	$(T_x,$	
	$\left. 1 \right\}$	ε^*	$-\varepsilon$	−1	$-\varepsilon*$	ε	−1	$-\varepsilon^*$	ε	1	ε^*	$\left. -\varepsilon \right\}$	$T_y)$	
E_{2u}	$\left\{ 1 \right.$	$-\varepsilon^*$	$-\varepsilon$	1	$-\varepsilon^*$	$-\varepsilon$	−1	ε^*	ε	−1	ε^*	$\left. \varepsilon \right\}$		
	$\left. 1 \right\}$	$-\varepsilon$	$-\varepsilon^*$	1	$-\varepsilon$	$-\varepsilon^*$	−1	ε	ε^*	−1	ε	$\left. \varepsilon^* \right\}$		

8.12.5 The D_p groups

(19)

$D_2 \equiv V$	E	$C_2(z)$	$C'_2(y)$	$C''_2(x)$		
A	1	1	1	1		α_{xx}, α_{yy}, α_{zz}
B_1	1	−1	1	−1	T_y; R_y	α_{xz}^{y}
B_2	1	1	−1	−1	T_z; R_z	α_{xy}^{z}
B_3	1	−1	−1	1	T_x; R_x	α_{yz}^{x}

(20)

D_3	E	$2C_3$	$3C_2$		
A_1	1	1	1		$\alpha_{xx} + \alpha_{yy}$, α_{zz}
A_2	1	1	−1	T_z; R_z	
E	2	−1	0	(T_x, T_y); (R_x, R_y)	$\left(\alpha_{xx}^{x} - \alpha_{yy}^{x}; \overline{\alpha}_{xy}^{y}\right)$, $\left(\alpha_{yz}^{x}; \overline{\alpha}_{xz}^{y}\right)$

(21)

D_4	E	$2C_4$	$C_4^2 = C''_2$	$2C_2$	$2C'_2$		
A_1	1	1	1	1	1		$\alpha_{xx} + \alpha_{yy}$, α_{zz}
A_2	1	1	1	−1	−1	T_z; R_z	
B_1	1	−1	1	1	−1		$\alpha_{xx} - \alpha_{yy}$
B_2	1	−1	1	−1	1		α_{xy}
E	2	0	−2	0	0	(T_x, T_y); (R_x, R_y)	$\left(\alpha_{yz}^{x};\ \overline{\alpha}_{zx}^{y}\right)$

(22)

D_6	E	$2C_6$	$2C_3$	C''_2	$3C_2$	$3C'_2$		
A_1	1	1	1	1	1	1		$\alpha_{xx} + \alpha_{yy}, \alpha_{zz}$
A_2	1	1	1	1	−1	−1	$T_z; R_z$	
B_1	1	−1	1	−1	1	−1		
B_2	1	−1	1	−1	−1	1		
E_1	2	1	−1	−2	0	0	$(T_x, T_y); (R_x, R_y)$	$\left(\alpha_{yz}^{x}; \overline{\alpha}_{zx}^{y}\right)$
E_2	2	−1	−1	2	0	0		$(\alpha_{xx} - \alpha_{yy}; \alpha_{xy})$

8.12.6 The D_{pd} groups

(23)

$D_{2d} \equiv V_d$	E	$2S_4$	C_2	$2C'_2$	$2\sigma_d$		
A_1	1	1	1	1	1		$\alpha_{xx} + \alpha_{yy}, \alpha_{zz}$
A_2	1	1	1	−1	−1	R_z	
B_1	1	−1	1	1	−1		$\alpha_{xx} - \alpha_{yy}$
B_2	1	−1	1	−1	1	T_z	α_{xy}^{z}
E	2	0	−2	0	0	$(T_x, T_y); (R_x, R_y)$	$\left(\alpha_{yz}^{x}; \alpha_{zx}^{y}\right)$

(24)

D_{3d}	E	$2C_3$	$3C_2$	i	$2S_6$	$3\sigma_d$		
A_{1g}	1	1	1	1	1	1		$\alpha_{xx} + \alpha_{yy}, \alpha_{zz}$
A_{2g}	1	1	−1	1	1	−1	R_z	
E_g	2	−1	0	2	−1	0	(R_x, R_y)	$\left(\alpha_{xx}^{I} - \alpha_{yy}^{I}, \overline{\alpha}_{xy}^{II}\right), \left(\alpha_{yz}^{I}, \overline{\alpha}_{xz}^{II}\right)$
A_{1u}	1	1	1	−1	−1	−1		
A_{2u}	1	1	−1	−1	−1	1	T_z	
E_u	2	−1	0	−2	1	0	(T_x, T_y)	

(25)

D_{4d}	E	$2S_8$	$2C_4$	$2S_8^3$	C''_2	$4C_2$	$4\sigma_d$		
A_1	1	1	1	1	1	1	1		$\alpha_{xx} + \alpha_{yy}, \alpha_{zz}$
A_2	1	1	1	1	1	−1	−1	R_z	
B_1	1	−1	1	−1	1	1	−1		
B_2	1	−1	1	−1	1	−1	1	T_z	
E_1	2	$\sqrt{2}$	0	$-\sqrt{2}$	−2	0	0	(T_x, T_y)	
E_2	2	0	−2	0	2	0	0		$(\alpha_{xx} - \alpha_{yy}; \alpha_{xy})$
E_3	2	$-\sqrt{2}$	0	$\sqrt{2}$	−2	0	0	(R_x, R_y)	$(\alpha_{xz}; \alpha_{yz})$

8.12.7 The D_{ph} groups

(26)

$D_{2h} = V_h$	E	$C_2(z)$	$C'_2(y)$	$C''_2(x)$	i	$\sigma(xy)$	$\sigma''(zx)$	$\sigma'(yz)$		
A_g	1	1	1	1	1	1	1	1		$\alpha_{xx}, \alpha_{yy}, \alpha_{zz}$
B_{1g}	1	1	−1	−1	1	1	−1	−1	R_z	α_{xy}
B_{2g}	1	−1	1	−1	1	−1	1	−1	R_y	α_{xz}
B_{3g}	1	−1	−1	1	1	−1	−1	1	R_x	α_{yz}
A_u	1	1	1	1	−1	−1	−1	−1		
B_{1u}	1	1	−1	−1	−1	−1	1	1	T_z	
B_{2u}	1	−1	1	−1	−1	1	−1	1	T_y	
B_{3u}	1	−1	−1	1	−1	1	1	−1	T_x	

(27)

D_{3h}	E	$2C_3$	$3C_2$	σ_h	$2S_3$	$3\sigma_v$		
A'_1	1	1	1	1	1	1		$\alpha_{xx} + \alpha_{yy}, \alpha_{zz}$
A'_2	1	1	−1	1	1	−1	R_z	
E'	2	−1	0	2	−1	0	(T_x, T_y)	$\left(\alpha^y_{xx} - \alpha^y_{yy}; \alpha^x_{xy}\right)$
A''_1	1	1	1	−1	−1	−1		
A''_2	1	1	−1	−1	−1	1	T_z	
E''	2	−1	0	−2	1	0	(R_x, R_y)	$(\alpha_{yz}; \overline{\alpha}_{xz})$

(28)

D_{4h}	E	$2C_4$	C''_2	$2C_2$	$2C'_2$	i	$2S_4$	σ_h	$2\sigma_v$	$2\sigma_d$		
A_{1g}	1	1	1	1	1	1	1	1	1	1		$\alpha_{xx} + \alpha_{yy}, \alpha_{zz}$
A_{2g}	1	1	1	−1	−1	1	1	1	−1	−1	R_z	
B_{1g}	1	−1	1	1	−1	1	−1	1	1	−1		$\alpha_{xx} - \alpha_{yy}$
B_{2g}	1	−1	1	−1	1	1	−1	1	−1	1		α_{xy}
E_g	2	0	−2	0	0	2	0	−2	0	0	(R_x, R_y)	$(\alpha_{xz}; \alpha_{yz})$
A_{1u}	1	1	1	1	1	−1	−1	−1	−1	−1		
A_{2u}	1	1	1	−1	−1	−1	−1	−1	1	1	T_z	
B_{1u}	1	−1	1	1	−1	−1	1	−1	−1	1		
B_{2u}	1	−1	1	−1	1	−1	1	−1	1	−1		
E_u	2	0	−2	0	0	−2	0	2	0	0	(T_x, T_y)	

(29)

D_{5h}	E	$2C_5$	$2C_5^2$	$5C_2$	σ_h	$2S_5$	$2S_5^3$	$5\sigma_v$		
A'_1	1	1	1	1	1	1	1	1		$\alpha_{xx} + \alpha_{yy}, \alpha_{zz}$
A'_2	1	1	1	−1	1	1	1	−1	R_z	
E'_1	2	2 **cos** 72°	2 **cos** 144°	0	2	2 **cos** 72°	2 **cos** 144°	0	(T_x, T_y)	
E'_2	2	2 **cos** 144°	2 **cos** 72°	0	2	2 **cos** 144°	2 **cos** 72°	0		$(\alpha_{xx} - \alpha_{yy}; \alpha_{xy})$
A''_1	1	1	1	1	−1	−1	−1	−1		
A''_2	1	1	1	−1	−1	−1	−1	1	T_z	
E''_1	2	2 **cos** 72°	2 **cos** 144°	0	−2	−2 **cos** 72°	−2 **cos** 144°	0	(R_x, R_y)	$(\alpha_{xz}; \alpha_{yz})$
E''_2	2	2 **cos** 144°	2 **cos** 72°	0	−2	−2 **cos** 144°	−2 **cos** 72°	0		

(30)

D_{6h}	E	$2C_6$	$2C_3$	C''_2	$3C_2$	$3C'_2$	i	$2S_3$	$2S_6$	σ_h	$3\sigma_d$	$3\sigma_v$		
A_{1g}	1	1	1	1	1	1	1	1	1	1	1	1		$\alpha_{xx} + \alpha_{yy}, \alpha_{zz}$
A_{2g}	1	1	1	1	−1	−1	1	1	1	1	−1	−1	R_z	
B_{1g}	1	−1	1	−1	1	−1	1	−1	1	−1	1	−1		
B_{2g}	1	−1	1	−1	−1	1	1	−1	1	−1	−1	1		
E_{1g}	2	1	−1	−2	0	0	2	1	−1	−2	0	0	(R_x, R_y)	$(\alpha_{yz}; \bar{\alpha}_{xz})$
E_{2g}	2	−1	−1	2	0	0	2	−1	−1	2	0	0		$(\alpha_{xx} - \alpha_{yy}; \alpha_{xy})$
A_{1u}	1	1	1	1	1	1	−1	−1	−1	−1	−1	−1		
A_{2u}	1	1	1	1	−1	−1	−1	−1	−1	−1	1	1	T_z	
B_{1u}	1	−1	1	−1	1	−1	−1	1	−1	1	−1	1		
B_{2u}	1	−1	1	−1	−1	1	−1	1	−1	1	1	−1		
E_{1u}	2	1	−1	−2	0	0	−2	−1	1	2	0	0	(T_x, T_y)	
E_{2u}	2	−1	−1	2	0	0	−2	1	1	−2	0	0		

(31)

$D_{\infty h}$	E	$2C_\infty^\varphi$	...	$\infty\sigma_v$	i	$2S_\infty^\varphi$	...	∞C_2		
Σ_g^+	1	1	...	1	1	1	...	1		$\alpha_{xx} + \alpha_{yy}, \alpha_{zz}$
Σ_g^-	1	1	...	−1	1	1	...	−1	R_z	
Π_g	2	2 **cos** φ	...	0	2	2 **cos** φ	...	0	(R_x, R_y)	$(\alpha_{yz}; \bar{\alpha}_{xz})$
Δ_g	2	2 **cos** 2φ	...	0	2	2 **cos** 2φ	...	0		$(\alpha_{xx} - \alpha_{yy}; \alpha_{xy})$
⋮	⋮	⋮	⋮	⋮	⋮	⋮	⋮	⋮		
Σ_u^+	1	1	...	1	−1	−1	...	−1	T_z	
Σ_u^-	1	1	...	−1	−1	−1	...	1		
Π_u	2	2 **cos** φ	...	0	−2	−2 **cos** φ	...	0	(T_x, T_y)	
Δ_u	2	2 **cos** 2φ	...	0	−2	−2 **cos** 2φ	...	0		
⋮	⋮	⋮	⋮	⋮	⋮	⋮	⋮	⋮		

8.12.8 The cubic groups

(32)

T	E	$4C_3$	$4C_3^2$	$3C_2$		$\varepsilon \equiv e^{2\pi i/3}$
A	1	1	1	1		$\alpha_{xx} + \alpha_{yy} + \alpha_{zz}$
E	$\left\{\begin{matrix} 1 \\ 1 \end{matrix}\right.$	$\begin{matrix} \varepsilon \\ \varepsilon^* \end{matrix}$	$\begin{matrix} \varepsilon^* \\ \varepsilon \end{matrix}$	$\left.\begin{matrix} 1 \\ 1 \end{matrix}\right\}$		$(\alpha_{xx} + \alpha_{yy} - 2\alpha_{zz}; \sqrt{3}\alpha_{xx} - \sqrt{3}\alpha_{yy})$
F	3	0	0	−1	(T);(R)	$(\alpha^z_{xy}; \alpha^y_{xz}; \alpha^x_{yz})$

(33)

T_d	E	$8C_3$	$3C_2$	$6S_4$	$6\sigma_d$		
A_1	1	1	1	1	1		$\alpha_{xx} + \alpha_{yy} + \alpha_{zz}$
A_2	1	1	1	−1	−1		
E	2	−1	2	0	0		$(\alpha_{xx} + \alpha_{yy} - 2\alpha_{zz}; \sqrt{3}\alpha_{xx} - \sqrt{3}\alpha_{yy})$
F_1	3	0	−1	1	−1	(R)	
F_1	3	0	−1	−1	1	(T)	$(\alpha^z_{xy}; \alpha^y_{xz}; \alpha^x_{yz})$

(34)

T_h	E	$4C_3$	$4C_3^2$	$3C_2$	i	$4S_6$	$4S_6^2$	$3\sigma_h$		$\varepsilon \equiv e^{2\pi i/3}$
A_g	1	1	1	1	1	1	1	1		$\alpha_{xx} + \alpha_{yy} + \alpha_{zz}$
E_g	$\left\{\begin{matrix} 1 \\ 1 \end{matrix}\right.$	$\begin{matrix} \varepsilon \\ \varepsilon^* \end{matrix}$	$\begin{matrix} \varepsilon^* \\ \varepsilon \end{matrix}$	$\begin{matrix} 1 \\ 1 \end{matrix}$	$\begin{matrix} 1 \\ 1 \end{matrix}$	$\begin{matrix} \varepsilon \\ \varepsilon^* \end{matrix}$	$\begin{matrix} \varepsilon^* \\ \varepsilon \end{matrix}$	$\left.\begin{matrix} 1 \\ 1 \end{matrix}\right\}$		$(\alpha_{xx} + \alpha_{yy} - 2\alpha_{zz}; \sqrt{3}\alpha_{xx} - \sqrt{3}\alpha_{yy})$
F_g	3	0	0	−1	3	0	0	−1	(R)	$(\alpha_{xy}; \alpha_{xz}; \alpha_{yz})$
A_u	1	1	1	1	−1	−1	−1	−1		
E_u	$\left\{\begin{matrix} 1 \\ 1 \end{matrix}\right.$	$\begin{matrix} \varepsilon \\ \varepsilon^* \end{matrix}$	$\begin{matrix} \varepsilon^* \\ \varepsilon \end{matrix}$	$\begin{matrix} 1 \\ 1 \end{matrix}$	$\begin{matrix} -1 \\ -1 \end{matrix}$	$\begin{matrix} -\varepsilon \\ -\varepsilon^* \end{matrix}$	$\begin{matrix} -\varepsilon^* \\ -\varepsilon \end{matrix}$	$\left.\begin{matrix} -1 \\ -1 \end{matrix}\right\}$		
F_u	3	0	0	−1	−3	0	0	1	(T)	

(35)

O	E	$8C_3$	$3C''_2$	$6C_4$	$6C_2$		
A_1	1	1	1	1	1		$\alpha_{xx} + \alpha_{yy} + \alpha_{zz}$
A_2	1	1	1	−1	−1		
E	2	−1	2	0	0		$(\alpha_{xx} + \alpha_{yy} - 2\alpha_{zz}; \sqrt{3}\alpha_{xx} - \sqrt{3}\alpha_{yy})$
F_1	3	0	−1	1	−1	(T);(R)	
F_1	3	0	−1	−1	1		$(\alpha_{xy}; \alpha_{xz}; \alpha_{yz})$

(36)

O_h	E	$8C_3$	$3C''_2$	$6C_4$	$6C_2$	i	$8S_6$	$3\sigma_h$	$6S_4$	$6\sigma_d$		
A_{1g}	1	1	1	1	1	1	1	1	1	1		$\alpha_{xx} + \alpha_{yy} + \alpha_{zz}$
A_{2g}	1	1	1	−1	−1	1	1	1	−1	−1		
E_g	2	−1	2	0	0	2	−1	2	0	0		$(\alpha_{xx} + \alpha_{yy} - 2\alpha_{zz}; \sqrt{3}\alpha_{xx} - \sqrt{3}\alpha_{yy})$
F_{1g}	3	0	−1	1	−1	3	0	−1	1	−1	(R)	
F_{2g}	3	0	−1	−1	1	3	0	−1	−1	1		$(\alpha_{xy}; \alpha_{xz}; \alpha_{yz})$
A_{1u}	1	1	1	1	1	−1	−1	−1	−1	−1		
A_{2u}	1	1	1	−1	−1	−1	−1	−1	1	1		
E_u	2	−1	2	0	0	−2	1	−2	0	0		
F_{1u}	3	0	−1	1	−1	−3	0	1	−1	1	(T)	
F_{2u}	3	0	−1	−1	1	−3	0	1	1	−1		

Exercises for Chapter 8

8.1 As described in Section 6.6 (Figure 6.9), the $N - H$ bonds of the ammonia NH_3 molecule can be approximated as perpendicular to each other (see also Figures 8.7 and 8.8). The Cartesian coordinates of the comprising atoms may then be given as $N(0,0,0)$; $H(1,0,0)$Å; $H(0,1,0)$Å; and $H(0,0,1)$Å. The nitrogen atomic number is $A_N = 14$.

(a) Calculate the mass-center coordinates,

(b) Calculate the moment of inertia tensor relative to a Cartesian coordinate system whose origin is at the mass center, and whose axes are parallel to the former coordinate system.

(c) Express the geometric equations of the three A, B, C principal axes of inertia and their corresponding I_A, I_B, and I_C moments of inertia.

(d) Calculate the kinetic energy eigenvalues of the molecule for its pure rotational states; generalize Eq. (7.50) for that purpose.

8.2 Figure 8.13 presents a molecule-like structure that comprises three "atoms": one of mass M, the other two of mass m. All are free to move along a linear frictionless rod, yet the m atoms on opposite sides of M are bonded to the M atom by identical springs of equilibrium length L ($M - m$ bonds).

Figure 8.13 Geometric plot aimed at clarifying the "molecule" structure described in Exercise 8.2.

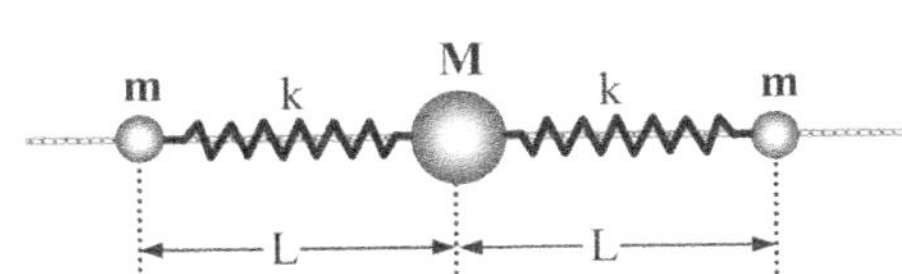

(a) Express the system potential matrix.
(b) Calculate the system normal vibrational frequencies and the normal coordinates.
(c) Plot the displacement vector for each of the normal vibrational modes.
(d) Assume that this is a model of the carbon dioxide CO_2 molecule, for which $M(C) = 12$ amu and $m(O) = 16$ amu. Given that the higher vibrational mode frequency is about 3800 cm^{-1}, estimate the effective spring constant k, and calculate the lower mode frequency.

8.3 The polarizability components of a gas molecule of a symmetric top structure are $\alpha_3 = 3.5 \times 10^{-24}$ cm^3 and $\alpha_1 = \alpha_2 = 1.25 \times 10^{-24}$cm^3. A gas comprising the molecules is compressed to 20 atm. at room temperature in a pressure cell. The gas is illuminated in "Y" direction through the cell windows using a 400 mW polarized laser beam of 1mm^2 cross-section and 514.5 nm wavelength. A 10 mm length of the beam crossing path is observed through a side window ("X" direction). Calculate the Rayleigh scattered light intensity at 20 cm distance from the side window at $Y(ZZ)X$ and $Y(XY)X$ Porto formations. Recall that the number N of participating molecules may be estimated using the ideal gas equation $PV = Nk_BT$, where P is the pressure, V is the illuminated gas volume, and k_B is the Boltzmann constant.

8.4 Analyze the vibrational modes of the following molecules by their Raman and infrared activity:
(a) Mono-deuterated ammonia NH_2D. The structure is identical to that of ammonia per Figures 6.9 and 8.7, with one of the hydrogen H atoms (protonium) replaced by deuterium D (whose nucleus comprises one proton and one neutron). The molecular structure is presented in Figure 8.14(a).
(b) Cis-1,2-dichloroethylene ($cis{-}1,2{-}ClHC{=}CClH$). Its structure is like that of ethylene per Figure 8.9, except that two of the hydrogen H atoms bonded to two different *close* carbon C atoms have been replaced by chlorine Cl atoms. The molecular structure is presented in Figure 8.14(b).
(c) Trans-1,2-dichloroethylene ($trans{-}1,2{-}ClHC{=}CClH$). Its structure is like that of ethylene per Figure 8.9, except that two of the hydrogen H atoms bonded to two different *far* carbon C atoms have been replaced by chlorine Cl atoms. The molecular structure is presented in Figure 8.14(c).
(d) Phosphorus pentafluoride PF_5. The phosphorus atom and three of the fluorine atoms form a common plane, where the fluorine atoms are vertices of an equilateral triangle with the phosphorus at its center. Two additional fluorine atoms reside oppositely above and below the plane, against the phosphorus atom. The molecular structure is presented in Figure 8.14(d).

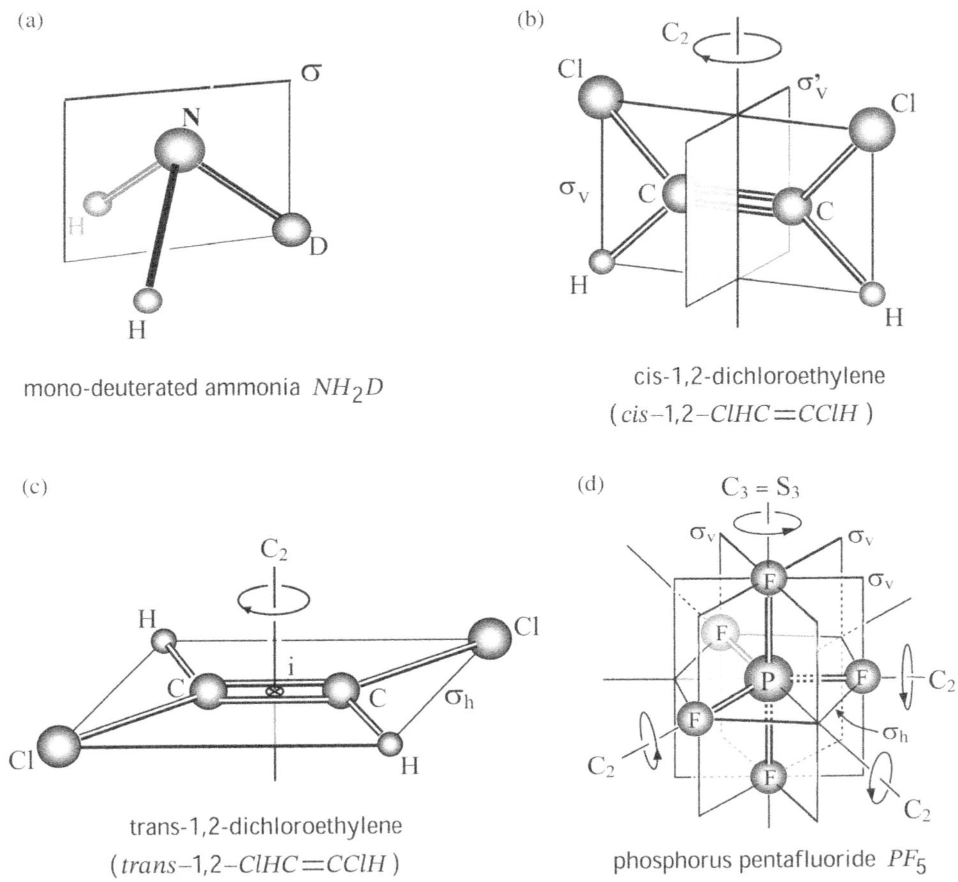

Figure 8.14 Spatial structure and symmetry elements of the molecules related to Exercise 8.4.

9 Crystalline Solids

9.1 Preamble

In Chapters 6 to 8, we described the essence of molecules and provided an analysis of their various properties. All of the provided examples involved a small number of comprising atoms – ten or fifteen at most – and their dimensions never exceeded several angstroms. The number of different atoms involved in the molecular construction was also limited to two or three. Fundamentally, however, as long as gravity effects may be neglected, molecules may comprise many billions of atoms, and their total dimensions may extend from several nanometers to tens, hundreds, and even many thousands of kilometers. Molecules of dimensions exceeding several thousand angstroms, that have skeletal structures that remain constant, are called **solids**. Liquids are comparable to solids in the sense that their interatomic distances are similar; however, they differ in that their skeletal structures change continually. In fact, every solid may change into a liquid under specific ranges of pressure and temperature. Observation in nature reveals that solids comprising only a small number of different atomic types tend to form special skeletal structures called **crystals**. The crystalline skeleton comprises a large number of identical molecules, whose dimensions are only several angstroms at most, arranged in an ordered, repetitive manner. Solids comprising a larger number of different atomic types usually lack such order; they are called **glasses**, or **amorphous solids**. Nevertheless, some materials comprising only a small number of different atomic types may form amorphous or glass structures. Likewise, many materials comprising a large number of different atomic types may form crystals under special conditions.

Crystal bonding, which is the effective potential energy that keeps each atom in an average constant position relative to other atoms, is similar to the molecular bonding discussed at length in Chapters 6 and 7, and specifically described in Sections 6.2 and 7.3.1.

The present chapter considers the properties of crystalline solids of macroscopic size (several micrometers or more).

9.2 Periodic Crystals

A simple case of a crystalline structure is the periodic crystal. In a periodic crystal, a periodic packaging of identical sets of atoms (or molecules) fill the space. The periodicity is described using a mathematical structure called a **lattice**. For mathematical simplicity, it is assumed that the crystal infinitely fills the entire space in all dimensions. The lattice is a set of points wherein the crystal appears the same from each lattice point; it is defined by the three **fundamental unit vectors** $\underline{a}$, $\underline{b}$, and $\underline{c}$, such that the crystal appears the same from any two points $\underline{r}$ and $\underline{r}'$, satisfying

$$\underline{r}' = \underline{r} + n_1\underline{a} + n_2\underline{b} + n_3\underline{c} \,, \tag{9.1}$$

Physics, Optics, and Spectroscopy of Materials, First Edition. Zeev Burshtein.

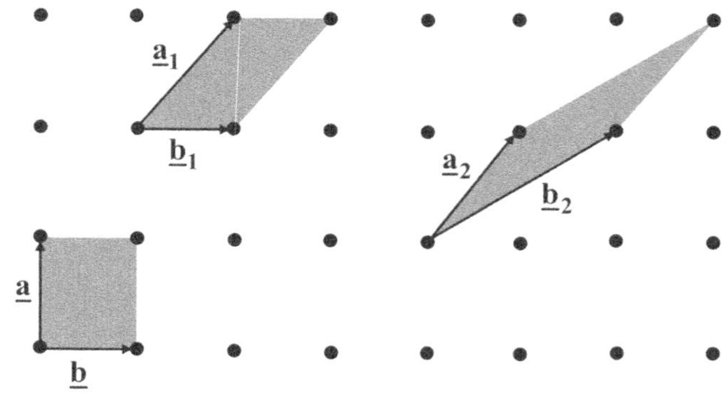

Figure 9.1 Schematic illustration of a two-dimensional lattice with three different selections of fundamental lattice vectors and the unit cells they generate (the shaded gray areas). The black small circles indicate the lattice points, and the different choices are given different subscript numbering.

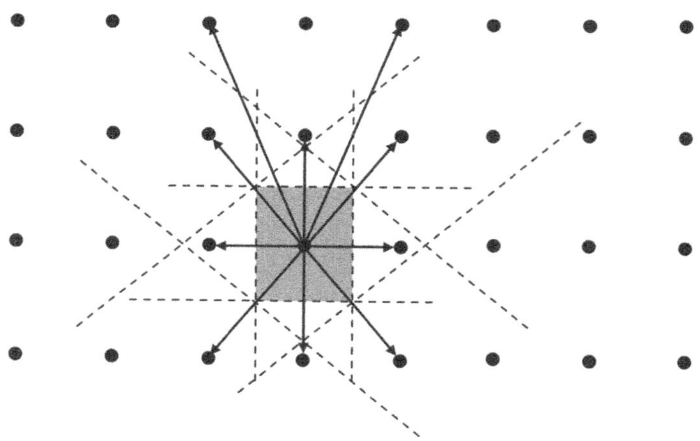

Figure 9.2 An illustration of the method of constructing the Wigner-Seitz unit cell. The dashed lines are projections on the paper plane of the perpendicular-slicing planes of lattice vectors, out-pointing from the lattice point selected as the origin.

where n_1, n_2, and n_3 are arbitrary integers. The set of points defined by $\underline{L} = n_1\underline{a} + n_2\underline{b} + n_3\underline{c}$ actually defines the crystal lattice. An $\underline{L}$ vector is called a **lattice vector** or **lattice displacement vector**. The three-dimensional parallelepiped based on the three fundamental vectors is called the **unit cell**. The lattice vectors and the unit cell are called **primitive** if all lattice points may be represented by Eq. (9.1). Otherwise, the unit cell is called **centered**. The aforementioned main underlying ideas are displayed in Figure 9.1, which actually shows a two-dimensional lattice. As demonstrated, it is possible to choose different combinations of fundamental vectors and unit cells for the same lattice (in fact, an infinite number of such different forms can be chosen). A unit cell exhibits the property that it is possible to fill the entire space with its translation using all lattice vectors.

Notably, the unit cells thus formed have identical volumes (or identical areas in the case of the two-dimensional lattice in the figure). It is also possible to generate unit cells that exhibit the same property, and are not necessarily structured as a spatial parallelepiped of the fundamental lattice vectors. One example is the **Wigner-Seitz unit cell**, which is formed by the following procedure: first, one of the lattice points is selected as the origin. Then, all out-pointing lattice vectors are sliced in two by adding infinite perpendicular planes. The smallest enclosed polyhedron is the desired unit cell. It is easy to verify, that its volume is identical to that of the parallelepiped unit cell, because it spans the entire lattice volume with no voids, and each cell is assigned to a single lattice point! Trivially, by its definition, the Wigner-Seitz unit cell is independent of the selection of fundamental lattice vectors.

The formation of a two-dimensional Wigner-Seitz unit cell is illustrated schematically in Figure 9.2.

The unit cell volume is given by the absolute value of a determinant, comprising as elements the Cartesian components of the fundamental $\underline{a}$, $\underline{b}$, and $\underline{c}$ vectors

$$V_C = |(\underline{a} \times \underline{b}\cdot\underline{c})| = \textbf{abs}\begin{vmatrix} a_x & a_y & a_z \\ b_x & b_y & b_z \\ c_x & c_y & c_z \end{vmatrix}. \tag{9.2}$$

The lattice is an abstract mathematical entity. The crystal itself is obtained by attaching an identical atom, or an identical stiff *collection of atoms*, to each lattice point. Such a collection is called an **atomic basis**. Like a unit cell construction, a polyatomic basis may be structured in infinite ways, but still yield the same crystal. The so-called "natural" polyatomic basis selection is the one included in the smallest possible volume sphere; we shall refer to it later. Quite often, the natural basis, or parts of it, are considered to exhibit properties similar to those of a separate molecule.

The attachment of an atomic basis to a lattice often reduces the lattice's symmetry. To demonstrate, we consider the two-dimensional square lattice shown in Figure 9.3, in which pairs of identical atoms, identically oriented, are attached to each lattice point, and the lattice point is positioned exactly in the middle of each pair. The lattice itself exhibits a fourfold rotational symmetry about the axes perpendicular to the paper plane when crossing the lattice points (C_4 symmetry; see Table 8.1). After the atomic basis is added, the crystal rotational symmetry about the same axes is only twofold (namely C_2). Thus, different crystals may exhibit different symmetry properties even while belonging to the same lattice.

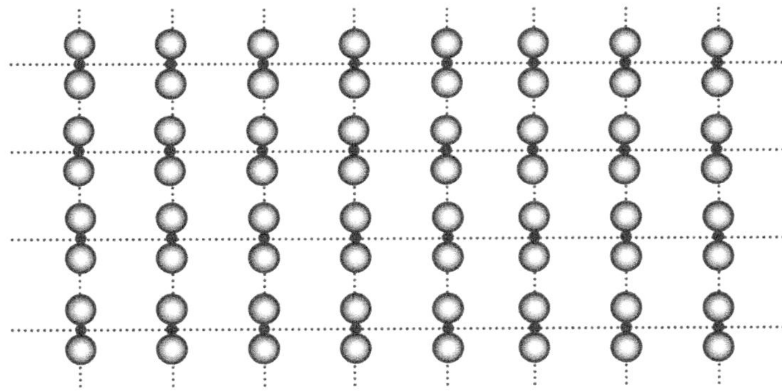

Figure 9.3 Schematic illustration of the rotational symmetry reduction of crystal lattices caused by the addition of an atomic basis. The two-dimensional lattice of the small full circles has a fourfold rotational symmetry about the axes perpendicular to the paper plane crossing the lattice points. This symmetry is reduced to only twofold rotational symmetry after the addition of an atomic basis comprising the two identical atoms (represented by balls) to each lattice point.

The various possible lattice shapes have been classified by Bravais into 14 different **families**, and subclassified as 7-crystalline **systems** that belong to similar point groups. The systems are defined by the relationship among the fundamental $\underline{a}$, $\underline{b}$, and $\underline{c}$ vectors lengths, and their corresponding α, β, γ inter angles. The $\underline{a}$, $\underline{b}$, and $\underline{c}$ vectors are obviously independent, **but are not necessarily always primitive**. The 14 Bravais lattices are thus represented by 7-crystalline systems, comprising a total of 7 primitive families and 7 centered families. In addition to the parallelepiped vertices, a centered unit cell contains other lattice points. We shall not attempt to prove the validity of Bravais' classification here, but instead will be satisfied by its systematic presentation. The crystalline systems and Bravais families are listed and plotted in Figure 9.4.

As in the case of molecules, the mathematical instrument used to present and analyze crystal symmetry is group theory. In group theory, the symmetry of the various lattice structures is described by 32 groups of point symmetry. Point group elements include inversion, mirror reflections, rotations (so-called "proper" rotations), and rotations followed by mirror reflections (so-called "improper" rotations) that leave at least a single point in space unaltered, and make the operated-upon object seem identical to the original structure. All of these groups are presented in Table 8.2 in an order common for molecules. Their classification according to the crystalline systems, as used in crystallography, is presented in Table 9.1.

When atomic bases are attached to the possible lattices and additional possible symmetry operations are included, a sum of 230 various symmetry groups are obtained, which are called **space-groups**. The added operations include lattice translations and mirror reflections, each followed by a translation that differs from any lattice translation (called a **glide-mirror**), and rotations, which are followed by a translation that differs from any lattice translation (called a **screw-axis**). Space-groups that *do not include* glide-mirrors or screw-axes are called **symmorphic**. Space-groups that include a glide-mirror and/or a screw-axis are called **nonsymmorphic**. All space groups are described in detail in Chapter 11.

Solids may appear in crystalline structures that exhibit disorder in one or more of the crystal symmetry sites; in polycrystalline structures where the crystalline order is maintained over limited ranges, such as several micrometers or a few hundred micrometers; in glass structures where the crystalline order is maintained over a range of only several unit cells; and in amorphous structures where the order is limited to a single unit cell, if at all. Macroscopically large structures exhibiting rotational symmetry inconsistent with a perfect periodic symmetry, such as the fivefold rotation C_5, were discovered in 1984. Although this discovery yielded a generalized definition of the crystalline concept, we shall not address that issue in the present book. Instead, our following discussions are limited to perfectly periodic crystals.

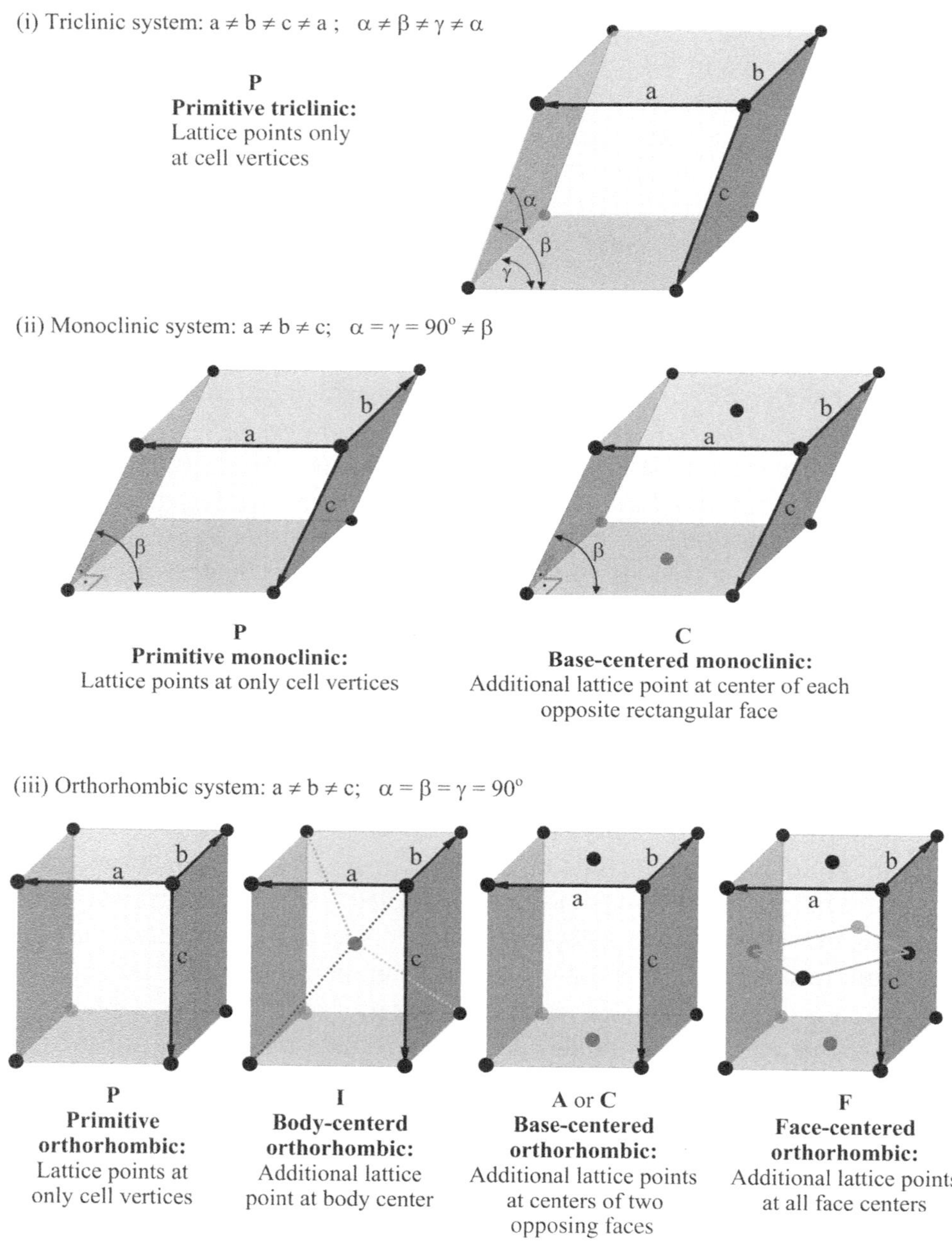

Figure 9.4 Bravais lattices.

(iv) Tetragonal system: $a = b \neq c$, $\alpha = \beta = \gamma = 90^{o}$.

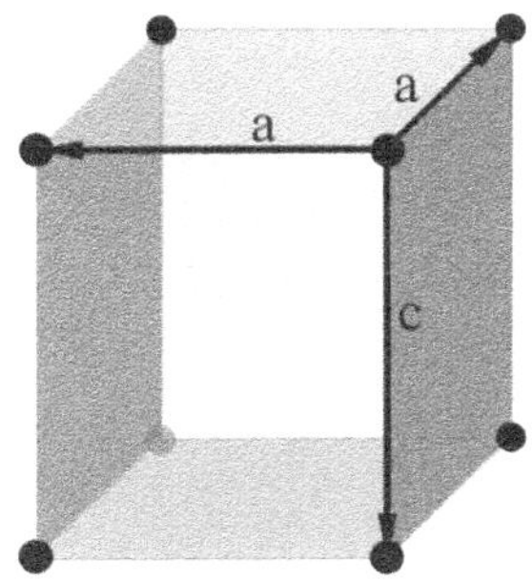

P
Primitive tetragonal:
Lattice points only at cell vertices

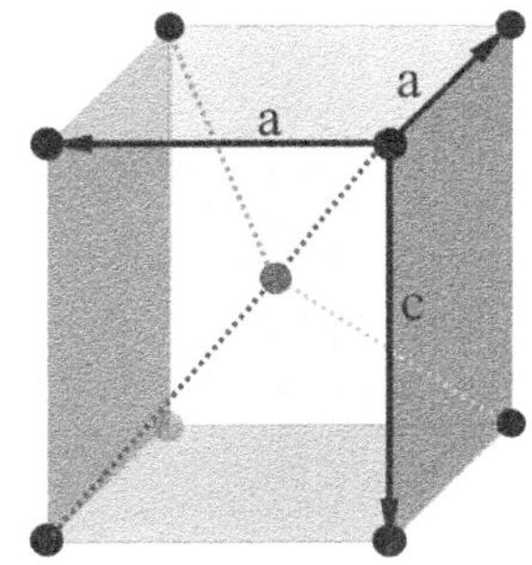

I
Body-centered tetragonal:
Additional lattice point at cell body center

(v) Trigonal (rhombohedral) system: $a = b = c$, $\alpha = \beta = \gamma < 120^{o}$, $\neq 90^{o}$ or 60°.

P or **R**
Primitive trigonal (rhombohedral):
Lattice points at only cell vertices

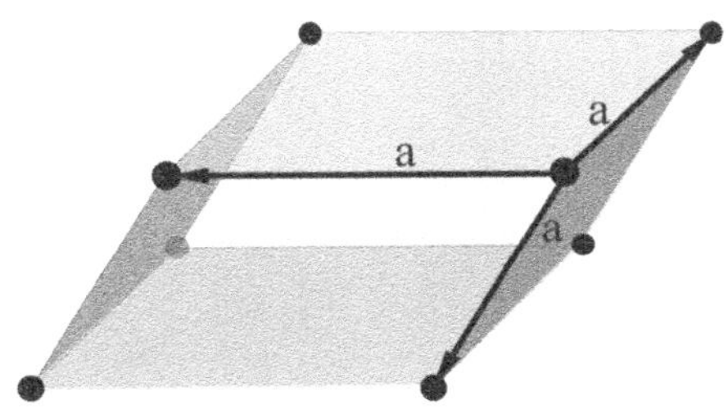

(vi) Hexagonal system: $a = b \neq c$; $\alpha = \beta = 90^{o}$, $\gamma = 60^{o}$

P
Primitive hexagonal:
Lattice points at only cell vertices

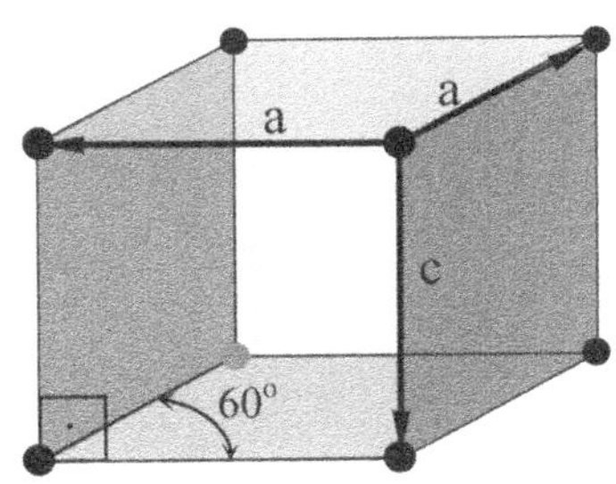

(vii) Cubic system $a = b = c$, $\alpha = \beta = \gamma = 90^{o}$.

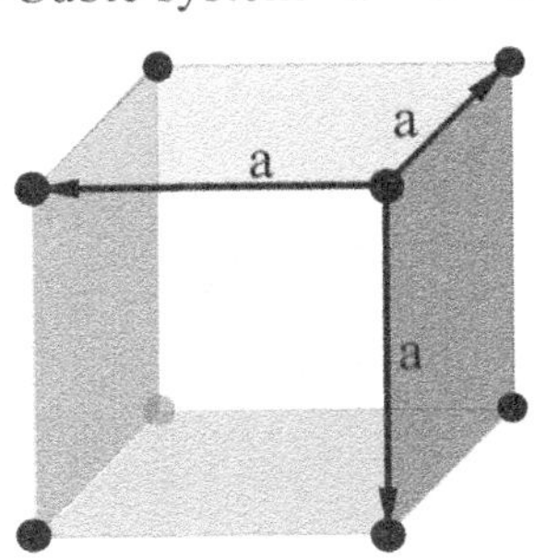

P
Primitive cubic:
Lattice points only at cell vertices

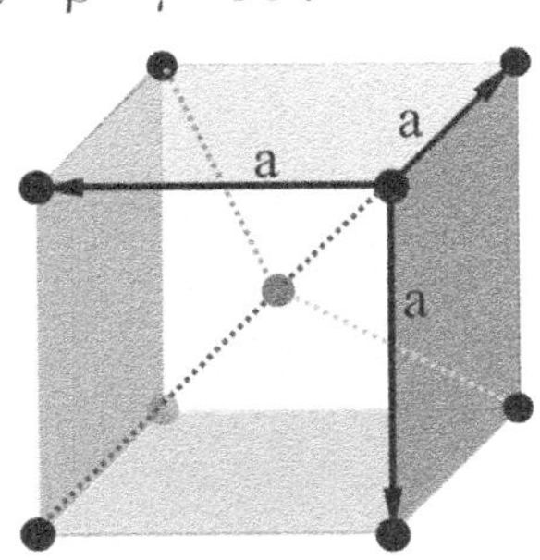

I
Body-centered cubic:
Additional lattice point at body center

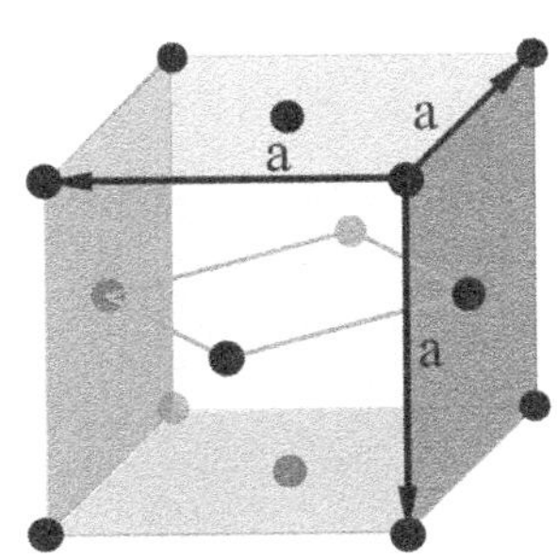

F
Face-centered cubic:
Additional lattice points at all face centers

Figure 9.4 (Continued)

Table 9.1 Descriptions of crystalline point group notations according to Shoenflies and Hermann-Mauguin. The alternative notations that are sometimes used are written in parentheses.

		Point group notation		
No.	**crystal system**	**Schoenflies**	**Hermann-Mauguin**	**No. in Table 8.2**
1	Triclinic	C_1	1	1
2		C_i	$\bar{1}$	6
3	Monoclinic	C_2	2	2
4		C_S	m	9
5		C_{2h}	2/m	15
6	Orthorhombic	$D_2(V)$	222	19
7		C_{2v}	mm2 (2mm,mm)	10
8		$D_{2h}(V_h)$	mmm (2/m 2/m 2/m)	26
9	Tetragonal	C_4	4	4
10		S_4	$\bar{4}$	7
11		C_{4h}	4/m	17
12		D_4	422	21
13		C_{4v}	4mm	12
14		$D_{2d}(V_d)$	$\bar{4}$ 2m	23
15		D_{4h}	4/mmm (4/m 2/m 2/m)	28
16	Trigonal	C_3	3	3
17		S_6 (C_{3i})	$\bar{3}$	8
18		D_3	32	20
19		C_{3v}	3m	11
20		D_{3d}	$\bar{3}$m ($\bar{3}$ 2/m)	24
21	Hexagonal	C_6	6	5
22		C_{3h} (S_3)	$\bar{6}$ (3/m)	16
23		C_{6h}	6/m	18
24		D_6	622 (62)	22
25		C_{6v}	6mm	13
26		D_{3h}	$\bar{6}m2$	27
27		D_{6h}	6/mmm (6/m 2/m 2/m)	30
28	Cubic	T	23	32
29		T_h	m3 (2/m $\bar{3}$)	34
30		O	432 (43)	35
31		T_d	$\bar{4}$ 3m	33
32		O_h	m$\bar{3}$m (4/m $\bar{3}$ 2/m)	36

9.3 Lattice-Vector and Lattice-Plane Orientations

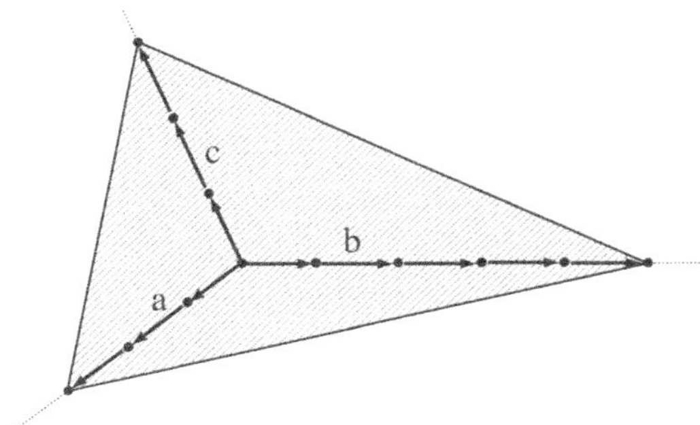

Figure 9.5 Demonstration of the definition method of the fundamental family of crystal planes. More details are provided in the text.

A **lattice-vector orientation** is the orientation of a lattice vector of the form $\underline{L} = h\underline{a} + k\underline{b} + \ell\underline{c}$, where h, k, and ℓ are specific integers that do not have a common factor; for brevity, the orientation is noted as $[h, k, \ell]$ (if, at first, the h, k, and ℓ numbers do have common factors, they are divided by the biggest one). A crystal (lattice) plane is a plane that contains lattice points. A fundamental plane is defined by three lattice points on vector tips oriented along fundamental lattice vectors, with the first at $p_1\underline{a}$, second at $p_2\underline{b}$, and third at $p_3\underline{c}$ (p_1, p_2, and p_3 integers), as demonstrated in Figure 9.5. A family of planes is the set containing *all lattice planes* parallel to the fundamental plane. Each family of planes contains *all* the lattice points, and each individual plane in a family forms a two-dimensional lattice, congruent with all of the others. In other words, any lattice vector belonging to the two-dimensional lattice of a single individual family of planes, is also a lattice vector in the two-dimensional lattices of all of the other families of planes.

In principle, it is possible to denote a family of planes using the three p_1, p_2, and p_3 quantities. Following Miller, however, the accepted notation is different, and defined as follows: let M be the smallest common product of p_1, p_2, and p_3. Then, the Miller indices (h, k, ℓ) are defined as

$$(h, k, \ell) \equiv \left(\frac{M}{p_1}, \frac{M}{p_2}, \frac{M}{p_3}\right). \tag{9.3}$$

For example, in the case plotted in Figure 9.5, $p_1 = 3$, $p_2 = 5$, and $p_3 = 3$. Thus $M = 15$. The Miller indices of the related family of planes are $(h, k, l) \equiv \left(\frac{15}{3}, \frac{15}{5}, \frac{15}{3}\right) = (5, 3, 5)$. In the special case of cubic unit cells, the $[h, k, \ell]$ vectors are always perpendicular to the (h, k, l) family of planes!

9.4 The Reciprocal Lattice

Every crystal lattice is associated with a private special lattice called a **reciprocal lattice**, which is defined by the former fundamental $\underline{a}$, $\underline{b}$, and $\underline{c}$ unit vectors as follows:

$$\boxed{\underline{A} \equiv 2\pi\frac{\underline{b}\times\underline{c}}{\underline{a}\times\underline{b}\cdot\underline{c}};\quad \underline{B} \equiv 2\pi\frac{\underline{c}\times\underline{a}}{\underline{b}\times\underline{c}\cdot\underline{a}};\quad \underline{C} \equiv 2\pi\frac{\underline{a}\times\underline{b}}{\underline{c}\times\underline{a}\cdot\underline{b}}}, \tag{9.4}$$

where $\underline{A}$, $\underline{B}$, and $\underline{C}$ are its fundamental vectors. The quantities appearing in the denominator of each expression are equal, and their absolute value equals the unit cell volume of the original lattice (Eq. [9.2]). When addressing both lattices, the original lattice is described as a **direct lattice**. The reciprocal lattice itself is identical for all possible selections of the direct cell fundamental vectors. The Wigner-Seitz unit cell of the reciprocal lattice is called the **first Brillouin zone**.

We now summarize several properties of reciprocal lattices:

1) Direct and crossed scalar products:

$$\underline{a}\cdot\underline{A} = \underline{b}\cdot\underline{B} = \underline{c}\cdot\underline{C} = 2\pi\ ;$$

$$\underline{a}\cdot\underline{B} = \underline{a}\cdot\underline{C} = \underline{b}\cdot\underline{A} = \underline{b}\cdot\underline{C} = \underline{c}\cdot\underline{A} = \underline{c}\cdot\underline{B} = 0\,.$$

The fundamental (unit) vectors of the reciprocal lattice are perpendicular to the corresponding direct lattice planes. In other words, $\underline{A}$ is perpendicular to the $(\underline{b},\underline{c})$ plane, $\underline{B}$ is perpendicular to the $(\underline{c},\underline{a})$ plane, and $\underline{C}$ is perpendicular to the $(\underline{a},\underline{b})$ plane.

2) Unit cell volume in the reciprocal lattice:

$$V_{rcp} = |(\underline{A}\times\underline{B}\cdot\underline{C})| = \frac{(2\pi)^3}{V_C}\,.$$

3) Cross-scalar products between direct and reciprocal unit vectors:
A reciprocal lattice vector: $\underline{G} = h\underline{A} + k\underline{B} + \ell\underline{C}$;
a direct lattice vector: $\underline{L} = n_1\underline{a} + n_2\underline{b} + n_3\underline{c}$.

Their scalar product: $\begin{cases} \underline{G}\cdot\underline{L} = (h\underline{A} + k\underline{B} + \ell\underline{C})\cdot(n_1\underline{a} + n_2\underline{b} + n_3\underline{c}) \\ \qquad = 2\pi(hn_1 + kn_2 + \ell n_3) = 2\pi n \end{cases}$, where n is an integer.

4) Distance between adjacent (h,k,ℓ) planes:
A reciprocal lattice vector: $\underline{G} = h\underline{A} + k\underline{B} + \ell\underline{C}$, (reduced (h,k,ℓ));
Two independent vectors in an (h,k,ℓ) plane: $\underline{v}_1 = \frac{M}{h}\underline{a} - \frac{M}{k}\underline{b}$; $\underline{v}_2 = \frac{M}{h}\underline{a} - \frac{M}{\ell}\underline{c}$.

Scalar products: $\begin{cases} \underline{G}\cdot\underline{v}_1 = (h\underline{A} + k\underline{B} + \ell\underline{C})\left(\frac{M}{h}\underline{a} - \frac{M}{k}\underline{b}\right) = M(\underline{A}\cdot\underline{a} - \underline{B}\cdot\underline{b}) = M(2\pi - 2\pi) = 0 \\ \underline{G}\cdot\underline{v}_2 = (h\underline{A} + k\underline{B} + \ell\underline{C})\left(\frac{M}{h}\underline{a} - \frac{M}{\ell}\underline{c}\right) = M(\underline{A}\cdot\underline{a} - \underline{C}\cdot\underline{c}) = M(2\pi - 2\pi) = 0 \end{cases}$;

thus the $\underline{G}(h,k,\ell)$ vector is perpendicular to the (h,k,ℓ) family of planes.
We further indicate, without proof, that $d_{h,k,\ell}$, the distance between adjacent (h,k,ℓ) planes, is given by:

$$\boxed{d_{h,k,\ell} = \frac{2\pi}{|\underline{G}(h,k,\ell)|}}\,. \tag{9.5}$$

5) The reciprocal lattice of a reciprocal lattice is again the original direct lattice.

9.5 Internal Crystalline Oscillations

9.5.1 Introduction

Similarly to molecules, the nuclei forming a crystal will oscillate about their equilibrium positions. However, the Born-Oppenheimer approximation (Section 7.2) is not simplistically applicable in the present case. Recall that the core of the Born-Oppenheimer approximation is the assumption that the electron wave function $\psi(\underline{R}_u,\underline{r}_i)$, where $\underline{R}_u$ is the position vector of the u nucleus and $\underline{r}_i$ is the position vector of the i electron, adjusts itself quasi-statically to the nuclei positions $\underline{R}_u$, and is independent of the nuclei velocities and their accelerations. Under those conditions, the wave function of the nuclei motions $\mathcal{S}(\underline{R}_u)$ could be calculated in an effective potential $E_{el}(\underline{R}_u)$, which is the eigenenergy of the electron state at a supposed (momentarily stiff) positions system $\underline{R}_u$ of the nuclei. For a macroscopically sized medium, however, the electron wave function cannot adjust itself to other nuclei position changes further than several thousand molecules. The reason is that the time period of a typical nuclear oscillation is of the order of 10^{-14} s, and the information of any position change cannot be delivered faster than the speed of light.

We must thus modify the "normal mode" term. It originally meant that all body parts oscillate at the same frequency and phase. We now define a vibrational normal mode of macroscopic crystals using an approximation method called **tight-binding approximation**. In a zero-order approximation, the natural atomic basis is considered similar to a free molecule, exhibiting internal vibrations as described in Chapters 7 and 8. This approach is not simplistically adequate for metal solids, where many electrons are shared by a large number of unit cells (so-called "free" electrons). It is, however, quite adequate for electrically insulating crystals (dielectric crystals). Considering the unit cell's translational and rotational motions, these concepts require modification, because of interactions with neighboring cells; return forces and force-moments transform these motions into oscillations about the equilibrium postures. The translations turn into translational vibrations; meanwhile, rotations transform into rotational oscillations called **librations**. All aforementioned motions generate oscillation waves that propagate through the crystal. These are the motions that define the crystal vibrational modes.

9.5.2 Hamiltonian and Dynamic Equations

We start by imagining the crystal as a collection of N oscillators that are not necessarily periodically organized, and may not exhibit any interactions. After turning on the coupling interactions, the collection exhibits exactly N different vibrational modes. We then assume that the lattice point positions remain the same for a limited-sized, yet macroscopic volume, when it is sliced off an ideal infinite crystal. Mathematically, it is obvious that none of the infinite crystal symmetries remain valid for the finite slice (except for the identity operation). We however assume, that the ideal symmetry properties are still a valid approximation. This means that a symmetry operation performed on a certain site, transforms other sites at a macroscopic distance into sites of identical properties. This assumption may not be strictly valid for sites close to the slice surface.

The Hamiltonian of each noninteracting oscillator located at the $\underline{x}$ lattice-point is given by $\mathcal{H}_{\underline{x}}^0 = \frac{1}{2}\left(\dot{q}_{\underline{x}}^{\,2} + \omega_0^{\,2} q_{\underline{x}}^{\,2}\right)$, where $q_{\underline{x}}$ is the generalized normal coordinate of the said oscillator, and ω_0 is its oscillation angular frequency; ω_0 is assumed as common for all oscillators (Eq. (8.13)). The coupling of each oscillator to its environment is assumed to belong to a perturbation potential of the form $V_{\underline{x}} = \frac{1}{2}\sum_{\underline{s}\neq 0} C_{\underline{s}}\left(q_{\underline{x}} - q_{\underline{x}+\underline{s}}\right)^2$, where $C_{\underline{s}}$ are coupling coefficients of the oscillator to a neighboring oscillator residing at $\underline{x} + \underline{s}$. This is a harmonic approximation for the perturbation coupling potential. It means that the oscillator's own returning-force potential is added with an additional contribution. It is the summed effect of simultaneous deviations from the equilibrium of normal coordinates of neighboring atomic bases that belong to the same oscillation mode. Coupling among different vibrational modes of the atomic basis is neglected! A different set of $C_{\underline{s}}$ coupling coefficients belongs to separate oscillation (vibrational) modes of the atomic basis (separate ω_0). Some of the coupling coefficients that belong to a specific mode may be negative; thus instead of increasing the internal return force, they reduce it, an effect called "softening". We further assume, that when the neighbor at distance $\underline{s}$ from the oscillator residing at $\underline{x}$ grows, the coupling coefficient's size $|C_{\underline{s}}|$ reduces steeply, practically vanishing beyond a certain distance from the $\underline{x}$ residing oscillator. That distance is typically of the order of only several unit-cell dimensions. The neglect of phase differences between the oscillator and neighbors' motions is thus well justified.

Generating theoretical calculations of the aforementioned coupling constants based on molecular orbitals assigned to the atomic basis is close to being a "mission impossible". To reduce the scope of the problem, we must resort to empirical models and approximations. We shall devote some more space to addressing the theoretical approach in Section 9.5.5. It should also be noted that the aforementioned expression for the perturbation potential is not valid for atomic bases near the sample surface, where the number of neighbors is fewer. The validity of our following treatment is thus limited to inner-volume positioned atomic bases. The number of surface-bases is, however, extremely small compared with the number of inner volume bases in samples of macroscopic dimensions.

The corrected Hamiltonian for each $\underline{x}$-positioned atomic basis is given by

$$\mathcal{H}_{\underline{x}} = \mathcal{H}_{\underline{x}}^0 + V_{\underline{x}} = \frac{1}{2}\left(\dot{q}_{\underline{x}}^{\,2} + \omega_0^{\,2} q_{\underline{x}}^{\,2}\right) + \frac{1}{2}\sum_{\underline{s}\neq 0} C_{\underline{s}}\left(q_{\underline{x}} - q_{\underline{x}+\underline{s}}\right)^2 . \tag{9.6}$$

The Hamiltonian of the entire vibrating crystal would then be

$$\mathcal{H} = \sum_{\underline{x}} \mathcal{H}_{\underline{x}} = \frac{1}{2}\sum_{\underline{x}}\left(\dot{q}_{\underline{x}}^{\,2} + \omega_0^{\,2} q_{\underline{x}}^{\,2}\right) + \frac{1}{2}\left[\frac{1}{2}\sum_{\underline{x}}\sum_{\underline{s}\neq 0} C_{\underline{s}}\left(q_{\underline{x}} - q_{\underline{x}+\underline{s}}\right)^2\right] . \tag{9.7}$$

The multiplication of the last summation by 1/2 stems from the fact that a quantity of the form $C_{\underline{s}}\left(q_{\underline{x}} - q_{\underline{x}+\underline{s}}\right)^2$ would appear twice in the summation (and thus once too often) when $\underline{x}$ assumes the $\underline{x} + \underline{s}$ value and correspondingly $\underline{s}$ assumes the $-\underline{s}$ value (recall that by Newton's third law, $C_{-\underline{s}} = C_{\underline{s}}$).

Hamilton's equation $\ddot{q} = -\partial\mathcal{H}/\partial q$ provides the dynamic equations for each lattice point oscillator

$$\ddot{q}_{\underline{x}} = -\omega_0^{\,2} q_{\underline{x}} - \sum_{\underline{s}\neq 0} C_{\underline{s}}\left(q_{\underline{x}} - q_{\underline{x}+\underline{s}}\right) ;$$

$$\boxed{\ddot{q}_{\underline{x}} = -\left(\omega_0^{\,2} + \sum\nolimits_{\underline{s}\neq 0} C_{\underline{s}}\right) q_{\underline{x}} + \sum\nolimits_{\underline{s}\neq 0} C_{\underline{s}} q_{\underline{x}+\underline{s}}} \; . \tag{9.8}$$

We seek a solution that is a normal mode of the entire crystal. The first requirement is that the vibrational frequencies of all the lattice oscillators be the same. We try a solution using a separation of the space and time variables of the form $q_{\underline{x}}(t) = \widetilde{q}(\underline{x})e^{-i\omega t}$. We set it in Eq. (9.8) and get

$$-\omega^2\widetilde{q}(\underline{x}) = -\left(\omega_0^{\,2} - \sum\nolimits_{\underline{s}\neq 0} C_{\underline{s}}\right)\widetilde{q}(\underline{x}) + \sum\nolimits_{\underline{s}\neq 0} C_{\underline{s}}\widetilde{q}(\underline{x}+\underline{s}) . \tag{9.9}$$

We then attempt to present the $\widetilde{q}(\underline{x})$ function by a Fourier series

$$\widetilde{q}(\underline{x}) = \frac{1}{\sqrt{N}}\sum\nolimits_{\underline{k}} \boldsymbol{A}_{\underline{k}} e^{i\underline{k}\cdot\underline{x}} , \tag{9.10}$$

where $\underline{k}$, the wave vector, is of reciprocal length dimensions and $\boldsymbol{A}_{\underline{k}}$ is the Fourier expansion coefficient (a $\hat{\boldsymbol{k}} \equiv \underline{k}/k$ symbol for the unity vector in the $\underline{k}$ direction is also occasionally used). The solution $q_{\underline{x}}(t) = \frac{1}{\sqrt{N}}\sum_{\underline{k}} \boldsymbol{A}_{\underline{k}} e^{-i(\omega t - \underline{k}\cdot\underline{x})}$ describes the sum of uncoupled plane-parallel waves. Each such separate $q_{\underline{x}}(t) = N^{-1/2}\boldsymbol{A}_{\underline{k}} e^{-i(\omega t - \underline{k}\cdot\underline{x})}$ plane-wave that belongs to a single $\underline{k}$ vector *defines* the normal crystal oscillation mode. The oscillators that belong to different $\underline{x}$ lattice points, whose total number is N, oscillate by the same frequency and amplitude, but at different phases. The number of possible (allowed) $\underline{k}$ vectors is also N. A periodically ordered macroscopic crystal always satisfies the following approximation:

$$\sum_{\underline{x}} e^{i(\underline{k}-\underline{k}\prime)\underline{x}} \cong N\,\delta_{\underline{k},\underline{k}'}\,. \tag{9.11}$$

In Section 9.6, we provide a justification for this above approximation. The Fourier coefficients $\boldsymbol{A}_{\underline{k}}$ are thus given by

$$\boldsymbol{A}_{\underline{k}} \cong \frac{1}{\sqrt{N}}\sum_{\underline{x}} \tilde{q}(\underline{x})e^{-i\underline{k}\cdot\underline{x}}\,. \tag{9.12}$$

We set the Fourier expansion of Eq. (9.10) in Eq. (9.9) and get

$$\sum_{\underline{k}} \boldsymbol{A}_{\underline{k}} e^{i\underline{k}\cdot\underline{x}}\left[-\omega^2+\omega_0{}^2+\sum_{\underline{s}\neq 0} C_{\underline{s}}-\sum_{\underline{s}\neq 0} C_{\underline{s}} e^{i\underline{k}\cdot\underline{s}}\right]=0\,. \tag{9.13}$$

It is possible to solve the above equation for each *pure* angular frequency ω belonging to a particular nonzero $\underline{k}$ vector, when all other $\boldsymbol{A}_{\underline{k}}$ wave amplitudes vanish.

One particular insight into this development should not be overlooked. Under quantization, a coherent-wave phonon energy $\hbar\omega$ is distributed over a huge number of oscillators; the energy thus belonging to each individual oscillator, as well as its oscillation amplitude, is tiny!

9.5.3 Allowed Wave-Number States and Their Density

Because of crystal symmetry properties, it is possible to represent all allowed $\underline{k}$-vector values within the first Brillouin zone's volume. The reason for this is that for two wave-vectors whose difference is a reciprocal lattice vector, the Fourier coefficient $\boldsymbol{A}_{\underline{k}}$ (the wave amplitude) is identical. For proof, we replace a wave vector $\underline{k}$ by $\underline{k}+\underline{G}$, where $\underline{G}$ is an arbitrary reciprocal lattice vector. According to Eq. (9.12),

$$\begin{aligned}\boldsymbol{A}_{\underline{k}+\underline{G}} &= \frac{1}{\sqrt{N}}\sum_{\underline{x}} \tilde{q}(\underline{x})e^{i(\underline{k}+\underline{G})\cdot\underline{x}} = \frac{1}{\sqrt{N}}\sum_{\underline{x}} \tilde{q}(\underline{x})e^{-i(\underline{k}+\underline{G})\cdot\underline{x}} = \frac{1}{\sqrt{N}}\sum_{\underline{x}} \tilde{q}(\underline{x})e^{-i\underline{k}\cdot\underline{x}}e^{-i\underline{G}\cdot\underline{x}}\\ &= \frac{1}{\sqrt{N}}\sum_{\underline{x}} \tilde{q}(\underline{x})e^{-i\underline{k}\cdot\underline{x}}e^{-i\cdot 2\pi n} = \frac{1}{\sqrt{N}}\sum_{\underline{x}} \tilde{q}(\underline{x})e^{-i\underline{k}\cdot\underline{x}} = \boldsymbol{A}_{\underline{k}}\,.\end{aligned}$$

The transition from the first to the second line follows the fact that $\underline{G}\cdot\underline{x}=2\pi n$ where n is an integer (Section 9.4.3). Because $e^{-i\cdot 2\pi n}=1$, then **Q.E.D.**

Any vector whose units are of inverse length may be represented by a sum of a vector from within the first Brillouin zone and a reciprocal lattice vector. Therefore **it is possible to enclose the vectors representing the normal crystalline oscillation modes within the first Brillouin zone**. Each such allowed $\underline{k}$ vector is represented by a point in the first Brillouin zone, located at the $\underline{k}$-vector tip. The zone is thus dotted by N such discrete points. The most simplistic approximation is to assume the dots to be evenly distributed throughout the first Brillouin zone's volume, irrespective of the crystal shape. In a macroscopic crystal, the density of the dots is very large. Therefore each thus-presented $\underline{k}$ vector is close to being perpendicular to a certain family of crystal planes. Their density is given by $N/V_{rcp}=N[V_C/(2\pi)^3]=V/(2\pi)^3$, where V is the crystal volume (Section 9.4.2). The $\underline{k}$ vector states density *per unit crystal volume* is given by

$$\boxed{\mathcal{D}(\underline{k})=3\frac{1}{(2\pi)^3}}\,. \tag{9.14}$$

The multiplication by 3 in Eq. (9.14) is required to account for the three independent vibrational modes per each $\underline{k}$ vector – one longitudinal and two transverse (see Section 9.5.4.1).

Waves propagating perpendicularly to the First Brillouin zone's flat boundary faces, whose $\underline{k}$ vector belongs to that surface, exhibit some special properties; namely, the inter-crystal planar distances they belong to are larger than those belonging to other $\underline{k}$ vectors in its vicinity (see Eq. [9.5]). It is possible to select at least three such independent vectors for any crystal. Their orientations are considered **major orientations** for the wave propagations. In all crystals, the three may be selected as perpendicular to the (100), (010), and (001) planes, according to the Bravais lattices classification (Figure 9.4). In high-symmetry crystals, they coincide with the [100], [010], and [001] lattice orientations.

For waves that propagate in a major orientation, it is possible to construct a model that provides accurate values for the $\underline{k}$ wave vectors. Let $\underline{k}$ be an allowed wave vector in a major orientation, and $\underline{G}$ be the shortest parallel reciprocal lattice vector ($\underline{G} \| \underline{k}$). Recall that a First Brillouin zone's flat side-boundary face is a plane perpendicular to $\underline{G}$ that bisects it. The distance between two adjacent crystalline planes that are perpendicular to $\underline{k}$ (as well as to $\underline{G}$) is $a \equiv 2\pi/G$. Note that in Eq. (9.5), the interplanar distance was instead noted by the letter d, as it usually is in crystallography. We define a vector $\underline{a}_{\underline{k}}$, whose length is a, that is co-oriented with $\underline{k}$ (when there is no risk for confusion, the $\underline{k}$ subscript can be omitted). We denote by $\mathcal{N}$ the number of such planes existing in the crystalline sample. In macroscopic crystals, the relation $N >> \mathcal{N} >> 1$ is always satisfied. We now make two approximations: for mathematical convenience, we assume $\mathcal{N}$ to be an even integer. Our second assumption relates to the drafting of boundary conditions for the vibrational q displacement variables; boundary conditions must exist in finite-sized crystals. The conditions we form are called **periodic boundary conditions**. For any lattice point $\underline{x}$,

$$\boxed{q_{\underline{x}} = q_{\underline{x} + \mathcal{N}\underline{a}}} \quad . \tag{9.15}$$

For a geometric feeling of the periodic boundary conditions of Eq. (9.15), imagine first that all said planes are of the same shape and area, and packed without side shifting. Such conditions may be realized only in regularly shaped crystals; we still assume those traits to be valid for all shapes of macroscopic crystals. We have to further imagine the chain of planes to be bent to form a closed ring *without distorting the interplanar distances*. Obviously, the periodic boundary conditions are a very rough approximation. We still adopt them for mathematical convenience, because, as we shall see, they straightforwardly provide separate registrations of oppositely propagating waves (for comparison, see Section 2.2, and especially the text related to Eq. [2.27]). Setting Eq. (9.15) in Eq. (9.12) yields

$$\boldsymbol{A}_{\underline{k}} e^{-i(\omega t - \underline{k}\cdot\underline{x})} = \boldsymbol{A}_{\underline{k}} e^{-i(\omega t - \underline{k}[\underline{x} + \mathcal{N}\underline{a}])} = \boldsymbol{A}_{\underline{k}} e^{-i(\omega t - \underline{k}\cdot\underline{x})} e^{ik\mathcal{N}a} .$$

This equality require that $e^{ik\mathcal{N}a} = 1$. In the development we used, the $\underline{k}$ and $\underline{a}$ vectors are co-oriented; therefore $\underline{k}\cdot\underline{a} = ka$. Then, k may assume the following values: $k = n(2\pi/\mathcal{N}a)$ for $n = 0, \pm 1, \pm 2, \cdots, \pm \frac{1}{2}\mathcal{N}$. There are thus $\mathcal{N} + 1$ different k values. It may be readily verified that $\underline{k}$ vectors belonging to $n = \pm\frac{1}{2}\mathcal{N}$ differ by a $\underline{G}$ vector of the reciprocal lattice; thus they represent the same oscillation mode. All other $\underline{k}$ vectors reside within the first Brillouin zone. The total number of the relevant $\underline{k}$ vectors is thus exactly $\mathcal{N}$, and they are evenly distributed along the major orientation.

The periodic boundary conditions are applicable *only for major orientations*. Attempts to apply them in other orientations perpendicular to crystalline planes yield $\underline{k}$ vectors outside of the first Brillouin zone. The reason for this is that the interplanar distances are smaller, and thus belong to $\underline{G}$ vectors of the reciprocal lattice that are too long. The even distribution in the first Brillouin zone for all other $\underline{k}$ vectors is satisfactory for all purposes.

For further demonstration of the meaning of different $\underline{k}$ vectors, we note that for $\underline{k}=0$, all oscillators vibrate at a common phase (Eq. [9.12]). For a first Brillouin zone edge in a major orientation, we have $\underline{k}=(\pi/a^2)\underline{a}=(\pi/a)\underline{u}$, where $\underline{u}$ is a unity vector in the major orientation. We consider next the phase changes between adjacent planes, which is given by $e^{-i\underline{k}\cdot\underline{a}}=e^{-i(\pi/a)\underline{u}\cdot\underline{a}}=e^{-i\pi}=-1$. This means that oscillators belonging to adjacent planes oscillate at opposite phases! A demonstration of phase changes in atomic bases vibrations for several different $\underline{k}$ vectors is provided in Figure 9.6.

9.5.4 Dispersion Curves

To find the frequencies belonging to each oscillation mode, we return to Eq. (9.1). The frequency belonging to each $\underline{k}$ vector obtains by solving the equation

$$-\omega^2+\omega_0{}^2+\sum\nolimits_{\underline{s}\neq 0}C_{\underline{s}}-\sum\nolimits_{\underline{s}\neq 0}C_{\underline{s}}e^{i\underline{k}\cdot\underline{s}}=0\ ,$$

whose solution is

$$\omega^2=\omega_0{}^2+\sum\nolimits_{\underline{s}\neq 0}C_{\underline{s}}-\sum\nolimits_{\underline{s}\neq 0}C_{\underline{s}}e^{i\underline{k}\cdot\underline{s}}. \tag{9.16}$$

For all practical purposes, it is sufficient to carry out the summations in Eq. (9.16) over a limited volume $\mathcal{R}$ encircling any lattice point $\underline{x}$, whose dimensions are of the order of several unit cells. In a more meticulous wording, it is assumed that the sums of all addends *outside* $\mathcal{R}$ are very small compared with the sums obtained for the lattice points $\underline{s}$ *inside* the $\mathcal{R}$ region. Recall that the coupling coefficients $C_{\underline{s}}$ are symmetric with respect to inversion about the $\underline{x}$ lattice point; namely $C_{\underline{s}}=C_{-\underline{s}}$; it is a direct conclusion of Newton's third law. It is thus possible to first divide $\mathcal{R}$ into two separate parts whose belonging lattice points are each an inversion of a lattice point of the other half. The summation may thus be carried out in steps, each time on a different half region. In particular,

$$\sum_{\underline{s}\neq 0}C_{\underline{s}}e^{-i\underline{k}\cdot\underline{s}}=\sum_{\underline{s}\in half\,\mathcal{R}}C_{\underline{s}}e^{-i\underline{k}\cdot\underline{s}}+\sum_{\underline{\sigma}=-\underline{s}}C_{\underline{\sigma}}e^{-i\underline{k}\cdot\underline{\sigma}}$$

$$=\sum_{\underline{s}\in half\,\mathcal{R}}\left\{C_{\underline{s}}e^{i\underline{k}\cdot\underline{s}}+C_{\underline{s}}e^{-i\underline{k}\cdot\underline{s}}\right\}=\sum_{\underline{s}\in half\,\mathcal{R}}2C_{\underline{s}}\cos\underline{k}\cdot\underline{s}\,,$$

where *half* $\mathcal{R}$ is one of the said separate parts. Similarly,

$$\sum_{\underline{s}\neq 0}C_{\underline{s}}=\sum_{\underline{s}\in half\,\mathcal{R}}2C_{\underline{s}}\,.$$

Thus we get

$$\omega^2(\underline{k})=\omega_0{}^2+\sum_{\underline{s}\in half\,\mathcal{R}}2C_{\underline{s}}(1-\cos\underline{k}\cdot\underline{s})\,. \tag{9.17}$$

The $\omega(\underline{k})$ function is called the **dispersion function**. If a single specific orientation is selected for the $\underline{k}$ vector, the dispersion function is symbolized as $\omega(k)$. If a single specific k_0 size is selected, the frequency dependence on the propagation orientation $\omega(k_0,\theta,\varphi)$ is called **spatial dispersion**.

We now classify the $\underline{s}$ crystal lattice points belonging to *half* $\mathcal{R}$ according to which lattice planes perpendicular to the $\underline{k}$ wave-vector they belong to. We assign a different numeral p to each plane

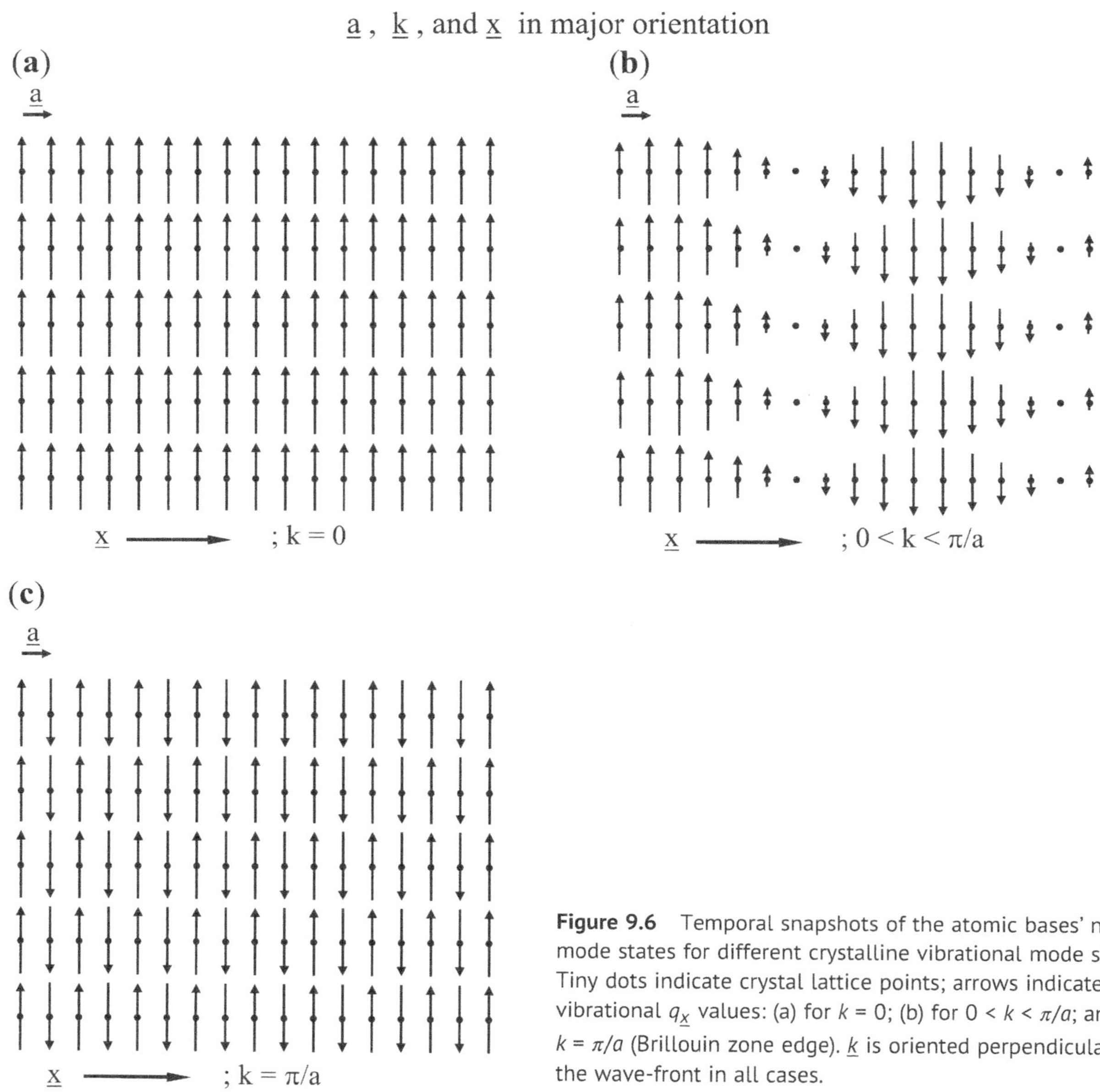

Figure 9.6 Temporal snapshots of the atomic bases' normal mode states for different crystalline vibrational mode states. Tiny dots indicate crystal lattice points; arrows indicate the vibrational $q_{\underline{x}}$ values: (a) for $k = 0$; (b) for $0 < k < \pi/a$; and (c) for $k = \pi/a$ (Brillouin zone edge). $\underline{k}$ is oriented perpendicularly to the wave-front in all cases.

according to its distance from the plane containing the origin ($\underline{x}$ lattice point). In other words, $p = 0$ marks the plane containing the $\underline{x}$ lattice point, $p = 1$ marks the closest parallel plane, and so on. Correspondingly, we include the p numbers as subscript labels to the $\underline{s}$ lattice points; namely $\underline{s}_p$ is a lattice point that belongs to the p plane. It is easy to verify that each $\underline{s}_p$ satisfies the relation $\underline{k}\cdot\underline{s}_p = pka$, where a is the interplanar distance along the $\underline{k}$ direction. Specifically, the scalar product $\underline{k}\cdot\underline{s}_p$ vanishes for $p = 0$. We apply the above nomenclature to the sums appearing in Eq. (9.17):

$$\sum_{\underline{s} \in half\, \mathcal{R}} 2C_{\underline{s}}(1 - \mathbf{cos}\, \underline{k}\cdot\underline{s}) = \sum_{all\ \underline{s}_p} 2C_{\underline{s}_p}(1 - \mathbf{cos}\, pka) = \sum_{p=1}^{\infty} 2\bar{C}_p(1 - \mathbf{cos}\, pka)\,,$$

where $\bar{C}_p = \sum_{\underline{s}_p; p=const} C_{\underline{s}_p}$. The $\bar{C}_p$ coefficient depends on the *orientation* of the $\underline{k}$ wave vector (but not on its size). For the sake of brevity we will not label this property in writing the following equations. The dispersion function (actually it square) is thus written as

$$\boxed{\omega^2(\underline{k}) = \omega_0{}^2 + \sum_{p=1}^{\infty} 2\bar{C}_p(1 - \mathbf{cos}\, pka)} \quad . \tag{9.18}$$

If all $\bar{C}_p$ coefficients are positive constants, $\omega^2(k)$ is a monotonically increasing function of k for small k values. Especially in the special cases where $\underline{k}$ is oriented in a major orientation, k changes between zero and π/a at the edge of the first Brillouin zone. The dispersion function correspondingly changes from a minimum $\omega_{\mathbf{min}}(0) = \omega_0$ to a maximum $\omega_{\mathbf{max}}{}^2 \equiv \omega^2\left(\frac{\pi}{a}\right) = \omega_0{}^2 + \sum_{p=1}^{\infty} 2\bar{C}_p(1 - \mathbf{cos}\, p\pi)$. Because $\mathbf{cos}\, p\pi = (-1)^p$, we get

$$\omega_{\mathbf{max}}{}^2 = \omega_0{}^2 + \sum_{p=1,3,\cdots} 4\bar{C}_p \,. \tag{9.19}$$

The sum of that infinite series is partial to the series for which we assumed convergence to a finite value in a previous paragraph (in the text following Eq. [9.16]). Therefore it also converges to a finite value. If, in contrast to the previous case, all $\bar{C}_p$ coefficients were to be negative constants, $\omega^2(k)$ would be a monotonically decreasing function of k between zero and π/a, and the terms "minimum" and "maximum" would be exchanged.

9.5.4.1 Acoustic Modes

Oscillation modes that are descendants of translational motions of the atomic bases belong to $\omega_0 = 0$, and are called **acoustic modes** or **acoustic branches**. Their dispersion curves are a special case of Eq. (9.18), namely

$$\boxed{\omega^2(k) = \sum_{p=1}^{\infty} 2\bar{C}_p(1 - \mathbf{cos}\, pka)} \quad , \tag{9.20}$$

and correspondingly also $\omega_{\mathbf{max}}{}^2 = \sum_{p=1,3,\cdots} 4\bar{C}_p$; (see Eq. [9.19]).

We analyze now the various properties of such modes. Recall that the dispersion function slope also describes the waves' group velocity $v_g \equiv \partial\omega/\partial k$. For its calculation, we differentiate both sides of Eq. (9.20)

$$2\omega \cdot d\omega = dk \cdot 2a \sum_{p=1}^{\infty} p\bar{C}_p\, \mathbf{sin}\, pka \quad \Rightarrow \quad \frac{d\omega}{dk} = \frac{a}{\omega} \sum_{p=1}^{\infty} p\bar{C}_p\, \mathbf{sin}\, pka \,.$$

For a major orientation in the upper limit of allowed k values (first Brillouin zone edge), we get

$$\left.\frac{d\omega}{dk}\right]_{k=\pi/a} = \frac{a}{\omega_{\mathbf{max}}} \sum_{p=1}^{\infty} p\bar{C}_p\, \mathbf{sin}\, p\pi = 0 \,. \tag{9.21}$$

Thus, in the π/a limit, the dispersion function attains a finite value with a vanishing slope (horizontal shape). In other words, the group velocity diminishes there. It is not an episodic effect related to the selected model; it is consistent with the fact that in the Brillouin zone edge limit, the oscillation modes belonging to $k = \pi/a$ and $k = -\pi/a$ are identical, because the k values differ by a reciprocal lattice vector! Separately, each describes a wave propagating in a direction opposite to the other. Recall, that combining two waves of identical amplitude and frequency that propagate in opposite orientations, creates a standing wave, whose group velocity is obviously zero. Importantly, the group velocities at a first Brillouin zone edge in off-major orientations usually do not vanish! We shall discuss that issue in more detail later.

To calculate the waves' group velocities in the long-waves limit ($k \to 0$), we rewrite Eq. (9.20) in the form

$$\omega^2(k) = 4\sum_{p=1}^{\infty} \overline{C}_p \,\mathbf{sin}^2 \frac{1}{2} pka . \tag{9.22}$$

We can now use the assumption that the coupling coefficients $C_{\underline{s}}$ turn small quite rapidly with growing $\underline{s}$, and practically diminish outside a certain volume surrounding each oscillator. Mathematically, we can assume that the sums of the two series, $\sum_{p=1}^{\infty} \overline{C}_p$, and $\sum_{p=1}^{\infty} p^2 \overline{C}_p$, converge to finite values. Namely, it is possible to stop each summation at a certain finite p_0, and the remainder will always be very small compared with the previous sum. Thus

$$\omega^2 \cong 4 \sum_{p=1}^{p_0} \overline{C}_p \,\mathbf{sin}^2 \frac{1}{2} pka \underset{k << 2/p_0 a}{\cong} k^2 a^2 \sum_{p=1}^{p_0} p^2 \overline{C}_p ;$$

$$\omega(k) \underset{k << 2/p_0 a}{\cong} k \cdot a \sqrt{\sum_{p=1}^{p_0} p^2 \overline{C}_p} . \tag{9.23}$$

The angular frequency vanishes for $k = 0$ (consistently with the underlying assumption), and grows linearly as a function of k for $k << \pi/a$. Frequencies in the very low range belong to the human acoustics, which donates its name to the entire branch. The acoustic wave propagation velocity, or actually the *sound velocity* in the matter, is given by the waves' group velocity $\mathrm{v}_g \equiv \partial\omega/\partial k$, which in this case is equal to the phase velocity $\mathrm{v}_\varphi \equiv \omega/k$:

$$\mathrm{v}_g = \mathrm{v}_\varphi = a \sqrt{\sum_{p=1}^{p_0} p^2 \overline{C}_p} . \tag{9.24}$$

It is customary to imagine the coupling coefficients among atomic bases as if the bases are connected by springs of $K_{\underline{s}}$ constants. Then, Eq. (9.24) would be written as

$$\mathrm{v}_g = \mathrm{v}_\varphi = a \sqrt{\frac{1}{M} \sum_{p=1}^{p_0} p^2 \overline{K}_p} , \tag{9.25}$$

where M is the mass of an atomic basis.

The acoustic oscillation modes are usually classified by the relationship between the motions' orientation of the M masses and the wave propagation motions. When the M masses' motions are co-oriented with the $\underline{k}$ wave-vector, the modes are called **longitudinal waves**. When the M masses' motions are orientated along a plane perpendicular to $\underline{k}$, the modes are called **transverse waves** or **shear waves**. The reason for the latter name is that the planes' motions describe a gliding (or shearing) of a crystal plane relative to the adjacent one. Each $\underline{k}$ vector thus exhibits three fundamentally independent modes: one longitudinal and two transverse.

Some caution, however, is required. The aforementioned classification is based on an assumption that the described longitudinal and transverse motions are all oscillation modes. Indeed, each $\underline{k}$ vector exhibits three fundamentally independent modes, but they are not always exactly longitudinal or transverse. Even in a cubic crystal, exact longitudinal or transverse modes obtain only for the special propagation orientations [100], [110], and [111]. Deviations in other orientations are, however, small. For simplicity and clarity, we shall not be meticulous about these differences; instead, we flatly assume each $\underline{k}$ vector to exhibit one longitudinal and two transverse independent modes.

Figure 9.7 demonstrates the oscillation modes. Crystal planes perpendicular to the $\underline{k}$ wave vector oscillate with an identical phase; the oscillations of adjacent planes of a interplanar distance along the wave propagation orientation differ by a constant phase.

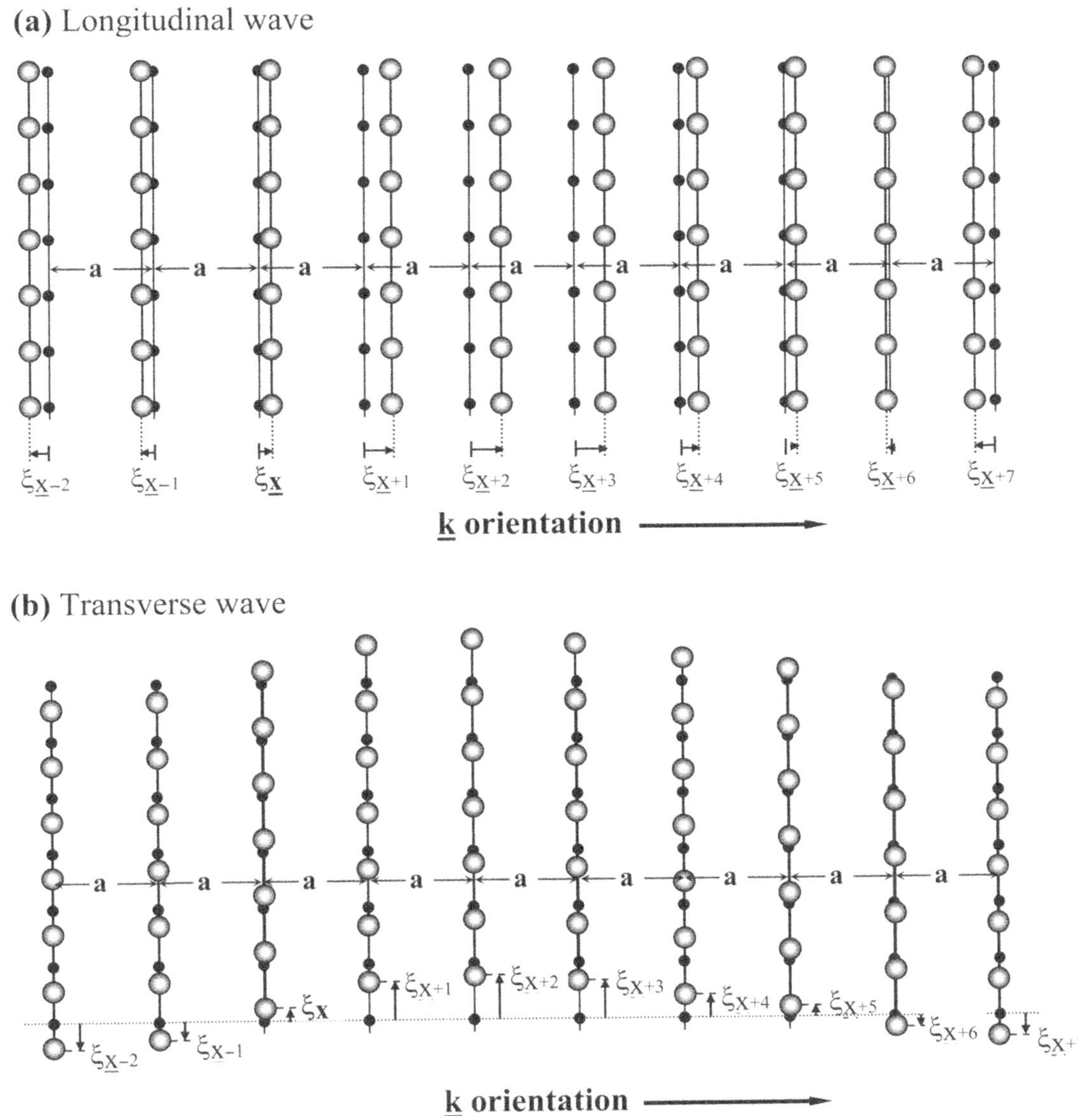

Figure 9.7 A schematic illustration of acoustic wave propagation in a crystal. (a) Longitudinal, (b) Transverse. Tiny full circles describe the equilibrium positions of the lattice points; the big balls describe their attached atomic bases. The $\xi_{\underline{x}+s}$ size deviations from equilibrium are grossly exaggerated; in reality, the deviations are very small compared with the interplanar distances. Recall that the actual deviations ξ are related to the normal coordinates q according to $q = \sqrt{M}\xi$, where M is the mass of an atomic basis.

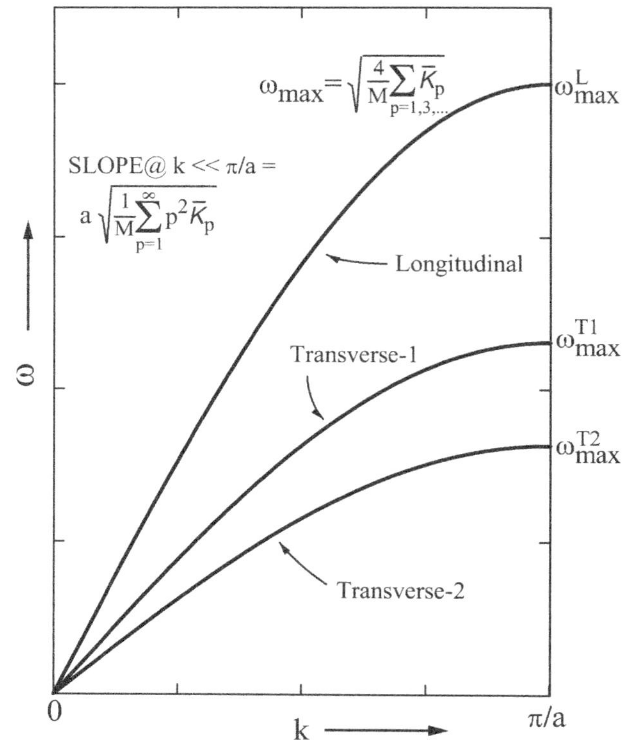

Figure 9.8 Schematic illustration of acoustic dispersion curves in a major orientation for the general case of longitudinal and transverse waves in a crystal of no degeneracy among branches. The Transverse-1 and Transverse-2 relate to planes shearing in two independent in-plane orientations.

Figure 9.8 provides a schematic plot of acoustic dispersion curves along a major orientation. The oscillation frequencies all start at zero for $k = 0$, grow (initially linearly with increasing k), and saturate at the first Brillouin zone edge. For the same wave-numbers, the transverse waves' frequencies are smaller than those of the longitudinal waves. This naturally connects to crystals macroscopic properties. In a longitudinal wave, the planes' displacements cause local changes of the crystal density, thus increasing the oscillation frequency. Shearing of a region by transverse waves leaves the local density practically unchanged; the return force, and hence the frequency, is considerably smaller.

Measurements of crystal dispersion curves are usually performed by studying inelastic neutron scattering (except for brief citations, this book will not deal further with neutron scattering). We shall see that once a dispersion curve for a major orientation is established, it allows a straightforward assessment of the different $\overline{C}_p$ values. To develop the needed expression, we multiply the two sides of Eq. (9.17) by $\mathbf{cos}\,rka$ for an integer $r \geq 1$. We then integrate the obtained expression over the wave-vector k variable in the $[0, +\pi/a]$ limits:

$$\int_0^{\pi/a} dk{\cdot}\omega^2(k)\,\mathbf{cos}\,rka = \sum_{p=1}^{\infty} 2\overline{C}_p \int_0^{\pi/a} dk(1 - \mathbf{cos}\,pka)\,\mathbf{cos}\,rka = -\pi\overline{C}_r/a\,.$$

The last transition on the righthand side is based on the fact that the integral factor in the sum vanishes, except for $p = r$, where its value is $(-\pi/2a)$. Thus we get

$$\overline{C}_p = -\frac{a}{\pi}\int_0^{\pi/a} dk{\cdot}\omega^2(k)\,\mathbf{cos}\,pka\,. \tag{9.26}$$

A similar expression is obtained for a nonmajor orientation when the upper-limit of the integral is selected at a point in the $\underline{k}$ space where the group velocity diminishes. Eq. (9.26) is valid not only for acoustic branches, but also for optical branches (see Sub-Section 9.5.4.2). In insulating (dielectric) crystals, the $\overline{C}_p$ values practically diminish for $p \gtrsim 3-4$. In metals, on the other hand, a long-range influence of up to 20 planes may be obtained!

Fundamentally, there is no need to calculate the dispersion curves in all of the crystallographic orientations. The first Brillouin zone satisfies all the point symmetries of the direct crystal lattice. It is thus possible to calculate the curves in a suitable slice of the first Brillouin zone, and apply it to the entire zone by performing the group symmetry operations. To demonstrate, we detail the symmetry properties of the first Brillouin zone of a primitive cubic crystal belonging to the O_h^1 space group (the crystal belongs to the O_h point group; see Table 9.1; detailed tables and use of the space-group symbols are also provided in Section 11.5). The first Brillouin zone is cubic, as plotted in Figure 9.9. The properties of points of special symmetry are indicated inside the cube body, and on its surfaces. These points are noted by capital letters, some Greek and some Roman, as follows:

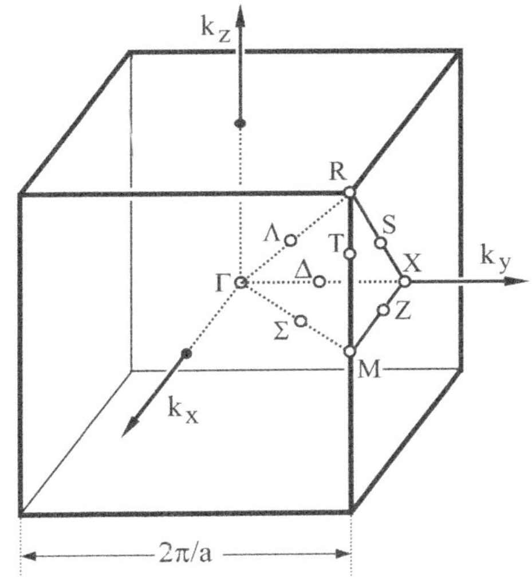

Figure 9.9 The first Brillouin zone of a primitive cubic crystal, with indications of points exhibiting special symmetries. The *a* quantity is the crystal lattice period. For details, see the text.

Γ. This is the point of the origin $k = 0$. The Brillouin zone exhibits the full O_h point symmetry relative to the Γ point.

Δ. This is a general point along the k_y axis. The first Brillouin zone exhibits a $C_{4\mathrm{v}}$ symmetry about this axis; the point is invariant (meaning it does not shift position) when k_y is the rotation axis and the (k_y, k_z) or alternatively (k_y, k_x) planes are σ_v symmetry plane.

Λ. This is a general point along the major diagonal axis of the Brillouin zone. The first Brillouin zone exhibits a $C_{3\mathrm{v}}$ rotational symmetry about this axis; a σ_v plane belonging to that symmetry is formed by that axis itself together with the k_x , k_y , or k_z axes.

Σ. This is a general point along the bisector of the angle between the k_x and k_y axes. The Brillouin zone symmetry about all Σ points is $C_{2\mathrm{v}}$; a σ_v plane belonging to that symmetry is formed by the axis itself, together with the k_z axis, or alternatively the $k_z = 0$ plane.

T. This is a general point situated along the cube edges. The point is invariant (meaning it maintains it position) under every mirror reflection in a plane containing its edge of residence and the Γ point. It transforms to an equivalent point under any operation of the $C_{4\mathrm{v}}$ subgroup of O_h, which preserves the k_z axis.

S. This is a general point situated along one of the diagonals of the cubic Brillouin zone's side faces. It stays invariant or transforms to an equivalent point under $C_{2\mathrm{v}}$ group operations; in the present case, the twofold rotational axis crosses the origin parallel to the diagonal containing the specific *S* point.

Z. This is a general point on the line obtained by cutting one of the Brillouin zone faces and the $k_z = 0$ plane. It stays invariant or transforms to an equivalent point under a $C_{2\mathrm{v}}$ group operations; in the present case, the twofold rotational axis is k_x, and the mirror symmetry plane is the $k_x = 0$ plane.

R. This is a point at the cube vertex. It stays invariant or transforms to an equivalent point by all operations of the O_h point symmetry relative to the Γ point.

X and *M*. The first point is located at a face center of the first Brillouin zone, and the second at an edge center. These points remain invariant or transform to equivalent points by operation of the D_{4h} point group. For *X*, the fourfold rotational axis crosses the *X* point itself; for *M*, the fourfold rotational axis is the k_z axis.

Even when the first Brillouin zone volume for which calculations of dispersion curves are required is reduced thanks to the crystal point symmetry, the remaining calculations are nontrivial. To calculate a dispersion curve along an arbitrary $\hat{\boldsymbol{k}}$ orientation using Eq. (9.19), it is necessary to use the interplanar distance $a_{\hat{k}}$ and the coupling coefficients $\overline{C}_p$ that belong to that orientation. For example, however, in an orientation of only slight deviation from a major orientation, the interplanar distance becomes tiny; correspondingly, the lattice points' density belonging to each such plane is also tiny. Calculation of the $\overline{C}_p$ coefficients using their formal definition by Eq. (9.18) then becomes impractical. There is a possibility to

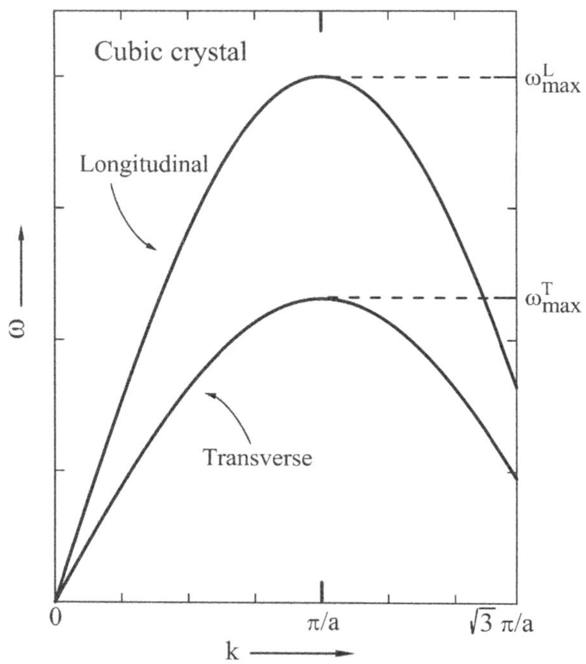

Figure 9.10 Schematic illustration of dispersion curves for longitudinal and transverse acoustic waves along the [111] orientation of a primitive cubic crystal of a lattice period. In this case, no degeneracy exists between longitudinal and transverse waves, but the transverse waves themselves are degenerate (meaning they exhibit no spatial dispersion) because of the cubic symmetry.

carry out calculations for a limited number of orientations where the lattice points' density of their belonging planes is reasonably large, allowing for sufficiently accurate estimation of their $\overline{C}_p$ coupling coefficients. By continuity considerations, it is expected that dispersion curves along orientations close to a calculated orientation are similar to the curve at the calculated orientation. Experiments absolutely justify that expectation! In other words, the dispersion curves behave as if the interplanar distance and their coupling coefficients are similar to those of the calculated curve.

In the following example, we demonstrate an extreme application of that approach. We assume that the dispersion curve belonging to an arbitrary orientation behaves as if the interplanar distance is an interpolation among those of their close *major orientation*, and the coupling constants are interpolations among the constants of their belonging orientations. Accordingly, we present the forms of dispersion curves calculated for acoustic oscillations in a primitive cubic lattice. The interplanes and coupling coefficients are identical for the three major orientations; thus all the interpolated parameters are also identical for all orientations. On the other hand, the upper-limit length of the $\underline{k}$ vectors will always be larger than π/a, where a is the lattice period. For example, the upper limit k length in the first Brillouin zone (of cubic shape; see Figure 9.9) in the orientation perpendicular to the (111) family of planes (the [111] crystal orientation) is $\sqrt{3}\,\pi/a$. The dispersion curves would be identical with those in the [001] direction up to π/a, where they would exhibit a frequency maximum, and then reduce as a function of k for $k > \pi/a$. That behavior is presented in Figure 9.10, and it is verified experimentally with good precision.

Interpolations are also applicable to the optical oscillation modes that will be presented in Sub-Section 9.5.4.2, as well as to crystals of lower symmetry, which are obviously more complex. We shall not elaborate on the issue further in the present book.

9.5.4.2 Optical Oscillation Modes

Oscillation modes for which $\omega_0 \neq 0$ are called **optical modes** or **optical branches**, and they are descendants of the vibrational and rotational motions of the atomic basis. The reason for their name is that some modes are excitable by absorption of visible or near-visible light. We turn to analyze the shape of their dispersion curves. Like with acoustic waves, a dispersion curve exhibits a finite value at the π/a limit, with a zero slope (becoming horizontal) irrespective of the value of ω_0. In other words, in a major orientation, the group velocity always diminishes at the edge of the first Brillouin zone.

For calculation of the group velocity at the long-wave limit ($k \rightarrow 0$), we convert Eq. (9.18) to read

$$\omega^2(k) = \omega_0{}^2 + 4\sum_{p=1}^{\infty} \overline{C}_p \,\mathbf{sin}^2 \frac{1}{2}pka\ . \tag{9.27}$$

Following the assumption that the coupling coefficients $C_{\underline{s}}$ reduce sharply and practically diminish beyond a certain distance from the oscillator, we may write (see also Eqs. [9.22] and [9.23] and related text)

$$\omega^2 \cong \omega_0^2 + 4\sum_{p=1}^{p_0} \overline{C}_p \,\mathbf{sin}^2\frac{1}{2}pka \underset{k \ll 1/p_0 a}{\cong} \omega_0^2 + k^2a^2\sum_{p=1}^{p_0} p^2\overline{C}_p \,. \tag{9.28}$$

Differentiation of both sides, as done in relation to the development of Eq. (9.21), yields

$$\frac{d\omega}{dk} \cong k\frac{a^2}{\omega_0}\sum_{p=1}^{p_0} p^2\overline{C}_p \quad\Rightarrow\quad \left.\frac{d\omega}{dk}\right]_{k=0} = 0\,. \tag{9.29}$$

The group velocity of the optical oscillations diminishes at the $k = 0$ limit as well. In other words, an optical dispersion curve is horizontal at both the $k = 0$ and $k = \pi/a$ ends.

The atomic basis may exhibit vibrational modes, producing an oscillating electrical dipole (see particularly Section 8.9 Case 3). Their descendant crystalline modes are known as **longitudinal optical modes** (abbreviated LO) when the wave-vector orientation coincides with the electrical dipole orientation. In high-symmetry crystals, the longitudinal orientation usually coincides with one of the major orientations. When the wave-vector orientation is perpendicular to the electrical dipole, the modes are called **transverse optical modes** (abbreviated TO). Then there is a full plane available for propagation of transverse optical modes. In high-symmetry crystals, it is usually possible to select two independent propagation orientations that coincide with major orientations. Crystalline modes belonging to a wave vector oriented in neither purely longitudinal nor purely transverse orientations, are called **mixed LO-TO modes**.

Each of the atomic basis *rotational* modes also generates longitudinal and transverse crystalline waves – the rotation axis defines the longitudinal orientation, and every perpendicular orientation defines a transverse orientation. If the atomic basis exhibits a permanent electrical dipole, the librational motion could generate an oscillating electrical dipole. In such cases, the dipole orientation defines the longitudinal one, and any perpendicular orientation is transverse.

Eq. (9.27) suggests that the dispersion curves of both longitudinal and transverse optical curves converge to a common angular frequency, ω_0 in the $k \to 0$ limit. Experiments show *a different behavior*: The longitudinal dispersion curve always converges to a frequency ω_0^L, which is normally higher than the transversely converged ω_0^T frequency. The difference $\omega_0^L - \omega_0^T$ is called **LO-TO splitting**. The transverse dispersion curves may also split (except in high-symmetry crystals). As for the Brillouin zone edge at $k = \pi/a$, in most cases, the longitudinal curve frequency is smaller than at $k \to 0$. A schematic plot of typical transverse and longitudinal optical dispersion curves obtained in a dielectric crystal is provided in Figure 9.11.

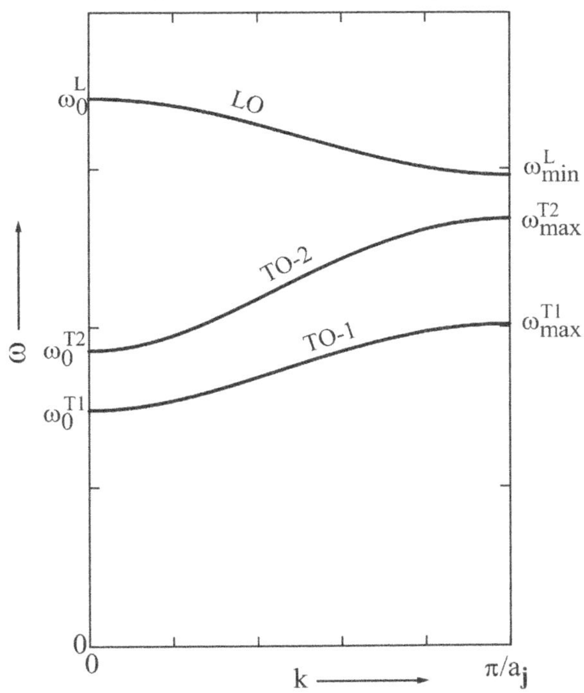

Figure 9.11 Schematic illustration of dispersion curves (angular frequency vs. wave-vector) for longitudinal (LO) and transverse (TO) waves that belong to a single mechanical vibration of the atomic basis. The wave-vector orientations are different for the different curves; the interplanar distance a_j may also be different for the different j major orientations. The longitudinal waves' dispersion curve convergence at $k = 0$ is normally larger than those of both transverse waves curves TO-1 and TO-2.

Figure 9.11 describes a reduction in an LO frequency ω as a function of wave-number k. It is a common effect, indicating that at least some of the coupling constants $\overline{C}_p$ for longitudinal waves are negative. In other words, the interaction among oscillators in the longitudinal orientation exhibits a considerable (perhaps anharmonic) softening component.

The LO-TO splitting is a general property of matter, but we shall present an interpretation for only dielectric materials. Properties of dielectric materials result from two independent contributions. One is polarization of the independent atoms caused by deflection of the electron clouds surrounding the atoms relative to their nuclei under the action of an external electric field. Quantum mechanically, the effect involves the participation of atomic virtual states (see the last paragraph in Section 8.5). This effect is termed **medium polarization**. The other factor is polarization caused by shifting of the atoms themselves (nuclei and electrons together) from their equilibrium positions under the action of the external electric field. For simplicity, we assume the crystal to have only a single polar vibrational species per each atomic basis (Figure 9.12). We further emphasize, that a $k \rightarrow 0$ limit means that the referred-to wavelengths are very long compared with the lattice period, but they are still very short compared with the crystal's dimensions.

We use basic electrostatic considerations. Consider a macroscopic-insulating crystal formed as a parallel plate, whose side faces are perpendicular to a major orientation, which is also the orientation of the vibrating dipole. Assume that an external electric field $\underline{\mathcal{E}}_{ext}$ is applied perpendicularly to the plate's side faces, such as by metal-plating the faces, and charging the obtained electrodes with equal electrical charges of opposite sign. $\underline{\mathcal{E}}_{ext}$ is defined as the electric field that would exist between the said electrodes if the volume between them would have been an empty space. Adding the crystal plate between the electrodes *without changing their charge*, polarizes the crystal. The effective electrical field then reduces, becoming $\underline{\mathcal{E}}_{eff} = \underline{\mathcal{E}}_{ext} - 4\pi\underline{\mathbf{P}}$, where $\underline{\mathbf{P}}$ is the induced material polarization (dipole moment per unit volume). Recalling the dielectric function (constant) definition $\varepsilon(\omega) \equiv \mathcal{E}_{ext}/\mathcal{E}_{eff}$, we get

$$\underline{\mathbf{P}}(0) = \frac{1}{4\pi}\left(1 - \frac{1}{\varepsilon(0)}\right)\underline{\mathcal{E}}_{ext} \,. \tag{9.30}$$

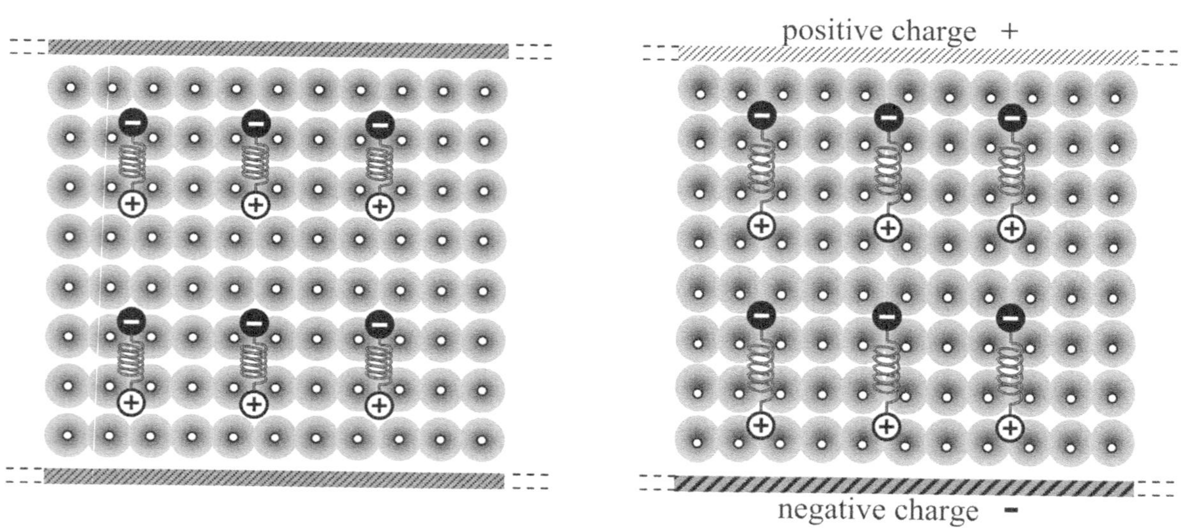

Figure 9.12 Material polarization caused by two contributions: a medium contribution described by electrical charges surrounding the nuclei, and a polar vibrational species contribution described as pairs of charges, positive and negative, bonded by springs. Both are affected by the application of an external electric field. Left side: Charge distribution without an external field. Right side: Charge distribution under an external field. The tiny white dots indicate the positively charged nuclei; the large gray circles indicate the negative electrical charge distribution.

The setting of $\omega = 0$ in Eq. (9.30) means that we are interested in a static electric field or in an alternating field of a relatively low frequency. We aim to understand the polar electric dipoles' contribution to the dielectric function. For that purpose, we have to subtract the medium polarizability from the total one. We thus advance a qualitative macroscopic empiric description of the refractive index frequency dependence (in Chapter 10, we provide an analysis based on microscopic grounds). Experimentally, far from resonance frequencies of electronic, vibrational, or rotational transitions, the material dielectric constant is larger than a unity size, and of a very moderate frequency dependence. These features are considered a result of the medium polarization, which is almost frequency independent. By that interpretation, if we measure the dielectric constant at a frequency high enough to be far from the material vibrational/librational frequencies, yet still small enough to be far from the material electronic transitions, the result would provide a fair assessment of its value at a static or near-static electric field. In most dielectrics, the desired range of measurements for that purpose belongs to the near infrared (wavelengths in the $1-5$ μm region).

Such a measurement is practically performed by measuring the refractive index dispersion in the visible and near-visible region. The refractive index $n(\omega)$ in optically transparent dielectrics is known to be real, because it is far below any electronic transition frequencies. The dielectric constant is the squared refractive index: $\varepsilon(\omega) = n^2(\omega)$. Empiric expressions describing the refractive index frequency (or wavelength) dependence in the *visible and near-visible range* exhibit changes of only several percentage points. Furthermore, extrapolation toward low frequencies (long waves) shows convergence to a finite value. The empiric expressions are called **Sellmeier equations**. A common and simple one has the form $(n^2 - 1)^{-1} = A - B/\lambda^2$, where A and B are empiric material constants, and λ is the wavelength of the electromagnetic radiation. It may be readily verified that in the $\lambda \to \infty$ limit (namely, $\omega \to 0$), the refractive index extrapolates to the finite value $n^2(\infty) = 1 + A^{-1}$. Mathematically, the "∞" indicator refers to the wavelength limit; it is, however, used to indicate that the *frequency range* from which the extrapolation is *made*, belongs to frequencies higher than those of the far infrared; most typical vibrational or librational frequencies belong to the very far infrared, and hence are not included in the analysis.

Figure 9.13 provides an example by analysis of the wavelength dependence of the extraordinary refractive index n_e of a sapphire (Al_2O_3) single crystal according to the aforementioned simple Sellmeier equation. In

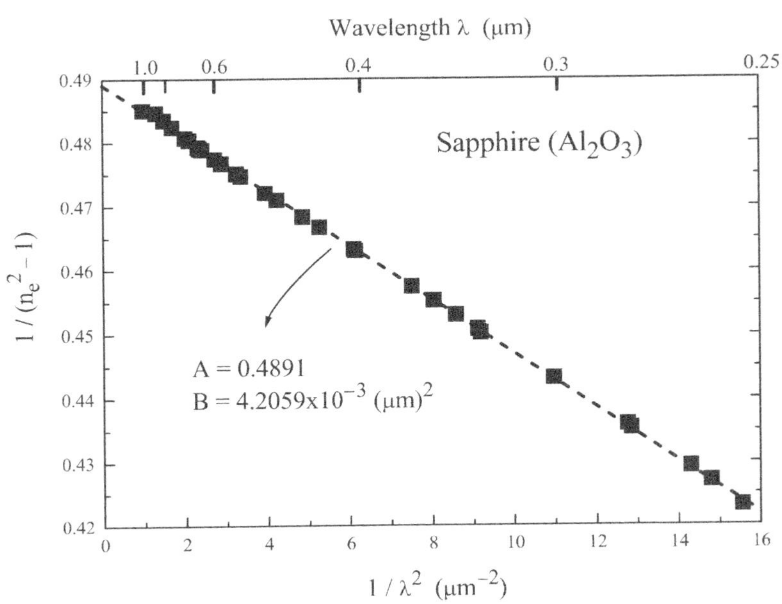

Figure 9.13 $1/(n_e^2 - 1)$ as function $1/\lambda^2$ for a sapphire single crystal. Solid squares – experimental results, dashed line – are a linear fit. The A and B fit parameters per the Sellmeier equation are inset in the figure frame.

the present case, an excellent fit is shown for the 1.0 – 0.25 μm spectral range, yielding $n_e^2(\infty) = \varepsilon(\infty) = 3.0446$. Recall that the "extraordinary" title refers to the light polarized parallel to the crystal optical axis; general optical properties of crystals in the visible and near-visible spectral regions are described in Section 15.7.

In view of the last discussion, we may regard the $\varepsilon(\infty)$ parameter obtained using Sellmeier's method as the *background refractive index* for the static or near-static medium polarization $\underline{\boldsymbol{P}}_B$. Following Eq. (9.30), we write

$$\underline{\boldsymbol{P}}_B = \frac{1}{4\pi}\left(1 - \frac{1}{\varepsilon(\infty)}\right)\underline{\mathcal{E}}_{ext}\,. \tag{9.31}$$

The polarization emanating from the contribution of the medium-induced dipoles may be written in the form $\underline{\boldsymbol{P}}_Q = \underline{\boldsymbol{Q}}/V_C$, where $\underline{\boldsymbol{Q}}$ is the induced dipole moment of each atomic basis, and V_C is the primitive unit cell volume (Eq. [9.2]). Thus we get

$$\underline{\boldsymbol{P}}_Q = \underline{\boldsymbol{P}} - \underline{\boldsymbol{P}}_B = \frac{\underline{\boldsymbol{Q}}}{V_C} = \frac{1}{4\pi}\left(\frac{1}{\varepsilon(\infty)} - \frac{1}{\varepsilon(0)}\right)\underline{\mathcal{E}}_{ext}\,. \tag{9.32}$$

It should be noted and emphasized, that both $\varepsilon(0)$ and $\varepsilon(\infty)$ are empiric parameters of the crystal.

We now consider the polarizing mechanism from a different perspective. The induced dipole moment Q is proportional to the generalized coordinate q of the atomic basis polarized vibrational species. It is thus possible to express it in the form $Q = \beta\, q$, where β is a proportionality constant. The q dimensions are $g^{1/2}$cm; it may thus be written as $q = \sqrt{\mu^*}\,\xi$, where ξ is a variable of length dimensions, to be addressed as "displacement," and μ^* is a constant of mass dimensions, to be addressed as "effective reduced mass". The meanings of the names stem from a model where the atomic basis comprises two stiff parts bonded by a spring of an effective K constant, vibrating one-dimensionally along their bond. Accordingly, we get $Q = \beta\sqrt{\mu^*}\xi \equiv e^*\xi$, where the $e^* \equiv \beta\sqrt{\mu^*}$ has electric-charge dimensions, and is thus addressed as an "effective charge" of the vibrating dipole. The model thus considers the two atomic basis parts as electrically charged, one with a positive ($+e^*$) charge, the other with a negative ($-e^*$) charge.

Application of an external static electric field $\underline{\mathcal{E}}_{ext}$, as described in the text preceding Eq. (9.31), causes a steady separation between the atomic basis parts, whereby the separating force $\mathcal{F} = e^*\mathcal{E}_{eff} = e^*\mathcal{E}_{ext}/\varepsilon(0)$ equals the spring return force $\mathcal{F} = K\xi$. Thus $\xi = (e^*/\varepsilon(0)K)\mathcal{E}_{ext}$ and $Q = e^*(e^*/\varepsilon(0)K)\mathcal{E}_{ext}$. Comparing the latter result with the expression given by Eq. (9.32), and extracting the effective spring constant, yields

$$K = \frac{4\pi e^{*2}}{V_C}\cdot\frac{\varepsilon(\infty)}{\varepsilon(0) - \varepsilon(\infty)}\,. \tag{9.33}$$

The term "electrostatics" indicates that if a polarized matter is sliced perpendicularly to the polarization orientation, any slice may be removed, and even metal plated on its flat surfaces, without the matter exhibiting any change of the inside electrical field strength. We now consider a parallel plate of matter exhibiting a constant polarization oriented perpendicularly to the plate surface, as shown in Figure 9.14a. We imagine that a tiny parallel plate is removed from inside the matter, thus forming a parallel-plate *void*, whose surfaces are also perpendicular to the polarization orientation. Such a case situation is indicated as a white region on the left side of Figure 9.14a. An electrical field $\underline{\mathcal{E}}_{ext}^{(1)} = 4\pi\underline{\boldsymbol{P}}$ then develops inside the void, because of the electrical surface charge exposed at its surface ends by the bulk material polarization. The same-sized electric field of opposite orientation $\underline{\mathcal{E}}_{ext}^{(2)} = -4\pi\underline{\boldsymbol{P}}$ exists on the removed tiny parallel plate, and equally so on the entire plate volume. When that plate is retuned to fill in the void, the total effective

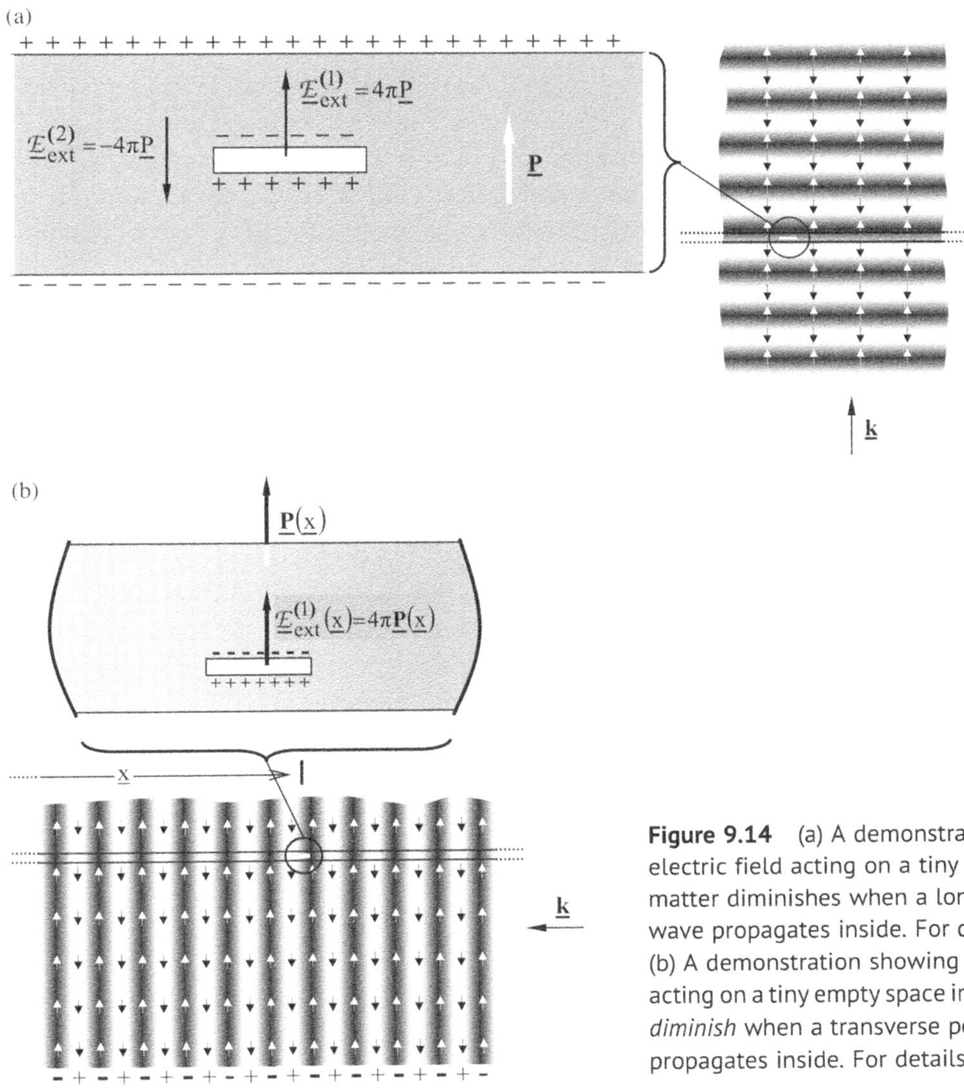

Figure 9.14 (a) A demonstration showing that an external electric field acting on a tiny empty space inside a dielectric matter diminishes when a longitudinal polarization plane wave propagates inside. For details – see the text. (b) A demonstration showing that an external electric field acting on a tiny empty space inside a dielectric matter *does not diminish* when a transverse polarization plane wave propagates inside. For details, see the text.

electric field it is subject to vanishes: $\underline{\mathcal{E}}_{ext} = \underline{\mathcal{E}}_{ext}^{(1)} + \underline{\mathcal{E}}_{ext}^{(2)} = 0$, which exists throughout the entire volume. Notably, that situation is valid despite the volume polarization!

The above consideration is applied to a plane-parallel longitudinal polarization wave. The described plate and void properties are attributed to a wave-front region as described on the right side of Figure 9.14a. This side demonstrates a temporal snapshot of a slice of matter carrying a propagating longitudinal polarization plane-parallel wave. A gray scale indicates the polarization strength and orientation: upward orientation for gray, downward for bright. The external electric field at any point is zero. Therefore, a return force acting between two parts of a vibrating dipole depends only on the effective spring, whose constant K has been evaluated by Eq. (9.33). Its relation to the dipole vibrational frequency is given by $K = \mu^*(\omega^L)^2$. Then

$$\left(\omega^L\right)^2 = \frac{4\pi e^{*2}}{\mu^* V_C} \cdot \frac{\varepsilon(\infty)}{\varepsilon(0)} \cdot \frac{\varepsilon(0)}{\varepsilon(0) - \varepsilon(\infty)} \cdot \tag{9.34}$$

Similar considerations are applied to a plane-parallel transverse polarization wave. The described plate and void properties are attributed to wave-front regions as described on the bottom of Figure 9.14b. It demonstrates a temporal snapshot of a slice of matter carrying a propagating longitudinal polarization plane-parallel wave. A gray scale indicates the polarization strength. The polarization orientation is indicated by arrows: upward orientation for gray, downward for bright. Like with the longitudinal propagating wave, we consider a sliced parallel plate perpendicular to the polarization orientation. In contrast, however, the slice surfaces are instead *parallel* to the wave propagation direction. The polarization now changes along the slice surfaces. A void is now produced inside the slice by removing a tiny parallel plate whose surfaces are parallel to those of the slice, while making sure that its long dimensions are very small compared with the polarization wavelength. We note by $\underline{x}$ the void position in the slice coordinates. As before, because of polarization exposed at its surfaces, the void is subject to an external electric field $\mathcal{E}_{ext}^{(1)} = 4\pi \boldsymbol{P}(\underline{x})$. In contrast to the longitudinal case, the influence of the far-plate surfaces now diminishes. The reason for this is that according to the wavelength, the electrical polarization reverses orientation periodically over the slice surface, and hence also the electrical charge there. For the void, it appears that the slice surface is not charged at all, namely $\mathcal{E}_{ext}^{(2)} \approx 0$. The external field in the void is thus only $\mathcal{E}_{ext} = \mathcal{E}_{ext}^{(1)} = 4\pi \boldsymbol{P}(\underline{x})$. On returning the removed tiny plate into the void while it is still $\boldsymbol{P}(\underline{x})$ polarized as its vicinity, it feels an external field of $\mathcal{E}_{ext} = 4\pi \boldsymbol{P}(\underline{x})$. That field is co-oriented with the polarization field, namely oppositely to the orientation of the force exerted by the K spring constant. It has a softening influence on the spring return field, causing a corresponding frequency lowering.

For a quantitative account, we can separate between the contributions of the medium and the polar oscillating dipoles. By Eq. (9.31), the medium contribution is given by $\boldsymbol{P}_B = \frac{1}{4\pi}\left(1 - \frac{1}{\varepsilon(\infty)}\right)\mathcal{E}_{ext}$. Setting $\mathcal{E}_{ext} = 4\pi \boldsymbol{P}$, we get $\boldsymbol{P}_B = \frac{1}{4\pi}\left(1 - \frac{1}{\varepsilon(\infty)}\right)\boldsymbol{P}$. The oscillating polar dipoles' contribution would be $\boldsymbol{P}_Q = \frac{e^*\xi}{V_C}$. The total polarization would be

$$\boldsymbol{P} = \boldsymbol{P}_B + \boldsymbol{P}_Q = \left(1 - \frac{1}{\varepsilon(\infty)}\right)\boldsymbol{P} + \frac{e^*\xi}{V_C}\ .$$

Solving for $\boldsymbol{P}$ from the two equations' sides. we get $\boldsymbol{P} = \varepsilon(\infty)(e^*\xi/V_C)$; thus,

$$\mathcal{E}_{ext} = 4\pi\varepsilon(\infty)\frac{e^*\xi}{V_C}\ . \tag{9.35}$$

The return force acting on the temporal dipole moment of a single oscillating $e^*\xi$ dipole is

$$\mathcal{F} = -K\xi + \frac{e^*\mathcal{E}_{ext}}{\varepsilon(0)} = -K\xi + 4\pi\frac{\varepsilon(\infty)}{\varepsilon(0)}\cdot\frac{e^{*2}}{V_C}\xi = -\left(K - 4\pi\frac{\varepsilon(\infty)}{\varepsilon(0)}\cdot\frac{e^{*2}}{V_C}\right)\xi\ .$$

Therefore, the effective spring constant acting on the oscillating dipole is $K^T = K - 4\pi\frac{\varepsilon(\infty)}{\varepsilon(0)}\cdot\frac{e^{*2}}{V_C}$. Setting the expression for K of Eq. (9.33), we get

$$K^T = \frac{e^{*2}}{V_C}\cdot 4\pi\frac{\varepsilon(\infty)}{\varepsilon(0)}\left(\frac{\varepsilon(0)}{\varepsilon(0)-\varepsilon(\infty)} - 1\right). \tag{9.36}$$

Similar to how we get Eq. (9.34) for longitudinal waves, for transverse waves we have

$$\left(\omega^T\right)^2 = \frac{e^{*2}}{\mu^* V_C}\cdot 4\pi\frac{\varepsilon(\infty)}{\varepsilon(0)}\left[\frac{\varepsilon(0)}{\varepsilon(0)-\varepsilon(\infty)} - 1\right]. \tag{9.37}$$

Dividing Eq. (9.34) by the last result (Eq. [9.37]), we get

$$\boxed{\left(\frac{\omega^L}{\omega^T}\right)^2 = \frac{\varepsilon(0)}{\varepsilon(\infty)}}. \tag{9.38}$$

The above simple relation between longitudinal and transverse wave frequencies in the $k \to 0$ limit is called a **Lyddane-Sachs-Teller** relation. Notably, in obtaining the ratio, the μ^*, e^*, and K quantities have been eliminated, and their knowledge is not relevant for using the said ratio. On the other hand, these particular quantities are necessary parameters for the shaping of models for evaluation of the frequencies themselves. We note further, that the model we used to describe the LO-TO splitting belongs to a family of physical effects that are called "cooperative", where, among other things, a cooperative behavior of a large number of identical physical bodies generates a feedback that influences each of the independent elements.

The above discussion seemingly suggests that the transverse waves' frequencies ω^T in the $k \to 0$ limit are independent of the $\underline{k}$ vector orientation. The real behavior is different, as schematically described in Figure 9.10. A spatial dispersion of transverse wave frequencies does exist in the $k \to 0$ limit. The physical reason is that the feedback that causes the softening of the dipoles' vibrations has a limited range, which is different for different waves' propagation directions.

The spatial dispersion has a considerable effect on electromagnetic propagation in dielectric matter in the far infrared region. For the visible and near-visible spectral regions, however, that effect is very small, and practically neglected. A fundamental assumption in uses of geometrical optics is that the refractive index depends solely on the beam polarization orientation, and not on its propagation direction. The optical properties of anisotropic crystals are described in Section 15.7; effects related to coupling between polar crystal vibrations and electromagnetic radiation are described in Chapter 10.

Electromagnetic waves (light) of frequency $\omega^T < \omega < \omega^L$ propagating outside the crystal in a direction that coincides with a major crystal orientation, cannot continue propagating into the crystal upon perpendicular incidence onto the surface. The reason for this is $\underline{k}$ vectors' orientation mismatch between the photons and same-frequency phonons; namely, the said phonons propagate only off perpendicularity to the surface. Therefore, a light-reflection band appears between ω^L and ω^T. Thus, an analysis of light absorption, transmission, and reflection measurements is sufficient to assess the ω^L and ω^T frequencies in the $k \to 0$ limit. That subject is addressed in Chapter 10, especially in Section 10.3. Inelastic neutron scattering measurements yield further information, covering the entire dispersion curves.

9.5.5 Theoretical Dispersion Curve Calculations – A Basic Approach

We provide now a swift summary of a basic approach to theoretical calculations of crystal vibrational waves' dispersion curves. The calculations include two basic steps: calculation of the vibrational frequencies of the atomic basis uncoupled oscillators, and a subsequent calculation of the coupling constants $C_{\underline{s}}$ (see Section 9.5.2).

Calculation of the oscillation modes and the vibrational and rotational frequencies is carried out by the procedures described in Chapter 8. The squared angular frequencies' matrix $\mathbf{\Omega}$ of the various modes (Eq. [8.10]) is obtained by diagonalizing the generalized force constants' $\mathbf{U}$ matrix defined in Eq. (8.8). In principle, they should be calculated from the position dependence of the overlap and exchange integrals of the ground-state atomic bases (see Section 6.2.3). In reality, such calculations are complicated, time consuming, and require the use of advanced numeric computing. To simplify and shorten the task, and reduce the number of variables, empiric dependences are sought, especially ones adjusted to uncoupled molecules, whose structure is similar to the atomic basis or to its parts. Empiric potentials like those of Morse, Hulburt-Hirschfelder, or Linett (Eqs. [7.11], [7.13], and [7.14]) are also used. To find the librational frequencies, we must first calculate the tensor of inertia (see Section 8.2, specifically the footnote at that Section third page). The three librational frequencies will be $\omega_0^A = \sqrt{\Omega_A/I_A}$, $\omega_0^B = \sqrt{\Omega_B/I_B}$, and $\omega_0^C = \sqrt{\Omega_C/I_C}$, where I_A , I_B , and I_C , are moments of inertia of the atomic basis about the principal axes A , B , and C of the tensor of inertia, and Ω_A , Ω_B , and Ω_C are return force moment constants for rotations about the A , B , and C principal axes, respectively.

All of the aforementioned approaches are based on breaking the forces into elemental contributions. Simplifications are obtained by adopting a semi-empiric approach. For example, in the case of librational motions, the relative motion of adjacent atomic bases rotating about parallel axes is, in fact, a shear motion. Thus, the return force moment constants may be derived from the macroscopic material elastic constants. It is possible to rely on force constants obtained by an analysis of transverse (shear) acoustic modes (see Section 9.5.4.1, especially Figure 9.8 and Eq. [9.26]). Except for the present citation, in the present book, we will not deal with material elastic properties.

For polar optical modes, it is necessary to consider the long-range interactions among the oscillating electrical dipoles. The dipoles are considered point-like, located at the centers of the atomic bases. When all dipoles are parallel, the contribution of the dipole/dipole interaction to the potential energy of the one located at $\underline{x}$ is given by

$$V_{\underline{x}}^{d-d} = \frac{1}{2}\sum\nolimits_{\underline{s}\neq 0}\left[\frac{e^{*2}}{\varepsilon(\infty)\,s^3}\right]\left\{1-3\cos^2(\underline{u},\underline{s})\right\}\xi_{\underline{x}}\xi_{\underline{x}+\underline{s}}\,, \tag{9.39}$$

where $\underline{u}$ is a unity vector in the dipole oscillation orientation and $\xi_{\underline{x}}$ and $\xi_{\underline{x}+\underline{s}}$ are the translations of all positive centers of the atomic bases from their belonging rest centers for the $\underline{x}$ and $\underline{x}+\underline{s}$ lattice points, respectively. Obviously, $\mu_{\underline{x}} \equiv e^*\xi_{\underline{x}}$ and $\mu_{\underline{x}+\underline{s}} \equiv e^*\xi_{\underline{x}+\underline{s}}$ are the dipole moments belonging to the $\underline{x}$ and $\underline{x}+\underline{s}$ atomic bases, respectively; the effective electrical charge e^* is an adjustable parameter. In addition, it is assumed that each of an atomic basis's parts, positive or negative, exhibits an independent polarizability emanating from the response of their electronic clouds to the action of an external electric field. Eq. (9.39) is a simplistic approximation, basically assuming that the dipoles are immersed in a uniform dielectric medium. The polarization of each of the said parts, or alternatively of the dielectric constant, are also adjustable parameters of the model. The model further includes short-ranged repulsions among parts of near atomic bases, included because of changes in the overlap integrals between electronic clouds (see

Section 6.2.2). Calculations of such integrals also involve the use of adjustable parameters. Models that are more complex include non-zero quadrupole moments, which oscillate in phase with the dipole ones.

Fair fits between theories and experiments are usually obtained when a sufficient number of adjustable parameters are used; sometimes more than 10 are needed. An important disadvantage of using a large number of adjustable parameters is that, often, the obtained fit values are inconsistent with their physical meaning. This can be interpreted as indicating that the different, seemingly unrelated fit parameters, are rather related in a form that has not been considered in the model construction.

9.5.6 Dispersion Curves and Specific Heats

In this section, we provide a limited description of the basic methods used in theoretical calculations of solids' specific heats. The subject is historically important because experimental measurements of specific heats' temperature dependences were used to strengthen the confidence in the validity of quantum theory, and its application to solids' internal oscillations. The calculation is based on the assumption that each oscillation mode of ω angular frequency could occupy a specific temperature-dependent number of $\hbar\omega$ energy. The occupation number is given by $\langle n(\omega)\rangle_T = \left(e^{\hbar\omega/k_BT} - 1\right)^{-1}$ (see Eq. [2.26]). The thermal energy stored in a unit crystal volume is given by

$$E_{Th} = \sum\nolimits_{\underline{k}} \langle n(\omega(\underline{k}))\rangle_T \; \omega(\underline{k}) \; . \tag{9.40}$$

It is possible to carry out such summations as integrals over thin layers of equifrequency ω. Eq. (9.40) then assumes the form

$$E_{Th} = \int_0^{\infty} \langle n(\omega)\rangle_T \cdot \hbar\omega \cdot D(\omega) \; d\omega \; , \tag{9.41}$$

where $D(\omega)$ is the density of states per unit angular frequency; namely, $D(\omega)\, d\omega$ is the number of states, whose frequency resides between ω and $\omega + d\omega$, per unit crystal volume. The basic requirements are then limited to calculation of the $D(\omega)$ function. In contrast to phonons in isotropic materials, the equifrequency surfaces are usually not spherical. The topological structure of these surfaces is quite complex, and the calculations often turn correspondingly complex even if the dispersion functions are well known. The most primitive approximations use a combination of the Debye model for the acoustical branches and Einstein's model for the optical branches. The Debye model assumes that the acoustical branches are identical for both transverse and longitudinal modes, exhibiting no spatial dispersion (dependence on wave propagation direction) and depend linearly on the wave vector's size (see Section 9.5.4.1). The dispersion curves are thus similar to those of electromagnetic waves (see Section 2.2). The true shape of the first Brillouin zone is further replaced by a sphere of the same state's density, and $3N$ total number per unit volume, where N is the number of unit cells in the same volume. The density of states then assumes the form (see Eq. [2.28])

$$D_D(\omega) \cong \frac{3\omega^2}{2\pi^2 \overline{v}_s{}^3} \quad for \quad 0 < \omega \leq \omega_D \equiv \overline{v}_s\left(6\pi^2 N\right)^{1/3} ,$$

where $\overline{v}_s$ is a representative value of the speed of sound in the crystal, which is a proper average between transverse and longitudinal waves.

In the Einstein model, each dispersion curve belonging to a κ mode is replaced by a constant angular frequency $\overline{\omega}_{0\kappa}$, which is an average over the frequencies belonging to the κ branch. In other words, the contribution of the optical branches to the density of states is given by

$$D_E(\omega) \cong N \sum_{\kappa=4}^{3K} \delta(\omega - \overline{\omega}_{0\kappa}) ,$$

where K is the number of atoms constituting the atomic basis attached to each lattice point, and $\kappa = 1, 2, 3$ relate to the three acoustic branches. The total density of states is given by

$$D(\omega) = D_D(\omega) + D_E(\omega) . \tag{9.42}$$

The specific heat per unit volume at constant crystal volume C_V (usually an experimentally obtained quantity) is given by

$$C_V \equiv \left.\frac{\partial E_{Th}}{\partial T}\right]_V . \tag{9.43}$$

Quite often, the calculated values of C_V exhibit reasonable accord with measured results. Improved accord is obtained by taking into account the various dispersions for assessing the density of states.

9.6 Appendix: Intermediate Calculation for Justifying Eq. (9.11)

We now perform a calculation that justifies the use of the approximation $\sum_{\underline{x}} e^{i\underline{k}\cdot\underline{x}} \cong N\delta_{\underline{k},0}$ for a macroscopic, periodically ordered crystal, as performed in section 9.5.2 (Eq. [9.11]). We write $\underline{x} = \nu_1\underline{a} + \nu_2\underline{b} + \nu_3\underline{c}$, where $\underline{a}$, $\underline{b}$, and $\underline{c}$ are the lattice fundamental vectors and ν_1, ν_2, and ν_3 are integers, describing a regularly shaped crystal. For each value of ν_1, ν_2, or ν_3 in such a crystal, the other values also belong to the lattice description. Thus,

$$\sum_{\underline{x}} e^{i\underline{k}\cdot\underline{x}} = \sum_{\nu_1,\nu_2,\nu_3} e^{i\,\underline{k}\cdot(\nu_1\underline{a} + \nu_2\underline{b} + \nu_3\underline{c})} = \left(\sum_{\nu_1} e^{i\nu_1\underline{k}\cdot\underline{a}}\right) \cdot \left(\sum_{\nu_2} e^{i\nu_2\underline{k}\cdot\underline{b}}\right) \cdot \left(\sum_{\nu_3} e^{i\nu_3\underline{k}\cdot\underline{c}}\right) .$$

We consider, for example, the first of the three factors on the rightside. It is a sum of a geometric series, whose ratio is $e^{i\,\underline{k}\cdot\underline{a}}$. Let its (finite) number of elements be $\mathcal{N}_1$. We get

$$\sum_{\nu_1} e^{i\,\nu_1\underline{k}\cdot\underline{a}} = \frac{e^{i\mathcal{N}_1\underline{k}\cdot\underline{a}} - 1}{e^{i\underline{k}\cdot\underline{a}} - 1} .$$

We denote the absolute value of the last expression by y. It is given by

$$y = \left| \frac{e^{i\mathcal{N}_1\underline{k}\cdot\underline{a}} - 1}{e^{i\underline{k}\cdot\underline{a}} - 1} \right| = \frac{\left| \sin\left(\frac{1}{2}\mathcal{N}_1\underline{k}\cdot\underline{a}\right) \right|}{\left| \sin\left(\frac{1}{2}\underline{k}\cdot\underline{a}\right) \right|} . \tag{9.44}$$

The above equation assumes an $\mathcal{N}_1$ value in the $\underline{k} = 0$ limit, and zeros at points satisfying $\mathcal{N}_1\underline{k}\cdot\underline{a} = n(2\pi)$, with n, the ordinal number of the zero points, being an integer. The values between zeros are larger than zero, but their heights reduce with increasing $\underline{k}\cdot\underline{a}$. When $\mathcal{N}_1 >> 1$, the width of the central peak around $\underline{k} = 0$ is $\sim 2\pi/\mathcal{N}_1 a$. The heights of the secondary peaks behave like $\sim (1/|n|)\ (\mathcal{N}_1/2\pi)$. For a graphical demonstration, we denote $z = \underline{k}\cdot\underline{a}$. In Figure 9.15, we plotted the function

Figure 9.15 The $y = \left|\sin \mathcal{N}_1 \cdot \frac{1}{2} z\right| / \left|\sin \frac{1}{2} z\right|$ versus z function for $\mathcal{N} = 100$.

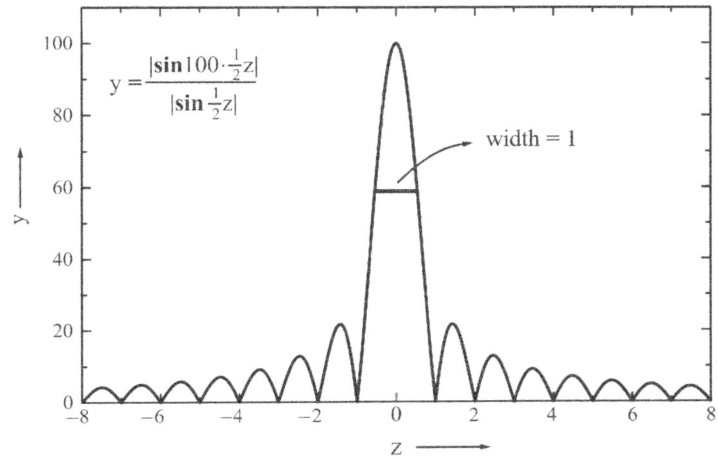

$y = \left|\sin \mathcal{N}_1 \cdot \frac{1}{2} z\right| / \left|\sin \frac{1}{2} z\right|$ for the $\mathcal{N}_1 = 100$ case. Notably, the zero points are obtained exactly for the $\underline{k}$ values allowed by the cyclic edge-conditions for the lattice vibrations expressed in section 9.5.3 (Eq. [9.15]). For those particular values, one gets exactly

$$\sum_{\nu_1} e^{i\nu_1 \underline{k}\cdot\underline{a}} = \frac{e^{i\mathcal{N}_1 \underline{k}\cdot\underline{a}} - 1}{e^{i\underline{k}\cdot\underline{a}} - 1} = \begin{cases} \mathcal{N}_1 & \text{for} \quad \underline{k} = 0 \\ 0 & \text{for} \quad \underline{k} \neq 0 \end{cases}, \tag{9.45}$$

and similar results for $\underline{b}$ and $\underline{c}$. In a regularly shaped crystal, $N = \mathcal{N}_1 \cdot \mathcal{N}_2 \cdot \mathcal{N}_3$. Therefore, Eq. (9.11) is accurate for all $\underline{k}$ values allowed by the cyclic edge conditions. For general $\underline{k}$ values, we may write

$$\sum_{\nu_1} e^{i\nu_1 \underline{k}\cdot\underline{a}} = \frac{e^{i\mathcal{N}_1 \underline{k}\cdot\underline{a}} - 1}{e^{i\underline{k}\cdot\underline{a}} - 1} \approx \begin{cases} \mathcal{N}_1 & \text{for} \quad |k| << 2\pi/\mathcal{N}_1 a \\ < \dfrac{1}{|n|}\dfrac{\mathcal{N}_1}{2\pi} & \text{for } |k| \approx n\dfrac{2\pi}{\mathcal{N}_1 a} \end{cases}. \tag{9.46}$$

Thus, for the general case, Eq. (9.11) is an approximation adopted for all allowed $\underline{k}$ values, independently of boundary conditions or the geometric structure of the crystal.

Exercises for Chapter 9

9.1 The Cartesian coordinates of the fundamental vectors for the primitive unit cell of a given crystal in angstrom (Å) units are:

$a(2.34, -10.14, 0)$; $b(0, 0, 6.72)$; $c(11.32, 5.28, 0)$.

(a) Plot the unit cell structure, and identify the crystalline system it belongs to.
(b) Calculate the unit cell volume and the volume of the first Brillouin zone.
(c) Calculate the fundamental vectors of the reciprocal lattice, plot its unit cell structure, and identify the crystalline system it belongs to.
(d) Calculate the $d_{7,7,12}$ distance between the adjacent (7, 7, 12) crystalline planes.

9.2 Imagine that aluminum *Al* exhibits a primitive cubic structure with a lattice period (constant) $a = 2.55$ Å. The return force constant between two adjacent atoms on shearing of two adjacent planes that are perpendicular to a crystal major axis is 1.15×10^4 dyn/cm; the return force constant between two adjacent atoms on compressing two such adjacent planes is 1.84×10^4 dyn/cm. Calculate numerically and plot graphically the different acoustic dispersion curves in aluminum; use them to calculate the speed of sound for longitudinal and transverse acoustic waves in the matter.

9.3 Prove Eq. (9.26), and especially the note about its relevance to optical modes and the way it may be adjusted for nonmajor orientations.

9.4 Consider a crystal belonging to a primitive tetragonal lattice whose lattice periods are $a = 5.6$ Å and $c = 6.9$ Å. The atomic base mass is 594 amu. Table 9.2 summarizes the return force constants for shear and compression between adjacent planes that are perpendicular to major orientations. Calculate numerically and plot graphically the different acoustic dispersion curves for the non-degenerate major propagation orientations [100] and [001]. Indicate the speed of sound (phase propagation velocity) per each mode.

Table 9.2 List of return force constants for shear and compression between different crystal planes.

$\underline{k}$ vector orientation	Distortion mode	Return force constant (10^4 dyn/cm)
[100]	Shear of (100) planes in [010] direction	1.4
	Shear of (100) planes in [001] direction	2.65
	Compression of (100) planes in [100] direction	6.05
[001]	Shear of (001) planes in [010] direction	0.8
	Compression of (001) planes in [001] direction	1.3

10 Dielectric Crystalline Solids

10.1 Light Propagation in a Dielectric Medium

The propagation of electromagnetic waves is derived from Maxwell's equations:

$$\begin{aligned} &(\mathrm{i}) \quad \textbf{rot}\ \underline{\mathcal{B}} = \frac{1}{c}\frac{\partial \underline{\mathcal{D}}}{\partial t} + \frac{4\pi}{c}\underline{J}_{ext} \\ &(\mathrm{ii}) \quad \textbf{div}\ \underline{\mathcal{D}} = 4\pi\rho_{ext} \\ &(\mathrm{iii}) \quad \textbf{rot}\ \underline{\mathcal{E}} = -\frac{1}{c}\frac{\partial \underline{\mathcal{B}}}{\partial t} \\ &(\mathrm{iv}) \quad \textbf{div}\ \underline{\mathcal{B}} = 0 \end{aligned} , \tag{10.1}$$

where $\underline{\mathcal{E}}$ is the electric field, $\underline{\mathcal{D}}$ is the electric displacement vector, $\underline{J}_{ext}$ is the "external" electric current density, ρ_{ext} is the "externally caused" volume density of the electric charge, $\underline{\mathcal{B}}$ is the magnetic induction vector, and c is the vacuum speed of light constant. The magnetic induction vector is often addressed as "the $\underline{\mathcal{B}}$ magnetic field" or "the magnetic flux density". The meaning of the "external" or "externally caused" adjective is that these quantities are not caused by $\underline{\mathcal{E}}$, $\underline{\mathcal{D}}$, or $\underline{\mathcal{B}}$, but are instead constraints imposed on the system. As a reminder, the physical meaning of the $\underline{\mathcal{E}}$ and $\underline{\mathcal{B}}$ fields is clarified by consideration of the expression for the force $\underline{\mathcal{F}}$ acting on a tiny point-like electric charge e moving at a velocity $\underline{\mathrm{v}}$ through the medium: namely, $\underline{\mathcal{F}} = e[\underline{\mathcal{E}} + (1/c)\underline{\mathrm{v}} \times \underline{\mathcal{B}}]$. As for the electric displacement $\underline{\mathcal{D}}$ vector, that quantity is obtained by the relation

$$\frac{\partial \underline{\mathcal{D}}}{\partial t} = \frac{\partial \underline{\mathcal{E}}}{\partial t} + 4\pi \underline{J} , \tag{10.2}$$

where $\underline{J}$ in Eq. (10.2) is the electric current density related to an actual motion of charge *caused* by the $\underline{\mathcal{E}}$ and $\underline{\mathcal{B}}$ fields. We further recall the relation $\underline{\mathcal{D}} = \underline{\mathcal{E}} + 4\pi\underline{P}$, where $\underline{P}$ is the electric polarization density vector of the medium.

Eqs. (10.1) and (10.2) assume the continuousness of all quantities appearing there, as is expected for a uniform medium. Such assumptions are also justified for a non-uniform medium for average values of the said quantities, where averaging is performed over a volume containing a large number of unit cells, whose dimensions are still very small compared with the electromagnetic beam wavelength. Alternatively presented, this is when the temporal resolution of the time scale is long compared with the time required for the wave propagation along many unit cells, but is still very short compared with the electromagnetic radiation time period.

We present now the solutions to Maxwell's equations for a dielectric medium. A characteristic feature of a dielectric medium is that all electrons are bound, and thus no free currents may develop. We further assume that the medium is electric-charge compensated (devoid of any space-charge), namely $\rho_{ext} = 0$. Maxwell's equations then reduce into

Physics, Optics, and Spectroscopy of Materials, First Edition. Zeev Burshtein.

$$\text{(i) } \boldsymbol{rot}\underline{\mathcal{B}} = \frac{1}{c}\frac{\partial \underline{\mathcal{D}}}{\partial t}\,; \text{ (ii) } \boldsymbol{div}\underline{\mathcal{D}} = 0\,; \text{ (iii) } \boldsymbol{rot}\underline{\mathcal{E}} = -\frac{1}{c}\frac{\partial \underline{\mathcal{B}}}{\partial t}\,; \text{ (iv) } \boldsymbol{div}\underline{\mathcal{B}} = 0\,. \tag{10.3}$$

We seek a solution of the form of a linearly polarized plane-parallel electromagnetic wave

$$\underline{\mathcal{E}} = \underline{\mathcal{E}}_0 e^{-i(\omega t - \underline{k}\cdot\underline{r})}\,; \; \underline{\mathcal{D}} = \underline{\mathcal{D}}_0 e^{-i(\omega t - \underline{k}\cdot\underline{r})}\,; \; \underline{\mathcal{B}} = \underline{\mathcal{B}}_0 e^{-i(\omega t - \underline{k}\cdot\underline{r})}\,, \tag{10.4}$$

where $\underline{r}$ is the position vector in the medium and $\underline{\mathcal{E}}_0$, $\underline{\mathcal{D}}_0$, and $\underline{\mathcal{B}}_0$ are the corresponding wave amplitudes that are constant over time and space. Such waves are considered **normal waves**. Attribution of a single frequency ω and a wave vector $\underline{k}$ for the propagating wave, actually means that a single Fourier component by time and space of $\underline{\mathcal{E}}$, $\underline{\mathcal{D}}$, and $\underline{\mathcal{B}}$ has been selected. Setting Eq. (10.4) in Eq. (10.3) yields the relations

$$\text{(i) } \underline{\mathcal{D}} = -\frac{c}{\omega}\underline{k}\times\underline{\mathcal{B}}\,; \text{ (ii) } \underline{k}\cdot\underline{\mathcal{D}} = 0\,; \text{ (iii) } \underline{\mathcal{B}} = \frac{c}{\omega}\underline{k}\times\underline{\mathcal{E}}\,; \text{ (iv) } \underline{k}\cdot\underline{\mathcal{B}} = 0\,. \tag{10.5}$$

For a demonstration, we prove in some detail the (10.5-i) relation. We set the relevant quantities of Eq. (10.4) in Eq. (10.3-i):

$$\boldsymbol{rot}\left[\underline{\mathcal{B}}_0 e^{-i(\omega t - \underline{k}\cdot\underline{r})}\right] = \frac{1}{c}\frac{\partial}{\partial t}\left[\underline{\mathcal{D}}_0 e^{-i(\omega t - \underline{k}\cdot\underline{r})}\right].$$

We then expand the left side according to the algebraic $\boldsymbol{rot}(\varphi\,\underline{A}) = \varphi\,\boldsymbol{rot}\underline{A} + \boldsymbol{grad}\varphi \times \underline{A}$ relation, where $\underline{A}$ is a three-dimensional vector and φ is a scalar function (see Section 10.5.1 item [xiv]); in our case, we choose $\varphi = e^{-i(\omega t - \underline{k}\cdot\underline{r})}$. To expand the above equation's right side, we start by the required time differentiation. We get

$$e^{-i(\omega t - \underline{k}\cdot\underline{r})}\cdot\boldsymbol{rot}\underline{\mathcal{B}}_0 + i\underline{k}\,e^{-i(\omega t - \underline{k}\cdot\underline{r})} \times \underline{\mathcal{B}}_0 = -\frac{i\omega}{c}\underline{\mathcal{D}}_0 e^{-i(\omega t - \underline{k}\cdot\underline{r})}\,.$$

The first addend on the left side vanishes because the ***rot*** operation on a spatially constant vector vanishes. Solving for $\underline{\mathcal{D}}$, we get $\underline{\mathcal{D}} = -\frac{c}{\omega}\,\underline{k}\times\underline{\mathcal{B}}$, **Q.E.D**.

A straightforward conclusion by Eqs. (10.5-ii) and (10.5-iv) is that the $\underline{\mathcal{D}}$ and $\underline{\mathcal{B}}$ vectors are both perpendicular to the $\underline{k}$ wave vector. The relation (10.5-i) indicates that $\underline{\mathcal{D}}$ and $\underline{\mathcal{B}}$ are also mutually perpendicular. On the other hand, the $\underline{\mathcal{E}}$ vector is not necessarily perpendicular to $\underline{k}$; in other words, the $\underline{\mathcal{D}}$ and $\underline{\mathcal{E}}$ vectors are not necessarily parallel, but the three vectors $\underline{\mathcal{D}}$, $\underline{\mathcal{E}}$, and $\underline{k}$ always reside on the same plane, because all three must be simultaneously perpendicular to $\underline{\mathcal{B}}$. Figure 10.1 presents schematically the orientation relations among the said vectors.

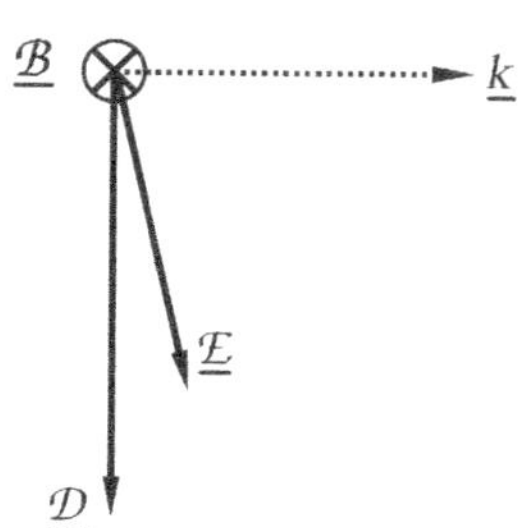

Figure 10.1 Orientation relations among magnetic and electric fields of linearly polarized plane-parallel electromagnetic fields propagating in a dielectric medium. The $\underline{\mathcal{D}}$ and $\underline{\mathcal{B}}$ fields are mutually perpendicular ($\underline{\mathcal{B}}$ oriented perpendicularly into the page) and both are perpendicular to the $\underline{k}$ wave vector. The $\underline{\mathcal{D}}$, $\underline{\mathcal{E}}$, and $\underline{k}$ vectors belong to a common plane (the page plane) perpendicularly to $\underline{\mathcal{B}}$.

A fundamental property of dielectric media is that the magnetic permeability tensor (μ) is diagonal, with all diagonal elements assuming a unity value; namely, $\mu_{ij} = \delta_{ij}$, where $i, j = x, y, z$ are the Cartesian coordinates. The $\underline{\mathcal{B}}$ magnetic induction is equal to the $\underline{\mathcal{H}}$ magnetic field. Thus, **hereinafter, we replace $\underline{\mathcal{B}}$ with $\underline{\mathcal{H}}$ in all equations**.

We now aim to derive the wave equations. We differentiate Eq. (10.3-i) by time to get $\textbf{\textit{rot}}\,\frac{\partial \underline{\mathcal{H}}}{\partial t} = \frac{1}{c}\frac{\partial^2 \underline{\mathcal{D}}}{\partial t^2}$. We set $\partial\underline{\mathcal{H}}/\partial t$ from Eq. (10.3-iii) to get

$$\textbf{\textit{rot rot}}\,\underline{\mathcal{E}} + \frac{1}{c^2}\frac{\partial^2 \underline{\mathcal{D}}}{\partial t^2} = 0 \; ; \tag{10.6}$$

$$\boxed{-\Delta\underline{\mathcal{E}} + \textbf{\textit{grad div}}\,\underline{\mathcal{E}} + \frac{1}{c^2}\frac{\partial^2 \underline{\mathcal{D}}}{\partial t^2} = 0} \; . \tag{10.7}$$

The $\Delta\underline{\mathcal{E}}$ expression is a vector, whose components are the result of Laplace operator Δ (see Section 3.2) acting separately on each component of the $\underline{\mathcal{E}}$ vector (see Section 10.5.1 items [v] and [vi]).

We set in Eq. (10.7) the assumed solution of a normal wave (a linearly polarized plane-parallel wave as in Eq. [(10.4)]). We get the relation

$$\underline{k}^2\underline{\mathcal{E}} - \underline{k}(\underline{k}\cdot\underline{\mathcal{E}}) - \frac{\omega^2}{c^2}\underline{\mathcal{D}} = 0 \; . \tag{10.8}$$

We now introduce the relation $\underline{\mathcal{D}} = \left(\varepsilon_{ij}(\omega, \underline{k})\right)\underline{\mathcal{E}}$, where $i, j = x, y, z$ mark the Cartesian coordinates. The quantities $\varepsilon_{ij}(\omega, \underline{k})$ are elements of the dielectric tensor. Inclusion of the $\underline{k}$ vector orientation as a variable of ε_{ij}, and not just its size k, allows for an analysis of the spatial dispersion effect in a similar way to those indicated in relation to lattice vibrations in Section 9.5.4. The dielectric tensor is symmetric; namely, its elements satisfy $\varepsilon_{ij}(\omega, \underline{k}) = \varepsilon_{ji}(\omega, \underline{k})$. Setting in Eq. (10.8), we get

$$\left(k^2\delta_{ij} - k_i k_j - \frac{\omega^2}{c^2}\varepsilon_{ij}((\omega, \underline{k}))\right)\underline{\mathcal{E}}_0 = 0 \; . \tag{10.9}$$

Eq. (10.9) exhibits a nontrivial solution $\underline{\mathcal{E}}_0 \neq 0$ only if the determinant of the tensor operating on $\underline{\mathcal{E}}_0$ in Eq. (10.9) vanishes; namely,

$$\left|k^2\delta_{ij} - k_i k_j - \frac{\omega^2}{c^2}\varepsilon_{ij}(\omega, \underline{k})\right| = 0 \; . \tag{10.10}$$

In principle, solutions for the set of equations presented in Eq. (10.10) provide relations between ω and $\underline{k}$, namely the dispersion curves $\omega(\underline{k})$ for the normal electromagnetic waves propagation in matter. At the present stage, we limit ourselves to considering specific cases, where the x, y, and z are principal axes of the dielectric tensor. Obviously, in such cases, the dielectric tensor is diagonal.

In a crystalline solid, the principal axes of the dielectric tensor $\varepsilon_{ij}(\omega, \underline{k})$ are determined by the crystalline symmetry. Table 10.1 classifies the tensor types according to the symmetry elements, and Bravais systems they belong to. The selected axis system in each case relates to the unit cell structures provided in Figure 9.4, which do not always coincide with the tensor principal axes. In using Table 10.1, the reader is also referred to Section 8.12, especially Tables 8.2 and 9.1. It should also be remembered that, in cases of degeneracy, degenerate axes may be replaced by an orthogonal conversion. For example, in all groups belonging to a trigonal crystal, the x and y Cartesian axes may be arbitrarily chosen on a plane perpendicular to the z axis.

Table 10.1 ε_{ij} tensors in Cartesian x, y, z coordinates relating to the fundamental unit-cell vectors of Bravais lattices and point symmetries of the crystal systems.

Crystal system	ε_{ij} tensor	Cartesian coordinates system
Triclinic	$\begin{pmatrix} \varepsilon_{xx} & \varepsilon_{xy} & \varepsilon_{xz} \\ \varepsilon_{xy} & \varepsilon_{yy} & \varepsilon_{yz} \\ \varepsilon_{xz} & \varepsilon_{yz} & \varepsilon_{zz} \end{pmatrix}$	The x, y, z, system may be chosen arbitrarily.
Monoclinic	$\begin{pmatrix} \varepsilon_{xx} & 0 & \varepsilon_{xz} \\ 0 & \varepsilon_{yy} & 0 \\ \varepsilon_{xz} & 0 & \varepsilon_{zz} \end{pmatrix}$	The y axis is parallel to the C_2 (or C_S) axis (the $\underline{b}$ axis in Figure 9.4ii).
Orthorhombic	$\begin{pmatrix} \varepsilon_{xx} & 0 & 0 \\ 0 & \varepsilon_{yy} & 0 \\ 0 & 0 & \varepsilon_{zz} \end{pmatrix}$	The x,y,z axes are co-oriented with the three C_2 axes ($\underline{a}$, $\underline{b}$, and $\underline{c}$ of Figure 9.4iii). In the C_{2v} group, x and y are perpendicular to the σ'_v and σ_v planes, respectively.
Tetragonal	$\begin{pmatrix} \varepsilon_{xx} & 0 & 0 \\ 0 & \varepsilon_{xx} & 0 \\ 0 & 0 & \varepsilon_{zz} \end{pmatrix}$	In all groups, the z axis is co-oriented with the C_4 or S_4 axes, which coincide with the $\underline{c}$ axis of Figure 9.4iv. In C_4, S_4, and C_{4h} groups, the x and y axes may be arbitrarily chosen in the plane perpendicular to z. In D_4, C_{4v}, and D_{2d} groups, the x and y axes are determined uniquely (consider $\underline{a}$ and $\underline{b}$ of Figure 9.4iii): in D_4, they are co-oriented with the two C_2 axes; in C_{4v}, they are positioned perpendicularly to the two σ_v planes; in D_{2d}, they are co-oriented with the two C'_2 axes.
Trigonal (rhombohedral)	$\begin{pmatrix} \varepsilon_{xx} & 0 & 0 \\ 0 & \varepsilon_{xx} & 0 \\ 0 & 0 & \varepsilon_{zz} \end{pmatrix}$	In all groups, the z axis is co-oriented with the C_3 axis ([111] orientation in the unit cell of Figure 9.4v). The x and y axes in the C_3 and S_6 groups may be arbitrarily chosen in a plane perpendicular to z. In D_3, C_{3v}, and D_{3d} groups, x and y axes are definitely determined: x is co-oriented with one of C_2's axes and y is perpendicular to the (z, x) plane.
Hexagonal	$\begin{pmatrix} \varepsilon_{xx} & 0 & 0 \\ 0 & \varepsilon_{xx} & 0 \\ 0 & 0 & \varepsilon_{zz} \end{pmatrix}$	In all groups, the z axis is co-oriented with the C_6 or C_3 axis (the $\underline{c}$ axis of the unit cell in Figure 9.4vi). The x and y axes in the C_6, C_{3h}, and C_{6h} groups may be arbitrarily chosen in a plane perpendicular to z. In D_6, D_{3h}, and D_{6h} groups, the x and y axes are uniquely determined: x is co-oriented with one of C_2's axes (the $\underline{a}$ vector in Figure 9.4vi) and y is perpendicular to the (z, x) plane. In the C_{6v} group, x is perpendicular to one of the σ_v planes ($\underline{a}$ vector in Figure 9.4vi) and y is perpendicular to the (z, x) plane.
Cubic	$\begin{pmatrix} \varepsilon & 0 & 0 \\ 0 & \varepsilon & 0 \\ 0 & 0 & \varepsilon \end{pmatrix}$	The x, y, z axes are co-oriented with the $\underline{a}$, $\underline{b}$, $\underline{c}$ unit-cell vectors of Figure 9.4vii. In T and T_h groups, they all are C_2 axes. In O, T_d, and O_h groups, they coincide with the C_4 axes.

After establishing a diagonal dielectric tensor, we further simplify the problem, and assume that only one of the $\underline{\mathcal{E}}_0$ electric field components is nonzero – $\mathcal{E}_{oi}$; in other words, we consider an eigenstate of the dielectric tensor. In that case, the electric field $\underline{\mathcal{E}}$ orientation coincides with that of the electric displacement $\underline{\mathcal{D}}$ vector; namely, it is also perpendicular to the $\underline{k}$ wave propagation orientation. It is easy to see (Eq. [(10.8)]) that the $\underline{k}(\underline{k}\cdot\underline{\mathcal{E}})$ addend vanishes, and Eq. (10.10) reduces to

$$k^2 - \frac{\omega^2}{c^2}\varepsilon_{ii}(\omega, \underline{k}) = 0 \; ;$$
$$\mathrm{v}_{\varphi i}(\omega, \underline{k}) \equiv \frac{\omega}{k} = \frac{c}{\sqrt{\varepsilon_{ii}(\omega, \underline{k})}} = \frac{c}{\widetilde{n}_i(\omega, \underline{k})} \, , \tag{10.11}$$

where $\mathrm{v}_{\varphi i}$ is the phase velocity of a normal electromagnetic wave of ω angular frequency, linearly polarized in the i principal axis of the dielectric tensor, and propagating in the $\underline{k}$ orientation inside the dielectric medium. The $\widetilde{n}_i(\omega, \underline{k}) \equiv \sqrt{\varepsilon_{i\,i}(\omega, \underline{k})}$ quantity is called the **complex refractive index** of the medium for the particular wave. Notably, for the visible spectral region, the fundamental dependence on the wave propagation $\underline{k}$ orientation of that quantity is *negligibly small.*

For a demonstration, we assume x to be the principal tensor orientation, and z to be the electromagnetic propagation direction. Eq. (10.4) then reduces to the form

$$\mathcal{E}_x = \mathcal{E}_{0x}\boldsymbol{exp}\left[-i\left(\omega t - \frac{\omega \widetilde{n}_x}{c} z\right)\right] . \tag{10.12}$$

Using Eq. (10.5-iii) we obtain the following expression for the magnetic field

$$\mathcal{H}_y = \frac{c}{\omega} k\mathcal{E}_x = \widetilde{n}_x \mathcal{E}_x = \widetilde{n}_x \mathcal{E}_{0x}\boldsymbol{exp}\left[-i\left(\omega t - \frac{\omega \widetilde{n}_x}{c} z\right)\right] . \tag{10.13}$$

As we shall see in Section 10.4, the dielectric function $\varepsilon_{xx}(\omega, \underline{k})$ of a dielectric medium may be complex. Correspondingly, the complex refractive index $\widetilde{n}_x(\omega, \underline{k}) = \sqrt{\varepsilon_{xx}(\omega, \underline{k})}$ would be a complex number. The refractive index is written as $\widetilde{n}_x(\omega, \underline{k}) \equiv n_x(\omega, \underline{k}) + i\,\kappa_x(\omega, \underline{k})$, where n_x , the real part of $\widetilde{n}_x$, is granted the name **refractive index**, and κ_x , the imaginary part of $\widetilde{n}_x$, is granted the name **extinction coefficient**. Setting the latter definitions in Eqs. (10.12) and (10.13), we get

$$\left.\begin{aligned} \mathcal{E}_x &= \mathcal{E}_{0x} e^{-i\omega t}\boldsymbol{exp}\left[i\frac{\omega n_x}{c} z\right]\boldsymbol{exp}\left[-\frac{\omega \kappa_x}{c} z\right] \\ \mathcal{H}_y &= \widetilde{n}_x \mathcal{E}_{0x} e^{-i\omega t}\boldsymbol{exp}\left[i\frac{\omega n_x}{c} z\right]\boldsymbol{exp}\left[-\frac{\omega \kappa_x}{c} z\right] \end{aligned}\right\} . \tag{10.14}$$

The factor $\widetilde{n}_x$ appearing on the right side of the expression for the magnetic field $\mathcal{H}_y$ in the second line of Eq. (10.14) is a complex number. It indicates the existence of a phase difference between the electric field oriented in the x-direction and the magnetic field oriented in the y-direction. That is entirely different than what is found in free space (vacuum), where the two fields are co-phased, peaking together and attaining zero values together (Eq. [1.2], Figure 1.1, and related text). Replacing the factor $\widetilde{n}_x$ with its real part n_x in the expression for $\mathcal{H}_y$, provides the component of $\mathcal{H}_y$ which is co-phased with $\mathcal{E}_x$; replacing it with the imaginary part $\kappa_x > 0$ (the extinction coefficient) provides the quarter-cycle advanced phase component of $\mathcal{H}_y$ relative to $\mathcal{E}_x$. In other words, in such a case, the magnetic field peak in a dielectric material appears before the peak of the electric field.

The latter observation is significant for calculating the light intensity (power per unit area) on crossing a plane perpendicular to the propagation direction. It is given by the real part of the Poynting vector $I(\omega, z, t) = \boldsymbol{Re}\left[\frac{c}{4\pi}\mathcal{E} \times \mathcal{H}\right]$ (see also Eq. [1.34]); for the particular present case of plane-parallel waves, we multiply the co-phased components of the electric and magnetic fields. After averaging them over a full time period (see Section 1.1, Eqs. [1.3]-[1.5], and related text), we get

$$I(\omega, z) = \left(\frac{c}{8\pi}\right) n_x \mathcal{E}_{0x}^{\;2} \cdot \boldsymbol{exp}\left[-\frac{2\omega\kappa_x}{c} z\right]. \tag{10.15}$$

According to the last result, if the extinction coefficient is positive ($\kappa_x(\omega, \underline{k}) > 0$), the beam intensity reduces as the beam progresses along the z orientation, with an absorption coefficient

$$\boxed{\alpha_x(\omega, \underline{k}) = \frac{2\omega\, \kappa_x(\omega, \underline{k})}{c}} \;. \tag{10.16}$$

It is possible to assume that as long as the material constituting atoms are in their ground state, the absorption coefficient $\alpha_x(\omega, \underline{k})$ is positive and, correspondingly, also $\kappa_x(\omega, \underline{k}) > 0$. The situation may overturn when the material contains excited electronic states. Under proper conditions, they may cause an increase in the beam intensity. We will not deal with that situation in the present chapter; the reader is referred to Section 2.5 for some more details.

We proceed now to present the specific connection between the refractive index parts and those of the dielectric function. The dielectric function is of the form $\varepsilon_x(\omega, k) = \varepsilon'_x(\omega, k) + i\varepsilon''_x(\omega, k)$, where $\varepsilon'_x(\omega, k)$ is the real part of the function and $\varepsilon''_x(\omega, k)$ is its imaginary part (we use a shorter subscript labeling of the tensors' components: ε_x instead of ε_{xx} , and similarly for all other parts).

According to Eq. (10.11), we have

$$\varepsilon'_x(\omega, k) + i\varepsilon''_x(\omega, k) = [n_x(\omega, k) + i\kappa_x(\omega, k)]^2;$$

$$\varepsilon'_x + i\varepsilon''_x = n_x^{\;2} - \kappa_x^{\;2} + i \cdot 2n_x\kappa_x \,.$$

Thus,

$$\boxed{\begin{aligned} \varepsilon'_x &= n_x^{\;2} - \kappa_x^{\;2} \\ \varepsilon''_x &= 2n_x\kappa_x \end{aligned}} \;. \tag{10.17}$$

And the inverse relations

$$\boxed{\begin{aligned} n_x^{\;2} &= \frac{1}{2}\sqrt{\varepsilon'^{\,2}_x + \varepsilon''^{\,2}_x} + \frac{1}{2}\varepsilon'_x \\ \kappa_x^{\;2} &= \frac{1}{2}\sqrt{\varepsilon'^{\,2}_x + \varepsilon''^{\,2}_x} - \frac{1}{2}\varepsilon'_x \end{aligned}} \;. \tag{10.18}$$

The existence of absorption coefficient in a medium may be assigned further physical meaning. Let us imagine the medium as a collection of identical electronic oscillators, where each belongs to a separate atomic basis of the crystal. We can further imagine that the oscillators are immersed in a uniform medium of constant real refractive index and dielectric constant (obviously satisfying $n^2(\infty) = \varepsilon(\infty)$ – see Section 9.5.4.2 and the text related to Figure 9.12). We assume that excitation of an individual oscillator by the electromagnetic field has no effect on the behavior of the other oscillators; a similar assumption is applied in Chapter 1 to atoms or molecules in a low-pressured gas (see Section 1.3). Such a model is known as the **independent-oscillators model**. We may correspondingly express the effective oscillator strength of each separate oscillating dipole f_{eff} using Eq. (2.59). Setting Eq. (10.16) in Eq. (2.59), and using the second relation in Eq. (10.17), we get

$$f_{eff}(k) \cong \frac{3m}{2\pi^2 e^2 \varepsilon(\infty)} \cdot \frac{1}{N} \int_0^\infty \omega\, \varepsilon''_x(\omega, k)\, d\omega \ , \qquad (10.19)$$

where m is the free electron rest mass, and N is the volume density of the crystal unit cells.

Equating a polar atomic basis oscillator with an electronic one allows us to compare the light absorption–causing excitation of an optical lattice vibration to the electronic optical transitions of the same resonance frequencies. The reader may justifiably argue that a more appropriate model would be to imagine an optical oscillation of an atomic basis as belonging to a two-parts molecule: positive and negative. The electric charge e^* of each part would be of the order of an elementary charge e, and the imagined molecule would vibrate like a diatomic molecule (see Section 7.3.2). Eq. (10.19) may become appropriate for describing that model's properties, if e^* is set instead of e, and the reduced mass μ^* of the vibrating atomic basis is set instead of m. The reduced mass may be assessed from the vibrational species description, as detailed in Section 8.10.

10.2 Light Transition from Vacuum into a Dielectric Medium

We consider a light beam in the form of a linearly polarized plane-parallel wave propagating in free space (zone I) perpendicularly to the plane surface of a dielectric medium (zone II). The x polarization orientation coincides with a principal orientation of the medium dielectric tensor. The z-noted propagation direction is obviously perpendicular to the medium surface. The $z = 0$ points define the surface plane, which acts as an interface between the medium and free space. We assume that the beam transmitted into the medium preserves its z propagation orientation and x polarization. We assume that a part of the beam is reflected, and propagates in the $-z$ direction while maintaining its x polarization. Figure 10.2 describes schematically the orientation relations among the electric and magnetic fields, and the propagation directions of the incident and reflected beams.

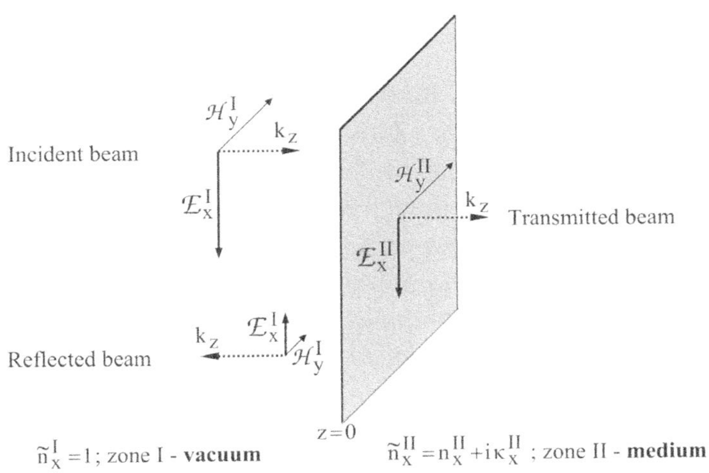

Figure 10.2 Relations among electric and magnetic field orientations, strengths, and propagation directions for a light beam propagating through vacuum, incident perpendicularly on the surface of a dielectric medium. The corresponding parameters are detailed for the incident, reflected, and transmitted beams.

In the free-space (vacuum) region, we accordingly have

$$\mathcal{E}_x(t, z<0) = \mathcal{E}^I_{0x} e^{-i(\omega t - kz)} + g\left(\tilde{n}^{II}_x\right) \mathcal{E}^I_{0x} e^{-i(\omega t + kz)} \; , \tag{10.20}$$

where $\mathcal{E}^I_{0x} e^{-i(\omega t - kz)}$ is the beam form that would exist in the free space *without the presence of the dielectric medium*. $g\left(\tilde{n}^{II}_x\right)$ is the reflectance factor of the electric fields; we assume that its value depends on the dielectric medium's (II) properties, as expressed by the different-than-a-unity refractive index, namely $g\left(\tilde{n}^{II}_x\right) \neq 1$. Recall that $g\left(\tilde{n}^{I}_x\right) = 1$ in vacuum. Using Eq. (10.5-iii), we get

$$\begin{aligned} \mathcal{H}_y(t, z<0) &= \frac{c}{\omega} k \mathcal{E}^I_{0x} e^{-i(\omega t - kz)} - g\left(\tilde{n}^{II}_x\right) \frac{c}{\omega} k \mathcal{E}^I_{0x} e^{-i(\omega t + kz)} \\ &= \tilde{n}^I_x \mathcal{E}^I_{0x} e^{-i(\omega t - kz)} - g\left(\tilde{n}^{II}_x\right) \tilde{n}^I_x \mathcal{E}^I_{0x} e^{-i(\omega t + kz)} \end{aligned} \; ;$$

$$\mathcal{H}_y(t, z<0) = \mathcal{E}^I_{0x} e^{-i(\omega t - kz)} - g\left(\tilde{n}^{II}_x\right) \tilde{n}^I_x \mathcal{E}^I_{0x} e^{-i(\omega t + kz)} \, . \tag{10.21}$$

As for the dielectric medium belonging to $z > 0$, it may be simply written as

$$\mathcal{E}_x(t, z>0) = \mathcal{E}^{II}_{0x} e^{i(\omega t - kz)}; \quad \mathcal{H}_y(t, z>0) = \tilde{n}^{II}_x \mathcal{E}^{II}_{0x} e^{i(\omega t - kz)} \, . \tag{10.22}$$

Because in the present case, both electric and magnetic fields are parallel to the interface plane, both are identical at the two interface sides. Therefore, at the $z = 0$ point, we have

Comparing electric fields: $\mathcal{E}^{II}_{0x} = \mathcal{E}^I_{0x} + g\left(\tilde{n}^{II}_x\right)\mathcal{E}^I_{0x} = \left[1 + g\left(\tilde{n}^{II}_x\right)\right]\mathcal{E}^I_{0x}$.

Comparing magnetic fields: $\tilde{n}^{II}_x \mathcal{E}^{II}_{0x} = \mathcal{E}^I_{0x} - g\left(\tilde{n}^{II}_x\right)\mathcal{E}^I_{0x} = \left[1 - g\left(\tilde{n}^{II}_x\right)\right]\mathcal{E}^I_{0x}$.

Therefore, we readily get

$$\boxed{\tilde{n}^{II}_x = \frac{1 - g\left(\tilde{n}^{II}_x\right)}{1 + g\left(\tilde{n}^{II}_x\right)}; \quad g\left(\tilde{n}^{II}_x\right) = -\frac{\tilde{n}^{II}_x - 1}{\tilde{n}^{II}_x + 1}} \, . \tag{10.23}$$

Both equations relate only to the dielectric medium parameters (zone II). Thus, we may save the "II" superscript notation in all equations, and write $\tilde{n}_x = n_x + i\kappa_x$, instead of $\tilde{n}^{II}_x = n^{II}_x + i\kappa^{II}_x$.

It is important to note the existence of a phase difference between the electric fields of the incident and reflected beams. In the particular case where the refractive index is real, the electric field reflectance factor $g = -(n_x - 1)/(n_x + 1)$ is indeed also real, but negative. Thus, at the vacuum side of the $z = 0$ plane, the phase of the reflected beam electric field is *retarded* by a half cycle compared with the incident beam. The phase retardation *reduces* by a φ angle for a complex refractive index. In other words, in that case, the reflected beam phase exhibits a relative *advancement*. Quantitatively, the phase advancement is obtained by expressing the g factor as follows:

$$g = -|g|e^{i\varphi} = -|g|\, \mathbf{cos}\, \varphi - i\, |g|\, \mathbf{sin}\, \varphi = -\frac{\left(n_x{}^2 - 1\right) + \kappa_x{}^2}{\left(n_x + 1\right)^2 + \kappa_x{}^2} - i\frac{2\kappa_x}{\left(n_x + 1\right)^2 + \kappa_x{}^2} \, . \tag{10.24}$$

The beam phase advancement thus satisfies

$$\tan \varphi = \frac{2\kappa_x}{(n_x{}^2 - 1) + \kappa_x{}^2} \, . \tag{10.25}$$

The intensities of the incident beam I_{inc}, the reflected beam I_{ref}, and the transmitted beam I_{trans} (power per unit area) at $z = 0$ are given by the average over a full time period of the corresponding Poynting vectors $(c/4\pi)|\underline{\mathcal{E}} \times \underline{\mathcal{H}}|$ (see also Eq. [(10.15)]):

$$\left.\begin{aligned} I_{inc} &= \frac{c}{8\pi}\left(\mathcal{E}^{I}_{0x}\right)^2 \\ I_{ref} &= \frac{c}{8\pi}\left(\mathcal{E}^{I}_{0x}\right)^2 \cdot | \, g(\tilde{n}_x) \, |^2 \\ I_{trans} &= \frac{c}{8\pi} n_x \left(\mathcal{E}^{II}_{0x}\right)^2 = \frac{c}{8\pi} n_x \left(\mathcal{E}^{I}_{0x}\right)^2 | \, 1 + g(\tilde{n}_x) \, |^2 \end{aligned}\right\} . \tag{10.26}$$

The transmittance $T \equiv I_{trans}/I_{inc}$ and reflectance $R \equiv I_{ref}/I_{inc}$ factors are simply calculated:

$$\boxed{R(\omega, \underline{k}) = \left|\frac{\tilde{n}_x - 1}{\tilde{n}_x + 1}\right|^2 ; \; T(\omega, \underline{k}) = \frac{4n_x}{|\tilde{n}_x + 1|^2}; \; R(\omega, \underline{k}) + T(\omega, \underline{k}) = 1} \, . \tag{10.27}$$

The $R + T = 1$ relation expresses the energy conservation law. It should be noticed that for two of the beams, the incident and the reflected, no phase difference exists between their electric and magnetic fields; this is a result of their belonging to a medium of a real refractive index (the vacuum). The transmitted beam, on the other hand, exhibits a phase difference, as discussed in relation to Eq. (10.15). We further note the relation $R = |g|^2$ or, alternatively, $|g| = \sqrt{R}$.

Given the reflectance R and the advancement phase φ allows one to calculate the refractive index n_x and the extinction coefficient κ_x by solving the relations appearing in Eq. (10.24). The result is

$$\boxed{\begin{aligned} n_x &= \frac{1 - R}{1 + R - 2\sqrt{R}\,\mathbf{cos}\,\varphi} \\ \kappa_x &= \frac{2\,\sqrt{R}\;\mathbf{sin}\,\varphi}{1 + R - 2\sqrt{R}\,\mathbf{cos}\,\varphi} \end{aligned}} \, . \tag{10.28}$$

We consider now in more detail the expressions for the reflectance and transmittance factors. The transmittance is given by

$$T = \frac{4n_x}{n_x{}^2 + 2n_x + 1 + \kappa_x{}^2} \, . \tag{10.29}$$

Obviously, the transmittance T diminishes and the reflectance correspondingly grows toward 100% in two different particular cases: when $n_x \to 0$ and simultaneously κ_x remains finite, and when $\kappa_x \to \infty$ but simultaneously n_x remains finite. A turning point occurs for $n_x = \kappa_x$, at which also $\varepsilon' = 0$ (Eq. [(10.18)]). When κ_x grows further ($\kappa_x > n_x$) the transmission reduces, and reflection grows. In spectral ranges over which that situation persists continuously, a "reflection band" occurs. It is a region where the real part of the dielectric function is negative ($\varepsilon' < 0$). In Section 10.4, we present a model of a dielectric function, with continuous ranges exhibiting $\varepsilon' < 0$.

10.3 Kramers-Kronig Relations

In the discussion presented in this section, we have to allow the angular frequency variable to assume mathematical complex values and not just positive values, as is the common understanding of the concept it represents. That generalization bears a physical meaning. On specifically writing $\omega = \omega' + i\omega''$ where ω' and ω'' are real, and setting it, for example, in the expression for the electric field of a normal electromagnetic wave of Eq. (10.4), we get

$$\underline{\mathcal{E}} = \underline{\mathcal{E}}_0 e^{-i(\omega t - \underline{k}\cdot\underline{r})} = \left[\underline{\mathcal{E}}_0 e^{\omega'' t}\right] e^{-i(\omega' t - \underline{k}\cdot\underline{r})} . \tag{10.30}$$

The result describes a wave whose amplitude grows with time for $\omega'' > 0$, and reduces with time for $\omega'' < 0$. Similarly, if the case $\omega' > 0$ relates to a wave propagating in a certain direction in space, the case $\omega' < 0$ describes the wave propagating in the opposite direction.

Let us consider a physical, time-dependent **driving factor** $P(t)$, acting on a physical system, and let $Q(t)$ be the time-dependent response of the physical system. We limit the analysis to a case where the response depends linearly on the driving factor $P(t)$, and their relation is scalar (not tensorial). We define a quantity $\hat{\beta}$, to be called the **response kernel,** by the relation

$$Q(t) = \int_{-\infty}^{t} \hat{\beta}\left(t', t\right) P(t') dt' . \tag{10.31}$$

The $\hat{\beta}(t', t)$ function actually describes the response $Q(t)$ when the driving factor $P(t')$ is a δ function of time around t. The write-up of the integral from $-\infty$ to t expresses a fundamental principle in the physical science called **causality**. The principle states that a result of any driving physical cause cannot precede that cause in time. We have activated the causality principle by assuming that the response at time t results from the driving factor application *up to* that point in time; in other words, the integration is performed only for the $t' \leq t$ duration.

If the physical properties of the responding system remain constant over the said time duration, the response kernel $\hat{\beta}$ depends only on the time separation $\tau \equiv t - t'$, namely

$$Q(t) = \int_{-\infty}^{t} \hat{\beta}\left(t - t'\right) P\left(t'\right) dt' = \int_{0}^{\infty} \hat{\beta}(\tau) P(t - \tau) d\tau . \tag{10.32}$$

We perform now a Fourier analysis of the above equation:

$$Q(t) = \frac{1}{\sqrt{2\pi}} \int_{-\infty}^{+\infty} \mathbf{Q}(\omega) e^{-i\omega t} d\omega = \frac{1}{\sqrt{2\pi}} \int_{0}^{\infty} d\tau\, \hat{\beta}(\tau) \int_{-\infty}^{\infty} \mathbf{P}(\omega) e^{-i\omega (t-\tau)} d\omega .$$

Exchanging the integration order on the right side, we get

$$\int_{-\infty}^{+\infty} d\omega \left[e^{-i\omega t} \mathbf{Q}(\omega)\right] = \int_{-\infty}^{\infty} d\omega \left[e^{-i\omega t} \mathbf{P}(\omega) \int_{0}^{\infty} d\tau\, \hat{\beta}(\tau)\, e^{i\omega\tau}\right] .$$

Consequently, $\mathbf{Q}(\omega) = \mathbf{P}(\omega)\int_0^{\infty} d\tau\, \hat{\beta}(\tau)e^{i\omega\tau} = \mathbf{P}(\omega)\widetilde{\beta}(\omega)$, where

$$\boxed{\widetilde{\beta}(\omega) \equiv \int_0^{\infty} d\tau\, \hat{\beta}(\tau)\, e^{i\omega\tau}} \quad . \tag{10.33}$$

The newly defined $\widetilde{\beta}(\omega)$ function is addressed as the **complex response function**. By its definition, it satisfies $\widetilde{\beta}^*(\omega) = \widetilde{\beta}(-\omega)$. A further property obtains by the following physical consideration: the response kernel $\hat{\beta}(\tau)$ should diminish if the driving factor acted for a very short time in the far past on a system that maintains thermal equilibrium (or a quasi-thermal equilibrium). This is simply because any physical system eventually reaches thermal equilibrium with its surroundings. In a mathematical expression, $\hat{\beta}(\tau \to \infty) = 0$. Therefore, the complex response function defined by Eq. (10.33) would not exhibit poles on the upper part of the complex $\omega = \omega' + i\omega''$ plane (namely for $\omega'' > 0$) or on the real axis $\omega'' = 0$. In mathematical terms, the $\widetilde{\beta}(\omega)$ function is regular in the said region. Importantly, in that region, it satisfies $\lim\limits_{|\omega| \to 0} \hat{\beta}(\omega) = 0$, and the zero limit is reached exponentially.

Because $\widetilde{\beta}(\omega)$ is *regular*, the function $\widetilde{\beta}(\omega)/(\omega - \varpi)$ where $\varpi > 0$ is *analytic*, with a single, first-order pole at the $\omega = \varpi$ point on the real ω axis. Under such conditions, one may apply the **Cauchy equation** (see Section 10.5.2)

$$\int_{-\infty}^{+\infty} \frac{\widetilde{\beta}(\omega)}{\omega - \varpi}\, d\omega = i\pi\, \widetilde{\beta}(\varpi)\ . \tag{10.34}$$

Expressing $\widetilde{\beta}$ in the form $\widetilde{\beta}(\omega) = \beta'(\omega) + i\beta''(\omega)$, where $\beta'(\omega)$ and $\beta''(\omega)$ are correspondingly the real and imaginary parts of the $\widetilde{\beta}(\omega)$ function, we get

$$\int_{-\infty}^{+\infty} \frac{\beta'(\omega)}{\omega - \varpi}\, d\omega + i\int_{-\infty}^{+\infty} \frac{\beta''(\omega)}{\omega - \varpi}\, d\omega = i\pi\left[\beta'(\varpi) + i\,\beta''(\varpi)\right] = -\pi\beta''(\varpi) + i\pi\,\beta'(\varpi)\,.$$

Comparing the real and imaginary parts of the two sides in the above equation, one gets♣

$$\left.\begin{aligned} &\text{(i)}\quad \beta''(\varpi) = -\frac{1}{\pi}\int_{-\infty}^{+\infty} \frac{\beta'(\omega)}{\omega - \varpi} d\omega \\ &\text{(ii)}\quad \beta'(\varpi) = \frac{1}{\pi}\int_{-\infty}^{+\infty} \frac{\beta''(\omega)}{\omega - \varpi} d\omega \end{aligned}\right\}. \tag{10.35}$$

Multiplying both numerator and denominator of the integration objects in Eq. (10.35-i) and Eq. (10.35-ii) by $(\omega + \varpi)$ yields

♣ If the real and imaginary parts of a complex function of a real variable exhibit the relations expressed in Eq. (10.35), they are called **Hilbert transforms** of each other.

$$\left.\begin{aligned}&\text{(i)}\ \beta''(\overline{\omega}) = -\frac{1}{\pi}\int_{-\infty}^{+\infty}\frac{\omega\beta'(\omega)}{\omega^2-\overline{\omega}^2}d\omega - \frac{\overline{\omega}}{\pi}\int_{-\infty}^{+\infty}\frac{\beta'(\omega)}{\omega^2-\overline{\omega}^2}d\omega\\&\text{(ii)}\ \beta'(\overline{\omega}) = \frac{1}{\pi}\int_{-\infty}^{+\infty}\frac{\omega\beta''(\omega)}{\omega^2-\overline{\omega}^2}d\omega + \frac{\overline{\omega}}{\pi}\int_{-\infty}^{+\infty}\frac{\beta''(\omega)}{\omega^2-\overline{\omega}^2}d\omega\end{aligned}\right\}. \tag{10.36}$$

Because $\widetilde{\beta}^*(\omega) = \widetilde{\beta}(-\omega)$, we immediately see that $\beta'(\omega) = \beta'(-\omega)$ and $\beta''(\omega) = -\beta''(-\omega)$. Breaking each integral in Eq. (10.36) into a sum of integrals, one between $-\infty$ and zero and a second one between zero and $+\infty$, and setting the relations among the different $\widetilde{\beta}$ components, we are left with

$$\boxed{\begin{aligned}&\text{(i)}\ \beta''(\overline{\omega}) = -\frac{2\overline{\omega}}{\pi}\int_{0}^{+\infty}\frac{\beta'(\omega)}{\omega^2-\overline{\omega}^2}d\omega\\&\text{(ii)}\ \beta'(\overline{\omega}) = \frac{2}{\pi}\int_{0}^{+\infty}\frac{\omega\beta''(\omega)}{\omega^2-\overline{\omega}^2}d\omega\end{aligned}}\ . \tag{10.37}$$

These relations between the real and imaginary parts of the complex response function are called **dispersion relations**, or more commonly, **Kramers-Kronig** relations.♣ Their physical significance is that the real and imaginary parts of the response function are not independent parameters of the medium. Given the dispersion function (ω dependence) of one, the other's dispersion function is determined unequivocally. This has a practical importance in the performing of physical experiments, because sometimes it is easier to measure one of them rather than the other. If the measurement is performed over a sufficiently broad spectral range, it is possible to calculate the other by the proper expression in Eq. (10.37). Physical effects for which the Kramers-Kronig relations are useful, are numerous, including electric circuits' impedances, anomalous X-ray fluorescence and scattering, and the particular issue of present interest – light reflection from the surface of a dielectric material.

Recall Eqs. (10.23) and (10.24), particularly the wave reflection factor $g(\omega) = \sqrt{R(\omega)}\,e^{i\varphi(\omega)}$. We consider now a new function

$$h(\omega) \equiv \boldsymbol{ln}\left[\frac{g(\omega)}{\sqrt{R(\overline{\omega})}}\right] = \boldsymbol{ln}\left[\frac{\sqrt{R(\omega)}}{\sqrt{R(\overline{\omega})}}\right] + i\,\varphi(\omega)\,. \tag{10.38}$$

Eq. (10.38) considers light incidence on a material surface, namely having positive frequencies $\omega > 0$. We generalize it artificially to negative frequencies $\omega < 0$ by defining $R(-\omega) = R(\omega)$, and $\varphi(-\omega) = -\varphi(\omega)$. We then use the Kramers-Kronig relations (Eq. [10.37-i]) to get

$$\boxed{\varphi(\overline{\omega}) = -\frac{\overline{\omega}}{\pi}\int_{0}^{\infty}\frac{\boldsymbol{ln}[R(\omega)/R(\overline{\omega})]}{\omega^2-\overline{\omega}^2}d\omega}\ . \tag{10.39}$$

♣ The Kramers-Kronig relation was presented in a paper published in 1927. For historic accuracy, the same relation between the real and imaginary parts of the complex refractive index was presented by Sellmeier already in 1871.

An analysis of optical reflectance and conclusions obtained thereof is called a **Kramers-Kronig analysis**. It usually requires the calculation of thousands of integrals and, therefore, the use of digital computing. In that respect, it is worthwhile noticing that the division of the $g(\omega)$ by $\sqrt{R(\varpi)}$ in the definition of the $h(\omega)$ function in Eq. (10.38) is not a necessity; the use of any constant value will suffice. That particular selection was done to ensure that the numerator of the integration object turns zero coincidently with the denominator zeros (at $\omega = \varpi$)! This allows for the use of L'Hôpital's theorem (separate differentiation of numerator and denominator in the ϖ limit). The object value at that point then converges to a finite value $\left.\frac{1}{2\varpi R(\varpi)} \cdot \frac{dR(\omega)}{d\omega}\right]_{\varpi}$. That mathematical "trick" assists the numerical integral calculation algorithms near $\omega = \varpi$.

> It is worthwhile emphasizing the order in which experiments, and experimental results analyses, need to be performed to acquire the material properties. Initially, the perpendicular optical reflectance $R(\omega)$ of a linearly polarized light should be measured over a continuous broad frequency range, where the light polarization is co-oriented with a principal axis of the dielectric tensor (say, the x axis). Then, use is made of Eq. (10.39) to calculate the phase advancement spectrum $\varphi(\omega)$. Then, Eq. (10.28) is used to calculate the refractive index $n_x(\omega)$ and the extinction coefficient $\kappa_x(\omega)$ spectrums. The extinction coefficient $\kappa_x(\omega)$ is further used to calculate the absorption coefficient spectrum $\alpha_x(\omega)$, using Eq. (10.16). The real and imaginary parts of the dielectric function, $\varepsilon'(\omega)$ and $\varepsilon''(\omega)$, respectively, are calculated using Eq. (10.17). Eventually, the effective oscillator strength of each absorption peak is calculated using Eq. (10.16) [for more on that matter, see also the following Eqs. (10.57) and (10.58)].

10.4 A Microscopic Model of the Dielectric Function

At the end of Section 10.1, we pointed out the possibility of imagining each polar optical oscillator (oscillating dipole) as an atomic basis comprising two stiff parts whose reduced mass is μ^* and the parts are oppositely charged by effective $\pm e^*$ charges. Accordingly, the deviation from equilibrium ξ of their inter-distance satisfies $\xi = q/\sqrt{\mu^*}$, where q is the generalized oscillation mode coordinate (see Eq. [9.32] and following text).

Consider now that the crystal is subject to the influence of an electromagnetic field, whose effective electric field inside the crystal is given by $\mathcal{E}(\omega, t) = \mathcal{E}_0 e^{-i\omega t}$, and the crystal exhibits only a single polar vibrational mode. Further assume that the electric field is co-oriented with the dipole. The oscillating field will generate a crystal wave, whose propagation direction coincides with that of the electromagnetic field. Because frequencies of crystal vibrations that are in resonance with electromagnetic radiation require the latter to belong to the infrared, the electromagnetic wavelength λ is very long compared with a crystal lattice a period. Therefore, the light wave vector $k = 2\pi/\lambda$ is very small compared with the dimensions of the first Brillouin-zone π/a.

We thus proceed to calculate the crystal response to the external driving electric field in the crystal $k \to 0$ wave-vector limit. For that purpose, we correct the equation of motion (Eq. [9.8]) by ignoring the $\sum_{\underline{s}} C_{\underline{s}}\left(q_{\underline{x}} - q_{\underline{x}+\underline{s}}\right)$ term; it is based on the assumption that for a $k \approx 0$ limit, generalized oscillation coordinates of near-atomic bases are practically identical (namely $q_{\underline{x}} - q_{\underline{x}+\underline{s}}$ is negligible for the relevant limited range of $\underline{s}$ bases). We, however, include the acceleration caused by the driving force that the oscillating electric field exerts on the two parts of the atomic basis. We further assume, that the oscillations of each atomic basis are damped with an effective constant damping factor γ . The equation of motion would be

$$\ddot{\xi}_{\underline{x}} + \gamma\dot{\xi}_{\underline{x}} + {\omega_0}^2\xi_{\underline{x}} = \frac{e^*}{\mu^*}\mathcal{E}_0\, e^{-i\omega t}. \tag{10.40}$$

Recall that the $2/\gamma$ quantity describes the time duration for which the oscillation amplitude reduces by an e^{-1} factor in the absence of a driving force (e here being the basis of the natural logarithm). We will not specifically address the damping causing mechanisms; most relate to inharmonicity of the binding potential. The steady-state solution of the equation is (see Eq. [1.32])

$$\xi(\omega, t) = \frac{(e^*/\mu^*)}{{\omega_0}^2 - \omega^2 - i\,\gamma\omega}\mathcal{E}_0 e^{-i\omega t}. \tag{10.41}$$

In writing Eq. (10.41), we have omitted the atomic basis position label $\underline{x}$ of ξ ; there is no loss of clarity, and we shall adhere to that practice of simplification where possible.

The presence of forced oscillation entails the oscillation of polarization density given by

$$P(\omega, t) = \frac{e^*\xi(\omega, t)}{V_C} = \frac{e^{*2}}{\mu^* V_C}\cdot\frac{1}{{\omega_0}^2 - \omega^2 - i\gamma\omega}\mathcal{E}_0 e^{-i\omega t}, \tag{10.42}$$

where V_C is the primitive unit cell volume.

We consider a specific case in which the orientations of the induced dipole $\underline{Q} = e^*\underline{\xi}$ and the medium polarization density $\underline{P}_B$ coincide with that of the driving electric field (namely, it is a principal orientation of the dielectric tensor). The electric field $\underline{\mathcal{E}}$ and displacement vector $\underline{\mathcal{D}}$ orientations also coincide. We may then write

$$\begin{aligned}\mathcal{D}(\omega, t) &\equiv \varepsilon(\omega)\mathcal{E}(\omega, t) = \varepsilon(\infty)\mathcal{E}(\omega, t) + 4\pi P(\omega, t)\\ &= \left[\varepsilon(\infty) + 4\pi\frac{e^{*2}}{\mu^* V_C}\cdot\frac{1}{{\omega_0}^2 - \omega^2 - i\gamma\omega}\right]\mathcal{E}_0 e^{-i\omega t},\end{aligned} \tag{10.43}$$

where $\varepsilon(\omega)$ is the **dielectric function**. The $\varepsilon(\infty)$ quantity is the dielectric constant of the matter surrounding the oscillating dipole, that is induced by much higher electronic resonance frequencies of the matter (see the detailed discussion in Section 9.5.4.2, especially in relation to Figures 9.11 and 9.12). Then, we get

$$\varepsilon(\omega) = \varepsilon(\infty) + 4\pi\frac{e^{*2}}{\mu^* V_C}\cdot\frac{1}{{\omega_0}^2 - \omega^2 - i\gamma\omega}. \tag{10.44}$$

For brevity, we denote $S \equiv 4\pi\cdot e^{*2}/\mu^* V_C$. We may then write

$$\varepsilon(\omega) = \varepsilon'(\omega) + i\varepsilon''(\omega) = \varepsilon(\infty) + \frac{S({\omega_0}^2 - \omega^2)}{({\omega_0}^2 - \omega^2)^2 + \gamma^2\omega^2} + i\frac{S\gamma\omega}{({\omega_0}^2 - \omega^2)^2 + \gamma^2\omega^2},$$

where $\varepsilon'(\omega)$ and $\varepsilon''(\omega)$ are, respectively, the real and imaginary parts of the dielectric function:

$$\left.\begin{aligned}\varepsilon'(\omega) &\equiv \varepsilon(\infty) + \frac{S({\omega_0}^2 - \omega^2)}{({\omega_0}^2 - \omega^2)^2 + \gamma^2\omega^2}\\ \varepsilon''(\omega) &\equiv \frac{S\gamma\omega}{({\omega_0}^2 - \omega^2)^2 + \gamma^2\omega^2}\end{aligned}\right\}. \tag{10.45}$$

We analyze now the frequency dependence of the dielectric function in the specific case where the damping factor vanishes, namely $\gamma = 0$. The imaginary part ε'' vanishes, and the real part ε' assumes the simple form $\varepsilon'(\omega) = \varepsilon(\infty) + S/(\omega_0^2 - \omega^2)$. That function exhibits a pole at the $\omega = \omega_0$ point. For $\omega < \omega_0$, the $\varepsilon'(\omega)$ function is positive and, in the $\omega \to 0$ limit, it converges to the value $\varepsilon'(0) = \varepsilon(\infty) + S/\omega_0^2$. For $\omega > \omega_0$, the $\varepsilon'(\omega)$ function is negative, with a huge absolute value that reduces upon frequency increase. Afterwards, the $\varepsilon'(\omega)$ function attains a zero value at the $\omega_1^2 = \omega_0^2 + S/\varepsilon(\infty)$ point. Closer to higher frequencies, the $\varepsilon'(\omega)$ function becomes positive again, finally converging asymptotically to the $\varepsilon(\infty)$ value. A continuous frequency range thus exists, where $\varepsilon'(\omega)$ is negative:

$$\omega_0 < \omega < \omega_1 \equiv \sqrt{\omega_0^2 + S/\varepsilon(\infty)}. \tag{10.46}$$

Experimental measurement of the ω_0 and ω_1 frequencies is rather simple. An electromagnetic wave cannot propagate in a medium where $\varepsilon'(\omega)$ is negative. Therefore, an electromagnetic wave incident on the medium surface will be fully reflected in the $\omega_0 < \omega < \omega_1$ range. Measuring the optical reflectance allows for a quite accurate determination of the reflection band edges, namely identification of the ω_0 and ω_1 frequencies. We shall discuss that issue later in more detail; see also Figure 10.3.

Solving for the two relations (i) $\omega_1^2 = \omega_0^2 + S/\varepsilon(\infty)$ and (ii) $\varepsilon'(0) - \varepsilon(\infty) = S/\omega_0^2$ yields

$$\left(\frac{\omega_1}{\omega_0}\right)^2 = \frac{\varepsilon(0)}{\varepsilon(\infty)}, \tag{10.47}$$

which is exactly the Lyddane-Sachs-Teller relation between the squares of the longitudinal and transverse ω^L and ω^T frequencies, respectively, of the same optical mode oscillation in the $k \to 0$ limit (Eq. [9.38]). Experimental measurements indeed show a very close resemblance between the ω_0 and ω_1 frequencies measured by optical reflection, and the ω^T and ω^L frequencies measured by Raman or neutron scattering. Still, on fundamental principle grounds, it is quite difficult to equate between the ω_0 and ω_1 pair and the ω^T and ω^L pair. The LO-TO splitting relates to dependence of the mode frequency on the crystal wave propagation orientation (see Section 9.5.4.2). Departure between the cases is theoretically revealed by considering the modes squared frequencies. By Eq. (10.46), one gets

$$\omega_1^2 - \omega_0^2 = \frac{S}{\varepsilon(\infty)}, \tag{10.48}$$

whereas by Eqs. (9.34) and (9.37), one gets

$$\left(\omega^L\right)^2 - \left(\omega^T\right)^2 = \frac{S}{\varepsilon(0)/\varepsilon(\infty)}. \tag{10.49}$$

Despite the above difference, and especially because of a multitude of experimental similarities, very often in the scientific literature, the ω^T symbol is used instead of ω_0, and ω^L instead of ω_1. This practice is partially justified by the argument that, in any case, the effective electric charge e^* and the reduced mass μ^* used to define the S quantity are adjustable parameters, that may be used differently in the application of different models.

We turn now to analyzing the case where the damping constant γ does not vanish. In the low frequencies limit, the $\varepsilon'(\omega)$ function converges to the $\varepsilon'(0) = \varepsilon(\infty) + S/\omega_0^2$, exactly as in the damping constant absence. The pole at ω_0 disappears; however, the $\varepsilon'(\omega)$ function attains a zero value at a point ω_n^T, which is very close to ω_0. Qualitatively, it is quite plausible because in the $\gamma \to 0$ limit, the function diverges to $+\infty$ at the left of ω_0, and to $-\infty$ at the right of ω_0. The connection between the two limits may be visualized as a straight vertical line crossing the frequency axis at ω_0. Near ω_0, the $\varepsilon'(\omega)$ function is negative for $\omega > \omega_0$ up to the point noted ω_n^L (slightly different than ω_1) where it becomes zero. It becomes positive for higher frequencies, eventually approaching $\varepsilon(\infty)$ asymptotically at very high frequencies.

The ω_n^T and ω_n^L are solutions of the $\varepsilon'(\omega) = 0$ equation. To simplify expressions we choose ω^2 as the unknown variable. We write $y = \omega^2$, recall that $\omega_1^2 - \omega_0^2 = S/\varepsilon(\infty)$, and receive the following equation for $\varepsilon'(\omega) = 0$:

$$\left(\omega_0^2 - y\right)^2 + \left[\omega_1^2 - \omega_0^2\right]\left(\omega_0^2 - y\right) + \gamma^2 y = 0. \tag{10.50}$$

We solve the above quadratic equation for $(\omega_0^2 - y)$ in the limit of sufficiently small γ. Thus, for finding ω_n^T, we set ω_0^2 in the last addend instead of y and get

$$\left(\omega_0^2 - y\right)^2 + \left[\omega_1^2 - \omega_0^2\right]\left(\omega_0^2 - y\right) + \gamma^2 \omega_0^2 = 0,$$

whose approximate solution for $(\omega_0^2 - y)$ would be $\left(\omega_0^2 - y\right) \cong -\dfrac{\gamma^2\omega_0^2}{\omega_1^2 - \omega_0^2}$ when $\gamma << \omega_1 - \omega_0$.

Recalling that $y = \left(\omega_n^T\right)^2$, we get

$$\left(\omega_n^T\right)^2 \cong \omega_0^2 + \frac{\omega_0^2\gamma^2}{\omega_1^2 - \omega_0^2} \cong \left[\omega_0 + \frac{1}{4}\frac{\gamma^2}{\omega_1 - \omega_0}\right]^2. \tag{10.51}$$

For finding ω_n^L we set ω_1^2 in the last addend of Eq. (10.50) instead of y, and get

$$\left(\omega_0^2 - y\right)^2 + \left[\omega_1^2 - \omega_0^2\right]\left(\omega_0^2 - y\right) + \gamma^2 \omega_1^2 = 0,$$

whose approximate solution for $(\omega_0^2 - y)$ when $\gamma << \omega_1 - \omega_0$ would be $\left(\omega_0^2 - y\right) \cong$

$-\left[\omega_1^2 - \omega_0^2\right] + \dfrac{\gamma^2\omega_1^2}{\omega_1^2 - \omega_0^2}$.

Recalling that now $y = \left(\omega_n^T\right)^2$, we get

$$\left(\omega_n^L\right)^2 \cong \omega_1^2 - \frac{\omega_1^2\gamma^2}{\omega_1^2 - \omega_0^2} \cong \left[\omega_1 - \frac{1}{4}\frac{\gamma^2}{\omega_1 - \omega_0}\right]^2. \tag{10.52}$$

Thus, the low-frequency zeroing of $\varepsilon'(\omega)$ is slightly higher than ω_0, the high-frequency zeroing of $\varepsilon'(\omega)$ is slightly lower than ω_1, and, for both cases, $\left|\omega_n^T - \omega_0\right| << \gamma$ and $\left|\omega_n^L - \omega_1\right| << \gamma$. That situation is very common in single crystals; in such cases, there is no practical distinction between ω_n^L and ω_1 and between ω_n^T and ω_0. In other words, for small or moderate damping constants, the reflection band of an electromagnetic radiation incident on crystal surfaces occurs practically between ω_n^L and ω_n^T.

When damping exists, the imaginary part $\varepsilon''(\omega)$ of the dielectric function never attains a zero value (Eq. [(10.45)]). Its frequency dependence exhibits a peak at the ω_0 point, whose height is $\varepsilon''(\omega_0) = S/\gamma\omega_0$. In the range where $\gamma << \omega_0$, the curve is symmetric about the peak, and its full width at half maximum, to an excellent approximation, is γ. Figure 10.3a shows an example of a dielectric function spectrum calculated using Eq. (10.45) for a moderate damping. The frequency axis scale is provided in reciprocal wavelength units (cm^{-1} in a cgs units system). It serves to demonstrate the common units used in the scientific literature in studies of crystal vibrational modes spectroscopy, and to indicate their order of magnitude. Recall that the actual angular frequency ω in rad/s units is obtained by multiplying the wave number $1/\lambda$ by $2\pi c$.

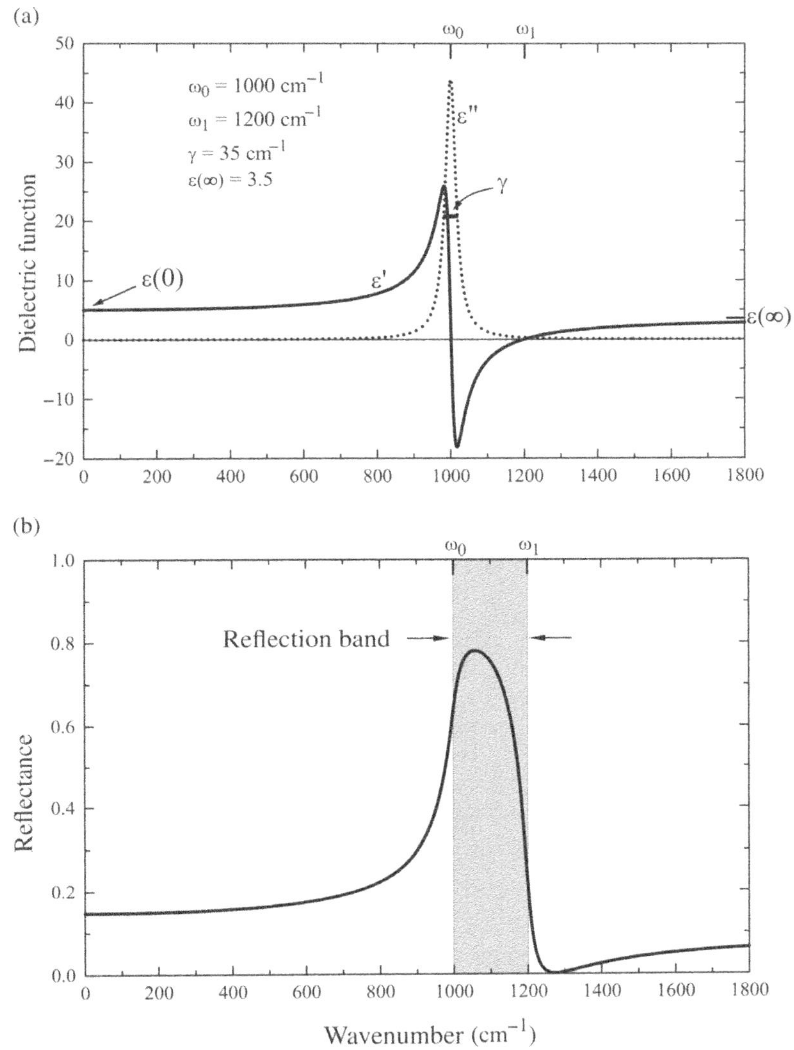

Figure 10.3 (a) The real ε' and imaginary ε'' parts of the dielectric function as functions of the incident light wavenumber per Eq. (10.45). Used parameters are inset on the left side of the figure frame. (b) The perpendicular optical reflection R spectrum from the crystal surface per Eqs. (10.45), (10.18), and (10.27).

Spectral measurements of the real and imaginary parts of the dielectric function provide a quite full characterization of the crystal vibrations in the $k \to 0$ limit, as well as the dielectric constants $\varepsilon(\infty)$ and $\varepsilon(0)$. Section 10.3 clearly indicates how these quantities may be obtained using spectral measurements of light reflection from a crystal surface. Figure 10.3b describes a perpendicular reflection spectrum $R(\omega)$ from the surface of the same sample, calculated using the dielectric function expressed in Eqs. (10.18) and (10.27). That spectrum exhibits a prominent reflection band, of which the crystal optical frequencies may be characterized to a first-order approximation. Observing the band from the high-frequency side, the reflection reduces to almost zero on approaching the longitudinal frequency ω_1 . The minimum position is approximately at $\omega_1 + 2\gamma$. Then, the reflection grows to form a band, whose width is approximately $\omega_1 - \omega_0$. The peak reflection occurs at approximately $\omega_0 + 2\gamma$. Closer to the lower frequencies, the reflection reduces, approaching asymptotically a constant positive value.

When the crystal contains several oscillators, they all respond to the incident electromagnetic radiation. The theoretical analysis is fundamentally based on an independent oscillators model, which means that the different oscillations are uncoupled. In such a case, the total polarization density would be a sum of elements like that of Eq. (10.42),

$$P(\omega,t)=\sum_j P_j(\omega,t)=\sum_j \frac{e_j{}^{*}\xi_j(\omega,t)}{V_C}=\left[\sum_j \frac{e_j{}^{*2}}{\mu^{*}{}_j V_C}\cdot\frac{1}{\omega_{0j}{}^2-\omega^2-i\gamma_j\omega}\right]\mathcal{E}_0 e^{-i\omega t}, \tag{10.53}$$

where j is an ordinal number of the oscillation mode. Usually, the elements are arranged in ascending order of their resonance frequencies. The dielectric function would then be (Eq. [(10.44)])

$$\varepsilon(\omega)=\varepsilon(\infty)+\sum_j \frac{4\pi e_j{}^{*2}}{\mu^{*}{}_j\cdot V_C}\cdot\frac{1}{\omega_{0j}{}^2-\omega^2-i\gamma_j\omega}. \tag{10.54}$$

The real ε' and imaginary ε'' parts of the dielectric function would be

$$\left.\begin{aligned}\varepsilon'(\omega)&\equiv\varepsilon(\infty)+\sum_j \frac{S_j\left(\omega_{0j}{}^2-\omega^2\right)}{\left(\omega_{0j}{}^2-\omega^2\right)^2+\gamma_j{}^2\omega^2}\\ \varepsilon''(\omega)&\equiv\sum_j \frac{S_j\gamma_j\omega}{\left(\omega_{0j}{}^2-\omega^2\right)^2+\gamma_j{}^2\omega^2}\end{aligned}\right\}, \tag{10.55}$$

where $S_j\equiv 4\pi e_j{}^{*2}/\mu_j{}^{*}\cdot V_C\equiv\ \varepsilon_j(\omega_{1j}{}^2-\omega_{0j}{}^2)$. The right-hand side of the latter relation is a definition of a "local" $\varepsilon(\infty)$ dielectric constant belonging to the j-numbered oscillation mode.

At the low-frequency limit, the $\varepsilon'(\omega)$ function converges to $\varepsilon(0)=\varepsilon(\infty)+\sum_j S_j/\omega_{0j}{}^2$, and at the high-frequency limit, it converges asymptotically to $\varepsilon(\infty)$. An important and interesting effect of the above calculations is that when the resonance frequencies are grossly separated, ε' assumes an almost constant value at frequencies between any two resonances, numbered ν and $\nu+1$, that are also sufficiently distant from both. That value would be $\varepsilon(\omega_{0,\nu}\ll\omega\ll\omega_{0,\nu+1})\ =\varepsilon(\infty)+\sum_{j\geq\nu+1} S_j/\omega_{0j}{}^2$. This means that the contribution of all oscillators whose ordinal number is larger than ν, namely $\varepsilon(\infty)+\sum_{j\geq\nu+1} S_j/\omega_{0j}{}^2$, play the role of a "quasi $\varepsilon(0)$" for the $\nu+1$ oscillator, and of a "quasi $\varepsilon(\infty)$" for the ν oscillator. This feature is approximately demonstrated in Figure 10.4a, which describes the spectral behavior of the real part $\varepsilon'(\omega)$ of the dielectric function. As in the case of the lone oscillator, the frequencies where $\varepsilon'(\omega)$ goes to zero describe well the ω_{0j} and ω_{1j} (namely, the ω_j^T and ω_j^L) of the j oscillator. In such cases, the Lyddane-Sachs-Teller relation (Eq. [9.38]) could be generalized to read

$$\frac{\varepsilon(0)}{\varepsilon(\infty)}\cong\prod_j\left(\frac{\omega_{1j}}{\omega_{0j}}\right)^2=\prod_j\left(\frac{\omega_j^L}{\omega_j^T}\right)^2. \tag{10.56}$$

It should be noted that the previous considerations are valid, and Eq. (10.56) reasonably accurate, *only* when the oscillators' frequencies are well separated; in other words, it is only valid when $(\omega_{0,j+1}-\omega_{0,j})$ is considerably large compared with $(\omega_{1j}-\omega_{0j})$ for all j. If that condition is not met, Eq. (10.56) is not at all a fair approximation, and the zero-positions of $\varepsilon'(\omega)$ fail to provide sufficiently accurate estimates of ω_{1j} and ω_{0j}.

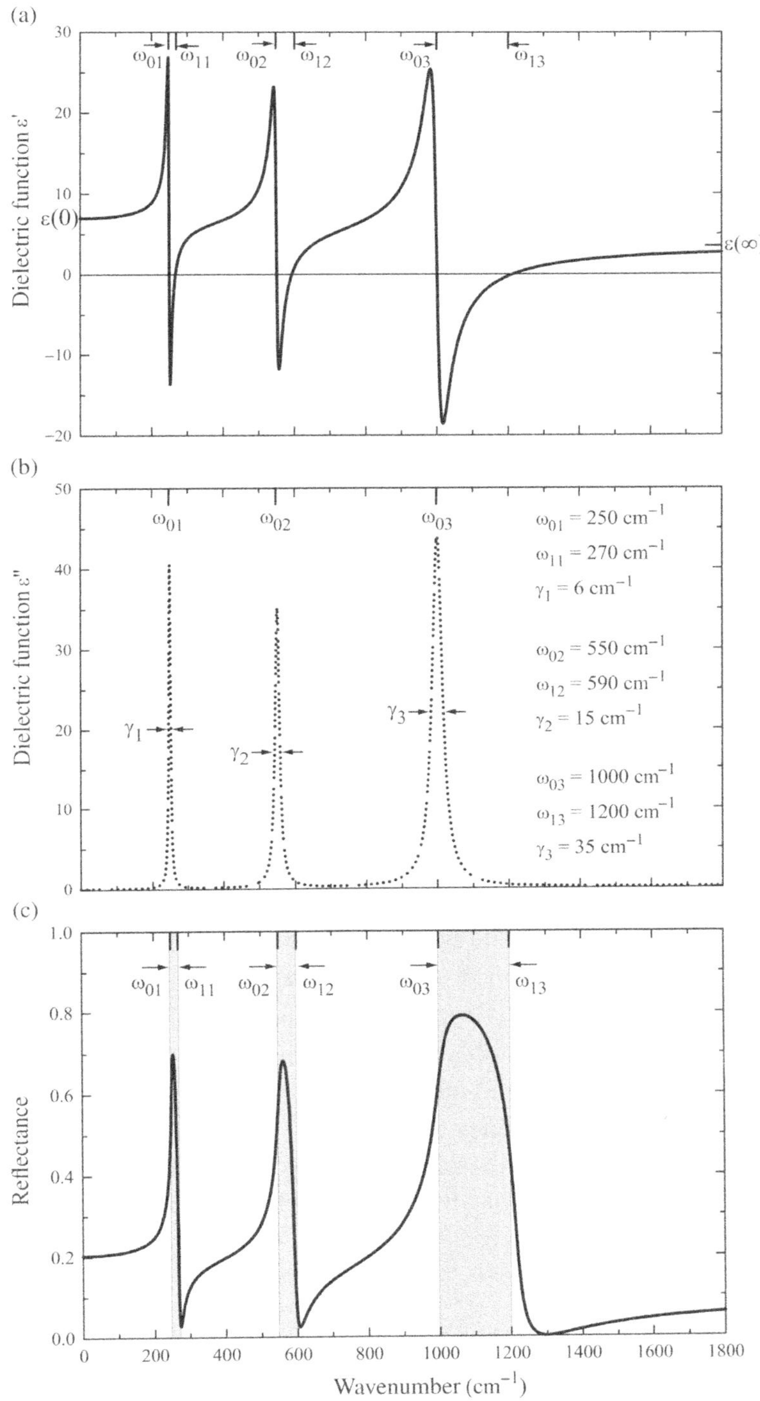

Figure 10.4 (a) Real part ε' and (b) imaginary part ε'' of the dielectric function as functions of the incident light wave number per Eq. (10.55), when the material exhibits three active optical oscillators. Used parameters are inset on the right side of the (b) figure frame. (c) The perpendicular optical reflection R spectrum from the crystal surface per Eqs. (10.45), (10.18), and (10.27).

Figures 10.4a and 10.4b present the dielectric function calculated per Eq. (10.55) for an intermediate case, where the oscillators' frequencies are separated, but not enough to exclude all inter-influences. The set parameters in Figure 10.4 determine the highest frequency mode, whose number 3 is chosen to be identical to those of the lone oscillator of Figure 10.3. Figure 10.4a shows the spectrum of the real part of the $\varepsilon'(\omega)$ dielectric function. The addition of the two new oscillators results in an increase in the low-frequency dielectric constant $\varepsilon(0)$ compared with its value in the case of a single oscillator: $\varepsilon(0) = 6.97$ in Figure 10.4a compared with $\varepsilon(0) = 5.04$ in Figure 10.3a. It is an example of a broad general effect, where the dielectric constant at the low-frequency limit is a result of the presence of all of the high-frequency polar oscillators. An additional effect is the influence of the low-frequency oscillators on the shape of the $\varepsilon'(\omega)$ spectrum in the high-frequency range. Their presence causes reduction of $\varepsilon'(\omega)$ in the high-frequency range. For example, the high-frequency zero-point of the third oscillator is larger than ω_{13}. In the case shown in Figure 10.4a, the zero value attainment actually occurs at $\omega_3^L = 1215\ \text{cm}^{-1}$, which is higher than the $\omega_{13} = 1200\ \text{cm}^{-1}$ parameter belonging to the no. 3 oscillator (Figure 10.4b). The general effect is that the presence of several oscillators with resonance frequencies close to that of a separate oscillator affects its longitudinal vibrational frequency.

The $\varepsilon''(\omega)$ function presented in Figure 10.4b displays a sum of separate peaks, where the peak position of the j- numbered oscillation is centered at ω_{0j}, its width is γ_j, and its height is $S_j/\gamma_j\omega_{0j}$. Inter-influence among various oscillation modes of close ω_{0j} frequencies is barely noticed. Thus, measurements of $\varepsilon''(\omega)$ provide quite accurate estimates of the S, γ, and ω_0 of each oscillator. To obtain an estimate of ω_1, a display of the function is needed; however, the value obtained is deviated upwards for oscillators belonging to the higher resonance frequencies. That difficulty is overcome by considering the said estimates as zero-order approximations, to be corrected until a best fit is obtained between the calculated model and the experimental results.

Figure 10.4c presents the $R(\omega)$ perpendicular reflectance spectrum. It is calculated using the dielectric functions, per Eqs. (10.18) and (10.27). The spectrum reveals three reflection bands, each situated approximately between ω_{0j} and ω_{1j}. Like in the case described in Figure 10.3, when viewing each band from the high-frequency aspect, a steep reduction in reflectance is seen on approaching the longitudinal frequency ω_{1j}, and the minimum position is approximately at $\omega_{1j} + 2\gamma_j$. Then, the reflectance increases, exhibiting the reflectance band. The band reflection peak is situated at approximately $\omega_{0j} + 2\gamma_j$. Closer to the low-frequency range, the reflectance gradually lessens until it sharpens in the vicinity of an adjacent reflection band belonging to a lower frequency.

We consider now the issue of the effective oscillator strength given by Eq. (10.19). Assume we have an experimental spectrum of the $\varepsilon''(\omega)$ imaginary part of the dielectric function (obtained, for example, by a Kramers-Kronig analysis of optical reflectance over a broad spectrum). The effective oscillator strength related to a spectral peak at ω_0 (see Figure 10.3a) is given by

$$\boxed{f_{eff} \cong \frac{3m}{2\pi^2 e^2 N} \cdot \frac{\omega_0\, \gamma\, \varepsilon''(\omega_0)}{\varepsilon(\infty)}} \,. \tag{10.57}$$

Alternatively, after observing an experimental spectrum of the $\varepsilon'(\omega)$ real part of the dielectric function, and provided that the reflection bands are well separated, the $\varepsilon'(\omega)$ function's zero-crossing points ω_1 and ω_0 may be obtained. According to Eq. (10.48), the effective oscillator strength may be expressed as

$$\boxed{f_{eff} \cong \frac{3m}{2\pi^2 e^2 N} \cdot \left(\omega_1^{\,2} - \omega_0^{\,2}\right)} \,. \tag{10.58}$$

10.5 A Reminder: Gradient, Divergence, Rotor, and the Cauchy Equation

10.5.1 Gradient, Divergence, and Rotor

The *gradient* ***grad*** of a scalar function $\varphi(x, y, z)$, where (x, y, z) are Cartesian coordinates of a point in space, is defined as

$$\boldsymbol{grad}\varphi(x,y,z) \equiv \underline{\nabla}\,\varphi(x,y,z) \equiv \left(\frac{\partial\varphi}{\partial x}, \frac{\partial\varphi}{\partial y}, \frac{\partial\varphi}{\partial z}\right). \tag{10.59}$$

The gradient operator generates a vector in space, where each of its $\partial\varphi/\partial x$, $\partial\varphi/\partial y$, and $\partial\varphi/\partial z$ components is a scalar function of (x, y, z). When the gradient operator is written in the $\underline{\nabla}$ short form, it is called a "**del**". A different write-up of the gradient (del) operator is $\underline{\nabla} \equiv \boldsymbol{i}\frac{\partial}{\partial x} + \boldsymbol{j}\frac{\partial}{\partial y} + \boldsymbol{k}\frac{\partial}{\partial z}$, where $\boldsymbol{i}, \boldsymbol{j}$, and $\boldsymbol{k}$ are unit vectors along the X, Y, and Z axes, respectively.

The *divergence* ***div*** of an $\underline{A}(x,y,z) \equiv \left(A_x(x,y,z),\ A_y(x,y,z),\ A_z(x,y,z)\right)$ vector is defined as

$$\boldsymbol{div}\underline{A}(x,y,z) \equiv \frac{\partial A_x}{\partial x} + \frac{\partial A_y}{\partial y} + \frac{\partial A_z}{\partial z}. \tag{10.60}$$

The divergence operator generates a scalar function of the (x, y, z) Cartesian coordinates. It may also be written as a scalar product: $\boldsymbol{div}\ \underline{A} = \underline{\nabla}\cdot\underline{A}$.

The *rotor* ***rot*** of an $\underline{A}(x,y,z)$ vector is defined as

$$\boldsymbol{rot}\,\underline{A}(x,y,z) \equiv \left(\frac{\partial A_z}{\partial y} - \frac{\partial A_y}{\partial z}, \frac{\partial A_x}{\partial z} - \frac{\partial A_z}{\partial x}, \frac{\partial A_y}{\partial x} - \frac{\partial A_x}{\partial y}\right). \tag{10.61}$$

The rotor operator generates a vector in space, where each of its components is a scalar function of (x, y, z). Quite often, the alternative ***curl*** symbol is used. It may also be written as a vector product: $\boldsymbol{rot}\,\underline{A} = \underline{\nabla} \times \underline{A}$.

The following are several theorems concerning the gradient, divergence, and rotor operators and their interrelations:

i) At any point on an equivalued surface $\varphi(x, y, z) = K$ of a constant K, the gradient vector is perpendicular to the surface at that point, and oriented toward the increased K valued surface.

ii) The derivative of the $\varphi(x, y, z)$ function in a general $\underline{s}$ direction is given by the scalar product

$$\frac{\partial\varphi(x,y,z)}{\partial s} = (\boldsymbol{grad}\varphi\cdot\underline{n}),$$

where $\underline{n} \equiv \underline{s}/s$ is a unit vector along the $\underline{s}$ orientation.

iii) The gradient operation on a product of two scalar functions yields

$$\boldsymbol{grad}(\varphi\,\psi) = \varphi\,\boldsymbol{grad}\psi + \psi\,\boldsymbol{grad}\varphi.$$

iv) The Φ_σ flux of an $\underline{A}(x,y,z)$ vector through a σ surface surrounding a closed spatial zone τ satisfies the relation

$$\Phi_\sigma \equiv \int_\sigma \underline{A}\cdot d\underline{\sigma} = \int_\tau \boldsymbol{div}\,\underline{A}\cdot d\underline{\tau},$$

where $d\underline{\sigma}$ is a vector whose length is identical to the size of the infinitesimal surface-area slice, and is oriented toward outside the τ zone.

v) The Laplace operator Δ (Laplacian) is defined via the gradient and divergence operators:

$$\Delta \equiv \textbf{\textit{div grad}}, \text{ namely } \Delta\varphi(x,y,z) \equiv \textbf{\textit{div grad}}\,\varphi(x,y,z) = \frac{\partial^2\varphi}{\partial x^2} + \frac{\partial^2\varphi}{\partial y^2} + \frac{\partial^2\varphi}{\partial z^2}.$$

The Laplacian of a scalar function is also a scalar function of the (x, y, z) coordinates. The Laplacian operation on an $\underline{A}$ vector is defined as a vector, whose components are results of the Laplace operation separately on the $\underline{A}$ vector components:

vi) $\Delta\underline{A} = (\Delta A_x, \Delta A_y, \Delta A_z)$.

vii) $\textbf{\textit{div}}(\varphi\underline{A}) = \varphi\,\textbf{\textit{div}}\underline{A} + \underline{A}\cdot\textbf{\textit{grad}}\,\varphi$.

viii) $\textbf{\textit{div}}(\underline{A}\times\underline{B}) = \underline{B}\cdot\textbf{\textit{rot}}\underline{A} - \underline{A}\cdot\textbf{\textit{rot}}\underline{B}$.

ix) **Stokes** theorem: The integral of an $\underline{A}$ vector along a closed path ℓ (**circulation**) equals the flux of $\textbf{\textit{rot}}\underline{A}$ through a σ surface, whose path ℓ encircles the surface at its edges

$$\oint_\ell \underline{A}\cdot d\underline{\ell} = \int_\sigma \textbf{\textit{rot}}\,\underline{A}\cdot d\underline{\sigma},$$

where the $d\underline{\sigma}$ vector orientation is in the advancement direction of a right-hand screw rotating along the integration orientation $d\underline{\ell}$.

x) **Theorem**: The rotor of a *true vector* field is an *axial vector* field and vice-versa; the rotor of an axial vector is a true vector. A true vector reverses its sign after inversion of the axes frame, and is thus often known as a "polar vector"; an axial vector preserves its sign after an axes frame inversion, and is thus often known as a "pseudo-vector". A true vector is demonstrated geometrically by drawing an arrowhead at *one* of the line edges; a pseudo vector is demonstrated geometrically by drawing oppositely oriented arrowheads at the *two* line edges.

xi) **Theorem**: A vector field $\underline{A}(x, y, z)$ is a gradient of a scalar function $\varphi(x.\,y.\,z)$ if and only if the rotor of the field vanishes identically, namely if $\textbf{\textit{rot}}\,\underline{A} \equiv 0$.

xii) $\textbf{\textit{div rot}}\,\underline{A} \equiv 0$.

xiii) $\textbf{\textit{rot rot}}\,\underline{A} = \textbf{\textit{grad div}}\underline{A} - \Delta\underline{A}$.

xiv) $\textbf{\textit{rot}}(\varphi\,\underline{A}) = \varphi\,\textbf{\textit{rot}}\,\underline{A} + \textbf{\textit{grad}}\varphi\times\underline{A}$.

xv) $\textbf{\textit{rot}}(\underline{A}\times\underline{B}) = \underline{A}\,\textbf{\textit{div}}\,\underline{B} - \underline{B}\,\textbf{\textit{div}}\,\underline{A} + (\underline{B}\cdot\textbf{\textit{grad}})\underline{A} - (\underline{A}\cdot\textbf{\textit{grad}})\underline{B}$.

10.5.2 Cauchy's Equation

A function on the complex-plane is called "regular" over a given, singly connected region D, if it exhibits no singular poles, which means that it is continuous and differentiable throughout the entire region.

A function on the complex-plane is called "analytic" over such D region if it is continuous and differentiable throughout the entire region, except for a finite number of points addressed as singularity points.

Let $R(z)$ be a regular function over a certain such region of the complex plane. We define an analytic function $f(z)$ over the same region according to

$$f(z) \equiv \frac{R(z)}{z - z_p} \,. \qquad (10.62)$$

The complex z_p point forms a single singularity of $f(z)$ in the D region in the form of a simple pole. Cauchy's equation states that a contour integral around z_p would be

$$\oint_{around\ z_p} f(z)dz = 2\pi\, i\, R(z_p)\,. \qquad (10.63)$$

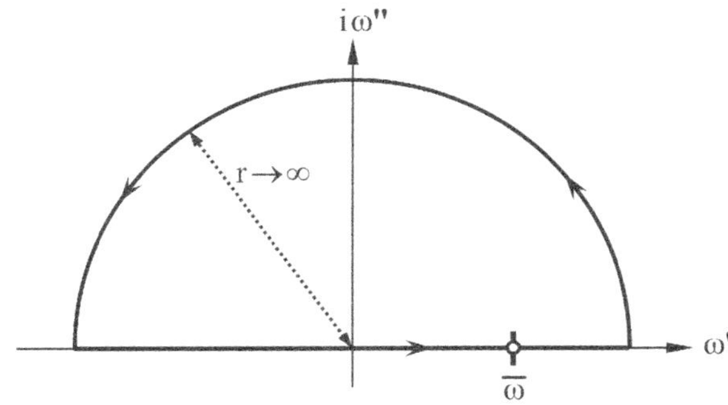

Figure 10.5 Integration contour for the integral appearing on the left-hand side of Eq. (10.34).

The contour integral vanishes when the z_p point is not included in the surrounded area; if z_p resides on the contour line itself, the integration result is half that given in Eq. (10.63), namely

$$\oint_{z_p\, on \circ} f(z)dz = \pi\, i\, R(z_p)\,. \qquad (10.64)$$

The $\widetilde{\beta}(\omega)/(\omega - \overline{\omega})$ function where $\overline{\omega} > 0$ is real, is the integration subject in Eq. (10.34); it is defined on the upper part of the complex plane. To finalize the integration, we define an integration contour between $-r$ and $+r$ on the real axis and, in addition, a half circle, whose center rests on the axis system origin. We then let r grow to infinity. The contour is shown in Figure 10.5. The integral part along the half circle diminishes in the said limit, leaving only the integral along the real axis between $-\infty$ and $+\infty$. The result, according to Eq. (10.64) is written in Eq. (10.34).

Exercises for Chapter 10

10.1 An electron moving in space at 300 km/s is under the action of a combined, mutually perpendicular static electric field of 10 kV/cm and a magnetic induction of 5.0 Tesla. The electron speed is perpendicular to the magnetic induction, while forming a 30° angle with the electric field. Calculate the effective force acting on the electron at that particular moment, and its acceleration.

10.2 A dielectric material whose refractive index is $n = 1.8$ at a frequency of 50 cm^{-1} also absorbs electromagnetic radiation at that frequency, with an absorption coefficient of 200 cm^{-1}. Calculate the extinction coefficient, and the phase difference between the electric and magnetic fields of the electromagnetic radiation of same frequency propagating inside the matter. Calculate the reflection and transmission factors of an electromagnetic radiation of same frequency incident perpendicularly to the material surface. What are the real and imaginary parts of the dielectric constant?

10.3 Plot the spectra of the refractive index $n(\omega)$ and extinction coefficient $\kappa(\omega)$ of the dielectric material, whose dielectric, and perpendicular reflectance functions, are plotted in Figure 10.3 for the same spectral range. Given that the unit-cells' density is 10^{22} cm^{-3}, calculate numerically the effective electronic oscillator strength that would generate the presented dielectric properties.

11 Crystalline Oscillation Species

11.1 Introduction

The calculation of crystalline oscillation species may be performed in a number of ways. One way is to consider the crystal as a gathering of identical atomic bases and, by a zero-order approximation, address each as an isolated molecule. In Section 8.5.1, we showed that it is possible to measure all of the elements of the polarizability tensor of a single molecule using light scattering in different Porto configurations from a group of molecules having an identical posture in space. The latter condition obviously exists for crystalline atomic bases. Accordingly, the oscillation modes of a single atomic basis are calculated according to the methods provided in Section 8.10 (see the comment concerning **Step 2** in Section 8.10), including rotation and translation modes. Recall that because of interactions with the surrounding material (see Section 9.5), translation and rotation motions of the atomic basis transform into "translational vibrations" and "librations", respectively, about the equilibrium posture. A subsequent step is to take into account the influence of the crystal symmetry. That approach is called **molecular site-group analysis**.

An alternative approach, which is mathematically much easier, is to consider the crystal as a gathering of many lattices, each comprising a single specific atom of the atomic basis. These atoms obviously exhibit both lattice translational symmetry and the unique site-symmetry existing for the site they occupy. In that approach, all calculated oscillation species are translational. This approach is called **nuclear-site group analysis**.

The method chosen in many spectroscopic studies is to combine the two methods; namely, the two separate calculations are performed, and a correlation map is then presented accordingly.

11.2 Crystalline Sites

A single isolated molecule obeys a point symmetry related to the mass center. When identical molecules (namely, identical atomic bases) are joined to form a crystal, however, special point-symmetry properties are generated at different positions in the unit cell, that do not necessarily coincide with those of the lattice points. These points are classified into sets, where each point belonging to a certain set has a vicinity of an identical form, and in that respect are equivalent. These sets are called **crystalline sites**. The set of symmetry operations that belong to each site is a symmetry group called a **site-group**. A site-group is always a sub-group of the crystalline point group, and is isomorphic with one of the 32 point groups allowed in crystalline structure (see Tables 9.1 and 8.2). If the location of one of the points belonging to a certain site is given, the other points may be found by performing all of the crystalline symmetry operations, which transform it successively into all others. The number of points belonging to any site is given by the ratio g_L/g_s, where g_L is the order of the crystalline point group and g_s is the order of the site group (recall that the "order" of any group is the number of different operations in the group. It may also be easily

Physics, Optics, and Spectroscopy of Materials, First Edition. Zeev Burshtein.

calculated using the group character table by summing the number of operations belonging to all similarity classes written in the top line of each character table). For example, the D_{3d} point group of the trigonal crystal (see Table 9.1) is $g_L = 12$ (see Table No. [24] in Section 8.12). Thus, when an atom resides in a C_1 symmetry site, there will be 12 identical atoms whose vicinities look the same, and it is possible to reach all of them by performing D_{3d} operations on any of the others; this is valid because $g_{C_1} = 1$. When an atom resides on a site of C_{2h} symmetry, there will be only three identical atoms residing in equivalent sites ((12/4) = 3); that is because $g_{C_{2h}} = 4$.

In certain cases, the position of a point belonging to a site is unique. Sites belonging to C_{pv}, C_{ph}, C_p, and C_S symmetries could exhibit one, two, or three degrees of freedom; namely, the site positions may be translated arbitrarily (but equally) along a unique line, arbitrarily (but equally) parallel to a unique plane, or just arbitrarily to any point of the unit cell, respectively. These properties are indicated in the (**i**) numbered set of tables provided in Section 11.5 below, and in a form to be described shortly.

11.3 Tabulation Method

Because there exists a finite number of crystalline point groups (32) and a finite number of crystalline space groups (230), and each space-group has a finite number of sites, it is possible to write-up all possible sites, and list their properties and the representations brought about by their occupation, over a limited number of pages. That is presented in Section 11.5 by the set of tables numbered (**i**). The set contains 32 tables, with each representing the data for one of the crystalline point groups. The title of each table specifies the ordinal number of the point group, its symbol by the Schoenflies notation, and, in parentheses, the symbol of the same group by the full/standard Hermann-Mauguin notation. For example, the number "7" point group is symbolized by C_{2v} in Schoenflies notation, and $mm2$ in the Hermann-Mauguin notation (Table 9.1). Importantly, concerning the Hermann-Mauguin notation, sometimes in the scientific literature, somewhat different notations are used, which depend on the naming choice of the fundamental unit cell vectors. The Schoenflies notation, on the other hand, is fundamentally unique.

Each table line belongs to a different space group. The far left column presents the ordinal group number among all space groups (a total of 230). The second from left column presents the space-group symbol; first by the Schoenflies notation, and then, in parentheses, by the Hermann-Mauguin notation. The Schoenflies notation is added by a numerical superscript that presents the space-group ordinal number among all space-groups belonging to the same point group. For example, the #97 space group among all 230 ones is symbolized as D_4^9; this means that it belongs to the D_4 point group and is numbered 9 among the 10 space groups belonging to D_4 (Table [**i**]-12). The same space-group is symbolized as $I422$ by the Hermann-Mauguin notation. The first letter indicates the centering type of the Bravais unit cell (Figure 9.4) according to the following convention:

"P" indicates a primitive Bravais cell; only one lattice point belongs to a primitive cell.
"A", "B", or "C" indicate a base-centered Bravais cell, which contains two lattice points.
"I" indicates a body-centered Bravais cell, which contains two lattice points.
"F" indicates a face-centered Bravais cell, which contains four lattice points.
"R" indicates a primitive rhombohedral Bravais cell, which contains one lattice point.

The number of lattice points in a centered unit cell is addressed as **lattice-points multiplicity**.

According to the above, the $I422$ space group of the last example belongs to a body-centered tetragonal Bravais lattice (Figure 9.4). The other letters and numbers in the space-group provide details on rotational and screw axes, and on reflection and glide plane mirrors. We will not elaborate further on those notations in the present book.

The third from left column details all of the sites existing per each space group. Each site symbol contains the symbol of the point group it belongs to; the number in parentheses on its right indicates the number of the same equivalent sites in the Bravais unit cell. On its left, it is given a name belonging to one of the letters of the English alphabet. This is called its **Wyckoff letter**. If the number of equivalent sites is infinite, the site symbol is enclosed in square parentheses, and the infinity "∞" symbol is added to its left. Consider for example the C_{2h}^6 space group whose general ordinal number is 15 (see Section 11.5, Table [**i**]-5). Its Hermann-Mauguin symbol is $C12/c1$. It is thus shown to belong to a base-centered monoclinic Bravais lattice. The group contains three site types:

- Sites of C_i symmetry, each comprising four equivalent points. The lattice contains four independent sites of the same symmetry, named consecutively by the Wyckoff "a", "b", "c", and "d" letters. Each site is unique with no degrees of freedom.
- A single site of C_2 symmetry comprising four equivalent points. The lattice contains a single such site of same symmetry, named by the Wyckoff "e" letter. The site is not unique; it has one degree of freedom and thus there are an infinite number of e-sites.
- A single site of C_1 symmetry comprising eight equivalent points. The lattice contains a single such site of same symmetry, named by the Wyckoff "f" letter. The site is not unique; it has three degree of freedom, and thus there are an infinite number of f-sites.

Notably, the sites' writing order from left to right exhibits a *descending* number of equivalent sites. The number of equivalent C_1 sites is largest (eight), and it reflects the multiplication between the C_{2h} point group order (four; see the character table [15] in Section 8.12) and the lattice points multiplicity (two), which results from Bravais cell centering. The C_2 and C_i orders are both two (character tables [2] and [6], respectively, in Section 8.12). Thus, each contains 4(=8/2) equivalent points. Apart from a very limited exception of partial exchanges (for example, the #22 D_2^7 space group), the Wyckoff lettering order is *ascending from right to left*, according to the English alphabet lettering order.

Notably, in counting the number of equivalent points for each site, the lattice-points' multiplicity from Bravais cell centering is taken into account. Thus, in base-centered or body-centered lattices, the number of equivalent sites is doubled compared with a primitive cell of same crystalline system. In face-centered lattices, the number of equivalent sites is fourfold larger compared with a primitive cell of the same crystalline system. As an example, consider the monoclinic C_2 system, whose properties are detailed in Table (**i**)-3. The number of C_1 sites in the #3 C_2^1 ($P121$) space group is two. On the other hand, the number of C_1 sites in the #5 C_2^3 ($C121$) space group is four. Still, in a primitive unit cell constructed on the basis of its centered cell, the number of equivalent sites returns to be identical to those of a primitive unit cell belonging to the same symmetry system.

When a crystalline site occupies an atom, this atom will exhibit three translational oscillation modes, the same as the number of any atom of three degrees of freedom. The number of independent oscillation modes will thus be three times the number of the equivalent sites. The vibrations' symmetry species are expressed by irreducible representations that are summarized in the Section 11.5-(**ii**) set of tables. The set contains 32 tables, each presenting the data of one of the crystalline point groups. The title of each table indicates the ordinal group number, then the group symbol by the Schoenflies notation and, in parentheses, the Hermann-Mauguin notation. Each table comprises two columns divided into lines. The first column details the sub-groups of the title group; each sub-group also indicates the site symmetry, and the right column of the same line details its irreducible representations (species). For example, we consider the representations stemming from occupation of the C_{2h} site of the C_{2h} point group (Table [**ii**]-5). The representations are $A_u \oplus 2B_u$. Recall that A and B representations are one-dimensional. Then, there are only three non-degenerate oscillation modes. That number is consistent with the number of C_{2h} sites, which is one for a primitive unit cell (see Section 11.5, Table [**i**]-5). On the other hand, representations of the C_2 site of the same point group are $A_g \oplus A_u \oplus 2B_g \oplus 2B_u$, a total of six non-degenerate oscillation modes. That number is consistent with the number of C_2 sites, which is two for a primitive unit cell (Table [**i**]-5). As a further example, consider the occupation of the S_4 site in the C_{4h} point group (Table [**ii**]-11). The representations are $A_u \oplus B_g \oplus E_g \oplus E_u$. Recall that E representations are two-dimensional (see Section 8.8 item [3]). Thus, there are a total of six oscillation modes, which is consistent with the number of S_4 sites, which is two for a primitive unit cell (Table [**i**]-11). Recall that E_g and E_u representations are each doubly degenerate.

When a point belonging to a certain site occupies a molecule, the translational modes are added with vibrational and librational motions. We first address the librations. There are three degrees of freedom for librations of any molecule. Thus, the number of independent librations obtained by occupation of a certain site is given by multiplying the number of equivalent sites by 3. The symmetry species of the librational modes are expressed by irreducible representations, which are detailed in the no. (**iii**) set of tables. The set contains 32 tables, and each presents the data for each of the crystalline point-groups. The title of each separate table indicates first the ordinal number of the point group, then the group symbol by the Schoenflies notation, and, in parentheses, the group symbol by the Hermann-Mauguin notation. The tables comprise two columns divided into lines. The first column details all sub-groups of the headed point group; each such sub-group also describes the symmetry of the site whose irreducible representations are listed in the second column of the same line. For example, consider the representations resulted by occupying the C_{2h} site belonging to the C_{2h} point group in Section 11.5 (Table [**iii**]-5). The representations are $A_g \oplus 2B_g$. Because A and B representations are one-dimensional (see Section 8.8 item [2]), there are three non-degenerate oscillation modes. That number is consistent with the number of C_{2h} sites, which is one for the primitive unit cell (Section 11.5, Table [**i**]-5). On the other hand, representations of the C_2 site of the same point group would be $A_g \oplus A_u \oplus 2B_g \oplus 2B_u$, with a total of six non-degenerate oscillation modes. That number is consistent with the number of C_2 sites, which is two for a primitive unit cell (Section 11.5 Table [**i**]-5). As a further example, consider the occupation of the S_4 site in the C_{4h} point group (Section 11.5, Table [**ii**]-11). The representations are $A_u \oplus B_g \oplus E_g \oplus E_u$. Recalling that E representations are two-dimensional (see Section 8.8 item [2]), there are a total of six oscillation modes, consistent with the number of S_4 sites, which is two for a primitive unit cell (Section 11.5, Table [**i**]-11). The E_g and E_u are each doubly degenerate.

There is an unequivocal relation between the symmetry species of any oscillation of a single atom or a molecule as seen in a particular occupying site (translation, rotation, or libration) and its expression by the symmetry of the entire unit cell. By definition, the symmetry on occupying the site is obviously the site symmetry. In particular, molecular oscillations may be calculated using the methods provided in Section 8.10. If the site comprises several equivalent points, the unit cell correspondingly contains the same number of identical atoms or molecules. The symmetry species of the entire unit cell are linked to those of the site symmetries. These links are summarized in the set of tables numbered (**iv**) in Section 11.5, which are called **correlation tables**. The set contains 32 tables, and each belongs to a different crystalline point group. Each table details the symmetry species of the sites belonging to the particular crystalline symmetry species. The title of each table indicates first the ordinal number of the crystalline point group, then the group symbol by the Schoenflies notation, and finally, in parentheses, the group symbol by the Hermann-Mauguin notation. The group symbol by the Schoenflies notation is written again at the top left table cell. The first column below it lists all the group species (the irreducible representations; see the character tables in Section 8.12). Further below that, in the bottom left cell, the number (1) is written; that is the number of equivalent points in a site whose symmetry is identical to the *primitive* unit cell symmetry. The top line lists all sub-groups of the cell symmetry (in certain cases, a sub-group is listed more than once, such as when there are different choices of its symmetry elements). Against each sub-group, the bottom line indicates, in parentheses, the number of equivalent points belonging to the top sub-group in a *primitive* unit cell of the parent point group (see Section 11.5 [**i**] tables).

To clarify the meaning of the above sorting, we analyze an example. Consider the B_g oscillation species of the C_{2h} point symmetry (Section 11.5 Table [**iv**]-5). If that species is formed by molecules occupying the C_i site, then, according to the table, each of the two occupying molecules oscillates at a "personal" A_g symmetry. As an additional example, consider the E_u oscillation of the D_{4h} crystalline point group (Section 11.5 Table [**iv**]-15). If it is formed by oscillations of eight molecules occupying the C_i site points, each molecule in itself oscillates by one of two non-degenerate A_u modes (identically for all eight molecules). If it is formed by oscillations of four molecules occupying the C_{2h} site points, each molecule in itself oscillates by either A_u or B_u modes.

An additional important use of a correlation table is for the description of symmetry changes of molecular oscillations resulting from the distortion of the molecular shape, which causes reduction of the molecular symmetry. For example, consider the carbon tetrachloride CCl_4 molecule (Figure 8.11) vibrating by the $\nu_3(F_2)$ mode (Eq. [8.42] and following text). If the four vertical edges of the cube enclosing the molecule (Figure 8.11) are equally shortened or lengthened to form a square-based box, while the carbon atom maintains its position at the mass center, the molecular symmetry reduces to S_4. According to Table (**iv**)-31, the F_2 representation (species) changes into $B \oplus E$. Namely, the ν_3 mode frequency splits in two: $\nu_3(B)$ and $\nu_3(E)$. The F_2 modes are called "parents" and the B and E modes are called "daughters" (if the distortion is tiny, the $\nu_3(B)$ and $\nu_3(E)$ frequencies coincide with the $\nu_3(F_2)$ one). If instead, two of the $C-Cl$ bonds are equally shortened, the symmetry reduces to C_2. According to Section 11.5 Table (**iv**)-31, the F_2 representation (species) changes into $A \oplus 2B$. Namely, the ν_3 mode frequency splits in three: $\nu_3(A)$, $\nu_3(B)$ and $\nu_3'(B)$.

11.4 Calculation of Crystalline Oscillation Species – An Example

As an example, we present a calculation of the oscillation species of the cadmium tungstate $CdWO_4$ single crystal. It is primitive monoclinic, belonging to the #13 C_{2h}^4 $(P12/c1)$ space group, whose atomic basis comprises two $CdWO_4$ formula units. The unit cell constants are $a = 5.02$ Å, $b = 5.85$ Å, $c = 5.07$ Å, and $\beta = 91°30'$. The fractional position coordinates of the atoms constituting the atomic basis are detailed in Table 11.1.

Two equivalent cadmium Cd atoms occupy an e-site of C_2 symmetry, and two equivalent tungsten W atoms occupy an f-site of C_2 symmetry. There are two sets of oxygen atoms: four equivalent O_I oxygen atoms occupy a g-site of C_1 symmetry, and four other equivalent O_{II} oxygen atoms occupy a different g-site of C_1 symmetry. The resulting natural atomic basis of the crystal is shown in Figure 11.1.

Table 11.1 Fractional coordinates of atoms constituting the $CdWO_4$ crystal atomic basis.

	Cd	W	O_I	O_{II}
x/a	0.500	0.000	0.210	0.240
y/b	0.692	0.180	0.100	0.380
z/c	0.250	0.250	0.960	0.380

We start the oscillation modes analysis by using nuclear-site group analysis. According to Section 11.5 Table (**ii**)-5, the cadmium Cd and tungsten W atoms each contribute six modes: $A_g \oplus A_u \oplus 2B_g \oplus 2B_u$. Each oxygen set contributes

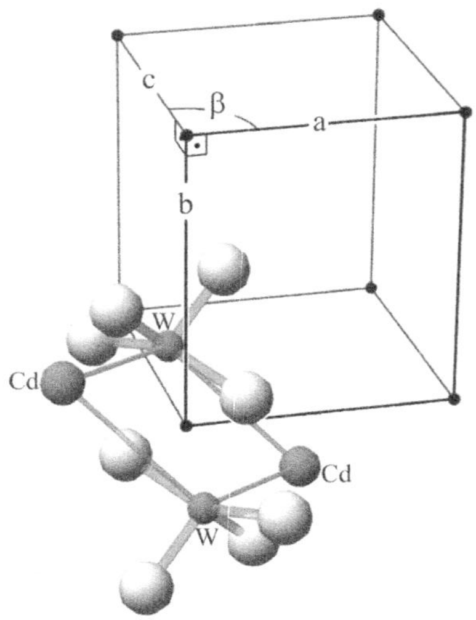

Figure 11.1 Natural atomic basis of the cadmium-tungstate $CdWO_4$ crystal unit cell. Large spheres symbolize oxygen O atoms. Adjacent tungsten-oxygen $W-O$ bonds are indicated by thick rods. Bonds forming the atomic basis skeletal structure (see text) are indicated by thinner rods.

12 oscillation modes: $3A_g \oplus 3A_u \oplus 3B_g \oplus 3B_u$. The total number of modes is 36: $8A_g \oplus 8A_u \oplus 10B_g \oplus 10B_u$. The unit-cell symmetry is obviously C_{2h}. According to the said table, the translational modes of the entire unit cell (those forming the acoustic waves in Section 9.5.4-[**i**]) are $A_u \oplus 2B_u$. The remaining, $8A_g \oplus 7A_u \oplus 10B_g \oplus 8B_u$, are thus the optical modes involved in forming the crystalline optical waves (Section 9.5.4-[**ii**]). By Table (**iii**)-5 of Section 11.5, we know that of these, $A_g \oplus 2B_g$, describe the unit cell librational modes. The A_g and B_g modes, a total of 18, are Raman active but infrared inactive. The A_u and B_u modes, a total of 15, are infrared active but Raman inactive (see Section 8.12 Table [15]).

To achieve some insight concerning the various modes, we seek their parents by collecting several, relatively strongly bonded atoms, to form separate quasi-molecular compounds. Relating to Figure 11.1, we separate each $CdWO_4$ formula unit into two type ions: Cd^{2+} and $WO_4{}^{2-}$. As before, the Cd^{2+} cation occupies a C_2 symmetry site. The tungsten W atom resides at the mass center of the $WO_4{}^{2-}$ anion; that anion thus exhibits the same C_2 symmetry as the tungsten atom. We assign a T_d-symmetry to its parent anion. In reality, however, that anion looks like a distorted tetrahedron: the lengths of two of the $W-O$ bonds are 1.79 Å, and the other two are 1.89 Å. There are two $W-O$ inter-bond angles of 94°, two of 105°, one of 99°, and one of 151°; in a regular tetrahedron, on the other hand, all bond-lengths are equal, and all inter-bond angles are 109°28′. The oscillation modes of the free $WO_4{}^{2-}$ parent ion, including translation and rotation, are calculated by the methods provided in Section 8.10; the result is $\Gamma = A_1 \oplus E \oplus F_1 \oplus 3F_2$, where translation species are $1F_2$, and rotation species are $1F_1$. The general anion oscillation-form reads

$$\Gamma_{WO_4{}^{2-}} = (\nu_1)1A_1 \oplus (\nu_2)1E \oplus (lib)1F_1 \oplus (trans, \nu_3, \nu_4)3F_2 \,. \tag{11.1}$$

Because the said ion occupies a C_2 symmetry site, and not a T_d one, the symmetry reduces, and the species frequencies split. Based on Section 11.5 Table (**iv**)-31, the A_1 species transforms into an A, the E species splits as $2A$, the F_1 species splits into $A \oplus 2B$, and the F_2 species splits the same way. These transformations are summarized in a "correlation map" shown in Figure 11.2.

Consider now the entire atomic basis, which includes two $WO_4{}^{2-}$ tungstate anions and two Cd^{2+} cations. We ignore now any libration or internal vibration of the $WO_4{}^{2-}$ compound. In other words, we imagine it to "shrink" to a point-size in its mass center, located in the C_2 symmetry site. We consider the two such $WO_4{}^{2-}$ compounds and the two Cd^{2+} cations, all occupying C_2 symmetry sites, as a so-called "skeleton" of the atomic basis. The six translational motions of the two tungstate anions ("trans" in Figure 11.2) couple to form the oscillations of the said skeleton. The skeleton is shaped like a parallelogram; two opposite vertices occupy tungstate anions. The two others occupy Cd^{2+} cations. Such skeleton, when free in space, exhibits a C_{2h} symmetry. Its motions calculated by the methods of Section 8.10 (translations and librations included) are given by $\Gamma = 4A_g \oplus 2B_g \oplus 2A_u \oplus 4B_u$. The translational motions ("trans") are $A_u \oplus 2B_u$; the librational motions ("Lib") are $A_g \oplus 2B_g$. The remaining vibrational modes are $3A_g \oplus A_u \oplus 2B_u$. That analysis is written in the form

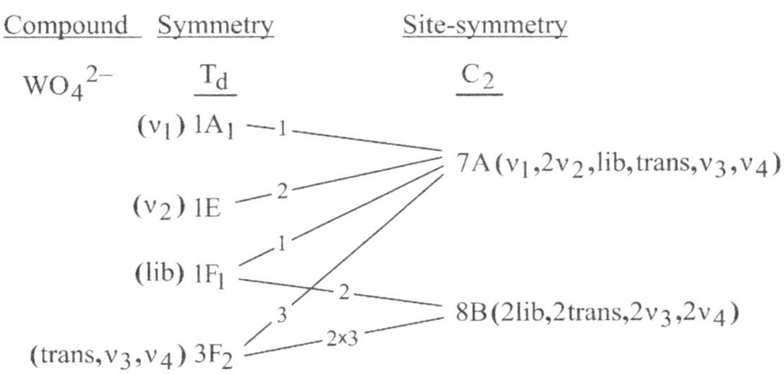

Figure 11.2 Correlation map between oscillation species of a WO_4^{2-} anion of T_d parent symmetry to a daughter species of a twisted anion of a C_2 symmetry.

$$\Gamma_{Skeleton} = \left\{ \begin{array}{rl} (f_1, f_2, f_3, Lib_1) & 4A_g \\ \oplus(f_4, acous_1) & 2A_u \\ \oplus(Lib_2 Lib_3) & 2B_g \\ \oplus(f_5, f_6, acous_2, acous_3) & 4B_u \end{array} \right. . \tag{11.2}$$

The skeletal vibrational modes are denoted by f_1, f_2, f_3, f_4, f_5, and f_6 in Eq. (11.2). These notations are specific for the present need; ν_1, ν_2, ν_3, and ν_4, on the other hand, are accepted notations in the scientific literature; see Section 8.11.3. Translational modes are denoted "acous", not "trans", because they are the ones that form the crystal acoustical waves (Section 9.5.4.1).

Figure 11.3 portrays the skeletal internal vibrational f_1, f_2, f_3, f_4, f_5, and f_6 motions (see methods and example calculations in Sections 8.10 and 8.11). The patterns' observation makes it possible to guess the frequency order among the species. The f_1 mode probably exhibits the highest frequency because it shows a periodic swelling and shrinking of the skeletal dimensions, and hence a large return force by the vicinity. The f_2 mode probably exhibits a smaller frequency because the skeleton grows and shrinks simultaneously only in one orientation and hence is subject to a smaller response by the vicinity. The f_3 mode exhibits no dimensional change during oscillation. It may be described as two oppositely phased librations – one of the $WO_4^{2-} - WO_4^{2-}$ pair, the other of the $Cd^{2+} - Cd^{2+}$ pair – both about an axis that crosses the mass center perpendicularly to the skeletal plane. This involves only shear strains, which entail only small return forces. Comparing the f_5 and f_6 modes, the translation vectors in f_5 are closer to the $Cd^{2+} - WO_4^{2-}$ bonding. They are thus likely to exhibit a larger frequency. As for f_4, the situation is not that clear, because of conflicting considerations. On the one hand, the mode describes a kind of bending similar to that of the free CO_2 molecule, whose frequency (667 cm^{-1}) is considerably smaller that the molecular stretching modes that are similar to f_5 and f_6 (1337 cm^{-1} and 2349 cm^{-1}). On the other hand, the motion involves deviation from the skeletal plane and hence is subject to a return reaction force by the vicinity. It may be reasonably concluded that the f_4 frequency is smaller than f_5 and f_6 but not by much. In many cases, qualitative considerations as detailed above are justified by theoretical calculations, as well as by experiment.

Because the skeleton resides in a site of C_i symmetry (the mass center on the lattice point), and not of a C_{2h} symmetry, the symmetry reduces, inducing change in the oscillation symmetry species. According to Table (**iv**)-5, the A_g and A_u species remain unchanged. B_g, however, changes into A_g, and the B_u changes into A_u. Figure 11.4 shows the relevant correlation table. The skeletal librations are indicated as "Lib". For brevity, the skeletal translations are indicated on the right side as "acs" instead of "acous".

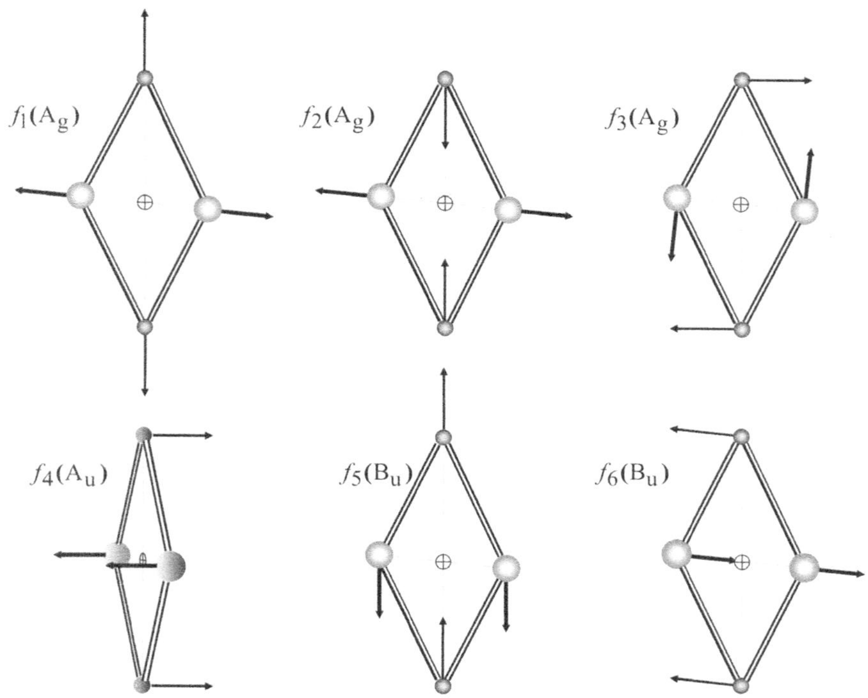

Figure 11.3 Translation vectors in the f_1, f_2, f_3, f_4, f_5, and f_6 vibrational oscillations of the $CdWO_4$ crystal skeleton components. A large ball describes a WO_4^{2-} anion, and a small one describes a Cd^{2+} cation. In the $f_4(A_u)$ species, the skeletal plane is almost perpendicular to the page plane; in all others, it lays parallel to the page plane.

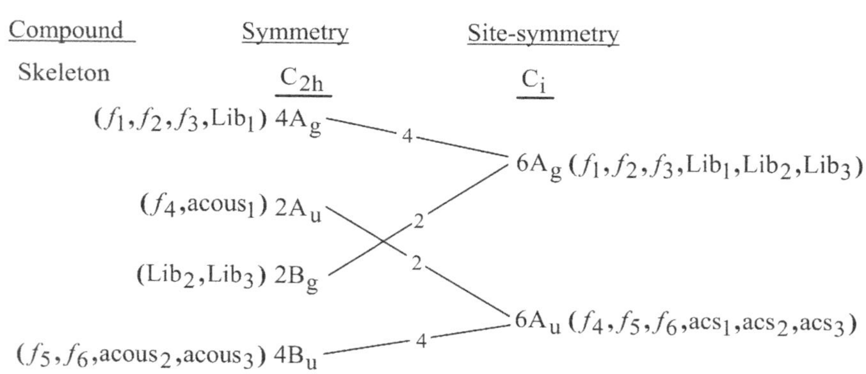

Figure 11.4 Correlation map between oscillation species of the cadmium tungstate crystal skeleton $(WO_4^{2-}[Cd^{2+}])_2$ of C_{2h} parent symmetry to daughter species where the skeleton resides in a site of C_i symmetry.

In Figure 11.5, we provide an integrated correlation map for all above-described oscillation modes of the cadmium tungstate $CdWO_4$ single crystal.

The symmetry linking of the crystal oscillation modes to those of the tungstate WO_4^{2-} ion and the unit-cell skeleton, sometimes allows for assessing of their frequencies. For example, oscillation modes originating in the tungstate WO_4^{2-} ion vibrations are bound to exhibit higher frequencies than those of the unit cell skeleton. It is customary to quote the following frequencies as representatives: $\nu_1 = 930\ \text{cm}^{-1}$;

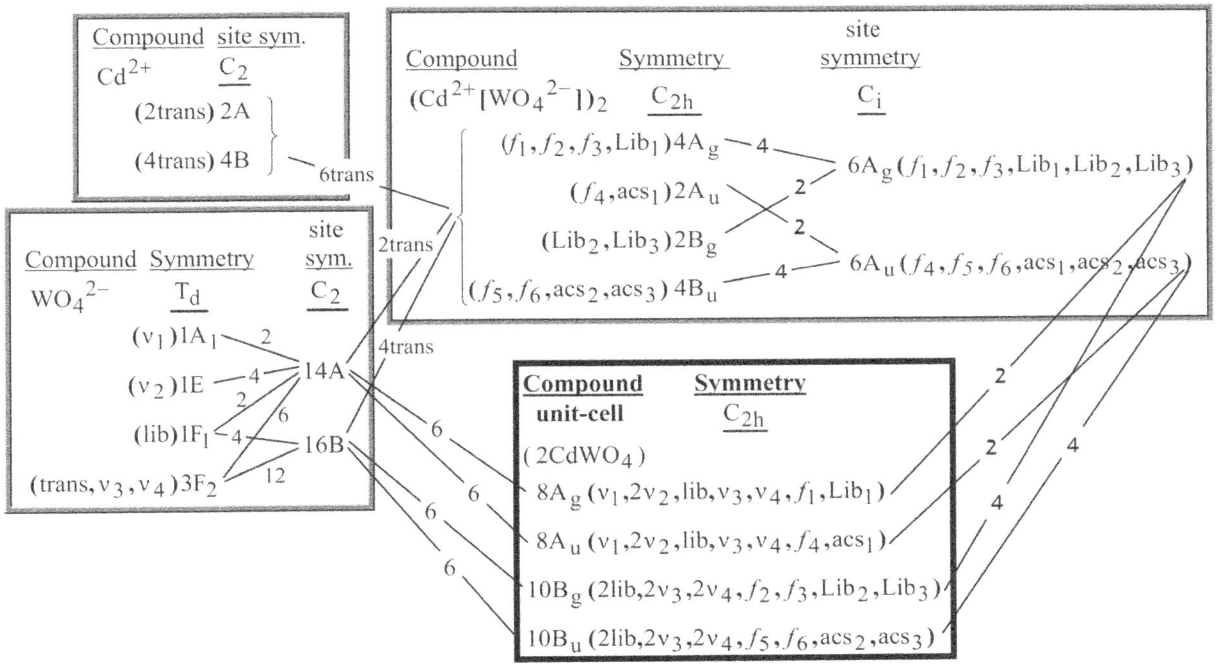

Figure 11.5 Correlation table linking the oscillation modes of the *CdWO*$_4$ single crystal to "independent" oscillation modes of the tungstate *WO*$_4^{2-}$ anions and the atomic basis skeleton belonging to each lattice point. For details, see the text.

$\nu_2 = 325\ \text{cm}^{-1}$; $\nu_3 = 840\ \text{cm}^{-1}$; and $\nu_4 = 325\ \text{cm}^{-1}$. These values were measured for the tungstate anion in an aqueous solution, and are likely to remain close in other compounds or solutions. The librational frequencies are also expected to be higher than those of the unit cell skeleton, because their moments of inertia are considerably smaller. Except for the above qualitative assessments, correlation tables do not relate directly to the frequency's size; detailed theoretical calculations and experiments are required for that purpose.

A quite comprehensive image of crystal vibrational states obtains by a combination of experimental Raman scattering with reflection and/or absorption measurements of infrared light. The character table of the C_{2h} group (Section 8.12, Table [15]) reveals that Raman scattering in (*XX*), (*YY*), (*ZZ*), or (*XY*) Porto configurations (Section 8.5.1) reflect vibrations of A_g species; Raman scattering in (*XZ*) or (*YZ*) Porto configurations reflect vibrations of B_g species. Absorption or perpendicular reflection of infrared light polarized in the *Z* direction (the $\underline{b}$ unit cell axis) reflects frequencies of the A_u species. Absorption or perpendicular reflection of infrared light polarized in any orientation on the *XY* plane reflects frequencies of the B_u species. Perpendicular reflection of infrared light off crystal surfaces parallel to specific crystallographic plane reveals the LO-TO splitting of each vibrational mode (Sections 9.5.4.2, 10.2, 10.3, and 10.4). Comparing reflections in $X(ZZ)\overline{X}$ and $Y(ZZ)\overline{Y}$ configurations would especially reveal the spatial dispersion of transverse optical TO waves belonging to the A_u species near $k = 0$ between the *X* and *Y* propagation orientations, respectively. Comparing reflections in $Y(XX)\overline{Y}$ and $Z(XX)\overline{Z}$ configurations would reveal the spatial dispersion of transverse optical TO waves belonging to the B_u species near $k = 0$ between the *Y* and *Z* propagation orientations, respectively.

11.5 Tabulation of Crystalline Space Group Properties

Tables (i): Sites existing for each space-group. Symmorphic groups are emphasized by a gray background indicating their ordinal number.

(**i**)-1. C_1 (1)

No.	Space-group	Sites
1	$C_1^1(P)$	$\infty[C_1(1)]$

(**i**)-2. C_i $(\overline{1})$

No.	Space-group	Sites
2	C_i^1 $(P\overline{1})$	$\infty[iC_1(2)] + (h + g + f + e + d + c + b + a)C_i(1)$

(**i**)-3. C_2 (2)

No.	Space-group	Sites
3	C_2^1 $(P121)$	$\infty[eC_1(2)] + \infty[(d + c + b + a)C_2(1)]$
4	C_2^2 $(P12_11)$	$\infty[aC_1(2)]$
5	C_2^3 $(C121)$	$\infty[cC_1(4)] + \infty[(b + a)C_2(2)]$

(**i**)-4. C_S (m)

No.	Space-group	Sites
6	C_S^1 $(P1m1)$	$\infty[cC_1(2)] + \infty[(b + a)C_S(1)]$
7	C_S^2 $(P1c1)$	$\infty[aC_1(2)]$
8	C_S^3 $(C1m1)$	$\infty[bC_1(4)] + \infty[aC_S(2)]$
9	C_S^4 $(C1c1)$	$\infty[aC_1(4)]$

(**i**)-5. C_{2h} $(2/m)$

No.	Space-group	Sites
10	C_{2h}^1 $(P12/m1)$	$\infty[oC_1(4)] + \infty[(n + m)C_S(2)] + \infty[(\ell + k + j + i)C_2(2)]$ $+ (h + g + f + e + d + c + b + a)C_{2h}(1)$
11	C_{2h}^2 $(P12_1/m1)$	$\infty[fC_1(4)] + \infty[eC_S(2)] + (d + c + b + a)C_i(2)$
12	C_{2h}^3 $(C12/m\ 1)$	$\infty[jC_1(8)] + \infty[iC_S(4)] + \infty[(h + g)C_2(4)] + (f + e)C_i(4)$ $+ (d + c + b + a)C_{2h}(2)$
13	C_{2h}^4 $(P12/c\ 1)$	$\infty[gC_1(4)] + \infty[(f + e)C_2(2)] + (d + c + b + a)C_i(2)$
14	C_{2h}^5 $(P12_1/c1)$	$\infty[eC_1(4)] + (d + c + b + a)C_i(2)$
15	C_{2h}^6 $(C12/c1)$	$\infty[fC_1(8)] + \infty[eC_2(4)] + (d + c + b + a)C_i(4)$

(**i**)-6. D_2 (222)

No.	Space-group	Sites
16	D_2^1 ($P222$)	$\infty[uC_1(4)] + \infty[(t+s+r+q)C_2{}^z(2)] + \infty[(p+o+n+m)C_2{}^y(2)] + \infty[(\ell+k+j+i)C_2{}^x(2)] + (h+g+f+e+d+c+b+a)D_2(1)$
17	D_2^2 ($P222_1$)	$\infty[eC_1(4)]\ \infty[(d+c)C_2{}^y(2)] + \infty[(b+a)C_2{}^x(2)]$
18	D_2^3 ($P2_12_12$)	$\infty[cC_1(4)] + \infty[(b+a)C_2{}^z(2)]$
19	D_2^4 ($P2_12_12_1$)	$\infty[aC_1(4)]$
20	D_2^5 ($C222_1$)	$\infty[cC_1(8)] + \infty[bC_2{}^y(4)] + \infty[aC_2{}^x(4)]$
21	D_2^6 ($C222$)	$\infty[\ell C_1(8)] + \infty[(k+j+i)C_2{}^z(4)] + \infty[(h+g)C_2{}^y(4)] + \infty[(f+e)C_2{}^x(4)] + (d+c+b+a)D_2(2)$
22	D_2^7 ($F222$)	$\infty[kC_1(16)] + \infty[(h+g)C_2{}^z(8)] + \infty[(i+f)C_2{}^y(8)] + \infty[(j+e)C_2{}^x(8)] + (d+c+b+a)D_2(4)$
23	D_2^8 ($I222$)	$\infty[kC_1(8)] + \infty[(j+i)C_2{}^z(4)] + \infty[(h+g)C_2{}^y(4)] + \infty[(f+e)C_2{}^x(4)] + (d+c+b+a)D_2(2)$
24	D_2^9 ($I2_12_12_1$)	$\infty[dC_1(8)] + \infty[cC_2{}^z(4)] + \infty[bC_2{}^y(4)] + \infty[aC_2{}^x(4)]$

(**i**)-7. C_{2v} ($mm2$)

No.	Space-group	Sites
25	C_{2v}^1 ($Pmm2$)	$\infty[iC_1(4)] + \infty[(h+g)C_S{}^{yz}(2)] + \infty[(f+e)C_S{}^{xz}(2)] + \infty[(d+c+b+a)C_{2v}(1)]$
26	C_{2v}^2 ($Pmc2_1$)	$\infty[cC_1(4)] + \infty[(b+a)C_S{}^{yz}(2)]$
27	C_{2v}^3 ($Pcc2$)	$\infty[eC_1(4)] + \infty[(d+c+b+a)C_2(2)]$
28	C_{2v}^4 ($Pma2$)	$\infty[dC_1(4)] + \infty[cC_S{}^{yz}(2)] + \infty[(b+a)C_2(2)]$
29	C_{2v}^5 ($Pca2_1$)	$\infty[aC_1(4)]$
30	C_{2v}^6 ($Pnc2$)	$\infty[cC_1(4)] + \infty[(b+a)C_2(2)]$
31	C_{2v}^7 ($Pmn2_1$)	$\infty[bC_1(4)] + \infty[aC_S{}^{yz}(2)]$
32	C_{2v}^8 ($Pba2$)	$\infty[cC_1(4)] + \infty[(b+a)C_2(2)]$
33	C_{2v}^9 ($Pna2_1$)	$\infty[aC_1(4)]$
34	C_{2v}^{10} ($Pnn2$)	$\infty[cC_1(4)] + \infty[(b+a)C_2(2)]$
35	C_{2v}^{11} ($Cmm2$)	$\infty[fC_1(8)] + \infty[eC_S{}^{yz}(4)] + \infty[dC_S{}^{xz}(4)] + \infty[cC_2(4)] + \infty[(b+a)C_{2v}(2)]$
36	C_{2v}^{12} ($Cmc2_1$)	$\infty[bC_1(8)] + \infty[aC_S{}^{yz}(4)]$
37	C_{2v}^{13} ($Ccc2$)	$\infty[dC_1(8)] + \infty[(c+b+a)C_2(4)]$

(**i**)-7. C_{2v} $(mm2)$ ...continued

No.	Space-group	Sites
38	C_{2v}^{14} $(Amm2)$	$\infty[fC_1(8)] + \infty[(e+d)C_S{}^{yz}(4)] + \infty[cC_S{}^{xz}(4)] + \infty[(b+a)C_{2v}(2)]$
39	C_{2v}^{15} $(Aem2)$	$\infty[dC_1(8)] + \infty[(cC_S{}^{xz}(4)] + \infty[(a+b)C_2(4)]$
40	C_{2v}^{16} $(Ama2)$	$\infty[cC_1(8)] + \infty[(bC_S{}^{yz}(4)] + \infty[aC_2(4)]$
41	C_{2v}^{17} $(Aea2)$	$\infty[bC_1(8)] + \infty[aC_2(4)]$
42	C_{2v}^{18} $(Fmm2)$	$\infty[eC_1(16)] + \infty[dC_S{}^{xz}(8)] + \infty[cC_S{}^{yz}(8)] + \infty[bC_2(8)] + \infty[aC_{2v}(4)]$
43	C_{2v}^{19} $(Fdd2)$	$\infty[bC_1(16)] + \infty[aC_2(8)]$
44	C_{2v}^{20} $(I\,mm2)$	$\infty[eC_1(8)] + \infty[dC_S{}^{yz}(4)] + \infty[cC_S{}^{xz}(4)] + \infty[(b+a)C_{2v}(2)]$
45	C_{2v}^{21} $(I\,ba2)$	$\infty[cC_1(8)] + \infty[(b+a)C_2(4)]$
46	C_{2v}^{22} $(I\,ma2)$	$\infty[cC_1(8)] + \infty[bC_S{}^{yz}(4)] + \infty[aC_2(4)]$

(**i**)-8. D_{2h} (mmm)

No.	Space-group	Sites
47	D_{2h}^{1} $(P2/m2/m2/m)$	$\infty[aC_1(8)] + \infty[(z+y)C_S{}^{xy}(4)] + \infty[(x+w)C_S{}^{xz}(4)] + \infty[(v+u)C_S{}^{yz}(4)]$ $+ \infty[(t+s+r+q)C_{2v}{}^{z}(2)] + \infty[(p+o+n+m)C_{2v}{}^{y}(2)]$ $+ \infty[(\ell+k+j+i)C_{2v}{}^{x}(2)] + (h+g+f+e+d+c+b+a)D_{2h}(1)$
48	D_{2h}^{2} $(P2/n2/n2/n)$	$\infty[mC_1(8)] + \infty[(\ell+k)C_2{}^{z}(4)] + \infty[(j+i)C_2{}^{y}(4)] + \infty[(h+g)C_2{}^{x}(4)]$ $+ (f+e)C_i(4) + (d+c+b+a)D_2(2)$
49	D_{2h}^{3} $(P\ 2/c\ 2/c\ 2/m)$	$\infty[rC_1(8)] + \infty[qC_S{}^{xy}(4)] + \infty[(p+o+n+m)C_2{}^{z}(4)] + \infty[(\ell+k)C_2{}^{y}(4)]$ $+ \infty[(j+i)C_2{}^{x}(4)] + (h+g+f+e)D_2(2) + (d+c+b+a)C_{2h}{}^{z}(2)$
50	D_{2h}^{4} $(P2/b2/a2/n)$	$\infty[mC_1(8)] + \infty[(\ell+k)C_2{}^{z}(4)] + \infty[(j+i)C_2{}^{y}(4)] + \infty[(h+g)C_2{}^{x}(4)]$ $+ (f+e)C_i(4) + (d+c+b+a)D_2(2)$
51	D_{2h}^{5} $(P2_1/m2/m2/a)$	$\infty[\ell C_1(8)] + \infty[kC_S{}^{yz}(4)] + \infty(j+i)C_S{}^{xz}(4)] + \infty[(h+g)C_2{}^{y}(4)]$ $+ \infty[(f+e)C_{2v}{}^{z}(2)] + (d+c+b+a)C_{2h}{}^{y}(2)$
52	D_{2h}^{6} $(P2/n2_1/n2/a)$	$\infty[eC_1(8)] + \infty[dC_2{}^{x}(4)] + \infty[cC_2{}^{z}(4)] + (b+a)C_i(4)$
53	D_{2h}^{7} $(P2/m2/n2_1/a)$	$\infty[iC_1(8)] + \infty[hC_S{}^{yz}(4)] + \infty[gC_2{}^{y}(4)]$ $+ \infty[(f+e)C_2{}^{x}(4)] + (d+c+b+a)C_{2h}{}^{x}(2)$
54	D_{2h}^{8} $(P2_1/c2/c2/a)$	$\infty[fC_1(8)] + \infty[(e+d)C_2{}^{z}(4)] + \infty[cC_2{}^{y}(4)] + (b+a)C_i(4)]$
55	D_{2h}^{9} $(P2_1/b2_1/a2/m)$	$\infty[iC_1(8)] + \infty[(h+g)C_S{}^{xy}(4)] + \infty[(f+e)C_2{}^{z}(4)] + (d+c+b+a)C_{2h}{}^{z}(2)$
56	D_{2h}^{10} $(P2_1/c2_1/c2/n)$	$\infty[eC_1(8)] + \infty[(d+c)C_2{}^{z}(4)] + (b+a)C_i(4)$
57	D_{2h}^{11} $(P2/b2_1/c2_1/m)$	$\infty[eC_1(8)] + \infty[dC_S{}^{xy}(4)] + \infty[cC_2{}^{x}(4)] + (b+a)C_i(4)$

(**i**)-8. D_{2h} (mmm) ...continued

No.	Space-group	Sites
58	D_{2h}^{12} $(P2_1/n2_1/n2/m)$	$\infty[hC_1(8)] + \infty[gC_S{}^{xy}(4)] + \infty[(f+e)C_2{}^z(4)] + (d+c+b+a)C_{2h}{}^z(2)$
59	D_{2h}^{13} $(P2_1/m2_1/m2/n)$	$\infty[gC_1(8)] + \infty[fC_S{}^{xz}(4)] + \infty[eC_S{}^{yz}(4)] + (d+c)C_i(4) + \infty[(b+a)C_{2v}{}^z(2)]$
60	D_{2h}^{14} $(P2_1/b2/c2_1/n)$	$\infty[dC_1(8)] + \infty[cC_2{}^y(4)] + (b+a)C_i(4)$
61	D_{2h}^{15} $(P2_1/b2_1/c2_1/a)$	$\infty[cC_1(8)] + (b+a)C_i(4)$
62	D_{2h}^{16} $(P2_1/n2_1/m2_1/a)$	$\infty[dC_1(8)] + \infty[cC_S{}^{xz}(4)] + (b+a)C_i(4)$
63	D_{2h}^{17} $(C2/m2/c2_1/m)$	$\infty[hC_1(16)] + \infty[gC_S{}^{xy}(8)] + \infty[fC_S{}^{yz}(8)] + \infty[eC_2{}^x(8)]$ $+ dC_i(8) + \infty[cC_{2v}{}^y(4)] + (b+a)C_{2h}{}^x(4)$
64	D_{2h}^{18} $(C2/m2/c2_1/e)$	$\infty[gC_1(16)] + \infty[fC_S{}^{yz}(8)] + \infty[eC_2{}^y(8)] + \infty[dC_2{}^x(8)]$ $+ cC_i(8) + (b+a)C_{2h}{}^x(4)$
65	D_{2h}^{19} $(C2/m2/m2/m)$	$\infty[rC_1(16)] + \infty[(q+p)C_S{}^{xy}(8)] + \infty[oC_S{}^{xz}(8)] + \infty[nC_S{}^{yz}(8)]$ $+ \infty[mC_2{}^z(8)] + \infty[(\ell+k)C_{2v}{}^z(4)] + \infty[(j+i)C_{2v}{}^y(4)]$ $+ \infty[(h+g)C_{2v}{}^x(4)] + (f+e)C_{2h}{}^z(4) + (d+c+b+a)D_{2h}(2)$
66	D_{2h}^{20} $(C2/c2/c2/m)$	$\infty[mC_1(16)] + \infty[\ell C_S{}^{xy}(8)] + \infty[(k+j+i)C_2{}^z(8)] + \infty[hC_2{}^y(8)]$ $+ \infty[gC_2{}^x(8)] + (f+e+d+c)C_{2h}{}^z(4) + (b+a)D_2(4)$
67	D_{2h}^{21} $(C2/m2/m2/e)$	$\infty[oC_1\ (16)] + \infty[nC_S{}^{xz}(8)] + \infty[mC_S{}^{yz}(8)] + \infty[\ell C_2{}^z(8)]$ $+ \infty[(k+j)C_2{}^y(8)] + \infty[(i+h)C_2{}^x(8)] + \infty[gC_{2v}{}^z(4)]$ $+ (f+e)C_{2h}{}^y(4) + (d+c)C_{2h}{}^x(4) + (b+a)D_2(4)$
68	D_{2h}^{22} $(C2/c2/c2/e)$	$\infty[iC_1(16)] + \infty[(h+g)C_2{}^z(8)] + \infty[fC_2{}^y(8)] + \infty[eC_2{}^x(8)]$ $+ (d+c)C_i(8) + (b+a)D_2(4)$
69	D_{2h}^{23} $(F2/m2/m2/m)$	$\infty[pC_1(32)] + \infty[oC_S{}^{xy}(16)] + \infty[nC_S{}^{xz}(16)] + \infty[mC_S{}^{yz}(16)] + \infty[\ell C_2{}^x(16)]$ $+ \infty[kC_2{}^y(16)] + \infty[jC_2{}^z(16)] + \infty[iC_{2v}{}^z(8)] + \infty[hC_{2v}{}^y(8)] + \infty[gC_{2v}{}^x(8)]$ $+ fD_2(8) + eC_{2h}{}^z(8) + dC_{2h}{}^y(8) + cC_{2h}{}^x(8) + (b+a)D_{2h}(4)$
70	D_{2h}^{24} $(F2/d2/d2/d)$	$\infty[hC_1(32)] + \infty[gC_2{}^z(16)] + \infty[fC_2{}^y(16)] + \infty[eC_2{}^x(16)]$ $+ (d+c)C_i(16) + (b+a)D_2(8)$
71	D_{2h}^{25} $(I2/m2/m2/m)$	$\infty[oC_1(16)] + \infty[nC_S{}^{xy}(8)] + \infty[mC_S{}^{xz}(8)] + \infty[\ell C_S{}^{yz}(8)]$ $+ kC_i(8) + \infty[(j+i)C_{2v}{}^z(4)] + \infty[(h+g)C_{2v}{}^y(4)]$ $+ [(f+e)C_{2v}{}^x(4)] + (d+c+b+a)D_{2h}(2)$
72	D_{2h}^{26} $(I2/b2/a2/m)$	$\infty[kC_1(16)] + \infty[jC_S{}^{xy}(8)] + \infty[(i+h)C_2{}^z(8)] + \infty[gC_2{}^y(8)]$ $+ \infty[(fC_2{}^x(8)] + eC_i(8) + (d+c)C_{2h}{}^z(4) + (b+a)D_2(4)$
73	D_{2h}^{27} $(I2_1/b2_1/c2_1/a)$	$\infty[fC_1(16)] + \infty[eC_2{}^z(8)] + \infty[dC_2{}^y(8)] + \infty[cC_2{}^x(8)] + (b+a)C_i(8)$
74	D_{2h}^{28} $(I2_1/m2_1/m2_1/a)$	$\infty[jC_1(16)] + \infty[iC_S{}^{xz}(8)] + \infty[hC_S{}^{yz}(8)] + \infty[gC_2{}^y(8)]$ $+ \infty[fC_2{}^x(8)] + \infty[eC_{2v}{}^z(4)] + (d+c)C_{2h}{}^y(4) + (b+a)C_{2h}{}^x(4)$

(**i**)-9. C_4 (4)

No.	Space-group	Sites
75	C_4^1 ($P4$)	$\infty[dC_1(4)] + \infty[cC_2(2)] + \infty[(b+a)C_4(1)]$
76	C_4^2 ($P4_1$)	$\infty[aC_1(4)]$
77	C_4^3 ($P4_2$)	$\infty[dC_1(4)] + \infty[(c+b+a)C_2(2)]$
78	C_4^4 ($P4_3$)	$\infty[aC_1(4)]$
79	C_4^5 ($I4$)	$\infty[cC_1(8)] + \infty[bC_2(4)] + \infty[aC_4(2)]$
80	C_4^6 ($I4_1$)	$\infty[bC_1(8)] + \infty[aC_2(4)]$

(**i**)-10. S_4 ($\bar{4}$)

No.	Space-group	Sites
81	S_4^1 ($P\bar{4}$)	$\infty[hC_1(4)] + \infty[(g+f+e)C_2(2)] + (d+c+b+a)S_4(1)$
82	S_4^2 ($I\bar{4}$)	$\infty[gC_1(8)] + \infty[(f+e)C_2(4)] + (d+c+b+a)S_4(2)$

(**i**)-11. C_{4h} ($4/m$)

No.	Space-group	Sites
83	C_{4h}^1 ($P4/m$)	$\infty[\ell C_1(8)] + \infty[(k+j)C_S{}^{xy}(4)] + \infty[iC_2(4)] + \infty[(h+g)C_4(2)] + (f+e)C_{2h}(2)$ $+ (d+c+b+a)C_{4h}(1)$
84	C_{4h}^2 ($P4_2/m$)	$\infty[kC_1(8)] + \infty[jC_S{}^{xy}(4)] + \infty[(i+h+g)C_2(4)] + (f+e)S_4(2) + (d+c+b+a)C_{2h}(2)$
85	C_{4h}^3 ($P4/n$)	$\infty[gC_1(8)] + \infty[fC_2(4)] + (e+d)C_i(4) + \infty[cC_4(2)] + (b+a)S_4(2)$
86	C_{4h}^4 ($P4_2/n$)	$\infty[gC_1(8)] + \infty[(f+e)C_2(4)] + (d+c)C_i(4) + (b+a)S_4(2)$
87	C_{4h}^5 ($I4/m$)	$\infty[iC_1(16)] + \infty[hC_S{}^{xy}(8)] + \infty[gC_2(8)] + fC_i(8) + \infty[eC_4(4)] + dS_4(4)]$ $+ cC_{2h}(4) + (b+a)C_{4h}(2)$
88	C_{4h}^6 ($I4_1/a$)	$\infty[fC_1(16)] + \infty[eC_2(8)] + (d+c)C_i(8) + (b+a)S_4(4)$

(**i**)-12. D_4 (422)

No.	Space-group	Sites
89	D_4^1 ($P422$)	$\infty[pC_1(8)] + \infty[(o+n+m+\ell)C_2{}^x(4)] + \infty[(k+j)C_2'(4)] + \infty[iC_2{}^z(4)] + \infty[(h+g)C_4(2)]$ $+ (f+e)D_2(2) + (d+c+b+a)D_4(1)$
90	D_4^2 ($P42_12$)	$\infty[gC_1(8)] + \infty[(f+e)C_2'(4)] + \infty[dC_2{}^z(4)] + \infty[cC_4(2)] + (b+a)D_2'(2)$
91	D_4^3 ($P4_122$)	$\infty[dC_1(8)] + \infty[cC_2'(4)] + \infty[(b+a)C_2{}^y(4)]$
92	D_4^4 ($P4_12_12$)	$\infty[bC_1(8)] + \infty[aC_2'(4)]$

(i)-12. D_4 (422) ...continued

No.	Space-group	Sites
93	D_4^5 ($P4_222$)	$\infty\,[pC_1(8)] + \infty\,[(o+n)C_2'(4)] + \infty\,[(m+\ell+k+j)C_2{}^x(4)] + \infty\,[(i+h+g)C_2{}^z(4)] + (f+e)D_2'(2) + (d+c+b+a)D_2(2)$
94	D_4^6 ($P4_22_12$)	$\infty\,[gC_1(8)] + \infty\,[(f+e)C_2'(4)] + \infty\,[(d+c)C_2{}^z(4)] + (b+a)D_2'(2)$
95	D_4^7 ($P4_322$)	$\infty\,[dC_1(8)] + \infty\,[cC_2'(4)] + \infty\,[(b+a)C_2{}^y(4)]$
96	D_4^8 ($P4_32_12$)	$\infty\,[bC_1(8)] + \infty\,[aC_2'(4)]$
97	D_4^9 ($I422$)	$\infty\,[kC_1(16)] + \infty\,[(i+h)C_2{}^x(8)] + \infty\,[(j+g)C_2'(8)] + \infty\,[fC_2{}^z(8)] + \infty\,[eC_4\ (4)] + dD_2'(4) + cD_2(4) + (b+a)D_4(2)$
98	D_4^{10} ($I4_122$)	$\infty\,[gC_1(16)] + \infty\,[fC_2{}^x(8)] + \infty\,[(e+d)C_2'8)] + \infty\,[cC_2{}^z(8)] + (b+a)D_2'(4)$

(i)-13. C_{4v} $(4mm)$

No.	Space-group	Sites
99	C_{4v}^1 ($P4mm$)	$\infty[gC_1(8)] + \infty\,[(f+e)C_S{}^v(4)] + \infty\,[dC_S{}^d(4)] + \infty\,[cC_{2v}{}^v(2)] + \infty\,[(b+a)C_{4v}(1)]$
100	C_{4v}^2 ($P4bm$)	$\infty[dC_1(8)] + \infty\,[(cC_S{}^d(4)] + \infty\,[bC_{2v}{}^d(2)] + \infty\,[aC_4(2)]$
101	C_{4v}^3 ($P4_2cm$)	$\infty[eC_1(8)] + \infty\,[dC_S{}^d(4)] + \infty\,[cC_2(4)] + \infty\,[(b+a)C_{2v}{}^d(2)]$
102	C_{4v}^4 ($P4_2nm$)	$\infty[dC_1(8)] + \infty\,[cC_S{}^d(4)] + \infty\,[bC_2(4)] + \infty\,[aC_{2v}{}^d(2)]$
103	C_{4v}^5 ($P4cc$)	$\infty[dC_1(8)] + \infty\,[cC_2(4)] + \infty\,[(b+a)C_4(2)]$
104	C_{4v}^6 ($P4nc$)	$\infty[cC_1(8)] + \infty\,[bC_2(4)] + \infty\,[aC_4(2)]$
105	C_{4v}^7 ($P4_2mc$)	$\infty[fC_1(8)] + \infty\,[(e+d)C_S{}^v(4)] + \infty\,[(c+b+a)C_{2v}{}^v(2)]$
106	C_{4v}^8 ($P4_2bc$)	$\infty[cC_1(8)] + \infty\,[(b+a)C_2(4)]$
107	C_{4v}^9 ($I4mm$)	$\infty[eC_1(16)] + \infty\,[dC_S{}^v(8)] + \infty\,[cC_S{}^d(8)] + \infty\,[bC_{2v}{}^v(4)] + \infty\,[aC_{4v}(2)]$
108	C_{4v}^{10} ($I4cm$)	$\infty[dC_1(16)] + \infty\,[cC_S{}^d(8)] + \infty\,[bC_{2v}{}^d(4)] + \infty\,[aC_4(4)]$
109	C_{4v}^{11} ($I4_1md$)	$\infty[cC_1(16)] + \infty\,[bC_S{}^v(8)] + \infty\,[aC_{2v}{}^v(4)]$
110	C_{4v}^{12} ($I4_1cd$)	$\infty[bC_1(16)] + \infty\,[aC_2(8)]$

(i)-14. D_{2d} $(\bar{4}2m)$

No.	Space-group	Sites
111	D_{2d}^1 $(P\bar{4}2m)$	$\infty[oC_1(8)] + \infty[nC_S(4)] + \infty[mC_2^z(4)] + \infty[(\ell+k+j+i)C_2(4)] + \infty[(h+g)C_{2v}(2)] + (f+e)D_2(2) + (d+c+b+a)D_{2d}(1)$
112	D_{2d}^2 $(P\bar{4}2c)$	$\infty[nC_1(8)] + \infty[(m+\ell+k)C_2^z(4)] + \infty[(j+i+h+g)C_2(4)] + (f+e)S_4(2) + (d+c+b+a)D_2(2)$
113	D_{2d}^3 $(P\bar{4}2_1m)$	$\infty[fC_1(8)] + \infty[eC_S(4)] + \infty[dC_2^z(4)] + \infty[cC_{2v}(2)] + (a+b)S_4(2)$
114	D_{2d}^4 $(P\bar{4}2_1c)$	$\infty[eC_1(8)] + \infty[(d+c)C_2^z(4)] + (b+a)S_4(2)$
115	D_{2d}^5 $(P\bar{4}m2)$	$\infty[\ell C_1(8)] + \infty[(k+j)C_S(4)] + \infty[(i+h)C_2(4)] + \infty[(g+f+e)C_{2v}(2)] + (d+c+b+a)D_{2d}(1)$
116	D_{2d}^6 $(P\bar{4}c2)$	$\infty[jC_1(8)] + \infty[(i+h+g)C_2^z(4)] + \infty[(f+e)C_2(4)] + (d+c)S_4(2)] + (b+a)D_2(2)$
117	D_{2d}^7 $(P\bar{4}b2)$	$\infty[iC_1(8)] + \infty[(h+g)C_2(4)] + \infty[(f+e)C_2^z(4)] + (d+c)D_2(2) + (b+a)S_4(2)]$
118	D_{2d}^8 $(P\bar{4}n2)$	$\infty[iC_1(8)] + \infty[(g+f)C_2(4)] + \infty[(h+e)C_2^z(4)] + (d+c)D_2(2) + (b+a)S_4(2)]$
119	D_{2d}^9 $(I\bar{4}m2)$	$\infty[jC_1(16)] + \infty[iC_S^{xz}(8)] + \infty[(h+g)C_2(8)] + \infty[(f+e)C_{2v}(4)] + (d+c+b+a)D_{2d}(2)$
120	D_{2d}^{10} $(I\bar{4}c2)$	$\infty[iC_1(16)] + \infty[(g+f)C_2^z(8)] + \infty[(h+e)C_2(8)] + (c+b)S_4(4) + (d+a)D_2(4)$
121	D_{2d}^{11} $(I\bar{4}2m)$	$\infty[jC_1(16)] + \infty[iC_S(8)] + \infty[hC_2^z(8)] + \infty[(g+f)C_2(8)] + \infty[eC_{2v}(4)] + dS_4(4) + cD_2(4) + (b+a)D_{2d}(2)$
122	D_{2d}^{12} $(I\bar{4}2d)$	$\infty[eC_1(16)] + \infty[dC_2^x(8)] + \infty[C_2^z(8)] + (b+a)S_4(4)$

(i)-15. D_{4h} $(4/mmm)$

No.	Space-group	Sites
123	D_{4h}^1 $(P4/m2/m2/m)$	$\infty[uC_1(16)] + \infty[(t+s)C_S^v(8)] + \infty[rC_S^d(8)] + \infty[(q+p)C_S^h(8)] + \infty[(o+n+m+\ell)C_{2v}(4)] + \infty[(k+j)C'_{2v}(4)] + \infty[iC_{2v}^v(4)] + \infty[(h+g)C_{4v}(2)] + (f+e)D_{2h}(2)] + (d+c+b+a)D_{4h}(1)$
124	D_{4h}^2 $(P4/m2/c2/c)$	$\infty[nC_1(16)] + \infty[(mC_S^h(8)] + \infty[(\ell+k)C_2(8)] + \infty[jC'_2(8)] + \infty[iC_2^z(8)] + \infty[(h+g)C_4^z(4) + fD_2(4) + eC_{2h}(4) + (d+b)C_{4h}(2) + (c+a)D_4(2)$
125	D_{4h}^3 $(P4/n2/b2/m)$	$\infty[nC_1(16)] + \infty[mC_S^d(8)] + \infty[(\ell+k)C_2(8)] + \infty[(j+i)C'_2(8)] + \infty[hC_{2v}^d(4)] + \infty[gC_4(4)] + (f+e)C'_{2h}(4) + (d+c)D'_{2d}(2) + (b+a)D_4(2)$
126	D_{4h}^4 $(P4/n2/n2/c)$	$\infty[kC_1(16)] + \infty[(j+i)C_2^v(8)] + \infty[hC_2^d(8)] + gC_2^z(8) + fC_i(8) + \infty[eC_4(4)] + dS_4(4) + cD_2(4) + (b+a)D_4(2)$
127	D_{4h}^5 $(P4/m2_1/b2/m)$	$\infty[\ell C_1(16)] + \infty[kC_S^d(8)] + \infty[(j+i)C_S^h(8)] + \infty[(h+g)C'_{2v}(4)] + \infty[fC_{2v}^d(4)] + \infty[eC_4(4)] + (d+c)D'_{2h}(2) + (b+a)C_{4h}(2)$
128	D_{4h}^6 $(P4/m2_1/n2/c)$	$\infty[iC_1(16)] + \infty[hC_S^h(8)] + \infty[gC'_2(8)] + \infty[fC_2^z(8)] + \infty[eC_4(4)] + dD'_2(4) + cC_{2h}^z(4) + (b+a)C_{4h}(2)$

(i)-15. D_{4h} $(4/mmm)$ ...continued

No.	Space-group	Sites
129	D_{4h}^{7} $(P4/n2_1/m2/m)$	$\infty[kC_1(16)] + \infty[jC_S{}^{d}(8)] + \infty[iC_S{}^{\mathrm{v}}(8)] + \infty[(h+g)C'_2(8)] + \infty[fC_{2\mathrm{v}}{}^{\mathrm{v}}(4)]$ $+ (e+d)C'_{2h}(4) + \infty[cC_{4\mathrm{v}}(2)] + (b+a)D_{2d}(2)$
130	D_{4h}^{8} $(P4/n2_1/c2/c)$	$\infty[gC_1(16)] + \infty[fC'_2(8)] + \infty[eC_2{}^{z}(8)] + dC_i(8) + \infty[cC_4(4)] + bS_4(4) + aD'_2(4)$
131	D_{4h}^{9} $(P4_2/m2/m2/c)$	$\infty[rC_1(16)] + \infty[(qC_S{}^{h}(8)] + \infty[(p+o)C_S{}^{\mathrm{v}}(8)] + \infty[nC'_2(8)]$ $+ \infty[(m+\ell+k+j)C_{2\mathrm{v}}(4)] + \infty[(i+h+g)C_{2\mathrm{v}}{}^{\mathrm{v}}(4)] + (f+e)D_{2d}(2)] + (d+c+b+a)D_{2h}(2)]$
132	D_{4h}^{10} $(P4_2/m2/c2/m)$	$\infty[pC_1(16)] + \infty[(oC_S{}^{d}(8)] + \infty[nC_S{}^{h}(8)] + \infty[(m+\ell)C_2(8)] + \infty[kC_2{}^{z}(8)]$ $+ \infty[(j+i)C'_{2\mathrm{v}}(4)] + \infty[(h+g)C_{2\mathrm{v}}{}^{d}(4)] + fC_{2h}{}^{z}(4) + eD_2(4)] + (d+b)D'_{2d}(2) + (c+a)D'_{2h}(2)$
133	D_{4h}^{11} $(P4_2/n2/b2/c)$	$\infty[kC_1(16)] + \infty[jC'_2(8)] + \infty[(i+h)C_2(8)] + \infty[(g+f)C_2{}^{z}(8)] + eC_i(8) + dS_4(4)$ $+ cD'_2(4) + (b+a)D_2(4)$
134	D_{4h}^{12} $(P4_2/n2/n2/m)$	$\infty[nC_1(16)] + \infty[mC_S{}^{d}(8)] + \infty[(\ell+k)C'_2(8)] + \infty[(j+i)C_2(8)] + \infty[hC_2{}^{z}(8)]$ $+ \infty[gC_{2\mathrm{v}}{}^{d}(4)] + (f+e)C'_{2h}(4) + dD'_2(4) + cD_2(4)] + (b+a)D'_{2h}(2)]$
135	D_{4h}^{13} $(P4_2/m2_1/b2/c)$	$\infty[iC_1(16)] + \infty[hC_S{}^{h}(8)] + \infty[gC'_2(8)] + \infty[(f+e)C_2{}^{z}(8)] + dD'_2(4)$ $+ bS_4(4) + (c+a)C_{2h}{}^{z}(4)$
136	D_{4h}^{14} $(P4_2/m2_1/n2/m)$	$\infty[kC_1(16)] + \infty[(jC_S{}^{d}(8)] + \infty[iC_S{}^{h}(8)] + \infty[hC_2{}^{z}(8)] + \infty[(g+f)C'_{2\mathrm{v}}(4)]$ $+ \infty[eC_{2\mathrm{v}}{}^{d}(4)] + dS_4(4) + cC_{2h}{}^{z}(4) + (b+a)D'_{2d}(2)$
137	D_{4h}^{15} $(P4_2/n2_1/m2/c)$	$\infty[hC_1(16)] + \infty[(gC_S{}^{\mathrm{v}}(8)] + \infty[fC'_2(8)] + eC_i(8) + \infty[(d+c)C'_{2\mathrm{v}}(4)] + (b+a)D_{2d}(2)$
138	D_{4h}^{16} $(P4_2/n2_1/c2/m)$	$\infty[jC_1(16)] + \infty[(iC_S{}^{d}(8)] + \infty[(h+g)C'_2(8)] + \infty[fC_2{}^{z}(8)]$ $+ \infty[eC_{2\mathrm{v}}{}^{d}(4)] + (d+c)C'_{2h}(4) + bS_4(4) + aD'_2(4)$
139	D_{4h}^{17} $(I4/m2/m2/m)$	$\infty[oC_1(32)] + \infty[nC_S{}^{\mathrm{v}}(16)] + \infty[mC_S{}^{d}(16)] + \infty[\ell C_S{}^{h}(16)] + \infty[kC'_2(16)]$ $+ \infty[(j+i)C_{2\mathrm{v}}(8)] + \infty[hC'_{2\mathrm{v}}(8)] + \infty[gC_{2\mathrm{v}}{}^{\mathrm{v}}(8)] + fC'_{2h}(8) + \infty[eC_{4v}(4)]$ $+ dD_{2d}(4) + cD_{2h}(4) + (b+a)D_{4h}(2)$
140	D_{4h}^{18} $(I4/m2/c2/m)$	$\infty[mC_1(32)] + \infty[\ell C_S{}^{d}(16)] + \infty[kC_S{}^{h}(16)] + \infty[jC_2(16)] + \infty[iC'_2(16)]$ $+ \infty[hC'_{2\mathrm{v}}(8)] + \infty[gC_{2\mathrm{v}}{}^{d}(8)] + \infty[fC_4(8)] + eC'_{2h}(8) + dD'_{2h}(4)$ $+ cC_{4h}(4) + bD'_{2d}(4) + aD_4(4)$
141	D_{4h}^{19} $(I4_1/a2/m2/d)$	$\infty[iC_1(32)] + \infty[(hC_S(16)] + \infty[gC'_2(16)] + \infty[fC_2(16)] + \infty[eC_{2\mathrm{v}}{}^{\mathrm{v}}(8)]$ $+ (d+c)C_{2h}(8) + (b+a)D_{2d}(4)$
142	D_{4h}^{20} $(I4_1/a2/c2/d)$	$\infty[gC_1(32)] + \infty[fC'_2(16)] + \infty[eC_2(16)] + \infty[dC_2{}^{z}(16)] + cC_i(16)$ $+ bD'_2(8) + aS_4(8)$

(i)-16. C_3 (3)

No.	Space-group	Sites
143	C_3^{1} $(P3)$	$\infty[dC_1(3)] + \infty[(c+b+a)C_3(1)]$
144	C_3^{2} $(P3_1)$	$\infty[aC_1(3)]$
145	C_3^{3} $(P3_2)$	$\infty[aC_1(3)]$
146	C_3^{4} $(R3)$	$\infty[bC_1(3)] + \infty[aC_3(1)]$

(**i**)-17. S_6 ($\bar{3}$)

No.	Space-group	Sites
147	S_6^1 ($P\bar{3}$)	$\infty[gC_1(6)] + (f+e)C_i(3) + \infty[(d+c)C_3(2)] + (b+a)S_6(1)$
148	S_6^2 ($R\bar{3}$)	$\infty[fC_1(6)] + (d+e)C_i(3) + \infty[cC_3(2)] + (b+a)S_6(1)$

(**i**)-18. D_3 (32)

No.	Space-group	Sites
149	D_3^1 ($P312$)	$\infty[\ell C_1(6)] + \infty[(k+j)C_2(3)] + \infty[(i+h+g)C_3(2)] + (f+e+d+c+b+a)D_3(1)$
150	D_3^2 ($P321$)	$\infty[gC_1(6)] + \infty[(f+e)C_2(3)] + \infty[(d+c)C_3(2)] + (b+a)D_3(1)$
151	D_3^3 ($P3_112$)	$\infty[cC_1(6)] + \infty[(b+a)C_2(3)]$
152	D_3^4 ($P3_121$)	$\infty[cC_1(6)] + \infty[(b+a)C_2(3)]$
153	D_3^5 ($P3_212$)	$\infty[cC_1(6)] + \infty[(b+a)C_2(3)]$
154	D_3^6 ($P3_221$)	$\infty[cC_1(6)] + \infty[(b+a)C_2(3)]$
155	D_3^7 ($R32$)	$\infty[fC_1(6)] + \infty[(e+d)C_2(3)] + \infty[cC_3(2)] + (b+a)D_3(1)$

(**i**)-19. C_{3v} ($3m$)

No.	Space-group	Sites
156	C_{3v}^1 ($P3m1$)	$\infty[eC_1(6)] + \infty[dC_S(3)] + \infty[(c+b+a)C_{3v}(1)]$
157	C_{3v}^2 ($P31m$)	$\infty[dC_1(6)] + \infty[cC_S(3)] + \infty[bC_3(2)] + \infty[aC_{3v}(1)]$
158	C_{3v}^3 ($P3c1$)	$\infty[dC_1(6)] + \infty[(c+b+a)C_3(2)]$
159	C_{3v}^4 ($P31c$)	$\infty[cC_1(6)] + \infty[(b+a)C_3(2)]$
160	C_{3v}^5 ($R3m$)	$\infty[cC_1(6)] + \infty[bC_S(3)] + \infty[aC_{3v}(1)]$
161	C_{3v}^6 ($R3c$)	$\infty[bC_1(6)] + \infty[aC_3(2)]$

(**i**)-20. D_{3d} ($\bar{3}m$)

No.	Space-group	Sites
162	$D_{3d}^1(P\bar{3}12/m)$	$\infty[\ell C_1(12)] + \infty[kC_S(6)] + \infty[(j+i)C_2(6)] + \infty[hC_3(4)] + (g+f)C_{2h}(3) + \infty[eC_{3v}(2)]$ $+ (d+c)D_3(2) + (b+a)D_{3d}(1)$
163	D_{3d}^2 ($P\bar{3}12/c$)	$\infty[iC_1(12)] + \infty[hC_2(6)] + gC_i(6) + \infty[(f+e)C_3(4)] + bS_6(2) + (d+c+a)D_3(2)$
164	D_{3d}^3 ($P\bar{3}2/m1$)	$\infty[jC_1(12)] + \infty[iC_S(6)] + \infty[(h+g)C_2(6)] + (f+e)C_{2h}(3) + \infty[(d+c)C_{3v}(2)] + (b+a)D_{3d}(1)$
165	D_{3d}^4 ($P\bar{3}2/c1$)	$\infty[gC_1(12)] + \infty[fC_2(6)] + eC_i(6) + \infty[(d+c)C_3(4)] + bS_6(2) + aD_3(2)$

(i)-20. D_{3d} ($\bar{3}m$) ....continued

No.	Space-group	Sites
166	D_{3d}^5 ($R\bar{3}2/m$)	$\infty[iC_1(12)] + \infty[hC_S(6)] + \infty[(g+f)C_2(6)] + (e+d)C_{2h}(3) + \infty[cC_{3v}(2)] + (b+a)D_{3d}(1)$
167	D_{3d}^6 ($R\bar{3}2/c$)	$\infty[fC_1(12)] + \infty[eC_2(6)] + dC_i(6) + \infty[cC_3(4)] + bS_6(2) + aD_3(2)$

(i)-21. C_6 (6)

No.	Space-group	Sites
168	C_6^1 ($P6$)	$\infty[dC_1(6)] + \infty[cC_2(3)] + \infty[bC_3(2)] + \infty[aC_6(1)]$
169	C_6^2 ($P6_1$)	$\infty[aC_1(6)]$
170	C_6^3 ($P6_5$)	$\infty[aC_1(6)]$
171	C_6^4 ($P6_2$)	$\infty[cC_1(6)] + \infty[(b+a)C_2(3)]$
172	C_6^5 ($P6_4$)	$\infty[cC_1(6)] + \infty[(b+a)C_2(3)]$
173	C_6^6 ($P6_3$)	$\infty[cC_1(6)] + \infty[(b+a)C_3(2)]$

(i)-22. C_{3h} ($\bar{6}$)

No.	Space-group	Sites
174	C_{3h}^1 ($P\bar{6}$)	$\infty[\ell C_1(6)] + \infty[(k+j)C_S(3)] + \infty[(i+h+g)C_3(2)] + (f+e+d+c+b+a)C_{3h}(1)$

(i)-23. C_{6h} ($6/m$)

No.	Space-group	Sites
175	C_{6h}^1 ($P6/m$)	$\infty[\ell C_1(12)] + \infty[(k+j)C_S(6)] + \infty[iC_2(6)] + \infty[hC_3(4)] + (g+f)C_{2h}(3) + \infty[eC_6(2)] + (d+c)C_{3h}(2) + (b+a)C_{6h}(1)$
176	C_{6h}^2 ($P6_3/m$)	$\infty[iC_1(12)] + \infty[hC_S(6)] + gC_i(6) + \infty[(f+e)C_3(4)] + bS_6(2) + (d+c+a)C_{3h}(2)$

(i)-24. D_6 (622)

No.	Space-group	Sites
177	D_6^1 ($P622$)	$\infty[nC_1(12)] + \infty[(m+\ell)C_2'(6)] + \infty[(k+j)C_2(6)] + \infty[iC_2^z(6)] + \infty[hC_3(4)] + (g+f)D_2(3) + \infty[eC_6(2)] + (d+c)D_3'(2) + (b+a)D_6(1)$
178	D_6^2 ($P6_122$)	$\infty[cC_1(12)] + \infty[bC_2'(6)] + \infty[aC_2(6)]$
179	D_6^3 ($P6_522$)	$\infty[cC_1(12)] + \infty[bC_2'(6)] + \infty[aC_2(6)]$
180	D_6^4 ($P6_222$)	$\infty[kC_1(12)] + \infty[(j+i)C_2'(6)] + \infty[(h+g)C_2(6)] + \infty[(f+e)C_2^z(6)] + (d+c+b+a)D_2(3)$
181	D_6^5 ($P6_422$)	$\infty[kC_1(12)] + \infty[(j+i)C_2(6)] + \infty[(h+g)C_2'(6)] + \infty[(f+e)C_2^z(6)] + (d+c+b+a)D_2(3)$
182	D_6^6 ($P6_322$)	$\infty[iC_1(12)] + \infty[hC_2(6)] + \infty[gC_2'(6)] + \infty[(f+e)C_3(4)] + (d+c+b)D_3'(2) + aD_3(2)$

(**i**)-25. C_{6v} $(6mm)$

No.	Space-group	Sites
183	C_{6v}^1 $(P6mm)$	$\infty\,[fC_1(12)] + \infty\,[eC_S^d(6)] + \infty\,[dC_S^v(6)] + \infty\,[cC_{2v}(3)] + \infty\,[bC_{3v}^d(2)] + \infty\,[aC_{6v}(1)]$
184	C_{6v}^2 $(P6cc)$	$\infty[dC_1(12)] + \infty\,[cC_2(6)] + \infty\,[bC_3(4)] + \infty\,[aC_6(2)]$
185	C_{6v}^3 $(P6_3cm)$	$\infty\,[dC_1(12)] + \infty\,[cC_S^v(6)] + \infty\,[bC_3(4)] + \infty\,[aC_{3v}^v(2)]$
186	C_{6v}^4 $(P6_3mc)$	$\infty\,[dC_1(12)] + \infty\,[cC_S^d(6)] + \infty\,[(b+a)C_{3v}^d(2)]$

(**i**)-26. D_{3h} $(\bar{6}m2)$

No.	Space-group	Sites
187	D_{3h}^1 $(P\bar{6}m2)$	$\infty\,[oC_1(12)] + \infty\,[nC_S^v(6)] + \infty\,[(m+\ell)C_S^h(6)] + \infty\,[(k+j)C_{2v}(3)] + \infty\,[(i+h+g)C_{3v}(2)]$ $+ (f+e+d+c+b+a)D_{3h}(1)$
188	D_{3h}^2 $(P\bar{6}c2)$	$\infty\,[\ell C_1(12)] + \infty\,[kC_S^h(6)] + \infty\,[jC_2(6)] + \infty\,[(i+h+g)C_3(4)] + (f+d+b)C_{3h}(2)$ $+ (e+c+a)D_3(2)$
189	D_{3h}^3 $(P\bar{6}2m)$	$\infty\,[\ell C_1(12)] + \infty\,[(k+j)C_S^h(6)] + \infty\,[iC_S^v(6)] + \infty\,[hC_3(4)] + \infty\,[(g+f)C_{2v}(3)]$ $+ \infty\,[eC_{3v}(2)] + (d+c)C_{3h}(2) + (b+a)D_{3h}(1)$
190	D_{3h}^4 $(P\bar{6}2c)$	$\infty\,[iC_1(12)] + \infty\,[hC_S^h(6)] + \infty\,[gC_2(6)] + \infty\,[(f+e)C_3(4)] + (d+c+b)C_{3h}(2) + aD_3(2)$

(**i**)-27. D_{6h} $(6/mmm)$

No.	Space-group	Sites
191	D_{6h}^1 $(P6/m2/m2/m)$	$\infty\,[rC_1(24)] + \infty\,[(q+p)C_S^h(12)] + \infty\,[oC_S^d(12)] + \infty\,[nC_S^v(12)] + \infty\,[(m+\ell)C_{2v}'(6)]$ $+ \infty\,[(k+j)C_{2v}(6)] + \infty\,[iC_{2v}^z(6)] + \infty\,[hC_{3v}^d(4)] + (g+f)D_{2h}(3) + \infty\,[eC_{6v}(2)]$ $+ (d+c)D_{3h}'(2) + (b+a)D_{6h}(1)$
192	D_{6h}^2 $(P6/m2/c2/c)$	$\infty\,[mC_1(24)] + \infty\,[\ell C_S^h(12)] + \infty\,[kC_2'(12)] + \infty\,[jC_2(12)] + \infty\,[iC_2^z(12)] + \infty\,[hC_3(8)]$ $+ gC_{2h}^z(6) + fD_2(6) + \infty\,[eC_6(4)] + dC_{3h}(4) + cD_3'(4) + bC_{6h}(2) + aD_6(2)$
193	D_{6h}^3 $(P6_3/m2/c2/m)$	$\infty\,[\ell C_1(24)] + \infty\,[kC_S^v(12)] + \infty\,[jC_S^h(12)] + \infty\,[iC_2'(12)] + \infty\,[hC_3(8)] + \infty\,[gC_{2v}(6)]$ $+ fC_{2h}(6) + \infty\,[eC_{3v}^v(4)] + dD_3'(4) + cC_{3h}(4) + bD_{3d}'(2) + aD_{3h}(2)$
194	D_{6h}^4 $(P6_3/m2/m2/c)$	$\infty\,[\ell C_1(24)] + \infty\,[kC_S^d(12)] + \infty\,[jC_S^h(12)] + \infty\,[iC_2(12)] + \infty\,[hC_{2v}'(6)] + gC_{2h}(6)$ $+ \infty\,[(f+e)C_{3v}^d(4)] + (d+c+b)D_{3h}'(2) + aD_{3d}(2)$

(**i**)-28. T (23)

No.	Space-group	Sites
195	T^1 ($P23$)	$\infty[jC_1(12)] + \infty[(i+h+g+f)C_2(6)] + \infty[eC_3(4)] + (d+c)D_2(3) + (b+a)T(1)$
196	T^2 ($F23$)	$\infty[hC_1(48)] + \infty[(g+f)C_2(24)] + \infty[eC_3(16)] + (d+c+b+a)T(4)$
197	T^3 ($I23$)	$\infty[fC_1(24)] + \infty[(e+d)C_2(12)] + \infty[cC_3(8)] + bD_2(6) + aT(2)$
198	T^4 ($P2_13$)	$\infty[bC_1(12)] + \infty[aC_3(4)]$
199	T^5 ($I2_13$)	$\infty[cC_1(24)] + \infty[bC_2(12)] + \infty[aC_3(8)]$

(**i**)-29. T_h ($m3$)

No.	Space-group	Sites
200	T_h^1 ($P2/m\bar{3}$)	$\infty[\ell C_1(24)] + \infty[(k+j)C_S(12)] + \infty[iC_3(8)] + \infty[(h+g+f+e)C_{2v}(6)] + (d+c)D_{2h}(3) + (b+a)T_h(1)$
201	T_h^2 ($P2/n\bar{3}$)	$\infty[nC_1(24)] + \infty[(g+f)C_2(12)] + \infty[eC_3(8)] + dD_2(6)] + (c+b)S_6(4) + aT(2)$
202	T_h^3 ($F2/m\bar{3}$)	$\infty[iC_1(96)] + \infty[hC_S(48)] + \infty[gC_2(48)] + \infty[fC_3(32)] + \infty[eC_{2v}(24)] + dC_{2h}(24) + cT(8) + (b+a)T_h(4)$
203	T_h^4 ($F2/d\bar{3}$)	$\infty[gC_1(96)] + \infty[fC_2(48)] + \infty[eC_3(32)] + (d+c)S_6(16)] + (b+a)T(8)$
204	T_h^5 ($I2/m\bar{3}$)	$\infty[hC_1(48)] + \infty[gC_S(24)] + \infty[fC_3(16)] + \infty[(e+d)C_{2v}(12)] + cS_6(8) + bD_{2h}(6) + aT_h(2)$
205	T_h^6 ($P2_1/a\bar{3}$)	$\infty[dC_1(24)] + \infty[cC_3(8)] + (b+a)S_6(4)$
206	T_h^7 ($I2_1/a\bar{3}$)	$\infty[eC_1(48)] + \infty[dC_2(24)] + \infty[cC_3(16)] + (b+a)S_6(8)$

(**i**)-30. O (432)

No.	Space-group	Sites
207	O^1 ($P432$)	$\infty[kC_1(24)] + \infty[(j+i)C_2'(12)] + \infty[hC_2(12)] + \infty[gC_3(8)] + \infty[(f+e)C_4(6)] + (d+c)D_4(3) + (b+a)O(1)$
208	O^2 ($P4_232$)	$\infty[mC_1(24)] + \infty[(\ell+k)C_2'(12)] + \infty[(j+i+h)C_2(12)] + \infty[gC_3(8)] + (f+e)D_2'(6) + dD_2(6) + (c+b)D_3(4) + aT(2)$
209	O^3 ($F432$)	$\infty[jC_1(96)] + \infty[iC_2(48)] + \infty[(h+g)C_2'(48)] + \infty[fC_3(32)] + \infty[eC_4(24)] + dD_2'(24) + cT(8) + (b+a)O(4)$
210	O^4 ($F4_132$)	$\infty[hC_1(96)] + \infty[gC_2'(48)] + \infty[fC_2(48)] + \infty[eC_3(32)] + (d+c)D_3(16) + (b+a)T(8)$
211	O^5 ($I432$)	$\infty[jC_1(48)] + \infty[(i+h)C_2'(24)] + \infty[gC_2(24)] + \infty[fC_3(16)] + \infty[eC_4(12)] + dD_2'(12) + cD_3(8) + bD_4(6) + aO(2)$
212	O^6 ($P4_332$)	$\infty[eC_1(24)] + \infty[dC_2'(12)] + \infty[cC_3(8)] + (b+a)D_3(4)$
213	O^7 ($P4_132$)	$\infty[eC_1(24)] + \infty[dC_2'(12)] + \infty[cC_3(8)] + (b+a)D_3(4)$

(**i**)-30. O (432) ....continued

No.	Space-group	Sites
214	O^8 ($I432$)	$\infty[iC_1(48)] + \infty[(h+g)C_2'(24)] + \infty[fC_2(24)] + \infty[eC_3(16)] + (d+c)D_2'(12) + (b+a)D_3(8)$

(**i**)-31. T_d $(\bar{4}3m)$

No.	Space-group	Sites
215	T_d^1 ($P\bar{4}3m$)	$\infty[jC_1(24)] + \infty[iC_S(12)] + \infty[hC_2(12)] + \infty[(g+f)C_{2v}(6)] + \infty[eC_{3v}(4)]$ $+ (d+c)D_{2d}(3) + (b+a)T_d(1)$
216	T_d^2 ($F\bar{4}3m$)	$\infty[iC_1(96)] + \infty[hC_S(48)] + \infty[(g+f)C_{2v}(24)] + \infty[eC_{3v}(16)] + (d+c+b+a)T_d(4)$
217	T_d^3 ($I\bar{4}3m$)	$\infty[hC_1(48)] + \infty[gC_S(24)] + \infty[fC_2(24)] + \infty[(eC_{2v}(12)] + dD_4(12) + \infty[cC_{3v}(8)] + bD_{2d}(6) + aT_d(2)$
218	T_d^4 ($P\bar{4}3n$)	$\infty[iC_1(24)] + \infty[(h+g+f)C_2(12)] + \infty[(eC_3(8)] + (d+c)S_4(6) + bD_2(6) + aT(2)$
219	T_d^5 ($F\bar{4}3c$)	$\infty[hC_1(96)] + \infty[(g+f)C_2(48)] + \infty[(eC_3(32)] + (d+c)S_4(24) + (b+a)T(8)$
220	T_d^6 ($I\bar{4}3d$)	$\infty[eC_1(48)] + \infty[(dC_2(24)] + \infty[(cC_3(16)] + (b+a)S_4(12)$

(**i**)-32. O_h $(m\bar{3}m)$

No.	Space-group	Sites
221	O_h^1 ($P4/m\bar{3}2/m$)	$\infty[nC_1(48)] + \infty[mC_S^d(24)] + \infty[(\ell+k)C_S^h(24)] + \infty[(j+i)C_{2v}'(12)] + \infty[hC_{2v}^h(12)]$ $+ \infty[gC_{3v}(8)] + \infty[(f+e)C_{4v}(6)] + (d+c)D_{4h}(3) + (b+a)O_h(1)$
222	O_h^2 ($P4/n\bar{3}2/n$)	$\infty[iC_1(48)] + \infty[hC_2'(24)] + \infty[gC_2(24)] + \infty[fC_3(16)] + \infty[eC_4(12)] + dS_4(12)$ $+ cS_6(8) + bD_4(6) + aO(2)$
223	O_h^3 ($P4_2/m\bar{3}2/n$)	$\infty[\ell C_1(48)] + \infty[kC_S^h(24)] + \infty[jC_2'(24)] + \infty[iC_3(16)] + \infty[(h+g+f)C_{2v}^h(12)]$ $+ eD_3(8) + (d+c)D_{2d}'(6) + bD_{2h}(6) + aT_h(2)$
224	O_h^4 ($P4_2/n\bar{3}2/m$)	$\infty[\ell C_1(48)] + \infty[kC_S^d(24)] + \infty[(j+i)C_2'(24)] + \infty[hC_2(24)] + \infty[gC_{2v}^d(12)]$ $+ fD_2'(12) + \infty[eC_{3v}(8)] + dD_{2d}(6) + (c+b)D_{3d}(4) + aT_d(2)$
225	O_h^5 ($F4/m\bar{3}2/m$)	$\infty[\ell C_1(192)] + \infty[kC_S^d(96)] + \infty[jC_S^h(96)] + \infty[(i+h)C_{2v}'(48)] + \infty[gC_{2v}^d(48)]$ $+ \infty[fC_{3v}(32)] + \infty[eC_{4v}(24)] + dD_{2h}'(24) + cT_d(8) + (b+a)O_h(4)$
226	O_h^6 ($F4/m\bar{3}2/c$)	$\infty[jC_1(192)] + \infty[iC_S^h(96)] + \infty[hC_2'(96)] + \infty[gC_3(64)] + \infty[fC_4(48)] + \infty[eC_{2v}^h(48)]$ $+ dC_{4h}(24) + cD_{2d}'(24) + bT_h(8) + aO(8)$
227	O_h^7 ($F4_1/d\bar{3}2/m$)	$\infty[iC_1(192)] + \infty[hC_2'(96)] + \infty[gC_S^d(96)] + \infty[fC_{2v}^d(48)] + \infty[eC_{3v}(32)]$ $+ (d+c)D_{3d}(16) + (b+a)T_d(8)$
228	O_h^8 ($F4_1/d\bar{3}2/c$)	$\infty[hC_1(192)] + \infty[gC_2'(96)] + \infty[fC_2(96)] + \infty[eC_3(64)] + dS_4(48) + cS_6(32)$ $+ bD_3(32) + aT(16)$
229	O_h^9 ($I4/m\bar{3}2/m$)	$\infty[\ell C_1(96)] + \infty[kC_S^d(48)] + \infty[jC_S^h(48)] + \infty[iC_2'(48)] + \infty[hC_{2v}(24)] + \infty[gC_{2v}^h(24)]$ $+ \infty[fC_{3v}(16)] + \infty[eC_{4v}(12)] + dD_{2h}'(12) + cD_{3d}(8) + bD_{4h}(6) + aO_h(2)$
230	O_h^{10} ($I4_1/a\bar{3}2/d$)	$\infty[hC_1(96)] + \infty[gC_2'(48)] + \infty[fC_2(48)] + \infty[eC_3(32)] + dS_4(24) + cD_2'(24)$ $+ bD_3(16) + aS_6(16)$

Tables (ii): Irreducible representations describing the symmetry properties of translational motions related to each occupied crystalline site in a primitive unit cell of the different space groups.

(**ii**)-1. C_1 (1)

Site	Translation representations
C_1	$3A$

(**ii**)-2. C_i ($\bar{1}$)

Site	Translation representations
C_1	$3A_g \oplus 3A_u$
C_i	$3A_u$

(**ii**)-3. C_2 (2)

Site	Translation representations
C_1	$3A \oplus 3B$
C_2	$A \oplus 2B$

(**ii**)-4. C_S (m)

Site	Translation representations
C_1	$3A \oplus 3B$
C_S	$A \oplus 2B$

(**ii**)-5. C_{2h} ($2/m$)

Site	Translation representations
C_1	$3A_g \oplus 3A_u \oplus 3B_g \oplus 3B_u$
C_i	$3A_u \oplus 3B_u$
C_2	$A_g \oplus A_u \oplus 2B_g \oplus 2B_u$
C_S	$2A_g \oplus A_u \oplus B_g \oplus 2B_u$
C_{2h}	$A_u \oplus 2B_u$

(**ii**)-6. D_2 (222)

Site	Translation representations
C_1	$3A \oplus 3B_1 \oplus 3B_2 \oplus 3B_3$
C_2^z	$A \oplus B_1 \oplus 2B_2 \oplus 2B_3$
C_2^y	$A \oplus 2B_1 \oplus B_2 \oplus 2B_3$
C_2^x	$A \oplus 2B_1 \oplus 2B_2 \oplus B_3$
D_2	$B_1 \oplus B_2 \oplus B_3$

(**ii**)-7. C_{2v} ($mm2$)

Site	Translation representations
C_1	$3A_1 \oplus 3A_2 \oplus 3B_1 \oplus 3B_2$
C_2	$A_1 \oplus A_2 \oplus 2B_1 \oplus 2B_2$
C_S^{xz}	$2A_1 \oplus A_2 \oplus 2B_1 \oplus B_2$
C_S^{yz}	$2A_1 \oplus A_2 \oplus B_1 \oplus 2B_2$
C_{2v}	$A_1 \oplus B_1 \oplus B_2$

(**ii**)-8. D_{2h} (mmm)

Site	Translation representations
C_1	$3A_g \oplus 3A_u \oplus 3B_{1g} \oplus 3B_{1u} \oplus 3B_{2g} \oplus 3B_{2u} \oplus 3B_{3g} \oplus 3B_{3u}$
C_2^z	$A_g \oplus A_u \oplus B_{1g} \oplus B_{1u} \oplus 2B_{2g} \oplus 2B_{2u} \oplus 2B_{3g} \oplus 2B_{3u}$
C_2^y	$A_g \oplus A_u \oplus 2B_{1g} \oplus 2B_{1u} \oplus B_{2g} \oplus B_{2u} \oplus 2B_{3g} \oplus 2B_{3u}$
C_2^x	$A_g \oplus A_u \oplus 2B_{1g} \oplus 2B_{1u} \oplus B_{2g} \oplus B_{2u} \oplus 2B_{3g} \oplus 2B_{3u}$
C_S^{xy}	$2A_g \oplus A_u \oplus 2B_{1g} \oplus B_{1u} \oplus B_{2g} \oplus 2B_{2u} \oplus B_{3g} \oplus 2B_{3u}$
C_S^{xz}	$2A_g \oplus A_u \oplus B_{1g} \oplus 2B_{1u} \oplus 2B_{2g} \oplus B_{2u} \oplus B_{3g} \oplus 2B_{3u}$
C_S^{yz}	$2A_g \oplus A_u \oplus B_{1g} \oplus 2B_{1u} \oplus B_{2g} \oplus 2B_{2u} \oplus 2B_{3g} \oplus B_{3u}$
C_i	$3A_u \oplus 3B_{1u} \oplus 3B_{2u} \oplus 3B_{3u}$
C_{2v}^z	$A_g \oplus B_{1u} \oplus B_{2g} \oplus B_{2u} \oplus B_{3g} \oplus B_{3u}$
C_{2v}^y	$A_g \oplus B_{1g} \oplus B_{1u} \oplus B_{2u} \oplus B_{3g} \oplus B_{3u}$
C_{2v}^x	$A_g + B_{1g} + B_{1u} + B_{2g} + B_{2u} + B_{3u}$
D_2	$B_{1g} \oplus B_{1u} \oplus B_{2g} \oplus B_{2u} \oplus B_{3g} \oplus B_{3u}$
C_{2h}^z	$A_u \oplus B_{1u} \oplus 2B_{2u} \oplus 2B_{3u}$
C_{2h}^y	$A_u \oplus 2B_{1u} \oplus B_{2u} \oplus 2B_{3u}$
C_{2h}^x	$A_u \oplus 2B_{1u} \oplus 2B_{2u} \oplus B_{3u}$
D_{2h}	$B_{1u} \oplus B_{2u} \oplus B_{3u}$

(**ii**)-9. C_4 (4)

Site	Translation representations
C_1	$3A \oplus 3B \oplus 3E$
C_2	$A \oplus B \oplus 2E$
C_4	$A \oplus E$

(**ii**)-10. S_4 $(\overline{4})$

Site	Translation representations
C_1	$3A \oplus 3B \oplus 3E$
C_2	$A \oplus B \oplus 2E$
S_4	$B \oplus E$

(**ii**)-11. C_{4h} $(4/m)$

Site	Translation representations
C_1	$3A_g \oplus 3A_u \oplus 3B_g \oplus 3B_u \oplus 3E_g \oplus 3E_u$
C_i	$3A_u \oplus 3B_u \oplus 3E_u$
C_2	$A_g \oplus A_u \oplus B_g \oplus B_u \oplus 2E_g \oplus 2E_u$
C_S	$2A_g \oplus A_u \oplus 2B_g \oplus B_u \oplus E_g \oplus 2E_u$
C_{2h}	$A_u \oplus B_u \oplus 2E_u$
C_4	$A_g \oplus A_u \oplus E_g \oplus E_u$
S_4	$A_u \oplus B_g \oplus E_g \oplus E_u$
C_{4h}	$A_u \oplus E_u$

(**ii**)-12. D_4 (422)

Site	Translation representations
C_1	$3A_1 \oplus 3A_2 \oplus 3B_1 \oplus 3B_2 \oplus 6E$
$C_2{}^z$	$A_1 \oplus A_2 \oplus B_1 \oplus B_2 \oplus 4E$
C_2	$A_1 \oplus 2A_2 \oplus B_1 \oplus 2B_2 \oplus 3E$
C_2'	$A_1 \oplus 2A_2 \oplus 2B_1 \oplus B_2 \oplus 3E$
D_2	$A_2 \oplus B_2 \oplus 2E$
D_2'	$A_2 \oplus B_1 \oplus 2E$
C_4	$A_1 \oplus A_2 \oplus 2E$
D_4	$A_2 \oplus E$

(**ii**)-13. C_{4v} $(4mm)$

Site	Translation representations
C_1	$3A_1 \oplus 3A_2 \oplus 3B_1 \oplus 3B_2 \oplus 6E$
C_2	$A_1 \oplus A_2 \oplus B_1 \oplus B_2 \oplus 4E$

(**ii**)-13. C_{4v} $(4mm)$ ....continued

Site	Translation representations
$C_S{}^v$	$2A_1 \oplus A_2 \oplus 2B_1 \oplus B_2 \oplus 3E$
$C_S{}^d$	$2A_1 \oplus A_2 \oplus B_1 \oplus 2B_2 \oplus 3E$
$C_{2v}{}^v$	$A_1 \oplus B_1 \oplus 2E$
$C_{2v}{}^d$	$A_1 \oplus B_2 \oplus 2E$
C_4	$A_1 \oplus A_2 \oplus 2E$
C_{4v}	$A_1 \oplus E$

(**ii**)-14. D_{2d} $(\overline{4}2m)$

Site	Translation representations
C_1	$3A_1 \oplus 3A_2 \oplus 3B_1 \oplus 3B_2 \oplus 6E$
$C_2{}^z$	$A_1 \oplus A_2 \oplus B_1 \oplus B_2 \oplus 4E$
C_2	$A_1 \oplus 2A_2 \oplus B_1 \oplus 2B_2 \oplus 3E$
C_S	$2A_1 \oplus A_2 \oplus B_1 \oplus 2B_2 \oplus 3E$
D_2	$A_2 \oplus B_2 \oplus 2E$
C_{2v}	$A_1 \oplus B_2 \oplus 2E$
S_4	$B_1 \oplus B_2 \oplus 2E$
D_{2d}	$B_2 \oplus E$

(**ii**)-15. D_{4h} $(4/mmm)$

Site	Translation representations
C_1	$3A_{1g}\oplus 3A_{1u}\oplus 3A_{2g}\oplus 3A_{2u}\oplus 3B_{1g}$ $\oplus 3B_{1u}\oplus 3B_{2g}\oplus 3B_{2u}\oplus 6E_g\oplus 6E_u$
C_i	$3A_{1u} \oplus 3A_{2u} \oplus 3B_{1u} \oplus 3B_{2u} \oplus 6E_u$
$C_2{}^z$	$A_{1g}\oplus A_{1u}\oplus A_{2g}\oplus A_{2u}\oplus B_{1g}\oplus B_{1u}$ $\oplus B_{2g}\oplus B_{2u}\oplus 4E_g\oplus 4E_u$
C_2	$A_{1g}\oplus A_{1u}\oplus 2A_{2g}\oplus 2A_{2u}\oplus B_{1g}\oplus B_{1u}$ $\oplus 2B_{2g}\oplus 2B_{2u}\oplus 3E_g\oplus 3E_u$
C_2'	$A_{1g}\oplus A_{1u}\oplus 2A_{2g}\oplus 2A_{2u}\oplus 2B_{1g}\oplus 2B_{1u}$ $\oplus B_{2g}\oplus B_{2u}\oplus 3E_g\oplus 3E_u$
$C_S{}^h$	$A_{1g}\oplus 2A_{1u}\oplus A_{2g}\oplus 2A_{2u}\oplus B_{1g}\oplus 2B_{1u}$ $\oplus B_{2g}\oplus 2B_{2u}\oplus 4E_g\oplus 2E_u$
$C_S{}^v$	$2A_{1g}\oplus A_{1u}\oplus A_{2g}\oplus 2A_{2u}\oplus 2B_{1g}\oplus B_{1u}$ $\oplus B_{2g}\oplus 2B_{2u}\oplus 3E_g\oplus 3E_u$

(**ii**)-15. D_{4h} (4/*mmm*) ....continued

Site	Translation representations
$C_S{}^d$	$2A_{1g} \oplus A_{1u} \oplus A_{2g} \oplus 2A_{2u} \oplus B_{1g} \oplus 2B_{1u} \oplus 2B_{2g} \oplus B_{2u} \oplus 3E_g \oplus 3E_u$
$C_{2h}{}^z$	$A_{1u} \oplus A_{2u} \oplus B_{1u} \oplus B_{2u} \oplus 4E_u$
C_{2h}	$A_{1u} \oplus 2A_{2u} \oplus B_{1u} \oplus 2B_{2u} \oplus 3E_u$
C'_{2h}	$A_{1u} \oplus 2A_{2u} \oplus 2B_{1u} \oplus B_{2u} \oplus 3E_u$
D_2	$A_{2g} \oplus A_{2u} \oplus B_{2g} \oplus B_{2u} \oplus 2E_g \oplus 2E_u$
D'_2	$A_{2g} \oplus A_{2u} \oplus B_{1g} \oplus B_{1u} \oplus 2E_g \oplus 2E_u$
$C_{2v}{}^v$	$A_{1g} \oplus A_{2u} \oplus B_{1g} \oplus B_{1u} \oplus 2E_g \oplus 2E_u$
$C_{2v}{}^d$	$A_{1g} \oplus A_{2u} \oplus B_{1u} \oplus B_{2g} \oplus 2E_g \oplus 2E_u$
C_{2v}	$A_{1g} \oplus A_{2g} \oplus A_{2u} \oplus B_{1g} \oplus B_{2g} \oplus B_{2u} \oplus E_g \oplus 2E_u$
C'_{2v}	$A_{1g} \oplus A_{2g} \oplus A_{2u} \oplus B_{1g} \oplus B_{1u} \oplus B_{2g} \oplus E_g \oplus 2E_u$
C_4	$A_{1g} \oplus A_{1u} \oplus A_{2g} \oplus A_{2u} \oplus 2E_g \oplus 2E_u$
S_4	$A_{1u} \oplus A_{2u} \oplus B_{1g} \oplus B_{2g} \oplus 2E_g \oplus 2E_u$
D_{2h}	$A_{2u} \oplus B_{2u} \oplus 2E_u$
D'_{2h}	$A_{2u} \oplus B_{1u} \oplus 2E_u$
C_{4h}	$A_{1u} \oplus A_{2u} \oplus 2E_u$
D_4	$A_{2g} \oplus A_{2u} \oplus E_g \oplus E_u$
C_{4v}	$A_{1g} \oplus A_{2u} \oplus E_g \oplus E_u$
D_{2d}	$A_{2u} \oplus B_{1g} \oplus E_g \oplus E_u$
D'_{2d}	$A_{2u} \oplus B_{2g} \oplus E_g \oplus E_u$
D_{4h}	$A_{2u} \oplus E_u$

(**ii**)-16. C_3 (3)

Site	Translation representations
C_1	$3A \oplus 3E$
C_3	$A \oplus E$

(**ii**)-17. S_6 $(\bar{3})$

Site	Translation representations
C_1	$3A_g \oplus 3A_u \oplus 3E_g \oplus 3E_u$
C_i	$3A_u \oplus 3E_u$
C_3	$A_g \oplus A_u \oplus E_g \oplus E_u$
S_6	$A_u \oplus E_u$

(**ii**)-18. D_3 (32)

Site	Translation representations
C_1	$3A_1 \oplus 3A_2 \oplus 6E$
C_2	$A_1 \oplus 2A_2 \oplus 3E$
C_3	$A_1 \oplus A_2 \oplus 2E$
D_3	$A_2 \oplus E$

(**ii**)-19. C_{3v} (3*m*)

Site	Translation representations
C_1	$3A_1 \oplus 3A_2 \oplus 6E$
C_S	$2A_1 \oplus A_2 \oplus 3E$
C_3	$A_1 \oplus A_2 \oplus 2E$
C_{3v}	$A_1 \oplus E$

(**ii**)-20. D_{3d} $(\bar{3}m)$

Site	Translation representations
C_1	$3A_{1g} \oplus 3A_{1u} \oplus 3A_{2g} \oplus 3A_{2u} \oplus 6E_g \oplus 6E_u$
C_i	$3A_{1u} \oplus 3A_{2u} \oplus 6E_u$
C_2	$A_{1g} \oplus A_{1u} \oplus 2A_{2g} \oplus 2A_{2u} \oplus 3E_g \oplus 3E_u$
C_S	$2A_{1g} \oplus A_{1u} \oplus A_{2g} \oplus 2A_{2u} \oplus 3E_g \oplus 3E_u$
C_3	$A_{1g} \oplus A_{1u} \oplus A_{2g} \oplus A_{2u} \oplus 2E_g \oplus 2E_u$
C_{2h}	$A_{1u} \oplus 2A_{2u} \oplus 3E_u$
S_6	$A_{1u} \oplus A_{2u} \oplus 2E_u$
D_3	$A_{2g} \oplus A_{2u} \oplus E_g \oplus E_u$
C_{3v}	$A_{1g} \oplus A_{2u} \oplus E_g \oplus E_u$
D_{3d}	$A_{2u} \oplus E_u$

(**ii**)-21. C_6 (6)

Site	Translation representations
C_1	$3A \oplus 3B \oplus 3E_1 \oplus 3E_2$
C_2	$A \oplus 2B \oplus 2E_1 \oplus E_2$
C_3	$A \oplus B \oplus E_1 \oplus E_2$
C_6	$A \oplus E_1$

(ii)-22. C_{3h} $(\overline{6})$

Site	Translation representations
C_1	$3A' \oplus 3A'' \oplus 3E' \oplus 3E''$
C_S	$2A' \oplus A'' \oplus 2E' \oplus E''$
C_3	$A' \oplus A'' \oplus E' \oplus E'$
C_{3h}	$A' \oplus E'$

(ii)-23. C_{6h} $(6/m)$

Site	Translation representations
C_1	$3A_g \oplus 3A_u \oplus 3B_g \oplus 3B_u \oplus 3E_{1g} \oplus 3E_{1u}$ $\oplus 3E_{2g} \oplus 3E_{2u}$
C_i	$3A_u \oplus 3B_u \oplus 3E_{1u} \oplus 3E_{2u}$
C_2	$A_g \oplus A_u \oplus 2B_g \oplus 2B_u \oplus 2E_{1g} \oplus 2E_{1u}$ $\oplus E_{2g} \oplus E_{2u}$
C_S	$2A_g \oplus A_u \oplus B_g \oplus 2B_u \oplus E_{1g} \oplus 2E_{1u}$ $\oplus 2E_{2g} \oplus E_{2u}$
C_3	$A_g \oplus A_u \oplus B_g \oplus B_u \oplus E_{1g} \oplus E_{1u} \oplus E_{2g} \oplus E_{2u}$
C_{2h}	$A_u \oplus 2B_u \oplus 2E_{1u} \oplus E_{2u}$
S_6	$A_u \oplus B_u \oplus E_{1u} \oplus E_{2u}$
C_6	$A_g \oplus A_u \oplus E_{1g} \oplus E_{1u}$
C_{3h}	$A_u \oplus B_g \oplus E_{1u} \oplus E_{2g}$
C_{6h}	$A_u \oplus E_{1u}$

(ii)-24. D_6 (622)

Site	Translation representations
C_1	$3A_1 \oplus 3A_2 \oplus 3B_1 \oplus 3B_2 \oplus 6E_1 \oplus 6E_2$
$C_2{}^z$	$A_1 \oplus A_2 \oplus 2B_1 \oplus 2B_2 \oplus 4E_1 \oplus 2E_2$
C_2	$A_1 \oplus 2A_2 \oplus B_1 \oplus 2B_2 \oplus 3E_1 \oplus 3E_2$
C_2'	$A_1 \oplus 2A_2 \oplus 2B_1 \oplus B_2 \oplus 3E_1 \oplus 3E_2$
C_3	$A_1 \oplus A_2 \oplus B_1 \oplus B_2 \oplus 2E_1 \oplus 2E_2$
D_2	$A_2 \oplus B_1 \oplus B_2 \oplus 2E_1 \oplus E_2$
D_3	$A_2 \oplus B_2 \oplus E_1 \oplus E_2$
D_3'	$A_2 \oplus B_1 \oplus E_1 \oplus E_2$
C_6	$A_1 \oplus A_2 \oplus 2E_1$
D_6	$A_2 \oplus E_1$

(ii)-25. C_{6v} $(6mm)$

Site	Translation representations
C_1	$3A_1 \oplus 3A_2 \oplus 3B_1 \oplus 3B_2 \oplus 6E_1 \oplus 6E_2$
C_2	$A_1 \oplus A_2 \oplus 2B_1 \oplus 2B_2 \oplus 4E_1 \oplus 2E_2$
$C_S{}^v$	$2A_1 \oplus A_2 \oplus 2B_1 \oplus B_2 \oplus 6E_1 \oplus 6E_2$
$C_S{}^d$	$2A_1 \oplus A_2 \oplus B_1 \oplus 2B_2 \oplus 3E_1 \oplus 3E_2$
C_3	$A_1 \oplus A_2 \oplus B_1 \oplus B_2 \oplus 2E_1 \oplus 2E_2$
C_{2v}	$A_1 \oplus B_1 \oplus B_2 \oplus 2E_1 \oplus E_2$
$C_{3v}{}^v$	$A_1 \oplus B_1 \oplus E_1 \oplus E_2$
$C_{3v}{}^d$	$A_1 \oplus B_2 \oplus E_1 \oplus E_2$
C_6	$A_1 \oplus A_2 \oplus 2E_1$
C_{6v}	$A_1 \oplus E_1$

(ii)-26. D_{3h} $(\overline{6}m2)$

Site	Translation representations
C_1	$3A_1' \oplus 3A''{}_1 \oplus 3A_2' \oplus 3A''{}_2 \oplus 6E' \oplus 6E''$
C_2	$A_1' \oplus A''{}_1 \oplus 2A_2' \oplus 2A''{}_2 \oplus 3E' \oplus 3E''$
$C_S{}^h$	$2A_1' \oplus A''{}_1 \oplus 2A_2' \oplus A''{}_2 \oplus 4E' \oplus 2E'$
$C_S{}^v$	$2A_1' \oplus A_1'' \oplus A_2' \oplus 2A_2'' \oplus 3E' \oplus 3E''$
C_3	$A_1' \oplus A_1'' \oplus A_2' \oplus A_2'' \oplus 2E' \oplus 2E''$
C_{2v}	$A_1' \oplus A_2' \oplus A_2'' \oplus 2E' \oplus E''$
D_3	$A_2' \oplus A_2'' \oplus E' \oplus E''$
C_{3v}	$A_1' \oplus A_2'' \oplus E' \oplus E''$
C_{3h}	$\mathrm{A''}_1 \oplus \mathrm{A''}_2 \oplus 2\mathrm{E'}$
D_{3h}	$A''{}_2 \oplus E'$

(ii)-27. D_{6h} $(6/mmm)$

Site	Translation representations
C_1	$3A_{1g} \oplus 3A_{1u} \oplus 3A_{2g} \oplus 3A_{2u} \oplus 3B_{1g} \oplus 3B_{1u}$ $\oplus 3B_{2g} \oplus 3B_{2u} \oplus 6E_{1g} \oplus 6E_{1u} \oplus 6E_{2g} \oplus 6E_{2u}$
C_i	$3A_{1u} \oplus 3A_{2u} \oplus 3B_{1u} \oplus 3B_{2u} \oplus 6E_{1u} \oplus 6E_{2u}$
$C_2{}^z$	$\mathrm{A_{1g} \oplus A_{1u} \oplus A_{2g} \oplus A_{2u} \oplus 2B_{1g}}$ $\mathrm{\oplus 2B_{1u} \oplus 2B_{2g} \oplus 2B_{2u} \oplus 4E_{1g} \oplus 4E_{1u}}$ $\mathrm{\oplus 2E_{2g} \oplus 2E_{2u}}$
C_2	$A_{1g} \oplus A_{1u} \oplus 2A_{2g} \oplus 2A_{2u} \oplus B_{1g} \oplus B_{1u}$ $\oplus 2B_{2g} \oplus 2B_{2u} \oplus 3E_{1g} \oplus 3E_{1u} \oplus 3E_{2g} \oplus 3E_{2u}$

(ii)-27. D_{6h} (6/mmm) ....continued

Site	Translation representations
C'_2	$A_{1g} \oplus A_{1u} \oplus 2A_{2g} \oplus 2A_{2u} \oplus 2B_{1g} \oplus 2B_{1u} \oplus B_{2g} \oplus B_{2u} \oplus 3E_{1g} \oplus 3E_{1u} \oplus 3E_{2g} \oplus 3E_{2u}$
$C_S{}^h$	$2A_{1g} \oplus A_{1u} \oplus 2A_{2g} \oplus A_{2u} \oplus B_{1g} \oplus 2B_{1u} \oplus B_{2g} \oplus 2B_{2u} \oplus 2E_{1g} \oplus 4E_{1u} \oplus 4E_{2g} \oplus 2E_{2u}$
$C_S{}^v$	$2A_{1g} \oplus A_{1u} \oplus A_{2g} \oplus 2A_{2u} \oplus B_{1g} \oplus 2B_{1u} \oplus 2B_{2g} \oplus B_{2u} \oplus 3E_{1g} \oplus 3E_{1u} \oplus 3E_{2g} \oplus 3E_{2u}$
$C_S{}^d$	$2A_{1g} \oplus A_{1u} \oplus A_{2g} \oplus 2A_{2u} \oplus 2B_{1g} \oplus B_{1u} \oplus B_{2g} \oplus 2B_{2u} \oplus 3E_{1g} \oplus 3E_{1u} \oplus 3E_{2g} \oplus 3E_{2u}$
C_3	$A_{1g} \oplus A_{1u} \oplus A_{2g} \oplus A_{2u} \oplus B_{1g} \oplus B_{1u} \oplus B_{2g} \oplus B_{2u} \oplus 2E_{1g} \oplus 2E_{1u} \oplus 2E_{2g} \oplus 2E_{2u}$
$C_{2h}{}^z$	$A_{1u} \oplus A_{2u} \oplus 2B_{1u} \oplus 2B_{2u} \oplus 4E_{1u} \oplus 2E_{2u}$
C_{2h}	$A_{1u} \oplus 2A_{2u} \oplus B_{1u} \oplus 2B_{2u} \oplus 3E_{1u} \oplus 3E_{2u}$
C'_{2h}	$A_{1u} \oplus 2A_{2u} \oplus 2B_{1u} \oplus B_{2u} \oplus 3E_{1u} \oplus 3E_{2u}$
D_2	$A_{2g} \oplus A_{2u} \oplus B_{1g} \oplus B_{1u} \oplus B_{2g} \oplus B_{2u} \oplus 2E_{1g} \oplus 2E_{1u} \oplus E_{2g} \oplus E_{2u}$
$C_{2v}{}^z$	$A_{1g} \oplus A_{2u} \oplus B_{1g} \oplus B_{1u} \oplus B_{2g} \oplus B_{2u} \oplus 2E_{1g} \oplus 2E_{1u} \oplus E_{2g} \oplus E_{2u}$
C_{2v}	$A_{1g} \oplus A_{2g} \oplus A_{2u} \oplus B_{1u} \oplus B_{2g} \oplus B_{2u} \oplus E_{1g} \oplus 2E_{1u} \oplus 2E_{2g} \oplus E_{2u}$
C'_{2v}	$A_{1g} \oplus A_{2g} \oplus A_{2u} \oplus B_{1g} \oplus B_{1u} \oplus B_{2u} \oplus E_{1g} \oplus 2E_{1u} \oplus 2E_{2g} \oplus E_{2u}$
S_6	$A_{1u} \oplus A_{2u} \oplus B_{1u} \oplus B_{2u} \oplus 2E_{1u} \oplus 2E_{2u}$
D_3	$A_{2g} \oplus A_{2u} \oplus B_{2g} \oplus B_{2u} \oplus E_{1g} \oplus E_{1u} \oplus E_{2g} \oplus E_{2u}$
D'_3	$A_{2g} \oplus A_{2u} \oplus B_{1g} \oplus B_{1u} \oplus E_{1g} \oplus E_{1u} \oplus E_{2g} \oplus E_{2u}$
$C_{3v}{}^v$	$A_{1g} \oplus A_{2u} \oplus B_{1u} \oplus B_{2g} \oplus E_{1g} \oplus E_{1u} \oplus E_{2g} \oplus E_{2u}$
$C_{3v}{}^d$	$A_{1g} \oplus A_{2u} \oplus B_{1g} \oplus B_{2u} \oplus E_{1g} \oplus E_{1u} \oplus E_{2g} \oplus E_{2u}$
C_6	$A_{1g} \oplus A_{1u} \oplus A_{2g} \oplus A_{2u} \oplus 2E_{1g} \oplus 2E_{1u}$
C_{3h}	$A_{1u} \oplus A_{2u} \oplus B_{1g} \oplus B_{2g} \oplus 2E_{1u} \oplus 2E_{2g}$
D_{2h}	$A_{2u} \oplus B_{1u} \oplus B_{2u} \oplus 2E_{1u} \oplus E_{2u}$
D_{3d}	$A_{2u} \oplus B_{2u} \oplus E_{1u} \oplus E_{2u}$
D'_{3d}	$A_{2u} \oplus B_{1u} \oplus E_{1u} \oplus E_{2u}$
C_{6h}	$A_{1u} \oplus A_{2u} \oplus 2E_{1u}$

(ii)-27. D_{6h}(6/mmm) ....continued

Site	Translation representations
D_6	$A_{2g} \oplus A_{2u} \oplus E_{1g} \oplus E_{1u}$
C_{6v}	$A_{1g} \oplus A_{2u} \oplus E_{1g} \oplus E_{1u}$
D_{3h}	$A_{2u} \oplus B_{2g} \oplus E_{1u} \oplus E_{2g}$
D'_{3h}	$A_{2u} \oplus B_{1g} \oplus E_{1u} \oplus E_{2g}$
D_{6h}	$A_{2u} \oplus E_{1u}$

(ii)-28. T(23)

Site	Translation representations
C_1	$3A \oplus 3E \oplus 9F$
C_2	$A \oplus E \oplus 5F$
C_3	$A \oplus E \oplus 3F$
D_2	$3F$
T	F

(ii)-29. T_h (m3)

Site	Translation representations
C_1	$3A_g \oplus 3A_u \oplus 3E_g \oplus 3E_u \oplus 9F_g \oplus 9F_u$
C_i	$3A_u \oplus 3E_u \oplus 9F_u$
C_2	$A_g \oplus A_u \oplus E_g \oplus E_u \oplus 5F_g \oplus 5F_u$
C_S	$2A_g \oplus A_u \oplus 2E_g \oplus E_u \oplus 4F_g \oplus 5F_u$
C_3	$A_g \oplus A_u \oplus E_g \oplus E_u \oplus 3F_g \oplus 3F_u$
C_{2h}	$A_u \oplus E_u \oplus 5F_u$
D_2	$3F_g \oplus 3F_u$
C_{2v}	$A_g \oplus E_g \oplus 2F_g \oplus 3F_u$
S_6	$A_u \oplus E_u \oplus 3F_u$
D_{2h}	$3F_u$
T	$F_g \oplus F_u$
T_h	F_u

(ii)-30. O (432)

Site	Translation representations
C_1	$3A_1 \oplus 3A_2 \oplus 6E \oplus 9F_1 \oplus 9F_2$
C_2	$A_1 \oplus A_2 \oplus 2E \oplus 5F_1 \oplus 5F_2$
C'_2	$A_1 \oplus 2A_2 \oplus 3E \oplus 5F_1 \oplus 4F_2$
C_3	$A_1 \oplus A_2 \oplus 2E \oplus 3F_1 \oplus 3F_2$

(ii)-30. O (432) ....continued

Site	Translation representations
D_2	$3F_1 \oplus 3F_2$
D'_2	$A_2 \oplus E \oplus 3F_1 \oplus 2F_2$
C_4	$A_1 \oplus E \oplus 3F_1 \oplus 2F_2$
D_3	$A_2 \oplus E \oplus 2F_1 \oplus F_2$
D_4	$2F_1 \oplus F_2$
T	$F_1 \oplus F_2$
O	F_1

(ii)-31. T_d $(\bar{4}3m)$

Site	Translation representations
C_1	$3A_1 \oplus 3A_2 \oplus 6E \oplus 9F_1 \oplus 9F_2$
C_2	$A_1 \oplus A_2 \oplus 2E \oplus 5F_1 \oplus 5F_2$
C_S	$2A_1 \oplus A_2 \oplus 3E \oplus 4F_1 \oplus 5F_2$
C_3	$A_1 \oplus A_2 \oplus 2E \oplus 3F_1 \oplus 3F_2$
D_2	$3F_1 \oplus 3F_2$
C_{2v}	$A_1 \oplus E \oplus 2F_1 \oplus 3F_2$
S_4	$A_2 \oplus E \oplus 2F_1 \oplus 3F_2$
C_{3v}	$A_1 \oplus E \oplus F_1 \oplus 2F_2$
D_{2d}	$3F_2$
T	$F_1 \oplus F_2$
T_d	F_2

(ii)-32. O_h $(m\bar{3}m)$

Site	Translation representations
C_1	$3A_{1g}\oplus 3A_{1u}\oplus 3A_{2g}\oplus 3A_{2u}\oplus 6E_g\oplus 6E_u \oplus 9F_{1g}\oplus 9F_{1u}\oplus 9F_{2g}\oplus 9F_{2u}$
C_i	$3A_{1u} \oplus 3A_{2u} \oplus 6E_u \oplus 9F_{1u} \oplus 9F_{2u}$
C_2	$A_{1g}\oplus A_{1u}\oplus A_{2g}\oplus A_{2u}\oplus 2E_g\oplus 2E_u \oplus 5F_{1g}\oplus 5F_{1u}\oplus 5F_{2g}\oplus 5F_{2u}$
C'_2	$A_{1g}\oplus A_{1u}\oplus 2A_{2g}\oplus 2A_{2u}\oplus 3E_g\oplus 3E_u \oplus 5F_{1g}\oplus 5F_{1u}\oplus 4F_{2g}\oplus 4F_{2u}$
$C_S{}^h$	$2A_{1g}\oplus A_{1u}\oplus 2A_{2g}\oplus A_{2u}\oplus 4E_g\oplus 2E_u \oplus 4F_{1g}\oplus 5F_{1u}\oplus 4F_{2g}\oplus 5F_{2u}$
$C_S{}^d$	$2A_{1g}\oplus A_{1u}\oplus A_{2g}\oplus 2A_{2u}\oplus 3E_g\oplus 3E_u \oplus 4F_{1g}\oplus 5F_{1u}\oplus 5F_{2g}\oplus 4F_{2u}$
C_3	$A_{1g}\oplus A_{1u}\oplus A_{2g}\oplus A_{2u}\oplus 2E_g\oplus 2E_u \oplus 3F_{1g}\oplus 3F_{1u}\oplus 3F_{2g}\oplus 3F_{2u}$

(ii)-32. O_h $(m\bar{3}m)$ ....continued

Site	Translation representations
C_{2h}	$A_{1u} \oplus A_{2u} \oplus 2E_u \oplus 5F_{1u} \oplus 5F_{2u}$
C'_{2h}	$A_{1u} \oplus 2A_{2u} \oplus 3E_u \oplus 5F_{1u} \oplus 4F_{2u}$
D_2	$3F_{1g} \oplus 3F_{1u} \oplus 3F_{2g} \oplus 3F_{2u}$
D'_2	$A_{2g}\oplus A_{2u}\oplus E_g\oplus E_u\oplus 3F_{1g}\oplus 3F_{1u} \oplus 2F_{2g}\oplus 2F_{2u}$
$C_{2v}{}^h$	$A_{1g} \oplus A_{2g} \oplus 2E_g \oplus 2F_{1g} \oplus 3F_{1u} \oplus 2F_{2g} \oplus 3F_{2u}$
$C_{2v}{}^d$	$A_{1g}\oplus A_{2u}\oplus E_g\oplus E_u\oplus 2F_{1g}\oplus 3F_{1u} \oplus 3F_{2g}\oplus 2F_{2u}$
C'_{2v}	$A_{1g}\oplus A_{2g}\oplus A_{2u}\oplus 2E_g\oplus E_u\oplus 2F_{1g}\oplus 3F_{1u} \oplus 2F_{2g}\oplus 2F_{2u}$
C_4	$A_{1g}\oplus A_{1u}\oplus E_g\oplus E_u\oplus 3F_{1g}\oplus 3F_{1u}\oplus 2F_{2g} \oplus 2F_{2u}$
S_4	$A_{1u}\oplus A_{2g}\oplus E_g\oplus E_u\oplus 2F_{1g}\oplus 3F_{1u}\oplus 3F_{2g} \oplus 2F_{2u}$
S_6	$A_{1u} \oplus A_{2u} \oplus 2E_u \oplus 3F_{1u} \oplus 3F_{2u}$
D_3	$A_{2g}\oplus A_{2u}\oplus E_g\oplus E_u\oplus 2F_{1g}\oplus 2F_{1u} \oplus F_{2g}\oplus F_{2u}$
C_{3v}	$A_{1g}\oplus A_{2u}\oplus E_g\oplus E_u\oplus F_{1g}\oplus 2F_{1u} \oplus 2F_{2g}\oplus F_{2u}$
D_{2h}	$3F_{1u} \oplus 3F_{2u}$
D'_{2h}	$A_{2u} \oplus E_u \oplus 3F_{1u} \oplus 2F_{2u}$
C_{4h}	$A_{1u} \oplus E_u \oplus 3F_{1u} \oplus 2F_{2u}$
C_{4v}	$A_{1g} \oplus E_g \oplus F_{1g} \oplus 2F_{1u} \oplus F_{2g} \oplus F_{2u}$
D_4	$2F_{1g} \oplus 2F_{1u} \oplus F_{2g} \oplus F_{2u}$
D_{2d}	$F_{1g} \oplus 2F_{1u} \oplus 2F_{2g} \oplus F_{2u}$
D'_{2d}	$A_{2g} \oplus E_g \oplus F_{1g} \oplus 2F_{1u} \oplus F_{2g} \oplus F_{2u}$
D_{3d}	$A_{2u} \oplus E_u \oplus 2F_{1u} \oplus F_{2u}$
T	$F_{1g} \oplus F_{1u} \oplus F_{2g} \oplus F_{2u}$
D_{4h}	$2F_{1u} \oplus F_{2u}$
T_h	$F_{1u} \oplus F_{2u}$
O	$F_{1g} \oplus F_{1u}$
T_d	$F_{1u} \oplus F_{2g}$
O_h	F_{1u}

Tables (iii): Irreducible representations describing the librational symmetry properties of atom-collections, each occupying a crystalline site (with each collection considered as a molecule exhibiting the site symmetry) in primitive unit cells belonging to different space groups.

(iii)-1. C_1 (1)

Site	Librational representations
C_1	$3A$

(iii)-2. C_i ($\bar{1}$)

Site	Librational representations
C_1	$3A_g \oplus 3A_u$
C_i	$3A_g$

(iii)-3. C_2 (2)

Site	Librational representations
C_1	$3A \oplus 3B$
C_2	$A \oplus 2B$

(iii)-4. C_S (m)

Site	Librational representations
C_1	$3A' \oplus 3A''$
C_S	$A' \oplus 2A''$

(iii)-5. C_{2h} ($2/m$)

Site	Librational representations
C_1	$3A_g \oplus 3A_u \oplus 3B_g \oplus 3B_u$
C_i	$3A_u \oplus 3B_u$
C_2	$A_g \oplus A_u \oplus 2B_g \oplus 2B_u$
C_S	$A_g \oplus 2A_u \oplus 2B_g \oplus B_u$
C_{2h}	$A_g \oplus 2B_g$

(iii)-6. D_2 (222)

Site	Librational representations
C_1	$3A \oplus 3B_1 \oplus 3B_2 \oplus 3B_3$
$C_2{}^z$	$A \oplus B_1 \oplus 2B_2 \oplus 2B_3$
$C_2{}^y$	$A \oplus 2B_1 \oplus B_2 \oplus 2B_3$
$C_2{}^x$	$A \oplus 2B_1 \oplus 2B_2 \oplus B_3$
D_2	$B_1 \oplus B_2 \oplus B_3$

(iii)-7. C_{2v} ($mm2$)

Site	Librational representations
C_1	$3A_1 \oplus 3A_2 \oplus 3B_1 \oplus 3B_2$
C_2	$A_1 \oplus A_2 \oplus 2B_1 \oplus 2B_2$
$C_S{}^{xz}$	$A_1 \oplus 2A_2 \oplus B_1 \oplus 2B_2$
$C_S{}^{yz}$	$A_1 \oplus 2A_2 \oplus 2B_1 \oplus B_2$
C_{2v}	$A_2 \oplus B_1 \oplus B_2$

(iii)-8. D_{2h} (mmm)

Site	Librational representations
C_1	$3A_g \oplus 3A_u \oplus 3B_{1g} \oplus 3B_{1u} \oplus 3B_{2g} \oplus 3B_{2u} \oplus 3B_{3g} \oplus 3B_{3u}$
$C_2{}^z$	$A_g \oplus A_u \oplus B_{1g} \oplus B_{1u} \oplus 2B_{2g} \oplus 2B_{2u} \oplus 2B_{3g} \oplus 2B_{3u}$
$C_2{}^y$	$A_g \oplus A_u \oplus 2B_{1g} \oplus 2B_{1u} \oplus B_{2g} \oplus B_{2u} \oplus 2B_{3g} \oplus 2B_{3u}$
$C_2{}^x$	$A_g \oplus A_u \oplus 2B_{1g} \oplus 2B_{1u} \oplus 2B_{2g} \oplus 2B_{2u} \oplus B_{3g} \oplus B_{3u}$
$C_S{}^{xy}$	$A_g \oplus 2A_u \oplus B_{1g} \oplus 2B_{1u} \oplus 2B_{2g} \oplus B_{2u} \oplus 2B_{3g} \oplus B_{3u}$
$C_S{}^{xz}$	$A_g \oplus 2A_u \oplus 2B_{1g} \oplus B_{1u} \oplus B_{2g} \oplus 2B_{2u} \oplus 2B_{3g} \oplus B_{3u}$
$C_S{}^{yz}$	$A_g \oplus 2A_u \oplus 2B_{1g} \oplus B_{1u} \oplus 2B_{2g} \oplus B_{2u} \oplus B_{3g} \oplus 2B_{3u}$
C_i	$3A_g \oplus 3B_{1g} \oplus 3B_{2g} \oplus 3B_{3g}$
$C_{2v}{}^z$	$A_u \oplus B_{1g} \oplus B_{2g} \oplus B_{2u} \oplus B_{3g} \oplus B_{3u}$
$C_{2v}{}^y$	$A_u \oplus B_{1g} \oplus B_{1u} \oplus B_{2g} \oplus B_{3g} \oplus B_{3u}$
$C_{2v}{}^x$	$A_u \oplus B_{1g} \oplus B_{1u} \oplus B_{2g} \oplus B_{2u} \oplus B_{3g}$
D_2	$B_{1g} \oplus B_{1u} \oplus B_{2g} \oplus B_{2u} \oplus B_{3g} \oplus B_{3u}$
$C_{2h}{}^z$	$A_g \oplus B_{1g} \oplus 2B_{2g} \oplus 2B_{3g}$
$C_{2h}{}^y$	$A_g \oplus 2B_{1g} \oplus B_{2g} \oplus 2B_{3g}$
$C_{2h}{}^x$	$A_g \oplus 2B_{1g} \oplus 2B_{2g} \oplus B_{3g}$
D_{2h}	$B_{1g} \oplus B_{2g} \oplus B_{3g}$

(iii)-9. C_4 (4)

Site	Librational representations
C_1	$3A \oplus 3B \oplus 3E$
C_2	$A \oplus B \oplus 2E$
C_4	$A \oplus E$

(iii)-10. S_4 $(\bar{4})$

Site	Librational representations
C_1	$3A \oplus 3B \oplus 3E$
C_2	$A \oplus B \oplus 2E$
S_4	$B \oplus E$

(iii)-11. C_{4h} $(4/m)$

Site	Librational representations
C_1	$3A_g \oplus 3A_u \oplus 3B_g \oplus 3B_u \oplus 3E_g \oplus 3E_u$
C_i	$3A_g \oplus 3B_g \oplus 3E_g$
C_2	$A_g \oplus A_u \oplus B_g \oplus B_u \oplus 2E_g \oplus 2E_u$
C_S	$A_g \oplus 2A_u \oplus B_g \oplus 2B_u \oplus 2E_g \oplus E_u$
C_{2h}	$A_g \oplus B_g \oplus 2E_g$
C_4	$A_g \oplus A_u \oplus E_g \oplus E_u$
S_4	$A_g \oplus B_u \oplus E_g \oplus E_u$
C_{4h}	$B_g \oplus E_g$

(iii)-12. D_4 (422)

Site	Librational representations
C_1	$3A_1 \oplus 3A_2 \oplus 3B_1 \oplus 3B_2 \oplus 6E$
$C_2^{\ z}$	$A_1 \oplus A_2 \oplus B_1 \oplus B_2 \oplus 4E$
C_2	$A_1 \oplus 2A_2 \oplus B_1 \oplus 2B_2 \oplus 3E$
C_2'	$A_1 \oplus 2A_2 \oplus 2B_1 \oplus B_2 \oplus 3E$
D_2	$A_2 \oplus B_2 \oplus 2E$
D_2'	$A_2 \oplus B_1 \oplus 2E$
C_4	$A_1 \oplus A_2 \oplus 2E$
D_4	$A_2 \oplus E$

(iii)-13. C_{4v} $(4mm)$

Site	Librational representations
C_1	$3A_1 \oplus 3A_2 \oplus 3B_1 \oplus 3B_2 \oplus 6E$
C_2	$A_1 \oplus A_2 \oplus B_1 \oplus B_2 \oplus 4E$

(iii)-13. C_{4v} $(4mm)$ ...continued

Site	Librational representations
$C_S^{\ v}$	$A_1 \oplus 2A_2 \oplus B_1 \oplus 2B_2 \oplus 3E$
$C_S^{\ d}$	$A_1 \oplus 2A_2 \oplus 2B_1 \oplus B_2 \oplus 3E$
$C_{2v}^{\ v}$	$A_2 \oplus B_2 \oplus 2E$
$C_{2v}^{\ d}$	$A_2 \oplus B_1 \oplus 2E$
C_4	$A_1 \oplus A_2 \oplus 2E$
C_{4v}	$A_2 \oplus E$

(iii)-14. D_{2d} $(\bar{4}2m)$

Site	Librational representations
C_1	$3A_1 \oplus 3A_2 \oplus 3B_1 \oplus 3B_2 \oplus 6E$
$C_2^{\ z}$	$A_1 \oplus A_2 \oplus B_1 \oplus B_2 \oplus 4E$
C_2	$A_1 \oplus 2A_2 \oplus B_1 \oplus 2B_2 \oplus 3E$
C_S	$A_1 \oplus 2A_2 \oplus 2B_1 \oplus B_2 \oplus 3E$
D_2	$A_2 \oplus B_2 \oplus 2E$
C_{2v}	$A_2 \oplus B_1 \oplus 2E$
S_4	$A_1 \oplus A_2 \oplus 2E$
D_{2d}	$A_2 \oplus E$

(iii)-15. D_{4h} $(4/mmm)$

Site	Librational representations
C_1	$3A_{1g} \oplus 3A_{1u} \oplus 3A_{2g} \oplus 3A_{2u} \oplus 3B_{1g} \oplus 3B_{1u} \oplus 3B_{2g} \oplus 3B_{2u} \oplus 6E_g \oplus 6E_u$
C_i	$3A_{1g} \oplus 3A_{2g} \oplus 3B_{1g} \oplus 3B_{2g} \oplus 6E_g$
$C_2^{\ z}$	$A_{1g} \oplus A_{1u} \oplus A_{2g} \oplus A_{2u} \oplus B_{1g} \oplus B_{1u} \oplus B_{2g} \oplus B_{2u} \oplus 4E_g \oplus 4E_u$
C_2	$A_{1g} \oplus A_{1u} \oplus 2A_{2g} \oplus 2A_{2u} \oplus B_{1g} \oplus B_{1u} \oplus 2B_{2g} \oplus 2B_{2u} \oplus 3E_g \oplus 3E_u$
C_2'	$A_{1g} \oplus A_{1u} \oplus 2A_{2g} \oplus 2A_{2u} \oplus 2B_{1g} \oplus 2B_{1u} \oplus B_{2g} \oplus B_{2u} \oplus 3E_g \oplus 3E_u$
$C_S^{\ h}$	$A_{1g} \oplus 2A_{1u} \oplus A_{2g} \oplus 2A_{2u} \oplus B_{1g} \oplus 2B_{1u} \oplus B_{2g} \oplus 2B_{2u} \oplus 4E_g \oplus 2E_u$
$C_S^{\ v}$	$A_{1g} \oplus 2A_{1u} \oplus 2A_{2g} \oplus A_{2u} \oplus B_{1g} \oplus 2B_{1u} \oplus 2B_{2g} \oplus B_{2u} \oplus 3E_g \oplus 3E_u$

(iii)-15. D_{4h} (4/*mmm*) ....continued

Site	Librational representations
$C_S{}^d$	$A_{1g}\oplus 2A_{1u}\oplus 2A_{2g}\oplus A_{2u}\oplus 2B_{1g}\oplus B_{1u}$ $\oplus B_{2g}\oplus 2B_{2u}\oplus 3E_g\oplus 3E_u$
$C_{2h}{}^z$	$A_{1g} \oplus A_{2g} \oplus B_{1g} \oplus B_{2g} \oplus 4E_g$
C_{2h}	$A_{1g} \oplus 2A_{2g} \oplus B_{1g} \oplus 2B_{2g} \oplus 3E_g$
C'_{2h}	$A_{1g} \oplus 2A_{2g} \oplus 2B_{1g} \oplus B_{2g} \oplus 3E_g$
D_2	$A_{2g} \oplus A_{2u} \oplus B_{2g} \oplus B_{2u} \oplus 2E_g \oplus 2E_u$
D'_2	$A_{2g} \oplus A_{2u} \oplus B_{1g} \oplus B_{1u} \oplus 2E_g \oplus 2E_u$
$C_{2v}{}^v$	$A_{1u} \oplus A_{2g} \oplus B_{1u} \oplus B_{2g} \oplus 2E_g \oplus 2E_u$
$C_{2v}{}^d$	$A_{1u} \oplus A_{2g} \oplus B_{1g} \oplus B_{2u} \oplus 2E_g \oplus 2E_u$
C_{2v}	$A_{1g} \oplus A_{2g} \oplus A_{2u} \oplus B_{1u} \oplus B_{2g} \oplus B_{2u} \oplus 2E_g \oplus E_u$
C'_{2v}	$A_{1u} \oplus A_{2g} \oplus A_{2u} \oplus B_{1g} \oplus B_{1u} \oplus B_{2u} \oplus 2E_g \oplus E_u$
C_4	$A_{1g} \oplus A_{1u} \oplus A_{2g} \oplus A_{2u} \oplus 2E_g \oplus 2E_u$
S_4	$A_{1g} \oplus A_{2g} \oplus B_{1u} \oplus B_{2u} \oplus 2E_g \oplus 2E_u$
D_{2h}	$A_{2g} \oplus B_{2g} \oplus 2E_g$
D'_{2h}	$A_{2g} \oplus B_{1g} \oplus 2E_g$
C_{4h}	$A_{1g} \oplus A_{2g} \oplus 2E_g$
D_4	$A_{2g} \oplus A_{2u} \oplus E_g \oplus E_u$
C_{4v}	$A_{1u} \oplus A_{2g} \oplus E_g \oplus E_u$
D_{2d}	$A_{2g} \oplus B_{1u} \oplus E_g \oplus E_u$
D'_{2d}	$A_{2g} \oplus B_{2u} \oplus E_g \oplus E_u$
D_{4h}	$A_{2g} \oplus E_g$

(iii)-16. C_3 (3)

Site	Librational representations
C_1	$3A \oplus 3E$
C_3	$A \oplus E$

(iii)-17. S_6 $(\bar{3})$

Site	Librational representations
C_1	$3A_g \oplus 3A_u \oplus 3E_g \oplus 3E_u$
C_i	$3A_g \oplus 3E_g$
C_3	$A_g \oplus A_u \oplus E_g \oplus E_u$
S_6	$A_g \oplus E_g$

(iii)-18. D_3 (32)

Site	Librational representations
C_1	$3A_1 \oplus 3A_2 \oplus 6E$
C_2	$A_1 \oplus 2A_2 \oplus 3E$
C_3	$A_1 \oplus A_2 \oplus 2E$
D_3	$A_2 \oplus E$

(iii)-19. C_{3v} (3*m*)

Site	Librational representations
C_1	$3A_1 \oplus 3A_2 \oplus 6E$
C_S	$A_1 \oplus 2A_2 \oplus 3E$
C_3	$A_1 \oplus A_2 \oplus 2E$
C_{3v}	$A_2 \oplus E$

(iii)-20. D_{3d} $(\bar{3}m)$

Site	Librational representations
C_1	$3A_{1g} \oplus 3A_{1u} \oplus 3A_{2g} \oplus 3A_{2u} \oplus 6E_g \oplus 6E_u$
C_i	$3A_{1g} \oplus 3A_{2g} \oplus 6E_g$
C_2	$A_{1g} \oplus A_{1u} \oplus 2A_{2g} \oplus 2A_{2u} \oplus 3E_g \oplus 3E_u$
C_S	$A_{1g} \oplus 2A_{1u} \oplus 2A_{2g} \oplus A_{2u} \oplus 3E_g \oplus 3E_u$
C_3	$A_{1g} \oplus A_{1u} \oplus A_{2g} \oplus A_{2u} \oplus 2E_g \oplus 2E_u$
C_{2h}	$A_{1g} \oplus 2A_{2g} \oplus 3E_g$
S_6	$A_{1g} \oplus A_{2g} \oplus 2E_g$
D_3	$A_{2g} \oplus A_{2u} \oplus E_g \oplus E_u$
C_{3v}	$A_{1u} \oplus A_{2g} \oplus E_g \oplus E_u$
D_{3d}	$A_{2g} \oplus E_g$

(iii)-21. C_6 (6)

Site	Librational representations
C_1	$3A \oplus 3B \oplus 3E_1 \oplus 3E_2$
C_2	$A \oplus 2B \oplus 2E_1 \oplus E_2$
C_3	$A \oplus B \oplus E_1 \oplus E_2$
C_6	$A \oplus E_1$

(iii)-22. C_{3h} $(\overline{6})$

Site	Librational representations
C_1	$3A' \oplus 3A'' \oplus 3E' \oplus 3E''$
C_S	$A' \oplus 2A'' \oplus E' \oplus 2E''$
C_3	$A' \oplus A'' \oplus E' \oplus E''$
C_{3h}	$A' \oplus E''$

(iii)-23. C_{6h} $(6/m)$

Site	Librational representations
C_1	$3A_g \oplus 3A_u \oplus 3B_g \oplus 3B_u \oplus 3E_{1g} \oplus 3E_{1u} \oplus 3E_{2g} \oplus 3E_{2u}$
C_i	$3A_g \oplus 3B_g \oplus 3E_{1g} \oplus 3E_{2g}$
C_2	$A_g \oplus A_u \oplus 2B_g \oplus 2B_u \oplus 2E_{1g} \oplus 2E_{1u} \oplus E_{2g} \oplus E_{2u}$
C_S	$A_g \oplus 2A_u \oplus 2B_g \oplus B_u \oplus 2E_{1g} \oplus E_{1u} \oplus E_{2g} \oplus 2E_{2u}$
C_3	$A_g \oplus A_u \oplus B_g \oplus B_u \oplus E_{1g} \oplus E_{1u} \oplus E_{2g} \oplus E_{2u}$
C_{2h}	$A_g \oplus 2B_g \oplus 2E_{1g} \oplus E_{2g}$
S_6	$A_g \oplus B_g \oplus E_{1g} \oplus E_{2g}$
C_6	$A_g \oplus A_u \oplus E_{1g} \oplus E_{1u}$
C_{3h}	$A_g \oplus B_u \oplus E_{1g} \oplus E_{2u}$
C_{6h}	$A_g \oplus E_{1g}$

(iii)-24. D_6 (622)

Site	Librational representations
C_1	$3A_1 \oplus 3A_2 \oplus 3B_1 \oplus 3B_2 \oplus 6E_1 \oplus 6E_2$
C_2^z	$A_1 \oplus A_2 \oplus 2B_1 \oplus 2B_2 \oplus 4E_1 \oplus 2E_2$
C_2	$A_1 \oplus 2A_2 \oplus B_1 \oplus 2B_2 \oplus 3E_1 \oplus 3E_2$
C_2'	$A_1 \oplus 2A_2 \oplus 2B_1 \oplus B_2 \oplus 3E_1 \oplus 3E_2$
C_3	$A_1 \oplus A_2 \oplus B_1 \oplus B_2 \oplus 2E_1 \oplus 2E_2$
D_2	$A_2 \oplus B_1 \oplus B_2 \oplus 2E_1 \oplus + E_2$
D_3	$A_2 \oplus B_2 \oplus E_1 \oplus E_2$
D_3'	$A_2 \oplus B_1 \oplus E_1 \oplus E_2$
C_6	$A_1 \oplus A_2 \oplus 2E_1$
D_6	$A_2 \oplus E_1$

(iii)-25. C_{6v} $(6mm)$

Site	Librational representations
C_1	$3A_1 \oplus 3A_2 \oplus 3B_1 \oplus 3B_2 \oplus 6E_1 \oplus 6E_2$
C_2	$A_1 \oplus A_2 \oplus 2B_1 \oplus 2B_2 \oplus 4E_1 \oplus 2E_2$
$C_S{}^v$	$A_1 \oplus 2A_2 \oplus B_1 \oplus 2B_2 \oplus 3E_1 \oplus 3E_2$
$C_S{}^d$	$A_1 \oplus 2A_2 \oplus 2B_1 \oplus B_2 \oplus 3E_1 \oplus 3E_2$
C_3	$A_1 \oplus A_2 \oplus B_1 \oplus B_2 \oplus 2E_1 \oplus 2E_2$
C_{2v}	$A_2 \oplus B_1 \oplus B_2 \oplus 2E_1 \oplus E_2$
$C_{3v}{}^v$	$A_2 \oplus B_2 \oplus E_1 \oplus E_2$
$C_{3v}{}^d$	$A_2 \oplus B_1 \oplus E_1 \oplus E_2$
C_6	$A_1 \oplus A_2 \oplus 2E_1$
C_{6v}	$A_2 \oplus E_1$

(iii)-26. D_{3h} $(\overline{6}m2)$

Site	Librational representations
C_1	$3A_1' \oplus 3A_1'' \oplus 3A_2' \oplus 3A_2'' \oplus 6E' \oplus 6E''$
C_2	$A_1' \oplus A_1'' \oplus 2A_2' \oplus 2A_2'' \oplus 3E' \oplus 3E''$
$C_S{}^h$	$A_1' \oplus 2A_1'' \oplus A_2' \oplus 2A_2'' \oplus 2E' \oplus 4E''$
$C_S{}^v$	$A_1' \oplus 2A_1'' \oplus 2A_2' \oplus A_2'' \oplus 3E' \oplus 3E''$
C_3	$A_1' \oplus A_1'' \oplus A_2' \oplus A_2'' \oplus 2E' \oplus 2E''$
C_{2v}	$A_1'' + A_2' \oplus A_2'' \oplus E' \oplus 2E''$
D_3	$A_2' \oplus A_2'' \oplus E' \oplus E''$
C_{3v}	$A_1'' \oplus A_2' \oplus E' \oplus E''$
C_{3h}	$A_1'' \oplus A_2' \oplus 2E''$
D_{3h}	$A_2' \oplus E''$

(iii)-27. D_{6h} $(6/mmm)$

Site	Librational representations
C_1	$3A_{1g} \oplus 3A_{1u} \oplus 3A_{2g} \oplus 3A_{2u} \oplus 3B_{1g} \oplus 3B_{1u} \oplus 3B_{2g} \oplus 3B_{2u} \oplus 6E_{1g} \oplus 6E_{1u} \oplus 6E_{2g} \oplus 6E_{2u}$
C_i	$3A_{1g} \oplus 3A_{2g} \oplus 3B_{1g} \oplus 3B_{2g} \oplus 6E_{1g} \oplus 6E_{2g}$
C_2^z	$A_{1g} \oplus A_{1u} \oplus A_{2g} \oplus A_{2u} \oplus 2B_{1g} \oplus 2B_{1u} \oplus 2B_{2g} \oplus 2B_{2u} \oplus 4E_{1g} \oplus 4E_{1u} \oplus 2E_{2g} \oplus 2E_{2u}$
C_2	$A_{1g} \oplus A_{1u} \oplus 2A_{2g} \oplus 2A_{2u} \oplus B_{1g} \oplus B_{1u} \oplus 2B_{2g} \oplus 2B_{2u} \oplus 3E_{1g} \oplus 3E_{1u} \oplus 3E_{2g} \oplus 3E_{2u}$

(iii)-27. D_{6h} (6/*mmm*) ....continued...

Site	Librational representations
C'_2	$A_{1g}\oplus A_{1u}\oplus 2A_{2g}\oplus 2A_{2u}\oplus 2B_{1g}\oplus 2B_{1u}\oplus B_{2g}\oplus B_{2u}\oplus 3E_{1g}\oplus 3E_{1u}\oplus 3E_{2g}\oplus 3E_{2u}$
$C_S{}^h$	$A_{1g}\oplus 2A_{1u}\oplus A_{2g}\oplus 2A_{2u}\oplus 2B_{1g}\oplus B_{1u}\oplus 2B_{2g}\oplus B_{2u}\oplus 4E_{1g}\oplus 2E_{1u}\oplus 2E_{2g}\oplus 4E_{2u}$
$C_S{}^v$	$A_{1g}\oplus 2A_{1u}\oplus 2A_{2g}\oplus A_{2u}\oplus 2B_{1g}\oplus B_{1u}\oplus B_{2g}\oplus 2B_{2u}\oplus 3E_{1g}\oplus 3E_{1u}\oplus 3E_{2g}\oplus 3E_{2u}$
$C_S{}^d$	$A_{1g}\oplus 2A_{1u}\oplus 2A_{2g}\oplus A_{2u}\oplus B_{1g}\oplus 2B_{1u}\oplus 2B_{2g}\oplus B_{2u}\oplus 3E_{1g}\oplus 3E_{1u}\oplus 3E_{2g}\oplus 3E_{2u}$
C_3	$A_{1g}\oplus A_{1u}\oplus A_{2g}\oplus A_{2u}\oplus B_{1g}\oplus B_{1u}\oplus B_{2g}\oplus B_{2u}\oplus 2E_{1g}\oplus 2E_{1u}\oplus 2E_{2g}\oplus 2E_{2u}$
$C_{2h}{}^z$	$A_{1g} \oplus A_{2g} \oplus 2B_{1g} \oplus 2B_{2g} \oplus 4E_{1g} \oplus 2E_{2g}$
C_{2h}	$A_{1u} \oplus 2A_{2u} \oplus B_{1g} \oplus 2B_{2g} \oplus 3E_{1g} \oplus 3E_{2g}$
C'_{2h}	$A_{1g} \oplus 2A_{2g} \oplus 2B_{1g} \oplus B_{2g} \oplus 3E_{1g} \oplus 3E_{2g}$
D_2	$A_{2g}\oplus A_{2u}\oplus B_{1g}\oplus B_{1u}\oplus B_{2g}\oplus B_{2u}\oplus 2E_{1g}\oplus 2E_{1u}\oplus E_{2g}\oplus E_{2u}$
$C_{2v}{}^z$	$A_{1u}\oplus A_{2g}\oplus B_{1g}\oplus B_{1u}\oplus B_{2g}\oplus B_{2u}\oplus 2E_{1g}\oplus 2E_{1u}\oplus E_{2g}\oplus E_{2u}$
C_{2v}	$A_{1u}\oplus A_{2g}\oplus A_{2u}\oplus B_{1u}\oplus B_{2g}\oplus B_{2u}\oplus 2E_{1g}\oplus E_{1u}\oplus E_{2g}\oplus 2E_{2u}$
C'_{2v}	$A_{1u}\oplus A_{2g}\oplus A_{2u}\oplus B_{1g}\oplus B_{1u}\oplus B_{2g}\oplus 2E_{1g}\oplus E_{1u}\oplus E_{2g}\oplus 2E_{2u}$
S_6	$A_{1g} \oplus A_{2g} \oplus B_{1g} \oplus B_{2g} \oplus 2E_{1g} \oplus 2E_{2g}$
D_3	$A_{2g}\oplus A_{2u}\oplus B_{2g}\oplus B_{2u}\oplus E_{1g}\oplus E_{1u}\oplus E_{2g}\oplus E_{2u}$
D'_3	$A_{2g}\oplus A_{2u}\oplus B_{1g}\oplus B_{1u}\oplus E_{1g}\oplus E_{1u}\oplus E_{2g}\oplus E_{2u}$
$C_{3v}{}^v$	$A_{1u} \oplus A_{2g} \oplus B_{1g} \oplus B_{2u} \oplus E_{1g} \oplus E_{1u} \oplus E_{2g} \oplus E_{2u}$
$C_{3v}{}^d$	$A_{1u} \oplus A_{2g} \oplus B_{1u} \oplus B_{2g} \oplus E_{1g} \oplus E_{1u} \oplus E_{2g} \oplus E_{2u}$
C_6	$A_{1g} \oplus A_{1u} \oplus A_{2g} \oplus A_{2u} \oplus 2E_{1g} \oplus 2E_{1u}$
C_{3h}	$A_{1g} \oplus A_{2g} \oplus B_{1u} \oplus B_{2u} \oplus 2E_{1g} \oplus 2E_{2u}$
D_{2h}	$A_{2g} \oplus B_{1g} \oplus B_{2g} \oplus 2E_{1g} \oplus E_{2g}$
D_{3d}	$A_{2g} \oplus B_{2g} \oplus E_{1g} \oplus E_{2g}$
D'_{3d}	$A_{2g} \oplus B_{1g} \oplus E_{1g} \oplus E_{2g}$
C_{6h}	$A_{1g} \oplus A_{2g} \oplus 2E_{1g}$

(iii)-27. D_{6h} (6/*mmm*) ....continued

Site	Librational representations
D_6	$A_{2g} \oplus A_{2u} \oplus E_{1g} \oplus E_{1u}$
C_{6v}	$A_{1u} \oplus A_{2g} \oplus E_{1g} \oplus E_{1u}$
D_{3h}	$A_{2g} \oplus B_{2u} \oplus E_{1g} \oplus E_{2u}$
D'_{3h}	$A_{2g} \oplus B_{1u} \oplus E_{1g} \oplus E_{2u}$
D_{6h}	$A_{2g} \oplus E_{1g}$

(iii)-28. T (23)

Site	Librational representations
C_1	$3A \oplus 3E \oplus 9F$
C_2	$A \oplus E \oplus 5F$
C_3	$A \oplus E \oplus 3F$
D_2	$3F$
T	F

(iii)-29. T_h ($m3$)

Site	Librational representations
C_1	$3A_g \oplus 3A_u \oplus 3E_g \oplus 3E_u \oplus 9F_g \oplus 9F_u$
C_i	$3A_g \oplus 3E_g \oplus 9F_g$
C_2	$A_g \oplus A_u \oplus E_g \oplus E_u \oplus 5F_g \oplus 5F_u$
C_S	$A_g \oplus 2A_u \oplus E_g \oplus 2E_u \oplus 5F_g \oplus 4F_u$
C_3	$A_g \oplus A_u \oplus E_g \oplus E_u \oplus 3F_g \oplus 3F_u$
C_{2h}	$A_g \oplus E_g \oplus 5F_g$
D_2	$3F_g \oplus 3F_u$
C_{2v}	$A_u \oplus E_u \oplus 3F_g \oplus 2F_u$
S_6	$A_g \oplus E_g \oplus 3F_g$
D_{2h}	$3F_g$
T	$F_g \oplus F_u$
T_h	F_g

(iii)-30. O (432)

Site	Librational representations
C_1	$3A_1 \oplus 3A_2 \oplus 6E \oplus 9F_1 \oplus 9F_2$
C_2	$A_1 \oplus A_2 \oplus 2E \oplus 5F_1 \oplus 5F_2$
C'_2	$A_1 \oplus 2A_2 \oplus 3E \oplus 5F_1 \oplus 4F_2$
C_3	$A_1 \oplus A_2 \oplus 2E \oplus 3F_1 \oplus 3F_2$

(iii)-30. O (432) ....continued

Site	Librational representations
D_2	$3F_1 \oplus 3F_2$
D'_2	$A_2 \oplus E \oplus 3F_1 \oplus 2F_2$
C_4	$A_1 \oplus E \oplus 3F_1 \oplus 2F_2$
D_3	$A_2 \oplus E \oplus 2F_1 \oplus F_2$
D_4	$2F_1 \oplus F_2$
T	$F_1 \oplus F_2$
O	F_1

(iii)-31. T_d $(\bar{4}3m)$

Site	Librational representations
C_1	$3A_1 \oplus 3A_2 \oplus 6E \oplus 9F_1 \oplus 9F_2$
C_2	$A_1 \oplus A_2 \oplus 2E \oplus 5F_1 \oplus 5F_2$
C_S	$A_1 \oplus 2A_2 \oplus 3E \oplus 5F_1 \oplus 4F_2$
C_3	$A_1 \oplus A_2 \oplus 2E \oplus 3F_1 \oplus 3F_2$
D_2	$3F_1 \oplus 3F_2$
C_{2v}	$A_2 \oplus E \oplus 3F_1 \oplus 2F_2$
S_4	$A_1 \oplus E \oplus 3F_1 \oplus 2F_2$
C_{3v}	$A_2 \oplus E \oplus 2F_1 \oplus F_2$
D_{2d}	$2F_1 \oplus F_2$
T	$F_1 \oplus F_2$
T_d	F_1

(iii)-32. O_h $(m\bar{3}m)$

Site	Librational representations
C_1	$3A_{1g} \oplus 3A_{1u} \oplus 3A_{2g} \oplus 3A_{2u} \oplus 6E_g \oplus 6E_u \oplus 9F_{1g} \oplus 9F_{1u} \oplus 9F_{2g} \oplus 9F_{2u}$
C_i	$3A_{1g} \oplus 3A_{2g} \oplus 6E_g \oplus 9F_{1g} \oplus 9F_{2g}$
C_2	$A_{1g} \oplus A_{1u} \oplus A_{2g} \oplus A_{2u} \oplus 2E_g \oplus 2E_u \oplus 5F_{1g} \oplus 5F_{1u} \oplus 5F_{2g} \oplus 5F_{2u}$
C'_2	$A_{1g} \oplus A_{1u} \oplus 2A_{2g} \oplus 2A_{2u} \oplus 3E_g \oplus 3E_u \oplus 5F_{1g} \oplus 5F_{1u} \oplus 4F_{2g} \oplus 4F_{2u}$
$C_S{}^h$	$A_{1g} \oplus 2A_{1u} \oplus A_{2g} \oplus 2A_{2u} \oplus 2E_g \oplus 4E_u \oplus 5F_{1g} \oplus 4F_{1u} \oplus 5F_{2g} \oplus 4F_{2u}$
$C_S{}^d$	$A_{1g} \oplus 2A_{1u} \oplus 2A_{2g} \oplus A_{2u} \oplus 3E_g \oplus 3E_u \oplus 5F_{1g} \oplus 4F_{1u} \oplus 4F_{2g} \oplus 5F_{2u}$
C_3	$A_{1g} \oplus A_{1u} \oplus A_{2g} \oplus A_{2u} \oplus 2E_g \oplus 2E_u \oplus 3F_{1g} \oplus 3F_{1u} \oplus 3F_{2g} \oplus 3F_{2u}$

(iii)-32. O_h $(m\bar{3}m)$ ....continued

Site	Librational representations
C_{2h}	$A_{1g} \oplus A_{2g} \oplus 2E_g \oplus 5F_{1g} \oplus 5F_{2g}$
C'_{2h}	$A_{1g} \oplus 2A_{2g} \oplus 3E_g \oplus 5F_{1g} \oplus 4F_{2g}$
D_2	$3F_{1g} \oplus 3F_{1u} \oplus 3F_{2g} \oplus 3F_{2u}$
D'_2	$A_{2g} \oplus A_{2u} \oplus E_g \oplus E_u \oplus 3F_{1g} \oplus 3F_{1u} \oplus 2F_{2g} \oplus 2F_{2u}$
$C_{2v}{}^h$	$A_{1u} \oplus A_{2u} \oplus 2E_u \oplus 3F_{1g} \oplus 2F_{1u} \oplus 3F_{2g} \oplus 2F_{2u}$
$C_{2v}{}^d$	$A_{1g} \oplus A_{2u} \oplus E_g \oplus E_u \oplus 3F_{1g} \oplus 2F_{1u} \oplus 2F_{2g} \oplus 3F_{2u}$
C'_{2v}	$A_{1u} \oplus A_{2g} \oplus A_{2u} \oplus 2E_g \oplus 2E_u \oplus 3F_{1g} \oplus 2F_{1u} \oplus 2F_{2g} \oplus 2F_{2u}$
C_4	$A_{1g} \oplus A_{1u} \oplus E_g \oplus E_u \oplus 3F_{1g} \oplus 3F_{1u} \oplus 2F_{2g} \oplus 2F_{2u}$
S_4	$A_{1g} \oplus A_{2u} \oplus E_g \oplus E_u \oplus 3F_{1g} \oplus 2F_{1u} \oplus 2F_{2g} \oplus 3F_{2u}$
S_6	$A_{1g} \oplus A_{2g} \oplus 2E_g \oplus 3F_{1g} \oplus 3F_{2g}$
D_3	$A_{2g} \oplus A_{2u} \oplus E_g \oplus E_u \oplus 2F_{1g} \oplus 2F_{1u} \oplus F_{2g} \oplus F_{2u}$
C_{3v}	$A_{1u} \oplus A_{2u} \oplus E_g \oplus E_u \oplus 2F_{1g} \oplus F_{1u} \oplus F_{2g} \oplus 2F_{2u}$
D_{2h}	$3F_{1g} \oplus 3F_{2g}$
D'_{2h}	$A_{2g} \oplus E_g \oplus 3F_{1g} \oplus 2F_{2g}$
C_{4h}	$A_{1g} \oplus E_g \oplus 3F_{1g} \oplus 2F_{2g}$
C_{4v}	$A_{1u} \oplus E_u \oplus 2F_{1g} \oplus F_{1u} \oplus F_{2g} \oplus F_{2u}$
D_4	$2F_{1g} \oplus 2F_{1u} \oplus F_{2g} \oplus F_{2u}$
D_{2d}	$2F_{1g} \oplus F_{1u} \oplus F_{2g} \oplus 2F_{2u}$
D'_{2d}	$A_{2u} \oplus E_u \oplus 2F_{1g} \oplus F_{1u} \oplus F_{2g} \oplus F_{2u}$
D_{3d}	$A_{2g} \oplus E_g \oplus 2F_{1g} \oplus F_{2g}$
T	$F_{1g} \oplus F_{1u} \oplus F_{2g} \oplus F_{2u}$
D_{4h}	$2F_{1g} \oplus F_{2g}$
T_h	$F_{1g} \oplus F_{2g}$
O	$F_{1g} \oplus F_{1u}$
T_d	$F_{1g} \oplus F_{2u}$
O_h	F_{1g}

Tables (iv): Correlation tables relating the irreducible representations of each crystalline site in a certain space group to the irreducible representations of its sub-groups.

(**iv**)-1. C_1 (1)

C_1	C_1
A	A
(1)	(1)

(**iv**)-2. C_i $(\bar{1})$

C_i	C_1
A_g	A
A_u	A
(1)	(2)

(**iv**)-3. C_2 (2)

C_2	C_1
A	A
B	A
(1)	(2)

(**iv**)-4. C_S (m)

C_S	C_1
A'	A
A''	A
(1)	(2)

(**iv**)-5. C_{2h} (2/m)

C_{2h}	C_1	C_i	C_2	C_S
A_g	A	A_g	A	A'
B_g	A	A_g	B	A''
A_u	A	A_u	A	A''
B_u	A	A_u	B	A'
(1)	(4)	(2)	(2)	(2)

(**iv**)-6. D_2 (222)

D_2	C_1	$C_2^z(C_2^z)$	$C_2^y(C_2^y)$	$C_2^x(C_2^x)$
A	A	A	A	A
B_1	A	A	B	B
B_2	A	B	A	B
B_3	A	B	B	A
(1)	(4)	(2)	(2)	(2)

(**iv**)-7. C_{2v} ($mm2$)

C_{2v}	C_1	C_2	$C_S^{xz}(\sigma_v^{xz})$	$C_S^{yz}(\sigma_v^{yz})$
A_1	A	A	A'	A'
B_1	A	B	A'	A''
A_2	A	A	A''	A''
B_2	A	B	A''	A'
(1)	(4)	(2)	(2)	(2)

(**iv**)-8. D_{2h} (mmm)

D_{2h}	C_1	$C_2^z(C_2^z)$	$C_2^y(C_2^y)$	$C_2^x(C_2^x)$	$C_S^{xy}(\sigma^{xy})$	$C_S^{xz}(\sigma^{xz})$	$C_S^{yz}(\sigma^{yz})$	C_i
A_g	A	A	A	A	A'	A'	A'	A_g
A_u	A	A	A	A	A''	A''	A''	A_u
B_{1g}	A	A	B	B	A'	A''	A''	A_g
B_{1u}	A	A	B	B	A''	A'	A'	A_u
B_{2g}	A	B	A	B	A''	A'	A''	A_g
B_{2u}	A	B	A	B	A'	A''	A'	A_u
B_{3g}	A	B	B	A	A''	A''	A'	A_g
B_{3u}	A	B	B	A	A'	A'	A''	A_u
(1)	(8)	(4)	(4)	(4)	(4)	(4)	(4)	(4)

(**iv**)-8. D_{2h} (mmm) ...continued

D_{2h}	C_{2v}^{z} $(C_2^z + \sigma^{xz} + \sigma^{yz})$	C_{2v}^{y} $(C_2^y + \sigma^{xy} + \sigma^{yz})$	C_{2v}^{x} $(C_2^x + \sigma^{xy} + \sigma^{xz})$	D_2	C_{2h}^{z} $(C_2^z + \sigma^{xy})$	C_{2h}^{y} $(C_2^y + \sigma^{xz})$	C_{2h}^{x} $(C_2^x + \sigma^{yz})$
A_g	A_1	A_1	A_1	A	A_g	A_g	A_g
A_u	A_2	A_2	A_2	A	A_u	A_u	A_u
B_{1g}	A_2	B_2	B_1	B_1	A_g	B_g	B_g
B_{1u}	A_1	B_1	B_2	B_1	A_u	B_u	B_u
B_{2g}	B_1	A_2	B_2	B_2	B_g	A_g	B_g
B_{2u}	B_2	A_1	B_1	B_2	B_u	A_u	B_u
B_{3g}	B_2	B_1	A_2	B_3	B_g	B_g	A_g
B_{3u}	B_1	B_2	A_1	B_3	B_u	B_u	A_u
(1)	(2)	(2)	(2)	(2)	(2)	(2)	(2)

(**iv**)-9. C_4 (4)

C_4	C_1	C_2
A	A	A
B	A	A
E	$2A$	$2B$
(1)	(4)	(2)

(**iv**)-10. S_4 ($\bar{4}$)

S_4	C_1	C_2
A	A	A
B	A	A
E	$2A$	$2B$
(1)	(4)	(2)

(**iv**)-11. C_{4h} ($4/m$)

C_{4h}	C_1	C_i	C_2	C_S	C_{2h}	C_4	S_4
A_g	A	A_g	A	A'	A_g	A	A
A_u	A	A_u	A	A''	A_u	A	B
B_g	A	A_g	A	A'	A_g	B	B
B_u	A	A_u	A	A''	A_u	B	A
E_g	$2A$	$2A_g$	$2B$	$2A''$	$2B_g$	E	E
E_u	$2A$	$2A_u$	$2B$	$2A'$	$2B_u$	E	E
(1)	(8)	(4)	(4)	(4)	(2)	(2)	(2)

(**iv**)-12. D_4 (422)

D_4	C_1	$C_2^z(C_2^z)$	$C_2(C_2)$	$C_2'(C_2')$	D_2 $(C_2^z + 2C_2)$	D_2' $(C_2^z + 2C_2')$	C_4
A_1	A	A	A	A	A	A	A
A_2	A	A	B	B	B_1	B_1	A
B_1	A	A	A	B	A	B_1	B
B_2	A	A	B	A	B_1	A	B
E	$2A$	$2B$	$A \oplus B$	$A \oplus B$	$B_2 \oplus B_3$	$B_2 \oplus B_3$	E
(1)	(8)	(4)	(4)	(4)	(2)	(2)	(2)

(**iv**)-13. C_{4v} (4mm)

C_{4v}	C_1	C_2	$C_S{}^{v}(\sigma_v)$	$C_S{}^{d}(\sigma_d)$	$C_{2v}{}^{v}(\sigma_v)$	$C_{2v}{}^{d}(\sigma_d)$	C_4
A_1	A	A	A'	A'	A_1	A_1	A
A_2	A	A	A''	A''	A_2	A_2	A
B_1	A	A	A'	A''	A_1	A_2	B
B_2	A	A	A''	A'	A_2	A_1	B
E	$2A$	$2B$	$A' \oplus A''$	$A' \oplus A''$	$B_1 \oplus B_2$	$B_1 \oplus B_2$	E
(1)	(8)	(4)	(4)	(4)	(2)	(2)	(2)

(**iv**)-14. D_{2d} $(\bar{4}2m)$ $C_2 \rightarrow C_2(z)$

D_{2d}	C_1	$C_2^z(C_2^z)$	$C_2(C_2)$	C_S	D_2	C_{2v}	S_4
A_1	A	A	A	A'	A	A_1	A
A_2	A	A	B	A''	B_1	A_2	A
B_1	A	A	A	A''	A	A_2	B
B_2	A	A	B	A'	B_1	A_1	B
E	$2A$	$2B$	$A \oplus B$	$A' \oplus A''$	$B_2 \oplus B_3$	$B_1 \oplus B_2$	E
(1)	(8)	(4)	(4)	(4)	(2)	(2)	(2)

(**iv**)-15. D_{4h} (4/mmm)

D_{4h}	C_1	C_i	$C_2^z(C_2^z)$	$C_2(C_2)$	$C_2'(C_2')$	$C_S{}^{h}(\sigma^h)$	$C_S{}^{v}(\sigma^v)$	$C_S{}^{d}(\sigma^d)$	$C_{2h}{}^{z}$ $(C_2{}^{z} + \sigma_h)$	C_{2h} $(C_2 + \sigma_v)$
A_{1g}	A	A_g	A	A	A	A'	A'	A'	A_g	A_g
A_{1u}	A	A_u	A	A	A	A''	A''	A''	A_u	A_u
A_{2g}	A	A_g	A	B	B	A'	A''	A''	A_g	B_g
A_{2u}	A	A_u	A	B	B	A''	A'	A'	A_u	B_u
B_{1g}	A	A_g	A	A	B	A'	A'	A''	A_g	B_g
B_{1u}	A	A_u	A	A	B	A''	A''	A'	A_u	B_u
B_{2g}	A	A_g	A	B	A	A'	A''	A'	A_g	A_g
B_{2u}	A	A_u	A	B	A	A''	A'	A''	A_u	A_u
E_g	$2A$	$2A_g$	$2B$	$A \oplus B$	$A \oplus B$	$2A''$	$A' \oplus A''$	$A' \oplus A''$	$2B_g$	$A_g + B_g$
E_u	$2A$	$2A_u$	$2B$	$A \oplus B$	$A \oplus B$	$2A'$	$A' \oplus A''$	$A' \oplus A''$	$2B_u$	$A_u + B_u$
(1)	(16)	(8)	(8)	(8)	(8)	(8)	(8)	(8)	(4)	(4)

(**iv**)-15. D_{4h} (4/*mmm*) .....continued.

D_{4h}	C'_{2h} $(C'_2 + \sigma_d)$	D_2 $(C_2{}^z + 2C_2)$	D'_2 $(C_2{}^z + 2C'_2)$	$C_{2v}{}^{v}$ $(C_2{}^z + 2\sigma_v)$	$C_{2v}{}^{d}$ $(C_2{}^z + 2\sigma_d)$	C_{2v} $(C_2 + \sigma_h + \sigma_d)$	C'_{2v} $(C'_2 + \sigma_h + \sigma_d)$	C_4
A_{1g}	A_g	A	A	A_1	A_1	A_1	A_1	A
A_{1u}	A_u	A	A	A_2	A_2	A_2	A_2	A
A_{2g}	B_g	B_1	B_1	A_2	A_2	B_1	B_1	A
A_{2u}	B_u	B_1	B_1	A_1	A_1	B_2	B_2	A
B_{1g}	B_g	A	B_1	A_1	A_2	A_1	B_1	B
B_{1u}	B_u	A	B_1	A_2	A_1	A_2	B_2	B
B_{2g}	A_g	B_1	A	A_2	A_1	B_1	A_1	B
B_{2u}	A_u	B_1	A	A_1	A_2	B_2	A_2	B
E_g	$A_g \oplus B_g$	$B_2 \oplus B_3$	$B_2 \oplus B_3$	$B_1 \oplus B_2$	$B_1 \oplus B_2$	$A_2 \oplus B_2$	$A_2 \oplus B_2$	E
E_u	$A_u \oplus B_u$	$B_2 \oplus B_3$	$B_2 \oplus B_3$	$B_1 \oplus B_2$	$B_1 \oplus B_2$	$A_1 \oplus B_1$	$A_1 + B_1$	E
(1)	(4)	(4)	(4)	(4)	(4)	(4)	(4)	(4)

(**iv**)-15. D_{4h} (4/*mmm*) ...continued

$C''_2 \to C'_2$ (over D_{2d}) $\quad$ $C'_2 \to C'_2$ (over D'_{2d})

D_{4h}	S_4	D_{2h} $(C_2{}^z + 2C_2)$	D'_{2h} $(C_2{}^z + 2C'_2)$	C_{4h}	D_4	C_{4v}	D_{2d} $(C_2{}^z + 2C_2 + \sigma_d)$	D'_{2d} $(C_2{}^z + 2C'_2 + \sigma_v)$
A_{1g}	A	A_g	A_g	A_g	A_1	A_1	A_1	A_1
A_{1u}	B	A_u	A_u	A_u	A_1	A_2	B_1	B_1
A_{2g}	A	B_{1g}	B_{1g}	A_g	A_2	A_2	A_2	A_2
A_{2u}	B	B_{1u}	B_{1u}	A_u	A_2	A_1	B_2	B_2
B_{1g}	B	A_g	B_{1g}	B_g	B_1	B_1	B_2	B_1
B_{1u}	A	A_u	B_{1u}	B_u	B_2	B_2	A_2	A_1
B_{2g}	B	B_{1g}	A_g	B_g	B_2	B_2	B_1	B_2
B_{2u}	A	B_{1u}	A_u	B_u	B_2	B_1	A_1	A_2
E_g	E	$B_{2g} \oplus B_{3g}$	$B_{2g} \oplus B_{3g}$	E_g	E	E	E	E
E_u	E	$B_{2u} \oplus B_{3u}$	$B_{2u} \oplus B_{3u}$	E_u	E	E	E	E
(1)	(4)	(2)	(2)	(2)	(2)	(2)	(2)	(2)

(**iv**)-16. C_3 (3)

C_3	C_1
A	A
E	$2A$
(1)	(3)

(**iv**)-17. S_6 $(\bar{3})$

S_6	C_1	C_i	C_3
A_g	A	A_g	A
A_u	A	A_u	A
E_g	$2A$	$2A_g$	E
E_u	$2A$	$2A_u$	E
(1)	(6)	(3)	(2)

(**iv**)-18. D_3 (32)

D_3	C_1	C_2	C_3
A_1	A	A	A
A_2	A	B	A
E	$2A$	$A \oplus B$	E
(1)	(6)	(3)	(2)

(**iv**)-19. C_{3v} $(3m)$

C_{3v}	C_1	C_S	C_3
A_1	A	A'	A
A_2	A	A''	A
E	$2A$	$A' \oplus A''$	E
(1)	(6)	(3)	(2)

(**iv**)-20. D_{3d} $(\bar{3}m)$

D_{3d}	C_1	C_i	C_2	C_S	C_3	C_{2h}	S_6	D_3	C_{3v}
A_{1g}	A	A_g	A	A'	A	A_g	A_g	A_1	A_1
A_{1u}	A	A_u	A	A''	A	A_u	A_u	A_1	A_2
A_{2g}	A	A_g	B	A''	A	B_g	A_g	A_2	A_2
A_{2u}	A	A_u	B	A'	A	B_u	A_u	A_2	A_1
E_g	$2A$	$2A_g$	$A \oplus B$	$A' \oplus A''$	E	$A_g \oplus B_g$	E_g	E	E
E_u	$2A$	$2A_u$	$A \oplus B$	$A' \oplus A''$	E	$A_g \oplus B_u$	E_u	E	E
(1)	(12)	(6)	(6)	(6)	(4)	(3)	(2)	(2)	(2)

(**iv**)-21. C_6 (6)

C_6	C_1	C_2	C_3
A	A	A	A
B	A	B	A
E_1	$2A$	$2B$	E
E_2	$2A$	$2A$	E
(1)	(6)	(3)	(2)

(**iv**)-22. C_{3h} $(\bar{6})$

C_{3h}	C_1	C_S	C_3
A'	A	A'	A
A''	A	A''	A
E'	$2A$	$2A'$	E
E''	$2A$	$2A''$	E
(1)	(6)	(3)	(2)

(**iv**)-23. C_{6h} $(6/m)$

C_{6h}	C_1	C_i	C_2	C_S	C_3	C_{2h}	S_6	C_6	C_{3h}
A_g	A	A_g	A	A'	A	A_g	A_g	A	A'
A_u	A	A_u	A	A''	A	A_u	A_u	A	A''
B_g	A	A_g	B	A''	A	B_g	A_g	B	A''
B_u	A	A_u	B	A'	A	B_u	A_u	B	A'
E_{1g}	$2A$	$2A_g$	$2B$	$2A''$	E	$2B_g$	E_g	E_1	E''
E_{1u}	$2A$	$2A_u$	$2B$	$2A'$	E	$2B_u$	E_u	E_1	E'
E_{2g}	$2A$	$2A_g$	$2A$	$2A'$	E	$2A_g$	E_g	E_2	E'
E_{2u}	$2A$	$2A_u$	$2A$	$2A''$	E	$2A_u$	E_u	E_2	E''
(1)	(12)	(6)	(6)	(6)	(4)	(3)	(2)	(2)	(2)

(**iv**)-24. D_6 (622)

D_6	C_1	$C_2^z(C_2^z)$	$C_2(C_2)$	$C_2'(C_2')$	C_3	D_2	$D_3(3C_2)$	$D_3'(3C_2')$	C_6
A_1	A	A	A	A	A	A	A_1	A_1	A
A_2	A	A	B	B	A	B_1	A_2	A_2	A
B_1	A	B	A	B	A	B_2	A_1	A_2	B
B_2	A	B	B	A	A	B_3	A_2	A_1	B
E_1	$2A$	$2B$	$A \oplus B$	$A \oplus B$	E	$B_2 \oplus B_3$	E	E	E_1
E_2	$2A$	$2A$	$A \oplus B$	$A \oplus B$	E	$A \oplus B$	E	E	E_2
(1)	(12)	(6)	(6)	(6)	(4)	(3)	(2)	(2)	(2)

(**iv**)-25. C_{6v} $(6mm)$

$\sigma_v \rightarrow \sigma(zx)$

C_{6v}	C_1	C_2	$C_S{}^v(\sigma_v)$	$C_S{}^d(\sigma_d)$	C_3	C_{2v}	$C_{3v}{}^v(\sigma_v)$	$C_{3v}{}^d(\sigma_d)$	C_6
A_1	A	A	A'	A'	A	A_1	A_1	A_1	A
A_2	A	A	A''	A''	A	A_2	A_2	A_2	A
B_1	A	B	A'	A''	A	B_1	A_1	A_2	B
B_2	A	B	A''	A'	A	B_2	A_2	A_1	B
E_1	$2A$	$2B$	$A' \oplus A''$	$A' \oplus A''$	E	$B_1 \oplus B_2$	E	E	E_1
E_2	$2A$	$2A$	$A' \oplus A''$	$A' \oplus A''$	E	$A_1 \oplus A_2$	E	E	E_2
(1)	(12)	(6)	(6)	(6)	(4)	(3)	(2)	(2)	(2)

(**iv**)-26. D_{3h} $(\bar{6}m2)$ $\sigma_h \rightarrow \sigma'_v(yz)$

D_{3h}	C_1	C_2	$C_S{}^h(\sigma_h)$	$C_S{}^v(\sigma_v)$	C_3	C_{2v}	D_3	C_{3v}	C_{3h}
A'_1	A	A	A'	A'	A	A_1	A_1	A_1	A'
A''_1	A	A	A''	A''	A	A_2	A_1	A_2	A''
A'_2	A	B	A'	A''	A	B_2	A_2	A_2	A'
A''_2	A	B	A''	A'	A	B_1	A_2	A_1	A''
E'	$2A$	$A \oplus B$	$2A'$	$A' \oplus A''$	E	$A_1 \oplus B_2$	E	E	E'
E''	$2A$	$A \oplus B$	$2A''$	$A' \oplus A''$	E	$A_2 \oplus B_1$	E	E	E''
(1)	(12)	(6)	(6)	(6)	(4)	(3)	(2)	(2)	(2)

(**iv**)-27. D_{6h} $(6/mmm)$

D_{6h}	C_1	C_i	$C_2^z(C_2^z)$	$C_2(C_2)$	$C'_2(C'_2)$	$C_S{}^h(\sigma_h)$	$C_S{}^v(\sigma_v)$	$C_S{}^d(\sigma_d)$	C_3	$C_{2h}{}^z$ $(C_2{}^z + \sigma_h)$
A_{1g}	A	A_g	A	A	A	A'	A'	A'	A	A_g
A_{1u}	A	A_u	A	A	A	A''	A''	A''	A	A_u
A_{2g}	A	A_g	A	B	B	A'	A''	A''	A	A_g
A_{2u}	A	A_u	A	B	B	A''	A'	A'	A	A_u
B_{1g}	A	A_g	B	A	B	A''	A''	A'	A	B_g
B_{1u}	A	A_u	B	A	B	A'	A'	A''	A	B_u
B_{2g}	A	A_g	B	B	A	A''	A'	A''	A	B_g
B_{2u}	A	A_u	B	B	A	A'	A''	A'	A	B_u
E_{1g}	$2A$	$2A_g$	$2B$	$A \oplus B$	$A \oplus B$	$2A''$	$A' \oplus A''$	$A' \oplus A''$	E	$2B_g$
E_{1u}	$2A$	$2A_u$	$2B$	$A \oplus B$	$A \oplus B$	$2A'$	$A' \oplus A''$	$A' \oplus A''$	E	$2B_u$
E_{2g}	$2A$	$2A_g$	$2A$	$A \oplus B$	$A \oplus B$	$2A'$	$A' \oplus A''$	$A' \oplus A''$	E	$2A_g$
E_{2u}	$2A$	$2A_u$	$2A$	$A \oplus B$	$A \oplus B$	$2A''$	$A' \oplus A''$	$A' \oplus A''$	E	$2A_u$
(1)	(24)	(12)	(12)	(12)	(12)	(12)	(12)	(12)	(8)	(6)

(**iv**)-27. D_{6h} (6/*mmm*) ... continued ...

D_{6h}	C_{2h} $(C_2+\sigma_d)$	C'_{2h} $(C'_2+\sigma_v)$	D_2	$C_{2v}{}^z$ $(C_2{}^z+\sigma_v+\sigma_d)$	C_{2v} $(C_2+\sigma_h+\sigma_d)$	C'_{2v} $(C'_2+\sigma_v+\sigma_d)$	S_6	$D_3(3C_2)$	$D'_3(3C_2)$
A_{1g}	A_g	A_g	A	A_1	A_1	A_1	A_g	A_1	A_1
A_{1u}	A_u	A_u	A	A_2	A_2	A_2	A_u	A_1	A_1
A_{2g}	B_g	B_g	B_1	A_2	B_1	B_1	A_g	A_2	A_2
A_{2u}	B_u	B_u	B_1	A_1	B_2	B_2	A_u	A_2	A_2
B_{1g}	A_g	B_g	B_2	B_1	A_2	B_2	A_g	A_1	A_2
B_{1u}	A_u	B_u	B_2	B_2	A_1	B_1	A_u	A_1	A_2
B_{2g}	B_g	A_g	B_3	B_2	B_2	A_2	A_g	A_2	A_1
B_{2u}	B_u	A_u	B_3	B_1	B_1	A_1	A_u	A_2	A_1
E_{1g}	$A_g\oplus B_g$	$A_g\oplus B_g$	$B_2\oplus B_3$	$B_1\oplus B_2$	$A_2\oplus B_2$	$A_2\oplus B_2$	E_g	E	E
E_{1u}	$A_u\oplus B_u$	$A_u\oplus B_u$	$B_2\oplus B_3$	$B_1\oplus B_2$	$A_1\oplus B_1$	$A_1\oplus B_1$	E_u	E	E
E_{2g}	$A_g\oplus B_g$	$A_g\oplus B_g$	$A\oplus B_1$	$A_1\oplus A_2$	$A_1\oplus B_1$	$A_1\oplus B_1$	E_g	E	E
E_{2u}	$A_u\oplus B_u$	$A_u\oplus B_u$	$A\oplus B_1$	$A_1\oplus A_2$	$A_2\oplus B_2$	$A_2\oplus B_2$	E_u	E	E
(1)	(6)	(6)	(6)	(6)	(6)	(6)	(4)	(4)	(4)

(**iv**)-27. D_{6h} (6/*mmm*) ...continued

$\sigma_h \to \sigma(xy)$
$\sigma_v \to \sigma(yz)$

D_{6h}	$C_{3v}{}^v(\sigma_v)$	$C_{3v}{}^d(\sigma_d)$	C_6	C_{3h}	D_{2h}	D_{3d} $(3C_2+3\sigma_d)$	D'_{3d} $(3C'_2+3\sigma_d)$	C_{6h}	D_6	C_{6v}
A_{1g}	A_1	A_1	A	A'	A_g	A_{1g}	A_{1g}	A_g	A_1	A_1
A_{1u}	A_2	A_2	A	A''	A_u	A_{1u}	A_{1u}	A_u	A_1	A_2
A_{2g}	A_2	A_2	A	A'	B_{1g}	A_{2g}	A_{2g}	A_g	A_2	A_2
A_{2u}	A_1	A_1	A	A''	B_{1u}	A_{2u}	A_{2u}	A_u	A_2	A_1
B_{1g}	A_2	A_1	B	A''	B_{2g}	A_{1g}	A_{2g}	B_g	B_1	B_2
B_{1u}	A_1	A_2	B	A'	B_{2u}	A_{1u}	A_{2u}	B_u	B_1	B_1
B_{2g}	A_1	A_2	B	A''	B_{3g}	A_{2g}	A_{1g}	B_g	B_2	B_1
B_{2u}	A_2	A_1	B	A'	B_{3u}	A_{2u}	A_{1u}	B_u	B_2	B_2
E_{1g}	E	E	E_1	E''	$B_{2g}\oplus B_{3g}$	E_g	E_g	E_{1g}	E_1	E_1
E_{1u}	E	E	E_1	E'	$B_{2u}\oplus B_{3u}$	E_u	E_u	E_{1u}	E_1	E_1
E_{2g}	E	E	E_2	E'	$A_g\oplus B_{1g}$	E_g	E_g	E_{2g}	E_2	E_2
E_{2u}	E	E	E_2	E''	$A_u\oplus B_{1u}$	E_u	E_u	E_{2u}	E_2	E_2
(1)	(4)	(4)	(4)	(4)	(3)	(2)	(2)	(2)	(2)	(2)

(**iv**)-27. D_{6h} $(6/mmm)$ ...continued

D_{6h}	D_{3h} $(3C_2 + 3\sigma_h)$	D'_{3h} $(3C'_2 + 3\sigma_d)$
A_{1g}	A'_1	A'_1
A_{1u}	A''_1	A''_1
A_{12}	A'_2	A'_2
A_{2u}	A''_2	A''_2
B_{1g}	A''_1	A''_2
B_{1u}	A'_1	A'_2
B_{2g}	A''_2	A''_1
B_{2u}	A'_2	A'_1
E_{1g}	E''	E''
E_{1u}	E'	E'
E_{2g}	E'	E'
E_{2u}	E''	E''
(1)	(2)	(2)

(**iv**)-28. T (23)

T	C_1	C_2	C_3	D_2
A	A	A	A	A
E	$2A$	$2A$	E	$2A$
F	$3A$	$A \oplus 2B$	$A \oplus E$	$B_1 \oplus B_2 \oplus B_3$
(1)	(12)	(6)	(4)	(3)

(**iv**)-29. T_h $(m3)$

T_h	C_1	C_i	C_2	C_S	C_3	C_{2h}	D_2	C_{2v}	S_6	D_{2h}	T
A_g	A	A_g	A	A'	A	A_g	A	A_1	A_g	A_g	A
A_u	A	A_u	A	A''	A	A_u	A	A_2	A_u	A_u	A
E_g	$2A$	$2A_g$	$2A$	$2A'$	E	$2A_g$	$2A$	$2A_1$	E_g	$2A_g$	E
E_u	$2A$	$2A_u$	$2A$	$2A''$	E	$2A_u$	$2A$	$2A_2$	E_u	$2A_u$	E
F_g	$3A$	$3A_g$	$A \oplus 2B$	$A' \oplus 2A''$	$A \oplus E$	$A_g \oplus 2B_g$	$B_1 \oplus B_2 \oplus B_3$	$A_2 \oplus B_1 \oplus B_2$	$A_g \oplus E_g$	$B_{1g} \oplus B_{2g} \oplus B_{3g}$	F
F_u	$3A$	$3A_u$	$A \oplus 2B$	$2A' \oplus A''$	$A \oplus E$	$A_u \oplus 2B_u$	$B_1 \oplus B_2 \oplus B_3$	$A_1 \oplus B_1 \oplus B_2$	$A_u \oplus E_u$	$B_{1u} \oplus B_{2u} \oplus B_{3u}$	F
(1)	(24)	(12)	(12)	(4)	(8)	(6)	(6)	(6)	(4)	(3)	(2)

(**iv**)-30. O (432)

O	C_1	$C_2(C_4^2)$	$C'_2(C_2)$	C_3	$D_2(3C_4^2)$	D'_2 $(C_4^2 + 2C_2)$	C_4	D_3	D_4	T
A_1	A	A	A	A	A	A	A	A_1	A_1	A
A_2	A	A	A	A	A	B_1	B	A_2	B_1	A
E	$2A$	$2A$	$2A$	E	$2A$	$A \oplus B_1$	$A \oplus B$	E	$A_1 \oplus B_1$	E
F_1	$3A$	$A \oplus 2B$	$A \oplus 2B$	$A \oplus E$	$B_1 \oplus B_2 \oplus B_3$	$B_1 \oplus B_2 \oplus B_3$	$A \oplus E$	$A_2 \oplus E$	$A_2 \oplus E$	F
F_2	$3A$	$A \oplus 2B$	$A \oplus 2B$	$A \oplus E$	$B_1 \oplus B_2 \oplus B_3$	$A \oplus B_2 \oplus B_3$	$B \oplus E$	$A_1 \oplus E$	$B_2 \oplus E$	F
(1)	(24)	(12)	(12)	(8)	(6)	(6)	(6)	(4)	(3)	(2)

(**iv**)-31. T_d $(\bar{4}3m)$

T_d	C_1	C_2	C_S	C_3	D_2	C_{2v}	S_4	C_{3v}	D_{2d}	T
A_1	A	A	A'	A	A	A_1	A	A_1	A_1	A
A_2	A	A	A''	A	A	A_2	B	A_2	B_1	A
E	$2A$	$2A$	$A' \oplus A''$	E	$2A$	$A_1 \oplus A_2$	$A \oplus B$	E	$A_1 \oplus B_1$	E
F_1	$3A$	$A \oplus 2B$	$A' \oplus 2A''$	$A \oplus E$	$B_1 \oplus B_2 \oplus B_3$	$A_2 \oplus B_1 \oplus B_2$	$A \oplus E$	$A_2 \oplus E$	$A_2 \oplus E$	F
F_2	$3A$	$A \oplus 2B$	$2A' \oplus A''$	$A \oplus E$	$B_1 \oplus B_2 \oplus B_3$	$A_1 \oplus B_1 \oplus B_2$	$B \oplus E$	$A_1 \oplus E$	$B_2 \oplus E$	F
(1)	(24)	(12)	(12)	(8)	(6)	(6)	(6)	(4)	(3)	(2)

(**iv**)-32. O_h $(m\bar{3}m)$

O_h	C_1	C_i	$C_2(C_4^2)$	$C_2'(C_2)$	$C_S{}^h(\sigma_h)$	$C_S{}^d(\sigma_d)$	C_3	C_{2h} $(C_4^2 + \sigma_h)$	C_{2h}' $(C_2 + \sigma_d)$	$D_2(3C_4^2)$
A_{1g}	A	A_g	A	A	A'	A'	A	A_g	A_g	A
A_{1u}	A	A_u	A	A	A''	A''	A	A_u	A_u	A
A_{2g}	A	A_g	A	B	A'	A''	A	A_g	B_g	A
A_{2u}	A	A_u	A	B	A''	A'	A	A_u	B_u	A
E_g	$2A$	$2A_g$	$2A$	$A \oplus B$	$2A'$	$A' \oplus A''$	E	$2A_g$	$A_g \oplus B_g$	$2A$
E_u	$2A$	$2A_u$	$2A$	$A \oplus B$	$2A''$	$A' \oplus A''$	E	$2A_u$	$A_u \oplus B_u$	$2A$
F_{1g}	$3A$	$3A_g$	$A \oplus 2B$	$A \oplus 2B$	$A' \oplus 2A''$	$A' \oplus 2A''$	$A \oplus E$	$A_g \oplus 2B_g$	$A_g \oplus 2B_g$	$B_1 \oplus B_2 \oplus B_3$
F_{1u}	$3A$	$3A_u$	$A \oplus 2B$	$A \oplus 2B$	$2A' \oplus A''$	$2A' \oplus A''$	$B \oplus E$	$A_u \oplus 2B_u$	$A_u \oplus 2B_u$	$B_1 \oplus B_2 \oplus B_3$
F_{2g}	$3A$	$3A_g$	$A \oplus 2B$	$2A + B$	$A' \oplus 2A''$	$2A' \oplus A''$	$A \oplus E$	$A_g \oplus 2B_g$	$2A_g \oplus B_g$	$B_1 \oplus B_2 \oplus B_3$
F_{2u}	$3A$	$3A_u$	$A \oplus 2B$	$2A + B$	$2A' \oplus A''$	$A' \oplus 2A''$	$B \oplus E$	$A_u \oplus 2B_u$	$2A_u \oplus B_u$	$B_1 \oplus B_2 \oplus B_3$
(1)	(48)	(24)	(24)	(24)	(24)	(24)	(16)	(12)	(12)	(12)

(**iv**)-32. O_h $(m\bar{3}m)$ ....continued...

O_h	D_2' $(C_4^2 + 2C_2)$	$C_{2v}{}^h$ $(C_4^2 + 2\sigma_h)$	$C_{2v}{}^d$ $(C_4^2 + 2\sigma_d)$	C_{2v}' $(C_2 + \sigma_h + \sigma_d)$	C_4	S_4	S_6	D_3
A_{1g}	A	A_1	A_1	A_1	A	A	A_g	A_1
A_{1u}	A	A_2	A_2	A_2	A	B	A_u	A_1
A_{2g}	B_1	A_1	A_2	B_1	B	B	A_g	A_2
A_{2u}	B_1	A_2	A_1	B_2	B	A	A_u	A_2
E_g	$A \oplus B_1$	$2A_1$	$A_1 \oplus A_2$	$A_1 \oplus B_1$	$A \oplus B$	$A \oplus B$	E_g	E
E_u	$A \oplus B_1$	$2A_2$	$A_1 \oplus A_2$	$A_2 \oplus B_2$	$A \oplus B$	$A \oplus B$	E_u	E
F_{1g}	$B_1 \oplus B_2 \oplus B_3$	$A_2 \oplus B_1 \oplus B_2$	$A_2 \oplus B_1 \oplus B_2$	$A_2 \oplus B_1 \oplus B_2$	$A \oplus E$	$A \oplus E$	$A_g \oplus E_g$	$A_2 \oplus E$
F_{1u}	$B_1 \oplus B_2 \oplus B_3$	$A_1 \oplus B_1 \oplus B_2$	$A_1 \oplus B_1 \oplus B_2$	$A_1 \oplus B_1 \oplus B_2$	$A \oplus E$	$B \oplus E$	$A_u \oplus E_u$	$A_2 \oplus E$
F_{2g}	$A \oplus B_2 \oplus B_3$	$A_2 \oplus B_1 \oplus B_2$	$A_1 \oplus B_1 \oplus B_2$	$A_1 \oplus A_2 \oplus B_2$	$B \oplus E$	$B \oplus E$	$A_g \oplus E_g$	$A_1 \oplus E$
F_{2u}	$A \oplus B_2 \oplus B_3$	$A_1 \oplus B_1 \oplus B_2$	$A_2 \oplus B_1 \oplus B_2$	$A_1 \oplus A_2 \oplus B_1$	$B \oplus E$	$A \oplus E$	$A_u \oplus E_u$	$A_1 \oplus E$
(1)	(12)	(12)	(12)	(12)	(12)	(12)	(8)	(8)

(**iv**)-32. O_h $(m\bar{3}m)$ ....continued...

O_h	C_{3v}	D_{2h} $(3C_4^2 + 3\sigma_h)$	D'_{2h} $(C_4^2 + 2C_2 + \sigma_h + \sigma_d)$	C_{4h}	C_{4v}	D_4	D_{2d} $(3C_4^2 + \sigma_d)$	D'_{2d} $(C_4^2 + 2C_2 + \sigma_h)$
A_{1g}	A_1	A_g	A_g	A_g	A_1	A_1	A_1	A_1
A_{1u}	A_2	A_u	A_u	A_u	A_2	A_1	B_1	B_1
A_{2g}	A_2	A_g	B_{1g}	B_g	B_1	B_1	B_1	B_2
A_{2u}	A_1	A_u	B_{1u}	B_u	B_2	B_1	A_1	A_2
E_g	E	$2A_g$	$A_g \oplus B_{1g}$	$A_g \oplus B_g$	$A_1 \oplus B_1$	$A_1 \oplus B_1$	$A_1 \oplus B_1$	$A_1 \oplus B_2$
E_u	E	$2A_u$	$A_u \oplus B_{1u}$	$A_u \oplus B_u$	$A_2 \oplus B_2$	$A_1 \oplus B_1$	$A_1 \oplus B_1$	$A_2 \oplus B_1$
F_{1g}	$A_2 \oplus E$	$B_{1g} \oplus B_{2g} \oplus B_{3g}$	$B_{1g} \oplus B_{2g} \oplus B_{3g}$	$A_g \oplus E_g$	$A_2 \oplus E$	$A_2 \oplus E$	$A_2 \oplus E$	$A_2 \oplus E$
F_{1u}	$A_1 \oplus E$	$B_{1u} \oplus B_{2u} \oplus B_{3u}$	$B_{1u} \oplus B_{2u} \oplus B_{3u}$	$A_u \oplus E_u$	$A_1 \oplus E$	$A_2 \oplus E$	$B_2 \oplus E$	$B_2 \oplus E$
F_{2g}	$A_1 \oplus E$	$B_{1g} \oplus B_{2g} \oplus B_{3g}$	$A_g \oplus B_{2g} \oplus B_{3g}$	$B_g \oplus E_g$	$B_2 \oplus E$	$B_2 \oplus E$	$B_2 \oplus E$	$B_1 \oplus E$
F_{2u}	$A_2 \oplus E$	$B_{1u} \oplus B_{2u} \oplus B_{3u}$	$A_u \oplus B_{2u} \oplus B_{3u}$	$B_u \oplus E_u$	$B_1 \oplus E$	$B_2 \oplus E$	$A_2 \oplus E$	$A_1 \oplus E$
(1)	(8)	(6)	(6)	(6)	(6)	(6)	(6)	(6)

(**iv**)-32. O_h $(m\bar{3}m)$ ....continued

O_h	D_{3d}	T	D_{4h}	T_h	O	T_d
A_{1g}	A_{1g}	A	A_{1g}	A_g	A_1	A_1
A_{1u}	A_{1u}	A	A_{1u}	A_u	A_1	A_2
A_{2g}	A_{2g}	A	B_{1g}	A_g	A_2	A_2
A_{2u}	A_{2u}	A	B_{1u}	A_u	A_2	A_1
E_g	E_g	E	$A_{1g} \oplus B_{1g}$	E_g	E	E
E_u	E_u	E	$A_{1u} \oplus B_{1u}$	E_u	E	E
F_{1g}	$A_{2g} \oplus E_g$	F	$A_{2g} \oplus E_g$	F_g	F_1	F_1
F_{1u}	$A_{2u} \oplus E_u$	F	$A_{2u} \oplus E_u$	F_u	F_1	F_2
F_{2g}	$A_{1g} \oplus E_g$	F	$B_{2g} \oplus E_g$	F_g	F_2	F_2
F_{2u}	$A_{1u} \oplus E_u$	F	$B_{2u} \oplus E_u$	F_u	F_2	F_1
(1)	(4)	(4)	(3)	(2)	(2)	(2)

Exercises for Chapter 11

11.1 For continuation and completion of the cadmium tungstate $CdWO_4$ oscillation modes study, extend the calculations of the atomic basis coordinates attached to each lattice point. Consider the following calculation steps:

(a) Change the values in the atomic fractional coordinates in Section 11.5 Table 11.1 by translating the oxygen O_I coordinates by $-\underline{c}$ and those of the cadmium Cd by $-\underline{b}$.

(b) Transform the fractional coordinates into full Cartesian coordinates. Choose the X axis orientation to coincide with the $\underline{a}$ axis, the Z axis orientation to coincide with the $\underline{b}$ axis, and the Y axis to complete the formation of a right-handed coordinate system.

(c) Calculate the coordinates of an additional O_I type oxygen atom (to be noted as O_I') by performing the C_2 symmetry operation transformation about an axis parallel to the Z axis crossing through the tungsten W atom coordinates. Use a similar means to obtain the oxygen O_{II}' coordinates.

(d) Calculate the $W - O_I$ and $W - O_{II}$ bond lengths and the $O_I - W - O_I$ and $O_{II} - W - O_{II}$ inter bond angles.

(e) It is given that the unit-cell origin is of a C_i site symmetry. Perform the inversion operation about that point to obtain the Cartesian coordinates of the second formula unit comprising the atomic basis of the unit cell. Its belonging atoms are labeled by an over-hanging bar.

(f) Calculate the $W - \overline{W}$ and $Cd - \overline{Cd}$ bond lengths and the angle between these bonds in the unit cell.

(g) Calculate the $Cd - W$ bond length and the $W - Cd - \overline{W}$ and $Cd - W - \overline{Cd}$ inter-bond angles.

(h) Calculate the tensor of inertia of the atomic basis about the origin, and establish its principal axes and the moments of inertia about these axes (see Section 8.2). Assume a simplistic model, by which the characteristic rotational angular frequency is inversely proportional to the square root of the moment of inertia. Indicate accordingly the prospected order among the possible unit cell librational frequencies in the crystal.

12 Atoms and Ions in Crystalline Sites

12.1 Introduction

There are many cases, where an atom or ion residing in a crystal, is encircled by a symmetric distribution of neighboring atoms or ions. In particular, such a situation occurs when a small number of atoms or ions are added to a crystal by replacing native majority atoms or ions that comprise the pure crystal. The replacing species are called **impurities** or **dopants**; the majority (native) atoms are known as **legitimate atoms**. Quite often, the presence of impurities grants unique properties to the crystal. In semiconductors, the material electrical conductivity may change by orders of magnitude; in transparent insulating crystals, dopants' presence may provide additional absorption bands. In the visible range, transparent materials assume special colors. A prominent example is the presence of triply ionized chromium Cr^{3+} ions replacing legitimate ions of the sapphire crystal. Sapphire is the Al_2O_3 crystal. Its pure form is fully transparent in the visible range. The triply ionized chromium Cr^{3+} ions in the sapphire are surrounded by six doubly ionized oxygen O^{2-} anions, forming an eight-faced (octahedral) spatial structure. A sapphire crystal containing those dopants is called a **ruby**; it is not fully transparent like the pure material, and it looks purple. Because of the combination of color and its original extraordinary mechanical strength, ruby crystals have been used as gemstones for thousands of years.

Changes in atomic states' structure of ions when they are embedded in a solid material, compared with their structure when they are free in space, may be simplistically attributed to two factors. First, the crystalline vicinity may be interpreted as a uniform medium, whose dielectric constant $\varepsilon(\infty)$ is larger than a unity. That case is similar to the one described in Section 10.4, where each electronic oscillator experiences a dielectric constant produced by all higher-frequency oscillators. Considering the present issue, penetration of the electronic clouds of the crystalline vicinity into the dopant-hosted site, increases the site dielectric constant. By basic electrostatic considerations, the attractive force between the dopant nucleus and between all its electrons becomes smaller, thus reducing all states' energy gaps. By that consideration solely, the energy order among the states remains unchanged compared with those of the free atom. The other factor causing changes in the states' structure is that the site symmetry is obviously lower than the free-space symmetry. The symmetry reduction causes both states' splitting and changes of their energetic order. In Sections 12.2 and 12.3, we address the electronic states' structure of free atoms and ions in more detail compared with their description in Chapter 5; in Sections 12.4 and 12.5, we address the states' splitting and their radiative transition probabilities.

12.2 Energy States of Alkali and Alkali-Like Atoms

Qualitative discussions of the states' energy schemes of alkali atoms (mono-valence many-electron atoms), including detailed presentation of lithium *Li* and sodium *Na* atoms, have already been presented in

Physics, Optics, and Spectroscopy of Materials, First Edition. Zeev Burshtein.

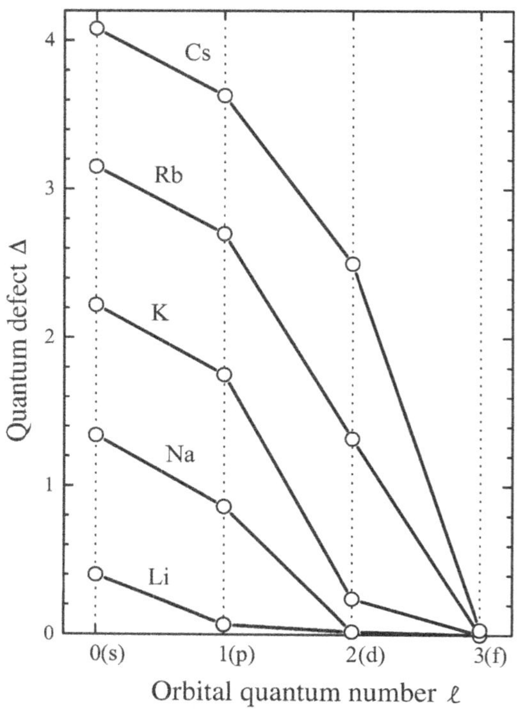

Figure 12.1 Quantum defect in alkali atoms as functions of the orbital angular momentum quantum number.

Section 5.5.1. It is convenient to provide an empiric presentation of their atomic energy schemes in a form that allows simple comparison with the states' schemes of the hydrogen and hydrogen-like atoms. By presenting the hydrogen and hydrogen-like atoms' energy in reciprocal wavelength units, we may write $\sigma_{n\ell} \equiv E_{n\ell}/hc = -R_{\infty}Z^2/n^2$, where the state energy $\sigma_{n\ell}$ in the cgs units' system is expressed in cm^{-1} units, $R_{\infty} = 109,737\ cm^{-1}$ is the Rydberg constant, n is the principal quantum number, ℓ is the orbital angular momentum quantum number, and Z is the atomic number (see Eq. [3.49]). The choice of presenting the energy in reciprocal wavelength units is very common in the field of atomic spectroscopy, and we adopt it for the following presentations. In generalizing the presentation to alkali and alkali-like metal atoms, the Z and n variables are considered adjustable parameters. In particular, it has been found that for alkali metals, fair approximation is obtained by setting $Z = 1$, and replacing n with $n^* \equiv n - \Delta_{n\ell}$, where the n^* quantity is a **corrected principal quantum number**. The energy is thus presented as

$$\boxed{\sigma_{n\ell} = -R_{\infty}/n^{*2}} \quad . \tag{12.1}$$

The $\Delta_{n\ell}$ quantity is often called a **quantum defect.** Experiments show that the quantum defect of a specific alkali atom is almost constant for the states' series of different n numbers that belong to the same orbital angular momentum quantum number ℓ. Figure 12.1 describes the experimental values of the quantum defect as function of ℓ for different alkali atoms. The symbols' sizes fully cover the experimental deviations among states of the same ℓ belonging to different n's. The most similar to a hydrogen atom is the lithium *Li* atom, whose maximal quantum defect (the 2*s* configuration) is 0.2. The most different is the cesium *Cs* atom, whose maximal quantum defect (the 6*s* configuration) is 4.08. Some empiric equations provide corrections to the figure, given values as function of the principal quantum number n.

Related to each alkali atom, there are ions that exhibit the same number of electrons. They are known as **alkali-like atoms**. For example, related to the sodium *Na* atom, whose electronic configuration is [*Ne*]3*s* (Table 5.5), there exist the magnesium Mg^+ ion, the doubly ionized aluminum Al^{2+}, the triply ionized silicon Si^{3+}, and so on. All noted ions exhibit 11 electrons orbiting about the nucleus; in other words, they are all **iso-electronic**. Their states' energies transform into

$$\boxed{\sigma_{n\ell} = -R_{\infty}(Z_i + 1)^2/n^{*2}} \quad , \tag{12.2}$$

where Z_i is the degree of ionization.

Experiment shows, as expected, that as the Z_i ionization degree increases, the energy scheme becomes more similar to the hydrogen atom one; in other words, the quantum defect Δ decreases toward zero for increasing Z_i. As for the alkali atoms, the quantum defect grows as a function of the orbital angular momentum quantum number ℓ. For a demonstration, the change of experimental quantum defect with the ionization degree for several sodium-like ions is presented in Figure 12.2.

Each of the energy states in alkali and alkali-like atoms is, in fact, a doublet; namely, it is split in two by the spin-orbit interaction according to the two possible spin states (see Section 4.5.2). The expression obtained for the doublet spin-orbit splitting of hydrogen and hydrogen-like atoms (Eq. [4.67]) is given as

$$\boxed{\Delta\sigma_{SL} = \alpha^2 R_\infty \frac{Z^4}{n^3\ell(\ell+1)}}, \tag{12.3}$$

where α is the fine-structure constant (see text related to Eq. [4.20]). In contrast to a general quantitative fit obtained by setting $Z = 1$ for the general energy scheme of alkali atoms (Eq. [12.1]), the doublet splitting by spin-orbit coupling in alkali-like atoms are orders of magnitude larger than the values obtained by setting $Z = 1$ in Eq. (12.3). Qualitatively, however, the general dependence on the n, ℓ, and Z parameters in Eq. (12.3) follows quite satisfactorily. Specifically, the splitting grows steeply with the atomic number Z, and is approximately inversely proportional to the third power of the principal atomic number n for the same Z atoms, and approximately inversely proportional to the $\ell(\ell + 1)$ product. A fine approximation may be obtained by considering Z as an adjustable parameter per each atom and per each series of states belonging to the same orbital angular momentum ℓ. The adjusted Z parameters usually rest between the actual atomic number of the particular atom and its ionization degree Z_i. The dependence on the n and ℓ quantum numbers is usually regular, yet prominent deviations occur in heavy atoms. For example, the doublet splitting in the d-series of the rubidium Rb atom for $n = 11$ is 6 times larger than for $n = 10$; the doublet splitting in the d-series of the cesium Cs atom for $n = 13$ is 8 times larger than for $n = 12$ (see Figure 12.3).

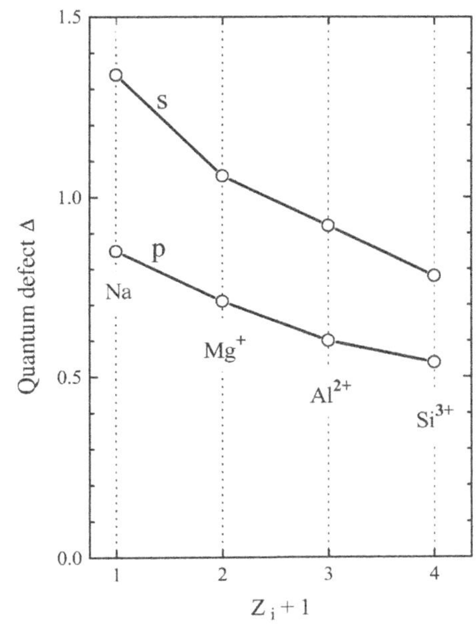

Figure 12.2 Quantum defect in equi-electrons' sodium-like ions as functions of the ionization degree.

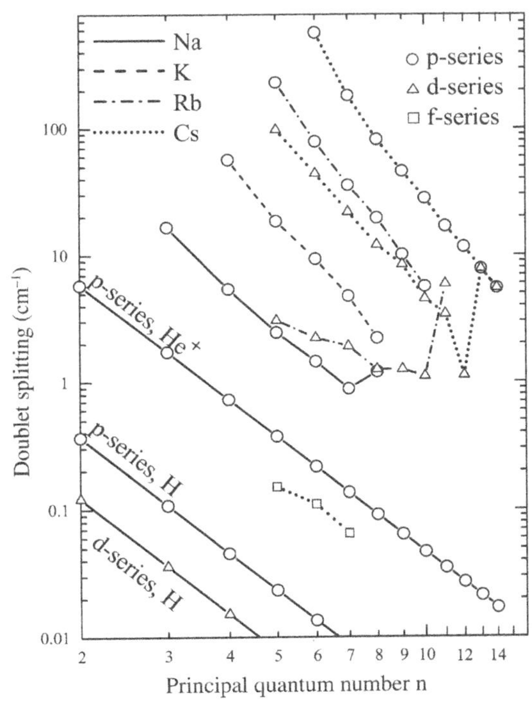

Figure 12.3 Doublet splitting of different states in the potassium *K*, sodium *Na*, rubidium *Rb*, and cesium *Cs* alkali atoms, as functions of the principal quantum number. Doublet splitting in the hydrogen *H* atom and the helium He^+ ion is shown for comparison.

12.3 Energy States of Many-Electron Atoms and Ions

In Chapter 5, we presented the underlying principles that describe the energy states of free, many-electron atoms. Identical principles are used to describe the energy states of free ions. In many cases, the energy states of different Terms may be presented using only a small number of parameters. In the present section, we outline the calculation methods in more detail, and present tables of simple expressions that allow us to form the energy schemes.

We recall that our theory refers to the independent electrons model, and limit our interest to Term energy calculations of valence electrons in atoms or ions, namely electrons in excess of those forming full shells or full sub-shells. For example, an observation of Table 5.5 reveals that the ground-state electronic configuration of the triply ionized neodymium Nd^{3+} contains three f-electrons in the shell belonging to the principal quantum number $n = 4$ (namely $4f^3$), all in addition to filled shells belonging to the $n = 1, 2, 3$ principal quantum numbers (like in the noble xenon Xe gas atom). The valence electrons' Hamiltonian would be given by

$$\mathbf{H} = \sum_{u=1}^{z}\left\{-\frac{\hbar^2}{2m}\underline{\nabla}_u^{\ 2} - \frac{[Z_i + 1 - \varsigma(r_u)]e^2}{r_u}\right\} + \sum_{u \neq v}\frac{e^2}{|\underline{r}_u - \underline{r}_v|}, \tag{12.4}$$

where z is the number of valence electrons, $\underline{r}_u$ for $u = 1, \ldots, z$ is the distance vector of the u electron from the nucleus, and $\varsigma(r)$ is the Hartree screening function (see Eq. [5.13] and related text). The zero-order Hamiltonian is selected as

$$\mathbf{H}_0 = \sum_{u=1}^{z}\mathbf{H}_{0u} = \sum_{u=1}^{z}\left\{-\frac{\hbar^2}{2m}\underline{\nabla}_u^{\ 2} - \frac{[Z_i + 1 - \varsigma(r_u)]e^2}{r_u}\right\}, \tag{12.5}$$

and the perturbation Hamiltonian as

$$\mathbf{V} = \sum_{u \neq v}\frac{e^2}{|\underline{r}_u - \underline{r}_v|}. \tag{12.6}$$

The zero-order Hamiltonian is a sum of several independent electrons, each belonging to a different u electron. The potential belonging to each is central. Thus, it is possible to perform variables separation, and represent the wave-function solution of each $\mathbf{H}_{0u}$ in the form $\psi_u(\underline{r}_u) = \boldsymbol{R}_{n_u\ell_u}(r_u)\boldsymbol{Y}_{\ell_u m_u}(\theta_u, \varphi_u)$. In the first stage of our considerations, we overlook the involvement of electronic spin states. The angular part of the solution, namely the $\boldsymbol{Y}_{\ell m}(\theta, \varphi) = \boldsymbol{\Theta}_\ell^m(\theta)\boldsymbol{\Phi}_m(\varphi)$ function, is one of the spherical harmonics described in Sections 3.3 and 3.4 (specifically in Table 3.3), like in the hydrogen atom. On the other hand, the $\boldsymbol{R}_{n\ell}(r)$ functions *differ* from the radial functions described there (specifically in Table 3.3). Their specific form depends on Hartree's screening function $\varsigma(r)$. In contrast to the hydrogen radial functions, energy degeneracy among states belonging to the same n principal quantum number of different orbital angular momentum ℓ numbers no longer exists. Selection of Hartree's screening function $\varsigma(r)$, or alternatively the form of the radial $\boldsymbol{R}_{n\ell}(r)$ function, is modified to exhibit maximal fits between calculated and experimentally observed states' energies.

The following steps require corrections to the states' energies according to the perturbation elements emanating from the electrostatic repulsion among the electrons (Eq. [12.6]). Such corrections should include considerations of exchange symmetry between each electron pairs (see Section 5.2.3). The correction calculations are lengthy and beyond the possible scope of the present book. We shall present only general principles underlying the calculations, and review results obtained for Terms belonging to different electronic configurations.

At this point, we need to get acquainted with a number of functions and mathematic developments needed for further discussions. For brevity, we refrain from presenting proofs for all developments and equations.

The Legendre polynomial $P_\ell(\xi)$ for an integer ℓ is defined for the ξ variable in the $-1 \leq \xi \leq +1$ range by the Rodrigues equation (text following Eq. [3.27])

$$P_\ell(\xi) \equiv P_\ell^0(\xi) = \frac{1}{2^\ell \cdot \ell!} \cdot \frac{d^\ell}{d\xi^\ell}\left(\xi^2 - 1\right)^\ell , \tag{12.7}$$

where $P_\ell^0(\xi)$ is the Legendre polynomial of $m = 0$ order (Eq. [3.27]). An important recurrence equation that allows for a convenient calculation of Legendre polynomials of any higher order is

$$P_{\ell+1}(\xi) = \frac{(2\ell+1)\cdot\xi\cdot P_\ell(\xi) - \ell P_{\ell-1}(\xi)}{\ell+1} . \tag{12.8}$$

Recall that a Legendre polynomial relates to the spherical harmonics $Y_{\ell m}(\theta, \varphi)$ (Table 3.2), according to

$$P_\ell(\cos\theta) = \left(\frac{4\pi}{2\ell+1}\right)^{1/2} Y_{\ell 0}(\theta, \varphi) . \tag{12.9}$$

In continuation of Table 3.2, in Table 12.1, we provide a list of $Y_{\ell m}(\theta, \varphi)$ spherical harmonics for $\ell = 4$, 5, 6, and the m values belonging to each ℓ. The reader will be requested to extend the table to include the $\ell = 7$ case, as an exercise.

Expansion of the $1/|\underline{r}_1 - \underline{r}_2|$ function into a power series, where $\underline{r}_1$ and $\underline{r}_2$ are vectors in a three-dimensional space, is important for theoretical calculations in general, and especially for the currently studied issue:

$$\frac{1}{|\underline{r}_1 - \underline{r}_2|} = \sum_{\ell=0}^{\infty} \frac{r_<^{\ell}}{r_>^{\ell+1}} P_\ell(\cos\upsilon) . \tag{12.10}$$

The $r_<$ and $r_>$ variables in Eq. (12.10) are the shortest and longest of the vectors pair $\underline{r}_1$ and $\underline{r}_2$, respectively, and υ is the angle between their orientations. The individual orientation angles of $\underline{r}_1$ and $\underline{r}_2$ in a Cartesian coordinate system are (θ_1, φ_1) and (θ_2, φ_2), respectively. The υ angle is given by

$$\cos\upsilon = \cos\theta_1 \cos\theta_2 + \sin\theta_1 \sin\theta_2 \cos(\varphi_1 - \varphi_2) , \tag{12.11}$$

An expression that satisfies the following relation, called the *addition theorem*, is

$$P_\ell(\cos\upsilon) = \frac{4\pi}{2\ell+1} \sum_{m=-\ell}^{\ell} Y_{\ell m}^*(\theta_1, \varphi_1) Y_{\ell m}(\theta_2, \varphi_2) . \tag{12.12}$$

Using Eq. (12.12), it is possible to write

$$V_{uv} = e^2 \sum_{\kappa=0}^{\infty} \left(\frac{r_<^{\kappa}}{r_>^{\kappa+1}}\right) P_\kappa(\cos\upsilon_{uv}) , \tag{12.13}$$

where $r_<$ and $r_>$ variables in Eq. (12.13) are, respectively, the shortest and longest of the r_u and r_v quantities, and υ_{uv} is the angle between $\underline{r}_u$ and $\underline{r}_v$. Using the addition theorem 12.12, we get

$$V_{uv} = e^2 \sum_{\kappa=0}^{\infty} \left(\frac{r_<^{\kappa}}{r_>^{\kappa+1}}\right) P_\kappa(\cos\upsilon_{uv}) \sum_{\mu=-\kappa}^{\kappa} \frac{4\pi}{2\kappa+1} Y_{\kappa\mu}^*(\theta_u, \varphi_u) Y_{\kappa\mu}(\theta_v, \varphi_v) . \tag{12.14}$$

Table 12.1 $Y_{\ell m}(\theta, \varphi)$ functions for $\ell = 4, 5, 6$ and the m values belonging to each ℓ. The corresponding spectroscopic g, h, i notations are written in parentheses. It is, in fact, a continuation of Table 3.2.

ℓ	m	$Y_{\ell m}(\theta, \varphi)$
4 (g)	0	$Y_{40} = \frac{3}{8}\sqrt{1/4\pi}\left(35\cos^4\theta - 30\cos^2\theta + 3\right)$
	±1	$Y_{4\pm1} = \mp\frac{3}{4}\sqrt{5/4\pi}\cdot\sin\theta\cos\theta\left(7\cos^2\theta - 3\right)\cdot e^{\pm i\varphi}$
	±2	$Y_{4\pm2} = \pm\frac{3}{8}\sqrt{5/2\pi}\cdot\sin^2\theta\left(7\cos^2\theta - 1\right)\cdot e^{\pm i2\varphi}$
	±3	$Y_{4\pm3} = \mp\frac{3}{4}\sqrt{35/4\pi}\cdot\sin^3\theta\cos\theta\cdot e^{\pm i3\varphi}$
	±4	$Y_{4\pm4} = \pm\frac{3}{16}\sqrt{35/2\pi}\cdot\sin^4\theta\cdot e^{\pm i4\varphi}$
5 (h)	0	$Y_{50} = \frac{1}{8}\sqrt{11/2\pi}\cdot\cos\theta\left(63\cos^4\theta - 70\cos^2\theta + 15\right)$
	±1	$Y_{5\pm1} = \mp\frac{1}{16}\sqrt{165/2\pi}\cdot\sin\theta\left(21\cos^4\theta - 14\cos^2\theta + 1\right)\cdot e^{\pm i\varphi}$
	±2	$Y_{5\pm2} = \pm\frac{1}{8}\sqrt{1155/2\pi}\cdot\sin^2\theta\cos\theta\left(3\cos^2\theta - 1\right)\cdot e^{\pm i2\varphi}$
	±3	$Y_{5\pm3} = \mp\frac{1}{16}\sqrt{385/4\pi}\cdot\sin^3\theta\left(9\cos^2\theta - 1\right)\cdot e^{\pm i3\varphi}$
	±4	$Y_{5\pm4} = \pm\frac{3}{16}\sqrt{385/2\pi}\cdot\sin^4\theta\cos\theta\cdot e^{\pm i4\varphi}$
	±5	$Y_{5\pm5} = \mp\frac{3}{16}\sqrt{77/4\pi}\cdot\sin^5\theta\cdot e^{\pm i5\varphi}$
6 (i)	0	$Y_{60} = \frac{1}{16}\sqrt{13/4\pi}\cdot\cos\theta\left(231\cos^6\theta - 315\cos^4\theta + 105\cos^2\theta - 5\right)$
	±1	$Y_{6\pm1} = \mp\frac{1}{16}\sqrt{273/2\pi}\cdot\sin\theta\cos\theta\left(33\cos^4\theta - 30\cos^2\theta + 5\right)\cdot e^{\pm i\varphi}$
	±2	$Y_{6\pm2} = \pm\frac{1}{32}\sqrt{1365/4\pi}\cdot\sin^2\theta\left(33\cos^4\theta - 18\cos^2\theta + 1\right)\cdot e^{\pm i2\varphi}$
	±3	$Y_{6\pm3} = \mp\frac{1}{16}\sqrt{455/4\pi}\cdot\sin^3\theta\cos\theta\left(11\cos^2\theta - 3\right)\cdot e^{\pm i3\varphi}$
	±4	$Y_{6\pm4} = \pm\frac{1}{16}\sqrt{91/4\pi}\cdot\sin^4\theta\left(11\cos^2\theta - 1\right)\cdot e^{\pm i4\varphi}$
	±5	$Y_{6\pm5} = \mp\frac{3}{16}\sqrt{1001/4\pi}\cdot\sin^5\theta\cos\theta\cdot e^{\pm i5\varphi}$
	±6	$Y_{6\pm6} = \pm\frac{1}{32}\sqrt{3003/4\pi}\cdot\sin^6\theta\cdot e^{\pm i6\varphi}$

Eq. (12.14) thus exhibits separation among the radiuses and angles of the u and v electrons for the perturbation Hamiltonian function $\boldsymbol{V}_{uv}$.

In the following calculations of the energy correction, we will not account for energy shifts and splitting caused by spin-orbit interactions. Spin states are obviously critical in determining the allowed Terms according to the Pauli Exclusion Principle (Table 5.4 and related text). Energy corrections are obtained by taking into account exchange symmetry among the electrons. The expressions contain addends of exchange integrals K_{ab} and Coulomb integrals C_{ab} of the form

$$K_{ab} = \int d^3\underline{r}_u d^3\underline{r}_v \boldsymbol{V}_{uv}\psi_a^*(\underline{r}_u)\psi_b(\underline{r}_u)\psi_b^*(\underline{r}_v)\psi_a(\underline{r}_v); \tag{12.15}$$

$$C_{ab} = \int d^3\underline{r}_u d^3\underline{r}_v \boldsymbol{V}_{uv}|\psi_a(\underline{r}_u)|^2|\psi_b(\underline{r}_v)|^2 , \tag{12.16}$$

where each of the a and b subscripts indicates a possible state, $|n_a\ell_a m_a\rangle$ or $|n_b\ell_b m_b\rangle$, of each of the independent electrons forming the (u, v) couple (see Section 5.2.3). Each such integral contains addends belonging to a different κ value. In the following, we limit our consideration to electronic configurations belonging to the same principal quantum number n (namely, $n_a = n_b$).

We consider first the exchange integral. An addend of the exchange integral J_{ab}^{κ} belonging to a specific κ would be

$$\begin{aligned} J_{ab}^{\kappa} = &\left\{ e^2 \int_0^\infty \int_0^\infty \frac{r_<^{\kappa}}{r_>^{\kappa+1}} \boldsymbol{R}_{n\ell_a}(r_u)\boldsymbol{R}_{n\ell_b}(r_u)\boldsymbol{R}_{n\ell_a}(r_v)\boldsymbol{R}_{n\ell_b}(r_v) r_u^2 r_v^2 dr_u dr_v \right. \\ &\cdot \frac{4\pi}{2\kappa+1} \sum_{\mu=-\kappa}^{\kappa} \left\{ \int_0^\pi \int_0^{2\pi} \boldsymbol{Y}_{\ell_a m_a}^*(\theta_u, \varphi_u)\, \boldsymbol{Y}_{\ell_b m_b}(\theta_u, \varphi_u) \boldsymbol{Y}_{\kappa\mu}^*(\theta_u, \varphi_u) \boldsymbol{\sin}\theta_u d\theta_u d\varphi_u \right. \\ &\left.\left. \cdot \int_0^\pi \int_0^{2\pi} \boldsymbol{Y}_{\ell_b m_b}^*(\theta_u, \varphi_u)\, \boldsymbol{Y}_{\ell_a m_a}(\theta_v, \varphi_v) \boldsymbol{Y}_{\kappa\mu}(\theta_v, \varphi_v) \boldsymbol{\sin}\theta_v d\theta_v d\varphi_v \right\}\right\}. \end{aligned} \tag{12.17}$$

The first factor in the integral depends only on the radial part of the states' functions; for brevity, we write

$$\begin{aligned} G_{ab}^{\kappa} &\equiv e^2 \int_0^\infty \int_0^\infty \frac{r_<^{\kappa}}{r_>^{\kappa+1}} \boldsymbol{R}_{n\ell_a}(r_u)\boldsymbol{R}_{n\ell_b}(r_u)\boldsymbol{R}_{n\ell_a}(r_v)\boldsymbol{R}_{n\ell_b}(r_v) r_u^2 r_v^2 dr_u dr_v \\ &= 2e^2 \int_0^\infty r_v^2 dr_v \int_0^{r_v} \frac{r_u^{\kappa}}{r_v^{\kappa+1}} \boldsymbol{R}_{n\ell_a}(r_u)\boldsymbol{R}_{n\ell_b}(r_u)\boldsymbol{R}_{n\ell_a}(r_v)\boldsymbol{R}_{n\ell_b}(r_v) r_u^2 dr_u \end{aligned} \tag{12.18}$$

Because the $\boldsymbol{R}_{n\ell}$ function is specific for each atom or ion, the G_{ab}^{κ} quantities have to be calculated specifically for each. The second factor in Eq. (12.17) depends only on the angular variables, and is thus common for all atoms and ions. When the specific spherical harmonics are inserted, then integrated over the φ azimuth angle, the integral vanishes unless $m_b = m_a$. Therefore, on summing over the μ variable, as required, there is only a single non-vanishing element, yet only provided that $\kappa \geq |m_b - m_a|$. Accordingly, the following quantity is defined as

$$c^{\kappa}(\ell_a m_a, \ell_b m_b) \equiv \sqrt{\frac{2}{2\kappa+1}} \int_0^\pi \boldsymbol{\Theta}_{\ell_a}^{m_a}(\theta)\boldsymbol{\Theta}_{\ell_b}^{m_b}(\theta)\boldsymbol{\Theta}_{\kappa}^{m_b - m_a}(\theta)\, \boldsymbol{\sin}\,\theta d\theta\ . \tag{12.19}$$

That quantity satisfies the relation $c^\kappa(\ell_a m_a, \ell_b m_b) = (-1)^{m_a - m_b} c^\kappa(\ell_b m_b, \ell_a m_a)$. It is possible to show that unless the following conditions are met,

$$\left.\begin{array}{l}\kappa + \ell_a + \ell_b = even \\ |\ell_a - \ell_b| \leq \kappa \leq \ell_a + \ell_b\end{array}\right\}, \tag{12.20}$$

the c^κ quantity vanishes identically.

Using all above short symbols, the exchange integral may be written in the following form:

$$K_{ab} = \sum_{\kappa = |m_b - m_a|}^{\infty} J_{ab}^{\kappa} = \sum_{\kappa = |m_b - m_a|}^{\infty} G_{ab}^{\kappa}\left[c^\kappa(\ell_a m_a, \ell_b m_b)\right]^2. \tag{12.21}$$

Using similar methods, the Coulomb integral may be written in the form

$$C_{ab} = \sum_{\kappa = |m_b - m_a|}^{\infty} C_{ab}^{\kappa} = \sum_{\kappa = |m_b - m_a|}^{\infty} F_{ab}^{\kappa}\left[c^\kappa(\ell_a m_a, \ell_a m_a)\right]\cdot\left[c^\kappa(\ell_b m_b, \ell_b m_b)\right], \tag{12.22}$$

where

$$F_{ab}^{\kappa} \equiv e^2 \int_0^\infty \int_0^\infty \frac{r_<^{\kappa}}{r_>^{\kappa+1}} \left|\boldsymbol{R}_{n\ell_a}(r_u)\right|^2 \left|\boldsymbol{R}_{n\ell_b}(r_v)\right|^2 r_u^2 r_v^2 dr_u dr_v . \tag{12.23}$$

The conditions expressed in Eq. (12.20) limit the number of elements that need to be calculated according to Eqs. (12.21) and (12.22) to a finite small number. In Section 12.7, we provide as an appendix a general expression that allows for the calculation of the c^κ coefficients. In practical issues, only a small finite number of non-zero elements exist for calculating each of the integrals, hence the c^κ values. It is therefore possible to summarize their numerical values in not-too-large tables. In Table 12.2, we list the numerical values of c^κ for *s*, *p*, *d*, and *f* electrons. The writing is intentionally presented in the form $c^\kappa = x_{ab}^\kappa\} \times 1/\sqrt{D_\kappa}$; for example, for the $\kappa = 3$, $\ell_a = 1(p)$, $m_a = \pm 1$, and $m_b = \pm 2$, the values are: $x_{ab}^\kappa = +\sqrt{3}$, and $D_\kappa = (7\sqrt{5})^2$ (see Table 12.2).

It is customary to use the x_{ab}^κ and D_κ quantities to simplify the write-up of the *F* and *G* quantities appearing in Eqs. (12.18) and (12.23). New "*F*" and "*G*" quantities are defined and labeled by κ subscripts instead of by superscripts, according to

$$F_\kappa(ab) \equiv F_{ab}^\kappa / D_\kappa \ ; \quad G_\kappa(ab) \equiv G_{ab}^\kappa / D_\kappa . \tag{12.24}$$

Using the newly defined *F* and *G* quantities, it is possible to write

$$\left.\begin{array}{l} K_{ab} = \displaystyle\sum_{\kappa = |m_b - m_a|}^{\infty} D_\kappa G_\kappa(ab)\left[x_{ab}^\kappa\right]^2 \\ C_{ab} = \displaystyle\sum_{\kappa = |m_b - m_a|}^{\infty} D_\kappa F_\kappa(ab)\left[x_{aa}^\kappa\right]\cdot\left[x_{bb}^\kappa\right] \end{array}\right\}. \tag{12.25}$$

The following Table 12.3 set of tables provides lists of energy correction expressions obtained for the various Terms generated by two or more electronic configurations belonging to the same principal quantum number *n*.

Table 12.2 Numerical values of $c^{\kappa}(\ell m, \ell' m')$; blank boxes mean "zero".

				$c^{\kappa}(\ell m, \ell' m')$		
ℓ	ℓ'	m	m'	$\kappa = 1$	3	5
$\ell + \ell' = odd$						
s	p	0	± 1	$-1/\sqrt{3}$		
		0	0	$+1/\sqrt{3}$		
s	f	0	± 3		$-1/\sqrt{7}$	
		0	± 2		$+1/\sqrt{7}$	
		0	± 1		$-1/\sqrt{7}$	
		0	0		$+1/\sqrt{7}$	
p	d	± 1	± 2	$-\sqrt{6}$	$+\sqrt{3}$	
		± 1	± 1	$+\sqrt{3}$	-3	
		± 1	0	-1	$+3\sqrt{2}$	
		0	± 2	0 $\Big\}\times 1/\sqrt{15}$	$+\sqrt{15}$ $\Big\}\times 1/7\sqrt{5}$	
		0	± 1	$-\sqrt{3}$	$-2\sqrt{6}$	
		0	0	$+2$	$+3\sqrt{3}$	
		± 1	∓ 2	0	$+3\sqrt{5}$	
		± 1	∓ 1	0	$-\sqrt{30}$	
d	f	± 2	± 3	$-\sqrt{15}$	$+\sqrt{10}$	-1
		± 2	± 2	$+\sqrt{5}$	$-2\sqrt{5}$	$+\sqrt{5}$
		± 2	± 1	-1	$-2\sqrt{6}$	$-\sqrt{15}$
		± 2	0	0	$-2\sqrt{5}$	$+\sqrt{35}$
		± 1	± 3	0	$+5$	$-\sqrt{7}$
		± 1	± 2	$-\sqrt{10}$	$-\sqrt{15}$	$+2\sqrt{6}$
		± 1	± 1	$+2\sqrt{2}$	$+\sqrt{2}$	$-5\sqrt{2}$
		± 1	0	$-\sqrt{3}$	$+\sqrt{2}$	$+4\sqrt{5}$
		0	± 3	0 $\Big\}\times 1/\sqrt{35}$	$+5$ $\Big\}\times 1/3\sqrt{35}$	$-2\sqrt{7}$ $\Big\}\times 1/6\sqrt{254}$
		0	± 2	0	0	$+3\sqrt{7}$
		0	± 1	$-\sqrt{6}$	-3	$-3\sqrt{10}$
		0	0	$+3$	$+4$	$+10$
		± 2	∓ 3	0	0	$-\sqrt{210}$
		± 2	∓ 2	0	0	$+3\sqrt{14}$
		± 2	∓ 1	0	$+\sqrt{10}$	$-\sqrt{70}$
		± 1	∓ 3	0	0	$-2\sqrt{21}$
		± 1	∓ 2	0	$+5$	$+4\sqrt{7}$
		± 1	∓ 1	0	$-\sqrt{15}$	$-\sqrt{105}$

Table 12.2 (Continued)

ℓ	ℓ'	m	m'	$\kappa = 0$	2	4
$\ell + \ell'$ = *even*						
s	s	0	0	$+1$		
s	d	0	± 2		$+1$	
		0	± 1		-1 } $\times 1/\sqrt{5}$	
		0	0		$+1$	
p	p	± 1	± 1	$+1$	-1	
		± 1	0	0	$+\sqrt{3}$	
		0	0	$+1$	$+2$ } $\times 1/5$	
		± 1	∓ 1	0	$-\sqrt{6}$	
p	f	± 1	± 3		$+3\sqrt{5}$	-1
		± 1	± 2		$-\sqrt{30}$	$+\sqrt{3}$
		± 1	± 1		$+3\sqrt{2}$	$-\sqrt{6}$
		± 1	0		-3	$+\sqrt{10}$
		0	± 3		0	$-\sqrt{7}$
		0	± 2		$+\sqrt{15}$ } $\times 1/5\sqrt{7}$	$+2\sqrt{3}$ } $\times 1/3\sqrt{21}$
		0	± 1		$-2\sqrt{6}$	$-\sqrt{15}$
		0	0		$+3\sqrt{3}$	$+4$
		± 1	∓ 3		0	$-2\sqrt{7}$
		± 1	∓ 2		0	$+\sqrt{21}$
		± 1	∓ 1		$+\sqrt{3}$	$-\sqrt{15}$
d	d	± 2	± 2	$+1$	-2	$+1$
		± 2	± 1	0	$+\sqrt{6}$	$-\sqrt{5}$
		± 2	0	0	-2	$+\sqrt{15}$
		± 1	± 1	$+1$	$+1$	-4
		± 1	0	0	$+1$ } $\times 1/7$	$+\sqrt{30}$ } $\times 1/21$
		0	0	$+1$	$+2$	$+6$
		± 2	∓ 2	0	0	$+\sqrt{70}$
		± 2	∓ 1	0	0	$-\sqrt{35}$
		± 1	∓ 1	0	$-\sqrt{6}$	$-2\sqrt{10}$

Table 12.2 (Continued)

ℓ	ℓ'	m	m'	$\kappa = 0$	2	4	6
$\ell + \ell'$ = even							
f	f	±3	±3	+1	−5	+3	−5
		±3	±2	0	+5	$-\sqrt{30}$	$+5\sqrt{7}$
		±3	±1	0	$-\sqrt{10}$	$+3\sqrt{6}$	$-10\sqrt{7}$
		±3	0	0	0	$-3\sqrt{7}$	$+10\sqrt{21}$
		±2	±2	+1	0	−7	+30
		±2	±1	0	$+\sqrt{15}$	$+4\sqrt{2}$	$-5\sqrt{105}$
		±2	0	0	$-2\sqrt{5}$	$-\sqrt{3}$	$+20\sqrt{14}$
		±1	±1	+1	+3	+1	−75
		±1	0	0	$+\sqrt{2}$	$+\sqrt{15}$	$+25\sqrt{14}$
		0	0	0	+4	+6	+100
		±3	∓3	0	0	0	$-10\sqrt{231}$
		±3	∓2	0	0	0	$+5\sqrt{462}$
		±3	∓1	0	0	$+\sqrt{42}$	$-5\sqrt{210}$
		±2	∓2	0	0	$\sqrt{70}$	$+30\sqrt{14}$
		±2	∓1	0	0	$-\sqrt{14}$	$-15\sqrt{42}$
		±1	∓1	0	$-2\sqrt{6}$	$-2\sqrt{10}$	$-10\sqrt{105}$
					× 1/15	× 1/33	× 1/429

Table 12.3(a) Energy corrections for p^q electronic configurations.

Configuration	Term	Energy	Configuration	Term	Energy
np^2	^1S	$F_0 + 10F_2$	np^4	^1S	$6F_0 + 10F_2$
	^1D	$F_0 + F_2$		^1D	$6F_0 + F_2$
	^3P	$F_0 - 5F_2$		^3P	$6F_0 - 5F_2$
np^3	^2P	$3F_0$			
	^2D	$3F_0 - 6F_2$			
	^4S	$3F_0 - 15F_2$			

Table 12.3(b) Energy corrections for d^q electronic configurations, presented by **Racah constants** defined as $A \equiv F_0 - 49F_4$, $B \equiv F_2 - 5F_4$, and $C \equiv 35F_4$.

Configuration	Term	Energy
nd^2	^1S	$A + 14B + 7C$
	^3P	$A + 7B$
	^1D	$A - 3B + 2C$
	^3F	$A - 8B$
	^1G	$A + 4B + 2C$

Configuration	Term	Energy
nd^3	^2P	$3A - 6B + 3C$
	^4P	$3A$
	^2D	$3A + 5B + 5C \pm (193B^2 + 8BC + 4C^2)^{1/2}$
	^2F	$3A + 9B + 3C$
	^4F	$3A - 15B$
	^2G	$3A - 11B + 3C$
	^2H	$3A - 6B + 3C$

Notably, the nd^3 configuration generates two energy states of a ^2D Term.

Configuration	Term	Energy	Term	Energy
nd^4	^1S	$6A + 10B + 10C \pm 2(193B^2 + 8BC + 4C^2)^{1/2}$	^3G	$6A - 12B + 4C$
	^3P	$6A - 5B + (11/2)C \pm \frac{1}{2}(912B^2 - 24BC + 9C^2)^{1/2}$	^3H	$6A - 17B + 4C$
	^1D	$6A + 9B + (15/2)C \pm \frac{3}{2}(144B^2 + 8BC + C^2)^{1/2}$	^1I	$6A - 15B + 6C$
	^3D	$6A - 5B + 4C$		
	^5D	$6A - 21B$		
	^1F	$6A + 6C$		
	^3F	$6A - 5B + (11/2)C \pm \frac{3}{2}(68B^2 + 4BC + C^2)^{1/2}$		
	^1G	$6A - 5B + (15/2)C \pm \frac{1}{2}(708B^2 - 12BC + 9C^2)^{1/2}$		

Notably, the nd^4 configuration generates two energy states for each of the ^1S, ^3P, ^1D, ^3F, and ^1G Terms.

Configuration	Term	Energy	Term	Energy
nd^5	^2S	$10A - 3B + 8C$	$^2F'$	$10A - 25B + 10C$
	^6S	$10A - 35B$	$^4F'$	$10A - 13B + 7C$
	^2P	$10A + 20B + 10C$	^2G	$10A - 13B + 8C$
	^4P	$10A - 28B + 7C$	$^2G'$	$10A + 3B + 10C$
	^2D	$10A - 3B + 11C \pm 3(57B^2 + 2BC + C^2)^{1/2}$	^4G	$10A - 25B + 5C$
	$^2D'$	$10A - 4B + 10C$	^2H	$10A - 22B + 10C$
	^4D	$10A - 18B + 5C$	^2I	$10A - 24B + 8C$
	^2F	$10A - 9B + 8C$		

Notably, the nd^5 configuration generates three energy states of a ^2D Term, and two energy states for each of the ^2F and ^2G Terms.

Table 12.3(c) Energy corrections for f^q electronic configurations, presented by Racah constants defined as $A \equiv F_0 - 21F_4 - 468F_6$, $B \equiv (5F_2 + 6F_4 - 91F_6)/5$, $C \equiv 7(F_4 - 6F_6)/5$, and $D \equiv 462F_6$.

Configuration	Term	Energy
nf^2	^1S	$A + 60B + 105C + 9D$
	^3P	$A + 45B$
	^1D	$A + 19B - 72C + 2D$
	^3F	$A - 10B$
	^1G	$A - 30B + 110C + 2D$
	^3H	$A - 25B$
	^1I	$A + 25B + 2D$

Configuration	Term	Energy
nf^3	^4S	$3A - 30B$
	^2P	$3A - 25B + 35C + 3D$
	^2D	$3A - 7B + (57/2)C + 3D \pm \frac{1}{2}(2176B^2 - 18096BC + 74529C^2)^{1/2}$
	^4D	$3A + 25B$
	^2F	$3A + 55B + (75/2)C + 6D \pm \frac{1}{2}\left[9700B^2 - 180B(45C + 2D) + 9(45C + 2D)^2\right]^{1/2}$
	^4F	$3A - 30B$
	^2G	$3A + 7B + (113/2)C + 3D \pm \frac{1}{2}(12676B^2 - 31676BC + 36169C^2)^{1/2}$
	^4G	$3A - 10B$
	^2H	$3A - 23B + (63/2)C + 3D \pm \frac{1}{2}(5056B^2 - 18816BC + 29169C^2)^{1/2}$
	^2I	$3A - 5B + 45C + 3D$
	^4I	$3A - 65B$
	^2K	$3A - 40B + 80C + 3D$
	^2L	$3A + 3D$

Notably, the nf^3 configuration generates two energy states for each of the ^2D, ^2F, ^2G, and ^2H Terms.

A comparison between detailed calculations and experimental results reveals a fair qualitative agreement, particularly related to the states' energy order and splitting order-of-magnitude, but with many deviations by specific details. Thus, the tables' listed energy constants are always considered as adjustable parameters. For a demonstration, in Figure 12.4a, we present graphically a list of Racah constants adjusted for the titanium Ti atom, and for a series of the equi-electrons Sc^+, Ti^{2+}, V^{3+}, Cr^{4+}, Mn^{5+}, Fe^{6+}, Co^{7+}, and Ni^{8+}; all ions contain 20 electrons orbiting about the nucleus (see Table 5.5), and the electronic configuration of the valence electrons is $3d^2$. It should be emphasized that the Terms for which the fit has been performed did not include the ^1S Term (see configuration nd^2 in Table 12.3b); the energy of that Term is considerably lower than the value expected by Racah equations in all atoms and ions. The fitted Racah constants generally increase with the degree of ionization. An exception occurs for the titanium *Ti* atom, most probably because of involvement of the two electrons occupying the $4s^2$ sub-shell (Table 5.5). For further demonstration, Figure 12.4b presents an energy scheme of the chromium Cr^{4+} ion shown in Figure 12.4a. The states' energies calculated for the ^3F, ^3P, ^1D, and ^1G Terms fairly coincide with the experimental results. In contrast, the ^1S Term state energy is higher by ~15,000 cm^{-1} compared with the experimental one. Similar deviations occur for all atoms referred to in Figure 12.4a.

Notably, the $\beta \equiv C/B$ ratio between corresponding Racah constants is quite similar among the material atoms addressed in Figure 12.4a. The mean value is $\beta_{mean} = 4.6$, with a standard deviation of about 10%. An

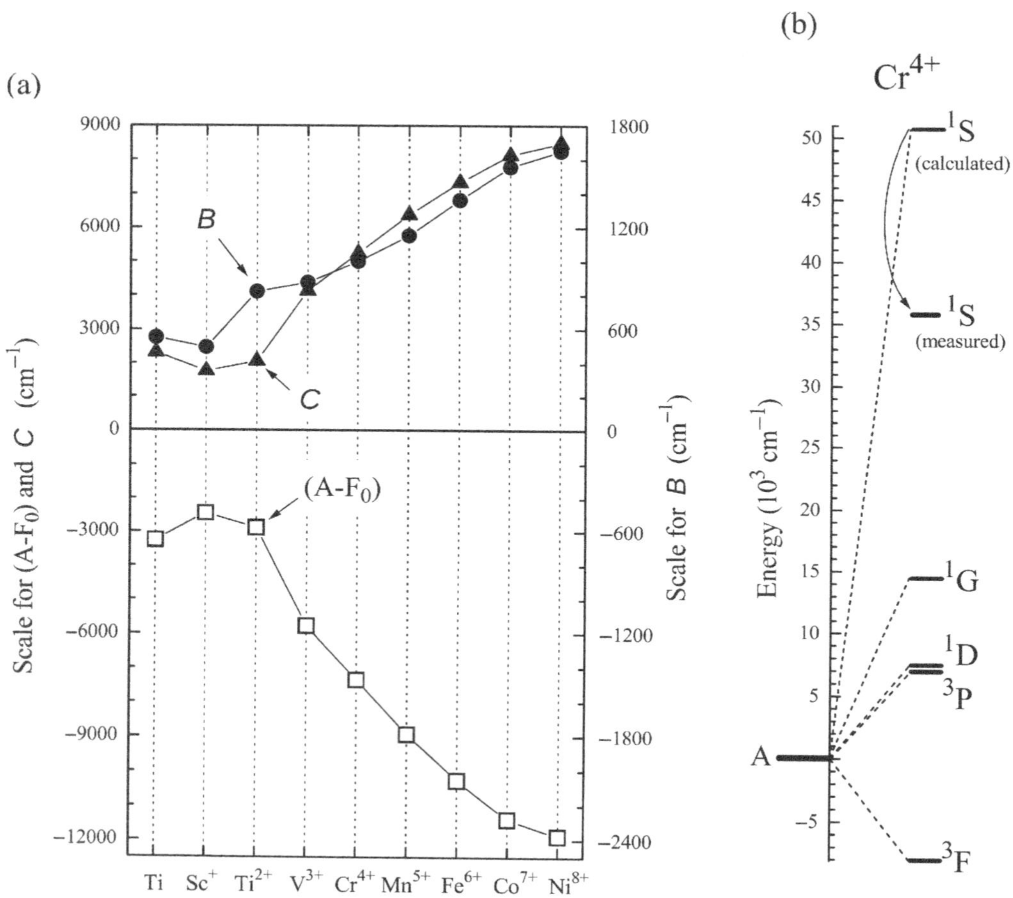

Figure 12.4 (a) Experimental values of Racah constants for a titanium *Ti* atom, and for ions of various atoms, all containing 20 electrons orbiting their nucleus. The valence electrons in the titanium atom and in all ions exhibit a $3d^2$ configuration. (b) Energy scheme of a fourfold ionized chromium atom Cr^{4+} calculated using the Racah constants of Figure 12.4 (a).

extreme small value, $\beta = 2.5$, is obtained for the Ti^{2+} ion; an extreme high value $\beta = 5.5$ is obtained for the Mn^{5+} ion. A similar behavior exists for Racah constants belonging to other sets of ions. For example, Figure 12.5 presents values of B and C Racah constants for different sets of atomic ions: the doubly ionized Ti^{2+}, V^{2+}, Cr^{2+}, Mn^{2+}, Fe^{2+}, Co^{2+}, and Ni^{2+}, and the triply ionized V^{3+}, Cr^{3+}, Mn^{3+}, Fe^{3+}, and Ni^{3+}. In the doubly ionized series of atoms, the electronic configurations are $3d^2$, $3d^3$, $3d^4$, $3d^5$, $3d^6$, $3d^7$, and $3d^8$, respectively. In the triply ionized series of atoms, the electronic configurations are $3d^2$, $3d^3$, $3d^4$, $3d^5$, $3d^6$, and $3d^7$, respectively (see also Table 5.5). These data are also indicated in Figure 12.5 next to the symbols showing the Racah constants values. The Racah constants are seen to grow with the atomic number Z; the C values grow by an average of 326 cm^{-1} per unit atomic number; the B values grow by an average of 53 cm^{-1} per unit atomic number. The $\beta = C/B$ ratio grows slightly with the atomic number, yet for practical purposes it is possible to assume it to be constant. In the series of values shown in Figure 12.5, the said ratio average is 4.55, with a 5% standard deviation.

We conclude this section by providing a schematic brief account of Terms splitting into multiplets because of spin/orbit coupling. Recall that the allowed total angular momentum quantum numbers J are $J = L + S,\ L + S - 1,\ \cdots\ |L - S|$ (Eq. [5.52]). In the example of Figure 12.4a, the ground state of all ions belongs to the ^3F Term (see specifically Fig 12.4b). In fact, that Term is a multiplet of sub-states split by the spin/orbit coupling. They are noted $^3F_2$, $^3F_3$, and $^3F_4$. According to Hund's laws (Eq. [5.64]-ii), the ground-state one is $^3F_2$.

Denoting a Term by the letter γ as a subscript, the shift $\Gamma_{\gamma,SLJ}$ of the J multiplet component belonging to a certain Term is given by

$$\Gamma_{\gamma,SLJ} = \xi_{\gamma,SL} \cdot \langle \underline{\mathbf{S}} \cdot \underline{\mathbf{L}} \rangle\ , \tag{12.26}$$

where $\xi_{\gamma,SL}$ is the spin/orbit interaction coefficient, and

$\langle \underline{\mathbf{S}} \cdot \underline{\mathbf{L}} \rangle = \frac{1}{2}[J(J+1) - S(S+1) - L(L+1)]$.

Direct detailed calculations of the spin/orbit interaction constants are successful in predicting their order of magnitude and trends of its dependence on the configuration parameters, the atomic number Z, and the Term parameters. Still, even so-called "direct" calculations are based on many empiric considerations; eventually, any result must be compared against experimental results of each specific case. For

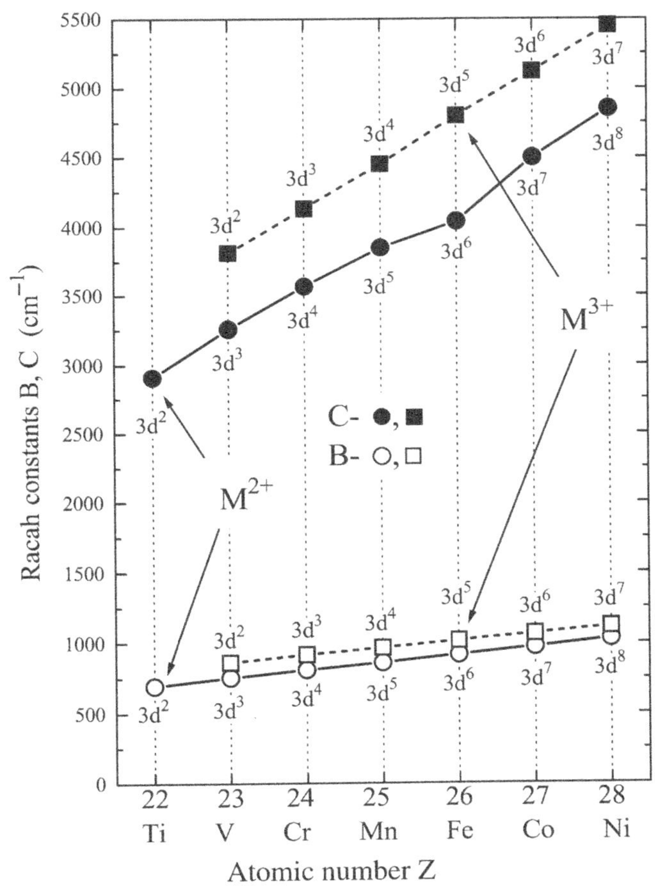

Figure 12.5 Experimental values of B and C Racah constants for series of doubly and triply ionized atoms of consecutive atomic numbers as functions of the atomic number. B values are indicated by open symbols and C ones by full symbols. Circles belong to the M^{2+} doubly ionized atoms; squares belong to the M^{3+} triply ionized ones.

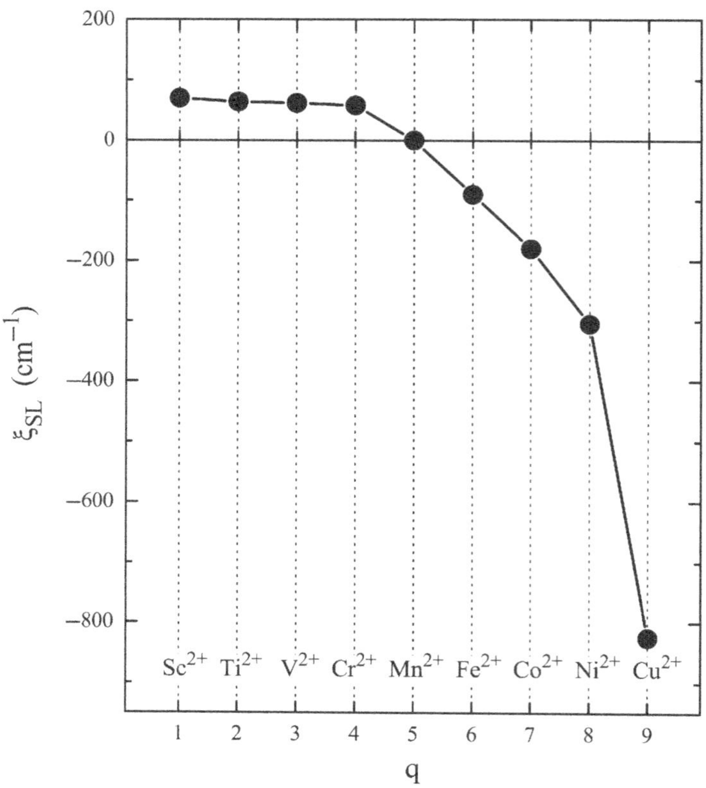

Figure 12.6 Experimental values of the spin/orbit coupling constant ξ_{SL} for doubly ionized atomic states of consecutive atomic numbers, all exhibiting a $3d^q$ electronic configuration, as a function of the number of electrons q forming the configuration.

demonstration, Figure 12.6 shows the experimental results of the spin/orbit interaction coefficients for the series of the doubly ionized atoms Sc^{2+}, Ti^{2+}, V^{2+}, Cr^{2+}, Mn^{2+}, Fe^{2+}, Co^{2+}, Ni^{2+}, and Cu^{2+}. All said ions belong to a $3d^q$ configuration, where q varies between 1 and 9. The results show that the spin/orbit coupling coefficient is positive, almost independent of q for less than half-filled shells, and turns negative for more than half-filled shells. Their absolute values increases with q. That behavior is typical of other cases as well.

12.4 Dopant Atoms or Ions in Crystalline Sites

In terminology used especially by chemists, a cluster of atoms, ions, or molecules that surround intimately a central atom or ion, is called "**ligand**". The ligand presence causes considerable spectroscopic change of the central atom/ion properties compared to its "free" state spectroscopy.

The said spectroscopic issue is addressed by two different methods. One is called the **crystal field** or **tight-binding** approach. The central atom/ion dopant is assumed to be surrounded by a charge distribution that forms a static electric field in the space it occupies. No covalent bonding is assumed between the atom/ion and its ligand-comprising atoms. The system-describing Hamiltonian is a combination of the $\mathcal{H}_F$ Hamiltonian of the free-atom electrons, and a perturbation potential $\boldsymbol{V}(\underline{r})$, where $\underline{r}$ is the distance vector of a dopant-belonging electron from its nucleus. Obviously, the perturbation potential depends on the position of the different dopant-belonging electrons $\underline{r}_j$, where j is an electron numerator. Thus,

$$\mathcal{H} = \mathcal{H}_F + \sum_j \boldsymbol{V}\left(\underline{r}_j\right). \tag{12.27}$$

The goal of the crystal field theory is to express the perturbation potential $\boldsymbol{V}(\underline{r})$ effect on the free-Hamiltonian eigenfunctions and eigenvalues, and assess the nature of the obtained differences compared with those of the free atom/ion.

In the second approach, the central dopant plus its ligand are considered to be a free isolated molecule. Accordingly, ionic and covalent bonds are also assumed between the dopant and the ligand-comprising members. In most cases, the calculations are performed by methods similar to those detailed in Chapter 6. That approach is called **ligand field theory**. Recall further that we have adopted a similar approach in describing the lattice vibrations in Chapter 9; there, in the zero-order approximation, the entire primitive unit cell was considered a free molecule.

Independently of the selected approach, symmetry considerations exhibit a crucial importance for the calculations. In particular, the ligand-hosting site-symmetry determines the degeneracy degree and splitting of each of the energy states dominated by the central atom/ion. Although the central atom/ion exhibits a full rotational symmetry when free, its symmetry reduces by residing in a crystalline site. In other words, the Hamiltonian containing the $\boldsymbol{V}(\underline{r})$ perturbation satisfies the site symmetry. States' degeneracy is then fully or partially lifted. Symmetry consideration alone could point out the existence of energy splitting, yet is unable to determine the splitting size. Like in the case of molecular vibrations (Chapter 8) and crystalline vibrations (Chapters 9 and 11), classification of energy states is performed using group theory.

The cause of states' splitting is assumed to be a static electric field. Therefore, the split states belonging to a certain parent state are often addressed as the field "Stark components" (see also Section 4.9.2). The "Stark" naming, however, might be misleading! The classical Stark effect considers the electric field as both static and constant across the subject atom/ion dimensions. In the crystal field effect, the considered electric field is indeed static; however, it is zero at the central atom/ion nucleus, and exhibits huge gradients across the central atom/ion dimensions!

12.4.1 The Full Rotation Group and its Representations

An atom or a free ion in space exhibits a full rotational symmetry; this means that any proper or improper rotation by any angle about any axis crossing the nucleus (mass center) brings the atom/ion to experience an identical environment. In other words, any rotation of order $p < \infty$, where p is real, belongs to the group. The group thus contains an infinite number of elements (Table 8.1). It is possible to describe any proper rotation of a rigid body in a three-dimensional space by a real orthogonal matrix whose determinant is +1. The group is called a "special orthogonal group", or "full rotation group", and denoted $SO[3]$. When an inversion operation is included, the obtained orthogonal group is denoted $O[3]$, and is often addressed as the "full rotation/inversion group". The determinant of a matrix describing an operation is +1 for a proper rotation, and −1 for an improper rotation (Table 8.1 and related text). In the following, we present without proof a number of equations, principles, and tables, all based on group theory.

We pause to explain the use of functions as representations of point symmetries. We consider a general $f(x, y, z)$ function, and form a point symmetry operation on its *coordinates system*. If no change is obtained in the *functional shape*, the function is regarded symmetric in relation to the said operation. As an example, consider the function $f \equiv x^2 + y^2 + z^2$. We perform the inversion operation i, whereby $x' = -x$, $y' = -y$, and $z' = -z$. Obviously, $f' = x'^2 + y'^2 + z'^2 = f$, and its shape in the new coordinates is identical, as in the original coordinates. Therefore, the function is addressed as symmetric in relation to inversion. If a function remains symmetric in relation to *all operations* belonging to a group, it is considered to be a *representation* of the group.

As a generalization, we refer to a more complex case. We consider a sub-space of functions spanned by a finite number of basis functions (namely, the set of all functions that are linear combinations of the basis functions). Assume that the group symmetry operations on any function belonging to the said sub-space leave it within the same sub-space, not necessarily preserving its shape (except for the identity operation). Such sub-space is an **invariant sub-space**. If the sub-space cannot be split into invariant sub-spaces of itself, it is considered an *irreducible representation* of the group. The sub-space order (namely, the number of its basis functions) defines the order of the representation.

A set of degenerate Hamiltonian eigenfunctions belonging to a certain symmetry group generates an invariant sub-space, which is a representation of the group. If the energy degeneracy is fundamental (not accidental), the representation is irreducible. If a single solution of the Hamiltonian has been established, the entire sub-space of functions it belongs to may be generated by operating on it with all the group symmetry operations.

For a demonstration, we present the relatively simple case of a hydrogen-like atom, comprising a single spinless electron (see Section 3.3). All energy states of the free atom that belong to the same n and ℓ quantum numbers, where n is the principal quantum number and ℓ is the orbital quantum number, are $(2\ell + 1)$-fold degenerate. The set of all $\boldsymbol{Y}_{\ell m}(\theta, \varphi)$ spherical harmonics, where m is the magnetic quantum number (Table 3.2), establish a complete set of functions that describe all orbital angular momentum states. The $\{\boldsymbol{Y}_{\ell m}\}$ sub-set of a constant ℓ spans a $2\ell + 1$ order invariant sub-space of the full rotation group, thus being an irreducible representation of the full rotation group. Any proper rotation of order p establishes a separate similarity class of the full rotation group. We present below, without proof, the expression describing the character belonging to the $\alpha = 2\pi/p$ rotation of the $\{\boldsymbol{Y}_{\ell m}\}$ irreducible representation of the special orthogonal $SO[3]$ rotation group:

$$\boxed{\chi_\alpha^\ell = \frac{\sin\left[\frac{1}{2}\alpha(2\ell + 1)\right]}{\sin\frac{1}{2}\alpha}} \quad . \tag{12.28}$$

As for the full rotation/inversion $O[3]$ group, the characters obtained using Eq. (12.28) are the same for proper or improper rotations for an even ℓ, and also for a proper rotation for an odd ℓ, but are of an opposite sign for an odd ℓ in an improper rotation.

It is now appropriate to extend the list of character tables of different symmetry groups presented in Section 8.12, and add, based on Eq. (12.28), a partial character table of the $SO[3]$ group for $\ell = 0, 1, 2, 3, 4, 5, 6, 7$ representations, namely for s, p, d, f, g, h, i, j ones, respectively. The similarity classes written in Table 12.4 belong to 1, 2, 3, 4, 6 order rotations possible for the crystalline space groups. As for all character tables, the column belonging to the identity operation expresses the degeneracy order of the representation; namely, it is $2\ell + 1$ in the present case.

When a hydrogen-like atom experiences perturbation by a potential of a lower symmetry, the $\{\boldsymbol{Y}_{\ell m}\}$ sub-spaces of a common ℓ split into partial sub-spaces of the lower-symmetry group. Basis vectors of the partial sub-spaces need not be one of the $\boldsymbol{Y}_{\ell m}$ functions, but they are always an appropriate linear combination of $\boldsymbol{Y}_{\ell m}$ functions. Each such sub-space is an irreducible representation of the lower symmetry group, and it is customary to present it by the same symbol (see Section 8.8). In Table 12.5, we present a correlation table relating each of the spherical harmonics listed in Table 12.4 to the irreducible representations of two groups: the cubic O_h group and the hexagonal D_{6h} group. It should serve as a "toolbox" for our following discussions. The table is, in fact, an addition for the (**iv**) series of tables presented in Section 11.5. It should be emphasized that O_h and D_{6h} are both sub-groups of $O[3]$; however, D_{6h} *is not* a subgroup of O_h.

Correlation to other symmetry groups may be derived using a combination of the following Table 12.5 with the correlation tables of Section 11.5 (**iv**).

Table 12.4 A partial character table of the $SO[3]$ full rotation group.

$SO[3]$	$E\ (\alpha = 0)$	$C_2\ (\alpha = 2\pi/2)$	$C_3\ (\alpha = 2\pi/3)$	$C_4\ (\alpha = 2\pi/4)$	$C_6\ (\alpha = 2\pi/6)$
$\ell = 0$ (s)	1	1	1	1	1
$\ell = 1$ (p)	3	−1	0	1	2
$\ell = 2$ (d)	5	1	−1	−1	1
$\ell = 3$ (f)	7	−1	1	−1	−1
$\ell = 4$ (g)	9	1	0	1	−2
$\ell = 5$ (h)	11	−1	−1	1	−1
$\ell = 6$ (i)	13	1	1	−1	1
$\ell = 7$ (j)	15	−1	0	−1	2

Table 12.5 Correlation table relating the irreducible representations of the full rotation-inversion $O[3]$ group of Table 12.4 to the irreducible representations of the cubic O_h and of the hexagonal D_{6h} groups. In electronic states' spectroscopy, many-electron Terms are indicated by capital letters; therefore, F-marked species of molecular groups are replaced by the T letter for electronic states' groups, to prevent confusion with the F-Term symbol.

$O[3]$	O_h	D_{6h}
$\ell = 0$ (s)	A_{1g}	A_{1g}
$\ell = 1$ (p)	T_{1u}	$A_{2u} \oplus E_{1u}$
$\ell = 2$ (d)	$E_g \oplus T_{2g}$	$A_{1g} \oplus E_{1g} \oplus E_{2g}$
$\ell = 3$ (f)	$A_{2u} \oplus T_{1u} \oplus T_{2u}$	$A_{2u} \oplus B_{1u} \oplus B_{2u} \oplus E_{1u} \oplus E_{2u}$
$\ell = 4$ (g)	$A_{1g} \oplus E_g \oplus T_{1g} \oplus T_{2g}$	$A_{1g} \oplus B_{1g} \oplus B_{2g} \oplus E_{1g} \oplus 2E_{2g}$
$\ell = 5$ (h)	$E_u \oplus 2T_{1u} \oplus T_{2u}$	$A_{2u} \oplus B_{1u} \oplus B_{2u} \oplus 2E_{1u} \oplus 2E_{2u}$
$\ell = 6$ (i)	$A_{1g} \oplus A_{2g} \oplus E_g \oplus T_{1g} \oplus 2T_{2g}$	$2A_{1g} \oplus A_{2g} \oplus B_{1g} \oplus B_{2g} \oplus 2E_{1g} \oplus 2E_{2g}$
$\ell = 7$ (j)	$A_{2u} \oplus E_u \oplus 2T_{1u} \oplus 2T_{2u}$	$A_{1u} \oplus 2A_{2u} \oplus B_{1u} \oplus B_{2u} \oplus 3E_{1u} \oplus 2E_{2u}$

To demonstrate the use of Table 12.5 and the correlation tables of Section 11.5, we show the splitting of the orbital *f* state in the hydrogen atom when placed in a site of D_2 point symmetry. We assume a succession of perturbations, where the atom is initially placed in the O_h site of a cubic crystal. Afterwards, the crystal is

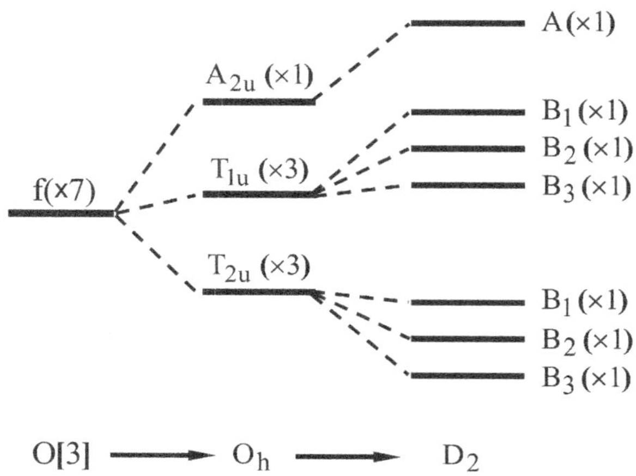

Figure 12.7 Correlation map, demonstrating calculated changes in the symmetry species of orbital angular momentum states in a hydrogen-like atom, starting with an f free atom state and ending differently when placed in a D_2 symmetry site. Each state degeneracy is indicated in parentheses. For details, see the text.

distorted, and the site symmetry reduces to D_2. By Table 12.5, the atomic state splits initially in three: $A_{2u} \oplus T_{1u} \oplus T_{2u}$. Afterwards, by Table (**iv**)-32 of Section 11.5, the A_{2u} state transforms into an A one, and the T_{1u} and T_{2u} states transform each into $B_1 \oplus B_2 \oplus B_3$ states. A correlation map describing the said changes and splitting is provided in Figure 12.7. The map appears similar to an energy scheme. Nevertheless, it should not be interpreted as such!

The similarity to an energy scheme may wrongly suggest that the level's height describes the energy order or, perhaps, their plotted separations relate to the size of the energy splitting. Transforming a correlation map into an energy scheme requires meticulous experiments and/or theoretical calculations. In the following, we present such a calculated example.

12.4.2 A Hydrogen-Like Atom in a Crystalline Perturbation Field

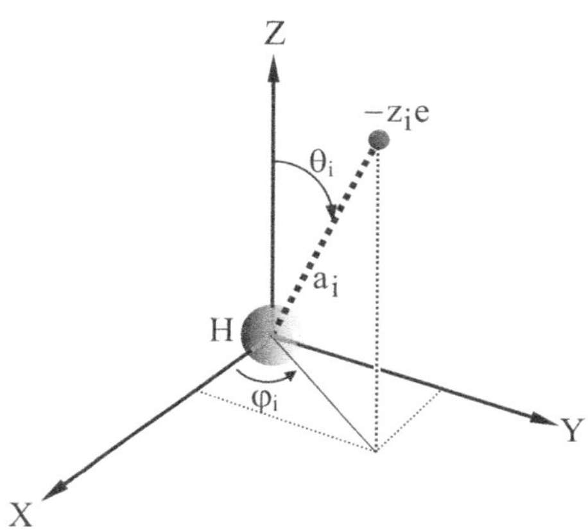

Figure 12.8 A hydrogen H atom (large gray ball) at Cartesian coordinates origin and a potential perturbation-forming charge $-z_ie$ (small dark ball) at a_i distance from the hydrogen nucleus. The angular θ_i and φ_i coordinates of the perturbing charge position are also indicated.

We imagine the electric field causing the perturbation as generated by a distribution of p point-like electric charges $-z_ie$ surrounding the atom, where e is the elementary charge and z_i for $i = 1, \ldots, p$ are unitless adjustable parameters. The z_i values' distribution is constrained to provide a zero electric field on the central atom nucleus at $\underline{r} = 0$. Cases of $z_i > 0$ indicate a negative charge; cases of $z_i < 0$ indicate a positive charge. The nucleus of the hydrogen-like atom itself, whose electric charge is $+Ze$ is assumed to be fixed at the coordinates system origin (Figure 12.8). The p surrounding charges distances from the origin are a_i.

The potential energy $\boldsymbol{V}_c$ perturbing the free electron revolution about the nucleus is given by

$$\boldsymbol{V}_c(\underline{r}) = \sum_{i=1}^{p} \frac{z_i e^2}{|\underline{R}_i - \underline{r}|}\,, \tag{12.29}$$

where $\underline{r}$ is the position vector of the single revolving electron of mass m, and $\underline{R}_i$ is the position vector of the i-th perturbing charge; obviously, $R_i = a_i$. Schrödinger's equation for the revolving electron would be (see Eq. [3.21])

$$\left(-\frac{\hbar^2}{2m}\underline{\nabla}^2 - \frac{Ze^2}{r} + \boldsymbol{V}_c(\underline{r})\right)\psi(\underline{r}) = E\psi(\underline{r})\,. \tag{12.30}$$

We further assume that the V_c potential establishes only a small perturbation of the free hydrogen atomic states, and that use of a static perturbation theory is adequate.

Using Eq. (12.13), we may readily write

$$V_c(\underline{r}) = \sum_{i=1}^{p} z_i e^2 \sum_{\kappa=0}^{\infty} \left(\frac{r_<^{\kappa}}{r_>^{\kappa+1}}\right) P_\kappa(\cos \upsilon_i) , \tag{12.31}$$

where $r_<$ and $r_>$ are the smaller and larger between the r and a quantities, and υ_i is the angle between $\underline{r}$ and $\underline{R}_i$. In the specific $a_i >> a_0$ case, where a_0 is the Bohr radius (see Eq. [3.38] and related text), it is possible to assume that $r < a_i$, and neglect the opposite case; Eq. (12.31) then reduces to

$$V_c(\underline{r}) = \sum_{i=1}^{p} \frac{z_i e^2}{a_i} \sum_{\kappa=0}^{\infty} \left(\frac{r}{a_i}\right)^{\kappa} P_\kappa(\cos \upsilon_i) = \sum_{\kappa=0}^{\infty} r^{\kappa} \sum_{i=1}^{p} \frac{z_i e^2}{a_i^{\kappa+1}} P_\kappa(\cos \upsilon_i) . \tag{12.32}$$

By Eq. (12.12), we get

$$P_\kappa(\cos \upsilon_i) = \frac{4\pi}{2\kappa+1} \sum_{\mu=-\kappa}^{\kappa} Y^*_{\kappa\mu}(\theta_i, \varphi_i) Y_{\kappa\mu}(\theta, \varphi) , \tag{12.33}$$

where (r, θ, φ) and $(a_i, \theta_i, \varphi_i)$ are the spherical coordinates of $\underline{r}$ and $\underline{R}_i$, respectively (Figures 3.3 and 12.8). Setting in Eq. (12.32), we get

$$\boxed{V_c(\underline{r}) = \sum_{\kappa=0}^{\infty} r^{\kappa} \sum_{\mu=-\kappa}^{\kappa} q_{\kappa\mu} \left(\frac{4\pi}{2\kappa+1}\right)^{1/2} Y_{\kappa\mu}(\theta, \varphi)} , \tag{12.34}$$

where

$$q_{\kappa\mu} = \left(\frac{4\pi}{2\kappa+1}\right)^{1/2} \sum_{i=1}^{p} \frac{z_i e^2}{a_i^{\kappa+1}} Y^*_{\kappa\mu}(\theta_i, \varphi_i) . \tag{12.35}$$

The $q_{\kappa\mu}$ coefficients depend on the size and distribution of the perturbing charges.

We aim to eventually calculate a state energy splitting using the static perturbation theory (see Section 2.1.1). The perturbation matrix elements are quantities of the form

$$\langle n\ell m|V_c|n'\ell' m'\rangle = \int \psi^*_{n\ell m}(\underline{r}) V_c(\underline{r}) \psi_{n'\ell' m'}(\underline{r}) d^3\underline{r} ,$$

where $\psi_{n\ell m}(\underline{r}) = R_{n\ell}(r) Y_{\ell m}(\theta, \varphi)$ is the wave-function solution of the unperturbed Hamiltonian (see Section 3.3, specifically Tables 3.2 and 3.3). We further assume that it is possible to neglect mixing among states belonging to different principal n quantum numbers. We then limit ourselves to consider only $n' = n$ cases, and the perturbation matrix under consideration would be of an $n^2 \times n^2$ order. By separation between the r radius and (θ, φ) angles in the $\psi_{n\ell m}(\underline{r})$ wave function, the perturbation elements comprise products of matrix elements of r powers by the $R_{n\ell}$ radial function, with matrix elements of the different $Y_{\kappa\mu}(\theta, \varphi)$ functions appearing in the expression for $V_c(\underline{r})$ (Eqs. [12.32], [12.33]) by the $Y_{\ell m}(\theta, \varphi)$ spherical harmonics:

$$\langle n\ell m|V_c|n\ell' m'\rangle = \sum_{\kappa=0}^{\infty} \langle n\ell|r^{\kappa}|n\ell'\rangle \sum_{\mu=-\kappa}^{\kappa} q_{\kappa\mu} \left(\frac{4\pi}{2\kappa+1}\right)^{1/2} \langle \ell m|Y_{\kappa\mu}(\theta, \varphi)|\ell' m'\rangle . \tag{12.36}$$

The matrix elements of the r powers would be given by

$$\langle n\ell|r^{\kappa}|n\ell'\rangle = \int_0^{\infty} \boldsymbol{R}^*_{n\ell}(r)\boldsymbol{R}_{n\ell'}(r)r^{\kappa+2}dr\,. \tag{12.37}$$

As to the angular part, using Eq. (12.20), we get

$$\left(\frac{4\pi}{2\kappa+1}\right)^{1/2}\langle \ell m|\boldsymbol{Y}_{\kappa\mu}(\theta,\varphi)|\ell' m'\rangle = c^{\kappa}(\ell m,\ell' m')\,, \tag{12.38}$$

then

$$\boxed{\langle n\ell m|\boldsymbol{V}_c|n\ell' m'\rangle = \sum_{\kappa=0}^{\infty}\langle n\ell|r^{\kappa}|n\ell'\rangle c^{\kappa}(\ell m,\ell' m')\sum_{\mu=-\kappa}^{\kappa} q_{\kappa\mu}}\,. \tag{12.39}$$

As a result of specific properties of the c^{κ} coefficients (Eq. [12.21] and Table 12.1), in most cases, the infinite sum written in Eq. (12.39) is, in fact, finite, containing a limited number of elements. A similar situation exists in relation to exchange and Coulomb integrals treated in section 12.3 (see also Eq. [12.37]). Table 12.6 summarizes calculated results of the $\langle n\ell|r^{\kappa}|n\ell'\rangle$ matrix elements for $n \le 4$ of different values of the κ power parameter.

12.4.3 Example: States Splitting in a Cubic Perturbation Field

As an example of a cubic crystal field of O_h symmetry, we consider it to be formed by six equal perturbation point charges ($z_i = z$, Figure 12.8), each located at a different face center of an imaginary cube surrounding a hydrogen H atom (Figure 12.9). The hydrogen atom itself is located at the cube center. The distance between the hydrogen atom nucleus and each of the perturbing charges is a.

We turn now to calculate the $q_{\kappa\mu}$ values for the said perturbation. The orientation (θ_i, φ_i) data of the different perturbation charges (Figure 12.8) are:

$$(\theta_1,\varphi_1) = \left(\frac{\pi}{2},\ 0\right);\quad (\theta_2,\varphi_2) = \left(\frac{\pi}{2},\ \frac{\pi}{2}\right);\quad (\theta_3,\varphi_3) = \left(\frac{\pi}{2},\ \pi\right)$$
$$(\theta_4,\varphi_4) = \left(\frac{\pi}{2},\ \frac{3\pi}{2}\right);\quad (\theta_5,\varphi_5) = (0,0);\quad (\theta_6,\varphi_6) = (\pi,\ 0)\,.$$

We set the latter data in Eq. (12.35), recall that $\boldsymbol{Y}^*_{\kappa\mu}(\theta,\varphi) = (-1)^{\mu}\boldsymbol{Y}_{\kappa,-\mu}(\theta,\varphi)$, and get

$$q_{\kappa 0} = \left(\frac{4\pi}{2\kappa+1}\right)^{1/2}\cdot\frac{ze^2}{a^{\kappa+1}}\left[4\boldsymbol{Y}_{\kappa 0}\left(\frac{\pi}{2},0\right) + \boldsymbol{Y}_{\kappa 0}(0,0) + \boldsymbol{Y}_{\kappa 0}(\pi,0)\right]\,; \tag{12.40 (a)}$$

for an even $\mu \neq 0$, we have

$$q_{\kappa\mu} = \left(\frac{4\pi}{2\kappa+1}\right)^{1/2}\cdot\frac{ze^2}{a^{\kappa+1}}\boldsymbol{Y}_{\kappa\mu}\left(\frac{\pi}{2},0\right)\left[1 + \boldsymbol{exp}\left(-i\mu\frac{\pi}{2}\right) + \boldsymbol{exp}(-i\mu\pi) + \boldsymbol{exp}\left(-i\mu\frac{3\pi}{2}\right)\right]\,; \tag{12.40 (b)}$$

for an odd μ, we have

$$q_{\kappa\mu} = 0\,. \tag{12.40 (c)}$$

Table 12.6 Numerical values of the $\langle n\ell|r^{\kappa}|n\ell'\rangle$ matrix elements for $n \leq 4$. Empty spaces are those where knowing the value is unnecessary because the $c^{\kappa}(\ell m, \ell' m')$ coefficients are zero (Table 12.1).

			$\langle n\ell|r^{\kappa}|n\ell'\rangle$ $([a_0/Z]^{\kappa})$			
n	ℓ	ℓ'	$\kappa = 1$	3	5	
$\ell + \ell' = odd$						
2	s	p	$-3\sqrt{3}$			
3	s	p	$-9\sqrt{2}$			
	p	d	$-9\sqrt{5}/2$	$-945\sqrt{5}$		
4	s	p	$-6\sqrt{15}$			
	s	f		$-1680\sqrt{35}$		
	p	d	$-12\sqrt{3}$	$-8400\sqrt{3}$		
	d	f	$-6\sqrt{7}$	$-3600\sqrt{7}$	$-2661120\sqrt{7}$	
n	ℓ	ℓ'	$\kappa = 0$	2	4	6
$\ell + \ell' = even$						
2	s	s	1			
	p	p	1	30		
3	s	s	1			
	s	d		45		
	p	p	1	180		
	d	d	1	126	25515	
4	s	s	1			
	s	d		$240\sqrt{5}$		
	p	p	1	600		
	p	f		$80\sqrt{21}$	$60480\sqrt{21}$	
	d	d	1	504	362880	
	f	f	1	360	190080	138378240

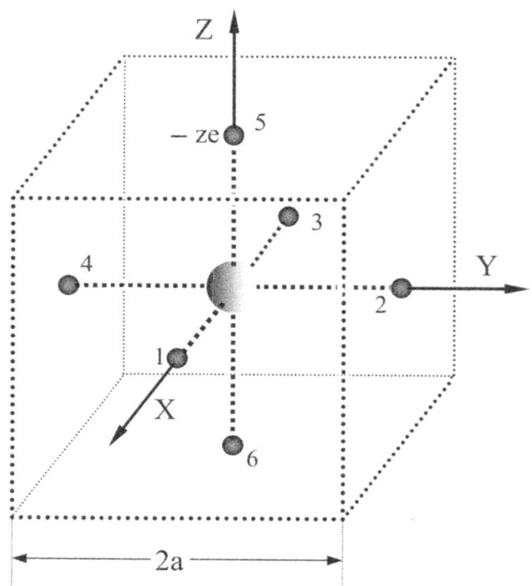

Figure 12.9 A hydrogen *H* atom (large gray ball) in a cube center surrounded by six equal −*ze* point charges (small dark balls), steadily located at the cube faces centers. The said charges are numbered consecutively from 1 to 6.

Calculation of energy splitting requires calculation of the perturbation matrix elements (Eq. [12.39]). Using Tables 3.2, 12.1, and 12.6, we get that all the matrix elements vanish except for sub-matrices belonging to $\ell = \ell'$. The physical meaning of that result is that in the perturbation type under the present study, there exists no mixing among different ℓ states, and each state splits independently of the others. The sub-matrix $\ell = \ell' = 0$ (namely, the *s* state) contains only a single element, whose value is $\langle ns0|\boldsymbol{V}_c|ns0\rangle = 6ze^2/a$. In the sub-matrix $\ell = \ell' = 1$ (namely, *p* states), the only non-zero elements are the diagonal ones $\langle np0|\boldsymbol{V}_c|np0\rangle = \langle np,\pm1|\boldsymbol{V}_c|np,\pm1\rangle = 6ze^2/a$. Nevertheless, when considering $\ell = \ell' = 2$ states, namely *d* states, off-diagonal non-zero elements also exist:

$$\left.\begin{aligned} \langle nd\pm2|\boldsymbol{V}_c|nd\pm2\rangle &= \mathcal{D}q + 6ze^2/a \\ \langle nd\pm1|\boldsymbol{V}_c|nd\pm1\rangle &= -4\mathcal{D}q + 6ze^2/a \\ \langle nd0|\boldsymbol{V}_c|nd0\rangle &= 6\mathcal{D}q + 6ze^2/a \\ \langle nd\pm2|\boldsymbol{V}_c|nd\mp2\rangle &= 5\mathcal{D}q \end{aligned}\right\}, \tag{12.41}$$

where

$$\boxed{\mathcal{D} \equiv \frac{35}{4}\frac{ze^2}{a^5};\quad q \equiv \frac{2}{105}\langle nd|r^4|nd\rangle}\ . \tag{12.42}$$

Notably, the $\mathcal{D}$ quantity expresses properties related to the perturbing potential, whereas the q quantity expresses properties of the central atom. The $\mathcal{D}q$ product is called the *splitting parameter* and has energy units.

We consider now the $\varepsilon^0 \equiv 6ze^2/a$ quantity. The physical reason is that ε^0 appears as an addend in all diagonal elements, and there only, which physically means an increase of all states energies by ε^0. For looking at solely the splitting, we deduct it from all diagonal elements (mathematically, it means consideration of the $\langle n\ell m|\boldsymbol{V}_c^0|n\ell'm'\rangle$ matrix, where $\boldsymbol{V}_c^0 = \boldsymbol{V}_c - \varepsilon^0 I$). All sub-matrices belonging to $\ell = \ell' = 0$ and $\ell = \ell' = 1$ are zero. It means that the *s* and *p* states do not split. The remaining sub-matrix is the one belonging to $\ell = \ell' = 2$ (namely, *d*-states), which reads

$$\begin{array}{c|ccccc} m\backslash m' & 2 & 1 & 0 & -1 & -2 \\ \hline 2 & \mathcal{D}q & 0 & 0 & 0 & 5\mathcal{D}q \\ 1 & 0 & -4\mathcal{D}q & 0 & 0 & 0 \\ 0 & 0 & 0 & 6\mathcal{D}q & 0 & 0 \\ -1 & 0 & 0 & 0 & -4\mathcal{D}q & 0 \\ -2 & 5\mathcal{D}q & 0 & 0 & 0 & \mathcal{D}q \end{array}\ . \tag{12.43}$$

The secular equation revealing the energy splitting would be

$$\begin{vmatrix} \mathcal{D}q-\varepsilon & 0 & 0 & 0 & 5\mathcal{D}q \\ 0 & -4\mathcal{D}q-\varepsilon & 0 & 0 & 0 \\ 0 & 0 & 6\mathcal{D}q-\varepsilon & 0 & 0 \\ 0 & 0 & 0 & -4\mathcal{D}q-\varepsilon & 0 \\ 5\mathcal{D}q & 0 & 0 & 0 & \mathcal{D}q-\varepsilon \end{vmatrix} = (-4\mathcal{D}q-\varepsilon)^3(6\mathcal{D}q-\varepsilon)^2 = 0. \tag{12.44}$$

Eq. (12.44) has three identical solutions $\varepsilon_T = -4\mathcal{D}q$, and two identical solutions $\varepsilon_E = 6\mathcal{D}q$. The T and E subscript labels indicate the obtained states' degeneracy: 3 and 2, respectively, according to the common species notation in group theory (see Section 8.8 and the caption for Table 12.5). Observation of Table 12.5 indeed indicates that the d-state under an O_h site symmetry splits according to $d \to E_g \oplus T_{2g}$. The last results are presented as an energy scheme in Figure 12.10. Concerning nomenclature, it is important to note the inclusion of the parent states' symmetry symbol s, p, or d for each split state; it is a recommended practice in spectroscopic literature. Also note two important properties related to the splitting size. One is the summed $10\mathcal{D}q$ splitting of the nd state; the other is that when split states are given their degeneracy weight, the split-states mass center is zero. Considering the correction preceding Eq. (12.42), the said center is in fact up-shifted by ε^0 above the original d-state position.

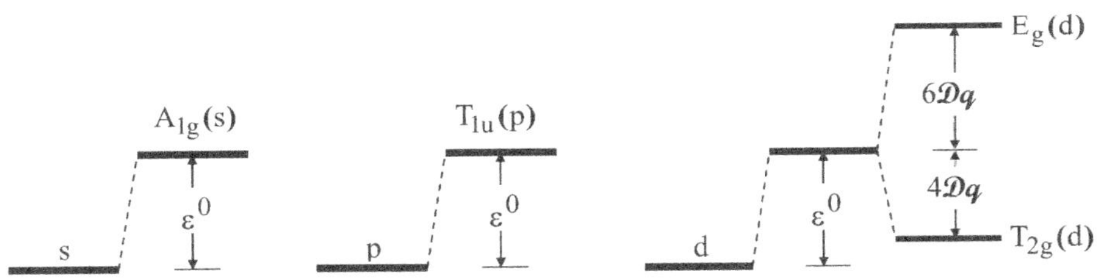

Figure 12.10 Schematic illustration of *ns*, *np*, and *nd* states' energy shifts and splitting in a hydrogen *H* atom under an O_h symmetry perturbation field. Recall that *d* states exist only for $n \geq 3$ shells.

To demonstrate an order of magnitude of prospected energy shifts and splitting, we present a numerical example for the $n = 3$ shell. Assume the perturbing charge to be $z = 0.1$, and the distances from the center to be $a = 10a_0$, where a_0 is the Bohr radius. The expression for the shift is $\varepsilon^0 = 6ze^2/a$. Accordingly, the given numerical values are

$$z = 0.1\,;\, e = 4.8 \times 10^{-10}\ \text{esu};\, a_0 = 0.529 \times 10^{-8}\,\text{cm}.$$

Setting and calculating:

$$\varepsilon^0 = \frac{6 \times 0.1 \times 4.8^2 \times 10^{-20}}{10 \times 0.529 \times 10^{-8}} = 2.61 \times 10^{-12}\ \text{erg} = \frac{2.61 \times 10^{-12}\ \text{erg}}{1.6 \times 10^{-12}\ \text{erg/eV}} = 1.631\ \text{eV}.$$

Turning to splitting calculation, using Eq. (12.42), we get

$$\mathcal{D}q = \frac{1}{6}\frac{ze^2}{a} \cdot \frac{1}{a^4}\langle 3d|r^4|3d\rangle = \frac{\varepsilon^0}{36} \cdot \frac{1}{a^4}\langle 3d|r^4|3d\rangle.$$

Using Table 12.6, we get $\langle 3d|r^4|3d\rangle = 25515{a_0}^4$. Setting it above, we get

$$\mathcal{D}q = \frac{\varepsilon^0}{36}\frac{1}{10^4{a_0}^4}\cdot 25515{a_0}^4 = \frac{2.5515}{36}\varepsilon^0 = \frac{2.5515}{36}\times 1.631\ \text{eV} = 0.116\ \text{eV}.$$

Recall now that the summed splitting is actually $10\mathcal{D}q = 1.16$ eV. That value is almost twice as large as the 0.66 eV separation between the unperturbed states of the $n = 3$ and $n = 4$ hydrogen atom shells (Eq. [3.38]). The above assessment clearly indicates that neglecting the mixing of higher lying unperturbed states is not always justified. Use of experimental results allows for the assessment of proper values for the fundamental a and z adjustable quantities.

The calculation just presented allows us to obtain quite readily an expression for a perturbation of same symmetry yet of a different nature. For example, we consider a case where the $-z_i e$ charges (Figures 12.8 and 12.9) are replaced by electric dipoles $\underline{\mu}_i$ of equal size ($|\underline{\mu}_i| = \mu$), each pointing from inside out. That system may be regarded as a combination of two, very close charge systems: the inner system of negative $-ze$ electric charges at an a distance like in Figure 12.9, and $+ze$ positive electric charges at $a + \delta a$ distance, such that $\delta a = \mu/ze$. A perturbation matrix element of that dipoles distribution will be the difference between the corresponding matrix element of the negative charge distribution and the matrix element of the positive charge distribution. Accordingly, it is possible to define a $\mathcal{D}^\mu$ quantity of analogous role as $\mathcal{D}$ in Eqs. 12.41 and 12.42, which is the difference between the two relevant $\mathcal{D}$'s:

$$\mathcal{D}^\mu = \frac{35}{4}\cdot\frac{ze^2}{a^5} - \frac{35}{4}\cdot\frac{ze^2}{(a+\delta a)^5} \underset{\delta a << a}{\cong} \frac{35}{4}\cdot\frac{5ze^2}{a^6}\cdot\delta a = \frac{175}{4}\cdot\frac{\mu e}{a^6}. \tag{12.45}$$

We turn now to calculate the wave functions of the split $3d$ states. By Eq. (2.8), these functions are linear combinations of the original, unperturbed functions, where the expansion coefficients are given by the eigenvectors of the perturbation matrix. The obtained matrix eigenvectors are

$$\frac{1}{\sqrt{2}}\begin{pmatrix}1\\0\\0\\0\\-1\end{pmatrix};\quad \begin{pmatrix}0\\1\\0\\0\\0\end{pmatrix};\quad \begin{pmatrix}0\\0\\0\\1\\0\end{pmatrix}$$ for the T_{2g} states, and $$\frac{1}{\sqrt{2}}\begin{pmatrix}1\\0\\0\\0\\1\end{pmatrix};\quad \begin{pmatrix}0\\0\\1\\0\\0\end{pmatrix}$$ for the E_g states. The eigenfunctions would then be

$$\begin{gathered}\Psi^{(1)}_{(d)T} = |3d-1\rangle;\quad \Psi^{(2)}_{(d)T} = |3d+1\rangle;\quad \Psi^{(3)}_{(d)T} = \frac{1}{\sqrt{2}}(|3d+2\rangle - |3d-2\rangle)\\ \Psi^{(1)}_{(d)E} = |3d0\rangle;\quad \Psi^{(2)}_{(d)E} = \frac{1}{\sqrt{2}}(|3d+2\rangle + |3d-2\rangle)\end{gathered} \tag{12.46}$$

All above functions are linearly independent and orthonormal. The three functions tagged by a "T" subscript span a three-dimensional invariant space under O_h symmetry transformations. It should be appreciated that it is possible to transform the $\Psi_{(d)T}$ basis functions of Eq. (12.46) into another orthonormal basis by performing a linear transformation via a unitary matrix, without changing the physical meaning. The two functions tagged by an E subscript span a two-dimensional invariant space under O_h symmetry transformations. Here as well, it is possible to transform the $\Psi_{(d)E}$ basis functions of Eq. (12.46) into another orthonormal basis functions by performing a linear transformation via a unitary matrix, without changing the physical meaning.

As for the s and p states wave functions, they are assumed to undergo no change at all by the perturbation, because for them, the entire perturbation effect is a shift of the entire energy scheme, with no change of separation. It is also important to note, that the radial parts of the wave functions remained unchanged; the only induced change occurred at the angular part.

The above splitting calculations exhaust all possible changes to an electron belonging to an $n = 3$ principal quantum number. For an electron belonging to $n = 4$, splitting of $4f$ states should also be calculated; the reader is requested to carry out such calculations as an exercise.

12.4.4 Tanabe-Sugano Diagrams

States of many-electrons atoms, whether hydrogen-like or not, differ from those of the hydrogen H atom in exhibiting no degeneracy among states of different orbital angular momentum, even when belonging to the same principal quantum number n. Accordingly, in most cases, it is possible to overlook the possibility of initial mixing among different states of the orbital angular momentum. Splitting calculations exhibit similarity with those of the hydrogen atom. In particular, considering the fourfold ionized Cr^{4+} chromium ion states provided in Figure 12.4, the ^4F ground-state Term would split in a O_h symmetry site exactly like the f state in the hydrogen atom; namely, $^3F \to {}^3A_{2u} \oplus {}^3T_{1u} \oplus {}^3T_{2u}$. The ^1G state would split according to $^1G \to {}^1A_{1g} \oplus {}^1E_g \oplus {}^1T_{1g} \oplus {}^1T_{2g}$ (Table 12.5). We emphasize again that in spectroscopic states' assignment practices, the parent state is also noted in parentheses; for example, the $^1T_{1g}$ that has split off the ^1G state is commonly written $^1T_{1g}(^1G)$ (see also Figure 12.11).

In the present book, we refrain from presenting detailed calculations of matrix elements belonging to ions under crystal field perturbations, whereby states' splitting are deduced. The calculations are analogous to the ones described in Sections 12.2 and 12.3 for the different Term energies of atoms and free ions. As expected, the said calculations involve Coulomb integrals C_{ab} and exchange integrals K_{ab} (Eqs. [12.25]) or, alternatively, the Racah constants A, B, C, and D they determine (Table 12.3) and, in addition, the splitting parameter $\mathcal{Dq}$. The $\mathcal{Dq}/B$ ratio is called *crystal field perturbation*. Plots describing the split states' energies for different ions as functions of the crystal field perturbation are known as **Tanabe-Sugano diagrams**. In Figure 12.11, we show as an example such a diagram for the V^{3+} triply ionized vanadium atom residing in a cubic crystal field perturbation of O_h symmetry (see also Figure 12.4a).

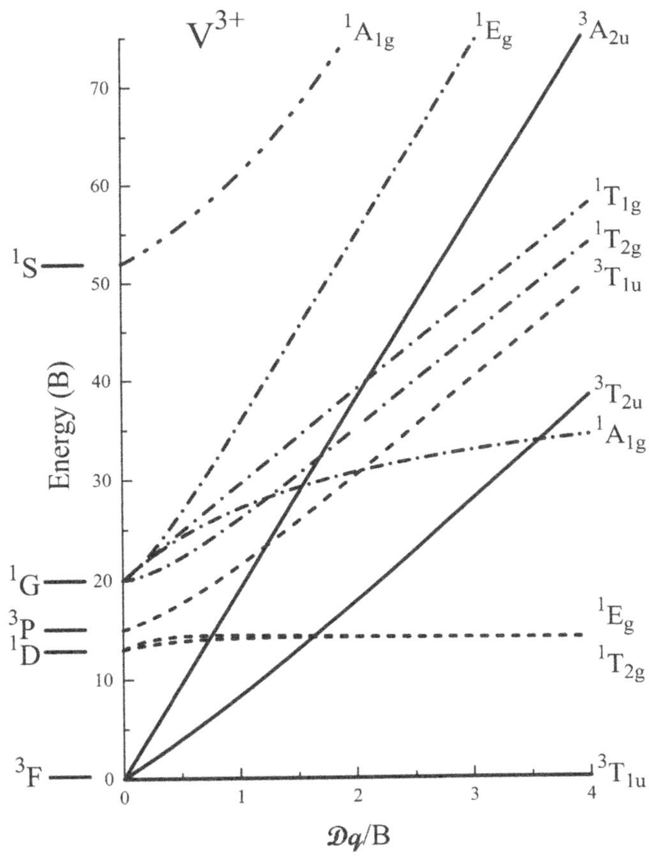

Figure 12.11 Energy separations between triply ionized V^{3+} vanadium ion states and the $^3T_{1u}(^3F)$ state split in a cubic O_h crystal field symmetry, as functions of the specific crystal field perturbation parameter $\mathcal{Dq}/B$. The Racah constant in that particular ion is $B = 860\ \text{cm}^{-1}$ (see Figure 12.5).

Several important issues should be recognized. Firstly, the ^3P state energy in the V^{3+} ion is higher than the ^1D state. The relation is inverted compared with the situation in the Cr^{4+} ion (Figure 12.4). Indeed, the said states in the two ions are relatively close. Secondly, the crystal field splitting under large field perturbations changes the states' energy order compared with the order among their parent ones. In Tanabe-Sugano diagrams, this effect expresses itself by curve crossings. Thirdly, we further emphasize the need to extend the states notations to include the parent Term. For example, there are two ^1E_g states in the Figure 12.11 diagram. The parent state is obvious in the diagram; however, to avoid confusion when optical transitions are indicated without a full diagram, the said states should be denoted as $^1E_g(^1G)$ and $^1E_g(^1D)$ for the ^1G and ^1D parents, respectively.

12.5 Transition Probabilities and Selection Rules

There are three factors that influence optical transition probabilities between different ion states when the ion occupies crystalline sites as previously described. One is optical transitions between the parent states (namely, between states of the original free ion). The second is the parity imposed on each split state by the crystal field. The third factor is crystal vibrations, which couple to the electronic states.

The discussion presented in the present section is basically qualitative. The influence of crystal vibrational states on the electronic states is similar to the influence of internal molecular vibrations on their electronic states in the original "stiff" state described in Chapter 7. Still, it is not exactly a direct generalization of the case described there to the currently discussed case of a "giant molecule"; the subject matter is not the entire molecular electronic states, but rather electronic states of so-called "foreign" atoms, which are a minority within the solid. It is thus assumed that the most prominent dopant ion interaction occurs with its surrounding ligand "as if the ligand is an isolated molecule". Fundamental approaches toward the latter case are valid with minor changes. Translational modes, which in free molecules exhibit practically no effect on the transition light emission (except for a small Doppler broadening), have an important part in the hosting crystal vibrations; they form quite a broad band of lattice acoustic motions. The internal vibrations are also broadened for similar reasons (see Chapter 9). Rotational modes turn into lattice librations. Thus, in application of the Franck-Condon principle (Section 7.5.4), each of the involved vibrational modes is already broadened because of interaction with the lattice acoustic modes of the host matrix.

In the discussion related to the Franck-Condon principle described in section 7.5.2, we obtained that the $\langle \underline{M} \rangle$ transition dipole between different electronic states is given by (see Eq. [7.45])

$$\langle \underline{M} \rangle = \left\langle \mathcal{S}^f | \mathcal{S}^i \right\rangle \left\langle \psi^f \middle| \underline{M}_e \middle| \psi^i \right\rangle , \tag{12.47}$$

where $\langle \psi^f | \underline{M}_e | \psi^i \rangle$ is the electronic transition dipole moment between the initial i state to the final f state and $\left\langle \mathcal{S}^f | \mathcal{S}^i \right\rangle$ is the Franck-Condon overlap between the oscillation wave functions of the nuclei in the initial i state and the final f state. If the electronic states of the central ion in the ligand are assumed yet unsplit, the first factor in the expression for a transition dipole moment belongs to the free ion. Those transitions abide by the selection rules summarized in Eq. (5.65):

$$\Delta L = \pm 1; \quad \Delta S = 0; \quad \Delta J = 0, \pm 1; \quad \Delta M = 0, \pm 1 , \tag{12.48}$$

except for the forbidden $J = 0 \rightarrow J' = 0$ transition, and the forbidden $\Delta M = 0$ transition if $\Delta J = 0$. In view of the dependence on the Franck-Condon integral, each transition is broadened, revealing the overlap structure between the ligand vibrational wave functions. At normal temperatures, absorption or emission transitions are initiated at the corresponding vibrational ground states; therefore, the absorption spectrum reveals the vibrational states' structure of the upper electronic state; the fluorescence spectrum, on the other hand, reveals the vibrational states' structure of the lower electronic state. The fluorescence spectrum will be shifted towards longer wavelengths (smaller energies) compared with the absorption spectrum, revealing the Stokes shift described in Section 7.5.5.

Transitions between different multiplet states belonging to the same Term are forbidden in a free atom (or a free ion) because they involve no change in the orbital angular momentum; namely, $\Delta L = 0$. Nevertheless, a crystal field, especially a low-symmetry one, allows for such a transition to some extent. The electronic transition moment $\underline{M}_e$ exhibits an odd symmetry. Therefore, the transition dipole moment $\langle \psi^f | \underline{M}_e | \psi^i \rangle$ between states of defined parity depends on both parities. Split states in a crystal field of a symmetry center (like the cubic O_h described earlier) are odd for u states, and even for g states. For lower symmetries, states are odd when belonging to T_x, T_y, T_z, or T species (by the nomenclature of character tables of Section 8.12), and are even for all others. Parity considerations allow for transitions between states of opposite parity. Therefore, some transitions between different multiplet states belonging to the same Term will be allowed, and others will be forbidden. In reality, relatively faint transitions are observed for many so-called "legally forbidden" electric dipole transitions. As to transitions forbidden by parity considerations, their occurrence is interpreted to result from imperfect symmetry of their inhabiting site, and sometimes to vibrations distorting the site symmetry for brief periods. Transitions may also be electric quadrupole moment allowed ($\Delta J = 0, +1, \pm 2$) or magnetic dipole allowed $\Delta L = 0, \pm 1, \pm 2$. Oscillator strengths measured for such transitions are of the order of $\boldsymbol{f} \approx 10^{-5} - 10^{-6}$, or even much smaller.

Properties of an odd state involved in allowed electric dipole transitions also indicate the transition dipole orientation. When a state symmetry is tagged as T_x, T_y, or T_z, light polarizations along the crystal X, Y, Z coordinate system axes are respectively allowed to involve in the transitions. If T tagged, all polarization states are allowed. States transforming like (T_x, T_y) or $(T_x, T_y, T_z) \equiv (T)$ are doubly or triply degenerate, respectively.

12.6 Spectroscopic Examples

We present several examples demonstrating effects related to the presence of different dopant ions in crystalline sites. The first is the fourfold ionized chromium Cr^{4+} ion embedded in $Y_3Al_5O_{12}$ crystal. The crystal is well known by its abbreviated *YAG* name, an acronym for "Yttrium-Aluminum-Garnet". Garnet is the name granted to a family of crystals belonging to the no. 230 $O_h^{10}\,(I4_1/a\bar{3}2/d)$ body-centered cubic space group (see Section 11.5, Table (**i**)-32). The centered unit cell contains 8 formula units. Aluminum Al atoms are triply ionized (Al^{3+}), inhabiting two different sites, an octahedrally coordinated "a" Wyckoff site of S_6 symmetry, and a tetrahedrally coordinated "d" Wyckoff site of S_4 symmetry. The yttrium Y atoms are also triply ionized (Y^{3+}), occupying a dodecahedrally coordinated "c" Wyckoff site of D_2 symmetry. The YAG crystal is used in several laser systems as a host for various ions. The Cr^{4+} ion in that crystal replaces the lattice aluminium Al^{3+} ion of the octahedrally coordinated Wyckoff "a" site, provided that a nearby lattice-yttrium Y^{3+} ion is replaced by a calcium Ca^{2+} ion for the purpose of charge compensation. That way, the charge of the $Cr^{4+} - Ca^{2+}$ dopant pair is the same as on the replaced lattice $Al^{3+} - Y^{3+}$ pair. The closest neighbours of the Cr^{4+} ion are six lattice oxygen ions, forming a near octahedral structure. It is not a regular octahedron of O_h symmetry like the one described in Figure 12.9, but a distorted one of only S_6 symmetry. It is quite likely that after addition of the $Cr^{4+} - Ca^{2+}$ pair, the symmetry drops further. The Cr^{4+} ion is still addressed as having a so-called "octahedral coordination", because the number of closest neighbors (6) remains unchanged. For first-order approximation splitting assignments, the symmetry is assumed as octahedral; this is not the highest O_h one, but the lower-symmetry O one. The latter differs from the O_h one by lacking an inversion symmetry (Table 8.2, and Section 8.12 Tables [35] and [36]).

Figure 12.12a shows a calculated energy scheme of the Cr^{4+} ion in the said crystal site and, on its side (b), a measured optical absorption spectrum. The energy scheme addresses the lower energies because for energies exceeding about 35,000 cm^{-1}, the host crystal fundamental absorption (lattice absorption) overshadows any possible transitions in the dopant ion. The singlet and triplet manifolds are indicated separately (left and right sides, respectively). Absorption transitions between the triplet ground-state $^3T_1(^3F)$ and the singlet states are forbidden in the free ion (Eq. [12.48]), thus expected to be rather faint. Absorption transitions are indicated within the triplet manifold: $^3T_1(^3F) \rightarrow {}^3T_2(^3F)$ and $^3T_1(^3F) \rightarrow {}^3T_2(^3P)$. Both are forbidden by the $\Delta L = \pm 1$ selection rule, yet are allowed because of the symmetry reduction in the occupying host site. Figure 12.12b shows the measured absorption spectrum in a YAG crystal hosting Cr^{4+} ions in the octahedrally coordinated Wyckoff "a" site. The spectrum reveals a complex absorption peak at about 0.5 μm, and an additional peak at about 0.26 μm whose wavelength is only slightly longer than the host lattice absorption threshold. The figure displays a resolution of the spectrum into component peaks. The component peaking at 20,750 cm^{-1} is assigned to the $^3T_1(^3F) \rightarrow {}^3T_2(^3F)$ transition; the component peaking at 35,900 cm^{-1} (close to the lattice absorption threshold) is assigned to the $^3T_1(^3F) \rightarrow {}^3T_1(^3P)$ transition. The origin of the components peaking at 24,600 cm^{-1} and 27,700 cm^{-1} is unclear; they might relate to transitions in the singlet manifold, or reflect different distances of the charge-compensating

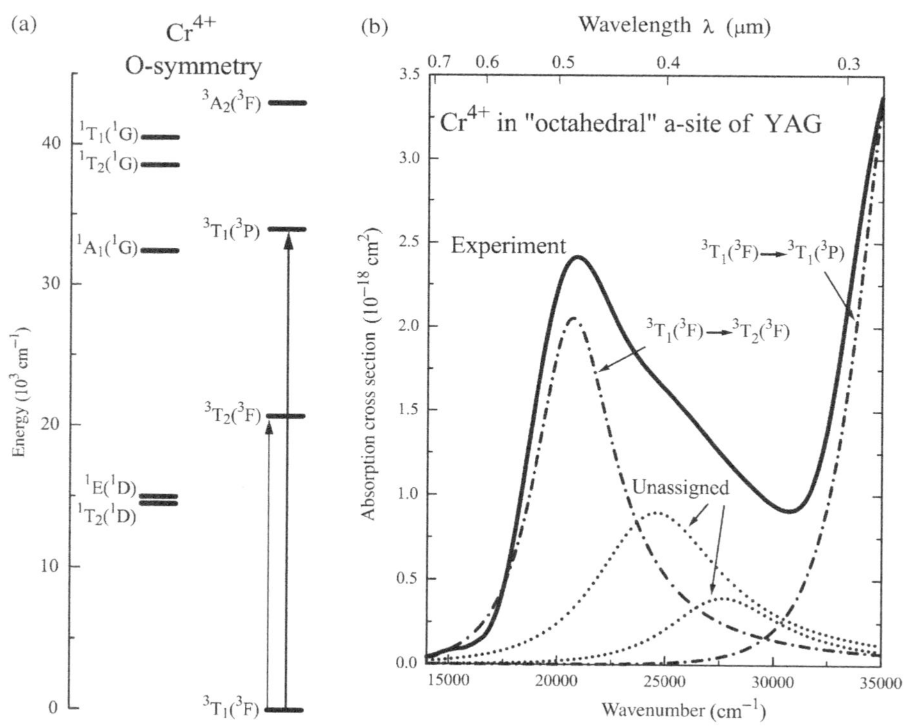

Figure 12.12 (a) Calculated energy scheme of the lower four-times ionized chromium Cr^{4+} ion replacing the Al^{3+} ion in the "a" Wyckoff site of a YAG crystal. (b) Absorption spectrum of the Cr^{4+} : *YAG* crystal in the 0.25 – 0.75 μm spectral range, where the Cr^{4+} ion resides in the octahedrally coordinated "a" site. For detailed discussion, see the text.

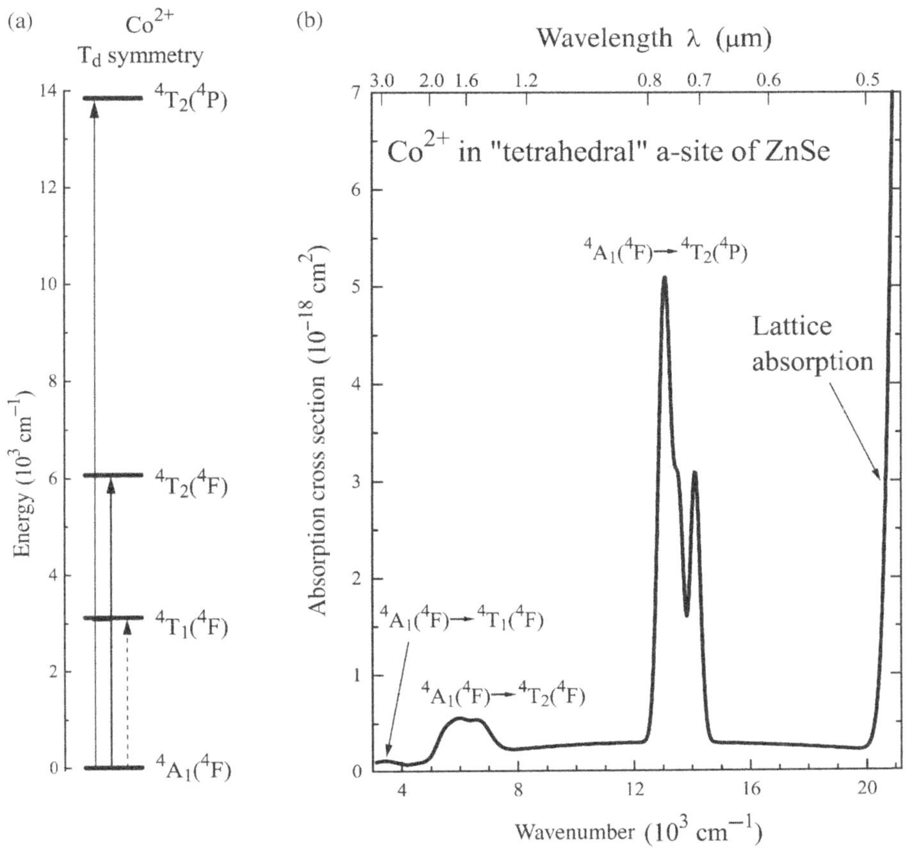

Figure 12.13 (a) Calculated lower states' energy scheme of doubly ionized cobalt Co^{2+} that replaces lattice Zn^{2+} ions in the "a" Wyckoff sites of the zinc-selenide *ZnSe* crystal. (b) Absorption spectrum in a Co^{2+}:*ZnSe* polycrystalline plate in the 0.4–3.2 μm range. For discussion, see the text.

calcium Ca^{2+} ions from the Cr^{4+} dopant ion, thus influencing the latter energy scheme. Quite notable is the strength of the $^3T_1(^3F) \to {}^3T_1(^3P)$ transition, which is supposed to be weak because it is forbidden in the free ion ($\Delta L = 2$). It might be explained by its proximity to the lattice absorption threshold, which induces states mixing. Such cases are referred to as "strength borrowing" between states.

The second example is the doubly ionized cobalt Co^{2+} ion hosted in a zinc-selenide *ZnSe* single crystal. The zinc selenide material is used for the manufacturing of optical devices in the infrared region because of its high optical transparency in that region. The *ZnSe* crystal is a face-centered cubic belonging to the no. 216 $T_d^2(F\bar{4}3m)$ space-group. Quite a number of crystals, called the "sphalerite family", belong to that space group. The centered unit cell comprises four formula units. The zinc *Zn* atoms are doubly ionized (Zn^{2+}), inhabiting an "a" Wyckoff site of T_d symmetry (see Section 11.5 Table (**i**)-31). A cobalt Co^{2+} ion replaces a lattice Zn^{2+} in its site. The closest neighbors to the Co^{2+} ion are four lattice selenium Se^{2-} anions, forming a tetrahedral structure such as the one described in Figures 8.11 and 8.12.

Figure 12.13a describes a calculated energy scheme of the Co^{2+} ion in the said crystal site. On its right side, it shows the absorption spectrum measured in a polycrystalline plate of the material (b). The energy scheme relates to the lower states energy of the Co^{2+} ion, because for energies exceeding ~20,200 cm^{-1}, the host lattice absorption overshadows any possible transitions in the dopant ion. Transitions between split states whose parent is ^4F, or between states of ^4F parent to those of a ^4P parent, are forbidden by a zero-order approximation, because they violate the $\Delta L = \pm 1$ selection rule (Eq. [12.48]). They, however, turn allowed by a higher-order approximation, because of the crystal field perturbation. Exact considerations require acquaintance with mathematical tools of group theory that have not been presented. We indicate, without proof, that no appreciable transition is expected between the $^4A_1(^4F)$ state and the $^4T_1(^4F)$ excited state; the reason is that operating on the $^4T_1$ state with the dipole transition operator, which is of a T_2 species symmetry, exhibits no component of an A_1 symmetry. On the other hand, the $^4A_1(^4F) \rightarrow {}^4T_2(^4F)$ and $^4A_1(^4F) \rightarrow {}^4T_2(^4P)$ transitions are allowed, because in these two cases, operating on a $^4T_2$ state with the dipole transition operator (which is of a T_2 species symmetry) does exhibit a component of an A_1 species. The reader is invited to verify the above statements by multiplying the relevant representations and reduce the result to its components using the T_d character table (see Section 8.12, Table [33]).

The experimental absorption spectrum is shown in Figure 12.13b, with transition assignments inset in the figure frame. Apparently, $^4A_1(^4F) \rightarrow {}^4T_2(^4P)$ and $^4A_1(^4F) \rightarrow {}^4T_2(^4F)$ transitions are split by spin-orbit interaction, the first obviously into three components, the latter most probably into four components, according to the possible resultant J states. Recall that the parent ^4P state is already split into $^4P_{1/2}$, $^4P_{3/2}$, and $^4P_{5/2}$ states at a decreasing energy order, and that the parent ^4F state is already split into $^4F_{3/2}$, $^4F_{5/2}$, $^4F_{7/2}$, and $^4F_{9/2}$ states at a decreasing energy order (see Table 5.4, Eq. [5.64], their related text, and Figure 12.6).

In principle, it should be possible to measure optical transitions between the ion excited states. Nevertheless, the initial state must be sufficiently populated to allow for the observation of transitions out of it (recall that a state "population" is the density of atoms/ions actually being at that state). Such a condition may be realized if the intended state's lifetime is sufficiently long. That condition is especially difficult to achieve for ions embedded in solids or liquids, where interaction with the surrounding matter induces non-radiative decay transitions. For the Co^{2+} ion residing in the $ZnSe$ matrix, the $^4T_1(^4F)$ state indeed exhibits a long lifetime (240 μs), but no absorption spectrum has yet been performed for it. Induced population of the $^4T_1(^4F)$ by absorption from the ground state is very inefficient, because the $^4A_1(^4F) \rightarrow {}^4T_1(^4F)$ transition is forbidden. Non-radiative decay transitions between states may compete with all radiative ones. A detailed discussion of non-radiative transitions in ions embedded in dielectric mediums is provided in Chapter 13.

12.7 Appendix: An Integral Over Three Multiplied Spherical Harmonics

$$\int_0^\pi \boldsymbol{\Theta}_\ell^m(\theta)\boldsymbol{\Theta}_{\ell'}^{m'}(\theta)\boldsymbol{\Theta}_\kappa^{m-m'}(\theta)\mathbf{sin}\theta d\theta = \frac{(-1)^{g-\ell+m'}(2g-2\ell')g!}{(g-\ell)!(g-\ell)!(2g+1)!}$$
$$\cdot\sqrt{\frac{(2\ell+1)(2\ell'+1)(2\kappa+1)(\kappa-m+m')!(\ell+m)!(\ell-m')!(\ell+m')!}{2(\kappa+m-m)!(\ell-m)!}},$$
$$\cdot\sum_t(-1)^t\frac{(\kappa+m-m'+t)!(\ell+\ell'-m+m'-t)!}{(\kappa-m+m'-t)!(\ell-\ell'+m-m'+t)!(\ell'+m'-t)!t!} \tag{12.49}$$

for $m \geq 0$, $m' \geq 0$, and $2g = \kappa + \ell + \ell'$, where, in summing over all t integers (third line), skipping over cases where the factorial operation is meaningless should be administered (the factorial of a negative integer is meaningless). Eq. (12.49) is known as the "Gaunt equation".

Exercises for Chapter 12

12.1 Extend Table 12.1 by including your calculated result of the spherical harmonics belonging to the $\ell = 7$ orbital quantum number. Make use of the following procedure:

(a) Use the Rodrigues equation (Eq. [12.7]) to calculate the $P_0(\xi)$ and $P_1(\xi)$ Legendre polynomials:

$$\boldsymbol{P}_\ell(\xi) = \frac{1}{2^\ell \cdot \ell!} \cdot \frac{d^\ell}{d\xi^\ell} \left(\xi^2 - 1\right)^\ell . \tag{12.50}$$

(b) Use the following recurrence properties of Legendre polynomials, and based on Eq. (12.50), calculate the $\boldsymbol{P}_2(\xi)$, $\boldsymbol{P}_3(\xi)$, $\boldsymbol{P}_4(\xi)$, $\boldsymbol{P}_5(\xi)$, $\boldsymbol{P}_6(\xi)$, and $\boldsymbol{P}_7(\xi)$ Legendre polynomials.

$$\boldsymbol{P}_{\ell+1}(\xi) = \frac{(2\ell + 1)\cdot\xi\cdot\boldsymbol{P}_\ell(\xi) - \ell\boldsymbol{P}_{\ell-1}(\xi)}{\ell + 1} . \tag{12.51}$$

(We present that equation without proof).

(c) Use Eq. (3.27), also given below as Eq. (12.52), to calculate the associated $\boldsymbol{P}_7^m(\theta, \varphi)$ Legendre polynomials for $m = 0, 1, \ldots, 7$:

$$\boldsymbol{P}_\ell^m(\xi) = \left(1 - \xi^2\right)^{m/2} \cdot \frac{d^m}{d\xi^m} \boldsymbol{P}_\ell(\xi) . \tag{12.52}$$

(d) Use the following Eq. (12.53) (see also Eqs. [3.31] and [3.33]) to calculate the spherical harmonics $\boldsymbol{Y}_7^m(\theta, \varphi)$ for $m = 0, \pm 1, \pm 2, \cdots, \pm 7$.

$$\boldsymbol{Y}_{\ell m}(\theta, \varphi) = (-1)^m \frac{1}{\sqrt{2\pi}} \left[\frac{2\ell + 1}{2} \cdot \frac{(\ell - m)!}{(\ell + m)!}\right]^{1/2} \boldsymbol{P}_\ell^m(\cos\theta)\, e^{im\varphi} . \tag{12.53}$$

12.2 Calculate the hydrogen $3d$ state splitting under the cubic perturbation described in Figure 12.9, where each $-ze$ perturbation charge is replaced by an outward pointing electric dipole, whose distance from the hydrogen nucleus is $10a_0$, and its strength is 2.54×10^{-20} $(\mathrm{erg} \cdot \mathrm{cm}^3)^{1/2}$.

12.3 Calculate the hydrogen $4f$ state splitting in a cubic O_h perturbation field generated by point charges. Also indicate the wave functions for each of the split states.

12.4 Plot an energy scheme describing the multiplets components split by the spin/orbit interactions for the cobalt Co^{2+} ion hosted in the zinc-selenide crystal (Figure 12.13a). In determining the shifts and splitting, use Eq. (12.26) and the given values of the spin/orbit coupling constants of Figure 12.6.

13 Non-Radiative and Mixed Decay Transitions♣

13.1 Non-Radiative Transitions Between Close Electronic States

This chapter considers ions embedded in a dielectric medium and the medium's influence on transitions between ionic excited states. The dominant decay mechanism between energetically close electronic levels in the ions is non-radiative. If the separation between two electronic levels is smaller than 0.05 eV (equivalent to 400 cm^{-1}), then the transition may exhibit resonance with acoustic phonons. This allows both the emission and absorption of acoustic phonons, namely the excitation or quenching of lattice vibrations, respectively. Such levels will be considered "very close". If the level separation ranges between approximately 0.05 and 0.25 eV, resonance with optical phonons may occur. Such a scheme will also allow both phonon absorption and emission transitions. Near room temperature, however, phonon absorptions exhibit smaller probabilities compared with phonon emissions. Electronic levels in this range are considered "close".

In the scientific literature, phonon energies and non-radiative transitions are often addressed by their angular frequency ω, whereas photon energies and radiative transitions are addressed by their frequency ν. We adhere here to this same practice.

13.1.1 Debye Approximation of Phonon Dispersion Curves

To facilitate calculations, we simplify the acoustic and optical phonon dispersion curves (see Section 9.5) in the spirit and according to the **Debye** approximation. One replaces the first Brillouin zone (FBZ) with a sphere of equal volume, which, like the FBZ, contains N points that represent the allowed wave vectors of the crystal vibrational states, where N is the volume density of crystal atomic bases (number of bases per unit crystal volume). The points are distributed uniformly in the said sphere at the same density as in the FBZ, namely at $\mathcal{D}(k) = 1/(2\pi)^3$ per unit crystal volume (see Eq. [9.14]). The Debye sphere radius (Debye wave vector) is then $k_D = (6\pi^2 N)^{1/3}$. For every acoustic branch, one assumes no spatial dispersion; in other words, the vibrational frequency depends only on the wave vector's length, and is independent of the wave propagation direction. The speed of sound v_s is assumed to be constant for all wave vectors, exhibiting the same value derived from the true dispersion curve at the long-wave limit ($k \to 0$). The phonon frequency dispersion function is then given by $\omega = v_s k$ throughout the entire Debye sphere.

For every optical branch, one assumes no spatial dispersion; the wave group velocity v_g is assumed to be constant, given by $v_g = \frac{a}{\pi}\left[\omega_{\boldsymbol{max}}^{opt} - \omega_{\boldsymbol{min}}^{opt}\right] = \frac{a}{\pi}\Delta_{opt}$, where $a = V_C^{1/3}$ is an average lattice constant, (V_C being the lattice primitive cell volume), π/a is approximately the first Brillouin zone edge in a principal

♣ Except for minimal additions and changes, the content of this chapter is taken from a paper previously published by the same author: Z. Burshtein, *Radiative, non-radiative, and mixed decay transitions of rare-earth ions in dielectric media*, SPIE Opt. Eng. **49**, 091005, 1-19 (2010).

Physics, Optics, and Spectroscopy of Materials, First Edition. Zeev Burshtein.

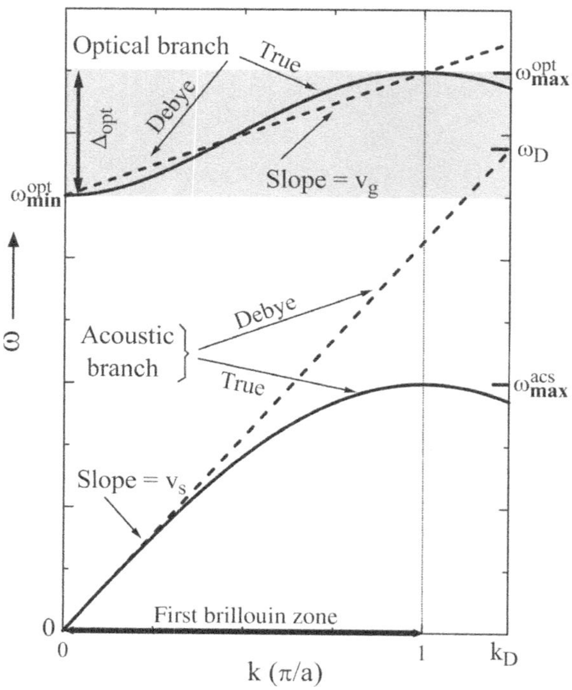

Figure 13.1 Demonstration of Debye approximations of acoustic and optical dispersion curves. For details, see text.

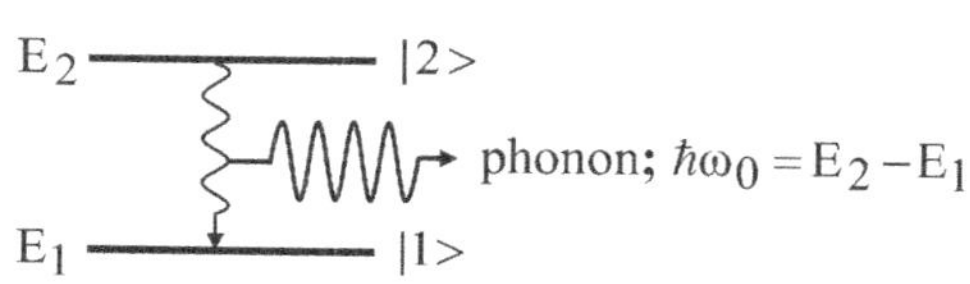

Figure 13.2 Energy scheme describing a phonon emission via a non-radiative transition between very close states.

direction, and ω^{opt}_{min} and ω^{opt}_{max} are the angular frequencies in the $k \to 0$ limit and at the Brillouin zone edge, respectively. Also, $\Delta_{opt} = \left|\omega^{opt}_{max} - \omega^{opt}_{min}\right|$ is the optical branch bandwidth. The above Debye approximations are illustrated schematically in Figure 13.1.

13.1.2 Non-Radiative Transitions Between Very Close Electronic States

In the following development, we take advantage of a similarity between the acoustic dispersion curves in the Debye approximation and the dispersion curves of the electromagnetic field. In Chapter 2, we developed an expression for the *radiative* lifetime of an excited state. In certain stages where the present development is analogous, we shall be satisfied with its short presentation. We encourage the reader to read the relevant developments in Chapter 2, especially Sections 2.3 to 2.5, as a means for improved comprehension.

We now further assume a full degeneracy between the longitudinal and transverse acoustic modes. Then, according the Debye approximation, the density of phonon frequencies along the frequency axis are given by (see Eq. [2.28] and Section 9.7.6)

$$D_D(\omega) \cong \begin{cases} \dfrac{3\omega^2}{2\pi^2 v_S{}^3} & for \quad 0 < \omega \le \omega_D \equiv v_s\left(6\pi^2 N\right)^{1/3} \\ 0 & for \quad \omega > \omega_D \end{cases}, \tag{13.1}$$

where ω_D is the Debye frequency. Typical Debye frequencies in crystals rarely exceed the order of 10^{14} rad/s , and in soft crystals, they are as low as 10^{12} rad/s . At thermal equilibrium, the volume density of thermal energy per unit angular frequency is given by (see Eq. [2.29])

$$\rho_\omega(\omega, T) = \frac{3\hbar\omega^3}{2\pi^2 v_S{}^3} \cdot \frac{1}{\boldsymbol{exp}\left(\hbar\omega/k_B T\right) - 1}. \tag{13.2}$$

Our following calculations concern transitions between two very close states of the embedded ion; a higher state $|2\rangle$ of E_2 energy, and a lower state $|1\rangle$ of E_1 energy. Figure 13.2 shows a fundamental energy scheme, expressing the process of a spontaneous phonon emission.

We assume the existence of coupling between acoustic vibrations and the valence electron revolving about the ion, such that when the ion's surrounding vicinity deviates from its equilibrium posture, a perturbing force is exerted on the revolving electron. Let the ion be positioned at $x = 0$, and the acoustic vibration normal coordinate be $q(x, t)$, where x indicates the equilibrium position coordinate of a crystal constituent atomic basis. Then we assume that the perturbing force exerted on the electron revolving in the ion is of the form $\varepsilon\, q'(0, t)$, where the prime indicates the spatial first derivative, and

ε is a coupling constant of dyn/g$^{1/2}$ units in the cgs system. The reason for relating the force to the deviation spatial derivative rather than to the deviation proper stems from the fact that a uniform translation of the ion and its vicinity would not change its posture relative to its surrounding; hence, no perturbing force is expected. The form assumed by the perturbing force is analogous to the action of an electric field on the electronic charge, where $q'(x, t)$ plays the role of the electric field and ε plays the role of the electronic charge. In analogy to Eq. (2.30), the perturbation Hamiltonian $\mathbf{H}'$ will thus become

$$\mathbf{H}' = \varepsilon(\underline{r}\cdot\underline{u})q'(0,t)\ , \tag{13.3}$$

where $\underline{r}$ is the average deviation vector of the valence electron revolving in the ion, and $\underline{u}$ is a unit vector in the force orientation. In analogy to Eq. (2.32), the probability per unit time for a stimulated transition from the $|2\rangle$ to the $|1\rangle$ state due to an incident acoustic wave is given by

$$W_{12} \cong \frac{\pi\varepsilon^2}{\hbar^2}|\langle 1|\underline{r}|2\rangle\cdot\underline{u}|^2\frac{d}{d\tau}\left|\mathbf{Q}'(\omega_0)\right|^2,$$

where τ is the time duration since the start of the wave incidence, $\langle 1|\underline{r}|2\rangle$ is the "transition radius" between the $|2\rangle$ and $|1\rangle$ states, $\omega_0 \equiv (E_2 - E_1)/\hbar$ is the transition angular frequency, and $\mathbf{Q}'(\omega)$ is the Fourier transform of $q'(x, t)$

$$\mathbf{Q}'(\omega) \equiv \frac{1}{\sqrt{2\pi}}\int_{-\infty}^{+\infty} q'(0,t)\ e^{-i\omega t}dt\ . \tag{13.4}$$

We recall that $q(x,t) \equiv \frac{1}{\sqrt{2\pi}}\int_{-\infty}^{+\infty}\mathbf{Q}(\omega)e^{i(\omega t - k(\omega)x)}d\omega$, where $\mathbf{Q}(\omega)$ is the Fourier transform of $q(x, t)$.

Therefore $q'(x,t) \equiv \frac{1}{\sqrt{2\pi}}\int_{-\infty}^{+\infty}\mathbf{Q}(\omega)(-ik(\omega))e^{i(\omega t - k(\omega)x)}d\omega$. We set $x = 0$, and the $k(\omega) = \omega/\mathrm{v}_s$ dispersion relation, and get

$$q'(0,t) \equiv \frac{1}{\sqrt{2\pi}}\int_{-\infty}^{+\infty}\frac{-i\omega}{\mathrm{v}_s}\mathbf{Q}(\omega)e^{i\omega t}d\omega\ . \tag{13.5}$$

Thus, we have $\mathbf{Q}'(\omega) = (-i\,\omega/\mathrm{v}_s)\,\mathbf{Q}(\omega)$ and $|\mathbf{Q}'(\omega_0)|^2 = (\omega_0/\mathrm{v}_s)^2|\mathbf{Q}(\omega_0)|^2$. Then we get

$$W_{12} \cong \frac{\pi\varepsilon^2\omega_0^2}{\hbar^2\mathrm{v}_s^2}|\langle 1|\underline{r}|2\rangle\cdot\underline{u}|^2\frac{d}{d\tau}|\mathbf{Q}(\omega_0)|^2\ . \tag{13.6}$$

The energy flux density carried by the acoustic waves' incident on the ion per unit angular frequency (energy per unit area per unit angular frequency) is given by $\mathbf{F}(\omega_0) = \frac{1}{2}\frac{\omega_0^2\mathrm{v}_s}{V_C}|\mathbf{Q}(\omega_0)|^2$, where V_C is the primitive unit cell volume (see Exercise 13.1). We set it in Eq. (13.6) and get

$$W_{12} \cong \frac{2\pi\varepsilon^2 V_C}{\hbar^2\mathrm{v}_s^3}|\langle 1|\underline{r}|2\rangle\cdot\underline{u}|^2\frac{d\mathbf{F}(\omega_0)}{d\tau}\ . \tag{13.7}$$

The $d\mathbf{F}(\omega_0)/d\tau$ quantity expresses the acoustic energy flux density per unit time per unit angular frequency (the acoustic waves' "intensity" per unit angular frequency). We further set $d\mathbf{F}(\omega_0)/d\tau = \mathrm{v}_s\rho_\omega(\omega_0)$, where ρ_ω is the volume density of acoustic waves' energy per unit angular frequency in the ion's vicinity, then average over all possible sound propagation orientations in space (see Eq. [2.34] and adjacent developments) and get

$$\boxed{W_{12} \cong \frac{2\pi\varepsilon^2 V_C}{3\hbar^2 \mathrm{v}_s{}^2} |\langle 1|\underline{r}|2\rangle|^2 \rho_\omega(\omega_0)} \quad . \tag{13.8}$$

As of this point, the development is strictly analogous to Einstein's development presented in Sections 2.3 and 2.4 for an atom exposed to an electromagnetic radiation field. The probability of the non-radiative decay transition per unit time (reciprocal of the spontaneous non-radiative transition lifetime τ_{nr}; also called the **decay rate constant**) is obtained by comparison with Eqs. (2.41) and (2.42) and their related text:

$$\tau_{nr}{}^{-1} = \frac{\varepsilon^2 \omega_0{}^3 V_C}{\pi\hbar \mathrm{v}_s{}^5} |\langle 1|\underline{r}|2\rangle|^2 . \tag{13.9}$$

The transition radius $\langle 1|\underline{r}|2\rangle$ that appears in Eq. (13.9) is, in fact, the electric dipole transition divided by the electron elementary charge e, namely $\langle 1|\underline{r}|2\rangle = \underline{\mu}/e$ (see Eq. [2.57] and related text). Its appearance in Eq. (13.9) establishes a linkage between the non-radiative and radiative lifetimes. By a first-order approximation, if the radiative transition is electric-dipole forbidden, the transition dipole diminishes. By an approach analogous to analysis of forbidden radiative transitions, we use the "oscillator strength" $\boldsymbol{f}$ concept (see Eq. [2.59] and related text) to assess the "allowed-ness" degree of the transition. It is readily seen that the effective squared radiative transition dipole μ^2 may be expressed as $\mu^2 = \boldsymbol{f}(e^2\hbar/m\omega_0)$, where e and m are the electron charge and mass, respectively. Thus,

$$\tau_{nr}{}^{-1} = \boldsymbol{f}\frac{\varepsilon^2 V_C \omega_0{}^2}{\pi m \mathrm{v}_s{}^5} . \tag{13.10}$$

Experiments show that for all frequencies smaller than Debye frequency $\omega_D = \mathrm{v}_s(6\pi^2 N)^{1/3}$ (see Eq. [13.1]), the non-radiative transition lifetimes are orders of magnitude shorter than the radiative lifetimes. A typical non-radiative lifetime order is $\tau_{nr} \sim 10^{-12}$ s, compared with a dipole-allowed radiative transition lifetime, which in the said frequency range is longer than the order of ~10^{-5} s. The critical parameter is obviously the non-radiative coupling parameter ε; the greater it is, the shorter would be the non-radiative lifetime. A further insight can be obtained by observing the ratio between the non-radiative and radiative lifetimes:

$$\frac{\tau_{nr}}{\tau_{rad}} = \left(\frac{\mathrm{v}_s}{c}\right)^3 \frac{2\pi e^2}{3\varepsilon^2 V_C/\mathrm{v}_s{}^2} . \tag{13.11}$$

The third power of the ~3×10^5 cm/s speed of sound to the ~3×10^{10} cm/s speed of light ratio $(\mathrm{v}_s/c)^3$~10^{-15} appearing in Eq. (13.11) plays an important, often dominant role in shortening the non-radiative lifetime compared with the radiative one.

The ε non-radiative coupling parameter itself may be expressed in terms of other material parameters. In achieving Eq. (9.25), we described a model by which each atomic basis is connected to its neighbors by a spring of an effective K constant. We extend the same idea by assuming that the "effective K constant" results from contributions of $\mathcal{N}_c$ identical effective springs, bonding each basis atom to its corresponding atom of the adjacent basis. We assume further that one of these springs bonds with the valence electron, which revolves in the dopant ion that resides in a specific atomic basis. Obviously, the said $\mathcal{N}_c$ quantity should be considered as an adjustable parameter, that is of the same order as the number of atoms comprising the atomic basis. Thus, the perturbing force $\mathcal{F}$ acting on the valence electron would be

$$\mathcal{F} \cong K(\xi_1 - \xi_0)/\mathcal{N}_c = K(q_1 - q_0)/M^{1/2}\mathcal{N}_c ,$$

where ξ_0 is the displacement of the atomic basis occupying the ion from its equilibrium position, and ξ_1 is the displacement of the neighboring atomic basis. M is the total atomic basis mass, and the q -quantities are the displacements' normal coordinates. We denote by a the average distance between the two bases (the lattice constant). We may then write

$$\mathcal{F} \cong \frac{Ka}{\mathcal{N}_c} \frac{(\xi_1 - \xi_0)}{a} \approx \frac{Ka}{\mathcal{N}_c M^{1/2}} \cdot q'(0,t) . \tag{13.12}$$

We thus conclude that $\varepsilon \approx Ka/M^{1/2}\mathcal{N}_c$. Based on Eq. (2.25), we conclude that $Ka \cong M{v_s}^2/a$. We further set $M = \rho_M a^3$, where ρ_M is the matter mass density (g/cm^3 units), and get

$$\varepsilon^2 \approx \rho_M a {v_s}^4 / {\mathcal{N}_c}^2 . \tag{13.13}$$

We set the latter result in Eq. (13.10), recalling that $V_C = a^3$, and get

$$\boxed{{\tau_{nr}}^{-1}(T=0) \approx \boldsymbol{f} \frac{\rho_M a^4 {\omega_0}^2}{\pi m v_S {\mathcal{N}_c}^2}} \; . \tag{13.14}$$

In the above equation, we have included the absolute temperature T as a specific variable of the process; recall that the physical process under consideration in Eq. (13.14) is a phonon spontaneous emission, equally occurring at all temperatures, including at $T = 0$. Notably, as long as the transition frequency is smaller than the Debye frequency, the non-radiative decay rate is proportional to the squared transition frequency ${\omega_0}^2$, and is inversely proportional to the speed of sound v_s. At a finite temperature, stimulated emissions by thermal phonons set in, causing an enhanced decay rate (see Eq. [2.44]):

$$\boxed{{\tau_{nr}}^{-1}(T) = {\tau_{nr}}^{-1}(0)\left[\frac{1}{\exp(\hbar\omega_0/k_B T) - 1} + 1\right]} \; . \tag{13.15}$$

The thermal *excitation rate* by phonon absorption between very close states is given by

$$W_{exc}^{1,2}(T) = {\tau_{nr}}^{-1}(0) \frac{1}{\exp(\hbar\omega_0/k_B T) - 1} \; . \tag{13.16}$$

As expected, the thermal excitation rate diminishes at the absolute zero temperature $T = 0$.

Figure 13.3 describes a typical case of non-radiative lifetime temperature dependence between two very close electronic states, whose separation (belonging to $\omega_0 = 4 \times 10^{13}$ rad/s) is equivalent to about 210 cm^{-1} (or 0.026 eV) . The dependence is presented for two different electronic oscillator strengths. The room-temperature (300*K*) non-radiative lifetime shortens to about 65% of its value at the low temperature limit. Notably, in the room-temperature vicinity, the shortening is an important factor up to the Debye frequency (actually, up to the acoustic

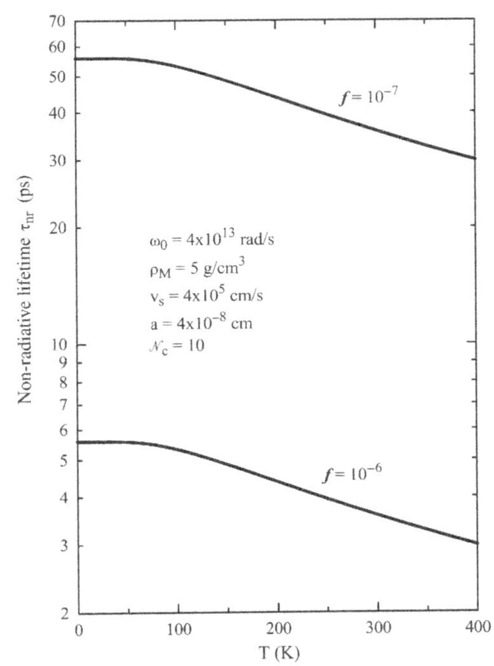

Figure 13.3 Semi-logarithmic plot of the non-radiative lifetime temperature dependence for two very close states' cases of different f electronic oscillator strengths. Transition frequency and material parameters are inset in the figure frame.

branch maximal frequency ω_{max}^{acs} ; see Figure 13.1). According to the presented model, the non-radiative decay mechanism through acoustic phonons would diminish for higher frequencies! No abrupt diminishing is to be expected, because the $|2\rangle$ and $|1\rangle$ electronic states themselves are broadened, and are sometimes as broad as the entire acoustic branch. That broadening provides a physical justification to overlook the infinite phonon states' density at the $\omega = \omega_{max}^{acs}$ point in the real dispersion curve, which would yield an infinite probability to the non-radiative transition at the $\omega_0 = \omega_{max}^{acs}$ point.

If the unit cell comprises several atoms, non-radiative decay mechanisms by optical phonons of higher frequencies would set in. That issue is the subject matter of the next subsection.

13.1.3 Non-Radiative Transitions Between Close Electronic States

We consider now a case where the energetic separation between the $|2\rangle$ and $|1\rangle$ electronic states (see Figure 13.2) equals the energy of an optical crystal oscillation. Hard crystals may exhibit optical oscillations of up to ~1600 cm^{-1} (~0.2 eV) and, in special cases, may go even higher. The highest angular frequency ω_{co} of an optical oscillation is called its **cutoff frequency**. The analysis is based on the Debye model described in Section 13.1. Under the Debye approximation, the energy surfaces (surfaces of equal energy) are spherical; thus, the frequency surfaces are, too. The $D(\omega)$ density of states along the angular frequency ω axis are defined through the relation $D(\omega)d\omega = \mathcal{D}(k)d^3\underline{k}$. We set $d^3\underline{k} = 4\pi k^2 dk$, and get

$$D(\omega) \cong \begin{cases} \dfrac{k^2(\omega)}{2\pi^2 v_g} & for \;\; 0 < k(\omega) \le k_D \equiv \left(6\pi^2 N\right)^{1/3} \\ 0 & for \;\; k(\omega) > k_D \end{cases}, \tag{13.17}$$

where $k(\omega) = \left(\omega - \omega_{min}^{opt}\right)/v_g$ is the reciprocal function to the $\omega(k)$ dispersion function in the Debye approximation described in Figure 13.1.

Under thermal equilibrium, the volume density of thermal energy per unit angular momentum (see Eq. [2.29]) is given by $\rho_\omega(\omega, T) = D(\omega)\langle \mathbf{n}(\omega)\rangle_T \hbar\omega$, where $\langle \mathbf{n}(\omega)\rangle_T$ is the phonon occupation number. Thus,

$$\rho_\omega(\omega, T) = \frac{k^2(\omega)\cdot\hbar\omega}{2\pi^2 v_g}\cdot\frac{1}{\mathbf{exp}\left(\hbar\omega/k_B T\right) - 1}, \tag{13.18}$$

for $k(\omega) \le k_D$.

We assume that the coupling between the optical oscillations and the valence electron revolving in the ion relates directly to the shape distortion of its ligand. For simplicity, we consider the atomic basis to play the role of the ligand. We assume further that the oscillation normal coordinate $q(t)$ exerts an external force on the revolving electron, whose size is $\eta \cdot q(t)$, where η is the coupling constant between the two oscillations, having a dyn/$g^{1/2}$cm units in the cgs system.

The mathematical structure of the expression for the force is analogous to the action of an electric field on the charged electron, where $q(t)$ plays the role of the electric field and η plays the role of the electric charge. In analogy to Eq. (2.30), the $\mathbf{H}'$ perturbation Hamiltonian on the ion electronic state would be

$$\mathbf{H}' = \eta(\underline{r}\cdot\underline{u})q(t), \tag{13.19}$$

where $\underline{r}$ is the average deviation vector of the valence electron revolving in the ion, and $\underline{u}$ is a unit vector in the force orientation. In analogy to Eq. (2.32), the probability per unit time for a stimulated transition from the $|2\rangle$ to the $|1\rangle$ state caused by an incident acoustic wave is given by

$$W_{12} \cong \frac{\pi\eta^2}{\hbar^2} \left|\langle 1|\underline{r}|2\rangle \cdot \underline{u}\right|^2 \frac{d}{d\tau} \left|\boldsymbol{Q}(\omega_0)\right|^2,$$

where τ is the time duration since the start of the wave-packet incidence, $\langle 1|\underline{r}|2\rangle$ is the "transition radius" between the $|2\rangle$ and $|1\rangle$ states, $\omega_0 \equiv (E_2 - E_1)/\hbar$ is the transition angular frequency, and $\boldsymbol{Q}(\omega)$ is the Fourier transform of $q(x, t)$:

$$\boldsymbol{Q}(\omega) \equiv \frac{1}{\sqrt{2\pi}} \int_{-\infty}^{+\infty} q(t)\, e^{-i\omega t} dt. \tag{13.20}$$

We assume that an optical oscillation mode of each atomic basis can be modeled by taking two identical $\frac{1}{2}M$ masses, where M is the mass of the entire basis, bonded by a spring, and oscillating one against the other (see Figure 7.2 and the text that follows Eq. [9.32]). Its reduced mass would then be $\mu^* = \frac{1}{4}M$. The energy flux density carried by the optical waves' incident on the ion per unit angular frequency (energy per unit area per unit angular frequency) is given by $\mathbf{F}(\omega_0) = \frac{1}{8} \frac{\omega_0{}^2 \mathrm{v}_s}{V_C} \left|\boldsymbol{Q}(\omega_0)\right|^2$. We set it in Eq. (13.19) and get

$$W_{12} \cong \frac{8\pi\eta^2 V_C}{\hbar^2 \mathrm{v}_g \omega_0{}^2} \left|\langle 1|\underline{r}|2\rangle \cdot \underline{u}\right|^2 \frac{d\mathbf{F}(\omega_0)}{d\tau}. \tag{13.21}$$

The $d\mathbf{F}(\omega_0)/d\tau$ quantity expresses the optical wave energy flux density per unit time per unit angular frequency (the optical waves' "intensity" per unit angular frequency). We set $d\mathbf{F}(\omega_0)/d\tau = \mathrm{v}_s \rho_\omega(\omega_0)$, where ρ_ω is the volume energy density of optical waves per unit angular frequency in the ion's vicinity, then average over all possible waves' propagation orientations in space (see Eqs. [2.34] and [13.8] and adjacent developments) and get

$$\boxed{W_{12} \cong \frac{8\pi\eta^2 V_C}{3\hbar^2 \omega_0{}^2} \left|\langle 1|\underline{r}|2\rangle\right|^2 \rho_\omega(\omega_0)}. \tag{13.22}$$

In a development analogous to Einstein's for an atom exposed to an electromagnetic radiation field presented in Sections 2.3 and 2.4, we obtain the non-radiative decay rate constant (reciprocal of spontaneous non-radiative transition lifetime τ_{nr}) as follows:

$$\tau_{nr}{}^{-1}(\omega_0) \cong \frac{4\pi^2\eta^2 \left(\omega_0 - \omega_{\boldsymbol{min}}^{opt}\right)^2}{3\hbar\omega_0 \Delta_{opt}{}^3} \left|\langle 1|\underline{r}|2\rangle\right|^2. \tag{13.23}$$

We recall further that the transition radius $\langle 1|\underline{r}|2\rangle$ is, in fact, the electric dipole transition divided by the electron elementary charge $\langle 1|\underline{r}|2\rangle = \underline{\mu}/e$ (see Eq. [2.57] and related text). The squared effective radiative transition dipole μ^2 satisfies the relation $\mu^2 = \boldsymbol{f}(e^2\hbar/m\omega_0)$, where $\boldsymbol{f}$ is the oscillator strength and e and m are the electron charge and mass, respectively. Thus, we get

$$\tau_{nr}{}^{-1}(\omega_0) \cong \boldsymbol{f} \frac{4\pi^2\eta^2 \left(\omega_0 - \omega_{\boldsymbol{min}}^{opt}\right)^2}{3m\omega_0{}^2 \Delta_{opt}{}^3}. \tag{13.24}$$

We relate now the η coupling constant to other material quantities. We extend the model which led to Eq. (9.33). We assume that the "effective spring" K constant that bonds the two atomic basis parts, which oscillate against each other in a polar oscillation, actually comprises $\frac{1}{2}\mathcal{N}_c$ identical springs, where $\mathcal{N}_c$ is the number of atoms constituting the atomic basis (in the following developed expressions, the $\mathcal{N}_c$ quantity may be considered as an adjustable parameter, which is of the order of the number of constituting atoms). We further assume that one of the springs binds with the valence electron of the dopant ion. The force acting on the electron during oscillation is given by $\mathcal{F} = (2K/\mathcal{N}_c)\,\xi = \left(4K/M^{1/2}\mathcal{N}_c\right)q$, therefore $\eta \approx 4K/M^{1/2}\mathcal{N}_c$. We use the expression for K appearing in the text that precedes Eq. (9.34): $K = \mu^*\,(\omega^L)^2$, where ω^L is the longitudinal propagation angular frequency of the wave belonging to the same optical oscillation (Section 9.5.4.3), and get $\eta \approx M^{1/2}\left(\omega^L\right)^2/\mathcal{N}_c$. Limiting ourselves to ω_0 frequencies within the optical band dispersion curve, we may use an approximation $\omega_0 = \omega^L$ in the expression for η, set it in Eq. (13.23), and get

$$\tau_{nr}{}^{-1}(\omega_0) \approx \boldsymbol{f}\,\frac{4\pi^2\rho_M a^3\omega_0{}^2\left(\omega_0 - \omega_{\boldsymbol{min}}^{opt}\right)^2}{3\mathcal{N}_c{}^2 m\Delta_{opt}{}^3}\,. \tag{13.25}$$

We have already noted that the electronic $|2\rangle$ and $|1\rangle$ states themselves (see Figure 13.2) are already broadened, sometimes to the extent of the entire optical phonon width. It thus suggests to present the $\tau_{rad}{}^{-1}(\omega_0)$ quantity as a simple average over the optical frequency band. We thus replace $\left(\omega_0 - \omega_{\boldsymbol{min}}^{opt}\right)^2$ by its average $\sim \frac{1}{3}\Delta_{opt}{}^2$ and get

$$\tau_{nr}{}^{-1}(\omega_0) \cong \boldsymbol{f}\,\frac{4\pi^2\rho_M a^3\omega_0{}^2}{9\mathcal{N}_c{}^2 m\Delta_{opt}}, \qquad for \quad \omega_{\boldsymbol{min}}^{opt} \lesssim \omega_0 \lesssim \omega_{\boldsymbol{max}}^{opt}\,. \tag{13.26}$$

In principle, it should be expected that the non-radiative transition probability would diminish in frequency regions belonging to gaps between the optical vibrational bands. In reality, such gaps rarely exist because of splitting between longitudinal and transverse oscillations (LO-TO splitting – see Section 9.5.4, especially Figure 9.11) and to spatial dispersion. In fact, frequency overlaps often exist among vibrational mode bands in addition to the above-mentioned broadening of the $|2\rangle$ and $|1\rangle$ states (see Figure 13.2). Recalling that $\mathcal{N}_c$ represents the number of atoms belonging to the atomic basis, the number of the crystal optical bands is $3\mathcal{N}_c - 3$. We thus simplify the case by assuming that the bands' dispersion spans the *entire* frequency range between $\omega_{\boldsymbol{max}}^{acs}$ up to ω_{co} (the maximal optical angular frequency). Thus, $\Delta_{opt} \approx \omega_{co}/(3\mathcal{N}_c - 3)$. We set in Eq. (13.26) and get

$$\boxed{\tau_{nr}{}^{-1}(\omega_0) \approx \boldsymbol{f}\,\frac{\mathcal{N}_c - 1}{\mathcal{N}_c{}^2}\cdot\frac{4\pi^2\rho_M a^3\omega_0{}^2}{3m\,\omega_{co}} \qquad for \quad \omega_0 \lesssim \omega_{co}}\,. \tag{13.27}$$

The one-phonon decay probability rate diminishes for $\omega_0 \gtrsim \omega_{co}$.

The expression related to the non-radiative decay lifetime provided in Eq. (13.27) belongs to a phonon spontaneous emission, valid at all temperatures, including the $T = 0$ absolute zero. At a finite T temperature, the effective decay lifetime shortens due to stimulated emissions according to Eq. (13.15). Nevertheless, for the room-temperature vicinity, the relation $\hbar\omega_0 >> k_B T$ is most often valid, and no considerable temperature dependence is expected.

We use Eq. (13.27) to assess a non-radiative lifetime for a representative case. Given data are: $\mathcal{N}_c = 10$; $\rho_M = 5\ \text{g/cm}^3$; $a = 4 \times 10^{-8}\text{cm}$; $\omega_0 = 10^{14}\,\text{rad/s}$; $\omega_{co} = 2 \times 10^{14}\,\text{rad/s}$; $\boldsymbol{f} = 10^{-7}$; and $m = 9.11 \times 10^{-28}\text{g}$. We set the data in the inverse of Eq. (13.27) and get

$$\tau_{nr} \approx \frac{10^2 \times 3 \times 9.11 \times 10^{-28} \times 2 \times 10^{14}}{10^{-7} \times 9 \times 4 \times 3.14^2 \times 5 \times 64 \times 10^{-24} \times 10^{28}} = 4.8 \times 10^{-13}\text{s}\,.$$

13.2 Radiative Transition Lifetime and Optical Absorption and Emission Spectra

We turn now to discuss the issue of transition dynamics in a complex state system, where some transitions are radiative and some non-radiative. A model of such system will be presented later on. Eq. (2.56) expressed the radiative transition lifetime between two homogeneously broadened states

$$\tau_{rad}^{-1} = \frac{8\pi n^2 \nu_0^2}{N_1 c^2} \int_0^{\infty} \alpha(\nu) d\nu , \tag{13.28}$$

where n is the medium refractive index, $\alpha(\nu)$ is the frequency ν dependent absorption coefficient from the ground-state to the upper state, ν_0 is the absorption peak frequency, N_1 is the density of ions at the ground-state, and c is the speed of light constant. The $\sigma_{abs}(\nu) \equiv \alpha(\nu)/N_1$ quantity is the absorption cross section. We may then write

$$\tau_{rad}^{-1} = \frac{8\pi n^2 \nu_0^2}{c^2} \int_0^{\infty} \sigma_{abs}(\nu) d\nu . \tag{13.29}$$

Eq. (13.29) assumes ideal symmetry between absorption and stimulated emission. The reason for presenting the lifetime on the basis of the absorption spectrum and not on the emission spectrum is that measuring a quantitative absorption spectrum is quite easy. On fundamental grounds, however, the radiative lifetime should be presented on the basis of the emission spectrum

$$\tau_{rad}^{-1} = \frac{8\pi n^2 \nu_0^2}{c^2} \int_0^{\infty} \sigma_{em}(\nu) d\nu , \tag{13.30}$$

where $\sigma_{em}(\nu)$ is the stimulated emission cross section.

In reality, radiative absorption and emission processes of ions residing in a material medium are not identical. The reason is that, usually, both ground and upper electronic states are each split, not necessarily identically, into a set of states called a **multiplet**. The upper multiplet is usually known as an **excited state**, and its component states as **Stark components**. If the lower multiplet stems from splitting of the lowest ionic state, its *lowest* Stark component is called the **ground-state**. If the separations between adjacent multiplet states are smaller than the largest (cutoff) phonon energy $\hbar\omega_{co}$, they are subject to very fast non-radiative transitions (see Eqs. [13.14], [13.16], and [13.27]) that yield preferred occupation of the lowest Stark components. The lowest multiplet components exhibit a nearly thermal equilibrium occupation distribution. If any component of the upper multiplet is excited (often by optical absorption), all multiplet states then assume a near thermal equilibrium distribution among themselves within several non-radiative lifetimes past excitation. That distribution is called **quasi-thermal distribution**. Thus, the radiative emission of the upper multiplet occurs mostly from its lowest component to all components of the lower multiplet. The said radiative transition may be observed when the involved states' separation exceeds several tenths of an electron-volt (eV).

Electronic states' splitting of an ion residing in a crystalline site is caused by both the crystal field (Stark splitting) and its coupling with optical oscillation modes of its surrounding ligand. In the model we present below, the couplings with the latter oscillations (vibrations, librations, and translations) are assumed to be relatively small, expressing themselves mostly by broadening of the crystal field split states.

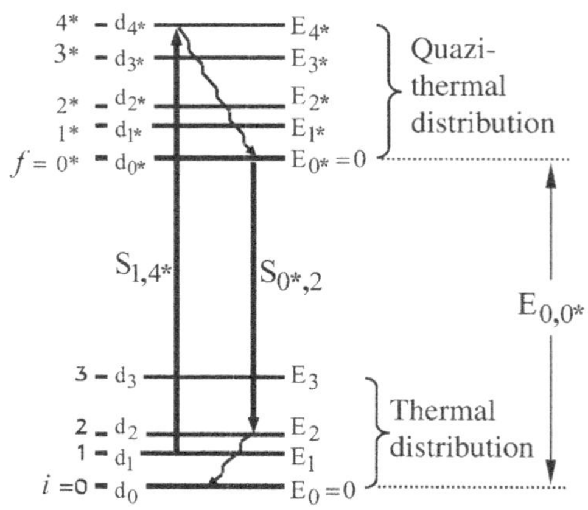

Figure 13.4 Model of the energy scheme and transitions used for calculating the radiative decay lifetime from an upper to a lower multiplet, where the energy states belonging to each multiplet have established a thermal equilibrium population distribution by both radiative and non-radiative transitions among themselves. Radiative transitions are indicated by straight arrows, and non-radiative ones by wiggled arrows.

The energy scheme is presented in Figure 13.4. Transitions between states are indicated, being radiative between different multiplet states and non-radiative within each multiplet. The $i = 0,\ 1,\ 2,\ \cdots$ labels indicate the different states of the lower multiplet at an increasing energy order; the $f = 0^*,\ 1^*,\ 2^*,\ \cdots$ labels indicate the different states of the upper multiplet at an increasing energy order. States' degeneracy in each multiplet are indicated by d_i and d_f, and their energy above the lowest one at each multiplet are indicated by E_i and E_f, respectively.

We designate by $\sigma^i_{abs}(\nu)$ the absorption cross section from a certain i state of the lower multiplet to the entire series of $f = 0^*,\ 1^*,\ 2^*,\ \cdots$ upper multiplet states. The spectrum may be written as $\sigma^i_{abs}(\nu) = \overline{S}^i g^i_{abs}(\nu)$, where $g^i_{abs}(\nu)$ is the absorption shape spectrum (see Eq. [2.47] and related text), and $\overline{S}^i$ (in cm^2/s units) is the said transition strength. We resolve the spectral shape into single-peak components (Gaussian, Lorentzian, or others; see Section 1.2) to be designated as $g^{if}_{abs}(\nu)$, such that $g^i_{abs}(\nu) = \sum_f g^{if}_{abs}(\nu)$. In fact, the peak positions of the resolved $g^{if}_{abs}(\nu)$ functions define the upper multiplet states energy. Each of the transitions belonging to the $g^{if}_{abs}(\nu)$ functions may exhibit a different width. We denote the corresponding width (standard deviation except for a Lorentzian) by w_{if}. We define $S_{if} \equiv \overline{S}_i \int_0^\infty g^{if}_{abs}(\nu)d\nu$ as the partial strength components. Obviously, $\overline{S}_i = \sum_f S_{if}$. Such quantities are indicated in Figure 13.4.

Analogous $\sigma^f_{em}(\nu)$, $g^f_{em}(\nu)$, $\overline{S}^f$, S_{fi} $g^{fi}_{em}(\nu)$, and w_{fi} quantities are defined for the stimulated emission. Because of the symmetry between absorption and stimulated emission, the following equalities exist: $S_{if} = S_{fi}$ and $w_{if} = w_{fi}$. These equalities between the said quantities are often described as "reciprocity principle".

The theoretical analysis we present below is based on the experimental reality that the absorption cross-section spectrum is easy to measure. The stimulated emission cross-section spectrum is difficult to measure; easy to measure is only the *spectral shape* of the spontaneous emission, namely the fluorescence spectral shape. The analysis shows how to obtain the stimulated emission cross-section spectrum from the measured data. The expression presenting the stimulated emission cross-section spectrum is known as the **Füchtbauer-Ladenburg** equation. The spectra allow also the assessment of the temperature dependence of the radiative lifetime by applying Eq. (13.28) to address the present issue.

As indicated, it is quite possible to measure the $\sigma_{abs}(T, \nu)$ absorption cross-section spectra and the $g_{em}(T, \nu)$ fluorescence shape spectra, where T is the absolute temperature. Initially, the said spectra are measured near the zero absolute temperature limit $T \cong 0$. All absorption transitions would then occur from the $i = 0$ ground-state to all $f = 0^*,\ 1^*,\ 2^*,\ \cdots$ upper multiplet states; all fluorescence emission transitions would then occur from the $f = 0^*$ excited multiplet state to all $i = 0,\ 1,\ 2,\ \cdots$ ground multiplet states. Resolution of the $\sigma_{abs}(0, \nu) \equiv \sigma^0_{abs}(\nu)$ and $g_{em}(0, \nu) \equiv g^0_{em}(\nu)$ into separate peaks defines the energetic E_f positions of the excited multiplet states, and the E_i positions of the ground multiplet states, respectively. The reciprocal decay lifetime from the $f = 0^*$ state to $i = 0,\ 1,\ 2,\ \cdots$ states would be given by

$$\tau_{rad}{}^{-1}(0) = \frac{8\pi n^2 \nu_{0,0^*}{}^2}{c^2} \int_0^\infty \sigma^0_{abs}(\nu)d\nu\,, \tag{13.31}$$

where $\nu_{0,0^*} = E_{0,0^*}/h$ is the absorption frequency from $i = 0$ to $f = 0^*$ (see Figure 13.3). Then we get

$$\sigma_{em}^{0^*}(\nu) = \frac{c^2}{8\pi n^2 \nu_{0,0^*}{}^2 \tau_{rad}(0)} \cdot g_{em}^{0^*}(\nu) = \overline{S}^{0^*} \cdot g_{em}^{0^*}(\nu)\,, \tag{13.32}$$

where

$$\overline{S}^{0^*} = \int_0^\infty \sigma_{abs}^{0}(\nu)d\nu\,. \tag{13.33}$$

Eq. (13.33) yields the $\overline{S}^{0^*}$ quantity as an experimental result, thus allowing calculation of the partial transition strengths as $S_{0^*,i} = \overline{S}^{0^*} \int_0^\infty g_{em}^{0^*,i}(\nu)d\nu$. Eq. (13.32) yields the emission cross-section spectrum for the low temperature limit. We shall detail below how to express the cross-section $\sigma_{em}(T, \nu)$ spectrum at any finite temperature T by using the previously obtained results.

A rise in temperature induces decay transitions from thermally excited *f* states of the upper multiplet to the series of the *i* ground multiplet states. It is *assumed*, that transitions initiated from a certain *f* state, duplicate those initiated from the lowest $f = 0^*$; however, they differ in three aspects:

(i) Each transition energy is by E_f larger than the lowest E_{0^*} state energy. Had it been the only consideration, we would write $g_{em}^{f}(\nu) = g_{em}^{0^*}\left(\nu - \frac{E_f}{h}\right)$. In reality, the transitions broaden with increasing temperature but only to a small extent. We overlook that broadening.

(ii) The S_{fi} emission strength parameter of an individual fluorescence component is given by

$$S_{fi} = S_{0^*,i} \cdot \frac{S_{0f}}{S_{0,0^*}}, \quad for \quad i = 0, 1, 2, \cdots. \tag{13.34}$$

The underlying physical reason is the assumption that the S_{fi} series of values for $i = 0, 1, 2, \cdots$ reflects the $S_{0^*,i}$ series of values, but is multiplied by the same amount by which S_{0f}, the *excitation* transition to the *f* state, is larger than $S_{0,0^*}$, the excitation transition to the 0^* state. It is readily verified that in the specific $f = 0^*$ case, Eq. (13.34) reconstructs what we already know. Thus,

$$\overline{S}^{f} = \frac{S_{0f}}{S_{0,0^*}} \overline{S}^{0^*}. \tag{13.35}$$

(iii) The width of the emission transitions depend on the initial *f* state. We assume that the series of $(w_{fi})^2$ variances for $i = 0, 1, 2, \cdots$ reflects the $(w_{0^*,i})^2$ series, just enlarged by the same amount that $(w_{0f})^2$ is greater than $(w_{0,0^*})^2$, namely

$$\left(w_{fi}\right)^2 = \left(w_{0^*,i}\right)^2 + \left[\left(w_{0f}\right)^2 - \left(w_{0,0^*}\right)^2\right]. \tag{13.36}$$

We thus need to revise the text of item **(i)** above, and write symbolically

$$g_{em}^{f}(\nu) = g_{em}^{0^*,wf}\left(\nu - \frac{E_f}{h}\right), \tag{13.37}$$

where the added "*wf*" superscript on the equation's right side indicates the inclusion of a separate width parameter to each $g_{em}^{0^*}$ component according to Eq. (13.36).

We are now properly prepared to calculate the stimulated emission spectra and radiative lifetime at all temperatures. The spectrum is a sum of all transition components from the *f* upper multiplet states to the ground multiplet states, where the separate contributions of the *f* states are weighed by their states' thermal occupancy

$$\begin{aligned}\sigma_{em}(T,\nu) &= \sum_f \frac{d_f \cdot \mathbf{exp}(-E_f/k_BT)}{Z_u(T)} \cdot \sigma_{em}^f\left(\nu - \frac{E_f}{h}\right) \\ &= \frac{1}{Z_u(T)} \sum_f d_f \cdot \mathbf{exp}(-E_f/k_BT) \cdot \bar{S}^f g_{em}^{0*,wf}\left(\nu - \frac{E_f}{h}\right) \\ &= \frac{1}{Z_u(T)} \cdot \frac{\bar{S}^{0*}}{S_{0,0*}} \sum_f d_f \cdot \mathbf{exp}(-E_f/k_BT) \cdot S_{0f}\, g_{em}^{0*,wf}\left(\nu - \frac{E_f}{h}\right),\end{aligned} \tag{13.38}$$

where $Z_u(T) \equiv \sum_f d_f \cdot \mathbf{exp}\left(-E_f/k_BT\right)$ is the thermodynamic distribution function of the upper multiplet states.

By reciprocity considerations, the absorption cross-section spectrum is given by

$$\sigma_{abs}(T,\nu) = \frac{1}{Z_\ell(T)} \cdot \frac{\bar{S}^{0^*}}{S_{0^*,0}} \sum_i d_i \cdot \mathbf{exp}\left(-E_i/k_BT\right) \cdot S_{0^*i}\, g_{abs}^{0,wi}\left(\nu - \frac{E_i}{h}\right), \tag{13.39}$$

where $Z_\ell(T) \equiv \sum_i d_i \cdot \mathbf{exp}\left(-E_i/k_BT\right)$ is the thermodynamic distribution function of the lower (ground) multiplet states.

The radiative lifetime is given by

$$\begin{aligned}\tau_{rad}^{-1}(T) &= \frac{8\pi n^2 \nu_{0,0^*}{}^2}{c^2} \int_0^\infty \sigma_{em}(T,\nu)d\nu\ ; \\ \tau_{rad}^{-1}(T) &= \tau_{rad}^{-1}(0) \cdot \frac{1}{Z_u(T)\, S_{0,0^*}} \sum_f S_{0f}\, d_f \cdot \mathbf{exp}\left(-E_f/k_BT\right).\end{aligned} \tag{13.40}$$

For demonstration, we present a detailed analysis of glass-embedded triply ionized Yb^{3+} ytterbium ions. The specific glass for which the spectra are presented is a silicate glass, comprising mainly silicon dioxide SiO_2, small additions of boron oxide B_2O_3, calcium oxide CaO, sodium oxide Na_2O, and potassium oxide K_2O. The Yb^{3+} spectral properties in other glasses, as well as in different crystals, are quite similar; the differences are expressed by small band shifts and states' broadening. The ion ground-state configuration is $[Xe]4f^{13}$ (see Table 5.5), and the ground-state Term is ^2F. Spin/orbit coupling splits the Term into $^2F_{7/2}$ and $^2F_{5/2}$ at an increasing energy order. The ytterbium ion is surrounded by six close oxygen O^{2-} anions, thus forming an octahedrally coordinated ligand, which may be approximated as forming a cubic O symmetry. The two spin-orbit split states are further split in three by the ligand field, according to $^2F \rightarrow {}^2T_1 \oplus {}^2T_2 \oplus {}^2A_2$ (see Table 12.5 and Table [**iv**]-32 in Section 11.5) at an increasing energy order.

Spin/orbit coupling in ytterbium Yb^{3+} ions (atomic number $Z = 70$) is considerably greater than a usual ligand-field splitting. The physical reason is that all the ion valence electrons belong to a low-lying shell (principal quantum number $n = 4$); therefore they revolve close to the nucleus, and the influence of the ligand field is relatively small. Further splitting is expected because the ligand symmetry is probably smaller due to glass site distortions. The glass disorder contributes a heterogeneous broadening of the electronic states. Additional broadening is caused by interactions with the glass vibrational modes.

Figures 13.5a, b, and c present the absorption and fluorescence emission spectra of Yb^{3+} ions in the said glass at a very low temperature, and their belonging energy scheme, respectively. The $(0, 0^*)$ absorption and $(0^*, 0)$ transmission exhibit very narrow peaks at $E_{0,0^*} = 10{,}260\ \text{cm}^{-1}$. The spectra exhibit $(0, f)$ absorption and $(0^*, i)$ emission structures, respectively. Spectral parameters obtained by the spectra analysis are inset in the figures' frames: radiative lifetime, state energies, strengths, and widths. The upper and lower multiplet energy schemes also include each split state's symmetry assignment, and degeneracy.

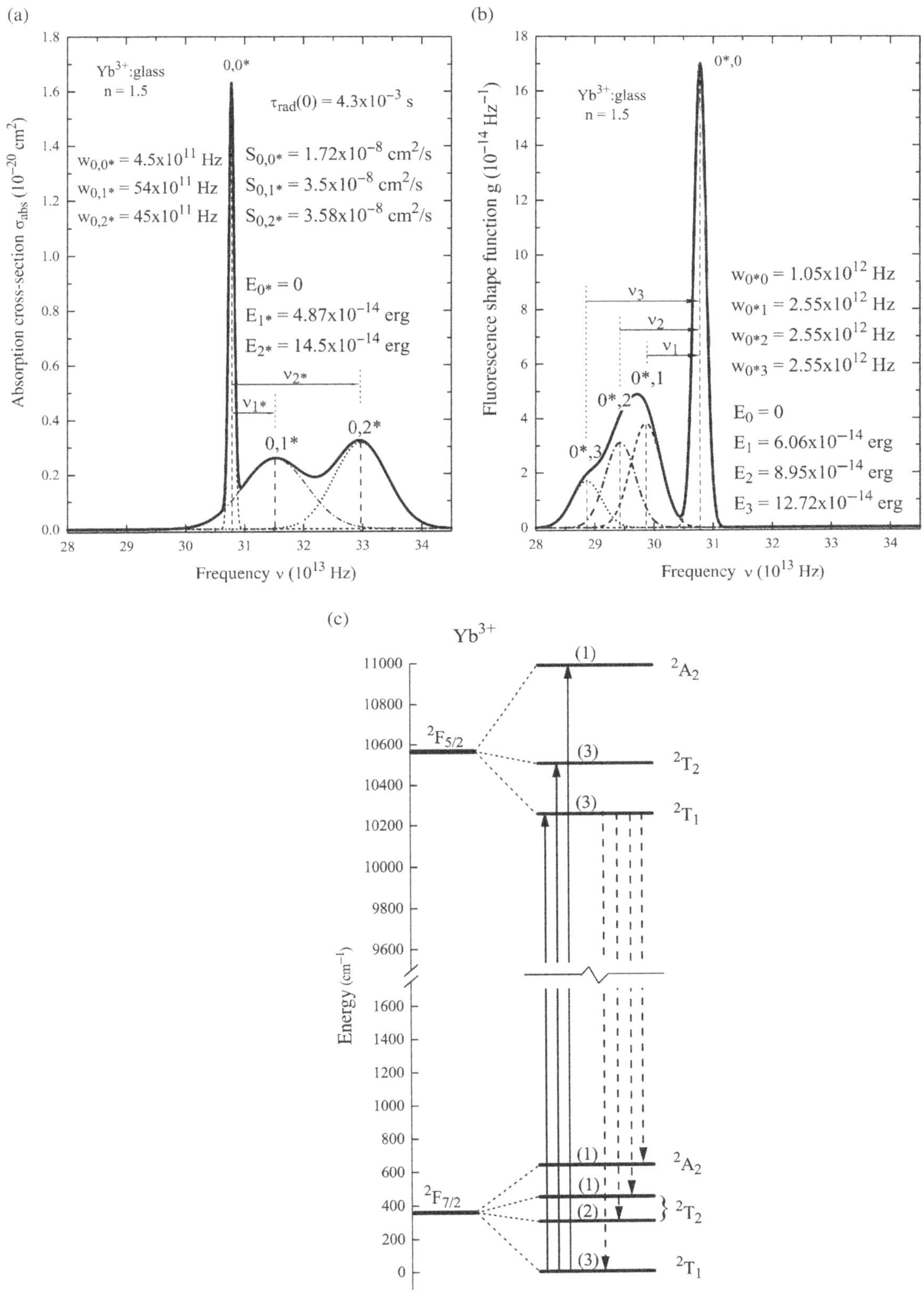

Figure 13.5 Absorption cross-section spectrum (a) and emission spectrum (b) of triply ionized ytterbium Yb^{3+} in a silicate-based glass at a low temperature, and its energy scheme (c). Transition parameters are indicated in the figure's frames. Each state degeneracy is indicated in parentheses within the energy scheme (c).

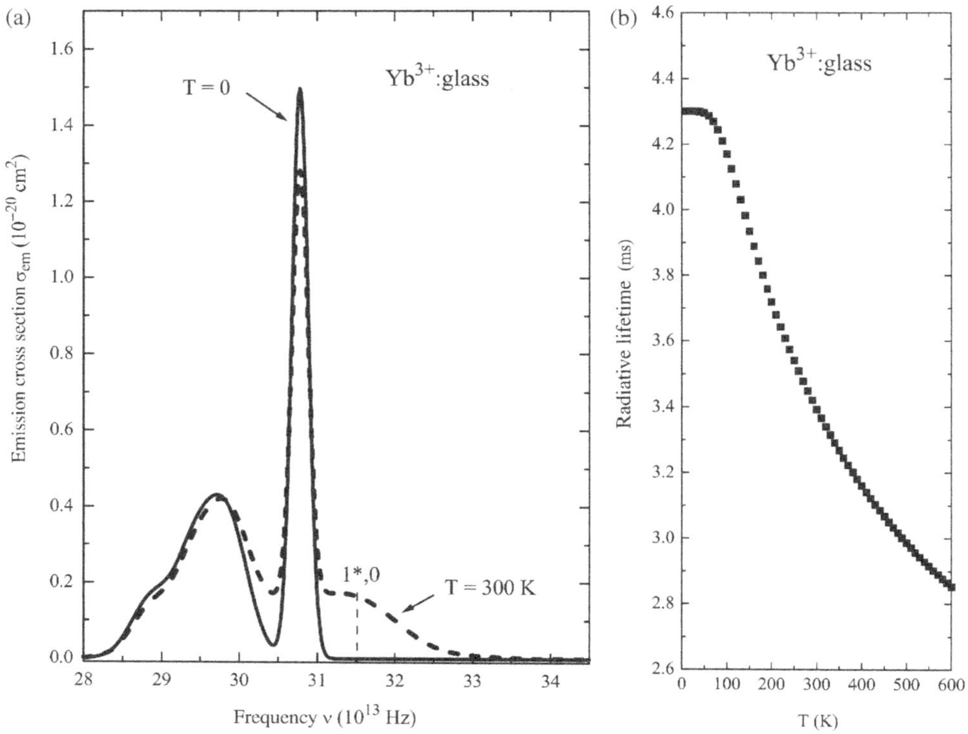

Figure 13.6 (a) Fluorescence cross-section emission spectrum calculated by Eq. 13.38 for a 300K temperature compared with the low temperature $T = 0$ limit. (b) Radiative lifetime temperature dependence calculated by Eq. 13.40.

The said parameters are used in calculating the emission spectrum at 300K compared with the low temperature one given by Eq. (13.38), and the radiative lifetime temperature dependence according to Eq. (13.40). The calculated results are presented in Figure 13.6. A reduction in the $(0^*, 0)$ transition is observed in Figure 13.6a compared with the low-temperature-limit spectrum, and with a simultaneous appearance of an additional, higher energy peak belonging to the $(1^*, 0)$ transition. The higher energy $(2^*, 0)$ transition would be revealed at a higher temperature; in principle, even higher-order transitions may be revealed as long as the hosting glass remains solid (does not melt). The radiative lifetime temperature dependence shortens upon increase of temperature, caused by enhanced $(1^*, i)$ radiative transitions on the expense of reduced $(0^*, i)$ transitions; note that the $(1^*, i)$ transitions are stronger than the $(0^*, i)$ ones, because $S_{0,\ 1^*} > S_{0,\ 0^*}$. Had the strengths ratio been inverted, a lengthening of the radiative lifetime would be effected by the increased temperature!

Inspection of Eq. (13.40) suggests that at the high temperature limit, the radiative lifetime approaches asymptotically a constant value, given by

$$\tau_{rad}{}^{-1}(\infty) = \tau_{rad}{}^{-1}(0)\cdot\frac{1}{S_{0,0^*}}\frac{\sum_f S_{0f}\, d_f}{\sum_f d_f}. \tag{13.41}$$

The last quantity relates to a special situation, where all higher-multiplet states participate equally in the radiative transitions to the lower multiplet states. In the specific case we use as an example, we get $\tau_{rad}(\infty) = 1.97$ ms.

13.3 Multi-Phonon Non-Radiative Transitions

A frequent occurrence in physical systems, like the one discussed in Section 13.2, involves the actually measured fluorescence lifetime being considerably shorter than the calculated radiative lifetime (Eq. [13.40]). It is particularly shorter than $\tau_{rad}(\infty)$ even if the distance between the electronic states is larger than the cutoff phonon energy $\hbar\omega_{co}$. Such effects are attributed to simultaneous non-radiative transitions involving a coherent emission of several phonons. Before addressing the process on its theoretical grounds, we provide a short description of the experimental methods used to calculate non-radiative transition rates between electronic states.

13.3.1 Principles and Methods in Experimental Measurement of Non-Radiative Lifetimes

We denote by N_u the number of excited ions in the upper multiplet. The rate equation governing the number of ions at any particular time t is given by

$$\frac{dN_u}{dt} = G - \frac{N_u}{\tau_{eff}}, \tag{13.42}$$

where $\tau_{eff}^{-1} = \tau_{rad}^{-1} + \tau_{nr}^{-1}$ is the effective decay rate of the excited ions states per unit time (the reciprocal lifetime) caused by the combined effects of the radiative and non-radiative decay mechanisms, and G is the excitation (generation) rate to the upper multiplet states (in ions/s or in abbreviated s^{-1} units). To observe the excited ions' concentration after the end of excitation, we set $G = 0$ in Eq. (13.42), whose solution is then

$$N_u(t) = N_u(0)\,\boldsymbol{exp}(-t/\tau_{eff}), \tag{13.43}$$

where $N_u(0)$ is the total number of the excited ions at a selected time $t = 0$ past excitation. As expected, the number of excited ions decays exponentially with time, with τ_{eff} being the decay time-constant. The photons' emission rate (in photons/s units) is given by $\tau_{rad}^{-1}N_u$. The measured fluorescence intensity $\boldsymbol{F}$ is proportional to the fluorescence photon emission rate, and is given by $\boldsymbol{F}(t) = C\tau_{rad}^{-1}N_u(t)$, where C is a dimensionless constant, whose value depends on the measuring conditions, and specifically on the efficiency of the emitted photons collection and the detector sensitivity. Namely,

$$\boldsymbol{F}(t) = C\tau_{rad}^{-1}N_u(0)\,\boldsymbol{exp}(-t/\tau_{eff}). \tag{13.44}$$

Thus, measuring the decay constant of the fluorescence intensity time trace provides the effective lifetime τ_{eff} without any need to know the actual values of C or τ_{rad}^{-1} constants. Measuring the absorption cross-section spectrum allows for the calculation of the radiative lifetime τ_{rad} through Eq. (13.29). The non-radiative decay probability per unit time τ_{nr}^{-1} (non-radiative decay rate) is then simply given by

$$\tau_{nr}^{-1} = \tau_{eff}^{-1} - \tau_{rad}^{-1}. \tag{13.45}$$

An alternative way to assess the non-radiative lifetime is by measuring the fluorescence quantum yield η, namely the ratio between the total number of emitted photons and the number of photons absorbed by a very short light pulse; the latter provides the $N_u(0)$ quantity. The total number of emitted photons is given by

$$N_{photons} = \int_0^\infty \tau_{rad}^{-1}N_u(t)dt = N_u(0)\cdot\left(\tau_{eff}/\tau_{rad}\right); \tag{13.46}$$

$$\eta \equiv \frac{N_{photons}}{N_u(0)} = \frac{\tau_{eff}}{\tau_{rad}}. \tag{13.47}$$

Using the latter result, and knowing the radiative lifetime τ_{rad}, allows us to calculate the effective lifetime τ_{eff}, and hence also τ_{nr}, by using Eq. (13.45), all without actually measuring the fluorescence decay constant.

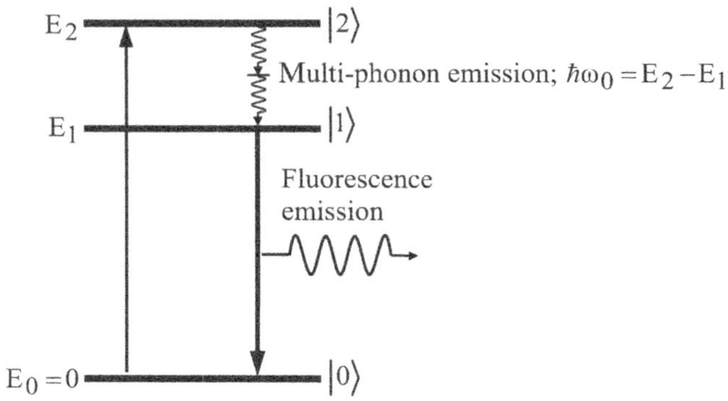

Figure 13.7 An energy states scheme that allows for the measurement of the multi-phonon non-radiative decay rate between two higher states, by measuring the fluorescence increase rate of the lower between the two latter ones. For details, see text.

Being acquainted with the state's energy scheme of each ion allows us to measure the non-radiative lifetimes between states whose radiative transitions efficiency is extremely small (namely, when radiative emission is practically unmeasurable). Figure 13.7 describes a fundamental energy scheme model that allows such a measurement. A very short light pulse excites ions from their $|0\rangle$ ground-state to the $|2\rangle$ excited state. The population generated at the $|2\rangle$ state decays by a predominantly non-radiative multi-phonon process to the $|1\rangle$ state of a very long radiative lifetime compared with the (non-radiative) transition from $|2\rangle$ to $|1\rangle$. As the $|1\rangle$ state population grows, the fluorescence out of that state gradually grows, thus reflecting the initial fluorescence *growth rate* in the $|1\rangle$ to $|0\rangle$ transition. It is a direct measure of the *decay rate* from $|2\rangle$ to $|1\rangle$, which is the measurement goal.

13.3.2 Theoretical Calculation of the Non-Radiative Lifetime

Seemingly, the probability for a non-radiative transition diminishes when the $\hbar\omega_0$ separation (gap) between the two electronic states is larger than $\hbar\omega_{co}$, where ω_{co} is the cutoff frequency of the medium optical oscillations. There exist a number of physical mechanisms, however, which enable such transitions. Two of them are anharmonicity of the electrons' motions about the nuclei, and anharmonicity of the atomic basis oscillations. As for the anharmonicity of the nucleus potential (the one binding the electrons to form atoms and ions), it also allows for multi-photon transitions, despite the fact that the electromagnetic field's oscillations are purely harmonic. Theoretically, n-photon transitions are calculated by using the n-th order perturbation theory; had the nucleus binding potential been perfectly harmonic, all perturbation elements of order larger than 1 would have vanished. In addressing the present problem, we calculate the multi-phonon transitions on the basis of the anharmonicity of the effective potential governing the atomic basis oscillations.

Recall that the atomic basis is identified as the ligand surrounding the fluorescing dopant ion. So far, we have addressed only non-radiative decay caused by single-phonon oscillation states; however, the atomic basis can oscillate at the same frequency yet of a larger amplitude – namely, at a state where the system inhabits several phonons. Such a situation is known as a "higher-order state". The above claim of "same frequency" for higher v-order oscillations is accurate only for a harmonic potential. In an anharmonic potential, the frequency changes as function of v. For example, in a Morse potential (Eq. [7.11]), the oscillation frequency decreases slightly as a function of v (see also Eq. [7.17]); however, for low orders, the change is rather small. In the following presented model, we assume that the configuration potential belonging to the largest oscillation frequency is given by the Morse potential (Eq. [7.11]) of the form

$$U(q) = D\left[1 - e^{-\beta q}\right]^2, \tag{13.48}$$

where q is the normal vibration configuration coordinate, and β and D are constants. The D quantity describes the dissociation energy of the ion ligand, and the characteristic vibrational frequency is given by $\omega_{vib} \cong (2\beta^2 D)^{1/2}$.

Because of the interaction between the vibrational oscillations of the atomic basis and its vicinity, a band of oscillation modes is obtained for each v-order frequency, as described in Section 9.5, especially in Section 9.5.4.2. A schematic potential for two electronic states is provided in Figure 13.8.

Figure 13.8 demonstrates a case of energy overlap between v′ = 0 zero-order vibration of the *higher* electronic state and a higher v-order vibration of the *lower* electronic state. Recall (see Section 7.5.2) that there always exists the probability for transition between vibrational states belonging to different electronic states, which depends critically on the Franck-Condon overlap integral (Eq. [7.47]). Because of the energy overlap, there also exists a probability for a non-radiative transition from the vibrational v′ = 0 state of the upper electronic state to the vibrational v state of the lower electronic state. That probability is, however, very small, because by the Franck-Condon principle, the exit point of the v′ = 0 state is quite distant from that of the v state (see Section 7.5.4, and especially Figure 7.11). We will not elaborate further on that particular mechanism, and now turn to discuss in detail a quite different one.

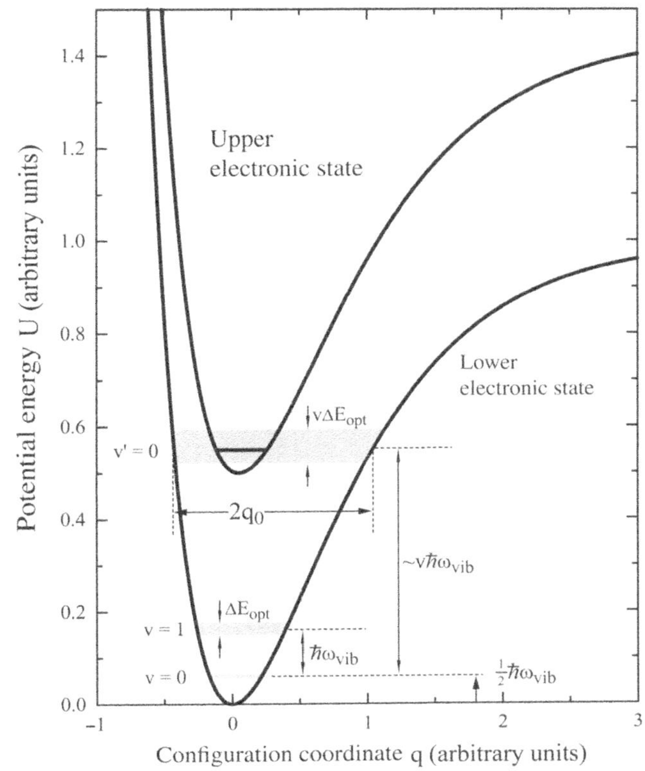

Figure 13.8 Schematic illustration of the configuration potentials for two ligand-surrounded fluorescing ion states. Vibrational states of the lower electronic state configuration potential are indicated by thick gray horizontal bars to demonstrate their broadening because of interaction with the surrounding matrix vicinity. The figure shows a case of energy overlap between a v′ = 0 zero-order vibration of the higher electronic state, and a higher v-order vibration of the lower electronic state.

The mechanism that we consider involves the force exerted by an atomic basis oscillating at a high vibrational order v on the valence electron of the hosted dopant ion. We first calculate the return force acting among the atomic basis parts and thus causing vibrations. We expand the configuration potential in a Taylor series of the q configuration coordinate about the $q = 0$ equilibrium position

$$U(q) = U(0) + \left.\frac{dU}{dq}\right]_0 q + \left.\frac{1}{2}\frac{d^2U}{dq^2}\right]_0 q^2 + \cdots + \frac{1}{(n+1)!}\left.\frac{d^{n+1}U}{dq^{n+1}}\right]_0 q^{n+1} + \cdots, \tag{13.49}$$

where $n = 1, 2, \cdots$. In the following, we denote $U^{(n)} \equiv d^nU/dq^n]_0$. Recall that the first derivative $U^{(1)}$ vanishes at the equilibrium position. The Return force between the two atomic basis parts is

$$-\frac{\mathcal{F}(q)}{\sqrt{\mu^*}} = \frac{dU(q)}{dq} = U^{(2)}q + \cdots + \frac{1}{n!}U^{(n+1)}q^n + \cdots, \tag{13.50}$$

where μ^* is the vibration-reduced effective mass. A simplistic model suggests that $\mu^* = M/4$, where M is the atomic basis mass (see Section 13.1.3). Despite the fact that oscillation frequencies are not identical at different vibrational orders, we may approximately write $q(t) \cong q_0 e^{i\omega_{vib}t}$, where in a classical approximation, $q_0 \cong (2\mathrm{v}\hbar/\omega_{vib})^{1/2}$ is the oscillation amplitude in the v-order vibration (Figure 13.8). Thus, we get

$$-\frac{\mathcal{F}(q(t))}{\sqrt{\mu^*}} \cong U^{(2)} q_0 e^{i\omega_{vib}t} + \cdots + \frac{1}{n!} U^{(n+1)} q_0^{\ n} e^{in\omega_{vib}t} + \cdots. \tag{13.51}$$

The acting force thus exhibits a component oscillating at an $n\omega_{vib}$ frequency in the $n + 1$ expansion order. Under resonance, namely when $\omega_0 = \mathrm{v}\omega_{vib}$, a stimulated emission is enabled, which involves v optical phonons. Such a transition is known as a **multi-phonon transition of order v**.

We now limit ourselves to discussion of that particular return force. Similarly to the discussion preceding Eq. (13.25), the force acting on the valence electron of the fluorescing ion would be $\frac{1}{2}\mathcal{N}_c$ times smaller than the one acting between the two parts of the atomic basis, where $\mathcal{N}_c$ is a number close to that of the different number of atoms comprising an atomic basis. Thus, the force component at a near ω_0 frequency, acting on the electron, would be

$$\mathcal{F}^{(\mathrm{v})}(t) \cong \frac{2\sqrt{\mu^*}}{\mathcal{N}_c \mathrm{v}!} U^{(\mathrm{v}+1)} [q(t)]^{\mathrm{v}}. \tag{13.52}$$

To enable a calculation of the transition probability using known methods and expressions, we convert the $[q(t)]^{\mathrm{v}}$ function into a different one: $C \cdot q_{\mathrm{v}}(t)$, that oscillates at a $\mathrm{v}\omega_{vib}$ angular frequency (same as $[q(t)]^{\mathrm{v}}$), and its amplitude $q_{\mathrm{v}0}$ is identical to the $[q(t)]^{\mathrm{v}}$ function amplitude; a proper selection of C should assure that property. Correspondingly, we choose the normal coordinate of a classical motion belonging to an imaginary configuration potential

$$\overline{U}_{\mathrm{v}}(q_{\mathrm{v}}) \equiv \frac{1}{2} \mathrm{v}^2 U^{(2)} q_{\mathrm{v}}^{\ 2}. \tag{13.53}$$

It may be readily verified that the characteristic frequency belonging to that potential is $\mathrm{v}\omega_{vib}$ - exactly the one coinciding with ω_0. The first-order vibration amplitude is $\mathrm{v}^{-1}(2\mathrm{v}\hbar/\omega_{vib})^{1/2}$. We then choose $C = \mathrm{v}(2\mathrm{v}\hbar/\omega_{vib})^{(\mathrm{v}-1)/2}$. It means performing a factor exchange in Eq. (13.52): $[q(t)]^{\mathrm{v}} \rightarrow \mathrm{v}\left(\frac{2\mathrm{v}\hbar}{\omega_{vib}}\right)^{(\mathrm{v}-1)/2} q_{\mathrm{v}}(t)$, thus getting

$$\mathcal{F}(t) \cong \frac{2\sqrt{\mu^*}}{\mathcal{N}_c \mathrm{v}!} U^{(\mathrm{v}+1)} \mathrm{v} \left(\frac{2\mathrm{v}\hbar}{\omega_{vib}}\right)^{(\mathrm{v}-1)/2} \cdot q_{\mathrm{v}}(t). \tag{13.54}$$

The coupling constant between the imaginary oscillator and the valence electron of the fluorescing ion is then (see Eqs. [13.3] and [13.19] and related text)

$$\eta \cong \frac{2\sqrt{\mu^*}}{\mathcal{N}_c \mathrm{v}!} U^{(\mathrm{v}+1)} \mathrm{v} \left(\frac{2\mathrm{v}\hbar}{\omega_{vib}}\right)^{(\mathrm{v}-1)/2}. \tag{13.55}$$

We further relate to the imaginary potential a coupling with the crystalline vicinity oscillations, which form a $\Delta_{opt}^{eff} = \mathrm{v}\Delta_{opt}$ frequency band, where $\Delta_{opt} = \Delta E_{opt}/\hbar$ is the bandwidth of the $\mathrm{v} = 1$ order of the original U configuration potential (Figure 13.8). $\mathrm{v}\Delta_{opt}$ is indeed the v-order phonon frequency bandwidth of the original configuration potential. We may now use the development performed in section 13.1.3, and generalize Eq. (13.23) to receive

$$\tau_{nr}^{-1}(\omega_0) \cong \boldsymbol{f}\frac{4\pi^2\eta^2\left(\omega_0 - \mathrm{v}\omega_{\boldsymbol{min}}^{opt}\right)^2}{3m\omega_0^2\Delta_{opt}^{eff\;3}}. \tag{13.56}$$

By considerations identical to those used in the text preceding Eq. (13.26), we replace the $\left(\omega_0 - \mathrm{v}\omega_{\boldsymbol{min}}^{opt}\right)^2$ expression by its average - $\frac{1}{3}\left(\Delta_{opt}^{eff}\right)^2$, set η according to Eq. (13.55), and also $\Delta_{opt}^{eff} = \mathrm{v}\Delta_{\mathrm{opt}}$, and get

$$\tau_{nr}^{-1}(\omega_0) \cong \boldsymbol{f}\frac{16\pi^2\mu^*\left[U^{(\mathrm{v}+1)}\right]^2\mathrm{v}(2\mathrm{v}\hbar/\omega_{\mathrm{vib}})^{\mathrm{v}-1}}{9\mathcal{N}_c^2(\mathrm{v}!)^2 m\omega_0^2\Delta_{opt}}. \tag{13.57}$$

A direct use of Eq. (13.48) yields

$$\left[U^{(\mathrm{v}+1)}\right]^2 = \frac{\omega_{vib}^{2(\mathrm{v}+1)}}{(2D)^{\mathrm{v}-1}}\left\{2^{\mathrm{v}}-1\right\}^2. \tag{13.58}$$

Setting it in Eq. (13.57), and in addition also $\mu^* \equiv M/4$, yields

$$\tau_{nr}^{-1}(\omega_0 = \mathrm{v}\omega_{\mathrm{vib}}) \cong \boldsymbol{f}\frac{4\pi^2 M\omega_{\mathrm{vib}}^2}{9\mathcal{N}_c^2 m\Delta_{opt}}\left(\frac{\hbar\omega_{\mathrm{vib}}}{D}\right)^{\mathrm{v}-1}\cdot\frac{\{2^{\mathrm{v}}-1\}^2\mathrm{v}^{\mathrm{v}-2}}{(\mathrm{v}!)^2}. \tag{13.59}$$

Following the consideration preceding Eq. (13.27), namely that $\Delta_{opt} \approx \omega_{\mathrm{vib}}/(3\mathcal{N}_c - 3)$, we get

$$\tau_{nr}^{-1}(\omega_0 = \mathrm{v}\omega_{\mathrm{vib}}) \cong \boldsymbol{f}\frac{\mathcal{N}_c - 1}{\mathcal{N}_c^2}\cdot\frac{4\pi^2 M\omega_{\mathrm{vib}}}{3m}\left(\frac{\hbar\omega_{\mathrm{vib}}}{D}\right)^{\mathrm{v}-1}\cdot\frac{\{2^{\mathrm{v}}-1\}^2\mathrm{v}^{\mathrm{v}-2}}{(\mathrm{v}!)^2}. \tag{13.60}$$

In the discussion preceding Eq. (13.27), we assumed that the optical branches' dispersion spans the entire frequency range from zero to the ω_{co} cutoff frequency. Under such a condition, there always exists a vibrational frequency $\omega_{vib} < \omega_{co}$ per each electronic transition frequency ω_0 that allows an integral number v for a multi-phonon transition. Eq. (13.60) would be thus valid on setting ω_0/v instead of ω_{vib}, with the properly related value for the D quantity. If D is preferably left unchanged, the non-radiative lifetime dependence on the transition energy is continuous. An alternative means for achieving continuousness is to replace ω_{vib} with ω_{co}, and to simultaneously allow the decay order v to assume continuous intermediate values between the integers; namely, we *define* $\mathrm{v} \equiv \omega_0/\omega_{\mathrm{co}}$. Thus, the final expression for the radiative decay rate would be

$$\boxed{\tau_{nr}^{-1}(\omega_0 = \mathrm{v}\omega_{\mathrm{vb}}) \cong \boldsymbol{f}\frac{\mathcal{N}_c - 1}{\mathcal{N}_c^2}\cdot\frac{4\pi^2 M\omega_{\mathrm{vib}}}{3m}\left(\frac{\hbar\omega_{\mathrm{vib}}}{D}\right)^{\mathrm{v}-1}\cdot\frac{\{2^{\mathrm{v}}-1\}^2\mathrm{v}^{\mathrm{v}-2}}{[\Gamma(\mathrm{v}+1)]^2}}, \tag{13.61}$$

for $\mathrm{v} > 1$.

As a reminder, the $\boldsymbol{\Gamma}(x)$ function appearing above in Eq. (13.61) and in the following Eq. (13.62) is an extension of the factorial (!) function for any real positive number. Its formal definition is $\boldsymbol{\Gamma}(x) \equiv \int_0^\infty e^{-\xi}\xi^{x-1}d\xi$; $\boldsymbol{\Gamma}(n) = (n-1)!$. For general analytic purposes, it is practical to use the excellent **Stirling** approximation $\boldsymbol{\Gamma}(x+1) = \sqrt{2\pi}\; x^{x+1/2}e^{-x}\left[1 + \frac{1}{12x} + \frac{1}{288x^2} + \cdots\right]$.

A direct calculation reveals that the function $\boldsymbol{B}(\mathrm{v}) \equiv \{2^{\mathrm{v}} - 1\}^2 \mathrm{v}^{\mathrm{v}-2}[\boldsymbol{\Gamma}(\mathrm{v}+1)]^{-2}$, which appears as a last factor in Eq. (13.61), is inscribed. It is invariably positive; its value increases continually from 1 to ~10 when v varies from 0 to 8, then reduces continually. The critical factor in determining the lifetime dependence of the non-radiative lifetime on v is rather its preceding factor $(\hbar\omega_{co}/D)^{\mathrm{v}}$. Because usually $\hbar\omega_{co} \ll D$, the multi-phonon non-radiative decay rate constant decreases exponentially with the number v of involved phonons. In other words, the non-radiative decay time lengthens as a function of v.

We now join together Eq. (13.61) with Eq. (13.27). We extend the $\mathrm{v} \equiv \omega_0/\omega_{co}$ definition to include the $\omega_0 < \omega_{co}$ case, and get

$$\tau_{nr}^{-1}(\omega_0 = \mathrm{v}\omega_{co}) \cong \begin{cases} f\dfrac{\mathcal{N}_c - 1}{\mathcal{N}_c^2} \cdot \dfrac{4\pi^2 M\omega_{co}}{3m}\mathrm{v}^2 & \textit{for } \mathrm{v} < 1 \\ f\dfrac{\mathcal{N}_c - 1}{\mathcal{N}_c^2} \cdot \dfrac{4\pi^2 M\omega_{co}}{3m}\left(\dfrac{\hbar\omega_{co}}{D}\right)^{\mathrm{v}-1} \cdot \dfrac{\{2^{\mathrm{v}} - 1\}^2 \mathrm{v}^{\mathrm{v}-2}}{[\boldsymbol{\Gamma}(\mathrm{v}+1)]^2} & \textit{for } \mathrm{v} > 1 \end{cases} \quad (13.62)$$

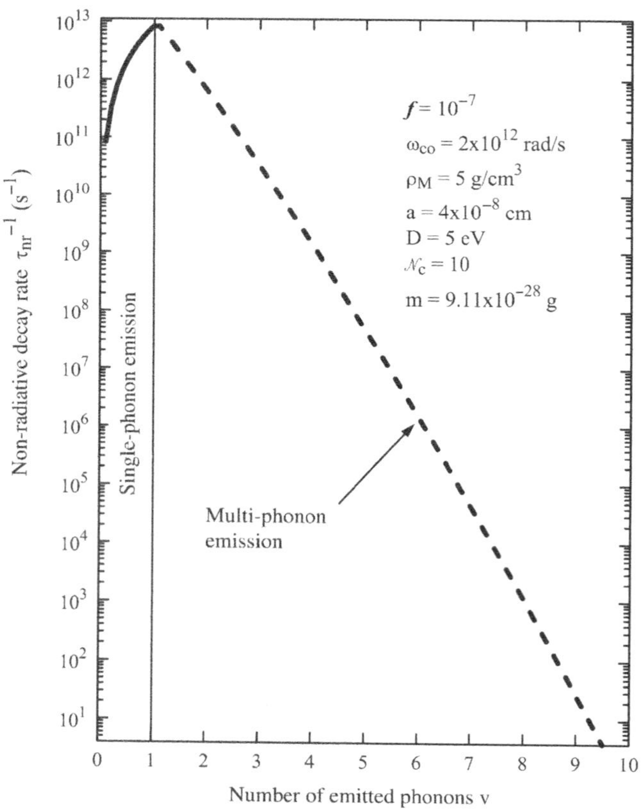

Figure 13.9 Semi-logarithmic plot of the non-radiative decay rate of an ion's excited state as a function of the number of cutoff frequency optical phonons emitted in the process for typical host parameters. Single phonon transitions are indicated by solid line, multi-phonon transitions – seen as a dashed line. Relevant physical and material parameters are indicated in the figure frame.

Notably, the τ_{nr}^{-1} non-radiative transition rate constants exhibit a common value at the v = 1 point, where both assume their maximal value as a function of v.

Following Eq. (13.62), in Figure 13.9 we describe the dependence of the non-radiative decay rate as function of v (the number of cutoff frequency phonons involved in the process). The process parameters are inset in the figure frame; note that the atomic basis mass parameter appearing in Eq. (13.62) may be simply obtained as $M = \rho_M a^3$, where ρ_M is the material mass density and a is the average unit-cell size. In fact, the v < 1 range describes a single phonon decay, and the set-there v describes the ratio between the transition energy and the cutoff phonon energy. It is clearly seen that the non-radiative decay rates are very fast for the single-phonon case, becoming faster with increasing frequency. In the multi-phonon decay process, on the other hand, the decay rates turn exponentially smaller with an increasing number of involved phonons. The practical probability to observe the radiative decay (fluorescence) of ions embedded in a solid matrix requires the non-radiative decay rate to be smaller than the radiative decay rate $\tau_{nr}^{-1} \lesssim \tau_{rad}^{-1}$. Eq. (13.62) indicates that a combination of a small optical cutoff frequency ω_{co} and high ligand dissociation energy D is favorable for observing fluorescence, even for small v values.

A practical empiric expression for multi-phonon transitions v > 1 is often used in the scientific literature, of the form

$$\boxed{\tau_{nr}^{-1} = K\,\boldsymbol{exp}\left\{-\alpha{\cdot}E_g\right\}}, \tag{13.63}$$

where K and α are experimentally adjustable parameters, and $E_g \equiv \hbar\omega_0 = \mathrm{v}\hbar\omega_{co}$ is to be addressed as an "energy gap." It may be readily shown that Eq. (13.62) may be transformed into that particular form by defining

$$K = \boldsymbol{f}\,\frac{\mathcal{N}_c - 1}{\mathcal{N}_c^{\,2}}\cdot\frac{4\pi^2 MD}{3\hbar m}\boldsymbol{B}(\mathrm{v}); \qquad \alpha = \frac{1}{\hbar\omega_{co}}\,\boldsymbol{ln}\left(\frac{D}{\hbar\omega_{co}}\right).$$

The K and α parameters are usually described as constants that characterize properties of the fluorescing ion and of the hosting matrix. Seemingly, the α parameter is independent of the dopant ion properties. Some dependence is, however expected, because the ion presence may slightly affect the vibrational modes of its vicinity (configuration potential and characteristic frequencies). As for the K constant, its value relates mostly to the dopant ion properties, mainly through the transition oscillator strength $\boldsymbol{f}$. It also depends on the transition vibrational order $\mathrm{v} = E_g/\hbar\omega_{co}$ through the $\boldsymbol{B}(\mathrm{v})$ function.

Figure 13.10 displays the experimental results of non-radiative decay rates measured at 80K for five triply ionized rare-earth atoms: europium Eu^{3+}, holmium Ho^{3+}, erbium Er^{3+}, neodymium Nd^{3+}, and thulium Tm^{3+}. The ions were hosted in an yttrium ortho-aluminate $YAlO_3$ single crystal. That crystal is orthorhombic, belonging to the #62 space-group D_{2h}^{16} $(P2_1/n\,2_1/m\,2_1/a)$. The unit cell is primitive, and each lattice point comprises four formula units. Thus, $N_c = 20$ and $M = \rho_M a^3 = 8.76 \times 10^{-22}$g. The four yttrium atoms occupy a Wyckoff c- site of a C_S symmetry, each surrounded by four close oxygen atoms. All said foreign ions replace yttrium atoms in their sites. The optical vibrational bands' cutoff frequency is $\omega_{co} \cong 750\ \mathrm{cm}^{-1}$.

The decay rates in Figure 13.10 are plotted as functions of the "energy gap", namely the separation energy between the initial state indicated in the figure frame and its closest lower energy state. Despite their belonging to ions of different materials, the results display a well-organized dependence, except for a considerable deviation for the initial $^5D_1$ state of europium Eu^{3+} ion, and some smaller deviations for the initial $^3F_4$ state of thulium Tm^{3+} and $^5I_7$ of holmium Ho^{3+} ions; as to the latter two cases, the experimental errors are also quite large. In view of the mostly well-organized dependence, it is reasonable to analyze the results according to

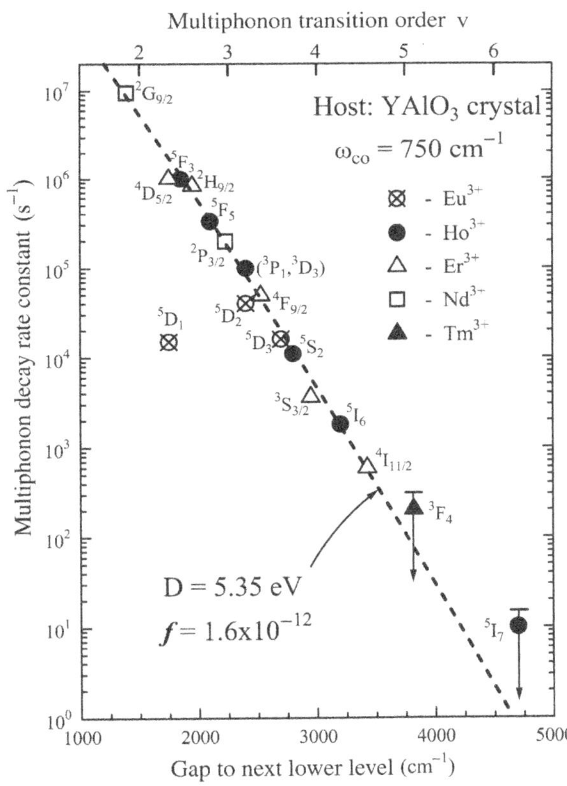

Figure 13.10 Semi-logarithmic plot of experimental non-radiative decay rates as a function of the energy gap between different transition-involved electronic states in triply ionized atoms of the europium Eu^{3+}, holmium Ho^{3+}, erbium Er^{3+}, neodymium Nd^{3+}, and thulium Tm^{3+} embedded in an yttrium ortho-aluminate $YAlO_3$ **single crystal**. The dashed line is the results fit to Eq. (13.62). The fit parameters are inset in the figure frame.

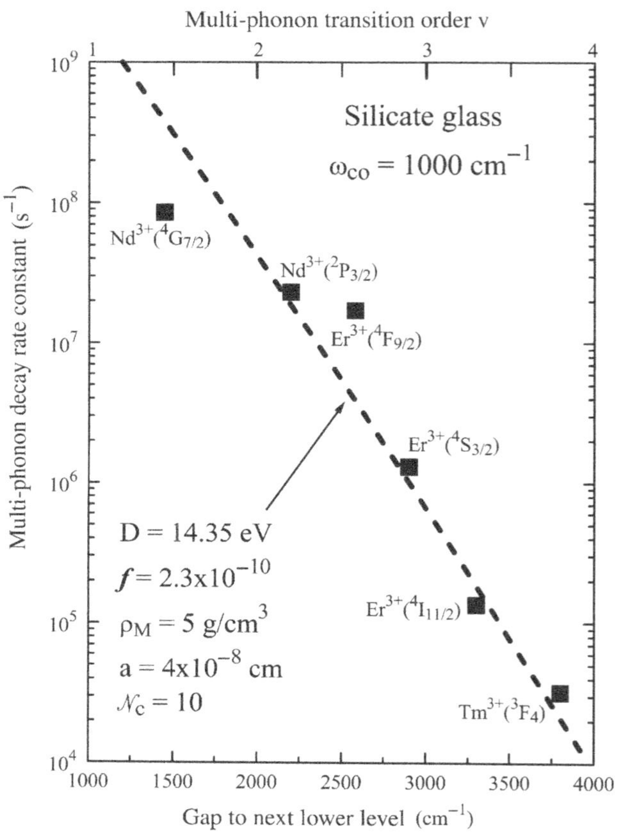

Figure 13.11 Semi-logarithmic plot of experimental non-radiative decay rates as a function of the energy gap between different electronic states in triply ionized atoms of the neodymium Nd^{3+}, erbium Er^{3+}, and thulium Tm^{3+} embedded in a silicate glass. The dashed line is the results fit to Eq. (13.62). The fit parameters are inset in the figure frame.

Eq. (13.62). The fit curve by that equation is plotted for $D = 5.35$ eV and $\boldsymbol{f} = 1.6 \times 10^{-12}$. Note that we use different energy units for different quantities: cm^{-1} for the transition energy, and electron-volt (eV) for the dissociation energy; that is a common usage in the scientific literature.

Scattering about the fit line in Figure 13.10 may be attributed to differences in the transitions oscillator strengths. The reasonably good fit for most data suggests that all indicated transitions exhibit oscillator strengths of a similar order – $\boldsymbol{f} \approx 1.6 \times 10^{-12}$. It should be emphasized that the D parameter may not mean strictly a "dissociation energy" of the free ion ligand floating in empty space; the ratio $\hbar\omega_{co}/D$ in the Morse potential (see Eq. [7.17]) expresses the potential anharmonicity measure; in a strictly harmonic potential, this ratio vanishes. Equations (13.62) and (13.63) show that when the "dissociation energy" D grows large compared with the $\hbar\omega_{co}$ phonons' cutoff energy, the $\tau_{nr}{}^{-1}$ non-radiative rate reduces steeply. Dielectric materials of such quality are desired as hosts for different ions when use of the ions' fluorescence is desired, especially in the infrared region.

Importantly, the theoretical conclusions obtained by modeling the non-radiative effects in a single crystal are valid for glasses as well. That is not surprising, because optical glass lattice oscillation bands are similar to those in the single-crystal phase, although considerably broadened. In Figure 13.11, we present experimental results of non-radiative decay rates measured in three triply-ionized atoms of the rare-earth elements neodymium Nd^{3+}, erbium Er^{3+}, and thulium Tm^{3+}. All ions were embedded in a silicate glass whose main component (67 mol.%) was silicon dioxide SiO_2, and smaller amounts, 15 mol.% and 18 mol.% sodium oxide Na_2O and barium oxide BaO, respectively. The decay rates are described as functions of the "energy gap", namely the transition energy from the initial state indicated in the figure to its closest lower electronic state. The cutoff optical oscillations in such a glass are $\omega_{co} \cong 1{,}000\ cm^{-1}$. Like in the case described in Figure 13.10, the results follow a well-organized dependence, despite belonging to ions of different materials. Indeed, there is a fundamental difficulty in defining a ligand surrounding the ion, especially since the host material is of a glassy phase. It is, however, possible to analyze the results according to Eq. (13.62). A fit line is accordingly presented in the figure for $D = 14.35$ eV, $\boldsymbol{f} = 2.3 \times 10^{-10}$, $\rho_M = 5\ g/cm^3$, $a = 4 \times 10^{-8}$ cm, and $\mathcal{N}_c = 10$. Here as well, results scattering about the fit line may be attributed to differences in the oscillator strengths of the different transitions. Still, because of the fit quality, it may be concluded that all oscillation strengths are of the same order of magnitude - $\boldsymbol{f} \approx 10^{-10}$.

In Figure 13.12, we provide an additional example that rather illustrates some limitations of the developed theoretical theory. We present the experimental results of non-radiative decay rates measured for different triply ionized ions belonging to three different rare earth elements: neodymium Nd^{3+}, holmium Ho^{3+}, and erbium Er^{3+}. The ions were embedded in a lithium-yttrium fluoride $LiYF_4$ single crystal. The crystal is body-centered tetragonal, belonging to the Scheelite family of crystals, all of the #88 C_{4h}^6 $(I4_1/a)$ space group (see Section 11.5 Table [**i**]-11). The cutoff frequency of optical phonons in the crystal is $\omega_{co} \cong 570\ \text{cm}^{-1}$. Each lattice point comprises two formula units. Thus, $\mathcal{N}_c = 12$, and $\rho_M a^3 = 5.75 \times 10^{-22}$ g. Each hosted ion replaces a "legitimate" lattice yttrium Y^{3+} ion in its Wyckoff a-site of S_4 symmetry. The decay rates are described as functions of their "energy gap", namely the transition energy from the indicated initial state to the closest lower electronic state.

The results for neodymium Nd^{3+} were obtained for a range of low energy gaps ($400 - 2250\ \text{cm}^{-1}$) and for the other ions for a range of higher energy gaps ($1850 - 5000\ \text{cm}^{-1}$), with some small overlap between the ranges. Observation of the energy gap dependence suggests a description as a sum of two quasi-exponential decays – one dominating the $400 - 2250\ \text{cm}^{-1}$ range, the other dominating the $1850 - 5000\ \text{cm}^{-1}$ range. The first may assume a fit to Eq. (13.62) with a $D = 45$ eV dissociation energy and $\boldsymbol{f} = 7 \times 10^{-9}$ oscillator strength; the second may assume a fit to Eq. (13.62) with grossly different parameters: $D = 0.75$ eV and $\boldsymbol{f} = 6 \times 10^{-14}$. Because especially the second range belongs to different ions, it is not possible to decide whether a particular ion presence is the one affecting the "dissociation energy" parameter, or that the observed dependence is intrinsic to the host material, and is relevant for all dopant ions. Because the ratio $\hbar\omega_{co}/D$ in the Morse potential expresses the potential anharmonicity, it is perhaps possible that a considerable deviation from harmonicity of the vibrational modes occurs at higher energies, namely that the Morse potential is not adequate to describe the true shape of the configuration potential at high energies. The analytic result may be expressed in changes of the D parameter of the Morse potential. On the other hand, it might be possible that a different mechanism is involved, like a transition to a higher vibrational state of the lower electronic state, as mentioned in the text that follows Figure 13.8.

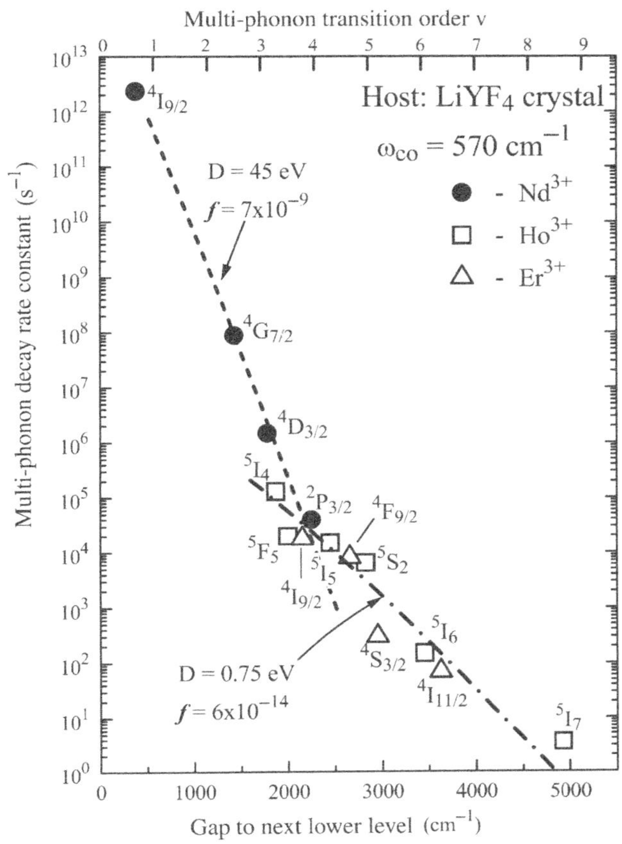

Figure 13.12 Semi-logarithmic plot of experimental non-radiative decay rates as functions of the energy gap between different electronic states in triply ionized atoms of the neodymium Nd^{3+}, holmium Ho^{3+}, and erbium Er^{3+}, embedded in a $LiYF_4$ **single crystal.** The dashed and dash-dotted lines are results fit to Eq. (13.62) in the lower and higher energy-gap regions, respectively. The fit parameters are inset in the figure frame. Other data are indicated in the text.

Experiments show that non-radiative transition rates are enhanced by increasing temperature. This effect is interpreted as being the result of enhanced stimulated phonon emissions due to thermal enhancement of the phonon field. Recall that the average energy of an oscillation at ω angular frequency is given by $\hbar\omega\left(\langle \boldsymbol{n}(\omega)\rangle_T + \frac{1}{2}\right)$, where $\langle \boldsymbol{n}(\omega)\rangle_T = [\boldsymbol{exp}(\hbar\omega/k_B T) - 1]^{-1}$ is the phonon occupation number (see Eqs. [2.26] and [5.27]). From a classical perspective, it means that the squared classical oscillation amplitude is proportional to $(\langle \boldsymbol{n}(\omega)\rangle_T + \frac{1}{2})$. On the other hand, the stimulated radiation rate is given by $\tau^{-1}(T=0)(\langle \boldsymbol{n}(\omega)\rangle_T + 1)$; (see Eqs. [2.44] and [13.15]). Thus, for the issue of stimulated decay, the squared classical amplitude relates to $(\langle \boldsymbol{n}(\omega)\rangle_T + 1)$, and should be considered as an "effective occupation number" for that purpose.

Eq. (13.52) shows that the force component due to atomic basis oscillations acting on the valence electron at the $v\omega_{co}$ frequency is proportional to the v-power of the classical oscillation. Thus, the probability for a multi-phonon transition would be proportional to the 2v- power of the classical amplitude, namely to the $v = E_g/\hbar\omega_{co}$ power of the generalized occupation number. The temperature-dependent non-radiative decay rate would thus be given by

$$\boxed{\tau_{nr}{}^{-1}(T) = \tau_{nr}{}^{-1}(0)\left[\frac{1}{\boldsymbol{exp}\,(\hbar\omega_{co}/k_B T) - 1} + 1\right]^{E_g/\hbar\omega_{co}}}, \tag{13.64}$$

where $\tau_{nr}{}^{-1}(0)$ is the multi-phonon non-radiative decay rate at the absolute zero temperature.

Figure 13.13 presents experimental results of non-radiative decay transitions as functions of temperature for states of three different rare-earth ions hosted in the $YAlO_3$ yttrium ortho-aluminate single crystal, which were also shown in Figure 13.10. The fit curves were made according to Eq. (13.64), all with a

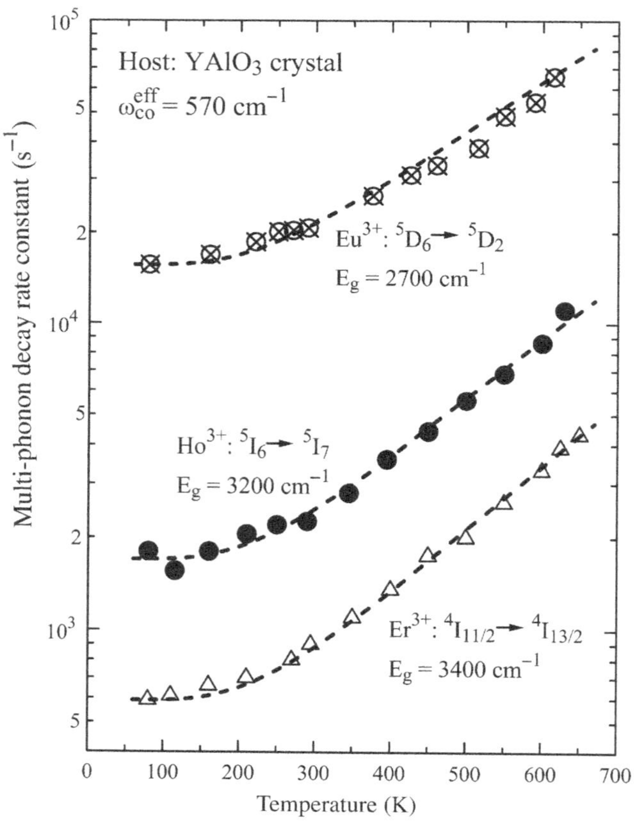

Figure 13.13 Semi-logarithmic plot of experimental non-radiative decay rates as functions of temperature for different states of ion hosts in an yttrium ortho aluminate $YAlO_3$ single crystal. The theoretical fits are calculated according to Eq. (13.64) for an ω_{co}^{eff} = 570 cm^{-1} effective cutoff frequency, instead of the known 750 cm^{-1} one used in Figure 13.10. The specific transitions and energy gaps are noted beside each curve.

$\omega_{co}^{eff} = 570\ \text{cm}^{-1}$ effective cutoff frequency, which is different from the known one of 750 cm^{-1} (also noted in Figure 13.10). That deviation may suggest that the effective cutoff frequency actually relates to the lattice vibration that exhibits the largest coupling with the valence electron of the dopant ion. If that particular frequency is used in fitting the results shown in Figure 13.10 by Eq. (13.62), the revised fit parameters would be $D = 1.45$ eV and $\boldsymbol{f} \approx 3.8 \times 10^{-12}$.

Recall now that the radiative lifetime and oscillator strength are reciprocally proportional. Using Eqs. (2.56) and (2.59) and defining $\nu_0 = E_g/2\pi\hbar$, it may be readily shown that

$$\boldsymbol{f} = \frac{3mc^3\hbar^2}{2e^2n^3E_g{}^2}\tau_{rad}{}^{-1}. \tag{13.65}$$

Setting in Eq. (13.62), one gets

$$\tau_{nr}{}^{-1}(\omega_0 = \text{v}\omega_{co}) \cong \begin{cases} \tau_{rad}{}^{-1}\cdot\dfrac{\hbar c}{e^2}\cdot\dfrac{\mathcal{N}_c-1}{\mathcal{N}_c{}^2}\cdot\dfrac{2\pi^2}{n^3}\dfrac{[Mc^2]\hbar\omega_{co}}{E_g{}^2}\text{v}^2 & for\ \text{v} < 1 \\ \tau_{rad}{}^{-1}\cdot\dfrac{\hbar c}{e^2}\cdot\dfrac{\mathcal{N}_c-1}{\mathcal{N}_c{}^2}\cdot\dfrac{2\pi^2}{n^3}\dfrac{[Mc^2]\hbar\omega_{co}}{E_g{}^2}\left(\dfrac{\hbar\omega_{co}}{D}\right)^{\text{v}-1}\boldsymbol{B}(\text{v}) & for\ \text{v} > 1 \end{cases}. \tag{13.66}$$

For brevity, we define $C_{nr} \equiv \frac{\hbar c}{e^2}\cdot\frac{\mathcal{N}_c-1}{\mathcal{N}_c{}^2}\cdot\frac{2\pi^2}{n^3}\frac{[Mc^2]\hbar\omega_{co}}{E_g{}^2}\left(\frac{\hbar\omega_{co}}{D}\right)^{\text{v-1}}\boldsymbol{B}(\text{v})$ for $\text{v} > 1$, to be addressed as the "non-radiative factor". It is seen to be inversely proportional to the squared energy gap. Furthermore, because always $\hbar\omega_{co} << D$, the non-radiative factor reduces steeply with increasing v.

To conclude this chapter, we point out an effect that causes fluorescence quenching of ions embedded in a dielectric medium through extrinsic properties, like structural faults of the host matrix, and/or impurities residing close to the fluorescing ion. An example is provided in Figure 13.14. Such entities may cause enhanced non-radiative transitions through the fault/impurity of electronic states, as illustrated in Figure 13.14. The decay processes may be very efficient if the said states' separations are relatively small, allowing non-radiative decay transitions, each involving only a small number of phonons. Practically, full quenching may be obtained if the fault/impurity densities are sufficiently high. The presence of structural faults/impurities may be expressed by a variety of effects; among them is rather a lengthening of the fluorescence lifetime by temporarily trapping the excitation energy at an extrinsic state of a long lifetime, and returning the energy after a while to the fluorescing ion via thermal activation.

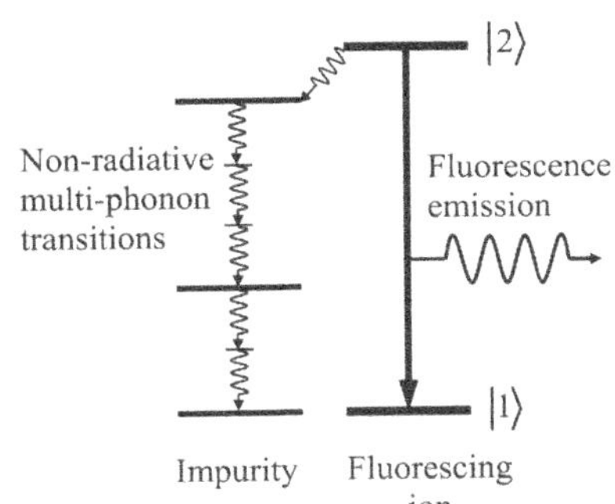

Figure 13.14 Energy scheme of a fluorescing ion and a nearby impurity, allowing fluorescence quenching due to fast, non-radiative multi-phonon transitions through the impurity energy levels. The radiative transition is marked by a solid arrow, the non-radiative ones by wiggled arrows.

Exercises for Chapter 13

13.1 The text preceding Eq. (13.7) presents the following expression for the energy flux per unit angular frequency incident on an atom, carried by acoustic waves (energy per unit area per unit angular frequency): $\mathbf{F}(\omega_0) = \frac{1}{2}\frac{\omega_0{}^2 v_S}{V_C}|\boldsymbol{Q}(\omega_0)|^2$, where V_C is the primitive unit cell volume. Prove this equation by presenting an expression analogous to the Poynting vector appropriate to the case of acoustic waves (see the text preceding Eq. [1.3]).

13.2 Eq. (13.15) describes the temperature dependence of the non-radiative decay rate. Show the expression structure in the $\hbar\omega_0 << k_B T$ limit (very high temperatures).

13.3 Calculate an expression for conditions where the radiative decay rate is large compared with the non-radiative rate. Use it to assess the limit of energy gaps for which it would be practical to observe the fluorescence of an ion embedded in the silicate glass, whose parameters are indicated in Figure 13.11.

14 Basic Acquaintance with the Laser and Its Components

14.1 General Description

A **laser** is an instrument that emits light based on the stimulated light emission process described in Section 2.4. The name "LASER" is an acronym of the responsible physical process "Light Amplification by Stimulated Emission of Radiation". The equipment contains a volume, producing intensity gain to crossing light over a limited spectral range. A positive gain is obtained by generating a population inversion between two electronic states. This generation process is called **pumping**. The most common means of pumping is by external illumination. When the intended amplifying medium is a gas, it may also be accomplished through electrical discharge or a chemical reaction. The geometrical shape of the amplifying medium is such that one of its dimensions, L_g, is longer than the others. As a result, the light amplification along that dimension, $\boldsymbol{exp}\ [\Gamma(\nu)L_g]$, where $\Gamma(\nu)$ is the gain at the ν frequency (see Eq. [2.54]), is orders of magnitude larger than along other dimensions. An optical **resonance cavity** is constructed for the long dimension by placing two parallel mirrors to reflect light propagating in that direction. That way, the effective length of the amplifying medium on the long dimension is doubled; each beam traverses twice, back and forth, until returning to the same place ($\ell_{eff} = 2L_g$). One of the mirrors, the **front mirror**, is only partially reflecting (hence partially transmitting), thus some of the amplified light is emitted out of the cavity; that is the **laser output**. For that reason, the front mirror is often referred to as the **output-coupling mirror**. The other mirror is called the **back mirror**. Figure 14.1 presents a schematic layout of a simple laser comprising the above indicated components: an amplifying medium described as a circular rod, a pump source presented as an electric discharge tube, and mirrors forming the optical cavity.

A positive effective gain is obtained in a frequency range where the $\Gamma(\nu)$ gain (Eq. [2.54]) is larger than a threshold value determined by the medium and cavity losses. There are various loss mechanisms that will be discussed later in relation to different contexts. The spectral output shape of the shown basic laser system is not identical to the $\Gamma(\nu)$ gain shape; it is narrower. It is approximately given by

$$I(\nu) = I_0 \cdot g(\nu) \cdot \boldsymbol{exp}\left[(\Gamma(\nu) - \alpha(\nu)) \cdot 2L_g\right], \tag{14.1}$$

where I_0 is a constant of intensity units (erg/cm$^2 \cdot$ s), $g(\nu)$ is the shape function of the spontaneous emission band (fluorescence), and $\alpha(\nu)$ is the frequency ν-dependent loss factor. In Section 14.7, we discuss the form of loss presentation in some more detail. As noted, during laser action, the relation $\Gamma(\nu) \geq \alpha(\nu)$ over a certain frequency range always exists. The variable in the exponential factor function of Eq. (14.1) is thus positive in that range. Gain calculation is not trivial, because the lasing process continually depletes the excited state's concentration, thus causing gain reduction. In the case of pulsed laser operations, the gain is time dependent, and thus the emission bandwidth is, too. Under continuous laser operations (CW – acronym of "continuous wave"), the excited state's depletion brings about a steady-state, where the gain is virtually identical with the losses at the lasing wavelength, an effect called **gain-clamping**. On the two

Physics, Optics, and Spectroscopy of Materials, First Edition. Zeev Burshtein.

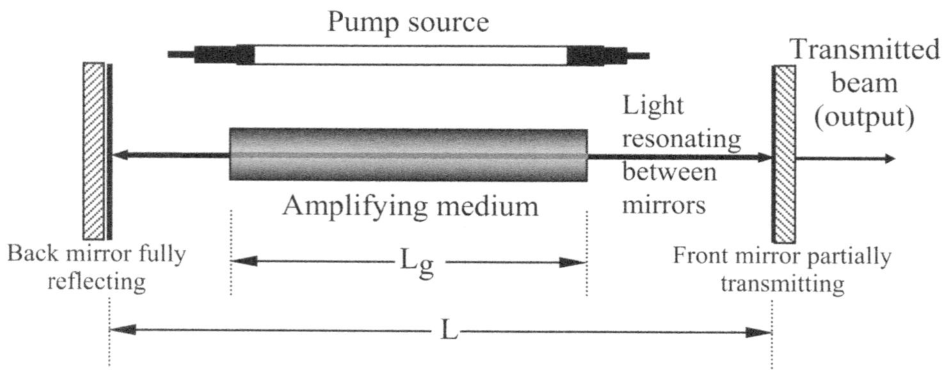

Figure 14.1 Schematic illustration of a basic laser system.

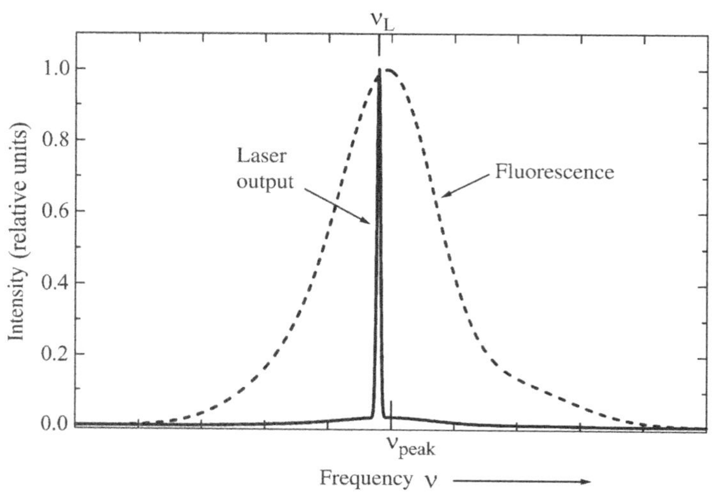

Figure 14.2 Schematic illustration of an amplifying medium fluorescence emission spectrum in a laser cavity and the laser output spectrum. The spectra peak values are normalized to a unity. The laser peak position ν_L is slightly shifted compared with the fluorescence peak position ν_{peak}, mostly because of some frequency dependence of the losses.

frequency sides of the lasing wavelength, the loss is larger than the gain. These conditions remain constant over time, as does the spectral width of the laser emission. We shall not elaborate further on that issue.

A schematic description of a usual laser emission spectrum compared with the fluorescence emission spectrum of the amplifying medium is presented in Figure 14.2. We shall see later, that it is possible to achieve a very narrow emission band by shortening the resonance cavity length L and adding frequency-selecting components within the cavity: a **prism**, a **reflection grating**, or an **etalon**.

Use of the simple basic cavity construction shown in Figure 14.1 is rarely satisfactory. A sophisticated construction, and the inclusion of different inside components, allows for the utilization of the resulting effects for various goals, like output wavelength tuning, polarization state selection, control of beam quality, and conversion of a CW output into a pulsed one. In the following section, we provide a survey of various such components, and analyze their functions.

14.2 The Optical Cavity

We consider first the function of a simple cavity comprising two plane mirrors with an L optical distance (Figure 14.1). Under steady-state conditions, standing electromagnetic waves exhibiting either junctions or nubs on the mirror's surfaces may exist for metallic or dielectric mirrors, respectively (dielectric mirrors are not a subject addressed in the present book). Allowed frequencies are given by $\nu_j = (c/2L)j$ for $j = 1,\ 2,\ 3,\ \cdots,\ \infty$, where c is the speed of light constant (see also Section 2.2). Waves belonging to those frequencies are called **longitudinal oscillation modes**. The difference between two adjacent frequencies is $\Delta\nu_j = \nu_{j+1} - \nu_j = c/2L$. In the optical region, and when the distance L between the mirrors is macroscopic, quite a large density of frequencies is obtained, namely a tiny difference between two adjacent allowed frequencies. As an example, we take $L = 15$ cm, and $c = 3 \times 10^{10}$ cm/s, and get $\Delta\nu_j = 3 \times 10^{10}/(2 \times 15) = 10^9$ Hz. That difference is indeed tiny compared with a typical ~5×10^{14} Hz frequency of a red light at 600 nm wavelength, which is itself small compared with the laser output bandwidth of a typical laser as described in Figure 14.2. A dye laser whose optical cavity comprises only mirrors could emit a beam whose spectral width is ~0.1 nm, namely $\delta\nu$~8×10^{10} Hz, meaning about 80 longitudinal modes. In other words, the laser output spectrum of Figure 14.2 is, in fact, an *envelope* of dozens of lines of much narrower width. If the laser cavity is designed sufficiently short, it is possible to impose lasing at a single longitudinal mode. The width of such a single mode is further addressed in section 14.5.2 during the discussion about the Fabry-Pérot etalon.

14.3 The Prism

Figure 14.3 presents the fundamental structure of a laser where a frequency-tuning triangular prism is inserted inside the resonance cavity. The prism resolves the crossing light into its component frequencies by changing their propagation orientation into different angles. The back mirror is placed perpendicularly to the frequency component desired for amplification. The other components are correspondingly reflected toward regions of lesser amplification. The effective gain length thus becomes frequency dependent: $\ell_{eff} = \ell_{eff}(\nu)$. There is then a drastic narrowing of the output beam frequency bandwidth, as well as the provision of an option for selecting different output central wavelengths (**tuning**) by rotating either the back mirror, or the prism, about an axis parallel to the prism axis (normal to the page). Tuning is possible within the gain spectral range, whose shape is governed by the $g(\nu)$ function (Eq. [2.54]), except for loss factors not yet detailed.

The reason for light dispersion on crossing the prism is the refractive index dependence on the light frequency. In Section 14.6, we elaborate further on light reflection and refraction in dielectric materials. Recall that we have already used some background acquaintance with the essence of a refractive index in developing Planck's equation for a black-body radiation (see Section 2.2), and Snell's refraction law in developing de Broglie's equation (see Section 3.1). The reader is also referred to Sections 10.1 and 10.2. In the following development, we convert expressions to use the wavelength instead of the frequency as the dispersion function variable. That usage method is a more common practice in related technological

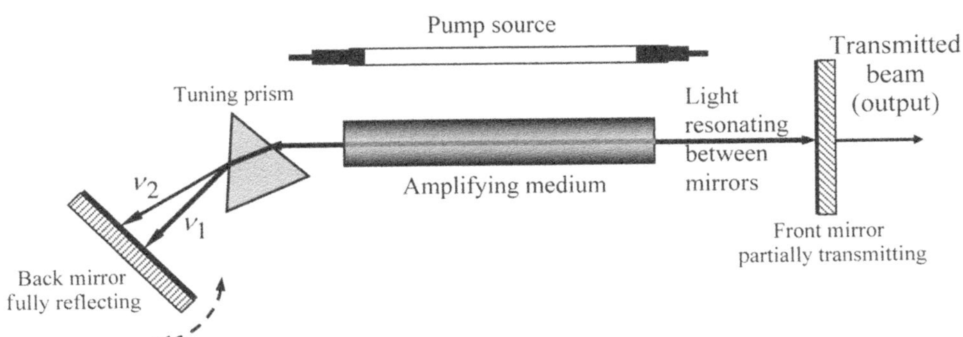

Figure 14.3 Schematic illustration of a laser construction with an added prism used for tuning the output frequency. Tuning is performed by rotating the back mirror about an axis perpendicular to the page plane. The amplified frequency is ν_1, which belongs to the beam reflected perpendicularly to the back mirror.

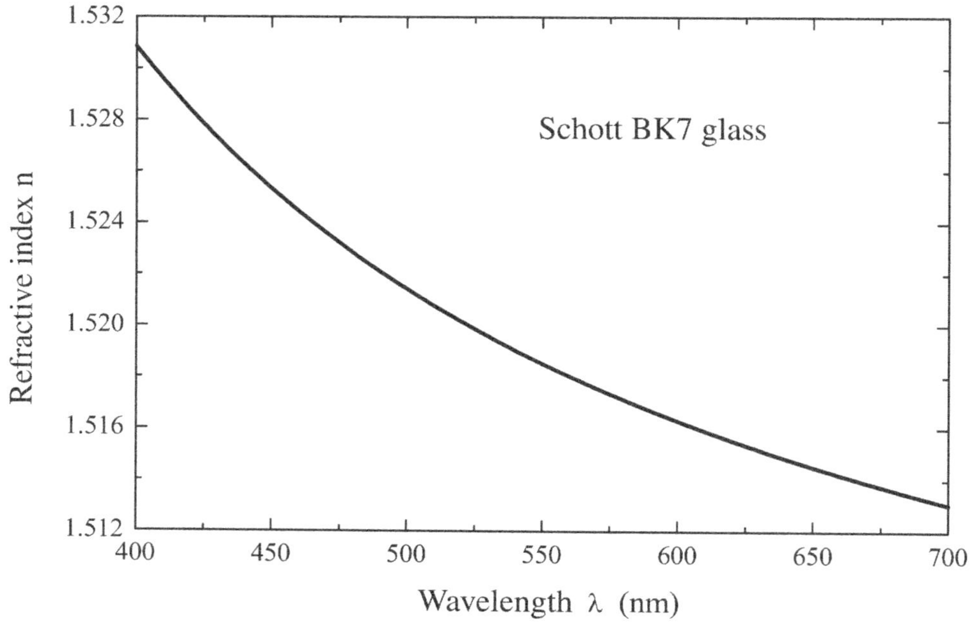

Figure 14.4 Refractive index versus wavelength in the commercial Schott BK7 glass.

literature. The refractive index n's dependence on the wavelength λ (dispersion equation; dispersion curve) is a material dependent property. In section 9.5.4.2, and specifically in Figure 9.13, we presented a dispersion curve analysis of the extraordinary refractive index in a sapphire crystal. Recall that technical equations describing dispersion curves are called **Sellmeier equations**. For example, the Sellmeier equation for the BK7 Schott-brand optical glass, which is very common for optical uses in the $0.35 - 2.0\ \mu\text{m}$ range, is given by

$$n^2 = 1 + \frac{B_1\lambda^2\,[\mu\text{m}]}{\lambda^2\,[\mu\text{m}] - C_1} + \frac{B_2\lambda^2\,[\mu\text{m}]}{\lambda^2\,[\mu\text{m}] - C_2} + \frac{B_3\lambda^2\,[\mu\text{m}]}{\lambda^2\,[\mu\text{m}] - C_3}, \tag{14.2}$$

where the "B" and "C" constants are

$$B_1 = 1.03961212; \quad B_2 = 2.31792344; \quad B_3 = 1.01046945$$
$$C_1 = 6.0006986 \times 10^{-3}; \quad C_2 = 2.00179144 \times 10^{-2}; \quad C_3 = 1.03560653 \times 10^{2}.$$

Figure 14.4 shows the dispersion function of the said glass in the visible region ($400 - 700\ \text{nm}$). The refractive index reduces with increasing wavelength, also exhibiting a reducing slope. The average slope in the visible range is $dn/d\lambda \equiv -7 \times 10^{-5}\ \text{nm}^{-1}$. Certain glasses may exhibit a local increase of refractive index with increasing wavelength. These cases are quite rare in the visible and near-visible infrared, but are more often apparent in the far infrared ($\lambda \approx 5{,}000 - 10{,}000\ \text{nm}$).

We now describe certain properties of prisms that are important for their function as tuning elements inside a laser resonance cavity.

14.3.1 A Prism Minimum Deviation Arrangement

We consider a special refraction case of a monochromatic light beam incident on a prism edge surface, where the incidence angle is equal to the exiting angle on the exit edge surface. That particular situation is

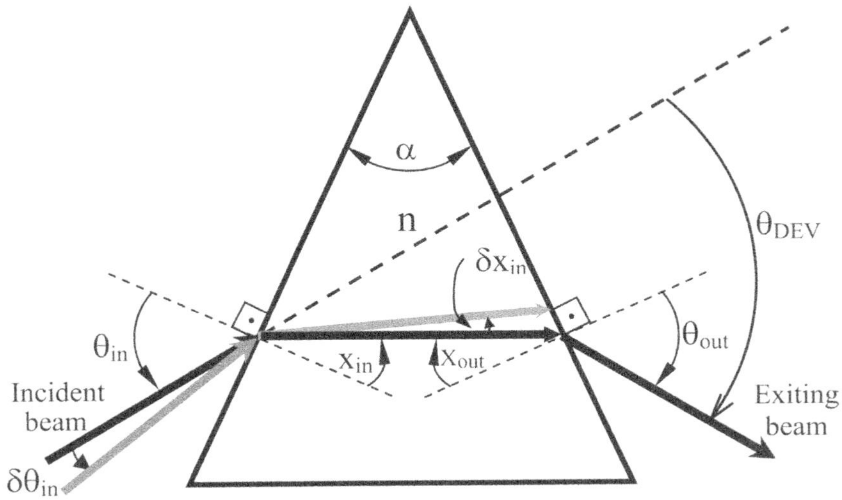

Figure 14.5 Schematic illustration of a beam trace on crossing a dielectric prism of n refractive index under symmetry between incidence and exit angles. These conditions also bring about the lowest deviation angle of the beam relative to its original orientation.

shown in Figure 14.5. It is a symmetric geometric case, where the beam crossing inside the prism portrays the basis of an isosceles triangle in relation to the prism vertex. It may be readily verified that the refraction angle of the incident beam x_{in} and the incidence angle of the beam inside the prism toward exiting x_{out}, both equal half the prism vertex angle; namely $x_{out} = x_{in} = \alpha/2$. By Snell's law, the incidence and refraction angles satisfy the relations

$$\boxed{n = \frac{\sin \theta_{in}}{\sin x_{in}} = \frac{\sin \theta_{out}}{\sin x_{out}} = \frac{\sin \theta_{in}}{\sin(\alpha/2)}} \,. \tag{14.3}$$

The deviation angle of the beam from its original propagation orientation θ_{DEV} is given by

$$\theta_{DEV} = (\theta_{in} - x_{in}) + (\theta_{out} - x_{out}) = (\theta_{in} + \theta_{out}) - \alpha \underset{\substack{Symmetric\\case}}{=} 2\theta_{in} - \alpha \,. \tag{14.4}$$

We may validate that the symmetric case conditions are also those that form the minimal beam deviation angle. Assume an infinitesimal $\delta\theta_{in}$ increase of the incidence angle; it may be achieved by an infinitesimal rotation of the prism. The transmitted refracted beam inside the prism increases by $\delta x_{in} = \frac{1}{n}\frac{\cos \theta_{in}}{\cos x_{in}}\delta\theta_{in}$. As a result, the in-prism incidence angle changes by $\delta x_{out} = -\delta x_{in}$. The exit angle thus changes by $\delta x_{out} = -\delta x_{in}$:

$$\delta\theta_{out} = \frac{n\cdot \cos x_{out}}{\cos \theta_{out}}\delta x_{out} = -\frac{n\cdot \cos x_{out}}{\cos \theta_{out}}\delta x_{in} \underset{\substack{Symmetric\\case}}{=} -\delta\theta_{in} \,. \tag{14.5}$$

Therefore, by Eq. (14.4), the change in the deviation angle satisfies $d\theta_{DEV}/d\theta_{in} = 0$. That means that the symmetric case is an extremum point of the deviation angle as a function of the incidence angle. In a more detailed analysis (second derivative), it may be shown that it is indeed a minimum.

The latter property suggests an experimental method of measuring the material refractive index: one measures the incidence angle θ_{in} that yields a minimum deviation, and uses Eq. (14.3). A further important and useful property of the symmetric case (namely of the minimum deviation) is that the geometric width of an incident beam remains the same on exiting the prism.

14.3.2 Light Dispersion in a Prism

Consider a polychromatic parallel beam incident on a dielectric prism postured to achieve a minimum deviation for one of the wavelength components λ_0. Because of the refractive index wavelength's dependence, different wavelengths refract at different angles on exiting the prism. This effect is called **dispersion**. We now calculate the wavelength dispersion of the different participating wavelengths relative to the angle deviation of λ_0, assuming them to be sufficiently close. The means would be Snell's equation (Eq. [14.3])

$$\sin x_{in} = \frac{\sin \theta_{in}}{n(\lambda)} . \tag{14.6}$$

The incidence angle θ_{in} is independent of the λ wavelength; the refracted beam propagation, however, changes. Therefore

$$\frac{dx_{in}}{d\lambda} = -\frac{\sin \theta_{in}}{n^2(\lambda_0)\cdot \cos x_{in}} \left.\frac{dn}{d\lambda}\right]_{\lambda_0} .$$

By Snell's equation, the exit angle is $\sin \theta_{out} = n(\lambda) \cdot \sin x_{out}$. In the present case, both incidence angle and refractive index change. Then

$$\frac{d\theta_{out}}{d\lambda} = \frac{1}{\cos \theta_{out}} \left[\sin x_{out} \left.\frac{dn}{d\lambda}\right]_{\lambda_0} + n(\lambda_0) \cos x_{out} \frac{dx_{out}}{d\lambda} \right] .$$

Setting the relations $\frac{dx_{out}}{d\lambda} = -\frac{dx_{in}}{d\lambda}$ and $\theta_{out} = \theta_{in}$, and the above result, we get that in the vicinity of the minimum deviation (Eq. [14.3]), the angular dispersion is

$$\boxed{\frac{d\theta_{out}}{d\lambda} = \frac{2 \sin \alpha/2}{\sqrt{1 - n^2(\lambda_0) \sin^2 \alpha/2}} \left.\frac{dn}{d\lambda}\right]_{\lambda_0}} . \tag{14.7}$$

Considering the last equation, it turns that for a BK7 glass whose refractive index dispersion curve is given in Figure 14.4, we get $d\theta_{out}/d\lambda]_{\lambda_0} \cong -7 \times 10^{-5}$ rad/nm $\cong 0.25$ arcmin/nm. If the practical gain range of the laser is about 50 nm, tuning the laser across the entire range requires a change of the back mirror angle by about 12 arcmin (Figure 14.3). To increase the angular resolution, sometimes a set of two consecutive prisms are used in the laser cavity.

14.3.3 Prism Wavelength Resolution

A prism could resolve between two $\delta\lambda$ differing colors if their refracted angles separated by $\delta\theta_R$ (calculated by Eq. [14.7]) are larger than the $\delta\theta_D \equiv \lambda/D$ natural angular beam divergence, where D is the cross-sectional width of the beam in the prism plane (see Figure 14.6). In Sections 15.2 and 15.6, we provide an in-depth discussion of the natural divergence of electromagnetic radiation. The "R" and "D" subscript labels for $\delta\theta$ in the above cases are acronyms for refraction and diffraction, respectively, as related to the responsible physical effects. Setting the $\delta\theta \equiv \delta\theta_R = \delta\theta_D \cong \lambda/D$ condition in Eq. (24.7), we get the following condition for the prism wavelength resolution:

$$\boxed{\delta\lambda \cong \left| \frac{\lambda_0}{2L\,\mathbf{sin}\,(\alpha/2)[dn/d\lambda]_{\lambda_0}} \right|}, \tag{14.8}$$

where $L = D/\,\mathbf{cos}\,\theta_{out}$ is the illuminated length of the prism edge (Figure 14.6).

Eq. (14.8) shows that the prism wavelength resolution increases (namely, $\delta\lambda$ decreases) with an increase of the prism dimensions (increase of L), and with increase of the $dn/d\lambda$ slope of the prism material. We consider a representative case: let $dn/d\lambda]_{\lambda_0} = 7 \times 10^{-5}\ \text{nm}^{-1} = 7 \times 10^2\ \text{cm}^{-1}$ (Figure 14.4 – a BK7 glass), $\alpha = 60°$, $L = 2$ cm, and $\lambda_0 = 500$ nm. Setting in Eq. (14.8) yields

$$\delta\lambda = \frac{5 \times 10^{-5}}{2 \times 2 \times \mathbf{sin}\ 30° \times 7 \times 10^2} \cong 0.36 \times 10^{-7}\ \text{cm} = 0.36\ \text{nm}.$$

In laser technology, the typical beam diameter within the amplifying medium is of the order of 0.3 – 0.5 mm. To make use of the full wavelength resolution of a tuning prism, such as calculated above, the beam needs to be broadened before crossing the prism. This is often done by adding a broadening telescope close to the prism, at the amplifying medium side. A schematic sketch of such a construction is provided in Figure 14.7. The presented telescope construction is called a **Galilean telescope**. It comprises a concave lens, which serves as the objective, and a convex lens, which serves as the eyepiece. The distance between the lenses is set to have a common focal point. A parallel plane beam crossing the objective lens exits the eyepiece lens plane-parallel and broader. To avoid secondary effects related to

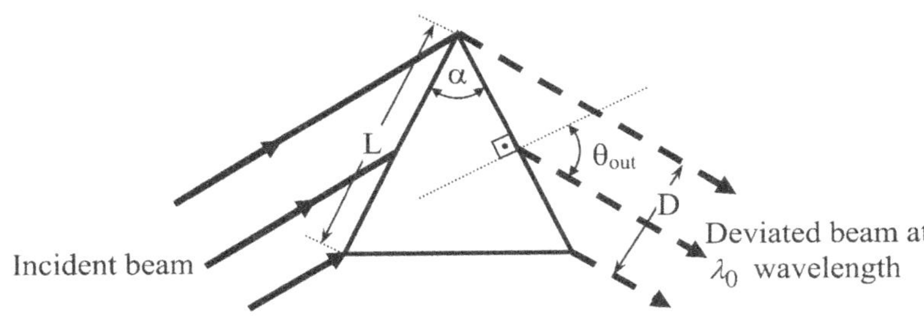

Figure 14.6 Sketch for assisting with the calculation of a prism wavelength resolution (see text).

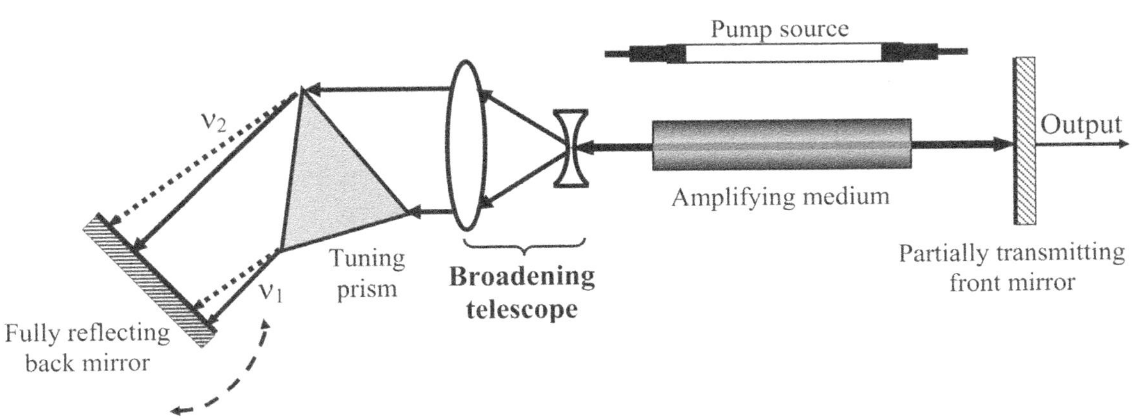

Figure 14.7 A laser with a wavelength tuning prism and an added beam broadening telescope to maximize the wavelength resolution.

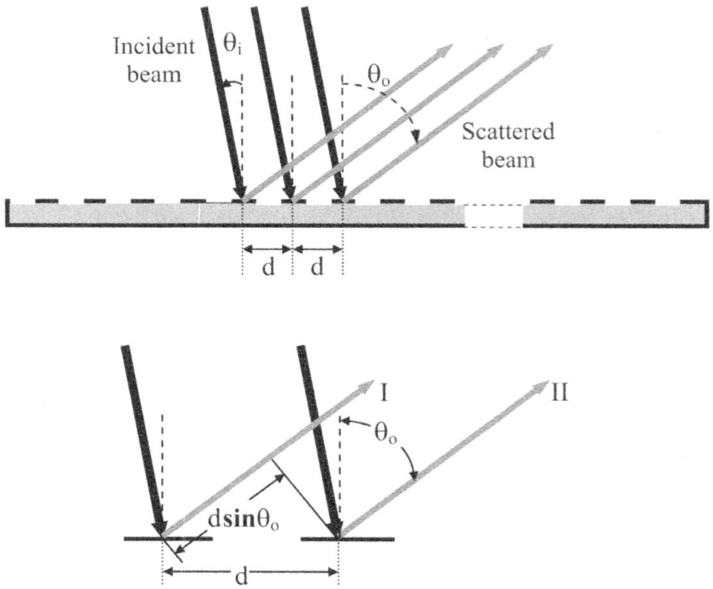

Figure 14.8 Sketch to assist in the calculation of the scattered light characteristics of a light beam incident on a reflection grating.

light dispersion in the lenses themselves, they may be replaced by an analogous spherical mirrors construction.

14.4 Reflection Grating

A reflection grating forms from a planar plate whose surface is sketched by alternating straight bands of high and low optical reflectivity zones (grooves), where d is the constant bands' period, and N is their total number. Figure 14.8 displays a schematic cross section of a grating structure.

14.4.1 Light Diffraction Off a Reflection Grating

Consider a monochromatic plane-parallel beam propagating toward a reflection grating at an angle θ_i relative to its plane normal, and the light is scattered at an angle θ_o. The path difference between two adjacent (incident plus reflected) waves – such as those marked "II" and "I" in Figure 14.8 – is given by

$$\Delta\ell = d\,\sin\,\theta_o - d\,\sin\,\theta_i\,. \tag{14.9}$$

The phase difference between the two waves will be

$$\delta = \frac{2\pi}{\lambda}\Delta\ell = \frac{2\pi\,d}{\lambda}(\sin\,\theta_o - \sin\,\theta_i)\,,$$

where λ is the beam wavelength.

We mark by $\mathcal{E}_p$ the complex amplitude of the beam scattered from the reflecting zone, whose ordinal number is p on counting from left to right. For simplicity, we assume the incident beam to be plane parallel and of a uniform wave front intensity distribution, and the width of each reflecting zone on the grating to be small compared with both the wavelength and grating period. Under such conditions, the complex amplitude of the scattered p component at the θ_o angle on the wave front is $\mathcal{E}_p = \mathcal{E}_0 e^{ip\delta}$. If the entire scattered beam is converged onto a point (a situation obtained at a lens focal point), the complex amplitude of the beam would be

$$\mathcal{E} = \sum_{p=0}^{N} \mathcal{E}_0 e^{ip\delta} = \mathcal{E}_0 \frac{e^{iN\delta} - 1}{e^{i\delta} - 1}\,.$$

The beam intensity would be given by (see text related to Eqs. [1.2] and [1.3])

$$I = \frac{c}{8\pi}\mathcal{E}^*\mathcal{E} = \frac{c}{8\pi}|\mathcal{E}_0|^2 \frac{\sin^2\left[\pi\frac{d}{\lambda}N(\sin\,\theta_o - \sin\,\theta_i)\right]}{\sin^2\left[\pi\frac{d}{\lambda}(\sin\,\theta_o - \sin\,\theta_i)\right]}\,.$$

It may be readily deduced that the light intensity dependence on the incidence and scattering angles exhibits a prominent peak for angles satisfying $\sin\left[\pi\frac{d}{\lambda}(\sin\theta_o - \sin\theta_i)\right] = 0$, namely for angles satisfying $\pi\frac{d}{\lambda}(\boldsymbol{sin}\,\theta_o - \boldsymbol{sin}\,\theta_i) = n\pi$, where $n = 0, 1, 2, \cdots$. The "n" quantity is called the **diffraction order**. The θ_o angle satisfies the relation

$$\boxed{d\,|\,\boldsymbol{sin}\theta_o - \boldsymbol{sin}\theta_i| = n\lambda}\ . \tag{14.10}$$

Eq. (14.10) is the **grating equation**, and it describes the light wavelength (color) dispersion upon reflection from the grating surface.

We assess now the angular width of a diffracted monochromatic light. Assume that the incidence angle θ_i remains constant, and the beam intensity is presented as function of $\delta\theta \equiv \theta - \theta_o$, where $\delta\theta << 1$, and θ_o satisfies the grating equation (Eq. [14.10]). We may then write

$$\frac{\boldsymbol{sin}^2[\pi\frac{d}{\lambda}N(\boldsymbol{sin}\theta - \boldsymbol{sin}\theta_i)]}{\boldsymbol{sin}^2[\pi\frac{d}{\lambda}(\boldsymbol{sin}\,\theta - \boldsymbol{sin}\theta_i)]} = \frac{\boldsymbol{sin}^2[\pi\frac{d}{\lambda}N(\boldsymbol{sin}\theta - \boldsymbol{sin}\theta_o + \boldsymbol{sin}\theta_o - \boldsymbol{sin}\theta_i)]}{\boldsymbol{sin}\theta^2[\pi\frac{d}{\lambda}(\boldsymbol{sin}\theta - \boldsymbol{sin}\theta_o + \boldsymbol{sin}\theta_o - \boldsymbol{sin}\theta_i)]}$$

$$= \frac{\boldsymbol{sin}^2[\pi\frac{d}{\lambda}N(\boldsymbol{sin}\theta - \boldsymbol{sin}\theta_o) + Nn\pi]}{\boldsymbol{sin}^2[\pi\frac{d}{\lambda}(\boldsymbol{sin}\theta - \boldsymbol{sin}\theta_o) + n\pi]} = \frac{\boldsymbol{sin}^2[\pi\frac{d}{\lambda}N(\boldsymbol{sin}\theta - \boldsymbol{sin}\theta_o)]}{\boldsymbol{sin}^2[\pi\frac{d}{\lambda}(\boldsymbol{sin}\theta - \boldsymbol{sin}\theta_o)]}$$

We define $z \equiv \pi\frac{d}{\lambda}(\boldsymbol{sin}\,\theta - \boldsymbol{sin}\,\theta_i) \cong \pi\frac{d}{\lambda}\,\boldsymbol{cos}\,\theta_o\delta\theta$. The beam intensity is then given by

$$I(z) = \frac{c}{8\pi}|\mathcal{E}_0|^2\frac{\boldsymbol{sin}^2 Nz}{\boldsymbol{sin}^2 z} \cong \frac{c}{8\pi}|\mathcal{E}_0|^2\frac{\boldsymbol{sin}^2\left(N\pi\frac{d}{\lambda}\,\boldsymbol{cos}\,\theta_o\cdot\delta\theta\right)}{\boldsymbol{sin}^2\left(\pi\frac{d}{\lambda}\,\boldsymbol{cos}\,\theta_o\cdot\delta\theta\right)}\ .$$

For large N values, the function $y = \boldsymbol{sin}^2Nz/\boldsymbol{sin}^2z$ peaks at $z = 0$, the peak height is N^2, its full width at half maximum is $\Delta z = \pi/N$, and it is tiny, practically zero, in all other regions (except in the vicinities of integral numbers of π). The integral over z is $N\pi$ (see also Figure 2.1).

For demonstration, in Figure 14.9 we compare two functions: $y_1 = \boldsymbol{sin}^2 10z/\boldsymbol{sin}^2 z$ and $y_2 = \boldsymbol{sin}^2 20z/\boldsymbol{sin}^2 z$. The figure demonstrates the peak height increase by a factor of 4 by increase of N from 10 to 20, and a corresponding width reduction by a factor of 2. The $\Delta\theta$ angular width of the reflected beam thus satisfies $(\pi d/\lambda)\,\boldsymbol{cos}\,\theta_o\Delta\theta = \Delta z = (\pi/N)$. Then

$$\boxed{\Delta\theta = \frac{\lambda}{Nd\,\boldsymbol{cos}\,\theta_o} = \frac{\lambda}{L\,\boldsymbol{cos}\,\theta_o}}\ , \tag{14.11}$$

where $L = Nd$ is the entire width of the illuminated grating part.

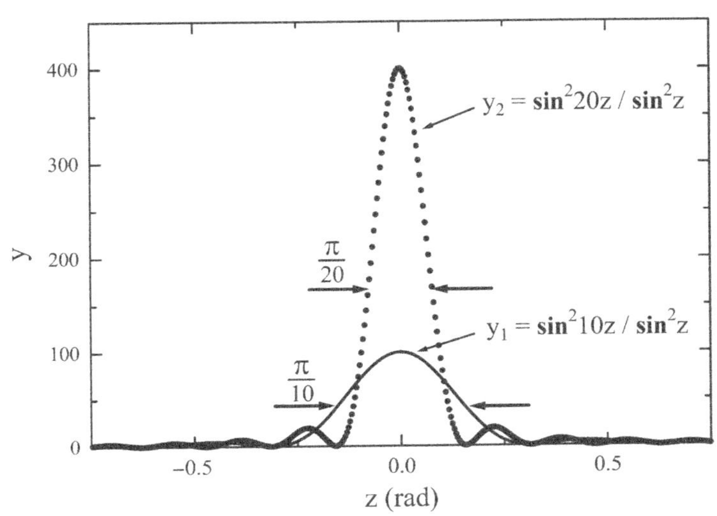

Figure 14.9 Plots of the $y_1 = \boldsymbol{sin}^2 10z/\boldsymbol{sin}^2 z$ and $y_2 = \boldsymbol{sin}^2 20z/\boldsymbol{sin}^2 z$ functions.

14.4.2 Wavelength Resolution of a Reflection Grating

A grating is capable of resolving between two wavelengths (colors) separated by $\delta\lambda$ if their angular diffraction orientation $\delta\theta_o = n \cdot \delta\lambda/d\mathbf{cos}\,\theta_o$ calculated by the grating equation (Eq. [14.10]) is equal to or greater than the angular width of each, $\Delta\theta = \lambda/L\,\mathbf{cos}\,\theta_o$, given in Eq. (14.11). The equality case limit defines the grating resolution limit. Then we get

$$\boxed{\delta\lambda = \frac{\lambda}{nN}}\ . \tag{14.12}$$

It is thus seen that the wavelength resolution capability increases (namely, $\delta\lambda$ decreases!) with the increasing number of reflecting bands, hence the illuminated grating area. If reflection orders larger than "1" are possible for the λ wavelength of the same grating, the resolution capability increases with the n reflection order. For demonstration, we consider a numerical example. Let $\lambda = 500$ nm $= 5 \times 10^{-5}$ cm (green light), $L = 2$ cm, $d = 2\lambda = 10^{-4}$ cm, $n = 1$. We then get

$$\delta\lambda = \frac{\lambda}{nN} = \frac{\lambda d}{nL} = \frac{5 \times 10^{-5} \times 10^{-4}}{1 \times 2} = 2.5 \times 10^{-9}\ \text{cm} = 2.5 \times 10^{-2}\ \text{nm}\,.$$

Thus, a typical resolution capability of a grating is approximately 15 times greater than a typical prism of an equal illuminated area (see Section 14.3.3). Like for the prism, for achieving a maximal wavelength resolution, the laser beam needs to be expanded when oriented toward the tuning grating. Figure 10.10 shows a possible integration of a reflection grating into a laser resonance cavity, in what is known as the **Littrow configuration**. The cavity includes a broadening telescope (**Galilean telescope**) like in Figure 14.7, but the back mirror is replaced by a reflection grating, whose surface normal is inclined relative to the cavity axis, in a form by which the grating bands (grooves) are perpendicular to the plane described by their surface normal and the cavity axis. The grating itself is rotatable about an axis parallel to the grating bands. In the specific Littrow configuration, the light beam will be amplified at the particular wavelength for which the reflected beam angle equals the incidence beam angle, but having an opposite sign (namely $\theta_o = -\theta_i$, hence $\mathbf{sin}\,\theta_o = -\mathbf{sin}\,\theta_i$). Setting the latter in the grating equation (Eq. [14.10]), we get the "wavelength tuning equation"

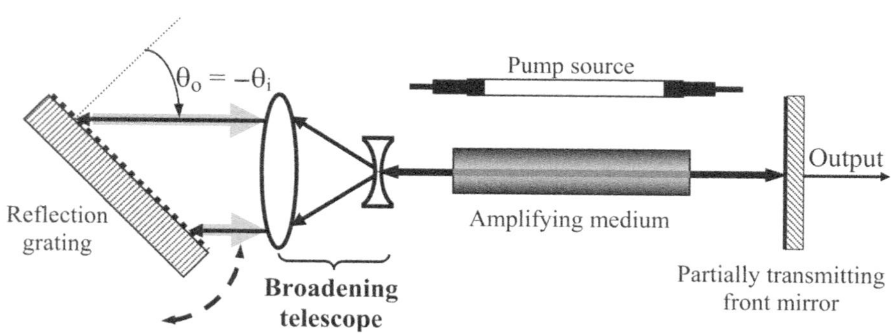

Figure 14.10 A laser with a wavelength tuning reflection grating replacing the back mirror (Littrow configuration), and a beam broadening telescope for maximizing the wavelength resolution. Grating rotation about an axis parallel to the grating bands (perpendicular to the page plane) changes the wavelength of the beam back-reflected into the amplifying medium.

$$\lambda = \frac{2d}{n} \sin \theta_i \,. \tag{14.13}$$

Setting a grating at the laser cavity end causes enhanced losses because a mirror may reflect almost a 100%, whereas a maximal grating reflection is approximately only 30% to 40%. To increase the grating reflection's efficiency, the grating high reflectivity zones are formed as planes tilted at an angle adequate for providing a mirror-like reflection for each of the wave components of a selected wavelength of interest. The reflectance yield increases for the selected wavelength and its vicinity, but it is of lower yield for other wavelengths. The particular wavelength selected for the said maximal reflection is called the **blaze wavelength**.

14.5 Fabry-Pérot Etalon

14.5.1 General Description and Fundamental Terms

The **Fabry-Pérot etalon** is a parallel plate of a uniform transparent material, usually a glass, whose faces are highly flat and parallel, of a refractive constant n and thickness L, within an ambience of a refractive index n'. We consider a plane parallel electromagnetic wave propagating at an angle θ' relative to the plate normal, as seen in Figure 14.11. The entering beam is refracted according to Snell's law, propagating inside at an angle θ. We consider the incident beam as a sum of partial small rays; two of them, $\mathcal{E}_0{}^{(0)}$ and $\mathcal{E}_0{}^{(1)}$, are sketched in Figure 14.11. We focus our interest on the wave exiting the etalon along the propagation path of $\mathcal{E}_0{}^{(0)}$, marked $\mathcal{E}_{T0}$. It results from the $\mathcal{E}_0{}^{(0)}$ ray that survived reflection, refraction, two interface transitions (the top and bottom surfaces), and bulk absorption. It is joined by a second wave, $\mathcal{E}_{T1}$, that also survived reflections at the top and bottom surfaces, and absorption due to a longer path, $2L/\cos\theta$, through the volume. That longer path is emphasized by a dashed line over the arrows describing the waves' propagation directions. By a simple trigonometric calculation, the additional optical path Δ is given by

$$\Delta = \frac{nL}{\cos\theta} \cos 2\theta + \frac{nL}{\cos\theta} = 2nL\cos\theta \,.$$

The impact of the added optical path is a phase lag. The phase lag δ of that wave compared with the first one would then be given by $\delta = (4\pi nL\cos\theta/\lambda_0)$, where λ_0 is the vacuum wavelength. The described two waves are joined by additional components belonging to the entire incident beam to the left of $\mathcal{E}_0{}^{(0)}$. Each exhibits a different amplitude, absorption, and phase lag, to be specified below. Interference of all exiting components generates the local exiting beam amplitude.

We denote by $\mathcal{E}_0{}^{(p)}$ the complex amplitudes of the wave components incident on the etalon plate, where the ordinal number p starts at 0 and grows toward the left direction. For simplicity, we assume that all amplitudes are equal, namely $\mathcal{E}_0{}^{(p)} = \mathcal{E}_0$ for all p. We denote as t_1 the amplitude transmission coefficient of the wave on crossing the top surface from n' to n refractive indices, and as t_2 the amplitude transmission coefficient of the wave on crossing the bottom surface from n to n' refractive indices. We denote as r_1 the wave amplitude

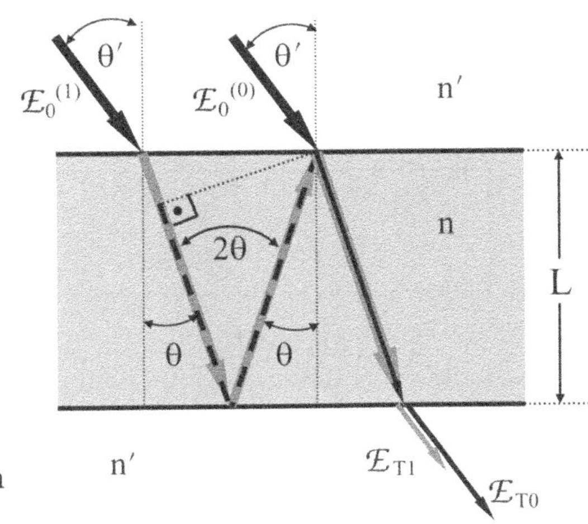

Figure 14.11 Sketch for assisting with the calculation of light characteristics transmitted through a Fabry-Pérot etalon.

reflection coefficient on crossing the top surface from n' to n refractive indices, and as t_2 the wave amplitude reflection coefficient on crossing the bottom surface from n to n' refractive indices. ${t_1}^2 = T_1$ is the intensity transmission through the top surface, and ${t_2}^2 = T_2$ is the intensity transmission through the bottom surface.

Similarly, ${r_1}^2 = R_1$ is the intensity reflection of the top surface, and ${r_2}^2 = R_2$ is the intensity reflection of the bottom surface. Obviously, by the energy conservation law, $R_1 + T_1 = R_2 + T_2 = 1$. It is possible to generate conditions, where the transmission and reflection coefficients of the two surfaces are different, by suitable surface coatings. Recalling specifically Section 10.2, if the imaginary part of the refractive index κ (the extinction coefficient) is very small compared with the real refractive index value n, the reflection phase lag on normal incidence may be ignored (see Section 10.2). The relation $\alpha = 2\omega\kappa/c$, where α is the medium absorption coefficient (Eq. [10.16]), shows that in the optical region it is sufficient to have $\alpha << 10^4\ \text{cm}^{-1}$. The same consideration is valid for non-normal incidence. In all practical cases $\alpha < 10^{-2}\ \text{cm}^{-1}$. It is therefore fully justified to ignore all phase changes on reflections and transmissions. Recall further that $\alpha/2$ is the amplitude absorption coefficient inside the plate. The amplitude of the transmitted component $\mathcal{E}_{Tp}$ is thus given by

$\mathcal{E}_{Tp} = \mathcal{E}_0 \cdot t_1 t_2\, \textbf{exp}\left(\frac{-\alpha L}{2\cos\theta}\right) \cdot {r_1}^p \cdot {r_2}^p \cdot \textbf{exp}\left(-p\frac{\alpha L}{\cos\theta}\right) \textbf{exp}(i \cdot p\delta)$. The amplitude of the exiting wave $\mathcal{E}_T$ would be the sum

$$\mathcal{E}_T = \sum_{p=0}^{\infty} \mathcal{E}_{Tp} = \mathcal{E}_0 t_1 t_2\, \textbf{exp}\left(\frac{-\alpha L}{2\cos\theta}\right) \sum_{p=0}^{\infty} {r_1}^p {r_2}^p\, \textbf{exp}\left(-p\frac{\alpha L}{\cos\theta}\right) \textbf{exp}\,(i \cdot p\delta)$$

$$\mathcal{E}_T = \frac{\mathcal{E}_0 t_1 t_2\, \textbf{exp}\left(\frac{-\alpha L}{2\cos\theta}\right)}{1 - r_1 r_2\, \textbf{exp}\left(-\frac{\alpha L}{\cos\theta} + i\delta\right)}$$

The ratio T_T between the intensity crossing the etalon to that incident is given by

$$\boxed{T_T = \frac{|\mathcal{E}_T|^2}{|\mathcal{E}_0|^2} = \frac{(1-R_1)(1-R_2)\,\textbf{exp}\,(-\alpha L/\cos\theta)}{\left(1-\sqrt{R_1R_2}\,\textbf{exp}\left(-\frac{\alpha L}{\cos\theta}\right)\right)^2 + 4\sqrt{R_1R_2}\,\textbf{exp}\left(-\frac{\alpha L}{\cos\theta}\right)\textbf{sin}^2\left(\frac{\delta}{2}\right)}}\ . \tag{14.14}$$

For brevity and clarity, we put forward some simplifying assumptions. We assume no absorption by the etalon material (namely, $\alpha = 0$), and that the etalon is symmetric (namely $R_1 = R_2 = R$). Eq. (14.14) then assumes the form

$$\boxed{T_T = \frac{(1-R)^2}{(1-R)^2 + 4R\,\textbf{sin}^2(\delta/2)}}\ . \tag{14.15}$$

Eq. (14.15) suggests that the transmission reaches 100% when the phase dependent denominator part vanishes, namely when $\delta = 2m\pi$, where m is an integer. The m quantity is called the "transmission order." Figure 14.12 below describes the optical transmission T_T's dependence on the phase lag δ for different surface reflectance R values. The transmission peaks turn narrower for growing reflectance. The lowest shown reflectance is typical of most commercial optical glasses, and the etalon transmission changes by approximately 15% between peak and bottom points.

The phase resolution capability of an etalon is defined as the width about a maximum transmission point where the transmissions drops to half its peak value. Such an evaluation is possible only for a sufficiently large surface reflectance. Under satisfactory conditions, the phase lags at half peak transmission satisfy the relation

$$(1-R)^2 = 4R\,\boldsymbol{\sin}^2\left(\frac{\delta_{1/2}-2m\pi}{2}\right).$$

For narrow transmission peaks, a property achieved when R is very close to a unity (Figure 14.12), the sine functions at a peak vicinity may be replaced by the phase variables themselves. Then we get

$$\left(\delta_{1/2}-2m\pi\right) = \pm\frac{1-R}{\sqrt{R}}$$
$$\Rightarrow \ \Delta\Phi_{1/2} = \frac{2(1-R)}{\sqrt{R}},$$

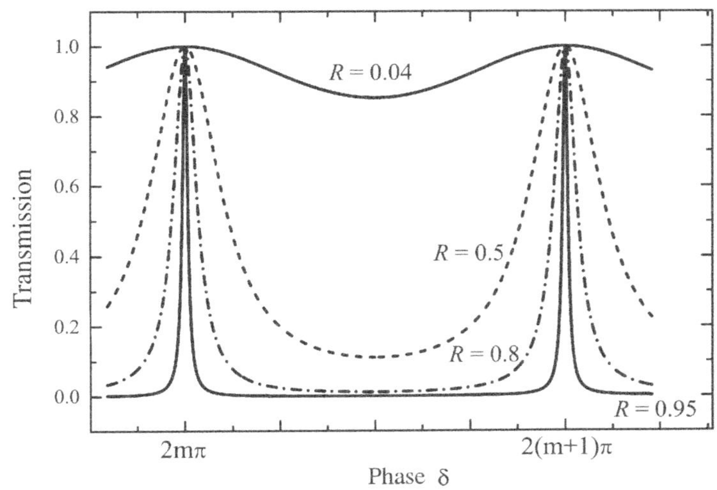

Figure 14.12 Optical transmission dependence of a symmetric, non-absorptive Fabry-Pérot etalon on the phase lag for several surface reflectance values. In the absence of absorption, the etalon reflectance is the complementary to a unity value.

where $\Delta\Phi_{1/2}$ is the transmission full phase width at half maximum. The phase separation between two transmission peaks is exactly 2π. Correspondingly, the etalon **finesse** is defined as the ratio between adjacent peaks phase separation and the full phase width at half maximum. For a symmetric, absorptions devoid etalon, the finesse is given by

$$\boxed{F \equiv \frac{2\pi}{\Delta\Phi_{1/2}} = \frac{\pi\sqrt{R}}{1-R}}. \tag{14.16}$$

14.5.2 The Etalon as an Optical Filter

We consider again the conditions for occurrence of a transmission peak of the etalon:

$$\delta = \frac{4\pi nL\,\boldsymbol{\cos}\,\theta}{\lambda_0} = 2m\pi \ \Rightarrow \ \frac{2nL\,\boldsymbol{\cos}\,\theta}{\lambda_0} = m, \quad \textit{where } m \textit{ is an integer}.$$

Since the phase lag is wavelength dependent, it is obvious that for a particular etalon of constant thickness and constant incidence angle, there would be a series of wavelengths for which a maximum transmission occurs. Recalling that $\nu = c/\lambda_0$, where c is the speed of light constant, we obtain the following series of wavelengths/frequencies of peaking transmissions:

$$\lambda_0^m = \frac{2nL\,\boldsymbol{\cos}\,\theta}{m}; \quad \nu_m = m\frac{c}{2nL\,\boldsymbol{\cos}\,\theta}. \tag{14.17}$$

Thus, the separation between two frequency peaks on the frequency axis is constant, but on the wavelength axis, the wavelength peaks become closer with increasing transmission order m. The range between two adjacent transmission peaking frequencies is called **free spectral range** (FSR). The reason for that naming will be clarified later on. The size of the free spectral range along the wavelength and frequency axes is

$$\Delta\lambda_{FSR} \equiv \lambda_0^m - \lambda_0^{m+1} = \frac{2nL\,\boldsymbol{\cos}\,\theta}{m(m+1)}; \quad \Delta\nu_{FSR} \equiv \nu_{m+1} - \nu_m = \frac{c}{2nL\,\boldsymbol{\cos}\,\theta}. \tag{14.18}$$

It may be readily seen, that the full peak widths at half maximum along the wavelength and frequency axes are given by

$$\boxed{\Delta\lambda_{1/2} = \frac{1}{m}\cdot\frac{\lambda_0}{F} = \left(\frac{m+1}{m}\right)\frac{\Delta\lambda_{FSR}}{F};\quad \Delta\nu_{1/2} = \frac{\Delta\nu_{FSR}}{F}}\ . \tag{14.19}$$

The last results demonstrate that the etalon may be used as a narrow-band optical filter, especially inside laser cavities. The spectral width of a laser beam is originally quite narrow due to gain effects and other included dispersive elements (see Eqs. [14.8] and [14.12] and related text). For further band narrowing, an etalon whose spectral transmission width at the desired wavelength is narrower than the spectral resolution of the other dispersive elements (a grating or a prism) is added inside the cavity. For example, we show a design of a BK7 made etalon, of a $n = 1.521$ refractive index (Figure 14.4), for a $\lambda_0 = 500$ nm $= 5 \times 10^{-5}$ cm wavelength, $F = 100$ finesse, and $\Delta\lambda_{1/2} = 2.5 \times 10^{-10}$ cm wavelength resolution (ten times better than the grating described in Section 14.4.2). The aimed at design parameters are the etalon surface's reflectance and thickness.

Initially, the surface's reflectance is calculated. The basic equation is $F = \dfrac{\pi\sqrt{R}}{1-R}$ (Eq. [14.16]). Knowing that R is close to a unity, we take $R \cong 1 - (\pi/F) = 1 - (3.14/100) = 0.969$ as an excellent approximation. To calculate the desired thickness, we first assess the transmission order m. We use the equation $\Delta\lambda_{1/2} = \lambda_0^m/mF$ (Eq. [14.19]). Developing and setting numerical given data we get $m = \dfrac{\lambda_0^m}{\Delta\lambda_{1/2}F} = \dfrac{5\times10^{-5}}{2.5\times10^{-10}\times100} = 2\times10^3$. Then we use the equation $\lambda_0^m = \dfrac{2nL\cos\theta}{m}$ to extract L. Assuming a normal incidence ($\cos\theta = 1$) and setting given numerical values, we get:

$$L = \frac{m\lambda_0^m}{2n} = \frac{2\times10^3\times5\times10^{-5}}{2\times1.521} = 3.2873\times10^{-2}\,\text{cm} \cong 0.3287\ \text{mm}\,.$$

There are several important issues to be addressed. Firstly, in contrast to the normal incidence assumption ($\theta = 0$), etalons under use are always placed with a slight inclination relative to the beam axis, with the goal of preventing undesired reflections along the cavity axis. That consideration should be included in the etalon thickness calculation. After that, the etalon free spectral range should be made broader than the laser effective gain (without the etalon) to prevent lasing development over more than a single wavelength. Use is made of the relation $\Delta\lambda_{FSR} = \dfrac{\lambda_0^m}{m+1} \cong \dfrac{5\times10^{-5}}{2000} = 2.5\times10^{-8}$ cm (Eqs. [14.18] and [14.17]). That range is larger than the width dictated in a laser cavity by using a reflection grating as described in Figure 14.10 and the related text. Figure 14.13 presents the structure of a laser that includes an etalon in addition to a reflection grating, with the goal of reducing the output line bandwidth. It is possible to

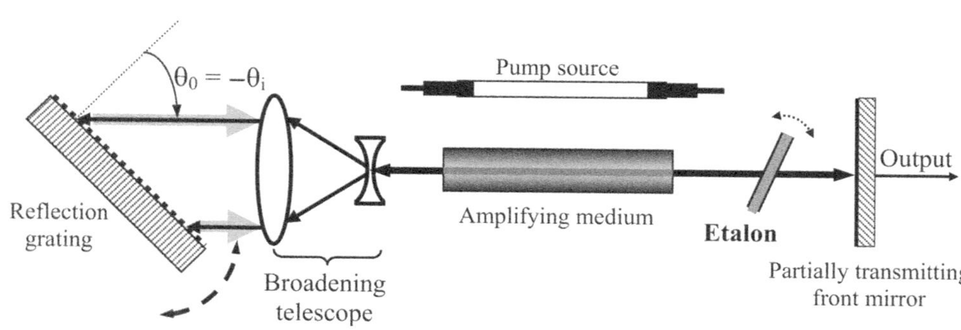

Figure 14.13 Laser construction with a reflection grating in Littrow configuration and an etalon, both integrated in the optical cavity for achieving a narrow-band color output.

tune the output over a small region of wavelengths permitted by the grating resolution, and by fine inclination changes of the etalon relative to the laser axis.

It should be noted that the laser cavity itself is actually an etalon, which under steady-state conditions supports a discrete set of wavelengths (see Section 14.2). The spectral width of each supported line is given by Eq. (14.19). The added etalon within the cavity selects a single line among the cavity-supported ones.

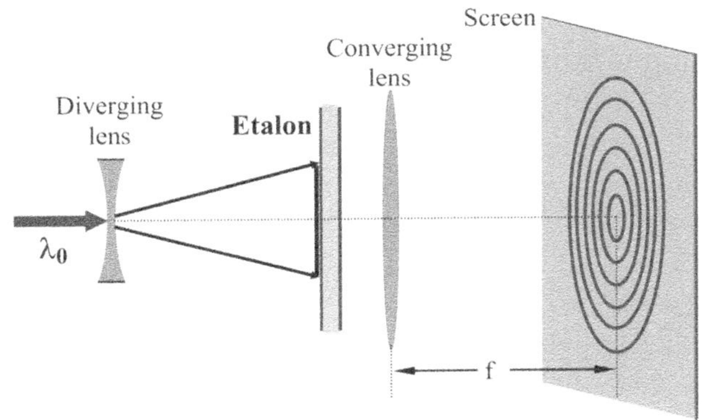

Figure 14.14 Schematic illustration of an etalon-based wavelength spectroscopy of laser beams. A diverging lens causes a plane-parallel beam to diverge on incidence on the etalon. A converging lens of f focal length placed after the etalon causes convergence of all wavelength components on a screen placed at the focal point, in a form of concentric circles. For details, see text.

14.5.3 The Etalon as a Spectrometer

Eq. (14.17) suggests that it is possible to tune the peak transmission frequency (wavelength) by changing either the incidence angle θ, or the etalon thickness L. These properties can be taken advantage of when using an etalon as a spectrometer.

14.5.3.1 A Solid Etalon

Consider a monochromatic light beam at λ_0 wavelength diverging at a relatively large angle incident on an etalon of a large finesse. Light transmission would occur at specific angles satisfying Eq. (14.17):

$$\boldsymbol{cos}\theta_m = m(\lambda_0/2nL)\,.$$

The exiting beam image on a distant screen describes a set of co-centered illuminated circles. The center is located on the beam axis, and the circles relate to the internal θ_m angles of the above equation. The most central circle belongs to the largest possible m order, $m_{max} = \boldsymbol{Int}(2nL/\lambda_0)$. The difference between internal angles belonging to successive internal reflections is

$$(\Delta\, \boldsymbol{cos}\theta)_{FSR} \equiv \boldsymbol{cos}\theta_{m+1} - \boldsymbol{cos}\theta_m = \lambda_0/2nL\,. \tag{14.20}$$

The latter quantity is known as the **angular free spectral range**. The difference between internal angles may be readily translated into differences between exiting angles using Snell's law.

Imagine now that the λ_0 wavelength is joined by additional close wavelengths. They will appear transmitted at θ angles close to θ_m. The corresponding wavelength change is given by $\lambda - \lambda_0 = \dfrac{2nL}{m}(\boldsymbol{cos}\theta - \boldsymbol{cos}\theta_m)$. Measuring the angular shift of the additional line (or lines) compared with the original λ_0 one provides an accurate value of the wavelength shift. A limit to a measurable wavelength shift is half the free spectral range; a larger shift generates a line resting closer to the next-order reference beam, and may be confused as belonging to it.

The size of the maximal measurable range is thus $|\lambda_{max} - \lambda_0| = \dfrac{2nL}{m}\cdot\dfrac{1}{2}\cdot\dfrac{\lambda_0}{2nL} = \dfrac{\lambda_0}{2m}$. A λ_0/m range on two sides of the original beam position belonging to the etalon internal angle reflection of $\lambda_0/2nL$ is then free for spectroscopic analysis. That is the reason for using the adjective "free" for the $(\Delta\, \boldsymbol{cos}\,\theta)_{FSR}$ quantity defined in Eq. (14.20) (see also Eqs. [14.18] and [14.19]). In a spectroscopic measurement, the free spectral range is calibrated using one of the internal reflection orders of an accurately known wavelength (preferably of the highest order). It is obtained by measuring the radiuses of the circles appearing on the screen (Figure 14.14). Then, one measures the radius shift generated by the studied beam.

14.5.3.2 A Scanning Etalon

A **scanning etalon** comprises two parallel mirrors, with air (refractive index $n = 1$) between. One mirror position is fixed, whereas the other is attached to a motor performing a slow continual scanning along its symmetry axis. Past a certain distance, the mirror returns to its start place, and a new scan follows. An axial Galilean telescope (broadening telescope) on one side of the etalon assures plane parallelism of the incident beam. An axial convex lens past the output concentrates the beam onto a light intensity detector.

Consider a monochromatic light of λ_0 wavelength incident on the etalon normally to its front surface. The light intensity on the detector as a function of the moving mirror scan-length displays a series of peaks similar to those described in Figure 14.12. According to Eq. (14.17), and setting $\cos\theta = 1$, the transmission peaks will correspond to scan lengths given by

$$L_m = m\frac{\lambda_0}{2n}.$$

The distance between two adjacent peaks relates to a scan length of

$$(\Delta L)_{FSR} \equiv L_{m+1} - L_m = \frac{\lambda_0}{2n}. \tag{14.21}$$

That quantity is called the **free spectral range**.

Imagine now that the λ_0 wavelength is joined by additional close wavelengths. They will appear as transmission peaks at scan lengths L close to L_m. The corresponding wavelength change is given by $\lambda - \lambda_0 = \frac{2n}{m}(L - L_m)$. Measuring the scan length shift of the additional line (or lines) compared with the original length belonging to λ_0 provides an accurate value of the wavelength shift.

A limit to a measurable wavelength shift is half the free spectral range; more than that generates a peak closer to the next order reference beam, and may be confused as belonging to it. The size of the maximal measurable range is thus $|\lambda_{max} - \lambda_0| = \frac{2nL}{m}\cdot\frac{1}{2}\cdot\frac{\lambda_0}{2nL} = \frac{\lambda_0}{2m}$. A λ_0/m range on two sides of the original peak position belonging to the $\lambda_0/2nL$ mirror scan-length is then free for spectroscopic analysis. That is the reason for using the adjective "free" for the $(\Delta L)_{FSR}$ quantity defined in Eq. (14.21) (see also Eqs. [14.18], [14.19], and [14.20]). In a spectroscopic measurement, the free spectral range about one of the accurately known original wavelength peaking orders (preferably the smallest order) is obtained by measuring the scan peaks distances. Then, the scan shift of the studied beam wavelength is measured according to that calibration.

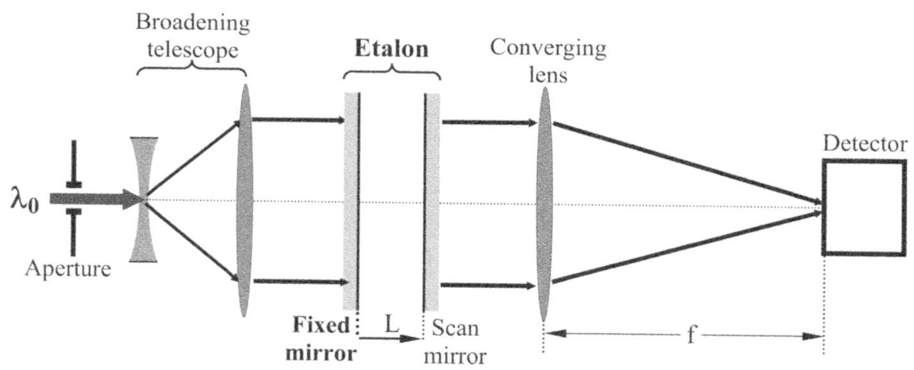

Figure 14.15 Schematic illustration of a scanning etalon used for wavelength spectroscopy of laser beams. A small aperture and a broadening telescope are used to obtain a broad plane-parallel beam incident on the etalon. A converging lens placed after the etalon scanning mirror concentrates its output on a light detector placed at the lens' focal point. For discussion and details, see text.

14.5.4 Etalon Transmission of Incoherent Light

For completeness of presentation, we reemphasize some details about the **light coherence** concept. That term addresses the relation among the electromagnetic radiation in different spaces and time zones. The extreme case of a fully coherent light exists when the light is monochromatic, of a well-defined time-independent polarization state, and of a constant intensity. If such a beam is split in two (such as by a partially transmitting mirror), then the optical path of one part is changed (such as by several mirror reflections) and rejoined. The interference result is obtained by adding the electric field values at each point, and the resulting beam exhibits the same time period. On the other hand, if the electromagnetic wave comprises successive wave packets of no steady relation between their phases and/or polarization states, the interference state would change over time if the optical path added to one of the split beams is longer than the typical length of the light wave packet. In fact, no interference pattern may be resolved. An interference pattern may still be resolved if the added path is very small compared with the typical length of a wave packet. The spatial length of a wave packet is called its **coherence length**. The coherence length of light emitted from a gas discharge tube, or an incandescent light bulb, could be as short as several nanometers; the coherence length of a laser-emitted CW beam, could be as long as tens, hundreds, or thousands of meters.

We consider now a case where the light incident on an etalon is incoherent. As a result, the assumption of interference between different order internal reflections inside the etalon is no more valid. It obviously depends on the etalon thickness, which in that case must be long compared with the beam coherence length. The multiple surface reflections must obviously be addressed, but summation should be made on the reflected *intensity* components (not on their amplitudes), and phase differences are of no importance. Thus, the exiting intensity component I_{Tp} belonging to the $\mathcal{E}_{Tp}$ amplitude (see Section 14.5.1) is given by $I_{Tp} = I_0 \cdot T_1 T_2 R_1{}^p R_2{}^p \cdot \mathbf{exp}\left(-\frac{\alpha L}{\mathbf{cos}\ \theta}\right) \cdot \mathbf{exp}\left(-p \cdot \frac{2\alpha L}{\mathbf{cos}\ \theta}\right)$. The exiting light intensity I_T is the sum

$$I_T = \sum_{p=0}^{\infty} I_{Tp} = T_1 T_2 I_0\, \mathbf{exp}\left(-\frac{\alpha L}{\mathbf{cos}\ \theta}\right) \sum_{p=0}^{\infty} R_1{}^p R_2{}^p\, \mathbf{exp}\left(-p \cdot \frac{2\alpha L}{\mathbf{cos}\ \theta}\right);$$

$$I_T = \frac{T_1 T_2 I_0\, \mathbf{exp}\left(-\frac{\alpha L}{\mathbf{cos}\ \theta}\right)}{1 - R_1 R_2\, \mathbf{exp}\left(-\frac{2\alpha L}{\mathbf{cos}\ \theta}\right)}.$$

The ratio T_T between the transmitted light intensity and the incident one is given by

$$\boxed{T_T = \frac{I_T}{I_0} = \frac{(1 - R_1)(1 - R_2)\, \mathbf{exp}\left(-\frac{\alpha L}{\mathbf{cos}\ \theta}\right)}{1 - R_1 R_2\, \mathbf{exp}\left(-\frac{2\alpha L}{\mathbf{cos}\ \theta}\right)}}. \tag{14.22}$$

The wavelength dependence of the transmission is smooth, and devoid of the dense peaking lines as obtained for a coherent beam. In uncoated materials, the transmission wavelength dependence actually reflects the wavelength dependence of the absorption coefficient α, because the wavelength dependence of the surface reflection coefficient R is very small for small θ angles of incidence.

14.6 Brewster Window and a Brewster Plate

14.6.1 Snell's Law and Fresnel Equations

In Section 10.2, we calculated the reflection and transmission characteristics of a special case of light, incident normally on the surface of a dielectric medium, where the medium may also exhibit light absorption. We now turn to address a more general case, where the light incidence

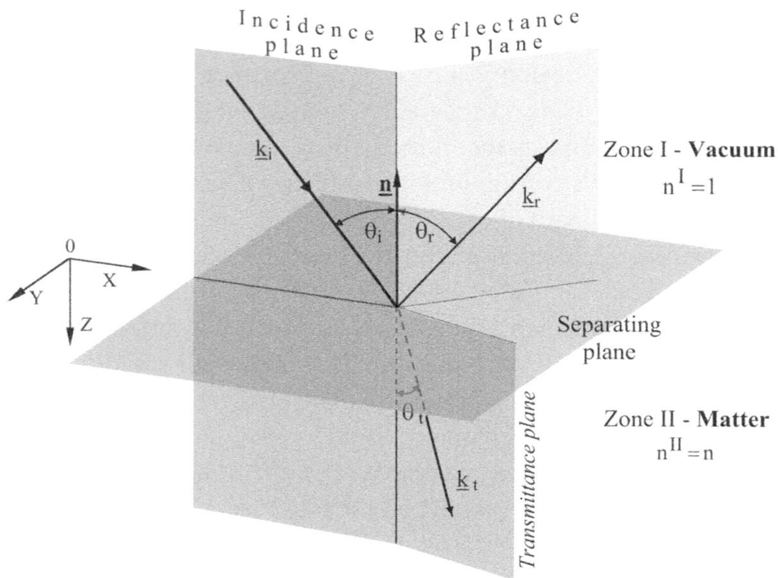

Figure 14.16 Definition of propagation orientations and angles for the incident, reflected, and transmitted beams. The figure assumes that the incidence, reflectance, and transmittance planes do not necessarily coincide. For detailed analysis, see text.

direction forms a non-zero angle relative to the surface normal. Bulk absorbance, however, is ignored. Figure 14.16 describes a plane-parallel beam incident at a non-zero angle on a surface dividing the space, acting as a border between the "I" and "II" dielectric media extending on each side from the border plane to infinity. The refractive indices are $n^I = 1$ (vacuum) and $n^{II} = n \neq 0$ (a material medium). All results are applicable to a case of two material media, by redefining $n = n^{II}/n^I$.

In the geometric description of the system provided in Figure 14.16, the Z axis is perpendicular to the separating $z = 0$ plane. A unity vector normal to the plane is marked by $\underline{\mathbf{n}}$ (it should not be confused as a refractive index, despite using the same letter). The X axis is selected along the crossing line between the plane separating the two media and the **incidence plane** (the plane generated by the $\underline{\mathbf{n}}$ vector and the incident beam wave vector $\underline{k}_i$). It is assumed that under steady-state conditions, electromagnetic wave propagation along $\underline{k}_i$ on the upper medium I side is simultaneously accompanied by a reflected beam along $\underline{k}_r$ in the upper medium I itself, and a transmitted beam along $\underline{k}_t$ in the lower medium II side (the matter). In all of the following expressions, the "i", "r", and "t"subscripts are used to refer to the incidence, reflection, and transmission processes, respectively. The reflection and transmission planes are defined analogously to the incidence plane definition. It is not assumed in advance that the three planes coincide. The electric field of the three plane-parallel waves are respectively given by

$$\left.\begin{aligned} \underline{E}_i(\underline{r},t) &= \underline{E}_{0i}\,\boldsymbol{exp}\left[-i(\omega_i t - \underline{k}_i\cdot\underline{r})\right] \\ \underline{E}_r(\underline{r},t) &= \underline{E}_{0r}\,\boldsymbol{exp}\left[-i(\omega_r t - \underline{k}_r\cdot\underline{r} + \varphi_r)\right] \\ \underline{E}_t(\underline{r},t) &= \underline{E}_{0t}\,\boldsymbol{exp}\left[-i(\omega_t t - \underline{k}_t\cdot\underline{r} + \varphi_t)\right] \end{aligned}\right\}, \tag{14.23}$$

where $\underline{r}$ is the position vector of the considered point in space. The different $\underline{E}_0$ electric field amplitudes are time-independent, $\underline{k}$ is the wave vector, ω is the angular frequency, and φ is the phase difference compared with the incident light phase; it need not necessarily be the same for the reflected and transmitted beams. In the Cartesian coordinates system chosen for Figure 14.16, the plane separation between the I and II media is $z = 0$.

According to the theory of electricity, the electric field component parallel to the separating surface transforms continuously between the two media sides ♣. The upper (*I*) side of the separating plane (the incident beam side) exhibits two contributions to the parallel electric field, one by the incident beam and the other by the reflected beam. The lower (*II*) side of the separating plane (the transmitted beam side) exhibits only a single contribution to the parallel electric field – the one belonging to the transmitted beam. The *size* of the parallel electric field component is given by the size of the vector product between the beam electric field vector and the plane normal $\underline{\mathbf{n}}$ vector (despite the product being rather perpendicular to it). The continuity requirement thus means that

$$\underline{\mathbf{n}} \times \underline{\mathcal{E}}_i + \underline{\mathbf{n}} \times \underline{\mathcal{E}}_r = \underline{\mathbf{n}} \times \underline{\mathcal{E}}_t \,. \tag{14.24}$$

It should again be emphasized that three vectors, $\underline{\mathbf{n}} \times \underline{\mathcal{E}}_i$, $\underline{\mathbf{n}} \times \underline{\mathcal{E}}_r$, and $\underline{\mathbf{n}} \times \underline{\mathcal{E}}_t$, are all parallel to the separating plane, while, in fact, being perpendicular to the corresponding electric field component. Using Eq. (14.24), we get

$$\underline{\mathbf{n}} \times \underline{\mathcal{E}}_{0i}\, \boldsymbol{exp}\left[-i(\omega_i t - \underline{k}_i\cdot\underline{r})\right] + \underline{\mathbf{n}} \times \underline{\mathcal{E}}_{0r}\, \boldsymbol{exp}\left[-i(\omega_r t - \underline{k}_r\cdot\underline{r} + \varphi_r)\right] = \underline{\mathbf{n}} \times \underline{\mathcal{E}}_{0t}\, \boldsymbol{exp}\left[-i(\omega_t t - \underline{k}_t\cdot\underline{r} + \varphi_t)\right].$$

The above relation must exist at any moment at any position on the $z = 0$ separation plane. In other words, the $\underline{\mathcal{E}}_i$, $\underline{\mathcal{E}}_r$, and $\underline{\mathcal{E}}_t$ electric field vectors must have the same functional dependence on time t and position $\underline{r}$ vector on the $z = 0$ plane. Namely,

$$(\omega_i t - \underline{k}_i\cdot\underline{r})_{z=0} = (\omega_r t - \underline{k}_r\cdot\underline{r} + \varphi_r)_{z=0} = (\omega_t t - \underline{k}_t\cdot\underline{r} + \varphi_t)_{z=0},$$

therefore $\omega_i = \omega_r = \omega_t = \omega$, as expected. Also,

$$(\underline{k}_i\cdot\underline{r})_{z=0} = (\underline{k}_r\cdot\underline{r} + \varphi_r)_{z=0} = (\underline{k}_t\cdot\underline{r} + \varphi_t)_{z=0}\,.$$

The left side equality in the last equation may be written as

$$[(\underline{k}_i - \underline{k}_r)\cdot\underline{r}]_{z=0} = \varphi_r\,. \tag{14.25}$$

This means that the projection of $\underline{r}$ on the $(\underline{k}_i - \underline{k}_r)$ vector is constant; namely, the edges of the $\underline{r}$ vector satisfying $z = 0$ describe a plane perpendicular to $(\underline{k}_i - \underline{k}_r)$. That plane must be the $z = 0$ plane. Then, the $(\underline{\mathrm{k}}_\mathrm{i} - \underline{\mathrm{k}}_\mathrm{r})$ vector is perpendicular to the $z = 0$ plane; namely, it is parallel to $\underline{\mathbf{n}}$.

A conclusion of the above is that the $\underline{k}_i$, $\underline{k}_r$, and $\underline{\mathbf{n}}$ vectors lay on the same plane; namely the incidence and reflectance planes coincide! It may be similarly concluded that the incidence and transmittance planes coincide. Thus, **the three vectors $\underline{k}_i$, $\underline{k}_r$, and $\underline{k}_t$ lay on the same plane**. A second conclusion stemming from the perpendicularity of $(\underline{k}_i - \underline{k}_r)$ to the separating plane is, that its vector product with $\underline{\mathbf{n}}$ vanishes: $\underline{\mathbf{n}} \times (\underline{k}_i - \underline{k}_r) = \underline{\mathbf{n}} \times \underline{k}_i - \underline{\mathbf{n}} \times \underline{k}_r = 0$. Similarly, $\underline{\mathbf{n}} \times (\underline{k}_i - \underline{k}_t) = \underline{\mathbf{n}} \times \underline{k}_i - \underline{\mathbf{n}} \times \underline{k}_t = 0$. Therefore

$$\left.\begin{aligned} k_i\, \boldsymbol{sin}\,\theta_i &= k_r\, \boldsymbol{sin}\,\theta_r \\ k_i\, \boldsymbol{sin}\,\theta_i &= k_t\, \boldsymbol{sin}\,\theta_t \end{aligned}\right\}. \tag{14.26}$$

Because the incident and reflected beams are of the same frequency, and propagate in the same refractive index medium, their corresponding wave vector's size is identical – $k_i = k_r$. Therefore, the above first relation yields $\boldsymbol{sin}\,\theta_i = \boldsymbol{sin}\,\theta_r$; namely, the **reflection angle is the same as the incidence angle**. On the

♣ In considering the two sides of a plane separating the two media, recall that, by electricity theory, the two sides' electric field $\mathcal{E}_{||}$ components parallel to the plane and the two sides' displacement electric field $\mathcal{D}_{\perp}$ components perpendicular to the plane, are identical ("transform continuously"). So are the magnetic field $\mathcal{H}_{||}$ components parallel to the plane and the magnetic induction field $\mathcal{B}_{\perp}$ components perpendicular to the plane.

other hand, the wave vector of the transmitted beam differs from that of the incident one; because they are of the same frequency, the wave vectors satisfy the relation $k_t = nk_i$. Thus, we get

$$\boxed{\frac{\sin\theta_i}{\sin\theta_t} = n}, \tag{14.27}$$

which is **Snell's refraction law** (see also the equation noted in the caption to Figure 3.1).

We turn now to calculating the amplitudes' relations among the incident, reflected, and transmitted light beams. In the following calculations, we shall see that these relations depend on the polarization state of the incident beam. In Figure 14.17, the polarization states are classified by two possible linear configurations: perpendicular to the incidence plane (called "S" polarization) or parallel to the incidence plane (called "P" polarization). Any non-polarized light may be resolved into the said two polarization states, permitting separate treatment of each. In a *uniform* dielectric medium, both electric field $\underline{\mathcal{E}}$ and adjacent magnetic field $\underline{\mathcal{H}}$ are perpendicular to the $\underline{k}$ wave vector orientation (see also Figure 10.1), and are mutually perpendicular. Figure 14.17 shows the different polarization states of the electric field $\underline{\mathcal{E}}$ and magnetic field $\underline{\mathcal{H}}$ in "S" and "P" configurations, for their incident, reflected, and transmitted light beams. The presentation is based on a (lucky) conjecture that the "S" or "P" configurations of the reflected and transmitted beams are the same as the incident beam.

We start by analyzing the "S" polarization (left side of Figure 14.15). The electric field $\mathcal{E}$ and magnetic field $\mathcal{H}$ in a uniform dielectric medium satisfy the relation $\mathcal{H} = n\mathcal{E}$, where n is the medium dielectric constant (a relation verifiable by using Eqs. [1.2] and [10.5] and Figures 1.1 and 10.1). We further use two boundary conditions: One is that the parallel electric field $\underline{\mathcal{E}}$'s components are identical on both sides of the separating plane (an already used condition); second is that the parallel magnetic field $\underline{\mathcal{H}}$'s components are identical on both sides of the separating plane (see footnote on the previous page). Thus,

$$\left.\begin{aligned} &\mathcal{E}_{0i} + \mathcal{E}_{0r} = \mathcal{E}_{0t} \\ &-\mathcal{E}_{0i}\cos\theta_i + \mathcal{E}_{0r}\cos\theta_r = -n\mathcal{E}_{0t}\cos\theta_t \end{aligned}\right\}.$$

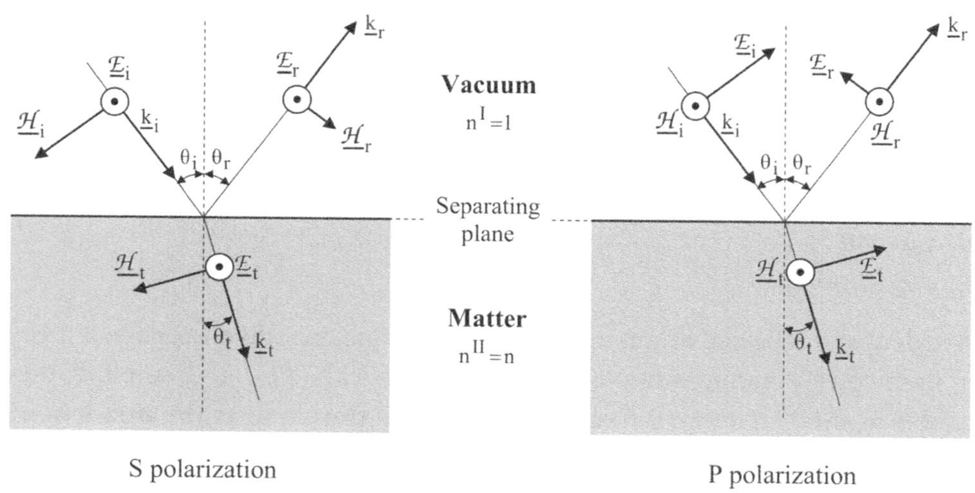

Figure 14.17 Description of electric and magnetic fields' orientations in "S" and "P" polarization configurations for the incident, reflected, and transmitted beams.

These relations allow us to obtain the ratios between electric field amplitudes as follows:

$$\left.\begin{aligned} r_S &\equiv \frac{\mathcal{E}_{0r}}{\mathcal{E}_{0i}} = \frac{\cos\theta_i - n\cos\theta_t}{\cos\theta_i + n\cos\theta_t} \\ t_S &\equiv \frac{\mathcal{E}_{0t}}{\mathcal{E}_{0i}} = \frac{2n\cos\theta_i}{\cos\theta_i + n\cos\theta_t} \end{aligned}\right\}, \tag{14.28}$$

where r_S and t_S are the amplitude's reflection and transmission factors, respectively, of incident light of "S" polarization configuration.

We turn now to a similar calculation for the "P" polarization configuration (right side of Figure 14.15). By continuity conditions for the transformation of the electric and magnetic fields that are parallel to the separating plane surface, we get

$$\left.\begin{aligned} &\mathcal{E}_{0i}\cos\theta_i - \mathcal{E}_{0r}\cos\theta_r = \mathcal{E}_{0t}\cos\theta_t \\ &\mathcal{E}_{0i} + \mathcal{E}_{0r} = -n\mathcal{E}_{0t} \end{aligned}\right\},$$

resulting in the following ratios between electric field amplitudes:

$$\left.\begin{aligned} r_P &\equiv \frac{\mathcal{E}_{0r}}{\mathcal{E}_{0i}} = \frac{n\cos\theta_i - \cos\theta_t}{n\cos\theta_i + \cos\theta_t} \\ t_P &\equiv \frac{\mathcal{E}_{0t}}{\mathcal{E}_{0i}} = \frac{2\cos\theta_i}{n\cos\theta_i + \cos\theta_t} \end{aligned}\right\}. \tag{14.29}$$

The energy power flux density carried by the oscillating electromagnetic field (power per unit area) is given by the Poynting vector

$$\boxed{\underline{S} = (c/4\pi)\underline{\mathcal{E}} \times \underline{\mathcal{H}}}\ . \tag{14.30}$$

After averaging over a single full period of the electromagnetic oscillation, the beam vacuum power density (intensity) is given by $I \equiv \langle S\rangle_{av} = (c/8\pi)\mathcal{E}_0{}^2$.

The incident beam intensity reflectance (reflection coefficient) is given by $R \equiv I_r/I_i = \mathcal{E}_{0r}{}^2/\mathcal{E}_{0i}{}^2$, whereas the transmittance (transmission coefficient) is given simply by $T \equiv I_t/I_i = 1 - R$. Using Eqs. (14.28) and (14.29), we get

$$\left.\begin{aligned} R_S &= \frac{\sin^2(\theta_i - \theta_t)}{\sin^2(\theta_i + \theta_t)}; \quad T_S = 1 - R_S \\ R_P &= \frac{\tan^2(\theta_i - \theta_t)}{\tan^2(\theta_i + \theta_t)}; \quad T_P = 1 - R_P \end{aligned}\right\}. \tag{14.31}$$

The above calculated reflections are called **Fresnel reflections**.

Figure 14.18 shows a graph of reflection coefficients' dependence on the incidence angle for beams of "S" and "P" polarization configurations, when the refractive index inside matter is $n = 1.5$. The two obtained curves coincide on two limits. One is at the zero incidence angle (namely, the perpendicular incidence); the reflection coefficient is given by $R(\theta_i = 0) = (n-1)^2/(n+1)^2$ (also obtainable by setting $\kappa = 0$ in Eq. [12.27]). The other is at near 90° incidence (**grazing incidence**); both reflection coefficients there converge to a unity. In the intermediate incidence angles, the reflection coefficient of the "S" polarization configuration grows continually with an increasing incidence angle. On the other hand, the reflection coefficient of the "P" polarization configuration exhibits a minimum of zero at a single specific angle! The

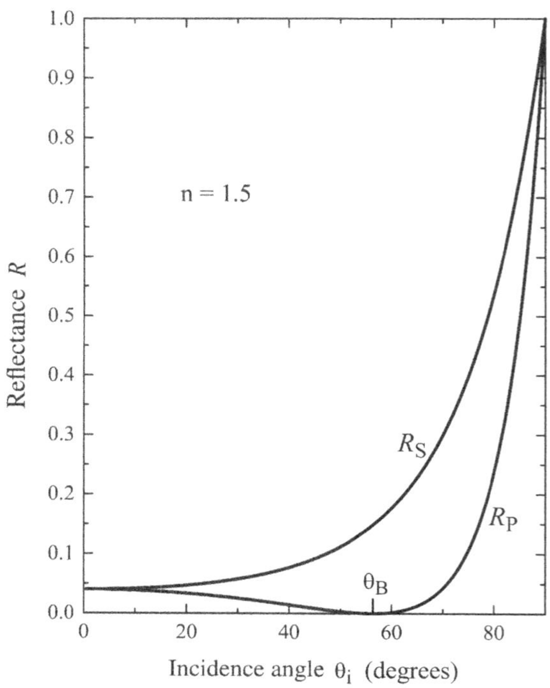

Figure 14.18 Reflection coefficient as the function of a beam incidence angle for "S" and "P" polarization configurations in a specific case, where the material refractive index is n = 1.5 (typical of a common commercial optical glass).

angle θ_B where that case occurs, is called a **Brewster angle**. Mathematically, this case occurs when $\theta_i + \theta_t = 90°$. Using Snell's refraction law for θ_t, we get

$$\tan \theta_B = n \,. \tag{14.32}$$

The last result bears a number of interesting and useful conclusions. For incidence of a non-polarized beam on a dielectric medium, the beam reflected at Brewster's angle is always totally "S" polarized. The phenomenon is quite useful for the generation of linearly polarized light. It is also useful for reduction, even close to elimination, of reflection losses of "P" polarized light on crossing between two dielectric media; this occurs when the beam incidence angle is arranged to be close to Brewster's angle.

14.6.2 Achieving Polarized Laser Emission

Thus far, described laser systems would emit light beams that are not necessarily polarized. The latter described effect of achieving a polarized beam upon reflection at Brewster's angle may be utilized to achieve a linearly polarized laser output. We describe two simple methods. One involves shaping the amplifying medium itself with boundary surfaces inclined at Brewster's angle relative to the medium axis (correspondingly addressed as **Brewster windows**); the other is the insertion of a **Brewster plate** inside the laser cavity. Figure 14.19 presents an example of a laser with an amplifying medium based on a solid rod of a refractive index n. The rod end surfaces are machined with end surfaces inclined at Brewster's angle $\theta_B = \tan^{-1} n$ relative to the beam propagation axis in the amplifying medium, where n is the medium refractive index. A beam propagating along the amplifying medium axis refracts at the boundary surface planes, forming a Brewster angle relative to the surface normal. The mirrors forming the laser cavity are perpendicular, each on its side, to the exiting beam. If, for example, the amplifying medium refractive index is n = 1.5, as in Figure 14.18, an

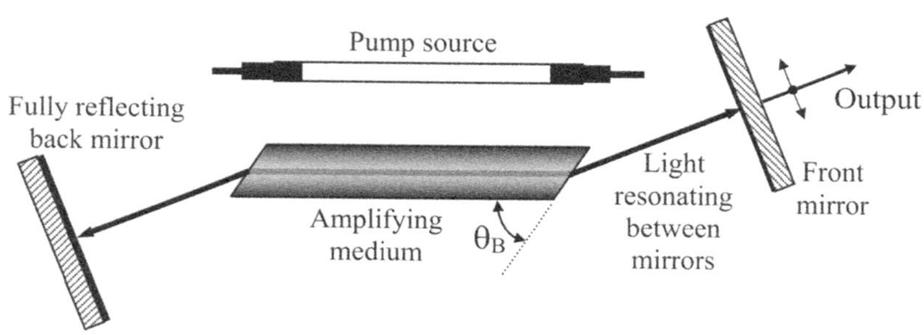

Figure 14.19 Schematic illustration of a typical laser construction emitting a polarized beam because of boundary surfaces of the amplifying medium that are inclined at Brewster's angle relative to the amplifying medium axis. For details – see text.

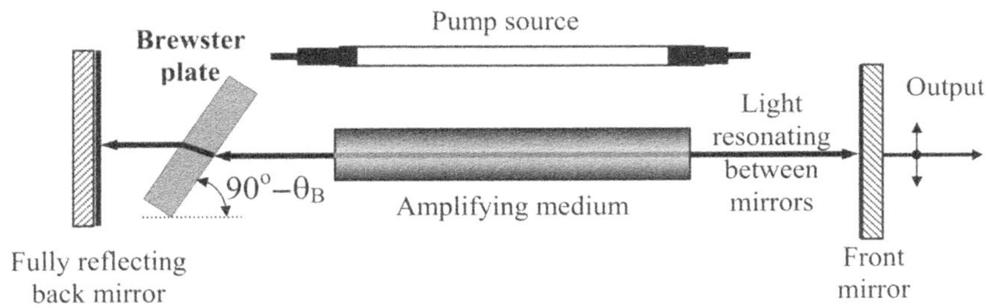

Figure 14.20 Schematic illustration of a typical laser construction emitting a linearly polarized beam because of the addition of a transparent plate inclined at Brewster's angle relative to the laser axis. For details, see text.

approximately 15% reflection occurs for the "S" polarization component (perpendicular to the page plane) upon each beam incidence on a surface whose one side is the medium and the other is a vacuum. The "P" polarization component, on the other hand, exhibits zero reflection (meaning it causes no loss). Each back and forth transition of the beam between the mirrors exhibits four such transitions; hence there are four 15% losses. Therefore amplification of the "P" polarization component is highly preferable. Practically, the exiting beam is fully "P" polarized (an orientation parallel to the page plane). The situation is demonstrated in Figure 14.19 by sketching a two-edged arrow on the output beam.

Figure 14.20 presents a different case, where a transparent parallel plate of a refractive index n is added into the laser cavity. The plate normal is inclined at Brewster's angle $\theta_B = \tan^{-1} n$ relative to the cavity axis. Such a plate is known as a **Brewster plate**. Similarly to the reasoning presented in relation to Figure 14.19, the plate transmits without loss the "P" beam polarization component (parallel to the page plane), yet causes considerable loss to any "S" beam component (perpendicular to the page plane). In that way as well, lasing in the "P" polarization orientation develops preferably, and the output beam is fully polarized parallel to the page plane, as demonstrated in Figure 14.20 by sketching a two-edged arrow on the output beam. It should be noted that in the construction described in Figure 14.19, the amplifying medium serves as a Brewster plate in addition to its role as a beam-amplifying medium.

Figures 14.3 and 14.7 above present laser resonance cavities, which include triangular prisms for wavelength output tuning. The vertex angle is usually selected such that the beam incidence angles from both sides (amplifying medium and back mirror) would be closest to its Brewster angle. In that way, losses related to transitions of a "P" polarized beam component are virtually eliminated, and the output beam is thus perfectly polarized in that direction. The added prism thus exhibits a double role: wavelength tuning, and beam polarization.

14.7 Loss Presentation in a Laser Cavity

In Section 14.1 we described the relationship between the output and loss spectra. Losses are classified as "distributed" and "lumped". Distributed losses usually belong to various light absorbing and scattering mechanisms in the amplifying medium, and are usually distributed uniformly along the light path in the amplifying medium. Lumped losses usually occur on various elements present inside the laser cavity, such as reflection losses on the cavity mirrors, on polarizing elements, and on wavelength tuning elements. To allow for the assessment of the relative importance among different type losses, lumped loss quantities are transformed to represent an equivalent (imaginary) distributed loss along the amplifying medium. Let $F(\nu)$ be the intensity remainder fraction of a light beam of ν frequency after undergoing a single, back and forth travel between the mirrors, without amplification yet including the lumped losses. The effective distributed loss $\alpha_F(\nu)$ is then

$$\alpha_F(\nu) \equiv -\frac{1}{2L_g} \ln F(\nu), \tag{14.33}$$

where L_g is the amplifying medium length. For example, if the reflection coefficients of the back and front mirrors in the basic laser cavity described in Figure 14.1 are R_1 and R_2, respectively, the effective distributed loss coefficient would be $\alpha_R(\nu) \equiv -\left(\frac{1}{2L_g}\right) \ln R_1R_2$; this is because the beam intensity reduces into an R_1 fraction upon reflection from the first mirror and an additional R_2 fraction upon reflection from the second mirror.

Exercises for Chapter 14

14.1 Consider a given simple laser system comprising two mirrors and an amplifying medium emitting a continuous wave (Figure 14.1). The amplifying medium is 5 cm long, and its fluorescence wavelength spectrum is Gaussian, peaking at 585 nm with a 35 nm full width at half maximum (FWHM). The cavity losses are 25 cm^{-1} throughout the entire spectral range.
(a) Express the gain function of the amplifying medium during lasing.
(b) Show a graph of the laser emission spectrum compared with its gain spectrum.

14.2 (a) The laser cavity described in Exercise 14.1 above is aided by a broadening telescope and a triangular, equilateral, BK7-made prism as in Figure 14.7, for the purpose of wavelength tuning. Calculate what should be the laser beam cross-section incident on the prism for the laser output spectral width to be less than 0.2 nm.
(b) The laser cavity described in Exercise 14.1 above is aided by a broadening telescope and a reflection grating of 600 grooves per millimeter as in Figure 14.10. What should be the grating inclination angle relative to the incident beam to achieve a 575 nm wavelength output. Calculate what should be the laser beam cross section on incidence on the grating for the laser output spectral width to be less than 0.2 nm.
(c) Design an etalon which, when properly added into the cavity of section (b) above, will narrow down the output beam spectral width to only 0.05 nm.

14.3 The central fluorescence emission of triply ionized titanium Ti^{3+} ions in a sapphire Al_2O_3 crystal is 750 nm. To achieve a linearly polarized output, the titanium sapphire Ti^{3+}:Al_2O_3 rod is machined with Brewster inclined edges as in Figure 14.19. Calculate the cutting angle of the rod edge. For a refractive index datum, consult Figure 9.13.

15 Transverse Optical Modes and Crystal Optics

15.1 Preamble

In all previous discussions, the laser beams were described as plane-parallel electromagnetic waves over the entire laser cavity, and quite often in the free space as well. As we shall see below, that description is not quite accurate for a beam of a finite cross section. To address the issue, we return to Maxwell's equations provided in Eq. (10.1). To simplify the mathematical treatment, we limit ourselves to a free-space medium (namely, an $n = 1$ refractive index). In that case, the magnetic field $\underline{\mathcal{H}}$ may replace the magnetic induction field $\underline{\mathcal{B}}$ in all equations, and the electric field $\underline{\mathcal{E}}$ may replace the electric displacement field $\underline{\mathcal{D}}$ in all equations. We then get

$$\left.\begin{aligned} &\text{(i)} && \boldsymbol{rot}\ \underline{\mathcal{H}} = \frac{1}{c}\frac{\partial \underline{\mathcal{E}}}{\partial t} \\ &\text{(ii)} && \boldsymbol{div}\ \underline{\mathcal{E}} = 0 \\ &\text{(iii)} && \boldsymbol{rot}\ \underline{\mathcal{E}} = -\frac{1}{c}\frac{\partial \underline{\mathcal{H}}}{\partial t} \\ &\text{(iv)} && \boldsymbol{div}\ \underline{\mathcal{H}} = 0 \end{aligned}\right\} . \tag{15.1}$$

A generalization of the equations for a uniform $n > 1$ dielectric medium is obtained by a simple replacement of the speed of light constant c by c/n, where n is the medium refractive index.

We now perform the rotor ***rot*** operation on Eq. (15.1-iii) above, and get

$$\boldsymbol{rot}\ \boldsymbol{rot}\,\underline{\mathcal{E}} = \boldsymbol{grad}\ \boldsymbol{div}\,\underline{\mathcal{E}} - \Delta\underline{\mathcal{E}} = -\frac{1}{c}\frac{\partial}{\partial t}\boldsymbol{rot}\,\underline{\mathcal{H}} = -\frac{1}{c^2}\frac{\partial^2 \underline{\mathcal{E}}}{\partial t^2} .$$

The first calculated step is based on the (xiii) mathematical relation indicated in Section 10.5.1, and the last one by setting ***rot*** $\underline{\mathcal{H}}$ according to Eq. (15.1-i) above. Because by Eq. (15.1-ii) $\boldsymbol{div}\underline{\mathcal{E}} = 0$, we are left with the relation

$$\Delta\underline{\mathcal{E}} = \frac{1}{c^2}\frac{\partial^2 \underline{\mathcal{E}}}{\partial t^2} . \tag{15.2}$$

The latter is a wave equation for the electric field components of the electromagnetic radiation. It is known to have solutions that describe the propagation of plane-parallel transverse electric field waves. We consider a specific solution describing a coherent monochromatic, x-polarized electromagnetic wave propagating in the z-direction, of the form $\mathcal{E}_x = \mathcal{E}_{x0}e^{-i(\omega t\,-\,kz)}$, where ω is the angular frequency and k is the wave vector. The complex amplitude $\mathcal{E}_{x0}$ quantity is time and position independent. We define a new quantity: $\Psi \equiv \mathcal{E}_{x0}e^{ikz}$; namely, $\mathcal{E}_x = \Psi\, e^{-i\omega t}$. Setting it in Eq. (15.2), and adding the $\omega = ck$ vacuum dispersion relation, we get the following equation for Ψ:

$$\boxed{\Delta\Psi = -k^2\Psi} . \tag{15.3}$$

Physics, Optics, and Spectroscopy of Materials, First Edition. Zeev Burshtein.

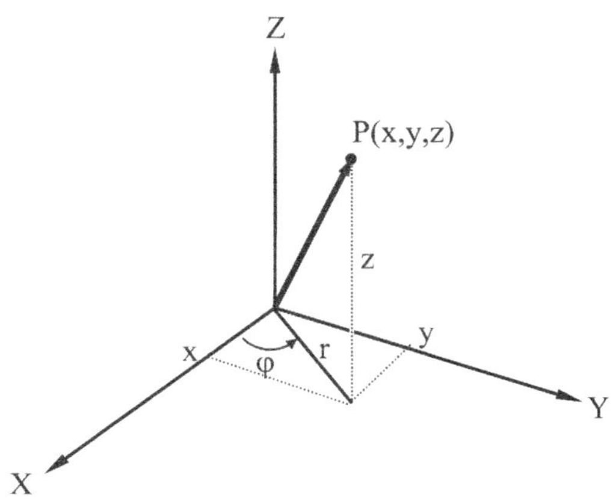

Figure 15.1 Cartesian and cylindrical coordinate systems. The (x, y, z) Cartesian coordinates of the P point are converted into (r, φ, z) in the cylindrical system.

Assume now that Eq. (15.3) establishes a good approximation for a case where the complex amplitude $\mathcal{E}_{x0}$ is not position independent, yet varies smoothly as function of the space (x, y, z) coordinates. Namely, $\Psi = \mathcal{E}_{x0}(x, y, z)\, e^{ikz}$. We set it in Eq. (15.3) and get

$$\boxed{\frac{\partial^2 \mathcal{E}_{x0}}{\partial x^2} + \frac{\partial^2 \mathcal{E}_{x0}}{\partial y^2} + 2ik\frac{\partial \mathcal{E}_{x0}}{\partial z} = 0} \ . \tag{15.4}$$

In writing of Eq. (15.4), we neglected the $\partial^2 \mathcal{E}_{x0}/\partial z^2$ addend; it would be justified if the variation of $\mathcal{E}_{x0}$ as a function of z is sufficiently moderate such that its second derivative according to z is extremely small compared with all other second derivatives. Interestingly, Eq. (15.4) appears analogous to a two-dimensional time-dependent Schrödinger equation, where the z axis variable plays the role of time t (Eq. 3.15). We now replace the $\mathcal{E}_{x0}$ symbol simply with $\mathcal{E}$.

In many cases, the amplifying mediums, as well as their outputs, exhibit a cylindrical symmetry. To express that symmetry, and to allow proper analysis, we convert Eq. (15.4) into the cylindrical (r, φ, z) coordinates defined in Figure 15.1. In Section 3.3, we described the initial steps relevant for conversion of Laplace operator Δ into spherical coordinates. The conversion steps in the present case are analogous. Because the mathematical procedure is quite long, we satisfy by presenting the end result. After conversion, Eq. (15.4) in the cylindrical coordinates reads

$$\boxed{\frac{1}{r}\frac{\partial}{\partial r}\left(r\frac{\partial \mathcal{E}}{\partial r}\right) + \frac{1}{r^2}\frac{\partial^2 \mathcal{E}}{\partial \varphi^2} + 2ik\frac{\partial \mathcal{E}}{\partial x} = 0} \ . \tag{15.5}$$

15.2 Transverse Single-Mode Gaussian Beam

We start by presenting a particular solution to Eq. (15.5), which is often called the **fundamental transverse mode**, and will be addressed hereinafter as a **transverse single-mode Gaussian beam**, denoted as TEM_{00}. The mathematical meaning of the notation will become clear later, on presenting the entire set of solutions. The particular solution will be of the form

$$\mathcal{E} = \mathbf{exp}\left[-i\left(P(z) + \frac{k}{2q(z)}r^2\right)\right] . \tag{15.6}$$

Apparently, the solution is independent of the azimuth φ angle. During further analysis, it will be apparent that the $P(z)$ quantity is a complex function, representing the variation of the beam phase on propagation along the Z axis, and the $q(z)$ quantity is a complex function representing the beam wave-front radius of curvature (nearly spherical close to the propagation axis) on propagation along the Z axis. Setting Eq. (15.6) in Eq. (15.5), we get that $q(z)$ and $P(z)$ need to satisfy the following conditions:

$$\frac{dq(z)}{dz} = -1 \ ; \qquad \frac{dP(z)}{dz} = \frac{i}{q(z)} . \tag{15.7}$$

Addressing the condition on the left, one receives a solution of the form $q(z) = q_0 - z$. The integration constant $q_0 \equiv q(z=0)$ obviously depends on selection of the $z = 0$ point along the beam propagation axis. Before addressing that needed selection, we resolve the $q(z)$ complex function into its real and imaginary parts as follows:

$$\frac{1}{q(z)} = \frac{1}{\mathcal{R}(z)} - i\frac{\lambda}{\pi w^2(z)}, \tag{15.8}$$

where $\lambda = 2\pi/k$ is the beam wavelength. We set a boundary condition as $\mathcal{R}(0) = \infty$ and denote $w(0) = w_0$. The $z = 0$ point is called the **waist position**, and w_0 is called the **waist width,** or the **beam waist radius**. Setting the above boundary condition in Eq. (15.8), we get $q_0 = i(\pi {w_0}^2/\lambda)$. Then

$$q(z) = i\frac{\pi {w_0}^2}{\lambda} - z \quad ; \qquad P(z) = -i\,\boldsymbol{ln}\left[\frac{q(z)}{C}\right], \tag{15.9}$$

where C is an integration constant. We set the expression for the $q(z)$ boundary condition of Eq. (15.9) into the left side of Eq. (15.8), equate between the real and imaginary parts belonging to the two sides, and get

$$\boxed{w(z) = w_0\left[1 + \left(\frac{\lambda z}{\pi {w_0}^2}\right)^2\right]^{1/2} \quad ; \qquad \mathcal{R}(z) = z\left[1 + \left(\frac{\pi {w_0}^2}{\lambda z}\right)^2\right]}. \tag{15.10}$$

Using a direct calculation, we may readily observe that the $w(z)$ quantity exhibits a w_0 minimum value at the $z = 0$ point. The $\mathcal{R}(z)$ quantity exhibits a $2\pi {w_0}^2/\lambda$ minimum value at the $z = \pi {w_0}^2/\lambda$ point.

The physical meaning of w and $\mathcal{R}$ quantities clarifies when Eq. (15.8) is set into Eq. (15.6). One gets

$$\mathcal{E} = C\left[\frac{1}{\mathcal{R}(z)} - i\frac{\lambda}{\pi w^2(z)}\right]\boldsymbol{exp}\left(-\frac{r^2}{w^2(z)}\right)\boldsymbol{exp}\left(-i\frac{\pi r^2}{\lambda \mathcal{R}(z)}\right). \tag{15.11}$$

For any z point, the $\boldsymbol{exp}(-r^2/w^2)$ factor determines the variation of the electric field as a function of the radial distance r from the beam propagation axis. The w quantity is known as the **beam radius**, and it shows the distance from the beam axis where the electric field reduces by a $1/e$ factor compared with its value on the axis itself, namely where the power density reduces by a $1/e^2$ factor. The beam intensity (power density) thus exhibits a Gaussian profile about the axis. That is the reason for using the "Gaussian" adjective to refer to the TEM_{00} beam mode (**transverse single-mode Gaussian beam** – see the beginning of the present section).

The last factor in Eq. (15.11), $\boldsymbol{exp}\left(-i\pi r^2/\lambda\mathcal{R}\right)$, describes the electric field phase variation as a function of r. In other words, the field phase at r distance from the axis *lags* behind its axial phase by $\pi r^2/\lambda\mathcal{R}$; namely, for a positive z point, the equi-phase point at r distance is located $(\lambda/2\pi)\cdot\left(\pi r^2/\lambda\mathcal{R}\right) = r^2/2\mathcal{R}$ *before* the z point. In the $r << \mathcal{R}$ limit ("paraxial limit"), the result describes a near circular curve of the $\mathcal{R}$ radius, centered on the beam axis at the $z - \mathcal{R}$ point. Geometrically, it means that $\mathcal{R}(z)$ provides the curvature radius of the wave-front crossing the axis at the z point.

Relying on Eq. (15.11), the beam intensity (power per unit area) would be

$$I(z,r) = Const\cdot\left[\frac{1}{\mathcal{R}^2(z)} + \frac{\lambda^2}{\pi^2 w^4(z)}\right]\boldsymbol{exp}\left(-\frac{2r^2}{w^2(z)}\right). \tag{15.12}$$

As expected, the beam density changes along the axis. However, the behavior of the $|1/q(z)|^2$ pre-exponential factor (square brackets in Eq. [15.12]) is such that the integrated energy flux on a transverse beam cross section fails to provide a constant value as a function of z (still being inscribed – neither zeroes nor diverges). That behavior violates the energy conservation law! The mathematical reason stems

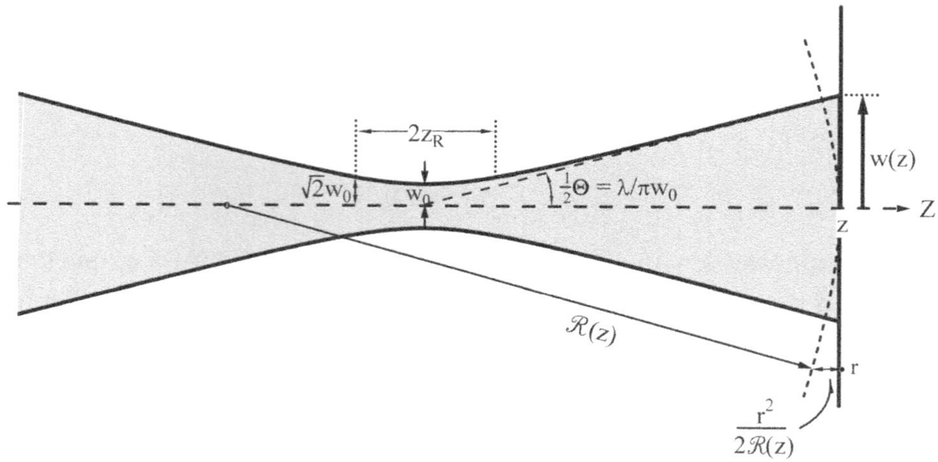

Figure 15.2 Fundamental parameters and propagation layout of a single-mode transverse Gaussian beam. For details, see the text.

from neglecting the electric field second derivative according to z in the development of Eq. (15.4). To comply with the energy conservation law, we modify Eq. (15.12) to the form

$$\boxed{I(z,r) = I_0 \frac{w_0^2}{w^2(z)} \boldsymbol{exp}\left(-\frac{2r^2}{w^2(z)}\right)}, \tag{15.13}$$

where I_0 is the beam intensity on the axis at $z = 0$ (the waist). That way, the energy flux (namely, the beam-integrated intensity at any transverse cross section) is constant, given by $I_0 \cdot 2\pi(w_0/2)^2$. The power fraction included within a radial cross section of $r = w$ radius is 86.5%. The fraction included within a radial cross section of $r = 2w$ radius is 99.9%.

Figure 15.2 shows a layout of the beam radius evolution per Eq. (15.10). In the large z value's limit $(z >> \pi w_0{}^2/\lambda)$, the tangent line describing the $w(z)$ function intersects the axis at the $z = 0$ point, forming a slope of $\frac{1}{2}\Theta = \lambda/\pi w_0$ relative to the beam propagation axis. Thus, the full-beam divergence angle at large z is given by

$$\Theta \equiv \lim_{z\to\infty} \frac{2w(z)}{z} = \frac{2\lambda}{\pi w_0} = 1.27\frac{\lambda}{(2w_0)}. \tag{15.14}$$

Apparently, at a large distance from the waist position, the beam radius grows linearly with distance, and its growth is inversely proportional to the beam waist w_0 . For example, a beam output from a helium-neon (He-Ne) gas laser of a $\lambda = 632.8$ nm wavelength and $w_0 = 1.0$ mm beam waist diverges by approximately 0.4 mrad. At a large distance, the beam appears as emitted out of a mathematical point. The beam wave-front appears spherical, with the sphere center at the waist position (Eq. [15.11]). Another quantity relevant to the Gaussian beam is the **Rayleigh range** parameter $z_R \equiv \pi w_0{}^2/\lambda$. It is the distance from each side of the waist position, where the beam radius increases from w_0 to $w(z_R) = \sqrt{2}w_0$. In the above example, the Rayleigh range is 500 cm.

The above development addressed a special case of an x- polarized beam. For a Gaussian beam, however, all results are valid for all polarization states in the (X, Y) plane.

15.3 Transverse Multi-Mode Beams

We now present the general solution to Eq. (15.5). For brevity, we indicate only the major development steps. The approach is similar to that presented in Section 3.3 for solving Schrödinger's equation of the hydrogen atom. A partial $\mathcal{E}(r,\varphi,z)=\boldsymbol{T}\left(\frac{r}{w}\right)\boldsymbol{\Phi}(\varphi)\boldsymbol{G}(r,z)$ variable separation is assumed, where $G(r,z)$ is the TEM_{00} function presented in the previous section (Eq. [15.6]). Solutions for the equations obtained separately for the $\boldsymbol{T}$ and $\boldsymbol{\Phi}$ functions are

$$\left.\begin{array}{ll}\Phi_m^+(\varphi)=\cos m\varphi;\quad \Phi_m^-(\varphi)=\sin m\varphi; & for\ \ m=integer>0\\ \Phi_m^{\pm}(\varphi)=1; & for\ \ m=0\\ T_p^m\left(\frac{r}{w}\right)=\left(\frac{\sqrt{2}r}{w}\right)^m \boldsymbol{L}_p^m\left(\frac{2r^2}{w^2}\right)\quad ; & for\ \ p=integer\geq 0\end{array}\right\}, \tag{15.15}$$

where the presented $\boldsymbol{L}_p^m(\zeta)$ functions of the ζ variable are the Laguerre associated polynomials presented in Eq. (3.39). For various uses, it is convenient to present those polynomials as

$$\boldsymbol{L}_p^m(\zeta)=\frac{[(m+p)!]^2}{p!m!}\boldsymbol{F}(-p|m+1|\zeta), \tag{15.16}$$

where $\boldsymbol{F}$ is the confluent hyper-geometric series of $p+1$ addends. The said series is described generally for any a and c real parameters as follows:

$$\boldsymbol{F}(a|c|\zeta)\equiv 1+\frac{a}{c}\zeta+\frac{a(a+1)}{2!c(c+1)}\zeta^2+\frac{a(a+1)(a+2)}{3!c(c+1)(c+2)}\zeta^3+\cdots. \tag{15.17}$$

Table 15.1 lists several Laguerre-associated polynomials of low p and m orders.

Table 15.1 Several Laguerre-associated polynomials of low p and m orders.

p	$\boldsymbol{L}_p^m(\zeta)$
0	$\boldsymbol{L}_0^0=1;\ \boldsymbol{L}_0^1=1;\ \boldsymbol{L}_0^2=2;\ \boldsymbol{L}_0^3=6;\ \boldsymbol{L}_0^m=m!;\ \cdots$
1	$\boldsymbol{L}_1^0=1-\zeta;\ \boldsymbol{L}_1^1=4-2\zeta;\ \boldsymbol{L}_1^2=18-6\zeta;\ \boldsymbol{L}_1^3=96-24\zeta;\ \cdots$
2	$\boldsymbol{L}_2^0=2-4\zeta+\zeta^2;\ \boldsymbol{L}_2^1=18-18\zeta+3\zeta^2;\ \boldsymbol{L}_2^2=144-96\zeta+12\zeta^2;\ \boldsymbol{L}_2^3=1200-600\zeta+60\zeta^2;\cdots$
3	$\boldsymbol{L}_3^0=6-18\zeta+9\zeta^2-\zeta^3;\ \boldsymbol{L}_3^1=96-144\zeta+48\zeta^2-4\zeta^3;\ \cdots$

The different solutions for $\mathcal{E}$ are usually denoted as TEM_{pm} as follows

$$TEM_{p,m\pm}(r,\varphi,z)\equiv\left(\frac{\sqrt{2}r}{w}\right)^m L_p^m\left(\frac{2r^2}{w^2}\right)\Phi_m^{\pm}(\varphi)\,G(r,z). \tag{15.18}$$

The *TEM* function notation is an acronym of the English term "transverse electromagnetic mode." The absolute square of the electric field provides the beam intensity on each transverse cross section. Introducing the $\rho\equiv 2r^2/w^2$ variable, the beam intensity of each transverse electromagnetic beam may be presented in the form

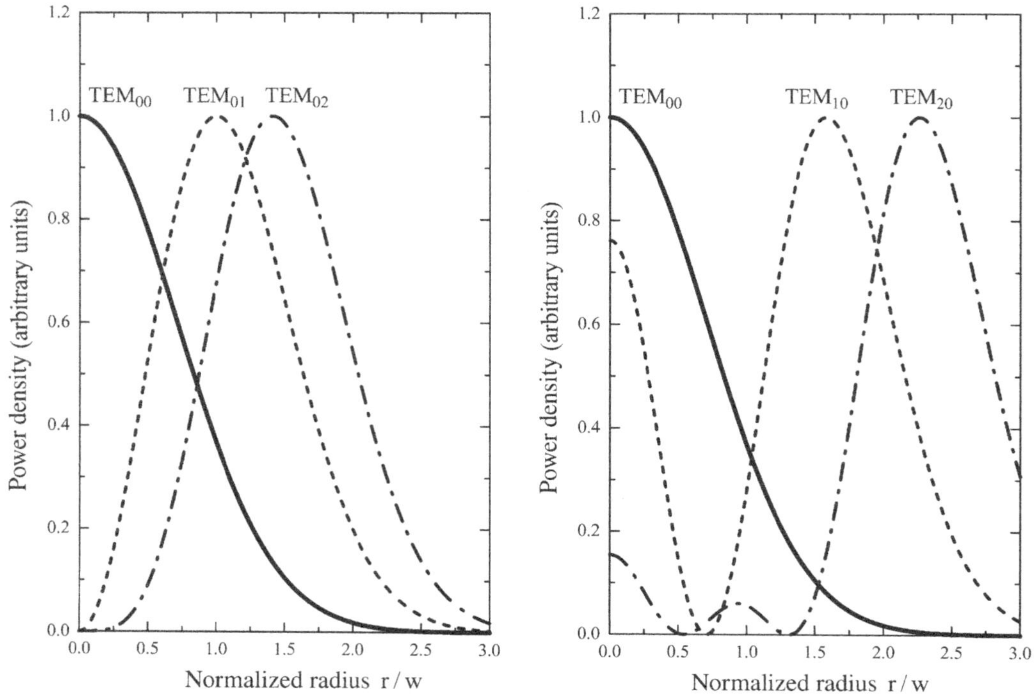

Figure 15.3 Intensity distributions as functions of the normalized radius *r/w* for TEM_{00}, TEM_{01}, and TEM_{02} transverse modes (figure on the left), and TEM_{00}, TEM_{10}, and TEM_{20} (figure on the right). All distributions are normalized to a unity peak value.

$$\boxed{I_{pm\pm}(\rho,\varphi,z) \equiv I_0^{pm}\rho^m\left[\boldsymbol{L}_p^m(\rho)\right]^2\left|\Phi_m^{\pm}(\varphi)\right|^2\boldsymbol{exp}(-\rho)}\quad, \tag{15.19}$$

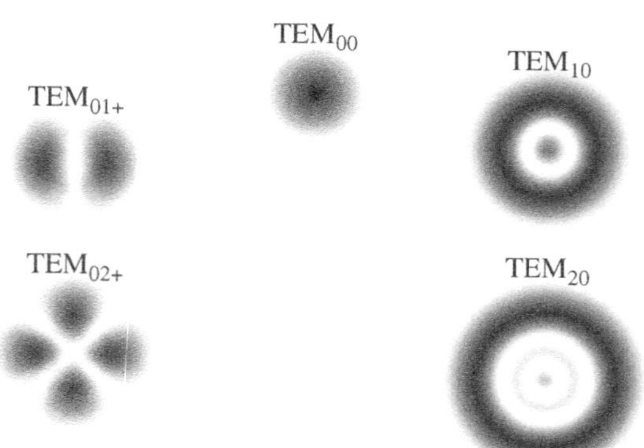

Figure 15.4 Negative brightness distribution images of the transverse modes, whose radial distributions are shown in Figure 15.3. The not-shown TEM_{01-} and TEM_{02-} images would exhibit the same shapes as TEM_{01+}, and TEM_{02+} ones, yet 90° and 45° rotated about their axes, respectively.

where I_0^{pm} is the integrated intensity (flux) contributed by the *pm*-order mode.

The intensity distribution (brightness) presented in Eq. (15.19) is a multiplication between a radial and angular parts. The angular part is constant, of a unity value, for the modes belonging to $m = 0$, namely for the TEM_{p0} functions. The produced images exhibit *p* dark rings of a common center, where each ring belongs to a zero value of the $\boldsymbol{L}_p^0(\rho)$ function. The radial intensity distribution decays with ρ due to the multiplying $\boldsymbol{exp}(-\rho)$ factor. The *pm* mode center would be bright for $m = 0$, but dark in all other cases because of the multiplying ρ^m factor. These modes exhibit $2m$ bright lobes. In Figure 15.3, we present the radial intensity distribution for several transverse modes, demonstrating the above description. Figure 15.4 presents brightness images on a gray scale after inclusion of the azimuth (φ-angle) dependence. The images are actually presented by their *negative* grayness; namely, the brightness is rather proportional to the darkness grade.

15.4 Selecting a Transverse Mode for a Laser Output

In principle, the transverse mode of a laser output is a result of boundary conditions determined by a combination of the spatial distribution of the population inversion in the amplifying medium and the cavity structure (especially the reflecting mirrors). If the mirror's design uses curved radiuses that coincide with the curvatures of the corresponding beam sides, the resonating beam will exhibit a waist radius w_0 corresponding to the said curvatures. In laser technology, care is taken to achieve maximal matching between the desired waist radius and the transverse pumping profile, with a goal of achieving maximal utilization of the pump energy.

The preferred transverse mode is the TEM_{00} because its divergence is smallest and its transverse distribution is the smoothest. To prevent the development of higher-order modes, spatial filtering means are used inside the cavity. The simplest means is an aperture that limits the reflected beam diameter on each mirror. Optimally, $4w(z)$ is the used diameter of such apertures, which allow use of 99.5% of that TEM_{00} mode energy. Figure 15.5 presents a schematic structure of a simple laser cavity, with concave mirrors and apertures, assuring an output of a single TEM_{00} transverse mode. The arrangement of two concave mirrors is appropriate when the cavity edges reside on two opposite sides of the beam waist. If both edges reside on the same side, one mirror will be concave, and the other convex.

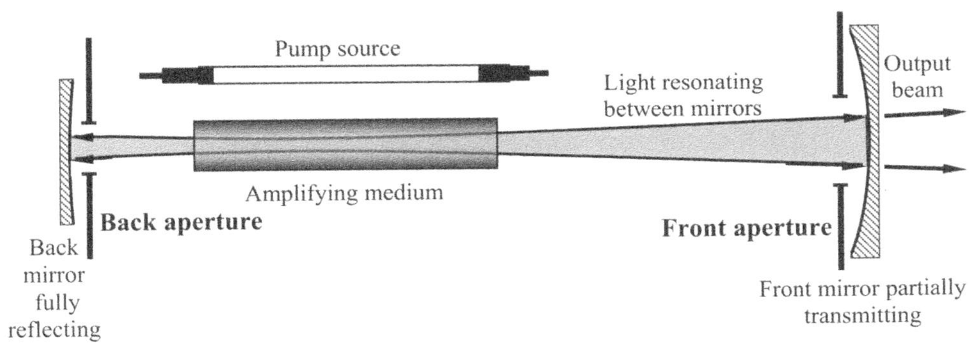

Figure 15.5 Schematic illustration of a typical laser system construction added with concave mirrors and apertures, intended to support and select the transverse TEM_{00} beam mode.

15.5 Lens Crossing of a Single-Mode Transverse Gaussian Beam

Lenses are a common means for manipulating the propagation nature of light in general, and laser beams specifically, through space. We assume that the reader is acquainted with the basic theory of geometrical optics. The lens equation, valid for thin lenses under paraxial (near-axis) conditions, is

$$\frac{1}{s} + \frac{1}{s'} = \frac{1}{f} \quad , \tag{15.20}$$

where s and s' are the object and image distances from the lens, respectively, and f is the lens focal length. The lens magnification m (ratio between the image and object linear sizes) is given by $m = s'/s$. These equations are, in fact approximations, obtained under the assumption that each beam wave-front curvature (reciprocal curvature radius) emanating from a point source on one side of the lens, reduces by $1/f$ after crossing toward the other lens's side. Also assumed is that the mediums on the two lens's sides are identical. We further recall that the focal length of a converging lens is positive, and the focal length of a diverging lens is negative.

The above relations are generalized for Gaussian TEM_{00} beams by defining the waist of a beam propagating toward the lens as the object, and the waist of the beam propagating away from the lens as the image. The fundamental underlying assumption is that on crossing a thin lens, the beam radius remains unchanged, but its curvature reduces by $1/f$. Using Eq. (15.10) and the z_R Rayleigh range definition $z_R \equiv \pi w_0^2/\lambda$ (see Section 15.2), one obtains the following relations:

$$\boxed{\begin{aligned} &(i)\quad \frac{1}{s + z_R^2/(s-f)} + \frac{1}{s'} = \frac{1}{f} \\ &(ii)\quad m \equiv \frac{w_0'}{w_0} = \left(\frac{z_R'}{z_R}\right)^{1/2} = \left\{[1-(s/f)]^2 + (z_R/f)^2\right\}^{-1/2} \end{aligned}}, \qquad (15.21)$$

where s and s' are the "object" and "image" distances from the lens, respectively; w_0 and w_0' are the object and image radiuses, respectively; and z_R and z_R' are the Rayleigh ranges of the object and image, respectively. The first equation – (15.21-i) – is called the **lens equation for Gaussian beams**. The second equation – (15.21-ii) – expresses the image magnification relative to the object.

We start by analyzing the significance of Eq. (15.21-i) above. We convert its form to read

$$\frac{1}{(s/f) + (z_R/f)^2/[(s/f)-1]} + \frac{1}{(s'/f)} = 1. \qquad (15.22)$$

It is obvious that Eq. (15.22) reduces into Eq. (15.20) in the $z_R << f$ limit. To emphasize rather the differences in ranges where $z_R \sim f$, we perform a different form of change, and write

$$\left(s'/f\right) = 1 + \frac{[(s/f)-1]}{[(s/f)-1]^2 + (z_R/f)^2}. \qquad (15.23)$$

Following Eq. (15.23), in Figure 15.6 we plotted the normalized axial image distance (s'/f) as a function of the normalized axial object distance (s/f) from a thin lens for different cases of the normalized Rayleigh range (z_R/f). Notably, the $(z_R/f) = 0$ case in Eq. (15.23) actually returns to the geometrical optics conditions (Eq. [15.20]). The appearance of a $(z_R/f) \neq 0$ parameter in Eq. (15.23) removes the mathematical pole at the $(s/f) = 1$ point. As well known, the pole meaning is that placing of a light source (object) at the focal point on one side of the lens results in a parallel beam propagating away from the lens. In other words, the image distance is infinite! That, in fact, is the *definition* of the focal distance.

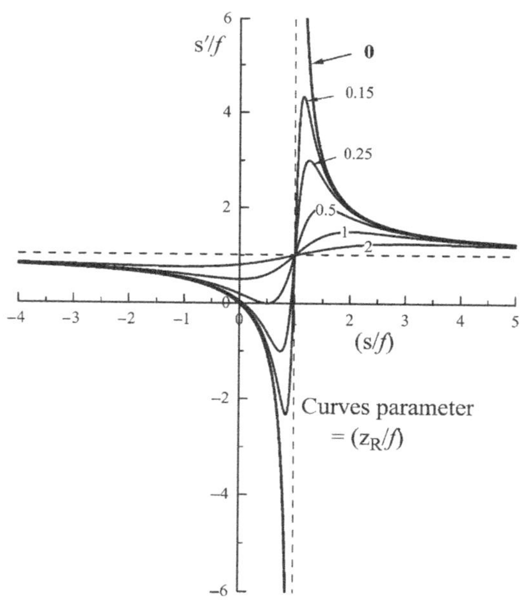

Figure 15.6 Normalized image distance (s'/f) of a single-mode transverse Gaussian beam as a function of the normalized object distance (s/f) from a thin lens, for different cases of the normalized Rayleigh range (z_R/f), according to Eq. 15.23.

A special effect observed in Figure 15.6 is that all curves belonging to $(z_R/f) \neq 0$ cross the common (1, 1) point. In other

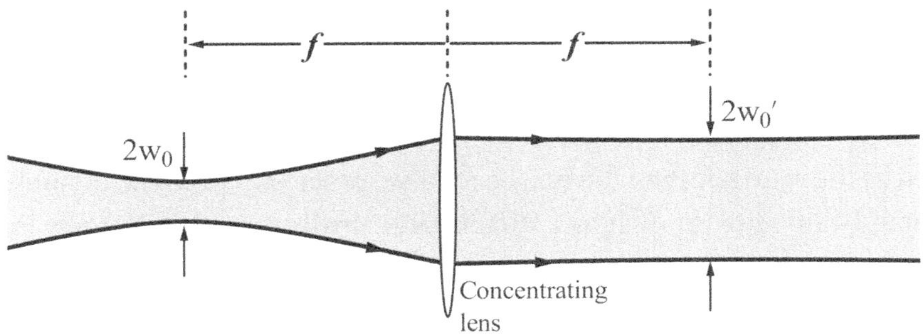

Figure 15.7 A propagation trace of a single-mode transverse Gaussian beam through a concentrating lens where the incident beam waist is located at the focal position before the lens. The incident beam Rayleigh range is a third of the focal length; thus, the transmitted beam waist is three times larger than the incident beam waist.

words, if the lens is placed so that its distance s from the waist position of a propagating beam equals the focal length f, the exiting beam will converge to create a waist at a distance f on the other side of the lens, irrespective of the Rayleigh range (z_R/f) size. Figure 15.7 schematically describes such a case. In the case of a negative lens, the common (1, 1) point relates to a situation where the incident beam exhibits an imaginary waist on the lens's back focal position, and the transmitted beam exhibits an imaginary waist on the lens's front focal position. It is further seen that for Gaussian beams, the image distance from the lens is always finite. It may be readily shown that the maximal image distance is $s'_{max} = f + f^2/2z_R$, and it is achieved for an object (waist) placed at an $s = f + z_R$ distance from the lens.

As for the magnification m given in Eq. (15.21-ii), Figure 15.8 presents the magnification as a function of the axial s/f normalized waist ("object") distance from the lens for different cases of the z_R/f normalized Rayleigh range. The maximal magnification obtains for $(s/f) = 1$, namely when the beam waist is situated at the focal point. The maximal magnification is (f/z_R).

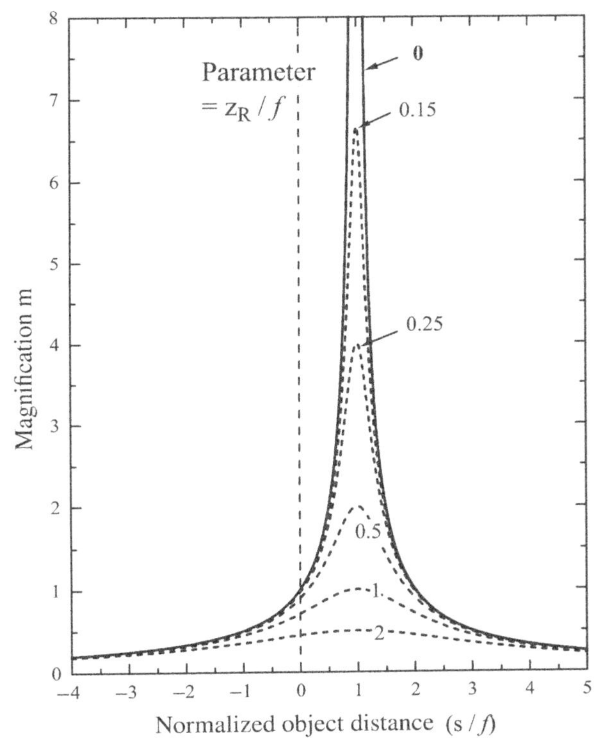

Figure 15.8 Magnification of the waist radius of a single-mode transverse Gaussian beam as a function of its normalized distance s/f for different cases of the normalized z_R/f Rayleigh range, according to Eq. 15.21.

15.6 Multi-Mode Transverse Gaussian Beams

As seen in Section 15.2, a TEM_{00} beam exhibits a Gaussian intensity distribution on any cross section perpendicular to the beam propagation direction. A reverse conclusion is not necessarily valid; it is possible that a beam exhibits a Gaussian intensity distribution at all of its cross sections without being a single mode. Imagine a combination of many nearly parallel TEM_{00} beams crossing the major axis at the common $z = 0$ point, all having the same λ wavelength and all exhibiting a w_0 waist at $z = 0$. Assume further that different beams are not mutually coherent, and have different propagation orientations. In particular, the propagation orientations are distributed evenly about the axis, but their flux reduces exponentially as a function of their inclination angle relative to the major axis. Under such conditions, the total intensity distribution is Gaussian on any perpendicular cross section! Such a **multi-mode beam**, however, differs from a single-mode one by exhibiting a greater divergence. Their **multi-mode quality figure** M^2 is defined as the ratio between the real beam divergence Θ_{real} and the theoretical divergence of a single-mode Gaussian beam Θ_{DL} (known as a **diffraction limited angle**). Namely (see Eq. [15.14]),

$$\Theta_{DL} = \frac{2\lambda}{\pi w_0}\,; \qquad \Theta_{real} = M^2 \frac{2\lambda}{\pi w_0}\,. \tag{15.24}$$

In continuation of Eq. (15.10), the divergence of a Gaussian multi-mode beam would be

$$w(z) = w_0 \left[1 + M^4 \left(\frac{\lambda z}{\pi {w_0}^2}\right)^2\right]^{1/2}, \tag{15.25}$$

where always $M^2 \geq 1$. The beam divergence may be correspondingly considered as if the Rayleigh range z_R reduces by an M^2 factor. Thus, transition through a lens of a multi-mode Gaussian beam becomes similar to geometrical (ray) optics as M^2 increases. A corresponding $K \equiv 1/M^2$ quantity is the **beam quality**. An important goal in laser technology is to achieve output laser beams of diffraction limited quality, namely of the highest $K = 1$ beam quality.

15.7 Crystal Optics

15.7.1 General Description

In Section 10.1, we applied Maxwell's equations for a dielectric medium while classifying the dielectric tensors $\varepsilon_{ij}(\omega, \underline{k})$ by the crystalline system they belong to (see Table 10.1). In following discussions it has been demonstrated, that the tensor elements may also be complex functions, possibly also depending on the $\underline{k}$ wave vector. Experiments prove that for sufficiently high electromagnetic frequencies compared with lattice vibrational frequencies, it may be safely assumed that all dielectric tensor elements are real, and independent of the $\underline{k}$ wave vector orientation. Our discussion of crystal optics is based on that approximation.

The dielectric tensor is symmetric, and thus can always be diagonalized, and the number of independent parameters is maximally three. The three principal $X,\ Y,\ Z$ axes of the dielectric tensor obviously relate to the crystalline symmetry. Table 10.1 addressed that relation. For the optical region, the addressed parameter is rather the "refractive index" and not the dielectric constant. Thus, $\varepsilon_{ii} = {n_i}^2$, where n_i is the "ii" element (component) of the optical refractive index; the "i" subscript label indicates one of the principal axes.

In the following section, we describe a geometric procedure that allows us to find the refractive index of light propagating in a known $\hat{\boldsymbol{k}}$ orientation in a crystalline medium, where $\hat{\boldsymbol{k}} \equiv \underline{k}/k$ is a unit vector in the $\underline{k}$ vector orientation. For brevity, we will not present a proof for the procedure.

A spatial ellipsoid is sketched according to the relation

$$\frac{x^2}{{n_x}^2} + \frac{y^2}{{n_y}^2} + \frac{z^2}{{n_z}^2} = 1\,, \tag{15.26}$$

where (x, y, z) are Cartesian coordinates co-oriented with the X, Y, Z principal axes of the dielectric tensor. The said ellipsoid is called the "index ellipsoid" or "indicatrix". Its crossing points with the axes are correspondingly (n_x, n_y, n_z). A plane crossing the origin O is sketched, oriented perpendicularly to $\hat{\boldsymbol{k}}$. Its crossing curve with the index ellipsoid describes an ellipse. The orientations of the long and short ellipse axes establish the allowed linear polarization orientations for a normal plane wave propagating in the $\hat{\boldsymbol{k}}$ direction. The full length of each axis is twice the refractive index for the same linear polarization: $2n_a$ and $2n_b$, respectively. The said procedure is demonstrated in Figure 15.9.

It should be emphasized that each said polarization orientation relates to the electric displacement field vector $\mathcal{D}$ and not the electric field vector $\mathcal{E}$. In an anisotropic dielectric medium, there is some small difference between the two, as detailed in Figure 10.1; we will address that issue further in section 15.7.3 below.

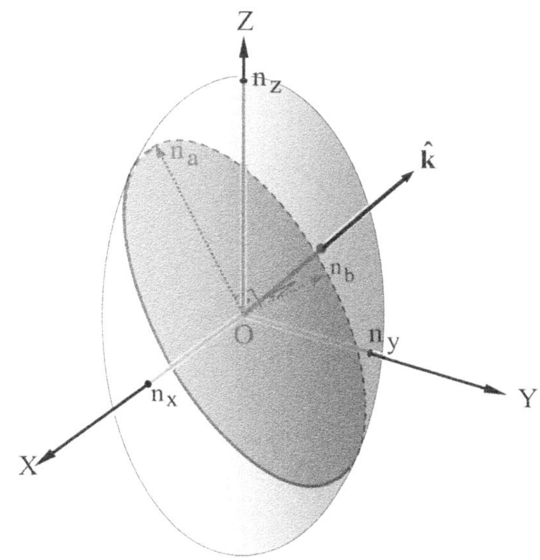

Figure 15.9 Demonstration of a refractive index ellipsoid and the geometric procedure that enables us to find the normal polarization states and their refractive indices for a beam propagating in a given orientation. For details, see text.

15.7.2 Uniaxial Crystals

The refractive indices of crystals belonging to the tetragonal, trigonal, and hexagonal systems satisfy the relation $n_x = n_y \neq n_z$ (see Table 10.1). The $n_o \equiv n_x = n_y$ quantity is called the **ordinary refractive index**; the quantity $n_e \equiv n_z$ is called the **extraordinary refractive index**. The Z axis is called the **optic axis**. It exhibits the important property that the refractive indices of any beam propagating in the optic axis direction are exactly n_o for all polarization states. Such crystals are called "uniaxial". When $n_e > n_o$, they are known as "positive uniaxial"; when $n_e < n_o$ they are known as "negative uniaxial". Their $\Delta n \equiv n_e - n_o$ difference is termed "refractive index anisotropy".

The mathematical treatment of light propagating in a uniaxial crystal is considerably simpler than that of a general case. First, it is possible to select either the X axis or the Y axis arbitrarily at any orientation in the plane perpendicular to the Z optic axis. Specifically, we may choose the Y axis in the plane generated by the Z axis and the $\hat{\boldsymbol{k}}$ vector, as presented in Figure 15.10. The figure demonstrates a case of a positive uniaxial crystal. The index ellipsoid is a rotation ellipsoid, symmetric about the Z axis, and the figure describes a cut in the (y, z) plane; θ is the angle between $\hat{\boldsymbol{k}}$ and Z. As demonstrated in Figure 15.10, the contour on a cut plane of the index ellipsoid that crosses the origin perpendicularly to the $\hat{\boldsymbol{k}}$ vector is an ellipse. Its long axis size is $2n_e(\theta)$, as indicated in Figure 15.10; its short axis size is $2n_o$, and it is perpendicular to Z on the origin O point. A beam propagating in the $\hat{\boldsymbol{k}}$ direction thus has two normal polarization states: one perpendicular to Z, of n_o refractive index, and known as "ordinary polarization", and the other perpendicular to $\hat{\boldsymbol{k}}$ on the (y, z) plane, of $n_e(\theta)$ refractive index, and known as "extraordinary polarization". The extraordinary index $n_e(\theta)$ depends on the θ angle, and it changes between n_o for $\theta = 0$ and n_e for $\theta = 90°$.

The validity of the following relations may be readily verified:

$$n_e^2(\theta) = z^2 + y^2; \qquad \frac{z}{n_e(\theta)} = \sin\theta,$$

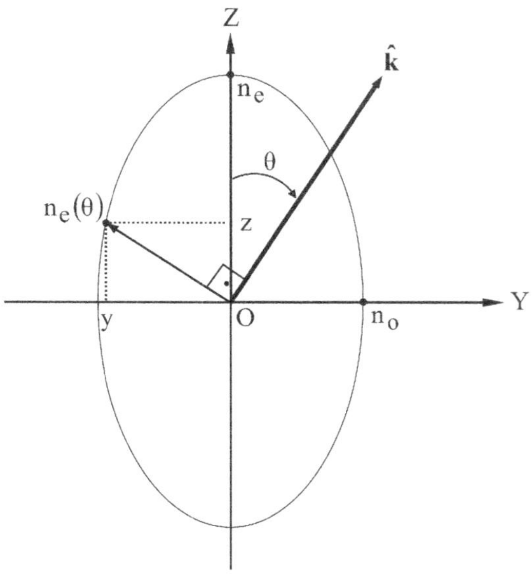

Figure 15.10 Cross section of the index ellipsoid of a uniaxial crystal on the (y,z) plane. $n_e(\theta)$ is the extraordinary refractive index of a plane-parallel beam propagating inside the crystal in the $\hat{k}$ direction.

in addition to the ellipse equation

$$\frac{y^2}{{n_o}^2} + \frac{z^2}{{n_e}^2} = 1 \,,$$

wherefrom we get

$$\boxed{\frac{1}{{n_e}^2(\theta)} = \frac{\cos^2\theta}{{n_o}^2} + \frac{\sin^2\theta}{{n_e}^2}} \,. \tag{15.27}$$

In the following, we present several effects related to light propagation in an anisotropic medium, and various resulting utilizations.

15.7.3 Walk-Off

Consider an optically polished plane-parallel plate of an anisotropic uniaxial material related to Figure 15.10, cut perpendicularly to the $\hat{k}$ light propagation orientation. Consider a beam of unpolarized light incident perpendicularly to the plate surface (Figure 15.11). It would split inside the plate into two polarization components, one X-polarized (perpendicularly to the page plane) belonging to the ordinary refractive index n_o, the second polarized in the page plane perpendicularly to X, belonging to the extraordinary refractive index n_e . Nevertheless, as noted in the text following Eq. (15.26), the said polarization components belong to the displacement vectors $\underline{\mathcal{D}}$ and not to the electric field $\underline{\mathcal{E}}$. Because the displacement vector's normal component transfers continuously from vacuum into matter (see also the footnote in the text of Section 14.6.1), the displacement vector's inside matter must be parallel to the surface, because the incident beam exhibits no component perpendicular to the surface. As for the ordinary beam component, the $\underline{\mathcal{E}}$ orientation coincides with that of $\underline{\mathcal{D}}$. Nevertheless, for the extraordinary component, the $\underline{\mathcal{E}}$ orientation differs from that of $\underline{\mathcal{D}}$ (see Figure 10.1).

At that point, it is important to clarify the "propagation orientation" concept. The $\underline{k}$ vector actually describes the propagation direction of the beam *phase*. The beam *energy*, however, propagates in the $\underline{S} = (c/4\pi)\underline{\mathcal{E}} \times \underline{\mathcal{H}}$ Poynting vector direction. Thus, the ordinary beam component propagates along the $\underline{k}$ vector orientation, but the extraordinary component propagates perpendicularly to the $\underline{\mathcal{E}}$ vector, with an orientation different from that of $\underline{k}$. This effect is called **walk-off**. It should be emphasized that phase propagation of the off-walking beam is identical with that of the ordinary beam, namely perpendicular to $\underline{k}$. On exiting the plate, the two beams return to propagate in a parallel manner, yet along different axes, each exhibiting a different polarization. Let ρ be the walk-off angle, and δ be the beams axes separation on exiting the plate. Figure 15.11 provides a schematic sketch of the said effect.

We now outline briefly the steps needed to calculate the walk-off angle δ. Let $\hat{\boldsymbol{d}}$, $\hat{\boldsymbol{e}}$, $\hat{\boldsymbol{h}}$, $\hat{\boldsymbol{s}}$, and $\hat{\boldsymbol{z}}$ be unit vectors in $\underline{\mathcal{D}}$, $\underline{\mathcal{E}}$, $\underline{\mathcal{H}}$, the Poynting vector $\underline{S}$, and the Z axis, respectively. The relation between $\underline{\mathcal{D}}$ and $\underline{\mathcal{E}}$ is given by $\underline{\mathcal{D}} = (\varepsilon)\underline{\mathcal{E}}$, where (ε) is the dielectric tensor, or alternatively by $\underline{\mathcal{E}} = (\varepsilon)^{-1}\underline{\mathcal{D}}$. Obviously,
$(\varepsilon)^{-1} = \begin{pmatrix} {n_o}^{-2} & 0 & 0 \\ 0 & {n_o}^{-2} & \\ 0 & & {n_e}^{-2} \end{pmatrix}$. Considering Figures 15.10 and 15.11, it is seen that $\hat{\boldsymbol{d}} = (0, -\boldsymbol{\cos}\theta, \boldsymbol{\sin}\theta)$.

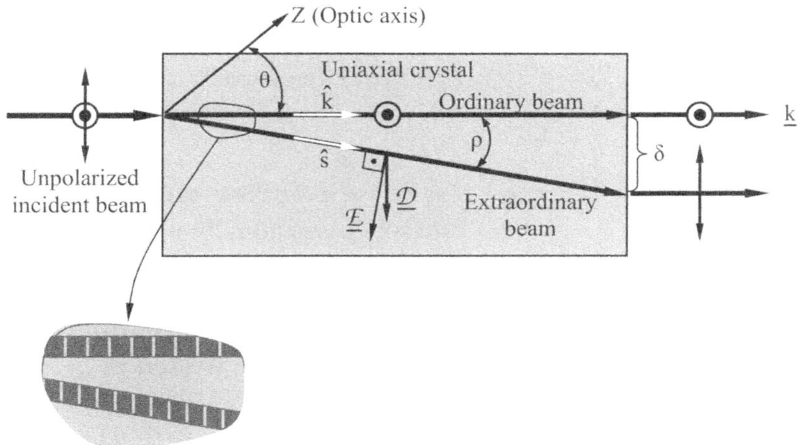

Figure 15.11 Schematic illustration of walk-off propagation inside matter of the extraordinary component of an unpolarized beam incident perpendicularly to a plane parallel plate of a uniaxial dielectric crystal, where its optic axis is neither perpendicular nor parallel to the plate planes. The enlarged sketch on the left bottom demonstrates that both beams' fronts (white lines) are perpendicular to $\hat{k}$; their different period demonstrates their different inner-matter wavelengths, because $n_e(\theta) \neq n_o$.

Use is made of the relation between $\mathcal{D}$ and $\mathcal{E}$ for calculating $\hat{\mathbf{e}}$. Also, $\hat{\boldsymbol{h}} = (1, 0, 0)$. Then $\hat{\mathbf{s}} = \hat{\mathbf{e}} \times \hat{\boldsymbol{h}}$ is calculated, hence $\boldsymbol{\cos}(\theta + \rho) = \hat{\mathbf{s}} \cdot \hat{\mathbf{z}}$, where $\hat{\mathbf{z}} = (0, 0, 1)$. Therefore,

$$\rho = \boldsymbol{arc\ cos}\left[\frac{\boldsymbol{cos}\,\theta}{n_o{}^2}\left(\frac{\boldsymbol{cos}^2\theta}{n_o{}^4} + \frac{\boldsymbol{sin}^2\theta}{n_e{}^4}\right)^{-1/2}\right] - \theta . \tag{15.28}$$

15.8 Retardation Plates

Consider an L thick parallel plate of a uniaxial dielectric crystal machined such that the Z optic axis is parallel to the plate plane (Figure 15.12). A plane-parallel beam of a linearly polarized monochromatic light of λ wavelength propagates in the Y direction incident perpendicularly to the plate surface plane, so that the polarization orientation is inclined by a φ angle relative to the optic axis. The figure describes the electric field state on the $y = 0$ plane at the $t = 0$ moment, where the electric field assumes its maximal $\mathcal{E}_0$ size. On entering the crystal, the beam separates into two normal oscillation components: one X polarized, whose amplitude is $\mathcal{E}_{x0} = \mathcal{E}_0\,\boldsymbol{sin}\,\varphi$, and the other Z polarized, whose amplitude is $\mathcal{E}_{y0} = \mathcal{E}_0\,\boldsymbol{cos}\,\varphi$. The phase propagation velocities v_ϕ of the two components are different; the X polarized component exhibits $v_\phi = c/n_o$, whereas the Z polarized component exhibits $v_\phi = c/n_e$. Using the dispersion relation $v_\phi = \omega/k$, where ω is the angular frequency and k is the wave number, we get $k^{(x)} = \frac{\omega}{c} n_o$, and $k^{(z)} = \frac{\omega}{c} n_e$, for wave numbers belonging to X and Z polarizations, respectively. Equations representing the two beams' propagations inside the crystal would be

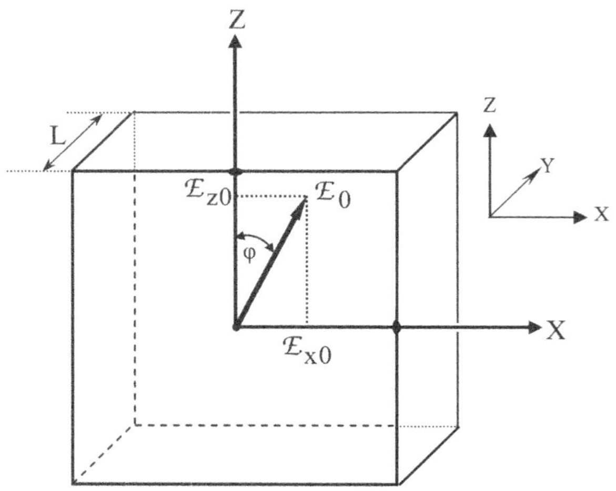

Figure 15.12 A dielectric uniaxial plane-parallel plate with a noted electric field and its components belonging to a monochromatic plane-wave light on incidence on the front plate surface. For discussion and details, see text.

$$\mathcal{E}_x(y,t) = \mathcal{E}_0 \boldsymbol{\sin\varphi}\ \boldsymbol{\cos}\,(\omega t - k_o y) = \mathcal{E}_0 \boldsymbol{\sin\varphi}\ \boldsymbol{\cos}\left(\omega t - \frac{\omega}{c} n_o y\right)$$
$$\mathcal{E}_z(y,t) = \mathcal{E}_0 \boldsymbol{\cos\varphi}\ \boldsymbol{\cos}\,(\omega t - k_e y) = \mathcal{E}_0 \boldsymbol{\cos\varphi}\ \boldsymbol{\cos}\left(\omega t - \frac{\omega}{c} n_e y\right) \quad (15.29)$$

A phase difference between the $\mathcal{E}_z$ and the $\mathcal{E}_x$ components gradually progresses, with a y distance from the front surface given by $\Delta\phi = \frac{\omega}{c}(n_e - n_o)y$. That phase difference is called **retardation**. Defining $t' = t - n_o y/c$, Eq. (15.29) may be written as

$$\mathcal{E}_x\left(y,t'\right) = \mathcal{E}_0\, \boldsymbol{\sin}\,\varphi\ \boldsymbol{\cos}\,\omega t'$$
$$\mathcal{E}_z\left(y,t'\right) = \mathcal{E}_0\, \boldsymbol{\cos}\,\varphi\ \boldsymbol{\cos}\left(\omega t' - \Delta\phi\right) \quad (15.30)$$

The initial linear electric field polarization obtained by summing its two components changes during the beam propagation into the crystal. We consider four special cases:

i) $\Delta\phi = \frac{\pi}{2} + 2\pi j$ with j an integer. Then $\boldsymbol{\cos}(\omega t' - \Delta\phi) = \boldsymbol{\sin}\ \omega t'$, resulting in the relation $\frac{\mathcal{E}_x^2(y,t')}{\mathcal{E}_0^2\, \boldsymbol{\sin}^2\varphi} + \frac{\mathcal{E}_z^2(y,t')}{\mathcal{E}_0^2\, \boldsymbol{\cos}^2\varphi} = 1$, an elliptic equation for the electric field components $\mathcal{E}_x$ and $\mathcal{E}_z$. This means that the electric field vector tip rotates continually along an elliptic circumference whose X-axis length is $2\mathcal{E}_0 \sin\varphi$ and Z-axis length is $2\mathcal{E}_0 \cos\varphi$. The position of the vector edge as a function of time during a full period is described in Figure 15.13. The rotation orientation along the circumference is counter clockwise. Because the beam propagates in the Y direction (into the page), its motion is similar to that of a left-handed screw. The polarization state is thus addressed as **left-handed elliptic polarization**. For $\varphi = 45°$, the two ellipse axes are equal (see Figures 15.12 and 15.13); the ellipse shape then degenerates into a simple circle, and the polarization state is known as **left-handed circular polarization**.

ii) $\Delta\phi = \pi + 2\pi j$ with j an integer. Then $\boldsymbol{\cos}(\omega t' - \Delta\phi) = -\,\boldsymbol{\cos}\omega t'$. Setting it in Eq. (15.30) yields $\mathcal{E}_x(y,t') = \mathcal{E}_0\ \boldsymbol{\sin\varphi\cos}\omega t'$; $\mathcal{E}_z(y,t') = -\ \mathcal{E}_0\ \boldsymbol{\cos\varphi\cos}\omega t'$. The result is a linearly polarized light, yet clockwise rotated by 180 °$- 2\varphi$ (in fact equivalent to a 2φ clockwise rotation). That condition is

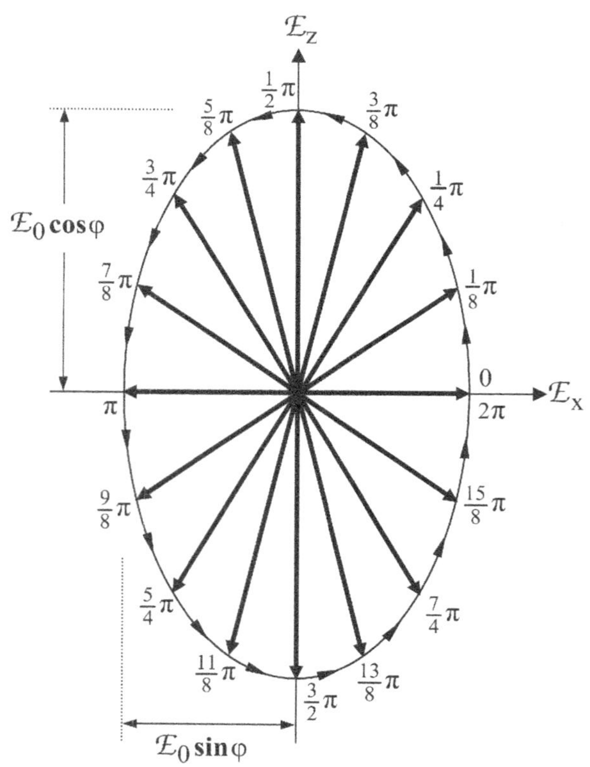

Figure 15.13 Tracing of the electric field state (thick arrow) as a function of time ($\omega t'$) over a full period of the electromagnetic field oscillation, when there is a $\pi/2$ phase retardation between its components. The $\omega t'$ time durations are indicated at the tips of the thick arrows that represent the temporary electric field.

demonstrated in Figure 15.14. In the special case where $\varphi = 45°$, the transmitted beam polarization is perpendicular to the incident beam polarization.

iii) $\Delta\phi = \frac{3}{2}\pi + 2\pi j$ with j as an integer. Then $\cos(\omega t' - \Delta\phi) = -\sin\omega t'$, a situation similar to case (i) above (namely, an elliptic polarization). In contrast, however, to case (i), the rotation orientation of the electric vector edge is rather clockwise, displaying a **right-handed elliptic polarization**. In the particular case of $\varphi = 45°$, it degenerates simply into a **right-handed circular polarization**.

iv) $\Delta\phi = 2\pi j$ with j as an integer. Then, $\cos(\omega t' - \Delta\phi) = \cos\omega t'$, and the polarization state of the transmitted beam returns to be linearly polarized as on incidence, namely in the original orientation.

A suitable selection of the plate thickness L and the inclination angle of its optic axis relative to the incident light polarization orientation (the φ angle), allows for the transformation of a beam polarization state from linear into elliptic or circular (left or right handed), as well as rotating a linear polarization from one orientation into a different one. It is easy to see that

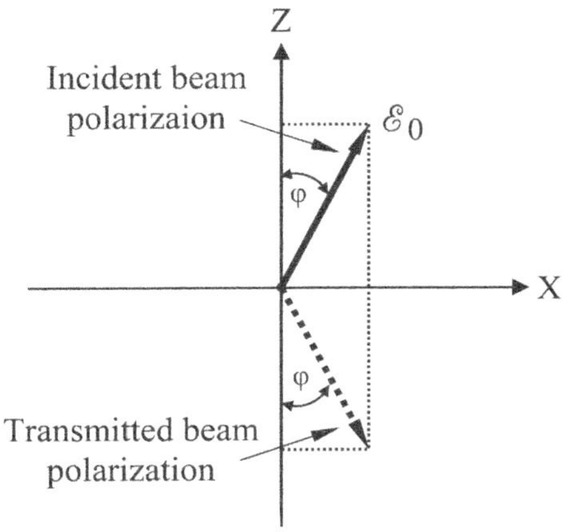

Figure 15.14 Comparison between states of the electric field amplitude of the incident beam (full arrow) and the transmitted beam amplitude (dashed arrow) after accumulation of a π phase retardation between their Z and X electric field components.

(a) A plate whose thickness is $L = \left(\frac{1}{4}\lambda + j\lambda\right)\frac{1}{n_e - n_o}$ or $L = \left(\frac{3}{4}\lambda + j\lambda\right)\frac{1}{n_e - n_o}$, where λ is the vacuum wavelength of the monochromatic incident beam, would transform the linear polarization state into right-handed elliptic or left-handed elliptic, respectively. The transmitted beam will be right-handed circular or left-handed circular, respectively, for $\varphi = 45°$. Such a retardation plate is called a **quarter-wave plate**, and j is called the "plate retardation order". Importantly, in the case of a negative uniaxial crystal ($n_e < n_o$), the right-handedness or left-handedness of the resulting elliptic or circular polarization states are exchanged.

(b) A plate whose thickness is $L = \left(\frac{1}{2}\lambda + j\lambda\right)\frac{1}{n_e - n_o}$ would transform a linear polarization state into a different linear polarization. The polarization orientation would rotate by 90° when $\varphi = 45°$. Such a retardation plate is called a **half-wave plate**; j is the "plate retardation order."

In all types of retardation plates, the polarization state of the transmitted beam is preserved if the orientation of the optic axis is parallel or perpendicular to the incident light polarization.

Retardation plates are often used in laser technology to manipulate, as needed, the polarization states of laser output beams.

Exercises for Chapter 15

15.1 An argon ion (Ar^+) gas laser emits a 100 mW green beam whose wavelength is 514.5 nm. The beam is a single-mode Gaussian, and its waist radius (located on the front mirror) is 0.5 mm.

(a) Calculate the beam Rayleigh range and its divergence angle.

(b) What would be the beam radius and its intensity (power per unit area) on the beam axis at 10 meters distance from the laser front mirror?
(c) A converging lens whose focal length is 1 meter is inserted on the beam path at a distance of 1 meter from the front mirror. Calculate the position and radius of the new waist.

15.2 (a) Perform a detailed calculation to prove Eq. (15.28).
(b) The ordinary and extraordinary refractive indices of a certain uniaxial crystal are $n_o = 1.7853$ and $n_e = 1.8135$, respectively, at the $\lambda = 632.8$ nm wavelength of the helium-neon (He-Ne) gas laser. The material is machined to the form of a 5 mm thick plane-parallel plate, with the optic axis inclined by 45° off the planes surfaces. The said unpolarized light beam is incident normally to the surface. Calculate the walk-off angle of the extraordinary beam component from the ordinary one, and their splitting upon exiting the plate counter side.
(c) Plan the exact thickness of a quarter-wave plate of the said material for the helium-neon laser wavelength; it is given, that for mechanical strength considerations, the plate must be thicker than 2.35 mm. Indicate the chosen order, and present a drawing of the plate, indicating the orientations of the optic axis and of a linearly polarized beam that the plate would use to change its linear polarization state into a circular one.

16 Pulsed High Power Lasers

16.1 Introduction

When laser cavity gain and loss are both constant over time, the laser output is also correspondingly constant. This is known as a continuous-wave (CW) laser. If either gain or loss are time dependent, light pulses are generated, which usually exhibit high peak intensities. It is possible to induce gain changes by changing the pump power. If pulsed pumping is used, the laser output is also pulsed. Such gain changes are called **gain switching**. One may also induce pulsed light output by causing temporal changes to the cavity losses. Such gain losses are called **Q-switching**. The "Q" letter is adopted from the initial letter of the term "quality factor", which is used to describe properties of electrical resonance circuits. The quality factor of an electrical resonance circuit is defined as 2π times the ratio between the energy stored in the circuit at resonance frequency and the energy lost (out radiated or heat converted) over a single oscillation period. In a laser cavity, the quality factor is defined analogously as 2π times the ratio between the electromagnetic energy stored in the cavity and the energy lost over a single back and forth photon transition between the cavity mirrors.

Mechanisms generating Q-switching are classified as "passive" or "active." In a passive mechanism, the cavity is added with a light absorbing element, whose absorption reduces temporarily because of population density changes of its absorbing states as a result of coupling with the cavity radiation field. In the active mechanism, the cavity is added with a loss-causing element that prevents any lasing, which, however, ceases to cause losses due to external triggering. In the present chapter, we discuss several methods and materials used in both passive and active Q-switching.

16.2 Passive Q-Switching Using a Saturable Light Absorber

16.2.1 Saturable Absorbers

Figure 16.1 presents a fundamental energy scheme of a "saturable absorbing" material. Under light incidence, the $|0\rangle$ ground-state absorbs photons of $h\nu$ energy, exciting the matter into the $|1\rangle$ state. The latter then decays very fast, mainly non-radiatively, to a slightly lower $|2\rangle$ state which is meta-stable, meaning it has a relatively long effective lifetime τ. Therefore, if the illumination intensity is sufficiently high, the atoms' population at the $|0\rangle$ state lowers as the population in the $|2\rangle$ state (**excited state**) correspondingly increases. If the light absorption cross section of the excited state σ_{es} is smaller than the absorption cross section of the ground-state σ_{gs} , the material effective absorption coefficient lowers. Increase of the light intensity decreases the absorption coefficient. Saturation of the absorption coefficient dependence on the light intensity is obtained when the first incident photons manage to transfer all ground-state atoms into their excited state.

Physics, Optics, and Spectroscopy of Materials, First Edition. Zeev Burshtein.

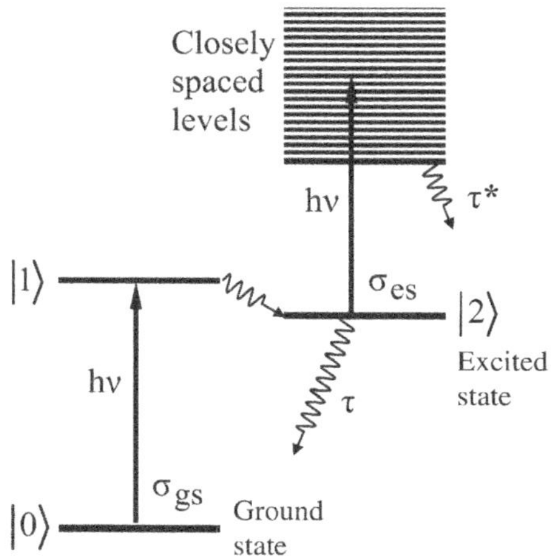

Figure 16.1 Typical energy scheme of a material suitable for performing as a saturable absorber. The decay rate from the excited |1⟩ state to the |2⟩ state is assumed to be very fast compared with the |2⟩ state itself τ^{-1}, and the light absorption cross section of the |2⟩ state is smaller than the absorption cross section of the |0⟩ ground-state (namely $\sigma_{es} < \sigma_{gs}$).

For theoretical analysis, we consider an L-thick parallel plate that contains absorbing ions, satisfying the energy scheme of Figure 16.1. A light pulse of $h\nu$ energy photons is incident perpendicularly to the $z = 0$ plate surface, then transmitted with reduced intensity at the $z = L$ surface (Figure 16.2). In the following section, we present a model where the photons are described as particles of a well-defined position at any moment. Let $I(z, t)$ be the photons' current density inside the plate (photons per unit area per unit time), and $E(z, t)$ be the photons' density (photons per unit volume) at a z distance from the front plate surface at time t. We follow the propagation of the photons' beam through the matter at the speed of light $dz/dt = c$. The rate equation determining the photons density is given by

$$\begin{aligned}\frac{dI(z,t)}{dz} &= \frac{dE(z,t)}{dt} = \frac{\partial}{\partial z}E(z,t)\cdot\frac{dz}{dt} + \frac{\partial}{\partial t}E(z,t)\\ &= c\cdot\frac{\partial}{\partial z}E(z,t) + \frac{\partial}{\partial t}E(z,t)\\ &= -\sigma_{gs}cN(z,t)E(z,t) - \sigma_{es}c[N_{tot} - N(z,t)]E(z,t) + \frac{\partial}{\partial t}E(z,t)\\ &= -\Delta\sigma\, cN(z,t)E(z,t) - \sigma_{es}\, cN_{tot}E(z,t) + \frac{\partial}{\partial t}E(z,t)\end{aligned}, \quad (16.1)$$

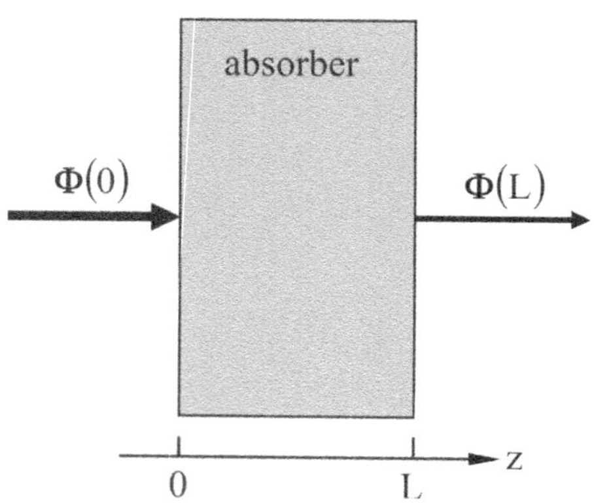

Figure 16.2 Schematic illustration of a light pulse of $\Phi(0)$ flux density (photons per unit area) incident on an absorber plate whose energy scheme is given in Figure 16.1, being transmitted at a lower density $\Phi(L)$. For details, see text.

where N_{tot} is the total concentration of light-absorbing ions in the sample, $N(z, t)$ is the concentration of absorbing ions remaining at the ground-state at time t at the z plane, and $\Delta\sigma \equiv \sigma_{gs} - \sigma_{es}$. The first addend on the third line in Eq. (16.1) describes the photons' concentration reduction rate resulting from absorption by the |0⟩ ground-state ions; the second addend describes the photons' concentration reduction rate resulting from absorption by the |2⟩ excited-state ions. The ions' concentration rate-determining equation for the different states is then

$$\frac{\partial}{\partial t}N(z,t) = -\sigma_{gs}cN(z,t)E(z,t) + \frac{N_{tot} - N(z,t)}{\tau}. \quad (16.2)$$

Eq. (16.2) expresses the growth rate of ion concentration at the ground-state. The first addend on the right side describes the concentration reduction rate resulting from photons' absorption, and the second one describes the concentration increasing rate resulting from decay of the |2⟩ excited-state ions. Both Eqs. (16.1) and (16.2) assume that the lifetime of the band of closely spaced excited states located above |2⟩ is short compared with any time scale where the photons' concentration or other states' concentration may vary. In other words, concentration changes of the |2⟩ state because of photons' absorption are tiny due to immediate back decay from the closely spaced higher levels.

16.2.1.1 Slow Saturable Absorber

The above **slow absorber** term relates to a case where the incident beam is a single light pulse, whose duration t_p is very short compared with the state's $|2\rangle$ decay time τ. It is then possible to neglect the second addend in Eq. (16.2). We then get $\frac{\partial N(z,t)}{N(z,t)} = -\sigma_{gs} cE(z,t)\partial t$. We integrate the equation from $t = -\infty$ until $t = t_p$ (which covers the entire pulse duration). We get

$$\int_{N_{tot}}^{N(z,t_p)} \frac{\partial N(z,t)}{N(z,t)} = -\int_{-\infty}^{t_p} \sigma_{gs} cE(z,t)\partial t = -\sigma_{gs}\boldsymbol{\Phi}(z)\,, \tag{16.3}$$

where $\boldsymbol{\Phi}(z) \equiv \int_{-\infty}^{t_p} cE(z,t)dt$ is the total light pulse flux density (photons per unit area) transmitted through a plane at z distance from the front surface inside the plate. Thus,

$$N(z) \equiv N(z,t_p) = N_{tot}\, \mathbf{exp}\left[-\sigma_{gs}\boldsymbol{\Phi}(z)\right], \tag{16.4}$$

where $N(z)$ is the remaining concentration of ground-state absorbing ions at the z plane immediately after the end of the incident light pulse.

Using the first and fourth lines of Eq. (16.1), we may write

$$\frac{\partial}{\partial z} cE(z,t) = -\Delta\sigma N(z,t)\, cE(z,t) - \sigma_{es} N_{tot}\, cE(z,t)\,. \tag{16.5}$$

Because the second addend in Eq. (16.2) is negligible, we convert Eq. (16.5) to the form

$$\frac{\partial}{\partial z} cE(z,t) = \frac{\Delta\sigma}{\sigma_{gs}} \cdot \frac{\partial N(z,t)}{\partial t} - \sigma_{es} N_{tot}\, cE(z,t)\,. \tag{16.6}$$

We perform a time integral on the latter equation from $t = -\infty$ to $t = t_p$ using Eq. (16.4):

$$\boxed{\frac{d}{dz}\boldsymbol{\Phi}(z) = N_{tot}\frac{\Delta\sigma}{\sigma_{gs}}\left[e^{-\sigma_{gs}\boldsymbol{\Phi}(z)} - 1\right] - N_{tot}\sigma_{es}\boldsymbol{\Phi}(z)}\,. \tag{16.7}$$

We consider now a special case of vanishing excited-state absorption, namely $\sigma_{es} = 0$. A saturable absorber of such a property is considered an "ideal saturable absorber". It is then possible to neglect the second addend in Eq. (16.7). The solution of Eq. (16.7) then yields

$$\boldsymbol{\Phi}(z) = \frac{1}{\sigma_{gs}}\, \boldsymbol{ln}\left\{1 + \left[e^{\sigma_{gs}\boldsymbol{\Phi}(0)} - 1\right] e^{-N_{tot}\sigma_{gs} z}\right\}. \tag{16.8}$$

Neglecting Fresnel surface reflections (see Section 14.6), and using the latter result, the optical transmission coefficient of an L thick plate $T_{FN} \equiv \boldsymbol{\Phi}(L)/\boldsymbol{\Phi}(0)$ is given by

$$\boxed{T_{FN} = \frac{1}{\sigma_{gs}\boldsymbol{\Phi}(0)}\, \mathbf{ln}\left\{1 + T_0\left[e^{\sigma_{gs}\boldsymbol{\Phi}(0)} - 1\right]\right\}; \quad T_0 \equiv e^{-N_{tot}\sigma_{gs} L}}\,. \tag{16.9}$$

Eq. (16.9) is known as the **Frantz-Nodvik** equation. The T_0 quantity expresses the transmission coefficient in the limit of very faint light pulses, satisfying $\sigma_{gs}\boldsymbol{\Phi}(0) << 1$. When the pulse intensity grows,

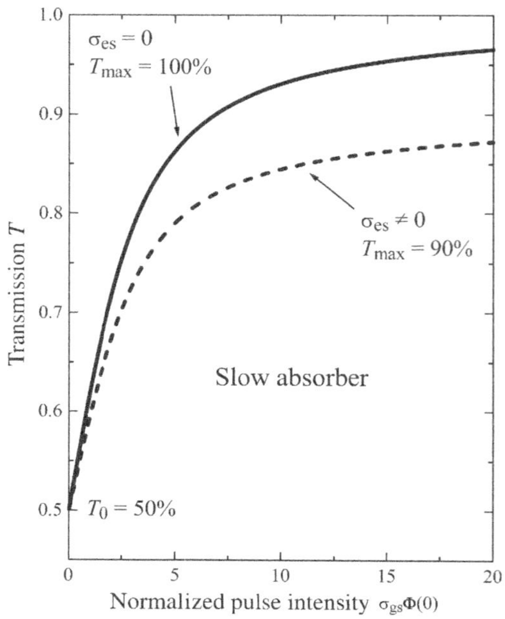

Figure 16.3 Optical transmission of a slow saturable absorber as a function of the normalized light pulse intensity $\sigma_{gs}\Phi(0)$ incident normally onto its front surface. Ideal absorber case ($\sigma_{es} = 0$) – solid line; Real absorber ($\sigma_{es} \neq 0$) –dashed line. Low light intensity transmission is 50% for both cases (Fresnel reflection losses overlooked).

the transmission correspondingly grows. Expanding the relation to a Taylor series about the $\sigma_{gs}\boldsymbol{\Phi}(0) = 0$ point up to the second order of the exponential and of the natural logarithm (neglecting all higher-order addends), we get that the increase in transmission is linear as a function of incident pulse intensity $\boldsymbol{\Phi}(0)$ as long as $\sigma_{gs}\boldsymbol{\Phi}(0) < 1$. The specific expression is

$$T_{FN} = T_0\left[1 + \frac{1}{2}(1 - T_0)\sigma_{gs}\boldsymbol{\Phi}(0)\right]. \tag{16.10}$$

In the opposite limit, namely for $\sigma_{gs}\boldsymbol{\Phi}(0) >> 1$, the transmission grows asymptotically toward a unity (namely toward $T_{max} \equiv e^{-N_{tot}\sigma_{es}L} = 100\%$). Figure 16.3 shows a plot of the said behavior for a specific case of $T_0 = 50\%$.

In reality, the excited state absorption cross section σ_{es} never vanishes, meaning the maximal transmission coefficient $T_{max} \equiv e^{-N_{tot}\sigma_{es}L} < 100\%$; maximal transmission is obtained when all ions occupy the excited $|2\rangle$ state. In that non-ideal case, the second addend in Eq. (16.7) may not be ignored. The equation is then transcendental (in other words, not solvable by an analytic solution). It is possible to present an *approximate* solution, which is actually a correction based on the Frantz-Nodvik expression as follows:

$$\boxed{T \cong T_0 + (T_{FN} - T_0)\,\frac{T_{max} - T_0}{1 - T_0}}\,. \tag{16.11}$$

Eq. (16.11) establishes an excellent approximation as long as $(T_{max} - T_0) \lesssim 0.3(1 - T_0)$. In that case as well, the transmission increases as a function of the light pulse intensity. Under low pulse intensities satisfying $\sigma_{gs}\boldsymbol{\Phi}(0) < 1$, the transmission assumes the approximate form

$$T \cong T_0\left[1 + \frac{1}{2}(1 - T_0)(\sigma_{gs} - \sigma_{es})\boldsymbol{\Phi}(0)\right]. \tag{16.12}$$

Under high pulse intensities, where $\sigma_{gs}\boldsymbol{\Phi}(0) >> 1$, the transmission grows steadily, asymptotically approaching the $T_{\max}$ value. Figure 16.3 shows a plot of the said behavior for two cases: One where $T_0 = 50\%$ and $T_{\max} = 100\%$ (the solid curve) and another one where $T_0 = 50\%$ and $T_{\max} = 90\%$ (the dashed curve).

16.2.1.2 Fast Saturable Absorber

Another extreme case of a saturable absorber occurs when the $|2\rangle$ excited-state lifetime τ (Figure 16.1) is very short compared with the pulse duration t_p. In that case, the last addend in Eq. (16.2) may not be ignored. It is possible to obtain an analytic solution to the rate equations under steady-state conditions, namely when both τ and τ^* are very short compared with any time scale variation of the incident light pulse intensity. Under the simplifying condition that the intensity during the light pulse duration is constant, the partial time derivative of $E(z, t)$ on the first line of Eq. (16.1) vanishes, and the equation assumes the form

$$\frac{dI(z)}{dz} = -\Delta\sigma N(z) I(z) - \sigma_{es} N_{tot} I(z), \tag{16.13}$$

where $I(z) = cE(z)$ is the light pulse intensity (photons per unit area per unit time) crossing the z plane inside the plate. Then Eq. (16.2) for the steady-state case assumes the form

$$\frac{\partial}{\partial t} N(z) = -\sigma_{gs} N(z) I(z) + \frac{N_{tot} - N(z)}{\tau} = 0. \tag{16.14}$$

Extracting $N(z)$, and setting the result in Eq. (16.13), yields the equation

$$\boxed{\frac{dI(z)}{dz} = -I(z)\left\{\frac{N_{tot}}{1 + \sigma_{gs}\tau I(z)}\Delta\sigma + N_{tot}\sigma_{es}\right\}}. \tag{16.15}$$

Observation of Eq. (16.15) suggests that in the limit of very small pulse intensities, the optical transmission is given by $T_0 = e^{-N_{tot}\sigma_{gs}L}$. Like for the slow saturable absorber, an approximate solution for sufficiently low pulse intensities satisfying $I(0)\tau(\sigma_{gs} - \sigma_{es}) \lesssim 0.3$ is given by

$$\boxed{T \cong T_0\left[1 + (1 - T_0)(\sigma_{gs} - \sigma_{es})\tau I(0)\right]}. \tag{16.16}$$

According to Eq. (16.16), the optical transmission for small pulse intensities starts at T_0, then grows linearly with increasing $I(0)$. By setting the opposite limit conditioning in Eq. (16.15), namely $\sigma_{gs}\tau\, I(z) >> 1$, the transmission converges asymptotically to $T_{max} = e^{-N_{tot}\sigma_{es}L}$. Thus, the initial transmission linear growth with $I(0)$ is moderated toward saturation at high intensities.

Eq. (16.15) does exhibit an analytic solution. We will not present the solving procedure, but instead just quote the result: $\frac{T_0}{T} = \left[\frac{I(L) + S}{I(0) + S}\right]^D$, where $T \equiv I(L)/I(0)$ is the optical transmission of an incident light pulse of $I(0)$ intensity, $T_0 = e^{-N_{tot}\sigma_{gs}L}$ is the optical transmission in the very low intensity limit, $D \equiv \Delta\sigma/\sigma_{es}$, and $S \equiv 1/\sigma_{gs}\tau$. Extracting $I(0)$ from the first relation yields

$$\boxed{\frac{I(0)}{S} = \frac{1 - (T_0/T)^{1/D}}{(T_0/T)^{1/D} - T}}. \tag{16.17}$$

Eq. (16.17) constitutes an implicit equation for T as a function of $I(0)$ (namely, it is impossible to present an algebraic solution for that relation). This expression is, however, very useful for curve fitting of experimental results, because one may readily swap between the $I(0)$ and T axes. As a demonstration, Figure 16.4 presents a transmission curve as a function of a normalized $\sigma_{gs}\tau\, I(0)$ light pulse intensity for two cases: one where $\sigma_{es} = 0.05\sigma_{gs}$, yielding $T_{max} = 96\%$ (full circles) and the second where $\sigma_{es} = 0.15\sigma_{gs}$, yielding $T_{max} = 90\%$ (open circles). In both cases, $T_0 = 50\%$.

16.2.1.3 Examples

An example of a slow saturable absorber is the calcium difluoride CaF_2 (fluorite) single crystal containing doubly ionized uranium U^{2+} atoms. The CaF_2 crystal is face-centered cubic belonging to the No. 225 $O_h^5(F\,4/m\,\bar{3}\,2/m)$ space group (see Section 11.5, Table [**i**]-32). The calcium Ca atoms occupy the $4a$ Wyckoff site, and the fluorine F atoms occupy the $8c$ sites. The uranium U^{2+} ion replaces the calcium atom in its 4*a* site, thus exhibiting an octahedral O_h symmetry. The replacing ion electronic configuration is $[Rn]5f^37s$ (Table 5.5), and the ground-state Term is $^5I_4$.

Figure 16.5a presents an energy scheme of the ion states, including transitions related to excitation in wavelengths in the vicinity of 1,500 nm (the mid-infrared region).

Because the crystal field states' splitting are of the same order as the spin-orbit splitting, the spectroscopic states' symmetry assignments are not unequivocal. The first absorption transition excites the ion from the $^5I_4$ multiplet states into the $^5I_6$ multiplet states. Fast non-radiative decay transitions send the ions into states belonging to a lower, $^5I_5$ multiplet, which then decay, mostly radiatively into the ground-state, with a typical 5 μs time-constant. While residing in the latter state (the $^5I_5$ multiplet), further absorption may be effected by excitation of $^5I_5$ states into the $^5I_7$ multiplet states. The system is thus similar to the schematic energy scheme described in Figure 16.1. Use of short light pulses compared with the 5 μs fluorescence decay time satisfies the slow absorber case conditions. Figure 16.5b presents experimental results of optical transmission as a function of pulsed light intensity, 30 ns long and 1,573 nm wavelength. The results reflect the behavior shown in Figure 16.3 and analyzed according to Eq. (16.11). The fit parameters are the ground-state and excited state's absorption cross sections (σ_{gs} and σ_{es}, respectively), and the N_{tot} concentration of the doubly ionized uranium atoms. The fit parameters are indicated in the figure frame.

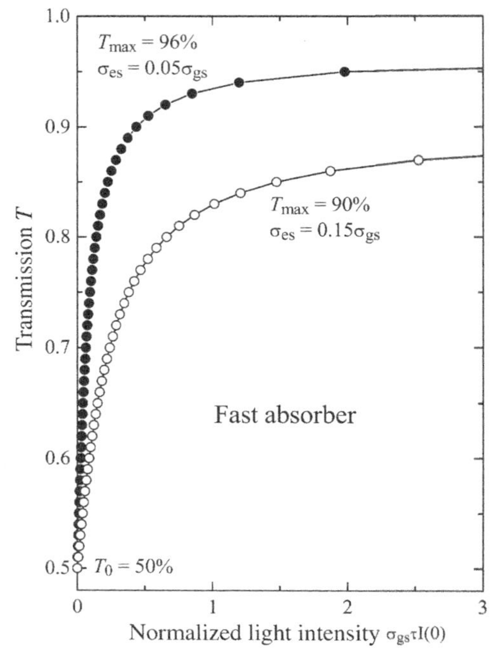

Figure 16.4 Optical transmission of a fast saturable absorber as a function of the normalized light pulse intensity $\sigma_{gs}\tau\, I(0)$ incident normally onto its front surface. Two cases of a relation between ground-states and excited states' absorption cross sections are demonstrated, where for both, the optical transmission in the low intensity limit is T_0 = 50%: in one case (full circles), $\sigma_{es} = 0.05\sigma_{gs}$ yielding $T_{max} = 96\%$; in the other case (open circles), $\sigma_{es} = 0.15\sigma_{gs}$, yielding $T_{max} = 90\%$ (Fresnel reflection losses overlooked).

An interesting example of a fast saturable absorber is a suspension of lead selenide *PbSe* nanocrystals (often abbreviated as "nc") in a liquid chloroform $CHCl_3$ medium. The lead selenide crystal is face-centered cubic belonging to the No. 225 O_h^5 $(F\,4/m\,\overline{3}\,2/m)$ space group (see Section 11.5, Table [**i**]-32). The centered unit cell contains four formula units. The lead *Pb* atoms occupy the 4*a* Wyckoff sites, and the selenium *Se* atoms occupy the 4*b* Wyckoff sites. The point symmetry of both sites is O_h. The lead selenide *PbSe* is a semiconductor, of an $E_g = 0.29$ eV direct energy gap (we assume here that the reader is acquainted with basic terms related to semiconductor physics). Both top valence band and bottom conduction band belong to the *L*-point of the first Brillouin zone, and are each fourfold degenerate. (The *L*-point is the center point on the large face of the first Brillouin zone belonging to a face-centered cubic unit cell; the Brillouin zone itself is a unit cell of a body centered cubic cell. Figure 9.9 presented an example of special symmetry points on a rather primitive cubic unit cell).

The effective masses of holes and electrons are both ~$0.05m_e$, where m_e is the free electron mass. Spatially close holes and electrons form a bound system, orbiting about each other quite similarly to a proton and electron that form the hydrogen atom. The Bohr radius for that motion, however, is very large, approximately 45 nm (only $a_0 = 0.0529$ nm of hydrogen – see Eq. [3.38] and related text). The reason relates to the small masses of the holes and electrons, and the large dielectric constant of that

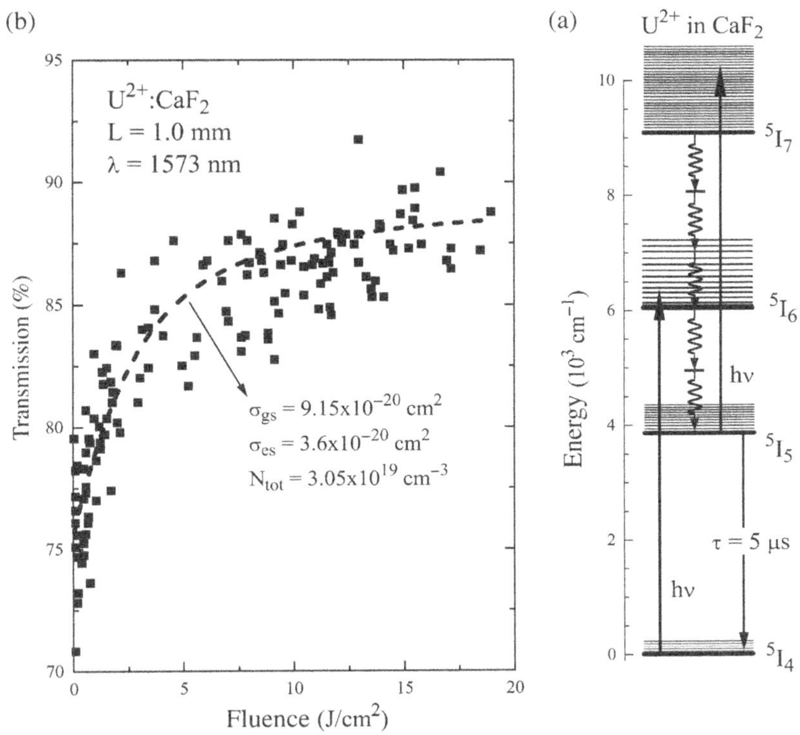

Figure 16.5 (a) Energy scheme of low-lying levels of a doubly ionized uranium U^{2+}, substituting the doubly ionized calcium Ca^{2+} in the 4*a* Wyckoff site of the calcium di-fluoride crystal. Solid arrows indicate possible radiative transitions, and wiggled arrows indicate non-radiative transitions. (b) Experimental results of optical transmission as a function of energy flux density of 30 ns long light pulses, $\lambda = 1573$ nm. Full squares are experimental results, and the dashed line is their theoretical fit to Eq. (16.11). Fit parameters are indicated inside the figure frame.

material – $\varepsilon_r \cong 25$. The system bound states are called **excitons**, and they exhibit a considerably smaller energy than the energy gap, residing between the conduction and valence bands. For considerations related to the origins of the valence bands' electronic orbitals, the lower state is assigned as 1*S* and the higher one as 1*P*.

When the nanocrystal's physical size reduces toward the Bohr radius, the energy state's scheme stretches; namely, the energy gap grows, together with the excitons' energy. Qualitatively, the effect is similar to the growing distances between states of an electron enclosed in a box upon the box dimensions reduction. It is then possible to tune the lead-selenide *PbSe* transition energies by selecting the crystal diameter. The particular example presented here refers to a 5.6 nm diameter, and the first excitonic transition energy is 0.75 eV. Its decay to the ground-state involves (mostly radiative) fluorescence emission of a τ_{rad}~100 ns lifetime. On using high pulse intensities for excitation, it is possible to generate several excitons within the same nanocrystal. Interactions between the excitons cause a fast non-radiative decay, whose effective time constant τ_{me} may be as short as several picoseconds (the "me" subscript label refers to **multi-excitonic**, which means that there are several interacting excitons in a single nanocrystal). The non-radiative decay mechanism is interpreted by an **Auger mechanism**; a decaying exciton delivers its energy to a near other exciton. The excess energy causes dissociation of the latter into a free hole and a free electron. Subsequently, the carriers' kinetic energy is dissipated non-radiatively, and the two carriers bond to re-establish an excitonic state. In summary, only a single exciton remains out of the two starting ones.

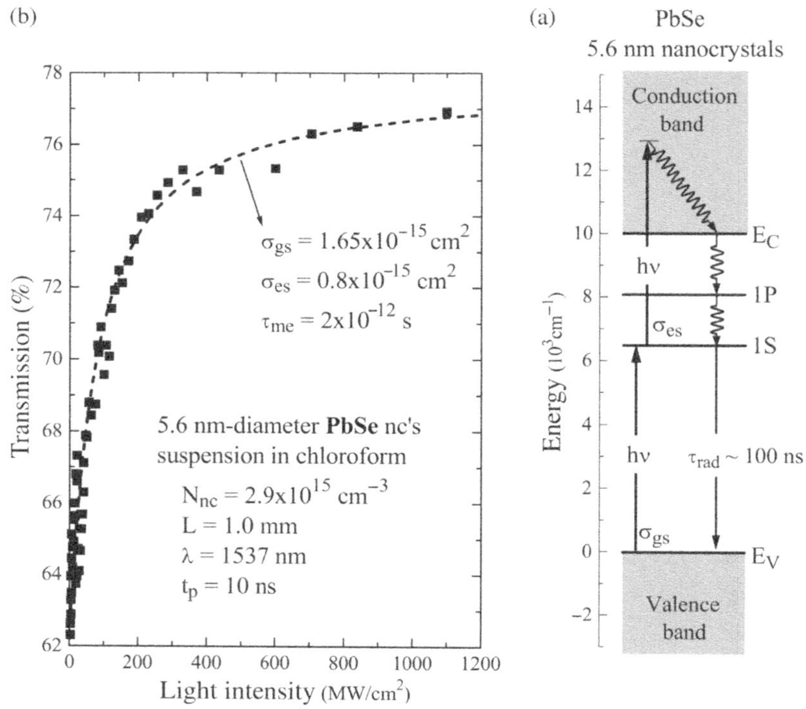

Figure 16.6 (a) Energy scheme of lead selenide *PbSe* nanocrystals of 5.6 nm diameter. Solid arrows indicate possible optical transitions; wiggled arrows indicate non-radiative transitions. (b) Optical transmission of the said nanocrystals suspension in liquid chloroform $CHCl_3$ as function of 1537 nm wavelength, ~10 ns long pulsed peak power. The absorbing medium thickness was 1.0 mm, and the nanocrystals density 2.9×10^{15} cm^{-3}. Full squares describe experimental results, the dashed line provide their fit to Eq. (16.16). Fit parameters are indicated inside the figure frame.

Figure 16.6a presents the energy scheme of a 5.6 nm diameter lead selenide *PbSe* nanocrystal suspended in a liquid chloroform $CHCl_3$ medium. The scheme represents optical absorption transitions for illumination at $\lambda = 1{,}537$ nm. A direct excitation of a 1*S* exciton is followed by a radiative decay back to the ground-state (fluorescence) at a very close wavelength (The Stokes shift is only $\sim$ 100 cm^{-1}). An exciton may also absorb a photon, and dissociate into a free electron-hole pair. Very fast non-radiative decay processes cause the charge carriers to lose their kinetic energies and bond back to form a 1*S* exciton. The system energy scheme and transitions are quite similar to the scheme described in Figure 16.1.

Seemingly, use of light pulses of the proper wavelength that are short compared with the 100 ns fluorescence lifetime of the 1S exciton, may be regarded to conform to a slow saturable absorber case. Nevertheless, very high light intensities are required to achieve saturation. Such intensities, however, also involve the excitation of several excitons on each *PbSe* nanocrystal, and the effective lifetimes shorten to the order of several picoseconds. Figure 16.6b presents the experimental results of optical transmission of nano-lead selenide *PbSe* crystals suspended in a chloroform $CHCl_3$ liquid at $N_{tot} \equiv N_{nc} = 2.9 \times 10^{15}$ cm^{-3} concentration as a function of $t_p = 10$ ns long light pulses at $\lambda = 1{,}537$ nm wavelength. The pulse length is very long compared with the multi-excitonic effective lifetime τ_{me}, thus satisfying conditions for a fast saturable absorber system. The results indeed behave in the form described in Figure 16.4, and are fit to Eq. (16.17). The fit parameters are the ground-state and excited state's absorption cross sections (σ_{gs} and σ_{es}, respectively) and the multi-excitonic effective lifetime τ_{me}. The fit parameters values are indicated in the figure frame.

16.2.2 Q-Switching Using a Saturable Absorber

As noted in Section 16.1, the addition of a saturable absorber into the laser cavity could help obtain short laser pulses due to the passive Q-switching occurring during pumping. Figure 16.7 demonstrates a development of the laser construction shown in Figure 14.20, which includes a Brewster plate in the cavity used to achieve a polarized output. In the present development, the Brewster plate is fabricated of a saturable absorber, whose absorption range covers the fluorescence spectral range of the amplifying medium. Thus, that single element combines two functions: linear polarization selection and Q-switching. Absorption of the amplifying medium fluorescence turns the plate gradually more transparent. The light resonating between the mirrors increases as well, and speeds up the transparency development. The population inversion also grows, because of the continuous pumping. At some moment, the radiation field inside the cavity exhibits a very fast growth that expresses itself by a corresponding growth of the laser output. Nevertheless, the latter growth quickly depletes the population inversion. The radiation field inside the cavity quickly diminishes, simultaneously with the laser output. Thus, in fact, the obtained output is a short light pulse, which is also polarized due to the saturable absorber plate inclination at Brewster's angle relative to the laser axis. If pumping steadily continues, the described process repeats itself and, after a while, new high power light pulses will be emitted, and so on. The described process is known as **repetitive Q-switching**.

Figure 16.8 shows a schematic comparison of output characteristics between a CW laser that contains the regular transparent Brewster plate described in Figure 14.20, and output characteristics obtained when the plate is replaced by a saturable absorber as described in Figure 16.7. The numerical values are typical of a common neodymium:YAG ($Nd^{3+}:Y_3Al_5O_{12}$) amplifying medium, and a chromium:YAG ($Cr^{4+}:Y_3Al_5O_{12}$) saturable absorber made Brewster plate. Under continuous pumping without a saturable absorber, the laser output is continuous as well, ~20 W at $\lambda = 1.064$ μm. After the Brewster plate assumes absorption saturability, the output displays a train of similar light pulses, each approximately 300 ns wide, nearly equally spaced by 35 μs in the presented example.

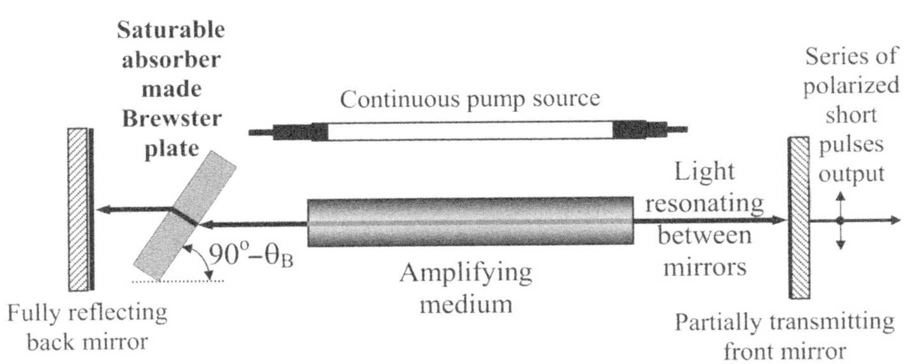

Figure 16.7 Schematic illustration of a typical laser construction emitting a long series of polarized pulses due to the addition of a saturable absorber plate inclined at Brewster's angle relative to the laser axis. For details, see the text.

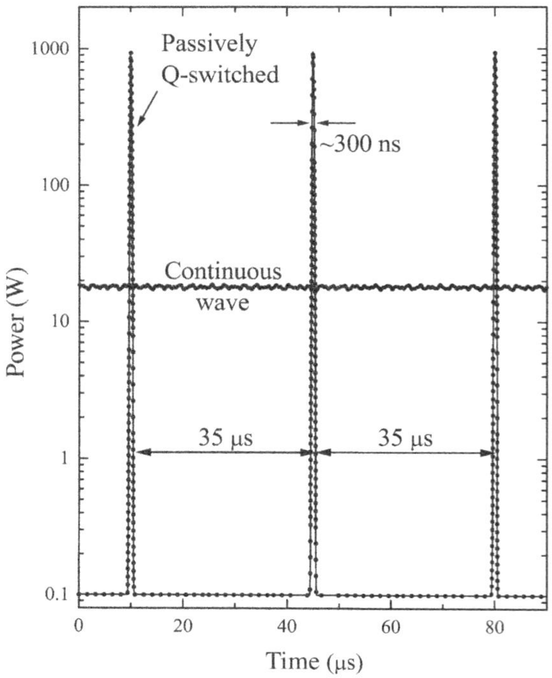

Figure 16.8 Time dependence semi-logarithmic plot of pulsed output of a continuously pumped laser that contains a saturable absorber in its cavity (beaded line) compared with the continuous output obtained without the absorber (solid line). The intensity values, the time separation between adjacent pulses, and their widths, are typical of a Nd:YAG CW laser, added with a YAG based saturable absorber containing fourfold ionized chromium atoms (Cr^{4+}:$Y_3Al_5O_{12}$).

For quantitative analysis of repetitive Q-switching, one needs to use rate equations that take into account pump rates, states' concentrations in both amplifying medium and saturable absorber, photon densities in the radiation field, transition rate constants, and losses. The rate equations are similar to Eqs. (16.1) and (16.2), which were used for the limited analysis of saturable absorption. Usually, the equations are simplified by assuming a constant distribution of states and photon densities over the entire cavity medium. Yet even then, computer-assisted calculations are needed for their solution, because of the large number of equations. We will not address that issue in our present book.

16.3 Active Q-Switching Using Electrooptic Crystals

16.3.1 The Electrooptic Effect

The **electrooptic effect** is the change in material refractive indices under an external electric field. In first-order approximation, no such effect occurs for centro-symmetric materials (those exhibiting inversion symmetry). As for materials lacking inversion symmetry, the effect is assumed to be proportional to the electric field. Recall that a centro-symmetric material appears identical under inversion symmetry about the center. Thus, if application of an electric field forms a change in the refractive indices $\underline{\Delta n} = (\alpha)\underline{\mathcal{E}}$, where $\underline{\Delta n}$ is the refractive index changes' vector and (α) is a 3×3 tensor, then on inversion, the same changes are effected, namely $\underline{\Delta n} = (\alpha)[-\underline{\mathcal{E}}]$. Therefore the (α) tensor must be identically zero. A crystal is centro-symmetric if the point symmetry of any of its sites is C_i, S_6, C_{2h}, C_{4h}, C_{6h}, D_{3d}, D_{2h}, D_{4h}, D_{6h}, T_h, or O_h (see Section 11.5, Table **[i]** and Table 8.2).

The electrooptic effect is described by expressing the changes in the constants representing the refractive indices ellipsoid (indicatrix). An ellipsoid equation generally satisfies the relation

$$A_1x^2 + A_2y^2 + A_3z^2 + 2A_4yz + 2A_5xz + 2A_6xy = 1\,. \tag{16.18}$$

On proper (diagonal) selection of the axes, the constants are $A_4 = A_5 = A_6 = 0$; $A_1 = \dfrac{1}{{n_x}^2}$; $A_2 = \dfrac{1}{{n_y}^2}$; and $A_3 = \dfrac{1}{{n_z}^2}$. On application of an external electric field, all A_i constants for $i = 1, \cdots, 6$ may change according to

$$\Delta A_i = \sum_{j=1}^{3} r_{ij}\mathcal{E}_j\,, \tag{16.19}$$

where $\mathcal{E}_j$ are the electric field components along the X, Y, and Z axes, and r_{ij} are characterizing constants of the specific crystal. Their units in the cgs system are cm/statV (in technology, the m/V units of the rationalized MKS (SI) system are more commonly used). They are called **electrooptic constants**. In a matrix form, Eq. (16.19) is written as

$$\begin{pmatrix} \Delta A_1 \\ \Delta A_2 \\ \Delta A_3 \\ \Delta A_4 \\ \Delta A_5 \\ \Delta A_6 \end{pmatrix} = \begin{pmatrix} r_{11} & r_{12} & r_{13} \\ r_{21} & r_{22} & r_{23} \\ r_{31} & r_{32} & r_{33} \\ r_{41} & r_{42} & r_{43} \\ r_{51} & r_{52} & r_{53} \\ r_{61} & r_{62} & r_{63} \end{pmatrix} \begin{pmatrix} \mathcal{E}_x \\ \mathcal{E}_y \\ \mathcal{E}_z \end{pmatrix} . \tag{16.20}$$

The above 6×3 matrix is called an "electrooptic constants tensor". We have already indicated, that in centro-symmetric crystals, all electrooptic constants vanish. Non-centro-symmetric crystals still exhibit some other symmetry properties; consequently, some constants vanish, some non-zero ones are identical to others, and some are of the same size but an opposite sign. Table 16.1 describes the fundamental structures of the electrooptic constants tensor for each of the non-centro-symmetric crystal systems. A symbols index is provided in the table caption. For example, the tensor belonging to the orthorhombic D_2 (222) system has only three non-zero elements: r_{41}, r_{52}, and r_{63}, of different size. The tensor belonging to the hexagonal $C_{3h}(\bar{6})$ system has six non-zero elements that are inter-related: $r_{21} = -r_{11}$, $r_{12} = -r_{22}$, $r_{61} = -2r_{22}$, and $r_{62} = -2r_{11}$.

Notably, the electrooptic constants tensor structure depends on the naming of the Cartesian coordinates relative to the crystalline symmetry elements. The table thus details some alternative coordinates' selection that are common in the scientific literature (of course always being principal axes of the dielectric tensor). It is also interesting to note that the electrooptic constants tensor that belongs to the $O(432)$ cubic system is identically zero despite being non-centro-symmetric!

Table 16.2 presents the electrooptic tensor values of several crystals of common use in laser technology, especially for controlled manipulation of a laser beam polarization that, among other purposes, allows for the initiation of Q-switching inside laser cavities, which will be described later on.

We now analyze in some more detail a specific option of using the di-hydrogen potassium phosphate KH_2PO_4 single crystal (often abbreviated as KDP) as a tunable retardation plate (see Section 15.8). The crystal is tetragonal, belonging to the D_{2d} $(\bar{4}2m)$ point group (see Section 11.5, Table [**i**]-14 and Tables 16.1 and 16.2). The structure of the electrooptic tensor is

$$\begin{pmatrix} 0 & 0 & 0 \\ 0 & 0 & 0 \\ 0 & 0 & 0 \\ r_{41} & 0 & 0 \\ 0 & r_{41} & 0 \\ 0 & 0 & r_{63} \end{pmatrix} . \tag{16.21}$$

Being tetragonal, the crystal is uniaxial. Thus, under an external electric field $\underline{\mathcal{E}} = \left(\mathcal{E}_x,\ \mathcal{E}_y,\ \mathcal{E}_z\right)$, the refractive indices ellipsoid assumes the form

$$\frac{x^2}{{n_o}^2} + \frac{y^2}{{n_o}^2} + \frac{z^2}{{n_e}^2} + 2r_{41}\mathcal{E}_x yz + 2r_{41}\mathcal{E}_y xz + 2r_{63}\mathcal{E}_z xy = 1 . \tag{16.22}$$

The appearance of the mixed xy, xz, and yz elements in the ellipsoid equation induces a gross orientation change of the principal axes of the ellipsoid tensor. The changed orientations obviously depend on the inducing external electric field $\mathcal{E}$. We consider the **special case** where the field is co-oriented with the original Z optic axis. The ellipsoid equation then reduces to

$$\frac{x^2+y^2}{n_o{}^2}+\frac{z^2}{n_e{}^2}+2r_{63}\mathcal{E}_Z xy=1\,. \tag{16.23}$$

Because the mixed variables addend does not involve the z variable, it is obvious that the Z axis remains a principal axis. Because of the symmetry between the roles of x and y, the new induced X and Y axes should

Table 16.1 Fundamental structures of electrooptic tensors for all point groups belonging to non-centro-symmetric crystal systems. A bold dot "•" symbolizes a zero element; a full circle "●" symbolizes a non-zero element. Two line-connected full circles " " symbolize two equal non-zero elements; two line-connected circles " ", one full and the other hollow, symbolize two non-zero, equally sized elements of opposite signs. A double hollow circle "◎" symbolizes a doubly sized element compared with its line/s-connected full circle, yet of the opposite sign.

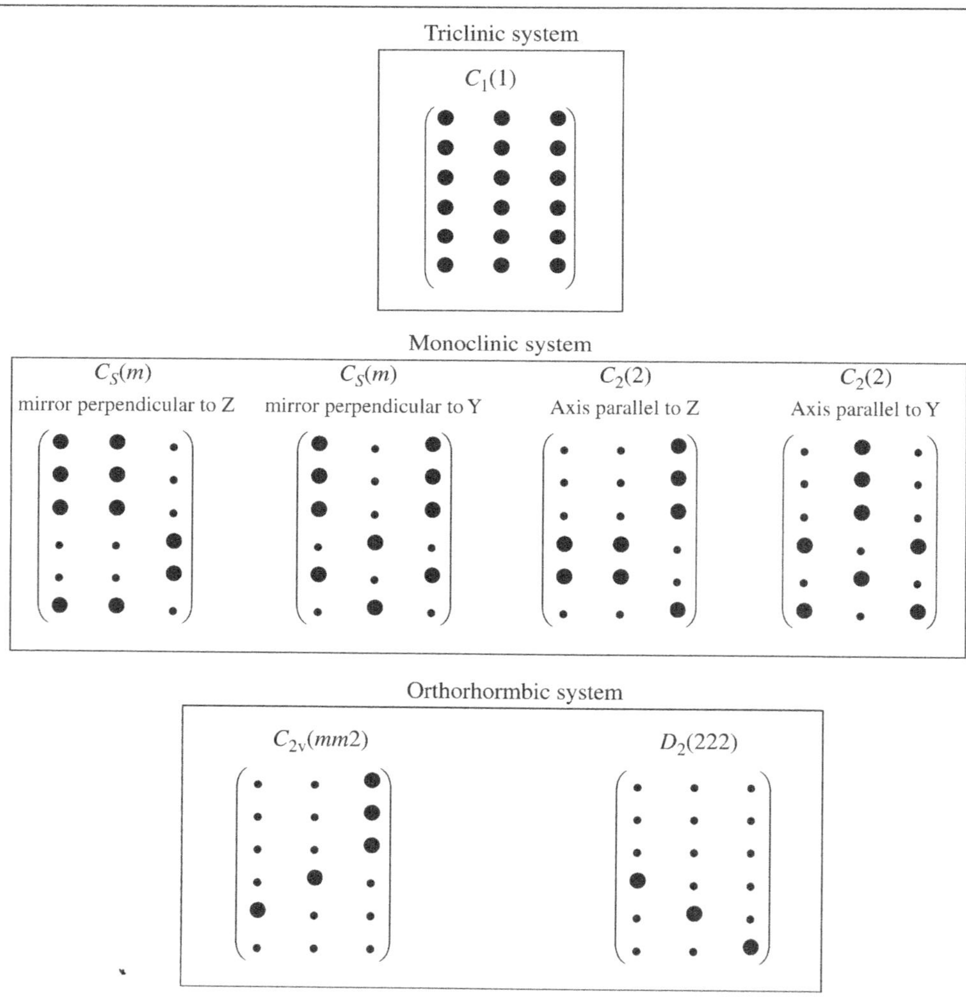

Table 16.1 (Continued)

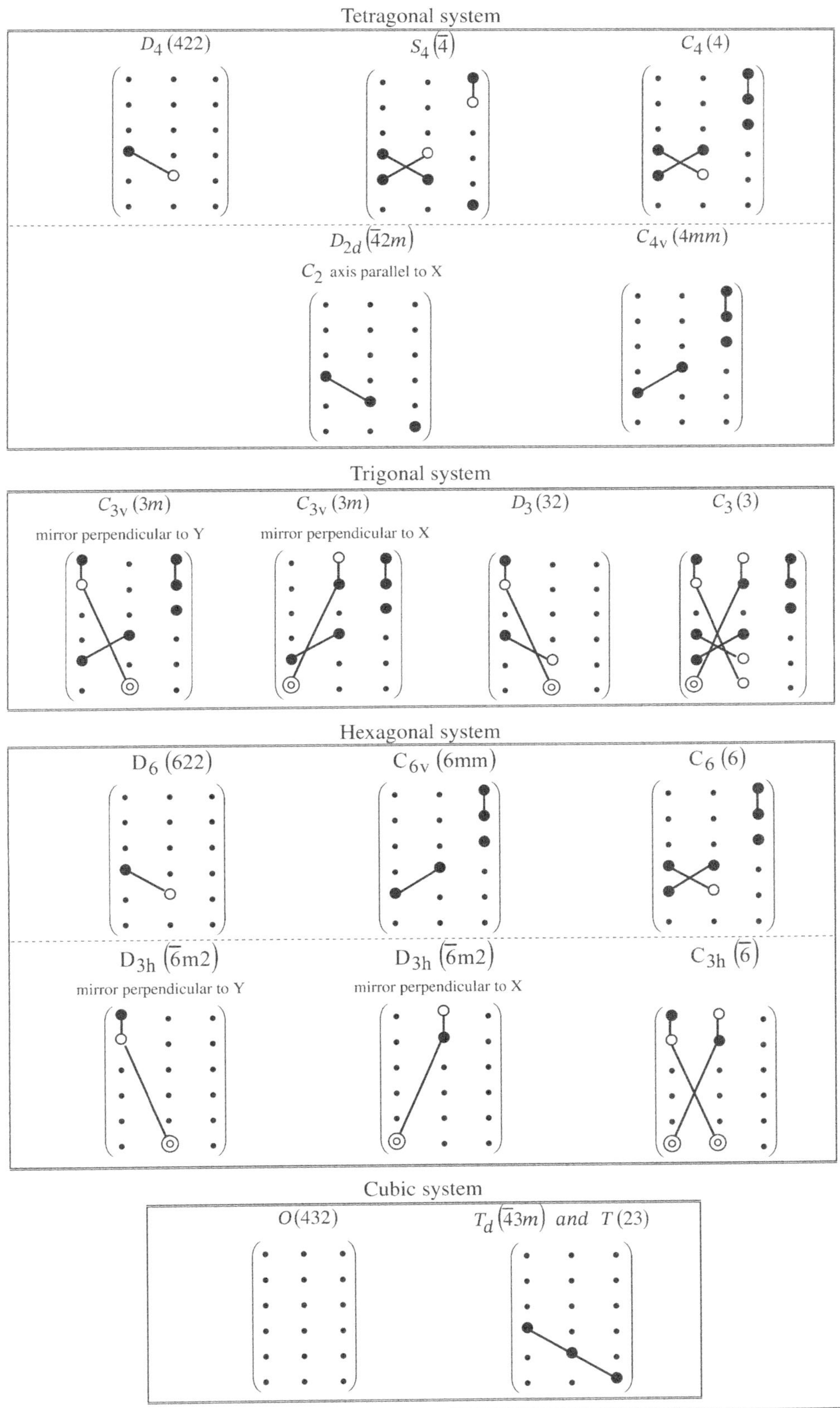

Table 16.2 Electrooptic tensor values of several useful crystals and related parameters (point group and refractive indices). Commonly used acronyms of the material names are indicated within parentheses in a couple of cases.

Material	Point group	Electrooptic constants (10^{-12} m/V)	Refractive indices
KH_2PO_4 (KDP)	$D_{2d}(\bar{4}2m)$	$r_{41} = 8.6$; $r_{63} = 10.6$	$n_o = 1.51$; $n_e = 1.47$
KD_2PO_4	$D_{2d}(\bar{4}2m)$	$r_{41} = 8.6$; $r_{63} = 23.6$	$n_o = 1.51$; $n_e = 1.47$
$NH_4H_2PO_4$ (ADP)	$D_{2d}(\bar{4}2m)$	$r_{41} = 28$; $r_{63} = 8.5$	$n_o = 1.52$; $n_e = 1.48$
ZnS	$T_d(\bar{4}3m)$	$r_{41} = 2.0$	$n_o = 1.97$
GaAs	$T_d(\bar{4}3m)$	$r_{41} = 1.6$	$n_o = 3.34$
ZnTe	$T_d(\bar{4}3m)$	$r_{41} = 3.9$	$n_o = 2.79$
CdTe	$T_d(\bar{4}3m)$	$r_{41} = 6.8$	$n_o = 2.6$
$LiNbO_3$	$C_{3v}(3m)$	$r_{13} = 8.6$; $r_{22} = 3.4$; $r_{33} = 30.8$; $r_{42} = 28$	$n_o = 2.29$; $n_e = 2.20$

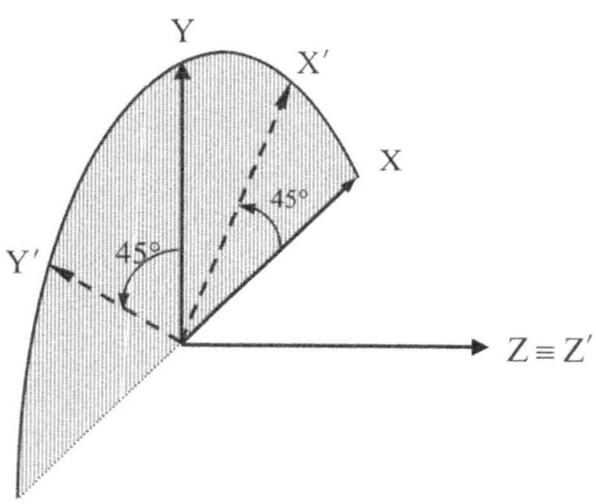

Figure 16.9 The principal Cartesian *XYZ* coordinate system of the dielectric tensor of a D_{2d} $(\bar{4}2m)$ symmetry crystal, relative to the principal Cartesian X' Y' Z' coordinate system obtained under a *Z*-oriented electric field.

be symmetric in relation to the original ones, namely rotated by 45° about the *Z* axis, as shown in Figure 16.9 below. Recall that in the D_{2d} $(\bar{4}2m)$ point symmetry, the *Z* axis coincides with the fourfold rotation axis of the tetragonal crystal, and the original *X* and *Y* axes coincide with the twofold rotation axes (see Table 10.1). Because the crystal is non-centro-symmetric, the *Z* axis is co-oriented with the [001] orientation, and not with the $[00\bar{1}]$ one.

Thus (see also Eq. [4.21] and Figure 4.2),

$$x = x' \cos 45^\circ - y' \sin 45^\circ$$
$$y = x' \sin 45^\circ + y' \cos 45^\circ \,.$$

Setting it in Eq. (6.23), one gets

$$\left(\frac{1}{{n_o}^2} + r_{63}\mathcal{E}_Z\right)x'^2 + \left(\frac{1}{{n_o}^2} - r_{63}\mathcal{E}_Z\right)y'^2 + \frac{z'^2}{{n_e}^2} = 1\,. \tag{16.24}$$

A simple calculation shows that the ellipsoid equation may be written as

$$\frac{x'^2}{n_{x'}^2} + \frac{y'^2}{n_{y'}^2} + \frac{z'^2}{n_e^2} = 1\,, \tag{16.25}$$

where, to an excellent approximation, $n_{x'} \cong n_o - n_o^3 r_{63}\mathcal{E}_Z$ and $n_{y'} \cong n_o + n_o^3 r_{63}\mathcal{E}_Z$, then $(n_{y'} - n_{x'}) \cong 2n_o^3 r_{63}\mathcal{E}_Z$. The Z axis thus ceases from being an optic axis. In other words, beams of different polarizations propagating in the Z direction now exhibit different dielectric constants. A parallel plate cut perpendicularly to the Z axis may now be used as a retardation plate (see Section 15.8). The phase difference between X' and Y' oriented polarization components of a light beam is given by $\Delta\varphi = (2\pi/\lambda)\,2n_o^3 r_{63}\mathcal{E}_Z L$, where L is the plate thickness. The plate may be used, for example, to convert a linearly polarized light into an elliptically or a circularly polarized one, or to rotate the polarization orientation by 90° about the Z axis, as described in Section 15.8.♣ A plate $L = \left(\frac{1}{2}\lambda + j\lambda\right)\left[2n_o^3 r_{63}\mathcal{E}_Z\right]^{-1}$ thick would perform as a half-wave plate and would rotate a beam polarization in the X or Y direction by 90°. A plate $L = \left(\frac{1}{4}\lambda + j\lambda\right)\left[2n_o^3 r_{63}\mathcal{E}_Z\right]^{-1}$ thick would perform as a quarter-wave plate and convert a beam polarized in the X or Y direction into a circularly polarized beam.

As a representative specific example, we consider the zero-order ($j = 0$) half-wave plate, and express the voltage as $V_Z^{(1/2)} = \mathcal{E}_Z L$. We then get that, **independently of the plate thickness**,

$$V_Z^{(1/2)} = \frac{\lambda}{4n_0^3 r_{63}}\,. \tag{16.26}$$

Under half of that voltage, the plate would function as a quarter-wave plate. The required voltages are very high – of the order of thousands volts. The given data for a $\lambda = 0.5$ μm beam wavelength, and a KDP crystal, are $r_{63} = 10.6 \times 10^{-12}$ m/V and $n_o = 1.51$ (Table 16.2). Setting in Eq. (16.26), we get

$$V_Z^{(1/2)} = \frac{5 \times 10^{-7}}{4 \times 1.51^3 \times 10.6 \times 10^{-12}} = 3425\ \text{V}\,.$$

To utilize the described effect, the two electrodes applied to the plate surface must be both **electrically conductive** and **optically transparent**!

16.3.2 Q-Switching Using an Electrooptic Crystal

The above described electrooptic effect suggests an opportunity to design an inside-cavity element, whose external activation would cause the laser to emit short high power pulses. Figure 16.10 describes such a modification of the laser construction shown in Figure 14.20. The goal there was the achievement of a linearly polarized output by addition of a Brewster plate into the laser cavity. In the present modification, the cavity is further added with the KDP (KH_2PO_4) electrooptic parallel plate, whose electrooptic properties are given in Table 16.2. The plate surface planes are perpendicular to the Z axis, which is co-oriented with the laser axis. Initially, the plate is biased by the specific high voltage that turns it into a quarter-wave plate (see the end of Section 16.1). A vertically polarized beam that crossed the Brewster plate must traverse the quarter-wave plate *twice* to return to the amplifying medium. Thus, that plate's action is equivalent to a half-wave plate! The beam "Y" polarization then changes into an "X" oriented polarization.

♣ Polarization conversion by use of the electrooptic effect is called **electrooptic modulation**. If the inducing external electric field is parallel to the beam propagation or perpendicular to it, the adjective **longitudinal** or **transverse** is correspondingly added to the name.

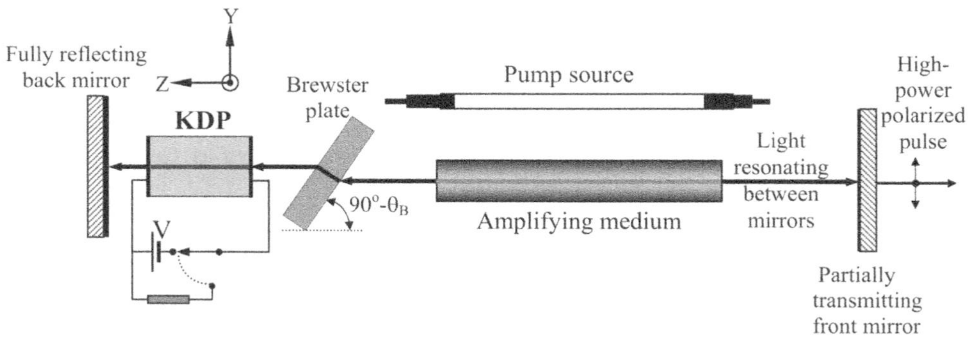

Figure 16.10 Schematic illustration of a typical laser construction that includes a Brewster polarizing plate, which emits powerful, linearly polarized short pulses, initiated by pulsed voltage variation applied over an electrooptic retardation plate. For details, see the text.

The Brewster plate it has to traverse for reaching the amplifying medium induces sufficient loss to prevent lasing. At a certain moment, the high external voltage is removed by disconnecting the voltage supplying circuitry, and subsequently short-circuiting the two electrodes through a small-valued resistor. In the figure, the switch is described as mechanical; in commercial systems, it is done electronically. The electrooptic plate ceases from being a retardation plate; no polarization change of the back-mirror reflected beam occurs; and lasing develops in the form of a short, high-powered pulse, which uses the entire energy accumulated in the amplifying medium.

As noted, the electrodes applied onto the KDP plate must be electrically conductive, optically transparent, and thin. A common material that satisfies that goal is the binary indium-tin oxide compound (indium oxide In_2O_3 [~90%] and tin oxide SnO_2 [~10%]). It is often abbreviated as ITO. Used layer thicknesses are of the order of several micrometers.

We should add several comments. The Brewster plate may be replaced by a different polarizer with similar results. It is also possible to use different voltage application configurations, where the electric field is perpendicular to the laser axis ("transverse" modulation); it saves the need to use transparent electrodes. KDP crystals are not the only usable option for Q-switching initiation; other electrooptic crystals are used at different spectral ranges, at different voltage application configurations, and in different intra-cavity combinations of polarizers and retardation plates.

External initiation of Q-switching by using an electrooptic crystal has many advantages but also a gross disadvantage related to the need for employing a very costly and mechanically heavy voltage supplier, with its entailed electronics. Where possible, passive Q-switching is preferred, as described in Section 16.2.2.

16.4 Mode-Locking

In the present section, we describe a way by which it is possible to achieve a very long series (train) of especially short light pulses via a mechanism called **mode locking**. Recall the discussion in Section 14.2 on properties of the optical cavity. We have concluded that the spectral output of a laser devoid of special frequency (wavelength) tuning elements usually comprises an envelope of tens or hundreds narrow lines, each being a longitudinal oscillation mode supported by the cavity. The supported frequencies are of equal separation: $\Delta\omega \equiv \omega_{j+1} - \omega_j = 2\pi c/2L$, where L is the distance between the cavity mirrors (Figure 14.1), and j is an ordinal number of the longitudinal frequency at an increasing frequency order. The spectral width of each line is given by Eq. (14.19). Figure 16.11 demonstrates a spectral output of a laser emitting red light of

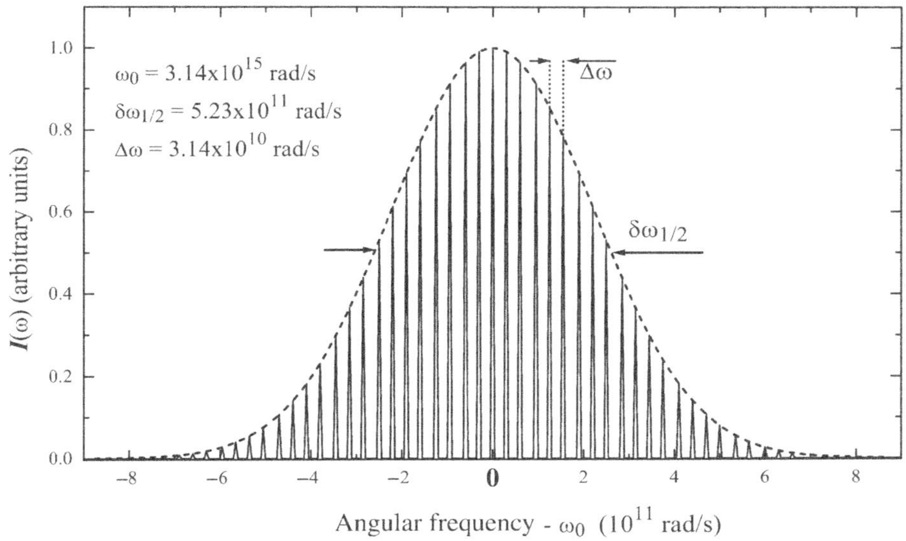

Figure 16.11 Magnified observation of a continuous wave laser spectral output (intensity per unit angular frequency as a function of the angular frequency), revealing the fine structure of the longitudinal electromagnetic radiation modes.

$\lambda = 600$ nm wavelength, belonging to a central 3.14×10^{15} rad/s angular frequency, of $\delta\lambda_{1/2} = 0.1$ nm full width at half maximum (equivalent to $\delta\omega_{1/2} = 5.23 \times 10^{11}$ rad/s). The cavity length is taken as $L = 3.0$ cm, yielding $\Delta\omega = 3.14 \times 10^{10}$ rad/s. The spectral width of the longitudinal modes, about $\Delta\omega = 3 \times 10^{9}$ rad/s, relates to an $F \cong 10$ finesse of the of the cavity (in fact, it performs as an etalon). For simplification, in the following analysis we assume that the longitudinal modes are infinitely narrow; namely, instead of using the frequency dependent complex electric field amplitude $\mathcal{E}_j(\omega)$ of the longitudinal mode j component, we define instead and use the quantity $\overline{\mathcal{E}}_j \equiv \int \mathcal{E}_j(\omega)d\omega$.

The above combination of longitudinal modes actually forms a wave group. The effective time dependence of electric field $\mathcal{E}(t)$ is written as

$$\mathcal{E}(t) = \sum_j \overline{\mathcal{E}}_j e^{i[(\omega_0 + j\Delta\omega)t + \varphi_j]}, \tag{16.27}$$

where $\omega_0 \equiv m\,\Delta\omega$ for an integer $m > 0$ is one of the longitudinal modes in the vicinity of the maximal gain, and φ_j is the phase of the j longitudinal mode (for simplicity hereafter, we avoid using the bar above the $\mathcal{E}$ symbol in all $\mathcal{E}_j$ parameters). The phase distribution is usually random, and might drift over time. Therefore the effective electric field amplitude, and hence the beam energy, will flicker over time. We consider first a special case where the φ_j phases are constant over time. We shall see that the electric field flickering exhibits a constant period $T = (2L/c) = (2\pi/\Delta\omega)$. For proof, we consider the expression

$$\begin{aligned}
\mathcal{E}(t+T) &= \sum_j \mathcal{E}_j\, \boldsymbol{exp}\left\{i\left[(\omega_0 + j\Delta\omega)(t+T) + \varphi_j\right]\right\} \\
&= \sum_j \mathcal{E}_j\, \boldsymbol{exp}\left\{i\left[(\omega_0 + j\Delta\omega)t + \varphi_j\right]\right\}\cdot \boldsymbol{exp}\left\{i\left[2\pi\left(\frac{\omega_0}{\Delta\omega} + j\right)\right]\right\} \\
&= \sum_j \mathcal{E}_j e^{i[(\omega_0 + j\Delta\omega)t + \varphi_j]}.e^{i[2\pi(m+j)]} = \mathcal{E}(t)
\end{aligned} \tag{16.28}$$

The final stage in the above development is based on the fact that since $m + j$ is an integer, then $e^{i[2\pi(m+j)]} = 1$ for each j.

The T flickering time period describes the back and forth photon transition time between the cavity mirrors. Because typical cavity lengths never exceed several tens of centimeters, the said time period is typically several nanoseconds. If the laser emission intensity is measured by a device of a considerably longer response time, the flickering will not be noticed!

A case of equal phases $\varphi_j = \varphi$ is called mode-phase-locking, or just **mode-locking**. To demonstrate, we assume a case where exactly N longitudinal modes exist over the entire spectral width, that all their amplitudes are equal $\mathcal{E}_j = \mathcal{E}_0$, and that N is odd. In other words, the longitudinal modes are non-zero for $-\frac{1}{2}(N-1) \leq j \leq +\frac{1}{2}(N-1)$, and zero otherwise. Then

$$\mathcal{E}(t) = \sum_{j=-(N-1)/2}^{(N-1)/2} \mathcal{E}_0 e^{i[(\omega_0 + j\Delta\omega)t + \varphi]} = \mathcal{E}_0 e^{i(\omega_0 t + \varphi)} \sum_{j=-(N-1)/2}^{(N-1)/2} e^{j[i\Delta\omega t]}.$$

The last multiplier on the right side is a sum of a geometric series, whose quotient is $e^{i\Delta\omega t}$. Carrying out the calculation, we get

$$\mathcal{E}(t) = \mathcal{E}_0 e^{i(\omega_0 t + \varphi)} \frac{\sin N \cdot \frac{1}{2}\Delta\omega t}{\sin \frac{1}{2}\Delta\omega t}. \tag{16.29}$$

Based on Eq. (10.15), the laser beam power would be given by

$$I(t) = \frac{c}{8\pi}|\mathcal{E}_0|^2 \frac{\sin^2 N \cdot \frac{1}{2}\Delta\omega t}{\sin^2 \frac{1}{2}\Delta\omega t}. \tag{16.30}$$

As expected, the $I(t)$ function is periodic, with a $T = (2L/c) = (2\pi/\Delta\omega)$ period. Over a short time length, around an integral number of a $2\pi/\omega$ time periods, a separate peak is obtained, of $N\frac{c}{8\pi}|\mathcal{E}_0|^2$ height and $\delta t_{1/2} \cong T/N$ half-maximum width. Thus, the periodic flickering of the laser output turns into a long ordered "train" of short pulses. Figure 16.12 shows the $I(t)$ function for a laser whose given properties

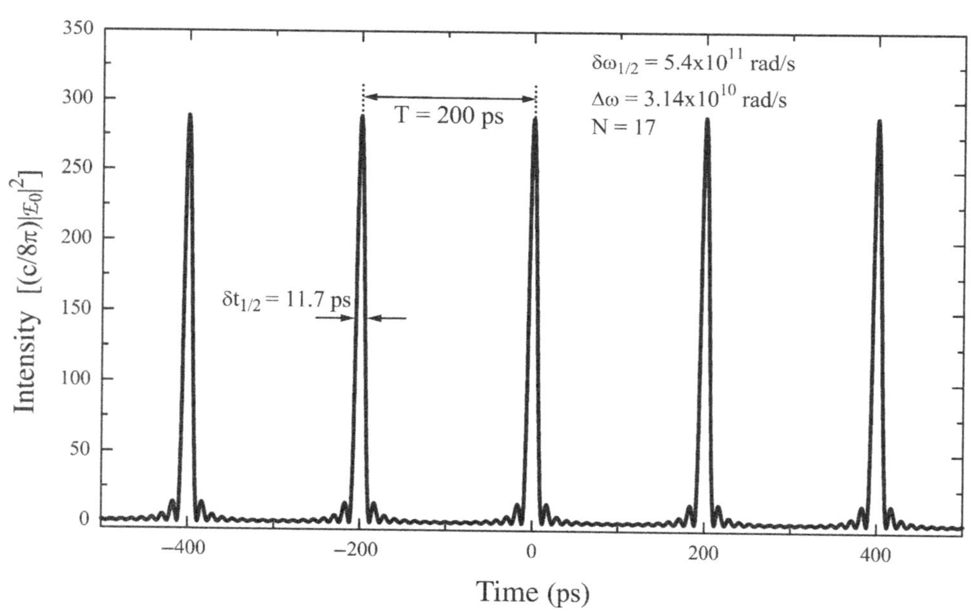

Figure 16.12 Time dependence of a laser output, whose operation parameters are given in relation to Figure 16.11, after phase-locking of all longitudinal modes.

are listed in the frame of Figure 16.11. Because $L = 3.0$ cm, then $T = 200$ ps (recall that 1 ps $\equiv 10^{-12}$ s) and $N = \delta\omega_{1/2}/\Delta\omega \cong 17$.

The reader is reminded that a function of the form $y = \sin^2 Nz/\sin^2 z$ appearing as a multiplier in Eq. (16.30), describing both the separate peaks and the repetitive nature, has also been discussed in relation to Figure 14.9. The latter figure and its related text consider the properties of a separate peak. An important property of a separate peak is the relation $\delta t_{1/2} = 2\pi/\delta\omega_{1/2}$, which, in fact, expresses Heisenberg's uncertainty principle (Eq. [1.19]). To generate separate pulses of the shortest possible widths, the spectral width of the laser output needs to be *as broad as possible*. An upper limit available for that goal is the fluorescence bandwidth, namely the spectral width of the *g*-function defined in Eq. (2.47), which appears also in various equations like Eq. (2.54), (2.58), and (14.1). The above equality relation is valid only under perfect mode locking. The relation should, in fact, be expressed in a form similar to Heisenberg's uncertainty principle:

$$\boxed{\delta t_{1/2} \cdot \delta\omega_{1/2} \geq 2\pi} . \tag{16.31}$$

The shortest separate pulses generated so far are about 8 fs wide (recall: 1 fs $\equiv 10^{-12}$ s); they have been obtained by utilizing the broad fluorescence of triply ionized titanium ions Ti^{3+} hosted in a sapphire crystal (sapphire is the single crystal form of aluminum oxide Al_2O_3). The fluorescence spectral width of those ions is about 200 nm, centered at an ~800 nm peak wavelength.

An imperative condition for obtaining the said short pulses' "train" is the occurrence of mode locking. Fast Q-switching in the laser cavity must be used for that purpose, of a time duration coinciding with the prospected $\delta t_{1/2}$ duration of each single mode-locked pulse repetitively, according to the prospected T period of the mode-locked pulses "train". It is extremely difficult to design an electronic system that would perform an externally initiated Q-switching (Figure 16.10) abided by that condition. Therefore, the usual chosen method is passive Q-switching, as described in Figure 16.7. A fast saturable absorber is used, whose lifetime τ (Figure 16.1) is shorter than the order of $\delta t_{1/2}$ (obviously also short compared with the prospected time T between successive pulses). The saturable absorber should be placed very close to one of the cavity mirrors, to ensure its transparency during a single mode-locked pulse transition back and forth between the Q-switching plate and the mirror. If, at a certain moment, mode locking appears spontaneously, the generated pulses accurately induce the required Q-switching, and mode locking continues uninterruptedly.

An interesting, very important mechanism that allows for the occurrence of a spontaneous mode locking, is the **optical Kerr effect**. The effect is a change in the refractive index n caused by the intensity I of a beam propagating through matter, according to

$$n(I) = n_0 + n_2 I , \tag{16.32}$$

where n_0 is the matter refractive index in the low light intensity limit, and n_2 is a positive coefficient called the **optical Kerr effect**. The "2" subscript is a reminder that the effect is of a second order in the radiation electric field dependence; namely, that the refractive index change is proportional to the squared electric field of the electromagnetic radiation. The Kerr effect exists in all matters. Regarding single crystals, it never vanishes for centro-symmetric crystals, as occurs for the electrooptic effect. Typical values of n_2 are of the order of 10^{-20} m^2/W; typical response times to the said refractive index change upon the electromagnetic radiation appearance are of the order of 2 – 5 fs in all materials. The Kerr effect belongs to a very broad family of "non-linear effects". Chapter 17 elaborates on several other non-linear effects used in laser technology.

When mode locking occurs, peak intensities of separate peaks are orders of magnitude greater than the average output power. These pulses thus induce considerable change of the refractive index in the amplifying medium according to Eq. (16.31). As described in Figure 15.5, laser cavities employ concave or convex mirrors, and apertures, to support the development of a preferable TEM_{00} Gaussian beam. The transverse Gaussian intensity necessarily causes a transverse distribution of the refractive index changes: a large change at the beam center, and a small change at its perimeter. The region where those changes occurred then functions as a kind of a converging lens, changing the shape of the beam wave front (**an effect called self-focusing**). The design of the laser cavity – mirrors' positions and curvature radiuses – takes into account the prospected self-focusing upon mode locking. In turn, under no mode locking, the design is not optimal for maximal output. Thus, on initiation of the laser operation, the output is of a continuous wave form, but less-then-optimal output. Nevertheless, if, at a certain moment, mode locking coincidently occurs, the laser output increases, and the mode locking form continues uninterruptedly.

Although the mode-locked laser produces a long train of short pulses, for various purposes it is possible to select separate pulses by a variety of means; such selection can, actually must, be done only outside the laser cavity.

In the present book, we remain satisfied with the qualitative presentation of the Kerr effect and the mode locking generation dynamics; we will not address the issues quantitatively.

Exercises for Chapter 16

16.1 The absorption cross section of ground-state ions in a saturable absorber at $\lambda = 1064$ nm wavelength is 6×10^{-19} cm^2 and of the excited state is 1.3×10^{-19} cm^2. The ions' concentration is 6.25×10^{18} cm^{-3} and their excited state lifetime is 2.2 μs. The material is machined as a 1.85 mm thick plane-parallel plate, and coated with an antireflection coating. Short, about 30 ns long light pulses of the said wavelength incident the plate perpendicularly to its surface planes. Calculate the plate optical transmission in the limits of very small and very high light intensities. What would be the optical transmissions for specific pulse energies of 0.2 J/cm^2 and 0.5 J/cm^2?

16.2 (a) An antireflection-coated parallel-plate formed crystal, 1.0 mm thick, contains ions absorbing at $\lambda = 1064$ nm, whose concentration is 9×10^{17} cm^{-3}. The optical transmission at low light intensities is 62%. Calculate the ground-state absorption coefficient of the ions.

(b) The ions' excited-state lifetime is 10 ns. The optical transmission grows to 68% for an incident $\lambda = 1064$ nm laser pulse, 150 ns long, of 20 mJ/cm^2 energy flux density. Calculate the absorption cross section of the ions' excited state.

16.3 The refractive indices of a KDP crystal at $\lambda = 1064$ nm are $n_o = 1.4942$ and $n_e = 1.4616$.

(a) Plan the accurate thickness for a quarter-wave plate of $j = 200$ order.

(b) The plate is machined to a box form, $d = 2$ mm thick along the crystal Z axis, as in Figure 16.13. Conductive electrodes are deposited on the faces perpendicular to the Z axis. Calculate the strength of the transverse voltage required for application between electrodes to convert it into a half-wave plate. What happens with an opposite voltage polarity?

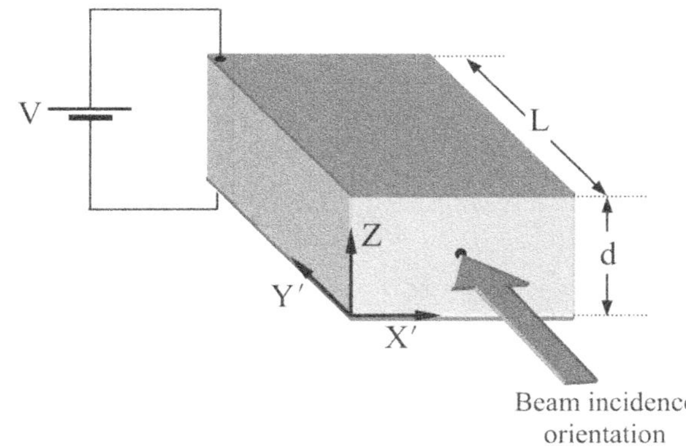

Figure 16.13 A KDP-based retardation plate used for conversion or modulation of an incident beam polarization state. The figure relates to Exercise 16.3.

16.4 The fluorescence emission of methanol-dissolved pyrromethene 567 is centered around the $\lambda = 546$ nm wavelength, and its full width at half maximum (FWHM) is 40 nm. Consider a laser whose amplifying medium is based on the said material, and whose cavity length is 25 cm. Calculate the minimal possible duration of a separate output light pulse under mode locking operation, and the time period between successive pulses.

17 Frequency Conversions of Laser Beams

17.1 Non-Linear Crystals

Mechanisms that enable the conversion of laser beam frequencies require the use of materials exhibiting non-linear properties, especially crystals. They enable the generation of diverse effects on laser beams. In the present chapter, we present the essence of non-linearity, and a small collection of effects demonstrating the phenomenon.

In Section 10.4, we presented a microscopic model of the dielectric function in dielectric crystals. The model imagined the polar crystal oscillation modes as emanating from polar oscillations of the atomic bases. Each atomic basis was described as comprising two rigid parts of μ^* reduced mass, carrying e^* -sized electrical charges of opposite signs. In the present chapter, we deal with wavelengths in the vicinity of the visible range, namely those at much higher frequencies than any crystal lattice oscillation resonances. Therefore, a crystal response to such frequencies involves oscillations of the electrons and not of their binding nuclei. We therefore convert the discussion in Section 10.4 into a model, whereby each atomic basis contains *electronic oscillators* of mass m and electric charge e, resonance frequencies ω_{0j}, and damping coefficients γ_j, where j is the ordinal number of each oscillator. For simplicity, we start by assuming the presence of only a single oscillator in an atomic basis. We additionally assume that the electron oscillation center exhibits a non-linear response to its deviation from the equilibrium position under an external electromagnetic field. The non-linearity expresses itself in the following equation of motion, which, in fact, is a modification of Eq. (10.40) that includes the said assumption:

$$\ddot{x} + \gamma\dot{x} + {\omega_0}^2 x + Dx^2 = \frac{e}{2m}\mathcal{E}_0\left(e^{-i\omega t} + e^{+i\omega t}\right), \tag{17.1}$$

where $\mathcal{E}(\omega, t) = \mathcal{E}_0 \cos \omega t = \frac{1}{2}\mathcal{E}_0\left(e^{-i\omega t} + e^{+i\omega t}\right)$ is the electric field component of the electromagnetic radiation field inside the crystal, x is the time-dependent electron oscillation center deviation from its equilibrium position, and D is a constant. The equation of motion presented in Eq. (17.1) derives from a $V(x) = -\frac{1}{2}m{\omega_0}^2x^2 - \frac{1}{3}mDx^3$ attractive potential. The non-linear $-\frac{1}{3}mDx^3$ element is an odd function; therefore D would vanish in a centro-symmetric crystal. Non-linear effects discussed later are thus possible only in non-centro-symmetric crystals!

We assume a solution for the differential equation of Eq. (17.1) as a periodic oscillation of x at an angular frequency ω, which is the same as that of the driving electric field. Recall that each periodic oscillation is presentable as a sum of harmonic oscillations belonging to frequencies that are an integer number of the period ω frequency. We thus attempt a solution of the form

$$x(t) = \frac{1}{2}\left(x_1 e^{-i\omega t} + x_2 e^{-2i\omega t} + c.c.\right), \tag{17.2}$$

where x_1 and x_2 are complex amplitudes of the oscillations at ω and 2ω, respectively, and "*c.c.*" is the complex conjugate of the former addends. We set it in Eq. (17.1) and get

Physics, Optics, and Spectroscopy of Materials, First Edition. Zeev Burshtein.

$$
\begin{aligned}
&\frac{1}{2}\omega^2\left[\left(-x_1e^{-i\omega t}-4x_2e^{-2i\omega t}\right)+c.c.\right]+\frac{1}{2}\omega\gamma\left[\left(-ix_1e^{-i\omega t}-2ix_2e^{-2i\omega t}\right)+c.c.\right]\\
&+\frac{1}{2}\omega_0^2\left[\left(x_1e^{-i\omega t}+x_2e^{-2i\omega t}\right)+c.c.\right]+\frac{1}{4}D\left[2|x_1|^2+2|x_2|^2\right]\\
&+\frac{1}{4}D\left[\left(2x_2x_1^*e^{-i\omega t}+x_1^2e^{-2i\omega t}+2x_1x_2e^{-3i\omega t}+x_2^2e^{-4i\omega t}\right)+c.c.\right]\\
&=\frac{e}{2m}\mathcal{E}_0\left(e^{-i\omega t}+c.c.\right).
\end{aligned}
\tag{17.3}
$$

The two equation sides may be equal at all times if the coefficients of $e^{-i\omega t}$ and its powers are equal at the two sides. We start by equating the coefficients of the first power $e^{-i\omega t}$, yet under a complementary assumption that $|Dx_2| \ll [(\omega_0^2-\omega^2)^2+\gamma^2\omega^2]^{1/2}$. We then get

$$
x_1 \cong \frac{e\mathcal{E}_0}{m}\frac{1}{(\omega_0^2-\omega^2)-i\gamma\omega}. \tag{17.4}
$$

The induced polarization density $\boldsymbol{P}(\omega, t)$ exhibits components oscillating at different frequencies. The ω frequency oscillating component, $\boldsymbol{P}^{(\omega)}(t)$, is given by (see also Eq. [10.42])

$$
\boldsymbol{P}^{(\omega)}(t)=\frac{ex_1}{2V_C}e^{-i\omega t}+c.c.=\frac{1}{2}\frac{e^2}{mV_C}\frac{1}{(\omega_0^2-\omega^2)-i\gamma\omega}\mathcal{E}_0e^{-i\omega t}+c.c.\,, \tag{17.5}
$$

where V_C is the crystal primitive unit cell volume. The $\chi_L^{(\omega)} \equiv \frac{e^2}{mV_C}\frac{1}{(\omega_0^2-\omega^2)-i\gamma\omega}$ quantity is called **dielectric susceptibility**; the "L" subscript in the symbol is the initial letter of the word "linear". In the present context, this means that the susceptibility allows the oscillator to respond by the same frequency as the driving force, and to exhibit the same resonance frequency as the free oscillator. In Section 10.4, we calculated (without naming it) the dielectric susceptibility of polar crystal vibrations devoid of non-linear effects (see Eq. [10.42]). For example, we could rewrite Eq. (10.44) as $\varepsilon(\omega)=\varepsilon(\infty)+4\pi\chi_L^{(\omega)}$. It is a common practice to define $N \equiv 1/V_C$; that quantity describes the volume density of electronic oscillators of ω_0 resonance frequency. We may then write

$$
\chi_L^{(\omega)} \equiv \frac{Ne^2}{m}\cdot\frac{1}{(\omega_0^2-\omega^2)-i\gamma\omega}. \tag{17.6}
$$

We now turn to calculate the polarization density component oscillating at a 2ω frequency. The right side of Eq. (17.3) has no such component. Therefore

$$
x_2=\frac{-Dx_1^2}{2(4\omega^2-\omega_0^2+2i\gamma\omega)}=\frac{De^2\mathcal{E}_0^2}{2m^2}\cdot\frac{1}{[(\omega_0^2-\omega^2)-i\gamma\omega]^2\left[\omega_0^2-(2\omega)^2-i\gamma\cdot 2\omega\right]}. \tag{17.7}
$$

Following Eq. (17.4), the $\boldsymbol{P}^{(2\omega)}(\omega, t)$ polarization density component oscillating at the 2ω frequency would be given by

$$
\begin{aligned}
\boldsymbol{P}^{(2\omega)}(t)&=\frac{ex_2}{2V_C}e^{-2i\omega t}+c.c.\\
&=\frac{1}{2}\frac{NDe^3}{2m^2}\frac{1}{[(\omega_0^2-\omega^2)-i\gamma\omega]^2\left[\omega_0^2-(2\omega)^2-i\gamma 2\omega\right]}\mathcal{E}_0^2e^{-2i\omega t}+c.c.
\end{aligned}
\tag{17.8}
$$

The $d_{NL}^{(2\omega)} \equiv \frac{NDe^3}{2m^2} \frac{1}{\left[\omega_0{}^2 - (2\omega)^2 - i\gamma 2\omega\right] \cdot [(\omega_0{}^2 - \omega^2) - i\gamma\omega]^2}$ quantity is called the **non-linear optical coefficient**, and its units in the cgs system are $\mathrm{cm}^{1/2}\mathrm{g}^{-1/2}\mathrm{s}$, or alternatively, cm/statV. The "*NL*" subscript in the quantity symbol is the initial letters of the term "non-linear", indicating that it relates to a component generating an oscillating polarization at a different (doubled) frequency compared with that of the electromagnetic field driving force. It may be readily shown that the non-linear optical coefficient may be expressed by the dielectric susceptibilities at ω and 2ω frequencies as follows:

$$\boxed{d_{NL}^{(2\omega)} \equiv \frac{mD\left[\chi_L^{(\omega)}\right]^2 \chi_L^{(2\omega)}}{2N^2e^3}} \,. \tag{17.9}$$

If both ω and 2ω frequencies are sufficiently far from resonance (the ω_0 frequency), namely if both $|\omega_0{}^2 - \omega^2| >> \gamma\omega$ and $|\omega_0{}^2 - (2\omega)^2| >> 2\gamma\omega$ conditions are satisfied, the non-linear optical coefficient is a real quantity to an excellent approximation.

As an intermediate comment we indicate that the left side of Eq. (17.3) shows that the non-linear response also contains a time-independent deviation component relative to the equilibrium point (namely, a component generating a constant polarization), as well as components oscillating at 3ω and 4ω. All these components are assumed to be sufficiently small to be safely ignored.

Following Eqs. (17.8) and (17.9), the polarization component oscillating at the 2ω frequency may be written as $\boldsymbol{P}^{(2\omega)}(t) = 2d_{NL}^{(2\omega)}[\mathcal{E}_0 \mathbf{cos}\,\omega t]\cdot[\mathcal{E}_0 \mathbf{cos}\,\omega t]$. In that form, the crystal appears as if it is subject to two *separate* radiation fields (in the present case of same ω frequency). Because of coupling between the separately linearly induced polarization fields, an entailing polarization field is produced, oscillating by the sum frequencies $2\omega = \omega + \omega$. Accordingly, we conjecture that the presented case is a special case of a much broader family of effects, whereby the crystal is exposed to two radiation fields of different ω_1 and ω_2 frequencies, and also of different "k" and "ℓ" polarization orientations. As a result of coupling between the two linearly induced polarizations, a new polarization state is formed, oscillating by the sum frequency $\omega_3 = \omega_1 + \omega_2$ at a j polarization orientation. We accordingly define a non-linear optical response tensor $d_{jk\ell}^{\omega_1+\omega_2}$ by the relation

$$\boxed{P_j^{\omega_1+\omega_2} = 2d_{jk\ell}^{\omega_1+\omega_2}\mathcal{E}_k^{\omega_1}\mathcal{E}_\ell^{\omega_2}} \,, \tag{17.10}$$

where $P_j^{\omega_1+\omega_2}$ is the complex amplitude of the j-oriented component of the induced polarization, oscillating by an $\omega_1 + \omega_2$ frequency; $\mathcal{E}_k^{\omega_1}$ is the complex amplitude of the k-oriented electric field component of ω_1 frequency; $\mathcal{E}_\ell^{\omega_2}$ is the complex amplitude of the ℓ-oriented electric field component of ω_2 frequency; and each of the $j,\ k,\ \ell$ may assume any of the $x,\ y,\ z$ orientations. Obviously, $x,\ y,\ z$ preferably coincide with the dielectric tensor principal axes. It is important to recall that in tensor symbols convention, the writing of identical labels such as in Eq. (17.10) means *summation* over all elements assuming the common label. We further recall the meaning of the complex amplitude; for example,

$$\mathcal{E}_k^{\omega_1}(t) = \boldsymbol{Re}\left(\mathcal{E}_k^{\omega_1}e^{-i\omega t}\right) = \frac{1}{2}\left(\mathcal{E}_k^{\omega_1}e^{-i\omega t} + c.c.\right), \tag{17.11}$$

and similarly for all other complex components appearing in Eq. (17.10).

In addition to a tensor that connects two optical fields to a polarization field oscillating by the *summed* frequencies, it is possible to define a $d_{jk\ell}^{\omega_1-\omega_2}$ tensor that connects them to a polarization field oscillating by the frequencies' *difference*; namely,

$$P_j^{\omega_1 - \omega_2} = 2d_{jk\ell}^{\omega_1 - \omega_2} \mathcal{E}_k^{\omega_1} \mathcal{E}_\ell^{-\omega_2}, \tag{17.12}$$

where $\mathcal{E}_\ell^{-\omega_2} \equiv \left(\mathcal{E}_\ell^{\omega_2}\right)^*$ (see Eq. [17.11]).

Physically, there should be no difference if we exchange between $\mathcal{E}_k^{\omega_1}$ and $\mathcal{E}_\ell^{\omega_2}$ in both (17.10) and (17.12) equations (if, for example, we write $\mathcal{E}_\ell^{\omega_2}\mathcal{E}_k^{\omega_1}$ instead of $\mathcal{E}_k^{\omega_1}\mathcal{E}_\ell^{\omega_2}$). Thus, we may replace any $k\ell$ label pairs of the $d_{jk\ell}^{\omega_1 + \omega_2}$ tensor with a single label; the convention for that purpose is:

$$xx \equiv 1; \; yy \equiv 2; \; zz \equiv 3; \; yz = zy \equiv 4; \; xz = zx \equiv 5; \; xy = yx \equiv 6\,.$$

Eq. (17.10) would then be written in the following reduced form:

$$\begin{pmatrix} P_x \\ P_y \\ P_z \end{pmatrix} = \begin{pmatrix} d_{11} & d_{12} & d_{13} & d_{14} & d_{15} & d_{16} \\ d_{21} & d_{22} & d_{23} & d_{24} & d_{25} & d_{26} \\ d_{31} & d_{32} & d_{33} & d_{34} & d_{35} & d_{36} \end{pmatrix} \begin{pmatrix} \mathcal{E}_x{}^2 \\ \mathcal{E}_y{}^2 \\ \mathcal{E}_z{}^2 \\ 2\mathcal{E}_z\mathcal{E}_y \\ 2\mathcal{E}_z\mathcal{E}_x \\ 2\mathcal{E}_x\mathcal{E}_y \end{pmatrix}. \tag{17.13}$$

The d elements' matrix of 3×6 order is called the **reduced tensor of non-linear optical coefficients**. We have already noted that in centro-symmetric crystals, all non-linear coefficients vanish. Non-centro-symmetric crystals still exhibit other symmetry properties; consequently, some constants vanish, some non-zero ones are identical to others, and some are of the same size but an opposite sign. Table 17.1 describes the fundamental structures of the electrooptic constants tensor for each of the non-centro-symmetric crystal systems, where x, y, and z, are principal axes of the dielectric tensor. The symbols index is provided in the table caption. For example, the tensor belonging to the orthorhombic D_2 (222) system has only three non-zero elements: d_{14}, d_{25}, and d_{36}, of different numerical values. The tensor belonging to the hexagonal C_{3h} $(\bar{6})$ system has six non-zero elements that are inter-related: $d_{12} = -d_{11}$, $d_{26} = -2d_{11}$, $d_{21} = -d_{22}$, and $d_{16} = -2d_{22}$.

The structure of the non-linear optical coefficients tensor also depends on the naming of the Cartesian coordinates relative to the crystalline symmetry elements. The table thus details some alternative names for the coordinates that are common in the scientific literature (of course always being principal axes of the dielectric tensor). It is also interesting to note that the electrooptic constants tensor that belongs to the O (432) cubic system is identically zero despite being non-centro-symmetric!

The reader should note that the structure of the matrices shown in Table 17.1 is, in fact, identical to the transposed matrices of the electrooptic tensors presented in Table 16.1. That identity stems from "selection rules" induced by the crystals' symmetry properties. It is, however, important to emphasize that the physical meaning of the d_{ij} optical non-linear coefficients is entirely different from those of the r_{ij} electrooptic coefficients.

Fundamentally, the optical non-linear coefficients are frequency dependent, which results from the frequency dependence of the dielectric susceptibilities at both ω and 2ω frequencies (Eq. [17.9]). The frequency range for which there exists a technological interest is where the crystal is transparent at both ω and 2ω frequencies, and both are quite far from any ω_0 resonance frequency. The frequency dependence of the relevant $d_{ij}^{(2\omega)}$ coefficients is then quite moderate. For most practical purposes, it is possible to present a single representative constant as adequate for all needed regions. Under such conditions, it is further possible to reduce the number of non-linear coefficients on the basis of an empiric rule called **Kleinman symmetry**. This rule states that coefficients whose extended labels are equal – irrespective of their labeling order – are equal. Thus, for example, $d_{xyy} \cong d_{yxy} \cong d_{yyx}$. The maximal number of independent coefficients reduces to only 10, and the reduced tensor looks like

$$\begin{pmatrix} \mathbf{d_{11}} & \mathbf{d_{12}} & \mathbf{d_{13}} & \mathbf{d_{14}} & \mathbf{d_{15}} & \mathbf{d_{16}} \\ d_{16} & \mathbf{d_{22}} & \mathbf{d_{23}} & \mathbf{d_{24}} & d_{14} & d_{12} \\ d_{15} & d_{24} & \mathbf{d_{33}} & d_{23} & d_{12} & d_{14} \end{pmatrix}. \tag{17.14}$$

The Kleinman independent elements are written with bold-faced letters.

Table 17.2 below presents experimental values for several non-linear optical coefficients of a number of materials commonly used in laser technology, especially for frequency conversions of laser outputs, an issue to be discussed later. Analysis of the experimental values reveals that differences among values of

Table 17.1 Fundamental structures of the non-linear electrooptic tensors for all point groups belonging to non-centro-symmetric crystal systems. A bold dot "•" symbolizes a zero element; a full circle "●" symbolizes a non-zero element. Two line-connected full circles "●–●" symbolize two equal non-zero elements; two line-connected circles "●–○", one full and the other hollow, symbolize two non-zero equally sized elements of opposite signs. A double hollow circle "◎" symbolizes a doubly sized element compared with its line-connected *full* circle, yet of the opposite sign.

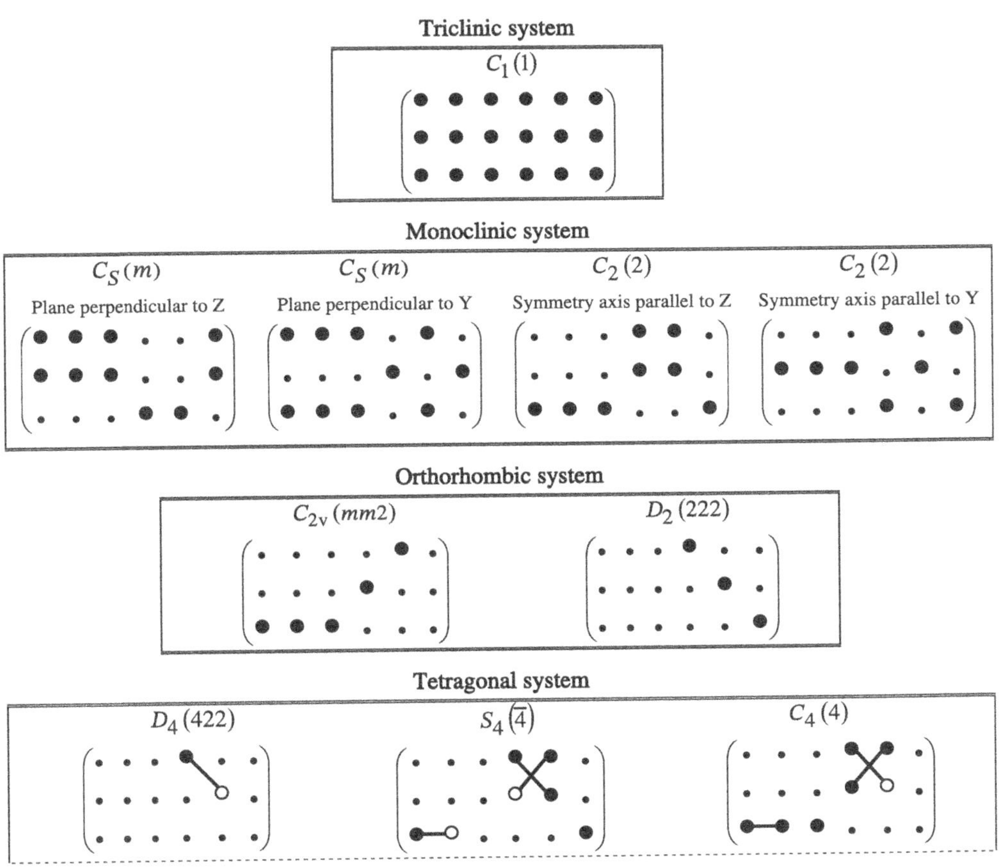

Table 17.1 (Continued)

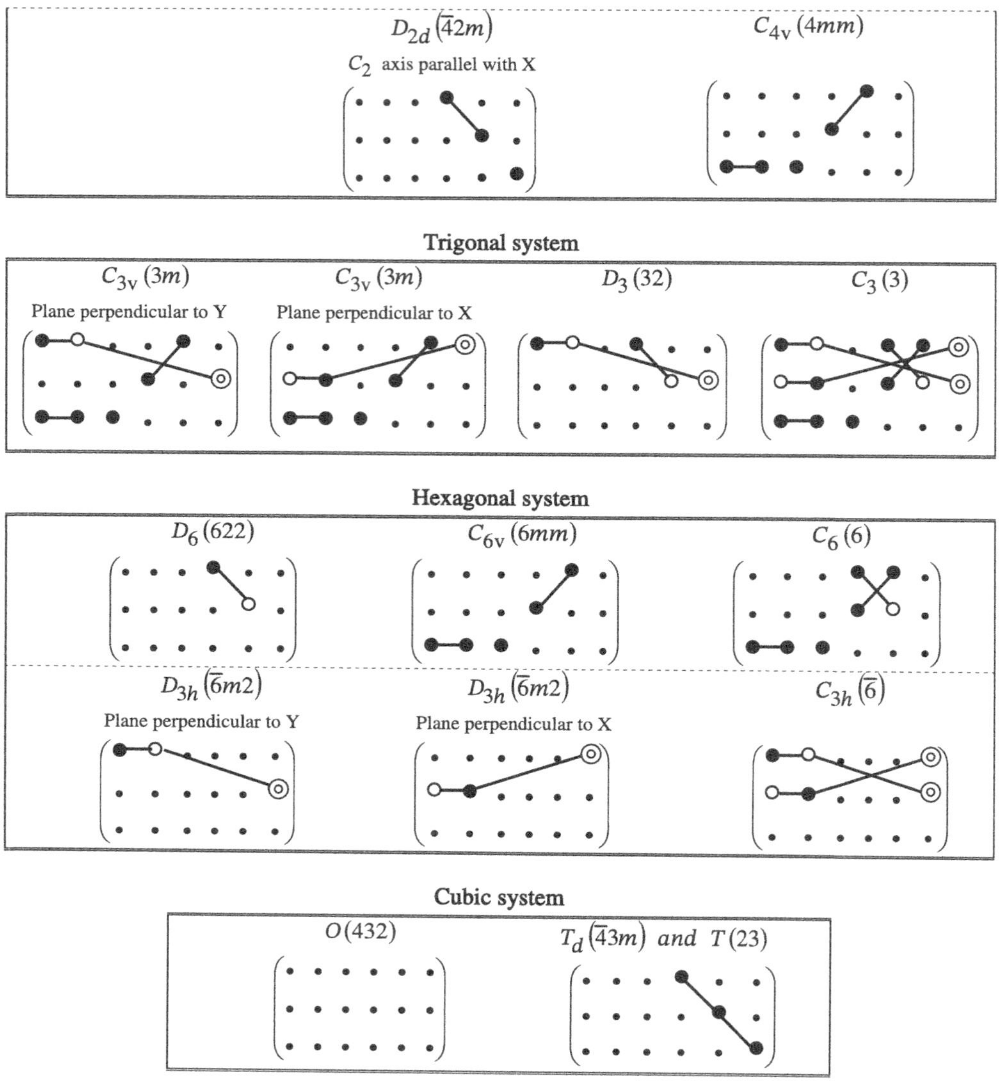

different materials are linked to differences in the values of the dielectric susceptibilities $\chi_L^{(\omega)}$ and $\chi_L^{(2\omega)}$. On the other hand, the $\delta^{(2\omega)} \equiv mD/2N^2e^3$ quantity that appears as a multiplier in Eq. (17.9) is quite similar in all materials – $\delta^{(2\omega)} \sim 2.5 \times 10^{-10}$ m/V, with about a 40% variance. Because the density of electronic oscillators N in all solid materials is of the same order of magnitude, the D quantity, which is the non-linear coefficient in the electronic oscillation (Eq. [17.9]), exhibits similar order values in all solid matters. The value obtained is $D \sim 10^{40}$ m^{-1}s^{-2}.

An underlying assumption in our former calculation was that each atomic basis of the crystal contains a single electronic oscillator of ω_0 frequency. That is obviously untrue because an atomic basis contains many electrons that contribute to the dielectric susceptibility and hence to the non-linear optical effects. By adopting a model of many independent oscillators, similar to the one used in Chapter 10 (specifically in Eq. [10.54]), we would have to write

$$\chi_L^{(\omega)} \equiv \frac{Ne^2}{m} \sum_p \frac{1}{(\omega_{0p}^2 - \omega^2) - i\gamma_p \omega}, \tag{17.15}$$

where p is the ordinal number of the electron rotating in the atomic basis, ω_{0p} is its resonance frequency, and γ_p is its oscillation damping factor. For numerical calculations, it is usually sufficient to include the effects of only the valence electrons; the resonance frequencies of ω_{0p} belonging to the core electrons are very high, and their contribution to the non-linearity of the dielectric susceptibility in the visible region is thus quite negligible.

Table 17.2 Compilation of non-linear optical tensor coefficients of several materials. The common abbreviation of a material's chemical formula is enclosed in parentheses to its side. Unlisted Tensor elements have not yet been measured; in some cases, they may be assessed using Kleinman's symmetry (Eq. [17.14]). Most values have been measured at $\lambda = 1.064$ μm, except for those indicated below the table margin. The choice of the dielectric *X, Y, Z* principal axes relative to the crystal *a, b, c* axes is detailed for the biaxial crystals.

Material	Point group	$d_{ij}^{(2\omega)}$ (10^{-12} m/V)	Dielectric axes
LiB_3O_5 (LBO)	C_{2v} $(mm2)$	$d_{31} = -0.67$; $d_{32} = 0.85$; $d_{33} = 0.04$	$X, Y, Z \leftrightarrow a, c, b$
KH_2PO_4 (KDP)	D_{2d} $(\bar{4}2m)$	$d_{36} = 0.39$; $d_{14} = 0.35$	
KD_2PO_4 (DKDP) ·	D_{2d} $(\bar{4}2m)$	$d_{36} = 0.37$; $d_{14} = 0.42$	
$NH_4H_2PO_4$ (ADP)	D_{2d} $(\bar{4}2m)$	$d_{36} = 0.47$	
$\beta - BaB_2O_4$ (BBO)	C_{3v} $(3m)$	$d_{22} = 2.3$; $d_{31} = -0.16$	
$LiIO_3$	C_6 (6)	$d_{31} = 4.4$; $d_{33} = 4.6$	
$KTiOPO_4$ (KTP)	C_{2v} $(mm2)$	$d_{31} = 1.4$; $d_{32} = 2.65$; $d_{33} = 10.7$	$X, Y, Z \leftrightarrow a, b, c$
$LiNbO_3$	C_{3v} $(3m)$	$d_{22} = 2.46$; $d_{31} = -4.64$; $d_{33} = -41.7$	
$KNbO_3$	C_{2v} $(mm2)$	$d_{31} = -11.9$; $d_{32} = -13.7$; $d_{33} = -20.6$	$X, Y, Z \leftrightarrow b, a, c$
$AgGaS_2$	D_{2d} $(\bar{4}2m)$	$d_{36} = 12.5^{\clubsuit}$	
$ZnGeP_2$	D_{2d} $(\bar{4}2m)$	$d_{36} = 69^{\clubsuit}$	

♣ Measured at $\lambda = 10.6$ μm

17.2 Electromagnetic Wave Propagation in a Non-Linear Crystal

17.2.1 Maxwell's Equations

A description of electromagnetic radiation propagation in any medium, including non-linear dielectric crystals, is obviously derived from Maxwell's equations (see Eq. [10.1]). In their present application, we elaborate in more detail on the electric polarization density vector $\underline{\boldsymbol{P}}$:

$$\begin{aligned}
&\text{(i)}\ \ \boldsymbol{rot}\ \underline{\mathcal{B}} = \frac{1}{c}\frac{\partial \underline{\mathcal{D}}}{\partial t} + \frac{4\pi}{c}\underline{\boldsymbol{J}}_{ext} = \frac{1}{c}\frac{\partial}{\partial t}(\underline{\mathcal{E}} + 4\pi\underline{\boldsymbol{P}}) + \frac{4\pi}{c}\underline{\boldsymbol{J}}_{ext}\\
&\text{(ii)}\ \ \boldsymbol{div}\ (\underline{\mathcal{E}} + 4\pi\underline{\boldsymbol{P}}) = 4\pi\rho_{ext}\\
&\text{(iii)}\ \ \boldsymbol{rot}\ \underline{\mathcal{E}} = -\frac{1}{c}\frac{\partial \underline{\mathcal{B}}}{\partial t}\\
&\text{(iv)}\ \ \boldsymbol{div}\ \underline{\mathcal{B}} = 0\,.
\end{aligned} \tag{17.16}$$

We further write $\underline{\boldsymbol{P}} = \chi_L\underline{\mathcal{E}} + \underline{\boldsymbol{P}}_{NL}$, where $\underline{\boldsymbol{P}}_{NL\,j}^{\omega_1+\omega_2} = 2d_{jk\ell}^{\omega_1+\omega_2}\mathcal{E}_k^{\omega_1}\mathcal{E}_\ell^{\omega_2}$ (see Eq. [17.10]). For simplicity, we assume that the dielectric susceptibility χ_L is a scalar quantity (namely, independent of the external electric field orientation). Because the material under consideration is a dielectric medium, all electrons are bonded, and no electrical currents may develop. Also, the magnetic permeability in a dielectric medium is a scalar of a unity size, thus the magnetic induction vector $\underline{\mathcal{B}}$ is the same as the magnetic field $\underline{\mathcal{H}}$. We further assume that the material is devoid of any space charge ($\rho = \rho_{ext} = 0$) and that no externally induced currents $\underline{\boldsymbol{J}}_{ext}$ are present. Under those conditions, the first listed Maxwell equation in Eq. (7.16) reduces to

$$\boldsymbol{rot}\ \underline{\mathcal{H}} = \frac{1}{c}\frac{\partial}{\partial t}[(1 + 4\pi\chi_L)\underline{\mathcal{E}}] + \frac{4\pi}{c}\frac{\partial}{\partial t}\underline{\boldsymbol{P}}_{NL}\,. \tag{17.17}$$

The $\varepsilon = 1 + 4\pi\chi_L$ quantity is the time-independent dielectric function. To achieve the electromagnetic wave equations, we differentiate Eq. (17.17) with respect to time and get $\boldsymbol{rot}\ \frac{\partial\underline{\mathcal{H}}}{\partial t} = \frac{\varepsilon}{c}\frac{\partial^2\underline{\mathcal{E}}}{\partial t^2} + \frac{4\pi}{c}\frac{\partial^2\underline{\boldsymbol{P}}_{NL}}{\partial t^2}$; we set $\partial\underline{\mathcal{H}}/\partial t$ according to Maxwell's Eq. (17.16)-iii). Recalling that $\varepsilon = n^2$, where n is the refractive index, we get

$$\boldsymbol{rot}\ \boldsymbol{rot}\ \underline{\mathcal{E}} + \frac{n^2}{c^2}\frac{\partial^2\underline{\mathcal{E}}}{\partial t^2} + \frac{4\pi}{c^2}\frac{\partial^2\underline{\boldsymbol{P}}_{NL}}{\partial t^2} = 0\,. \tag{17.18}$$

Using the mathematical relation provided in Section 10.5.1 item (xiii), we get

$$-\boldsymbol{\Delta}\underline{\mathcal{E}} + \boldsymbol{grad}\ \boldsymbol{div}\ \underline{\mathcal{E}} + \frac{n^2}{c^2}\frac{\partial^2\underline{\mathcal{E}}}{\partial t^2} + \frac{4\pi}{c^2}\frac{\partial^2\underline{\boldsymbol{P}}_{NL}}{\partial t^2} = 0\,. \tag{17.19}$$

Recall now that the $\boldsymbol{\Delta}\underline{\mathcal{E}}$ quantity is a *vector*, whose components are results of activating the Laplace operator $\boldsymbol{\Delta}$ (see Section 3.2) separately on each component of the $\underline{\mathcal{E}}$ vector. We further set $\boldsymbol{div}\ \underline{\mathcal{E}} = 0$ and get the following wave equation:

$$\boxed{\boldsymbol{\Delta}\underline{\mathcal{E}} + \frac{n^2}{c^2}\frac{\partial^2\underline{\mathcal{E}}}{\partial t^2} + \frac{4\pi}{c^2}\frac{\partial^2\underline{\boldsymbol{P}}_{NL}}{\partial t^2} = 0}\ . \tag{17.20}$$

17.2.2 Overlapping Beams of Different Frequencies Propagating in the Same Direction

We consider three overlapping different plane-wave beams, differently linearly-polarized, propagating in the crystal in some arbitrary Z direction, of ω_1, ω_2, and ω_3 frequencies:

$$\begin{aligned}
\mathcal{E}_\ell^{(\omega_1)}(z,t) &= \frac{1}{2}\left[\mathcal{E}_{1\ell}(z)e^{-i(\omega_1 t - k_1 z)} + c.c.\right]\\
\mathcal{E}_k^{(\omega_2)}(z,t) &= \frac{1}{2}\left[\mathcal{E}_{2k}(z)e^{-i(\omega_2 t - k_2 z)} + c.c.\right]\\
\mathcal{E}_j^{(\omega_3)}(z,t) &= \frac{1}{2}\left[\mathcal{E}_{3j}(z)e^{-i(\omega_3 t - k_3 z)} + c.c.\right].
\end{aligned} \tag{17.21}$$

The j, k, ℓ subscripts correspond to the electric field polarizations along some Cartesian axes on the plane perpendicular to Z, namely on the (X, Y) plane. We further assume that the three waves are coupled such that $\omega_3 = \omega_1 + \omega_2$.

As an example, we set the expression for the electric field ℓ component in Eq. (17.17). The partial derivatives of the Laplace $\boldsymbol{\Delta}$-operated electric field with respect to x and y vanish, due to the plane-parallel nature of the beams. As for the z derivative, we assume that the variation of the beam amplitude as a function of z is sufficiently moderate such that $\frac{\partial^2\mathcal{E}_{1i}}{\partial z^2} \ll k_1\frac{\partial\mathcal{E}_{1i}}{\partial z}$. We may then neglect the second z derivative of the field strength in all equations and write

$$\Delta \mathcal{E}_{\ell}^{(\omega_1)}(z,t) = -\frac{1}{2}\left[k_1{}^2 \mathcal{E}_{1\ell}(z) - 2ik_1 \frac{\partial \mathcal{E}_{1\ell}(z)}{\partial z}\right] e^{-i(\omega_1 t - k_1 z)} + c.c. \tag{17.22}$$

The complete wave equation would then become

$$\begin{aligned} &\left[ik_1 \frac{\partial \mathcal{E}_{1\ell}(z)}{\partial z} - \frac{k_1{}^2}{2}\partial \mathcal{E}_{1\ell}(z)\right] e^{-i(\omega_1 t - k_1 z)} + c.c. \\ &= -\frac{n_1{}^2}{c^2}\left[\frac{\partial \mathcal{E}_{1\ell}(z)}{\partial z} e^{-i(\omega_1 t - k_1 z)} + c.c.\right] + \frac{4\pi}{c^2}\frac{\partial^2 \mathbf{P}_{NL_\ell}^{\omega_1}(z,t)}{\partial t^2}. \end{aligned} \tag{17.23}$$

The crystal dispersion relation satisfies $n_1{}^2\omega_1{}^2/c^2 = k_1{}^2$. Recall that we have assigned a non-linear coupling among the three $\mathcal{E}_\ell^{(\omega_1)}$, $\mathcal{E}_k^{(\omega_2)}$, and $\mathcal{E}_j^{(\omega_3)}$ waves, satisfying $\omega_1 = \omega_3 - \omega_2$ (Eq. [17.12]). Thus,

$$ik_1 \frac{\partial \mathcal{E}_{1\ell}}{\partial z} e^{-i(\omega_1 t - k_1 z)} = -\frac{4\pi}{c^2}\omega_1{}^2 d_{jk\ell}\mathcal{E}_{3j} e^{-i(\omega_3 t - k_3 z)} \mathcal{E}_{2k}^* e^{i(\omega_2 t - k_2 z)}.$$

The $d_{jk\ell}$ tensor appearing in the above equation is the non-linear optical coefficients tensor for the $\omega_1 = \omega_3 - \omega_2$ coupling, which corresponds to the specific j, k, ℓ polarizations. After dividing by $ik_1 e^{-i(\omega_1 t - k_1 z)}$, setting $\omega_1 = \omega_3 - \omega_2$, and including the $k_1 = \omega_1 n_1/c$ dispersion relation, we get

$$\frac{\partial \mathcal{E}_{1\ell}}{\partial z} = i\omega_1 \frac{4\pi}{cn_1} d_{jk\ell}\mathcal{E}_{3j}\mathcal{E}_{2k}^* e^{i(k_3 - k_2 - k_1)z}, \tag{17.24}$$

similarly, for the other waves,

$$\frac{\partial \mathcal{E}^*{}_{2k}}{\partial z} = -i\omega_2 \frac{4\pi}{cn_2} d_{\ell jk}\mathcal{E}_{1\ell}\mathcal{E}_{3j}^* e^{i(k_1 - k_3 + k_2)z}, \tag{17.25}$$

$$\frac{\partial \mathcal{E}_{3j}}{\partial z} = i\omega_3 \frac{4\pi}{cn_3} d_{kj\ell}\mathcal{E}_{1\ell}\mathcal{E}_{2k} e^{i(k_1 + k_2 - k_3)z}. \tag{17.26}$$

17.2.3 Frequency Doubling

We turn to solving Eq. (17.26) for a specific case, where the two ω_1 and ω_2 are equal; namely, $\omega_3 = 2\omega_1$. We consider the simple case where the "1" and "2" beams are identical, actually meaning the case of a single beam; hence $\mathcal{E}_{2k} = \mathcal{E}_{1\ell}$. That specific case is called **frequency doubling** or, alternatively, **second harmonic generation** (SHG). The boundary condition for initial strength of the frequency-doubled beam is $\mathcal{E}_{3j}(z=0) = 0$. We further limit ourselves to consider only z values, where $\mathcal{E}_{1\ell}$ (obviously also $\mathcal{E}_{2k}$) remain almost unchanged. In other words, no considerable depletion of energy occurs for the $\mathcal{E}_{1\ell}$ beam. We define $\Delta k \equiv k_1 + k_2 - k_3$. Then $\Delta k = 2k_1^{(\ell)} - k_3^{(j)}$, where the upper superscript attached to each wave vector indicates the beam polarization orientation. Setting the said assumptions, and integrating Eq. (17.26) with respect to z yields

$$\mathcal{E}_{3j}(z) = \omega_3 \frac{4\pi}{cn_3} d_{\ell j\ell}\mathcal{E}_{1\ell}(0)\mathcal{E}_{1\ell}(0)\frac{e^{i\Delta k z} - 1}{\Delta k}. \tag{17.27}$$

The power density $I^{(\omega_3)}(z)$ of the beam developing at the $\omega_3 = 2\omega_1$ frequency is given by $\frac{cn_3}{8\pi}\left|\mathcal{E}_{3j}\right|^2$ (see Eq. [10.15] and the related text). Thus, we get

$$I^{(\omega_3)}(z) = {\omega_3}^2 \frac{2\pi}{cn_3}\left(d_{\ell j \ell}\right)^2 |\mathcal{E}_{1\ell}(0)|^2 |\mathcal{E}_{1\ell}(0)|^2 z^2 \frac{\sin^2(\Delta k\, z/2)}{(\Delta k\, z/2)^2},$$

$$I^{(\omega_3)}(z) = \left(d_{\ell j \ell}\right)^2 \frac{2(4\pi)^3 {\omega_3}^2}{c^3 {n_1}^2 n_3}\left[I_{1\ell}^{(\omega_1)}(0)\right]^2 z^2 \frac{\sin^2(\Delta k\, z/2)}{(\Delta k\, z/2)^2}. \tag{17.28}$$

For brevity, in the following text we write ω instead of ω_3. The frequency doubling efficiency, defined as $\eta_{SHG}(z) \equiv I^{(2\omega)}(z)/I_{1\ell}^{(\omega)}(0)$, is then given by

$$\boxed{\eta_{SHG}(z) = \left(d_{\ell j \ell}\right)^2 \frac{(8\pi)^3 \omega^2}{c^3 {n_1}^2 n_3} I_{1\ell}^{(\omega)}(0)\, z^2 \frac{\sin^2(\Delta k\, z/2)}{(\Delta k\, z/2)^2}}, \tag{17.29}$$

a relation valid as long as $\eta_{SHG}(z) << 1$.

The frequency doubling efficiency is proportional to the input beam power density $I_{1\ell}^{(\omega)}(0)$, and to the squared distance it progressed – z^2. A crucial factor in the frequency doubling efficiency is provided by the size of the last multiplying factor $y \equiv \sin^2(\Delta k\, z/2)/(\Delta k\, z/2)^2$. That function is described in detail in Figure 2.1, where the y-function variable is $x \equiv (\Delta k\, z/2)$. A maximal value of unity is obtained under **phase-matching** conditions, namely when $\Delta k = 0$. Conditions allowing phase matching are unique, and will be later discussed in detail. Under **phase mismatch** conditions, namely $\Delta k \neq 0$, the frequency doubling efficiency diminishes after a z propagation distance, satisfying $\Delta k\, z/2 = \pi$, and stays practically zero for longer distances. That limit distance $L_c = 2\pi/\Delta k$ is called **coherence length**; it describes the maximal propagation distance that permits the frequency doubling. For the purpose of enhanced frequency doubling, lengthening of the coherence length is imperative; practically, a full phase matching is required! Phase matching actually describes a case of linear momentum conservation, because the photon momentums' sum of the two beams of ω frequency equals the photon momentum of the doubled frequency beam: $k_\omega^{(\ell)} + k_\omega^{(\ell)} = k_{2\omega}^{(j)}$.

Proper phase matching means $2k_\omega^{(\ell)} = k_{2\omega}^{(j)}$. Recall that $k = 2\pi/\lambda = 2\pi n/\lambda_0$, where $\lambda_0 = 2\pi c/\omega$ is the beam wavelength on passing through vacuum. We thus obtain the following condition

$$\boxed{n_{2\omega}^{(j)} = n_\omega^{(\ell)}}, \tag{17.30}$$

where $n_{2\omega}^{(j)}$ and $n_\omega^{(\ell)}$ are the refractive indices of the beams whose frequencies are 2ω and ω in the "j" and "ℓ" polarizations, respectively. The required indices' equality clarifies an important physical meaning of the phase matching requirement: under phase matching, the "generating" beam at ω frequency propagates at the *same phase velocity* as the "generated" beam of the 2ω frequency. It apparently allows a continual amplification of the "generated" beam amplitude without destructive interferences.

Dielectric function's dispersion in dielectric materials would not allow the required indices equality (Eq. [17.30]) for the same ℓ and j polarizations; except for unique and special cases, the refractive index increases with frequency (wavelength shortening – see Figures 9.13 and 14.4) across the entire material's transparency region. On the other hand, refractive index anisotropy in non-cubic crystals allows the occurrence of an index equality between ω and 2ω belonging beams. Obviously, special propagation orientations and beam polarizations are required.

Figure 17.1 demonstrates the configuration arrangement chosen for a specific case. The used medium is the non-linear, uniaxial KH_2PO_4 (KDP) crystal (Table 17.2). The goal is to double the 1.064 μm wavelength output of a $Nd:YAG$ laser, and obtain a beam of half that wavelength – 0.532 μm (the two noted

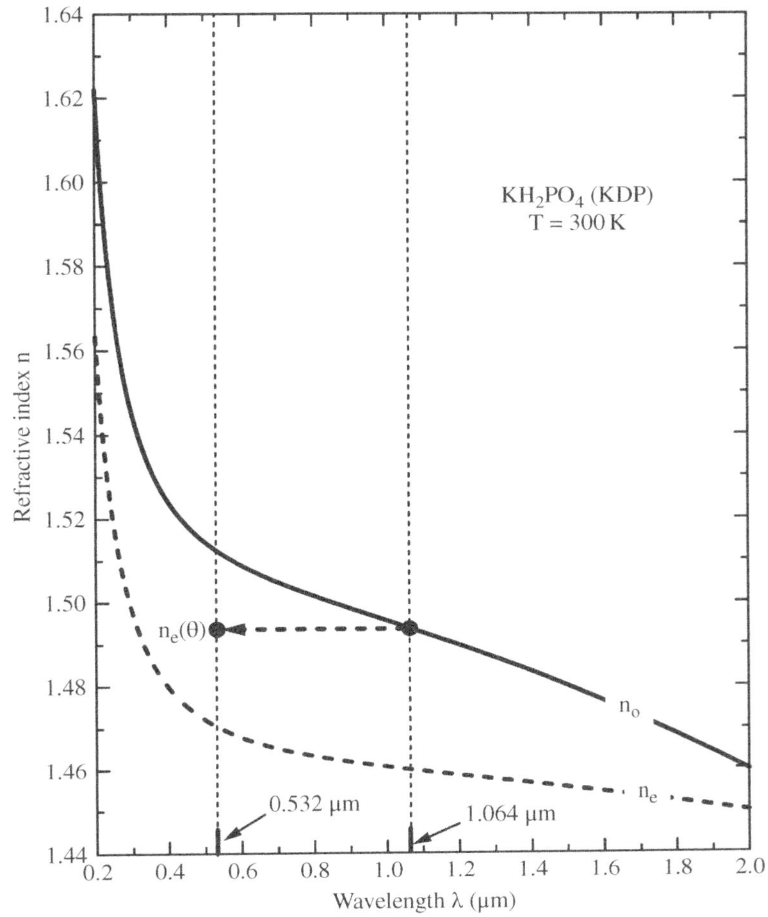

Figure 17.1 Ordinary and extraordinary refractive indices as functions of wavelength in the uniaxial KH_2PO_4 (KDP) single crystal, and a demonstration of phase matching configuration choice for frequency doubling of the 1.064 μm wavelength beam. The dashed arrow indicates the direction of frequency conversion. For detailed discussion, see text.

wavelengths are the vacuum wavelength ones). The known Sellmeier expressions for the KDP refractive indices are (wavelengths given in μm units)

$$n_o^2 = 2.259276 + \frac{13.00522\lambda^2[\mu m]}{\lambda^2[\mu m] - 400} + \frac{0.01008956}{\lambda^2[\mu m] - (77.26408)^{-1}};$$
$$n_e^2 = 2.132668 + \frac{3.2279924\lambda^2[\mu m]}{\lambda^2[\mu m] - 400} + \frac{0.008637494}{\lambda^2[\mu m] - (81.42631)^{-1}}. \quad (17.31)$$

Figure 17.1 correspondingly presents the KDP refractive indices versus wavelength dispersion curves. For frequency doubling of a 1.064 μm wavelength beam, one has an ordinary polarization of $n_o(1.064\ \mu m) = 1.4938$ refractive index. For the doubled frequency generated beam, one uses an extraordinary polarization of the *same* refractive index. It further requires selection of the appropriate θ propagation angle orientation, satisfying $n_e(\theta,\ 0.532\ \mu m) = 1.4938$ (see Section 15.7.2, especially Eq. [15.27]). The equation that needs to be solved for that purpose is

$$\frac{1}{\left(n_{\omega}^{(o)}\right)^2} = \frac{\cos^2\theta}{\left(n_{2\omega}^{(o)}\right)^2} + \frac{\sin^2\theta}{\left(n_{2\omega}^{(e)}\right)^2}, \tag{17.32}$$

whose general solution is

$$\sin^2\theta = \frac{\left(n_{\omega}^{(o)}\right)^{-2} - \left(n_{2\omega}^{(o)}\right)^{-2}}{\left(n_{2\omega}^{(e)}\right)^{-2} - \left(n_{2\omega}^{(o)}\right)^{-2}}. \tag{17.33}$$

Setting the following data: $n_{2\omega}^{(e)} = 1.4705$, $n_{2\omega}^{(o)} = 1.5124$ (Figure 17.1), and as before $n_{\omega}^{(o)} = 1.4938$, in Eq. (17.33), we get $\theta = 41°$.

Figure 17.2 below describes the aforementioned frequency doubling configuration. A beam of $\lambda = 1.064$ μm wavelength (infrared) propagates through a KDP crystal formed as a rectangular prism plate cut in the proper crystallographic orientations. It generates a beam of $\lambda = 0.532$ μm wavelength (green) that exits the plate. The remainder of the generating beam (whose power has been reduced) also exits the plate but is not indicated in the figure. The issue of its remaining intensity will be discussed later, in relation to the frequency doubling efficiency under conditions where its intensity reduces gradually due to energy transfer to the frequency-doubled beam.

We turn now to discuss the size of the non-linear optical coupling coefficient $d_{\ell j \ell}$. Using the tensor form of the non-linear optical coupling coefficient (space group D_{2d} $(\bar{4}2m)$ item in Table 17.1), we may write the induced $\underline{\boldsymbol{P}}^{(2\omega)}$ polarization vector in the form (Eq. [17.13])

$$\begin{aligned} P_x^{(2\omega)} &= 2d_{14}\mathcal{E}_z^{(\omega)}\mathcal{E}_y^{(\omega)} \\ P_y^{(2\omega)} &= 2d_{14}\mathcal{E}_z^{(\omega)}\mathcal{E}_x^{(\omega)} \\ P_z^{(2\omega)} &= 2d_{36}\mathcal{E}_x^{(\omega)}\mathcal{E}_y^{(\omega)}. \end{aligned} \tag{17.34}$$

Because the beam of ω frequency is set to be polarized perpendicularly to Z, namely $\mathcal{E}_z^{(\omega)} = 0$, the induced 2ω polarization components $P_x^{(2\omega)}$ and $P_y^{(2\omega)}$ in the X and Y directions, respectively, vanish. We further set $\mathcal{E}_x^{(\omega)} = \mathcal{E}_{1\ell}\sin\varphi$, and $\mathcal{E}_y^{(\omega)} = \mathcal{E}_{1\ell}\cos\varphi$ (see Figure 17.2), and get $P_z^{(2\omega)} = d_{36}\mathcal{E}_{1\ell}^{\,2}\sin 2\varphi$. The 2ω generated beam of continually increasing power (the phase-matched one) is oriented $90°-\theta$ off the Z axis.

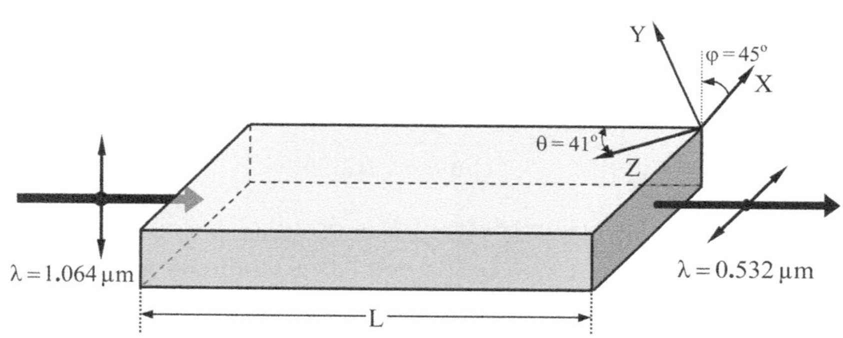

Figure 17.2 Configuration of a system based on a KDP (KH_2PO_4) rectangular prism plate set for frequency doubling of a $\lambda = 1.064$ μm wavelength laser beam. The choice of the θ and φ angles is aimed at phase matching and maximizing of the non-linear optical coupling $d_{\ell j \ell}$ coefficient, respectively. For a detailed discussion, see the text.

We should thus consider the polarization component in the polarization orientation of the continually increasing beam power, namely $P_z^{(2\omega)}\ \boldsymbol{sin\theta} = d_{36}\ \boldsymbol{sin\theta sin}2\varphi \cdot \mathcal{E}_{1\ell}^{\ 2}$. Correspondingly,

$$d_{eff} \equiv d_{\ell j\ell} = d_{36}\ \boldsymbol{sin\theta sin}2\varphi\,. \tag{17.35}$$

The maximal d_{eff} as a function of φ thus obtains for $\varphi = 45°$, as drafted in Figure 17.2.

Another interesting case would be the frequency doubling of a $\lambda = 0.517$ μm wavelength beam (the doubled frequency would belong to 0.2585 μm), where $\theta = 90°$. In that case, $d_{eff} = d_{36}$, and the polarization orientation of the frequency-doubled beam coincides with the Z-axis direction. It is thus impossible to find a configuration that allows for frequency doubling of any wavelength shorter than 0.517 μm. In other words, it is the **shortest wavelength** whose frequency could be doubled using a KH_2PO_4 (KDP) crystal.

Eq. (17.29) describing the frequency doubling efficiency is valid as long as the generated beam intensity is very small compared with the generating beam's one. We extend our discussion to a broader case, which takes into account a reduction in the generating beam intensity with a simultaneous increase in the generated beam intensity. We return to Eq. (17.26) and solve it for the phase matching conditions:

$$\frac{\partial \mathcal{E}_{3j}(z)}{\partial z} = i\omega_3 \frac{4\pi}{cn_3} d_{\ell j \ell} \mathcal{E}_{1\ell}(z)^2\,. \tag{17.36}$$

If we choose to present $\mathcal{E}_{1\ell}(0)$ as a real quantity, then the $\mathcal{E}_{3j}(0)\mathcal{E}_{2k}^*(0) = \mathcal{E}_{3j}(0)\mathcal{E}_{1\ell}(0)$ product in Eq. (17.24) will be imaginary; therefore the $\mathcal{E}_{1\ell}(z)$ function that solves Eq. (17.24) will be real. We may then write $|\mathcal{E}_{1\ell}(z)|^2$ instead of $\mathcal{E}_{1\ell}(z)^2$. We now define a new *real* quantity $\overline{\mathcal{E}}_{3j} \equiv -i\mathcal{E}_{3j}$. We recall further that the power densities of the different beams are given by $I_{1\ell}^{(\omega)}(z) = \frac{cn_1}{8\pi}|\mathcal{E}_{1\ell}(z)|^2$ and $I_{3j}^{(2\omega)}(z) = \frac{cn_3}{8\pi}|\mathcal{E}_{3j}(z)|^2 = \frac{cn_3}{8\pi}\overline{\mathcal{E}}_{3j}(z)^2$ (see Eq. [10.15] and related text). By energy conservation law, we get $|\mathcal{E}_{1\ell}(z)|^2 = |\mathcal{E}_{1\ell}(0)|^2 - \frac{n_3}{n_1}\overline{\mathcal{E}}_{3j}(z)^2$. We set it in Eq. (17.34) and get

$$\frac{\partial \overline{\mathcal{E}}_{3j}(z)}{\partial z} = \frac{8\pi\omega}{cn_1n_3} d_{\ell j \ell}\left[n_1\mathcal{E}_{1\ell}(0)^2 - \ n_3\overline{\mathcal{E}}_{3j}(z)^2\right]. \tag{17.37}$$

To simplify the expressions, we define a new parameter $A_{1\ell}(0) \equiv \sqrt{n_1}\,\mathcal{E}_{1\ell}(0)$ and a new variable $\overline{A}_{3j} \equiv \sqrt{n_3}\,\overline{\mathcal{E}}_{3j}$. The differential equation then assumes the form

$$\frac{\partial \overline{A}_{3j}(z)}{\partial z} = \frac{8\pi\omega\sqrt{n_3}}{cn_1n_3} d_{\ell j \ell}\left[A_{1\ell}(0)^2 - \ \overline{A}_{3j}(z)^2\right]. \tag{17.38}$$

For further simplification, we denote $y(z) = \overline{A}_{3j}(z)/A_{1\ell}(0)$ and get

$$\frac{\partial y}{\partial z} = \frac{8\pi\omega\sqrt{n_3}}{cn_1n_3} d_{\ell j \ell} A_{1\ell}(0)\left[1 - \ y(z)^2\right]. \tag{17.39}$$

As a next step in simplification, we denote $K = \frac{8\pi\omega\sqrt{n_3}}{cn_1n_3} d_{\ell j \ell}$ and change the equation variable from z into the new one $\xi = KA_{1\ell}(0)\,z$. The differential equation form then simplifies to read

$$\frac{\partial y(\xi)}{\partial \xi} = \left[1 - \ y(\xi)^2\right]. \tag{17.40}$$

The solution to that equation is $y(\xi) =$ ***tanh*** $\xi =$ ***tanh*** $[KA_{1\ell}(0)z]$ (the reader is referred to Section 17.4 to be reminded of the properties of the hyperbolic "trigonometric" functions). Because the $KA_{1\ell}(0)$ quantity is a constant, the solution may be represented as a function of z, namely $y^2(z) = \boldsymbol{tanh}^2[KA_{1\ell}(0)z]$. Reintroducing all of the above-defined relations into the latter solution, we get

$$I_{3j}^{(2\omega)}(z) = I_{1\ell}^{(\omega)}(0)\,\boldsymbol{tanh}^2\left\{\frac{(8\pi)^{3/2}\,\omega}{c^{3/2}n_1\sqrt{n_3}}d_{\ell j \ell}\left[I_{1\ell}^{(\omega)}(0)\right]^{1/2}z\right\}. \tag{17.41}$$

Finally, recalling that under phase matching $n_1 = n_3 \equiv n$, and denoting $d_{\ell j \ell} = d_{eff}$, we get

$$\boxed{\eta_{SHG} = \boldsymbol{tanh}^2\left\{\frac{(8\pi)^{3/2}\,\omega}{c^{3/2}n^{3/2}}d_{eff}\left[I_{1\ell}^{(\omega)}(0)\right]^{1/2}z\right\}}\ . \tag{17.42}$$

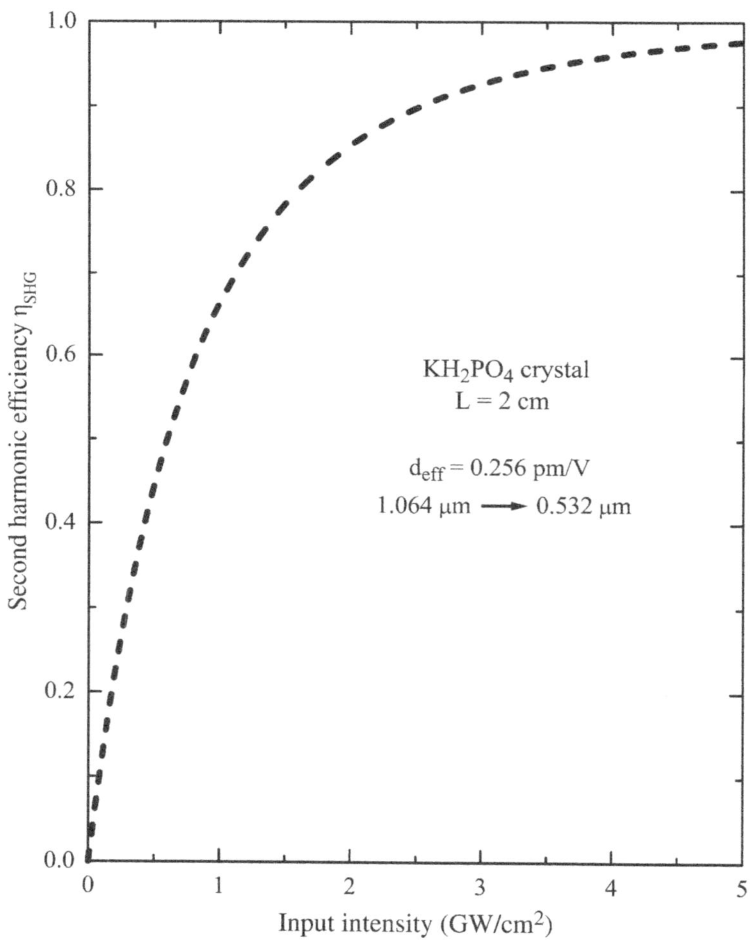

Figure 17.3 Theoretically calculated second harmonic (frequency doubling) generation efficiency by an incident 1.064 μm light beam as a function of the beam intensity per Eq. (17.42). The calculation assumed a 2-cm long KDP (KH_2PO_4) single crystal in the configuration shown in Figure 17.2.

Eq. (17.41) shows that at very large z distances, the $I^{(2\omega)}(z)$ doubled frequency beam intensity approaches asymptotically the initial value of the original $I^{(\omega)}(0)$ beam intensity. In other words, the frequency doubling intensity asymptotically approaches a unity. The initial efficiency is proportional to $I^{(\omega)}(0)$. For demonstration, in Figure 17.3, we present a theoretical curve describing the doubling frequency efficiency of a $\lambda = 1.064$ μm wavelength beam as a function of the initial (input) beam intensity using a 2 cm long KDP crystal according to the configuration of Figure 17.2.

In reality, it is quite difficult to achieve conditions for obtaining a close to a unity conversion efficiency. The main reason is walk-off of the frequency doubled beam relative to the incident beam, which gradually reduces the geometric overlap between the two (see Section 15.7.3, especially Figure 15.11). The trivial means for enhanced efficiency would be to increase the $I^{(\omega)}(0)$ incident beam intensity. A limitation for that approach relates to possible structural damage to the used crystal. Concentrating the beam aimed at increasing its local intensity would be helpful up to a limit where the doubled frequency beam walk-off reduces the two beams geometric overlap and hence the effective conversion efficiency. Other effects limiting the frequency doubling efficiency are intrinsic losses of the crystal, such as Rayleigh and Raman scattering, and extrinsic scattering because of impurities and structural defects.

The above presented configuration for frequencies' conversion is not the only possible one when using non-linear crystals. Configurations are usually labeled by three letters, which are either "o" for an ordinary polarized beam or "e" for an extraordinary polarized one. The label on the farthest left relates to the lower frequency beam; the label farthest right relates to the highest frequency beam. For example, the configuration described in Figures 17.1 and 17.2 for the KDP crystal (which is negative uniaxial) is "ooe". In a positive uniaxial crystal (where $n_o(\lambda) < n_e(\lambda)$), the chosen configuration would be "eeo", where the incident beam used for frequency doubling is extraordinarily polarized, and the frequency doubled beam is ordinarily polarized. Other configurations are chosen for cases of **sum-frequency generation** (SFG) or **difference frequency generation** (DFG). The most useful configurations relate to parallel beams' propagation; for various purposes, different configurations are selected for non-parallel propagating beams. The phase matching conditions are expressed in terms of linear photon momentum conservation: $\underline{k}_{\omega_1}^{(\ell)} + \underline{k}_{\omega_2}^{(k)} = \underline{k}_{\omega_1+\omega_2}^{(j)}$. In the present book, we will not elaborate further on the various configurations of frequency conversions.

17.3 Optical Parametric Oscillations

17.3.1 Forced Parametric Oscillations

In Chapter 1 we showed, that it is possible to force a mechanical oscillator, like an electron bonded to a positive center, to oscillate at a desired frequency by exerting an external driving force, oscillating at the said desired frequency. We shall see that an alternative way to enforce such oscillations is by adding an oscillating component to one of the system parameters. Such an external enforcement is called **parametric forcing**. As an example, we consider a mechanical oscillator, whose equation of motion is given by (see Eq. [1.10])

$$\ddot{x} + \Gamma\dot{x} + {\omega_0}^2 x = 0\ , \tag{17.43}$$

where x is the position displacement from equilibrium, Γ is the damping parameter, and ω_0 is the resonance frequency. As known, the solution of the differential equation describes a damped oscillation, whose amplitude decays over time with a characteristic $2/\Gamma$ decay time (see text related to Eq. [1.10]). We now imagine an external enforcement that modulates periodically the resonance frequency, such as a periodic modulation of the spring constant of a hanging mass, or a periodic moment of inertia modulation in a physical pendulum (an action executed by circus swing acrobats – see Exercise 17.2 at the end of this chapter). A similar effect may be obtained in an electrical resonance circuit comprising a resistor, an inductor, and a capacitor (RLC circuit) by a periodic modulation of the capacitance. We now express an equation of motion of parametrically forced mechanical oscillator in the form

$$\ddot{x} + \Gamma\dot{x} + \left({\omega_0}^2 - \Omega^2 \boldsymbol{\sin}\,\omega_p t\right) x = 0\ , \tag{17.44}$$

where ω_p is the angular frequency modulation of the squared oscillator resonance angular frequency, and $\Omega^2 \ll {\omega_0}^2$ is the modulation amplitude of the said squared resonance frequency. To convert the equation into a form similar to a simple forced oscillator, like in Eq. (1.28), we write

$$\ddot{x} + \Gamma\dot{x} + {\omega_0}^2 x = \left[\Omega^2 \boldsymbol{\sin}\,\omega_p t\right] x\ . \tag{17.45}$$

We attempt a solution of the form $x = \textbf{Re}\,[x_0 e^{\beta t} e^{i(\omega t + \varphi)}]$, where x_0 and β are real constants. Then $x = \frac{1}{2} x_0 \left\{ e^{i[(\omega - i\beta)t + \varphi]} + c.c. \right\}$. For $\beta < 0$ in the attempted solution, the oscillations decay with time; for $\beta > 0$, the oscillations grow as a function of time. Setting that attempted solution in Eq. (17.45), we get

$$- (\omega - i\beta)^2 e^{i[(\omega - i\beta)t + \varphi]} + i\Gamma(\omega - i\beta) e^{i[(\omega - i\beta)t + \varphi]} + \omega_0^{\,2} e^{i[(\omega - i\beta)t + \varphi]} + c.c\,.$$
$$= \Omega^2 \frac{e^{i\omega_p t} + e^{-i\omega_p t}}{2i} \left\{ e^{i[(\omega - i\beta)t + \varphi]} + c.c. \right\},$$

or

$$\left\{ -(\omega - i\beta)^2 + i\Gamma(\omega - i\beta) + \omega_0^{\,2} \right\} e^{i[\omega t + \varphi]} + \left\{ -(\omega + i\beta)^2 - i\Gamma(\omega + i\beta) + \omega_0^{\,2} \right\} e^{-i[\omega t + \varphi]}$$
$$= \frac{\Omega^2}{2i} \left\{ e^{i[(\omega_p - \omega)t - \varphi]} + e^{-i[(\omega_p - \omega)t - \varphi]} \right\} + \frac{\Omega^2}{2i} \left\{ e^{i[(\omega_p + \omega)t + \varphi]} + e^{-i[(\omega_p + \omega)t + \varphi]} \right\}. \tag{17.46}$$

Ignoring the right side addend displayed on a grey background, and equating parameters exhibiting identical phase and frequency on the two equation sides, we get the following relations, which establish a **resonance state**:

$$\boxed{\begin{array}{lll} \omega_p - \omega = \omega & \Rightarrow & \omega = \frac{1}{2}\omega_p \\ e^{-i\varphi} = e^{+i\varphi} & \Rightarrow & \varphi = 0 \quad or \quad \pi \end{array}}\,. \tag{17.47}$$

The right side addend of Eq. (17.46) (gray background) is an off-resonance contribution. Its presence indicates that the attempted solution is, in fact, just an approximation. The occurrence of identical addends on the equation's two sides indicates that the system would exhibit a component oscillating at half the forced modulation frequency, whose phase is coordinated with the modulation phase. Under conditions to be specified shortly, the said component amplitude increases with time. To see that, we compare between the coefficients of the corresponding exponential addends on the two sides of Eq. (17.46), and get

$$(\omega - i\beta)^2 - i\Gamma(\omega - i\beta) - \left(\omega_0^{\,2} + \frac{1}{2} i\Omega^2 \right) = 0\,.$$

Under conditions where $\Gamma^2 < < \Omega^2 < < \omega_0^{\,2}$, the above equation exhibits two solutions for β:

$$\beta_1 = i(\omega_0 - \omega) - \frac{1}{2}\Gamma + \frac{1}{4}\frac{\Omega^2}{\omega_0}; \qquad \beta_2 = -i(\omega_0 + \omega) - \frac{1}{2}\Gamma - \frac{1}{4}\frac{\Omega^2}{\omega_0}\,.$$

The β_2 solution describes a high frequency modulation, $(\omega_0 + \omega)$, that nonetheless causes a decay of the oscillation's amplitude. The β_1 solution, on the other hand, describes a low frequency modulation, $|\omega_0 - \omega|$, and, in cases where $\Omega^2 > 2\omega_0\Gamma$, also describes a *negative* damping, or a continual *amplification* of the oscillation's amplitude. In the particular case where $\omega = \frac{1}{2}\omega_p = \omega_0$ we get

$$\boxed{\beta = -\frac{1}{2}\Gamma + \frac{1}{4}\frac{\Omega^2}{\omega_0}}\,, \tag{17.48}$$

the meaning of which is the existence of amplitude gain with no entailed frequency modulation.

A point worth emphasizing is the need for having phase matching between the forced modulation and the amplitude growing oscillations. If, at any moment, there exist oscillations whose phases deviate from the required $\varphi = 0$ or $\varphi = \pi$ condition, their amplification will either turn smaller, or they will gradually reduce and eventually disappear.

The specific case of amplitude growing oscillations is called **parametric amplification**. The ω_p frequency is called the **pump frequency** and ω is called the **signal frequency**. As we shall see in Section 17.3.2, under non-linear coupling between light beams propagating through a material medium, there could be a case where the presence of a high frequency beam induces amplification of a lower frequency beam. Because of the similarity between the two effects, the case of such light beam amplification is called **parametric amplification**. That is the subject of our next section.

17.3.2 Optical Parametric Amplification

The **optical parametric amplification** is commonly abbreviated as **OPA**. The rate equations describing parametric amplifications of laser beams are special cases of Eqs. (17.24) – (17.26), with initiation boundary conditions $\mathcal{E}_{3j}(z) = \mathcal{E}_{3j}(0)$, $\mathcal{E}_{1\ell}(0) \neq 0$, and $\mathcal{E}_{2k}(0) = 0$. The beam belonging to $\mathcal{E}_{3j}$ whose frequency is $\omega_3 = \omega_1 + \omega_2$, is considered **pump beam**. The beam belonging to $\mathcal{E}_{1\ell}$ whose frequency is ω_1, is termed **signal beam**; the beam belonging to $\mathcal{E}_{2k}$ whose frequency is ω_2 is termed **idler beam**. Under an assumption of constant intensity of the $\mathcal{E}_{3j}$ beam, there are two differential equations requesting a solution:

$$\left.\begin{aligned} \frac{\partial \mathcal{E}_{1\ell}}{\partial z} &= i\omega_1 \frac{4\pi}{cn_1} d_{jk\ell}\mathcal{E}_{3j}\mathcal{E}_{2k}^{*} e^{i(k_3 - k_2 - k_1)z} && (i) \\ \frac{\partial \mathcal{E}_{2k}^{*}}{\partial z} &= -i\omega_2 \frac{4\pi}{cn_2} d_{\ell jk}\mathcal{E}_{1\ell}\mathcal{E}_{3j}^{*} e^{i(k_1 - k_3 + k_2)z} && (ii) \end{aligned}\right\}. \tag{17.49}$$

We now define $\Delta k = k_3 - (k_1 + k_2)$, and choose to represent $\mathcal{E}_{3j}(z) = \mathcal{E}_{3j}(0)$ as a real quantity (namely, satisfying $\mathcal{E}_{3j}^{*} = \mathcal{E}_{3j}$). For further simplicity, we adopt the Kleinman symmetry (Eq. [17.14] and related text) according to which $d_{jk\ell} = d_{\ell jk} \equiv d_{eff}$. Eq. (17.49) then reduces to

$$\left.\begin{aligned} \frac{\partial \mathcal{E}_{1\ell}}{\partial z} &= i\omega_1 \frac{4\pi}{cn_1} d_{eff}\mathcal{E}_{3j}(0)\mathcal{E}_{2k}^{*} e^{i\Delta k z} && (i) \\ \frac{\partial \mathcal{E}_{2k}^{*}}{\partial z} &= -i\omega_2 \frac{4\pi}{cn_2} d_{eff}\mathcal{E}_{3j}(0)\mathcal{E}_{1\ell} e^{-i\Delta k z} && (ii) \end{aligned}\right\}. \tag{17.50}$$

We turn now to solve Eqs. (17.50) under phase-matching conditions, namely $\Delta k = 0$, hence $e^{\pm i\Delta k z} = 1$. We differentiate Eq. (17.50) (i) with respect to z, and set on its right side the expression for $\partial \mathcal{E}_{2k}^{*}/\partial z$, of the left side of Eq. (17.50)-(ii). We perform a symmetric substitution in Eq. (17.50)-(ii) and get

$$\left.\begin{aligned} \frac{\partial^2 \mathcal{E}_{1\ell}}{\partial z^2} &= \frac{(4\pi)^2\omega_1\omega_2}{c^2 n_1 n_2} d_{eff}^{\,2}\,\mathcal{E}_{3j}(0)^2 \cdot \mathcal{E}_{1\ell} && (i) \\ \frac{\partial^2 \mathcal{E}_{2k}^{*}}{\partial z^2} &= \frac{(4\pi)^2\omega_1\omega_2}{c^2 n_1 n_2} d_{eff}^{\,2}\,\mathcal{E}_{3j}(0)^2 \mathcal{E}_{2k}^{*} && (ii) \end{aligned}\right\}. \tag{17.51}$$

As an example, we address first the solution of Eq. (17.51)-(i). The quantity $I_p = \frac{cn_3}{8\pi}\mathcal{E}_{3j}(0)^2$ is the pump beam intensity. We also denote $K = \frac{2(4\pi)^3\omega_1\omega_2}{c^3 n_1 n_2 n_3} d_{eff}^{\,2}$. The equation is then written as $\frac{\partial^2 \mathcal{E}_{1\ell}}{\partial z^2} = KI_p \cdot \mathcal{E}_{1\ell}$. We perform a variable change by defining $\xi = (KI_p)^{1/2} z$. The differential equation transforms into

$\frac{\partial^2 \mathcal{E}_{1\ell}}{\partial \xi^2} = \mathcal{E}_{1\ell}$, with $\mathcal{E}_{1\ell}(z) = A\,\boldsymbol{sinh}\left[(KI_p)^{1/2}\cdot z\right] + B\,\boldsymbol{cosh}\left[(KI_p)^{1/2}\cdot z\right]$ as a general solution (see Section 17.4 as a reminder of the properties of the hyperbolic "trigonometric" functions). Observation of Eq. (17.51)-(i) shows that under the said initial boundary conditions, the first derivative of $\mathcal{E}_{1\ell}$ vanishes at $z = 0$ (namely, $\partial \mathcal{E}_{1\ell}/\partial z]_{z=0} = 0$). Therefore the constants are $A = 0$ and $B = \mathcal{E}_{1\ell}(0)$. The solution is then $\mathcal{E}_{1\ell}(z) = \mathcal{E}_{1\ell}(0)\,\boldsymbol{cosh}\left[(KI_p)^{1/2}\cdot z\right]$. In a similar way we get $\mathcal{E}^*_{2k}(z) = -i\left(\frac{n_1\omega_2}{n_2\omega_1}\right)^{1/2}\mathcal{E}_{1\ell}(0)\,\boldsymbol{sinh}\left[(KI_p)^{1/2}\cdot z\right]$. Expressing the above results in terms of beam intensities, we get

$$\begin{aligned} I^{\omega_1}_{sig}(z) &= I^{\omega_1}_{sig}(0)\,\boldsymbol{cosh}^2\left[\frac{\sqrt{2}(4\pi)^{3/2}}{c^{3/2}} d_{eff}\sqrt{\frac{\omega_1\omega_2}{n_1n_2n_3}}\, I_p^{\ 1/2}\cdot z\right] \\ I^{\omega_2}_{idl}(z) &= \frac{\omega_2}{\omega_1} I^{\omega_1}_{sig}(0)\,\boldsymbol{sinh}^2\left[\frac{\sqrt{2}(4\pi)^{3/2}}{c^{3/2}} d_{eff}\sqrt{\frac{\omega_1\omega_2}{n_1n_2n_3}}\, I_p^{\ 1/2}\cdot z\right] \end{aligned} \quad . \tag{17.52}$$

In Eq. (17.52), we have attached a "sig" subscript to the beam intensity belonging to $\mathcal{E}_{1\ell}$, and an "idl" subscript to the beam intensity belonging to $\mathcal{E}^*_{2k}$ (referring them to the **signal** and **idler** terms, respectively). It is straightforwardly seen that the signal beam fed into the crystal with non-zero intensity at $z = 0$, grows exponentially with increasing distance into the crystal. The idler signal, assumed as totally absent at $z = 0$, also grows exponentially, yet that growth occurs **only if a non-zero signal is present at the crystal front surface**; namely, the presence of an $I^{\omega_1}_{sig}(0) \neq 0$, **is an imperative condition** for the existence of an idler signal as well. The gain itself is approximately proportional to the pump beam power I_p. It should be emphasized that Eq. (17.52) is valid only as long as $I_p(z)$ remains reasonably constant over the propagation length through the crystal. In fact, by energy conservation requirement, increase of the signal and idler intensities depletes the pump beam intensity. In the present frame we will not develop the specific equations that show the different beam intensities' variation profiles under such conditions.

We turn now to the means of achieving phase matching for the parametric amplifications, namely the simultaneous achievement of linear momentum conservation $k_1^{(\omega_1)} + k_2^{(\omega_2)} = k_3^{(\omega_3)}$, and of energy conservation $\omega_1 + \omega_2 = \omega_3$. Considering that each frequency belongs to a different refractive index, the above conditions summarize as

$$\begin{aligned} \omega_1 + \omega_2 &= \omega_3 \\ n_1\omega_1 + n_2\omega_2 &= n_3\omega_3\,. \end{aligned} \tag{17.53}$$

To demonstrate a solution, we consider a specific configuration that is reciprocal to the one used for frequency doubling presented in Figures 17.1 and 17.2. The pump beam of ω_3 frequency is the one belonging to the extraordinary polarization with an $n_3^{(e)}(\theta)$ refractive index, while both signal and idler beams are of an identical frequency, satisfying $\omega_1 = \omega_2 = \frac{1}{2}\omega_3$, with an ordinary polarization satisfying $n_1^{(o)}(\theta) = n_2^{(o)}(\theta) = n_3^{(e)}(\theta)$. That configuration is called **degenerate parametric amplification**, and may be readily represented in Figure 17.1 by just reversing the dashed vector indicating the frequency conversion orientation. Calculation of the θ angle by which the used crystal should be fabricated to achieve the said degenerate parametric amplification is given in Eq. (17.33). The described process is, in fact, a frequency *dividing* rather than the frequency doubling described in Figures 17.1 and 17.2.

Figure 17.4 describes the proper configuration for the said degenerate parametric amplification using a KDP crystal, and a $\lambda = 0.532$ μm pump beam. The crystal design and its crystallographic orientation are identical to those presented in Figure 17.2. The incident pump beam is extraordinarily polarized, and it is

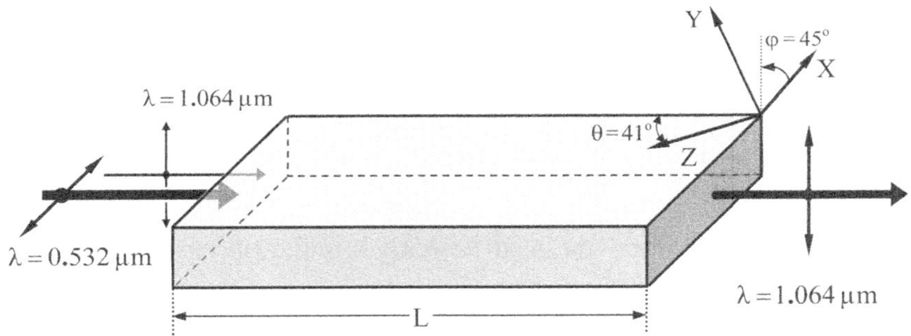

Figure 17.4 Configuration of a system based on a KDP (KH_2PO_4) rectangular prism plate fabricated for degenerate parametric amplification of a $\lambda = 1.064$ μm wavelength laser beam using a pump beam of $\lambda = 0.532$ μm wavelength. The $\lambda = 1.064$ μm wavelength beam sketched as entering the crystal from its left side, which is aimed at being amplified, should overlap the $\lambda = 0.532$ μm wavelength light beam. It is plotted at the side of the latter to prevent overcrowding. For discussion and details, see text.

joined by a $\lambda = 1.064$ μm wavelength beam, ordinarily polarized. In reality, the signal $\lambda = 1.064$ μm wavelength beam must overlap the pump beam, but is sketched to its side to prevent information overcrowding. The signal beam exits the crystal prism with an increased intensity. We have refrained from sketching the remainder of the pump beam (obviously of lowered intensity).

It is customary to define the following quantity as the **amplitude gain coefficient** (see Eq. [17.52])

$$g = \frac{\sqrt{2}(4\pi)^{3/2}}{c^{3/2}} d_{eff} \sqrt{\frac{\omega_1 \omega_2}{n_1 n_2 n_3}} I_p^{\ 1/2} . \tag{17.54}$$

For demonstration of the said gain order of magnitude, we consider a case where $I_p = 1\ \text{GW/cm}^2 = 10^{16}\ \text{erg/cm}^2 \cdot \text{s}$ at $\lambda_3 = 0.532 \times 10^{-4}$ cm, identical signal and idler wavelengths $\lambda_1 = \lambda_2 = 1.064 \times 10^{-4}$ cm, refractive indices $n_1 = n_2 = n_3 = 1.4938$ (in the KDP crystal), and $d_{eff} = 0.25 \times 10^{-12}$ m/V $= 7.5 \times 10^{-9}$ cm/statV (see Table 17.2 and Eq. (17.35)). Setting the above data in Eq. (17.54) yields $g = 3.11\ \text{cm}^{-1}$.

Consider now a case where the crystal prism crystallographic orientation deviates by a $\Delta\theta$ angle from the orientation adequate for the degenerate parametric amplification. This deviation entails a change in the n_3 refractive index, thus deteriorating the system phase matching. We seek a solution to the required phase matching, whereby the signal and idler frequencies change to become $\omega_1' = \omega_1 + \Delta\omega$, and $\omega_2' = \omega_2 - \Delta\omega$; the refractive indices would also change according to $n_1' = n_1 + \Delta n_1$ and $n_2' = n_2 + \Delta n_2$, where $\Delta\omega$ is the change in the signal frequency, Δn_1 is the change in the signal refractive index, and Δn_2 is the change in the idler refractive index. Following Eq. (17.53), the following relation must exist:

$$(n_1 + \Delta n_1)(\omega_1 + \Delta\omega) + (n_2 + \Delta n_2)(\omega_2 - \Delta\omega) = (n_3 + \Delta n_3)\omega_3 . \tag{17.55}$$

Under degeneracy, $\omega_1 = \omega_2 = \frac{1}{2}\omega_3$. For small $\Delta\theta$ changes, we may write $\Delta n_1 \cong \left.\frac{\partial n_0}{\partial \omega}\right]_{\omega_1} \Delta\omega$, and $\Delta n_2 \cong \left.\frac{\partial n_0}{\partial \omega}\right]_{\omega_1} (-\Delta\omega)$. Therefore, $\Delta n_2 = -\Delta n_1$. Setting these relations in Eq. (17.55) we get $\Delta\omega = \omega_1 \left(\frac{\Delta n_3}{\Delta n_1} - 1\right)$.
We set the relation for Δn_1 expressed in the previous line, and get a quadratic equation for $\Delta\omega$:

$$\Delta\omega^2 + \omega_1 \cdot \Delta\omega - \omega_1 \frac{\Delta n_3}{\partial n_o/\partial\omega]_{\omega_1}} = 0 .$$

We set the relation $\Delta n_3 = \partial n_e(\omega_3)/\partial\theta]_{\theta_d}\Delta\theta$ in the above equation, where θ_d is the angle that satisfied the degenerate amplification condition. Solution of the quadratic equation thus obtained is

$$\Delta\omega = -\frac{\omega_1}{2} + \frac{\omega_1}{2}\sqrt{1 + 4\frac{\partial n_e(\omega_3)/\partial\theta]_{\theta_d}}{\omega_1\cdot\partial n_o/\partial\omega]_{\omega_1}}\Delta\theta} .$$

In a range where the angular deviation $\Delta\theta$ is sufficiently small, and carrying out a Taylor expansion of the square root, $\Delta\omega$ may be approximated as

$$\Delta\omega \cong -\frac{\omega_1}{2} + \frac{\omega_1}{2}\left\{1 + \frac{2}{\omega_1}\frac{\partial n_e(\omega_3)/\partial\theta]_{\theta_d}}{\partial n_o/\partial\omega]_{\omega_1}}\Delta\theta - \frac{2}{\omega_1^2}\left(\frac{\partial n_e(\omega_3)/\partial\theta]_{\theta_d}}{\partial n_o/\partial\omega]_{\omega_1}}\right)^2(\Delta\theta)^2\right\}$$

$$\boxed{\Delta\omega \cong \left(\frac{\partial n_e(\omega_3)/\partial\theta]_{\theta_d}}{\partial n_o/\partial\omega]_{\omega_1}}\right)\Delta\theta - \frac{2}{\omega_1}\left(\frac{\partial n_e(\omega_3)/\partial\theta]_{\theta_d}}{\partial n_o/\partial\omega]_{\omega_1}}\right)^2(\Delta\theta)^2} . \tag{17.56}$$

Using Eq. (15.27), one obtains

$$\left.\frac{\partial n_e(\omega_3,\theta)}{\partial\theta}\right]_{\theta_d} = \frac{n_e^3(\omega_3,\theta_d)}{2}\,\mathbf{sin}2\theta_d\left(\frac{1}{n_o^2(\omega_3)} - \frac{1}{n_e^2(\omega_3)}\right) . \tag{17.57}$$

Normally, Sellmeier dispersion equations are represented as functions of wavelength λ. Then,

$$\left.\frac{\partial n_o}{\partial\omega}\right]_{\omega_1} = -\frac{\lambda_1^2}{2\pi c}\cdot\left.\frac{\partial n_o}{\partial\lambda}\right]_{\lambda_1} .$$

Using Sellmeier's equations for the KDP crystal provided in Eq. (17.31), we get $\partial n_e(\omega_3,\theta)/\partial\theta]_{\theta_d} = -0.0835\ \text{rad}^{-1}$ and $\partial n_o/\partial\omega]_{\omega_1} = 1.72\times10^{-17}$ s/rad for the specific parametric amplification described in Figure 17.4. We then use these values to present an empiric expression for the frequency shift of the signal versus the $\Delta\theta$ angular deviation of the pump beam propagation direction:

$$\Delta\omega\left[\frac{\text{rad}}{\text{s}}\right] \cong -4.855\times10^{15}\Delta\theta\ [\text{rad}] - 1.33\times10^{16}\Delta\theta^2\ [\text{rad}] , \tag{17.58}$$

also recall that $\Delta\omega_2 = -\Delta\omega$!

Based on Eq. (17.58), in Figure 17.5 we present the wavelengths of the signal and idler beams that provide phase matching with the pump beam as functions of the angular deviation of its propagation direction from the $\theta_d = 41°$ for which degenerate phase matching is obtained. Note that the mentioned propagations relate to orientations *inside the crystal*. It is possible to alter the propagation direction by changing the pump beam incidence angle on the front crystal prism surface (Snell's law – Eq. [14.27]). Thus, even if the crystal plate is cut for the degenerate phase matching, the dependence described in Figure 17.5 allows a corresponding wavelength tuning of both signal and idler beam amplification by simply rotating the prism described in Figure 17.4 about an axis perpendicular to its horizontal faces.

17.3.3 Optical Parametric Oscillations Based Laser

If an optical parametric oscillations amplifier is enclosed between mirrors of an optical resonance cavity, it is possible to obtain oscillations that are similar to those obtained in a regular laser by the mechanism of stimulated emission. Such a device is called the **Optical parametric oscillations laser**, abbreviated as

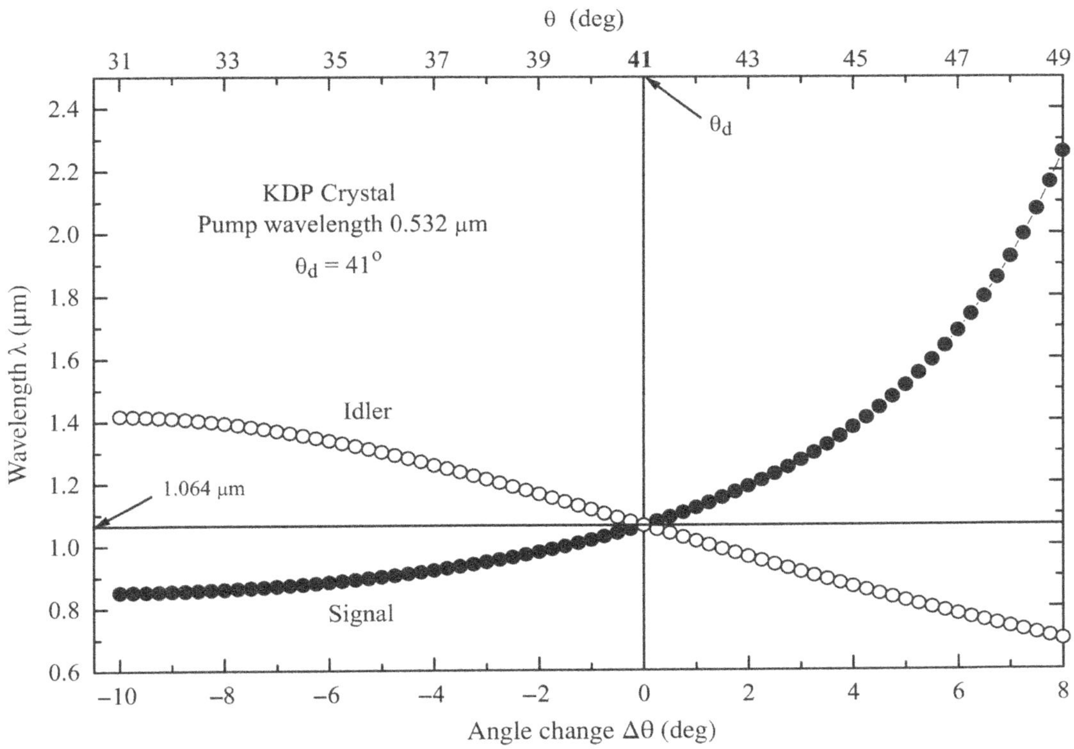

Figure 17.5 Wavelengths providing phase matching for signal and idler beams as functions of the angular deviation of the pump beam propagation direction relative to the $\theta_d = 41°$ orientation, which provides degenerate phase matching in a KDP crystal for the 0.532 μm pump beam wavelength (see Figure 17.4).

OPO-laser. Figure 17.6 displays a schematic construction of such a laser. The back mirror is coated by a special dielectric coating that provides maximal transparency to the pump beam of ω_3 frequency, and full reflectivity for the signal ω_1 and idler ω_2 frequencies, respectively. The pump beam is fed into the cavity through the back mirror. The front mirror is coated for maximal transparency for the pump beam of ω_3 frequency, and partial reflectance to the signal ω_1 and idler ω_2 frequencies. That construction ensures that the signal and idler beams resonate between the mirrors, thus undergoing further amplifications.

It is important to appreciate that the presence of a pump beam is not sufficient to initiate lasing. A necessary factor is the presence of electromagnetic radiation at a proper frequency (ω_1 or ω_2) oriented perpendicularly to the front surface. In most OPO systems, *thermal optical radiation* serves that purpose. It is fundamentally different from an ordinary laser system, where spontaneous emission from the upper laser state is the triggering factor. The beam crossing the front mirror contains the residual pump beam for the ω_3 frequency, as well as both signal and idler. An out-of-cavity frequency selective mirror complements the laser construction. It is inclined at 45° relative to the beams propagation direction; its dielectric coating allows for the transmission of the signal frequency and reflects fully the pump and idler frequencies, diverting them in a direction perpendicular to the cavity axis. The entire system thus produces a coherent light output of ω_1 frequency along the cavity axis direction. In many systems, it is possible to tune the output frequency by rotating the non-linear crystal plate described in Figure 17.6 about an axis perpendicular to the page plane.

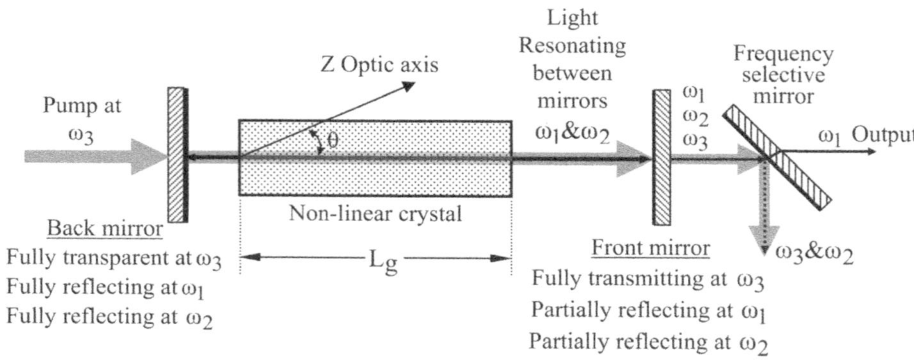

Figure 17.6 Schematic construction of a typical OPO laser.

17.4 A Reminder: Hyperbolic "Trigonometric" Functions

i) The hyperbolic sine ***sinh*** and hyperbolic cosine ***cosh*** functions of a real x variable are defined as

$$\boldsymbol{sinh}\ x \equiv \frac{e^x - e^{-x}}{2}; \qquad \boldsymbol{cosh}\ x \equiv \frac{e^x + e^{-x}}{2}.$$

The ***sinh*** x function is zero at $x = 0$. The function is positive for $x > 0$, continually growing to infinity for $x >> 1$. It also satisfies the relation ***sinh*** $(-x) = -$ ***sinh*** x.

The ***cosh*** x function is a unity 1 at $x = 0$. The function is positive, continually growing to infinity for $|x| >> 1$. It also satisfies the relation ***cosh*** $(-x) =$ ***cosh*** x.

It may be readily verified that the ***sinh*** x and ***cosh*** x functions satisfy the relation ***cosh***$^2 x -$ ***sinh***$^2 x = 1$. The functions' derivatives satisfy the relations $($***sinh*** $x)'' =$ ***sinh*** x and $($***cosh*** $x)'' =$ ***cosh*** x. Thus, a solution to a differential equation of the form $y'' = y$ is of the form $y = A$ ***sinh*** $x + B$ ***cosh*** x, where the specific numerical values of the A and B constants are determined by boundary conditions to the solution function, and to its derivatives.

ii) The hyperbolic tangent ***tanh*** function of a real x variable is defined as ***tanh*** $x \equiv \dfrac{e^x - e^{-x}}{e^x + e^{-x}}$. It is, in fact, the ratio between the hyperbolic sine and the hyperbolic cosine. It is zero for $x = 0$, positive for $x > 0$, grows linearly with x for $x << 1$, and then "saturates", asymptotically approaching a unity value 1 for $x >> 1$. It may be readily verified, that ***tanh*** $(-x) = -$ ***tanh*** x; its derivative satisfies $($***tanh*** $x)' = 1 -$ ***tanh***$^2 x$. Thus, a solution to a differential equation of the form $y' = 1 - y^2$ is $y =$ ***tanh*** x.

Exercises for Chapter 7

17.1 Calculate the phase mismatch Δk for intended frequency doubling of a 1.064 μm wavelength beam incident on a KDP single crystal plate at a $\theta = 0$ in an ordinary polarization (namely a situation where the polarizations of both incident and frequency doubled beams are ordinary; see Figure 17.2). For calculation of the refractive indices, use Eq. (17.31). Use the result to calculate the coherence length whose equation is given in the text that follows Eq. (17.29).

17.2 A circus swing acrobat performs acts on a swing comprising a horizontal bar attached to the ceiling by two side ropes. The acrobat hangs himself with both hands on the bar. His center of gravity is located 6 meters below the ceiling. If the acrobat stays still, making no willful motion, the swing oscillation amplitude reduces to $1/e$ of its initial value within 50 oscillations. Nevertheless, when the acrobat raises and lowers his gravity center at a proper rate, phase, and distance, he manages to preserve the swing oscillation amplitude.

(a) Calculate the natural resonance frequency of the swing plus still acrobat and its oscillations damping coefficient.

(b) At what rate, and by how much, does the acrobat need to raise and lower his gravity center to maintain a constant swinging amplitude.

17.3 The refractive index dispersion functions (Sellmeier equations) of the $ZnGeP_2$ single crystal are given by

$$n_o^2 = 4.47330 + \frac{5.26576\lambda^2[\mu m]}{\lambda^2[\mu m] - 0.13381} + \frac{1.49085\lambda^2[\mu m]}{\lambda^2[\mu m] - 662.55}\,;$$
$$n_e^2 = 4.63318 + \frac{5.34215\lambda^2[\mu m]}{\lambda^2[\mu m] - 0.14255} + \frac{1.45795\lambda^2[\mu m]}{\lambda^2[\mu m] - 662.55}\,, \tag{17.59}$$

where, as noted, the wavelength λ is expressed in μm units.

(a) Plot a graph of the wavelength dispersion curves in the 0.75–12 μm range in a way similar to that presented in Figure 17.1.

(b) Plan a crystal prism shape to be machined for parametric amplification of an electromagnetic beam of $\lambda = 8.0$ μm wavelength by using a pump beam of $\lambda_p = 4.0$ μm, in a way similar to that described in Figure 4. Calculate the amplitude gain coefficient for that particular case.

(c) Calculate the wavelengths of the signal and idler beams if the tilt angle of the pump beam propagation orientation relative to the optic axis is reduced by 2°.

18 Examples of Various Laser Systems

18.1 Introduction

In this chapter, we provide a survey of the various laser systems used for both industrial and research purposes. The goal is to demonstrate the diversity related to that scientific/technological subject on aspects of the amplifying medium state of matter, energy schemes, and emission wavelengths. Also, it allows us to demonstrate some of the challenges involved in the design of laser systems. The survey neither intends to be, nor can it be exhaustive of the subject; the current diversity of laser systems is huge!

The electronic states' structures of amplifying mediums utilized in laser systems are classified as "three level" or "four level". A schematic description of the said structures and the energy transitions involved in their functioning in laser systems is provided in Figure 18.1. In the three-level structure (Figure 18.1a), pumping excites the $|1\rangle$ ground state into the higher $|2\rangle$ state, which decays quickly and dominantly non-radiatively (Chapter 13) into the $|3\rangle$ state. The $|2\rangle$ state serves as the upper laser level. The radiative transition of state $|3\rangle$ down to state $|1\rangle$ is the lasing transition. In the four-level structure described in Figure 18.1b, the lower laser state marked $|4\rangle$ is higher than the ground state $|1\rangle$. A $|3\rangle \rightarrow |4\rangle$ transition emits a photon, then the $|4\rangle$ state decays quickly, dominantly non-radiatively, back into the $|1\rangle$ state, which again is available for further pumping.

The four-level structure has a great advantage over the three-level one. If the $|4\rangle$ state resides sufficiently higher than the $|1\rangle$ ground state, its occupation under thermal and quasi-thermal equilibrium is very small. The population inversion required for sufficient amplification is obtained immediately after the start of pumping, and is maintained as long as pumping continues.

18.2 Helium-Neon Laser

The amplifying medium in a helium/neon (HeNe) laser is a mixture of the helium *He* and neon *Ne* noble gases. The partial pressures of the two gases are usually 0.1 Torr *He* and 0.1 Torr *Ne*. (A reminder: $1\,\text{Torr} \equiv 1\,\text{mmHg} = (1/760)\,\text{Atm.} = 1335\,\text{dyn/cm}^2$). The active transitions during lasing actually take place in the neon *Ne* atom, whereas the presence of helium *He* serves to enhance the pumping efficiency. Pumping is performed by inducing electrical discharge through the gas mixture. The discharge generates a **plasma** state, which is a mixture of positive ions and (negative) free electrons. The electrons, and helium and neon ions, are accelerated by the electric field, thus generating excited states on impact with non-ionized atoms of the two kinds.

In Figure 18.2, we present a partial energy scheme of the two kinds of atoms participating in the pumping and lasing. The lower helium excited states are $2\,^1S$ of para-helium and $2\,^3S$ of ortho-helium (Figure 5.3). Both are meta-stable (in other words, exhibiting long radiative lifetimes). The radiative lifetime of the first one is $\sim 5\,\mu\text{s}$ and of the second one is $\sim 100\,\mu\text{s}$. The $2\,^1S$ helium state is close in energy to the neon state

Physics, Optics, and Spectroscopy of Materials, First Edition. Zeev Burshtein.

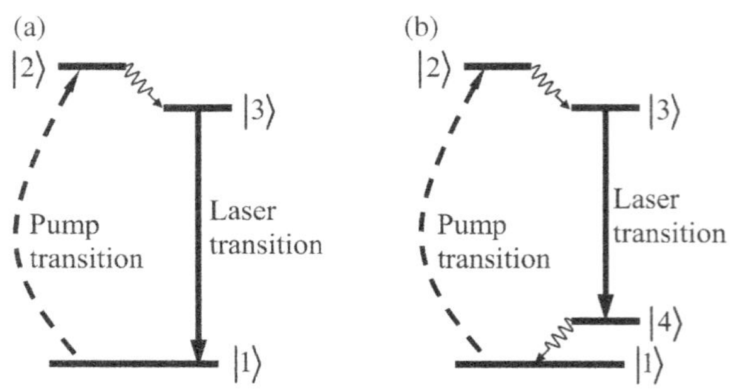

Figure 18.1 Energy schemes of amplifying mediums described as "three-level" (a) and "four-level" (b) systems. For discussion, see text.

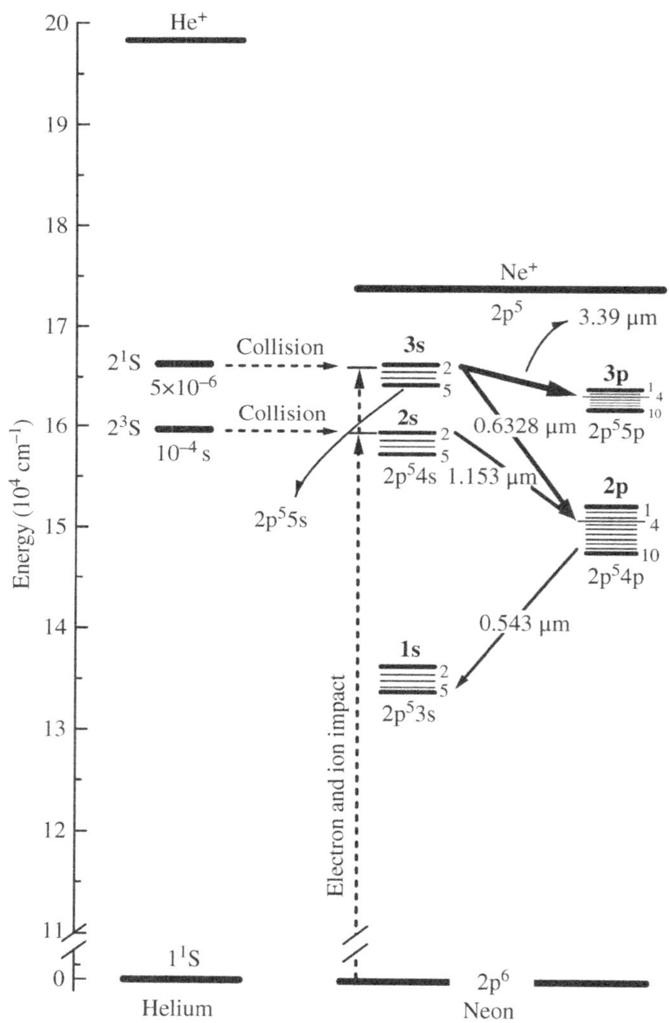

Figure 18.2 Partial energy states scheme of helium *He* and neon *Ne* participation in pumping and lasing in a HeNe laser. Except for energy state assignments, the figure indicates various excitation and fluorescence transitions usable for pumping and lasing.

marked 3*s*; the $2\,^3S$ helium state is very close in energy to the neon state marked 2*s*. Collisions between excited helium atoms and ground-state neon atoms induce energy transfer, and excitation of the neon atoms to the corresponding close states.

The neon *Ne* states of Figure 18.2 are classified as belonging to two manifolds – one generated by a single electron excited to *s* states, and the other generated when the electron is excited to *p* states. The states' assignments are provided by the electronic configuration, and added detailed naming after **Paschen**, rather than according to the **Russell-Saunders** coupling scheme (see Section 5.4.4).

The Paschen method aims at presenting the energy scheme in as similar as possible a form to that of the hydrogen *H* atom (see Figure 3.7). The naming of a single electron excited state assumes the form $\Delta n\,\ell$, where Δn is the *increase* in the principal quantum number compared with that of the ground-state, and ℓ is the orbital angular momentum quantum number. For example, the neon *Ne* ground-state configuration is [*He*] $2s^2 2p^6$ (see Table 5.5). When exciting an outer-shell electron into a 5*s* state, the generated configuration is [*He*] $2s^2 2p^5 5s$. The state is then marked as 3*s*, because the principal quantum number has increased by 3 (see Figure 18.2). The belonging multiplet states are marked by a $\Delta n\,\ell_q$ type symbol, where *q* is the state ordinal number, counting from higher to lower energies. For example, the lowest state of the 3*s* multiplet is marked $3s_5$. The dipole-allowed transitions presented on the scheme (exploited to achieve laser emissions) satisfy the $\Delta\ell = \pm 1$ selection rule; however, no justification exists to assign any symmetry, or any entailing selection rules, to the "neutral" ordinal *q* number. The main reasons for presenting the neon states by the Paschen method are that all spectroscopic studies of that atom have administered it, and the same practice continues in many atomic spectroscopic studies used for laser technology. Another important reason is that the Russell-Saunders and related Hund-rules (see Eq. [5.64]) fail to present properly the actual states order and selection rules of the neon *Ne* atom. It has been conjectured that a $j-\ell$ coupling (not discussed in our present book) would perhaps be more appropriate. The neon 1*s* state is meta-stable; its population draining back to the ground-state occurs predominantly by collisions with free electrons and with the gas container walls. Under sufficiently low pumping rates, the system exhibits the nature of a four-level system (see Figure 18.1).

Figure 18.3 schematically displays the construction of a Helium-neon laser, in fact demonstrating the fundamental ingredients of gas lasers in general. The gas mixture is contained inside a long glass tube, whose end windows are inclined at Brewster's angle relative to the tube axis (see Figures 14.19 and 14.20). A typical tube length is several tens of centimeters and its diameter is several millimeters. The optical windows need to be transparent to the intended output wavelength. A quartz-made tube is usually preferable, being stable up to ~ 1200°C, and relative to other glasses, it is of a very low thermal expansion coefficient. Each side of the inner tube includes a ring-shaped electrode, usually tantalum-made, connected to an electrical circuit that comprises a high voltage supply, a series resistor, and an on/off switch. The circuit ignites the electrical discharge at the needed high voltage, and supplies the constant current required after ignition at a lower voltage. The high voltage source is ~ 1500 V and the series resistance is ~ 100 kΩ. Before being switched on, the gas tube resistance is orders of magnitude larger than any other part of the circuitry. On the moment of on-switching, the voltage between the electrodes rises to almost the entire voltage supply output, initiating the gas discharge. The gas resistance inside the tube drops down to several kilo-ohms, the current increases, and the voltage between the electrodes drops down. By the given values of the voltage and circuit series resistance (see Figure 18.3), the series resistor secures that the current will never exceed 15 mA; in light of its role, the series resistor is called the "ballast resistor". Usually, the remaining steady discharge voltage between the electrodes is approximately 1000 V. In special cases, the cathode is heated externally up to several hundred degrees centigrade to activate electron thermionic emission; this reduces the chances for unintended termination of the electrical discharge.

As may be verified by observation of the system energy scheme, laser emission is achievable at separately four different wavelengths: 3.39 μm, 1.15 μm, 0.6328 μm, and 0.543 μm. Selection of one is performed through selection of the (Brewster) windows-forming materials, the front mirror–forming material, and the dielectric coatings on both back and front mirrors. Specifically, selection of a regular BK7-glass made Brewster window will entirely inhibit the possibility to achieve lasing in the 3.39 μm wavelength, because that material is optically opaque in the infrared. The most common use of a HeNe laser is the 0.6328 μm red wavelength. It is possible to achieve a stable continuous wave (CW) emission at high mono-chromaticity and beam quality (see Section 15.6), permitting various measurements and calibrations, especially

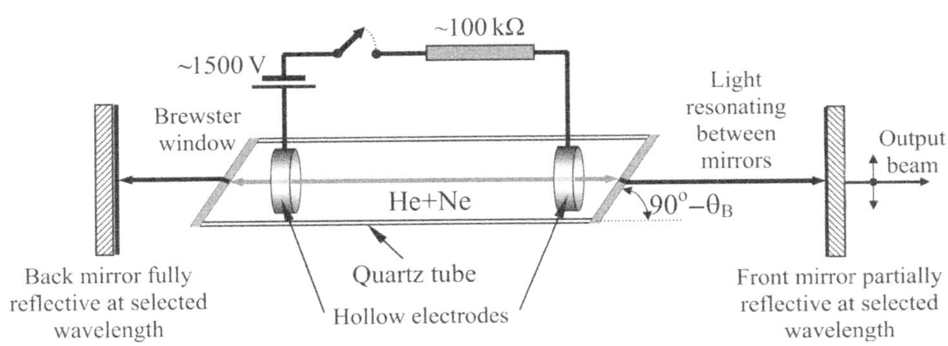

Figure 18.3 Schematic illustration of a typical structure of a gas laser like the HeNe laser. The windows are transparent to the desired wavelength, and inclined at Brewster's angle relative to the laser axis, thus enforcing a polarized output beam. For details, see the text.

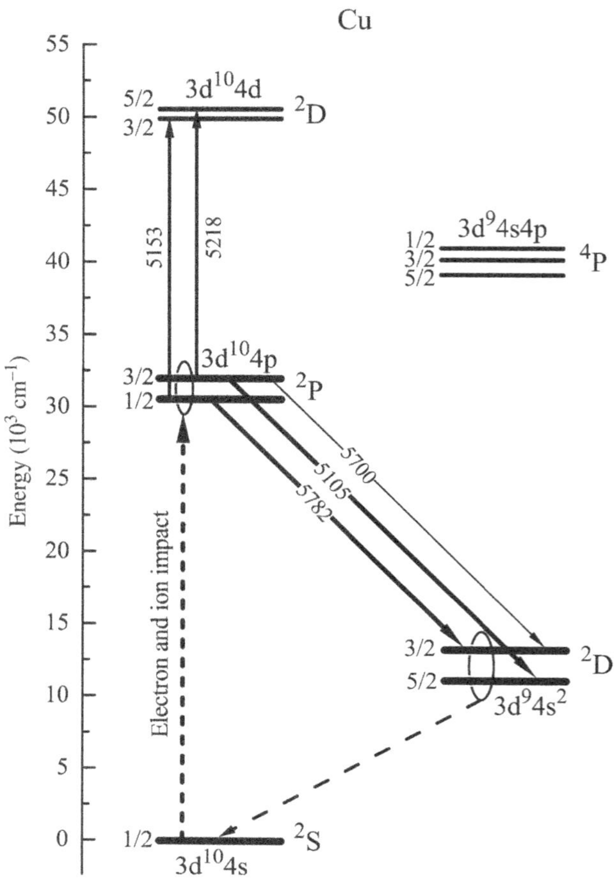

Figure 18.4 Partial energy scheme of the copper *Cu* atom states participating in the pumping and lasing of a copper vapor laser. The doublet splitting of the $3d^{10}4p(^2P)$ and $3d^{10}4d(^2D)$ levels are almost 10 times exaggerated, to demonstrate their existence (the lower of each is placed accurately on the energy scale). Excitation transitions used for pumping, radiative decay transitions used for lasing, and decay transition used for draining the lower laser state, are indicated by arrows. Transition wavelengths are provided in angstrom (Å) units.

calibration of wavelength measuring equipment. A typical power output is 0.5 – 5 mW. In special cases, outputs of up to 100 mW are achievable.

18.3 Copper Vapor Laser

A copper vapor laser takes advantage of radiative transitions between energy states of the copper *Cu* atom. Copper is solid at room temperature. Melting generates a considerable density of sublimed copper atoms. Various radiative and non-radiative transitions involve a mixed neon *Ne* gas plasma and ionized copper vapor atoms. The energy scheme of the participating copper vapor atoms is provided in Figure 18.4. The ground-state belongs to the $^2S_{1/2}$ Term of the $3d^{10}4s$ configuration (see Table 5.5). One of the presented energy manifolds is generated by excitation of a $4s$ electron into a $4p$ single electron configuration, generating $^2P_{1/2}$ and $^2P_{3/2}$ Terms; an excited $4d$ single electron configuration generates $^2D_{3/2}$ and $^2D_{5/2}$ Terms. The doublet splitting in the ^2P Terms is about 182 cm^{-1}; in the ^2D Terms, it is considerably smaller – about 61 cm^{-1} (see Eq. [12.3]). The ^2P Terms serve as upper laser states. Another multiplet involves a single electron state, created when a d electron of the $3d^{10}$ shell is excited to the $4s$ level; a $3d^94s^2$ configuration is thus established. Coupling between the two $4s$ electrons of opposite spin generates a state where the Term is actually determined by the $3d^9$ one-electron-missing shell, which is ^2D (see Table 5.4). The doublet splitting creates two Terms – $^2D_{5/2}$ and $^2D_{3/2}$ – where by Hund's law (see Eq. [5.64] **ii**), the latter is of a higher energy. The doublet splitting between the two is 1950 cm^{-1}, and both serve as lower laser states. Another single electron state forms when a d electron of the $3d^{10}$ shell is excited to form a $4p$ level, establishing a $3d^94s4p$ configuration. Its lower energy Term is ^4P, and by spin/orbit coupling, it generates $^4P_{5/2}$, $^4P_{3/2}$, and $^4P_{1/2}$ Terms.

Various radiative transitions are indicated by arrows on the level's energy scheme. Their relative strengths are demonstrated by the arrows' thickness. The lasing transitions are $^2P_{3/2} \rightarrow {}^2D_{1/2}$ at 510.5 nm wavelength (green) and $^2P_{1/2} \rightarrow {}^2D_{3/2}$ at 578.2 nm wavelength (yellow). Without the intended wavelength selection, the laser output comprises both wavelengths. The scheme does not indicate weak transitions to the $3d^94s4p$ configuration states or the strongest $^2P \rightarrow {}^2S$ radiative transition from the higher laser state to the ground-state. On the other hand, the scheme indicates the pump transition in the opposite direction. It takes place through the impact of free electrons carrying high kinetic energies with copper atoms at their ground-state. The said free electrons are constituents of the neon *Ne* and copper *Cu* plasma.

Usually, the system does not reach population inversion of the ^2P state compared with the ground-state; thus, no lasing is achievable for that transition. The transition radiative lifetime is approximately 7 ns. When the copper vapor atom density is high, an emitted photon by one excited atom is absorbed by a near ground-state atom, and the effective lifetime lengthens up to several tens of nanoseconds. That time allows the ^2P state to be used as an upper laser state relative to the ^2D state. The ^2D levels are meta-stable, which means they exhibit relatively long radiative lifetimes because their transitions to the ground-state are first-order forbidden. The radiative lifetimes of the $^2D_{5/2}$ and $^2D_{3/2}$ levels (serving as lower laser states) are 150 μs and 25 μs, respectively. Draining of those levels' population densities back to the ground-state is assisted by collisions with the ambient plasma free electrons and ions, as well as collisions with the container walls. Nevertheless, those mechanisms are insufficient to shorten their decay time considerably. On lasing, the lower laser state fills in quickly, causing the population inversion, hence lasing, to diminish within several tens of nanoseconds. For that reason, operation of the laser is performed cyclically: each cycle comprises an intense short pumping pulse, short lasing that switches off spontaneously, and eventually a longer waiting time aimed at allowing recombination between the copper ions and free electrons. It also allows their population to return to the ground-state. The laser output is thus pulsed repetitively.

Room temperature copper *Cu* metal is solid; its melting point is 1083°C. Vaporization starts after melting, establishing the copper atom population needed to participate in lasing. The vapor pressure rises steeply with temperature; for example, it is $\sim 10^{-3}$ Torr at 1150°C and $\sim$ 10 Torr at 1870°C. The optimal laser operation temperature is $\sim$ 1450°C, which is much lower than the copper boiling point of 2595°C. The vapor pressure at the optimal laser operation temperature is $\sim$ 1.0 Torr. The copper vapor atoms share the laser amplifying medium with a neon *Ne* gas at a pressure of several tens to several hundreds of Torr. Electrical discharge through the gas serves a double role – it heats the solid copper, and generates a plasma of free electrons and ions accelerated under the electric field, thereby pumping the copper atoms from the ground-state level to the upper laser level. The electrical discharge lasts maximally 20 – 50 ns, and completion of the lower laser state's draining and recombination of copper ions with free electrons lasts at least 200 – 250 μs.

A fundamental schematic structure of a copper vapor laser tube (quite adequate for other metal vapor lasers) is presented in Figure 18.5. The tube is usually made of compressed alumina (the aluminum oxide Al_2O_3) and is capable of withstanding high temperatures up to $\sim$ 1700°C. A typical tube length is of the order of 1m and a diameter of 5 cm. In addition to the construction displayed in Figure 18.5, the alumina laser tube is enclosed within a higher diameter porous alumina tube of a larger side-walls thickness, itself enclosed within a quartz (silicon dioxide SiO_2) tube, further wrapped by a metallic sleeve, cooled by running water. The laser tube side-windows are either quartz or BK7 glass plates, inclined at Brewster's

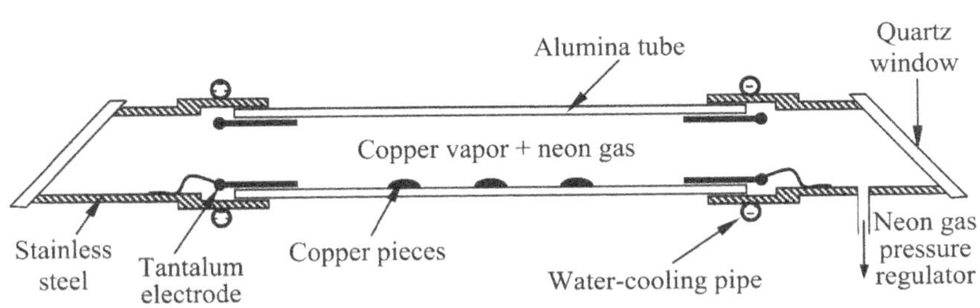

Figure 18.5 Schematic illustration of a copper vapor laser core construction. It is typical of any other metal vapor laser. For details, see text.

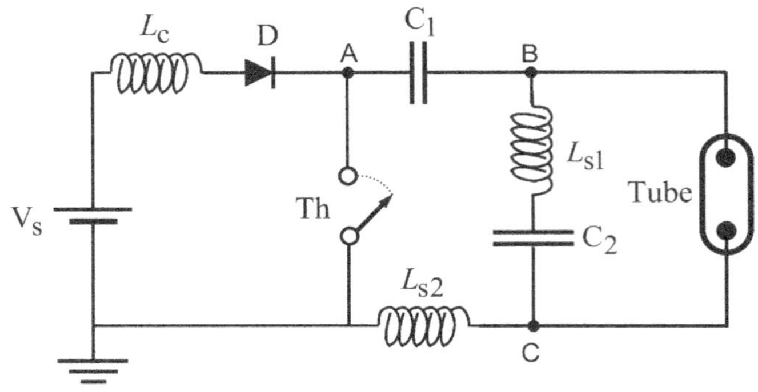

Figure 18.6 Fundamental layout of the pulse generation circuitry used for activating a copper vapour laser. Details are given in the text.

angle relative to the tube axis. The window holders are made of stainless steel, cooled by running water to protect the rubber seals present near different connections, aimed at preventing atmospheric gas leakage. Two hollow electrodes made of thin (0.15 mm) tantalum *Ta* are electrically connected to the window holders, consuming about 40% of the gross tube length. Several pieces of highly clean ("electrolytic" grade) copper, several grams each, are placed at equal distances between the electrodes. The copper/alumina system is highly inert – molten copper does not wet the alumina, nor does it form any reaction even under the high temperatures involved. The laser tube is connected to a rotational vacuum pump by a flexible pipe, whose exit tap is located on a window holder. Before any initial operation, the laser tube is pumped down to a residual vacuum of 5×10^{-3} Torr, after which the selected accompanying gas, usually neon *Ne*, is let in.

As noted, the electrical discharge in the accompanying gas is pulsed. A fundamental layout of a pulse generating circuitry is provided in Figure 18.6. An inductor L_c, a rectifying diode D, a capacitor C_1, a second inductor L_{s1}, a capacitor C_2, and a third inductor L_{s2} are connected in series at the said order to a high voltage source $V_S \cong 10,000$ V, of a grounded negative electrode. The laser tube "Tube" connects between the "B" and "C" points (parallel to the C_2 and L_{s1} series connection). The C_2 capacitance is very large compared with C_1. A fast "Th" switch connects or disconnects between the "A" point (midpoint between D and C_1) and ground. The particular switch is itself a gas tube called a "thyratron valve". It switches on and off electronically. Resistors present in the circuit exhibit only a minor role on the operation. In an open switch, the C_1 and C_2 capacitors get gradually charged at a rate determined mainly by the L_c inductance. At the end of charging, the "A" point voltage turns equal to the source voltage V_s, and the "B" point voltage is relatively small (it is, in fact, the voltage across the C_2 capacitor, whose capacitance is very large compared with C_1). The C_1 and C_2 capacitances are designed so that the "B" point voltage at that moment would be smaller than the discharge break through voltage of laser tube.

On the "Th" connection, the point "A" grounds within one or two nanoseconds. Because the capacitors are incapable of discharging so fast, the voltage on the "C" point immediately turns to be $-V_S$. On that time scale, the effective impedance♣ of the L_{s1} inductor (given by $2\pi L_{s1}/\tau_{Th}$ where τ_{Th} is the thyratron switching duration) is much larger that the C_1 capacitor impedance. Thus, the voltage division between the "A" and "C" points changes, and most voltage falls between the "B" and "C" points. That voltage is

♣ As a reminder: The "impedance" term relates to properties of resistors, capacitors, and inductors as complex quantities in circuits of alternating voltage/current (also presented as complex quantities). A resistor R impedance is $Z_R = R$, a capacitor C impedance is $Z_C = -i/\omega C$, and an inductor L impedance is $Z_L = i\omega L$, where ω is the angular frequency. An electrical circuit comprising capacitors and/or inductors under an alternating voltage usually exhibits phase differences between the currents and voltages.

sufficiently high to discharge through the laser tube. Within several nanoseconds, the tube resistance falls from being extremely high down to $5-25\,\Omega$, and the C_2 capacitor discharges within $20-50$ ns. The energy delivered to the gas filling the tube pumps the copper atoms up to the laser-excited state, as described previously. As discharge ends, the thyratron switch opens, and the capacitors' charging resumes towards a next pulse. An optimal pulsing frequency is of the order of 5 kHz (namely about 200 μs between successive pulses). Optimization relates to functioning parameters of the time scale required for draining the copper lower laser states, for recombination between the copper ions and the plasma electrons, and for maintaining a desired temperature inside the laser tube.

A considerable difficulty in maintaining a stable continual operation of a copper vapor laser relates to diffusion of copper vapor from the hot zone between the electrodes toward the cold tube edges. To reduce the diffusion rate, the neon pressure is raised up to 200 Torr. The laser may then operate continuously for several thousand hours. Purity of the accompanying (neon) gas is very important. The presence of regular atmospheric gases – oxygen O_2, nitrogen N_2, and carbon dioxide CO_2 results in laser output reduction due to relaxation mechanisms related to excitation of those gases states by the copper gas excited states, which competes with the copper pumping mechanisms. Atmospheric gases leak into the laser tube, especially because the neon gas pressure is lower than the atmospheric pressure. It is possible to reduce the leakage effects by either adding of proper selective adsorbing materials into the tube, or administering a tiny flow of neon gas through the tube during the laser operation, exchanging its full volume over several hundred hours of operation.

Continual discharging events over 10 to 15 minutes are needed until the temperature inside the tube reaches its target value, the copper pieces melt, and the copper vapor pressure assumes its needed value. Only then, lasing starts. As previously noted, the laser emits a beam that simultaneously comprises two wavelengths – 0.5105 μm (green) and 0.51782 μm (yellow) – at approximately 2:1 intensities ratio, respectively. A typical output (obviously dependent on the laser size and operation parameters) is $10-100$ W (namely $2-20$ mJ/pulse, respectively). Because of a typical large diameter of the amplifying medium relative to its length, the beam quality is rather poor, approximately $M^2 \approx 20-100$ (see section 15.6). The spectral width of each line is only slightly narrower than its fluorescence Doppler width – approximately 0.02 Å (Eq. [1.24]). The main use of copper vapor lasers is in laboratory experiments, especially for pumping of liquid dye lasers to be described in section 18.6. Attempts were made for usage in industrial materials cutting and medical operations, but they are yet insufficiently mature.

18.4 Hydrogen Fluoride Chemical Laser

The adjective "chemical" granted to a laser relates to the pumping process resulting from a chemical reaction in the amplifying medium. Most materials exploited that way are diatomic molecules of hydrogen H or deuterium D, with one of the halogens – fluorine F, chlorine Cl, bromine Br, or iodine I. The reactions where the compound forms of its constituting components are exothermic, namely they are "burning-like". A part of the emitted heat is given to the newly formed bond. Often, the probability of some exciting higher v-order vibrational states (see Chapter 7) is considerably greater, thus creating a population inversion among vibrational states. For example, consider the reaction between hydrogen H_2 and fluorine F_2 molecules. When mixed and thermally activated to induce dissociation of the fluorine molecule, a chain reaction between the gas mixture constituents occurs as follows:

$$\left.\begin{array}{llll} F + H_2 & \rightarrow & HF(\text{v}) + H; & E_a = 595\ \text{cm}^{-1};\ E_{ex} = 11{,}085\ \text{cm}^{-1} \\ H + F_2 & \rightarrow & HF(\text{v}) + F; & E_a = 840\ \text{cm}^{-1};\ E_{ex} = 34{,}240\ \text{cm}^{-1} \\ \vdots & & & \end{array}\right\}, \tag{18.1}$$

where $\text{v} = 1,\ 2,\ 3,\ \ldots$ is the vibrational order of the generated hydrogen fluoride molecule, E_a is the energy required for activating the process, and E_{ex} is the emitted heat. As to the thermal activation, recall that the rate constant for a chemical reaction of that kind is given by $k = \nu_0\, \boldsymbol{exp}\,(\text{-}E_a/k_B T)$, where the ν_0 quantity is called the "attempt to react rate", k_B is the Boltzmann constant, and T is the absolute temperature. In the process under consideration, $\nu_0 \approx 10^{14}\,\text{cm}^3/\text{mole}\cdot\text{s}$. The emitted thermal energy dissipates between the kinetic energy of the molecule and vibration/rotation motions.
It has been experimentally found, that the energy distribution among the vibrational modes may be approximated as ranging between states of $\frac{1}{3}E_{ex}$ and E_{ex} energy, peaking at about states of $\frac{2}{3}E_{ex}$ energy. Most of the remaining energy turns into a kinetic energy of the molecule, and only a small part into rotational energies. The first vibrational energy of hydrogen fluoride is $E_1 = 3961\ \text{cm}^{-1}$ and similar, slightly reducing gaps between higher energies. Table 18.1 summarizes the vibrational state energies of the hydrogen fluoride HF molecule, from $\text{v} = 1$ to $\text{v} = 10$, and for comparison also the vibrational states' energies of the other said molecules. All those molecules are polar; therefore, efficient dipole transitions are expected between their vibrational and rotational levels (see Section 7.3.5).

To elaborate further on the significance of the data provided in Table 18.1, Figure 18.7 presents the dependence of the vibrational energy of each molecule on the vibrational order. The dependences appear to

Table 18.1 Comparative summary among v = 1 − 10 vibrational states energies E_v of hydrogen H and deuterium D halogen compounds.

Vibrational order v	$E_\text{v}(\text{cm}^{-1})$						
	HF	**DF**	**HCl**	**DCl**	**HBr**	**DBr**	**HI**
1	3,961	2,906	2,885	2,091	2,557	1,840	2,228
2	7,751	5,722	5,669	4,127	5,026	3,634	4,379
3	11,372	8,446	8,348	6,114	7,404	5,383	6,450
4	14,831	11,083	10,926	8,044	9,691	7,086	8,436
5	18,130	13,633	13,405	9,922	11,888	8,744	10,338
6	21,273	16,099	15,786	11,748	13,993	10,352	12,157
7	24,261	18,484	18,071	13,521	16,008	11,920	13,888
8	27,097	20,782	20,261	15,242	17,935	13,440	15,536
9	29,780	23,003	22,356	16,910	19,768	14,913	17,095
10	32,483	-	24,356	18,526	21,528	16,340	18,565

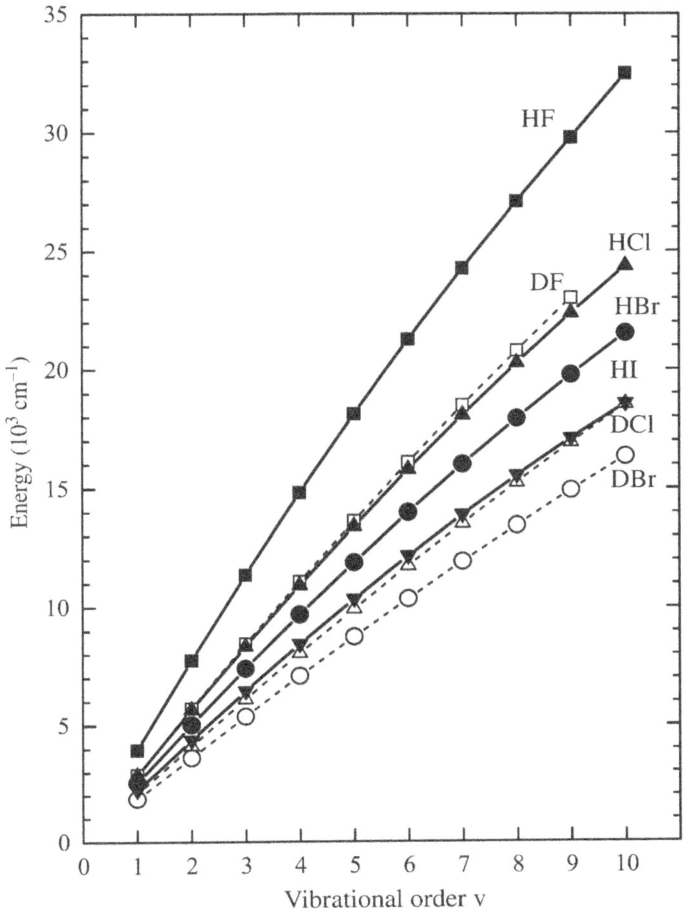

Figure 18.7 Vibrational energies of different hydrogen and deuterium halide molecules as functions of the vibrational order. Full symbols indicate hydrogen compounds, and open symbols indicate deuterium compounds. It is a graphic demonstration of the data provided in Table 18.1.

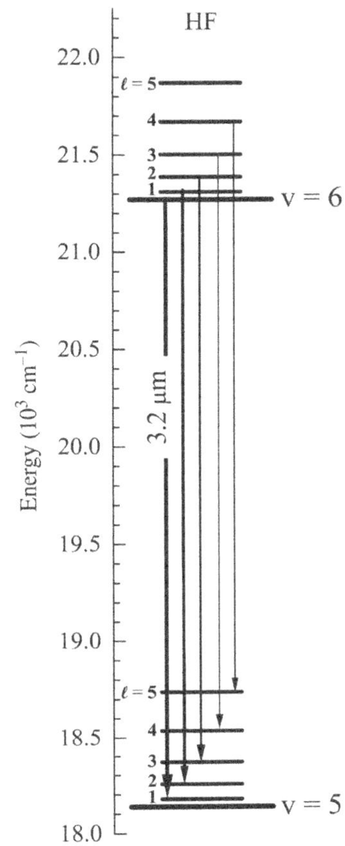

Figure 18.8 Energy scheme of vibration-rotation levels of the "5" and "6" hydrogen fluoride *HF* vibrational levels, and an indication of allowed emission transitions among the "R branch" states. The relative transition strength demonstrated by the arrows thickness is dominated by the thermal occupation distribution.

deviate from straight linearity, and the curves' slopes decrease gradually with growing order. It is quite expected because the attractive potential between the comprising atoms is not straightly harmonic (see Section 7.3). The vibrational frequencies turn smaller with increasing atomic masses, either by replacing the halogen or by replacing the hydrogen H with deuterium D.

In Chapter 7, we presented a detailed analysis of optical vibration/rotation transitions in diatomic molecules. In the following section, we concentrate on the functioning of the hydrogen fluoride HF molecular gas laser. Thus, in Figure 18.8, we present an energy scheme related to the "5" and "6" vibrational states of that molecule, including their associated rotational levels. Recall that a rotational energy is given by $\ell(\ell+1)\left[\hbar^2/2I_A\right]$, where $\ell = 0, 1, 2, \cdots$, $\hbar$ is the reduced Planck constant, and I_A is the molecule moment of inertia about an axis perpendicular to the molecular bond, crossing their mass center. In hydrogen fluoride, $\left[\hbar^2/2I_A\right] = 20\ \mathrm{cm}^{-1}$. The energy scheme indicates the allowed radiative emissions for $\Delta\ell = +1$ (a reminder: emission spectrum related to $\Delta\ell = +1$ reflects an absorption related to $\Delta\ell = -1$, "**P**-branch" – see Section 7.4.3; emission spectrum related to $\Delta\ell = -1$ reflects an absorption related to $\Delta\ell = +1$, the "**R**-branch"). Radiative transitions between states belonging to other close vibrational states are quite similar. The gain for the v = 6 → v = 5 transition peaks at about the 3.2 μm wavelength, and comprises several equi-distant lines of approximately $2\left[\hbar^2/2I_A\right] = 40\ \text{cm}^{-1}$ separation.

To achieve the desired reaction, hydrogen and fluorine gases must obviously be brought together. In CW lasers, it is customary to mix jetstreams of the two separate gases; mixing completion is brought about by diffusion. For improved mixing efficiency, a system of close narrow nostrils is used, where the hydrogen and fluorine flows are separate. A system for clearing out the resulted gaseous compounds and remnants is attached. The system construction accounts for the chemical aggressiveness and toxicity of the said materials. The pipes in contact with fluorine are made of a special nickel-copper alloy known by its commercial name **Monel**. The cleared-out gases are transferred through special absorption plates: an **activated charcoal** plate absorbs fluorine, a **porous alumina** Al_2O_3 absorbs hydrogen fluoride HF. A practical means to reduce the fluorine chemical aggressiveness is to use sulphur hexafluoride SF_6 gas as a fluorine source. That gas is highly inert at room temperature; specifically, it never reacts with hydrogen at that temperature. The reaction ignition is activated by an electrical discharge that dissociates the SF_6 compound, thus releasing fluorine, which immediately reacts with the present hydrogen (see Eq. [(18.1)]). Still, one of the SF_6 dissociation products is an S_2F_{10} molecular gas, which is very toxic.

Continuously operating lasers may generate beam powers of about 10 – 100 kW. Continuous output duration obviously requires a continuous supply of reaction gases. A beam spectrum usually contains several lines about the central 2.8 μm wavelength, belonging to transitions between different vibrational orders. Such lasers are used mainly for industrial cutting, but were also considered for military applications such as anti-rockets and artillery shells. Various laboratory lasers use an optical grating in a Littrow configuration (Figure 14.10) for wavelength tuning; they are used mainly for research purposes.

Figure 18.9 presents a version of a chemical hydrogen fluoride laser that produces a series of short pulses. The reaction cell is made of Monel. A common gas-mixture reservoir contains 89.5 wt.% He, 9.5 wt.% SF_6, and 1.0 wt.% H_2. Ignition of the chemical reactions is obtained by high-voltage electrical

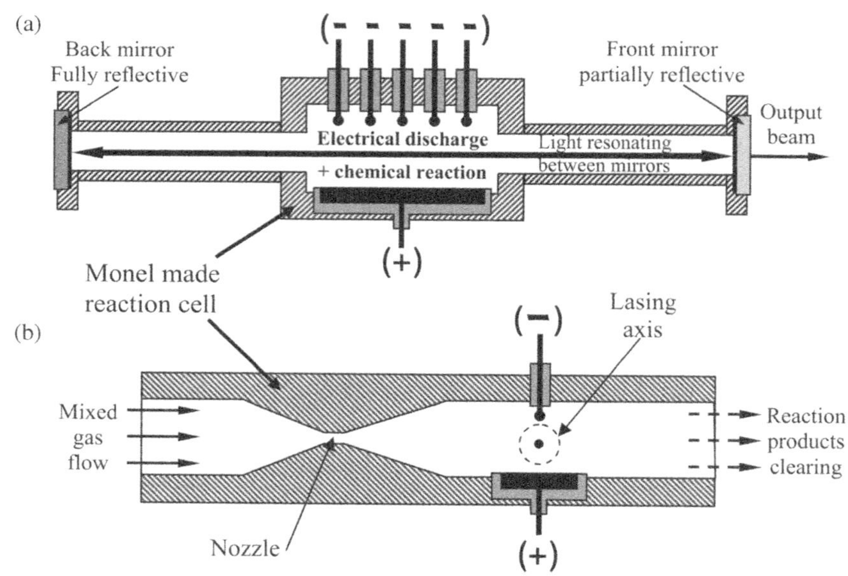

Figure 18.9 Schematic illustration of the structure of a pulsed chemical hydrogen fluoride HF laser. Electrical discharge pulses ignite the chemical reactions, thus causing its following lasing. (a) Cross section on the lasing axis. (b) Cross section on the gas flow axis, perpendicularly to the lasing axis. For details, see text.

discharges (see above). The cathode is a linear series of several stainless steel equi-distant ($1-2$ mm) pins. The anode is a flat aluminum table. The electrodes are electrically isolated from the cell body by Teflon coating. The reaction zone length in the lasing orientation is approximately 15 mm; its perpendicular cross-section area is approximately 5 cm^2. To prevent the hazard that the gases' "burning" position would advance *backward* from the reaction cell into the mixed gases reservoir, the gas mixture is enforced to flow through a narrow nozzle at a supersonic velocity (twice or three times faster than the speed of sound). Clearing of reaction products is done through **activated charcoal** and **porous alumina** filters into a large, constantly pumped vacuum container, maintaining a residual pressure of ~ 1 Torr. The optical cavity mirrors are each placed several millimeters away from the reaction cavity. The back mirror is an optically polished copper plate, dielectrically coated for chemical protection. Its reflectance is close to 100%. The front mirror is a zinc selenide *ZnSe* optically polished plate, which is transparent at the lasing wavelength range ($2.6-3.2\,\mu m$); it is dielectrically coated for ~85% reflectance. The laser cavity length is ~ 10 cm. The electrical circuit producing the voltage discharge pulses is similar to the one used in pumping the copper vapor laser (see Section 18.3 and Figure 18.6). The pulse rate is about $8-10$ kHz, and each discharge duration lasts about 10 ns. The voltage pulse heights are set to produce electrical fields of about 1000 V/cm between the electrodes. Laser pulses appear about 100 ns after the end of a discharge pulse, lasting for about 500 ns. The rate of inter-molecular clashes in the gas is relatively low; therefore the gas system never reaches full thermal equilibrium during the time period between successive pulses. The laser emission spectrum thus exhibits quite a number of different wavelengths around the central 2.8 μm wavelength. As a result of the broad distribution of state populations, lasing at different wavelengths occurs at different time delays relative to the discharge pulses. A typical integrated pulse output intensity is of the order of $0.1-1.0$ J/pulse.

18.5 Neodymium-YAG Laser

A most practical means for constructing a laser whose amplifying medium is solid is to use a glass or a crystal, where one of their sites hosts a dopant ion. Radiative transition between that ion's states are then used for lasing. The role of the glass or crystal is to fix the ions in their sites, assist in dissipating the heat generated during pumping and lasing, and enhance draining of the lower laser state population down to the ground-state by non-radiative transitions (see Chapter 13). Such solid lasers exhibit a clear advantage over gas lasers by the greater achievable density of the active component. For example, the density of neon *Ne* atoms in a helium-neon laser (see Section 18.2) never exceeds the order of $5 \times 10^{15}\ cm^{-3}$, whereas in the $Nd:YAG$ laser to be described below, the density of neodymium Nd^{3+} ions – the active ingredient in the lasing mechanism – is at least $10^{20}\ cm^{-3}$. Thus, the active medium size in the laser resonance cavity may be made very small compared to a gas laser.

The $Nd:YAG$ laser utilises triply ionized Nd^{3+} neodymium atoms hosted in relevant sites in a $Y_3Al_5O_{12}$ single crystal. The laser is also known by the YAG name, an acronym for "Yttrium Aluminium Garnet"; Garnet is the name of a family of crystals, all belonging to the same cubic body-centred space group (No. 230 O_h^{10} $(I4_1/a\bar{3}2/d)$ (see Section 11.5 Table [i]-32, and also Section 12.6). The (centered) unit cell comprises 8 formula units. A Nd^{3+} ion replaces a "legitimate" Y^{3+} lattice ion belonging to the Wyckoff c-site of dodecahedral coordination and a D_2 symmetry. A typically used concentration of neodymium ions is by replacement of 1% of the yttrium ions (denoted 1 at.%, to read "one atomic percent").

Figure 18.10 presents a partial energy scheme of the Nd^{3+} ion in the YAG crystal. The ion electronic configuration is $[Xe]4f^3$. Correspondingly, the ground state Term is determined by the three equivalent 4f

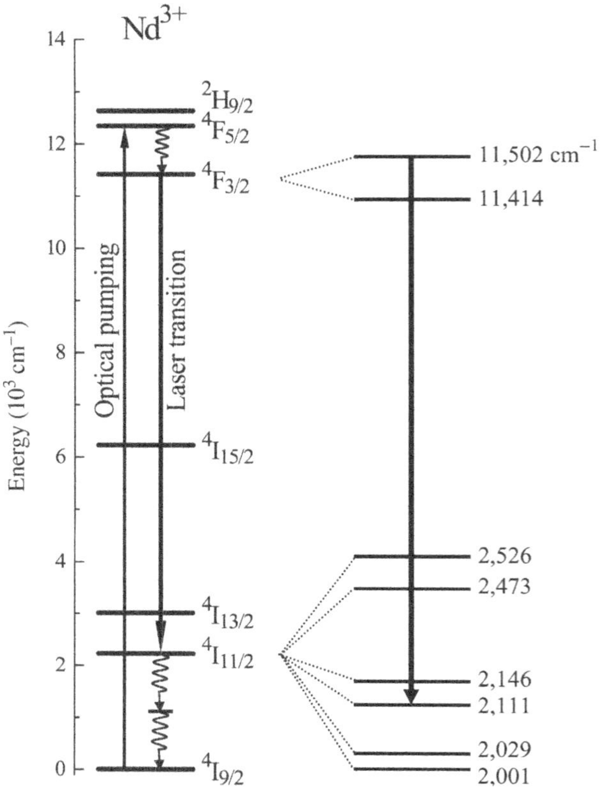

Figure 18.10 On the left, the energy scheme of the lower Nd^{3+} ion as dopant in the c-site of the $Y_3Al_5O_{12}$ single crystal. Solid arrows mark radiative transitions; wiggled arrows mark non-radiative ones. Crystal field splitting is detailed for the $^4F_{3/2}$ and $^4I_{11/2}$ states involved in the laser transitions. The energy of each split state is indicated on its right in cm^{-1} units. For detailed discussion, see text.

electrons to be $^4I_{9/2}$ (same as the praseodymium *Pr* atom – see Table 5.5). The multiplet belonging to the ^4I Term obviously comprises four states, $^4I_{9/2}$, $^4I_{11/2}$, $^4I_{13/2}$, and $^4I_{15/2}$, arranged in an ascending order of the resultant J angular momentum, because the $4f^3$ subshell is less than half full (Eq. [5.64-ii]). The $^4I_{11/2}$ state serves as the lower laser level. A different combination of the three 4f electron orbital angular momentums provides a resultant F orbital angular momentum, also of a resultant $S = 3/2$ spin, namely a ^4F Term. The energy of that Term is higher than the previously described ^4I one, consistent with Hund's rules (Eq. [5.64-i]). Two of its spectroscopic identified multiplet levels are indicated on the levels scheme: $^4F_{3/2}$ and $^4F_{5/2}$. The $^4F_{3/2}$ level serves as the higher laser level. Also indicated, is an additional level obtained by a different combination of the orbital angular momentums, providing a resultant H orbital angular momentum with an $S = 1/2$ resultant spin. The spectroscopic identified state component of its multiplet is $^2H_{9/2}$. The reader will be invited to calculate the other combinations of angular momentums that provide the other Terms presented in Figure 18.10. Each of those is split as a result of the crystal field (Table 12.5 and its related text). For brevity we refrain from presenting all those split states on the levels scheme. It should be emphasized that the spectroscopic studies performed so far do not suggest unequivocal identification of all split states. The exercises appearing at the end of this chapter request the reader to suggest assignments for all crystal-field split states appearing in the energy scheme of Figure 18.10.

In zero-order approximation, the $^4F_{3/2} \rightarrow {}^4I_{11/2}$ laser transmission indicated in the energy scheme is forbidden by failing to satisfy the conditions $\Delta L = \pm 1$ and $\Delta J = 0, \pm 1$ (see Eq. [5.65]). Its spectral width is 0.45 nm (full width at half maximum), and its radiative transition lifetime is $\tau_{rad} = 550\,\mu s$. The effective measured lifetime is only about $230\,\mu s$, a result of simultaneous non-radiative multi-phonon transitions (see Chapter 13). The first designed neodymium-YAG lasers were pumped optically by exposing the laser rod to light flashes emitted from a discharge tube, usually containing xenon *Xe* gas (Figure 14.1). Modern lasers are usually pumped by emission of diode-laser arrays, whose $\lambda = 810\,\mathrm{nm}$ output wavelength fits the ground-state $^4I_{9/2} \rightarrow {}^4F_{5/2}$ absorption transition; a detailed description of diode lasers is provided in Section 18.7. The $^4F_{5/2} \rightarrow {}^4F_{3/2}$ decay transition to the upper laser state is fast, and mostly non-radiative, as is the $^4I_{11/2} \rightarrow {}^4I_{9/2}$ draining transition of the lower laser state; both end within less than a nanosecond. Thus, the neodymium-YAG laser functions as a four-level system (see Section 18.1 and Figure 18.1).

In many laser systems, the amplifying medium is shaped as a long circular rod, typically of 5–10 mm diameter and 50–100 mm length. Because the amplifying medium is solid, some properties that are negligible in gases become apparent, especially the steady and transient thermal distortions that cause local refractive index changes. As a result, some distortions are effected in the laser output wave-fronts; hence, there is a reduction of the output beam quality (see Section 15.6). To minimize such effects, the rods are drilled out from selected sections of the crude crystal boule which exhibit minimal internal distortions. In shaping the rod, special care is devoted to achieving the best planarity of the front and end surfaces, and the best polishing and antireflection dielectric coating. Diode laser pumping is mostly performed through the rod cylindrical surface (**side pumping**). Therefore the cylindrical surface is also polished and antireflection coated, thus allowing for the concentration of the pump beams at the rod axis. This assists in achieving the goal of a TEM_{00} shape to the laser beam output (see Section 15.4). In side pumping by a discharge tube, often the cylindrical surface is roughened for two reasons; one is to disperse the pump light evenly inside the rod volume, and the second is to frustrate competing (parasitic) cross lasing between points on the cylindrical surface.

Figure 18.11 schematically displays a common construction of a diode-lasers-pumped $Nd:YAG$ laser. The $Nd:YAG$ rod is held inside a transparent tube with cooling water streaming through. The role of the cooling water is to dissipate the heat generated by the various non-radiative decay transitions during the laser operation (Figure 18.10). Pumping of the laser rod is obtained through the outer cylindrical envelope of the cooling water tube by diode laser arrays placed outside. Figure 18.11 describes four such arrays positioned at two opposite sides of the rod. Usually, more diode arrays are distributed at an odd circular symmetry around the rod axis. The goal is to provide the pumping distribution inside the rod the best possible circular symmetry, and simultaneously prevent the remnant pump radiation transmission from reaching diode arrays otherwise positioned on the opposite side; if it occurs, it might cause them damage. Because the $Nd:YAG$ lasing transition in itself is quite narrow (0.45 nm – see above), no wavelength-tuning components are needed. To obtain pulsed outputs, the resonance cavity is added using Q-switching devices (see Sections 16.1 and 16.2). The diode lasers themselves may provide either continuous or pulsed pumping. Pulsed $Nd:YAG$ lasers may reach outputs of 100–150 mJ/pulse, typically at a repetition rate of 10 Hz, where each pulse duration is about 10–30 ns.

Different versions of neodymium-YAG lasers have many different applications in industry and science, and are considered a "work horse" in laser technology. In industry, it is used for cutting and welding; in medicine, it is used for surgical purposes, including fine eye retina surgery; and in the military, it is used for range finding and target marking for homing munitions. In scientific research, use is made of the

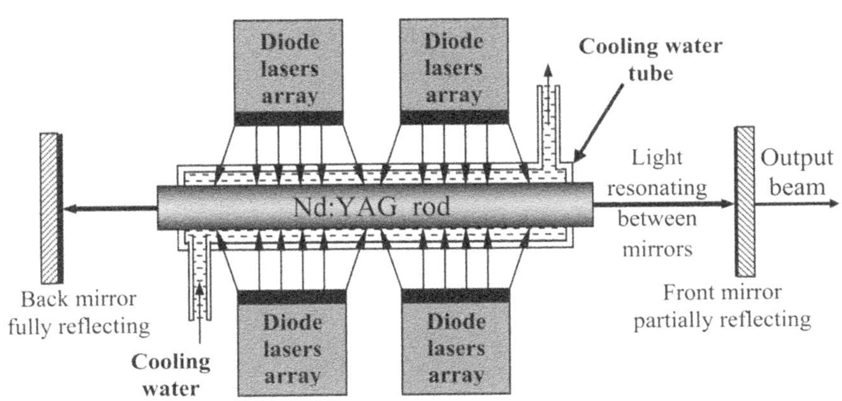

Figure 18.11 Schematic fundamental construction of a diode-lasers-pumped neodymium YAG laser. For details, see the text.

original wavelength output ($\lambda = 1.064\,\mu m$) for wavelength conversions (see Chapter 17) and pumping of other laser systems.

18.6 Dye Lasers

The name "dye lasers" refers to a very broad family of lasers whose amplifying medium is based on liquid-dissolved organic dyes. Typically, organic dye compound molecules comprise several benzene-like rings (six carbon atoms forming a regular hexagon – see Figure 6.20). The rings bond by sharing a side, or by sharing a σ bond at apexes, with in-ring carbon exchanges and the addition of bonded radicals. Table 18.2 provides a short list of such representative dyes. Besides the dye commercial/chemical name, the table shows its spatial construction, the used solvent/s, and the lasing wavelength range. For the benefit of readers who are not fluent in organic chemistry symbol conventions, we provide the gentle reminder that unless there is a different notation, each ring apex hosts a carbon *C* atom; unless an apex is indicated to bond with a specific atom or with another radical group, the apex carbon is bonded to a hydrogen *H* atom by a σ bonding. A dot above or below an atom symbol indicates the presence of a non-bonding valence electron; a short bar above or besides an atom symbol means it is a negative ion.

Polar dyes dissolve well in polar solvents like water, ethanol, and methanol. Non-polar dyes dissolve well in non-polar solvents; for example, the BASF-241 dye dissolves well in chloroform. which is non-polar, yet exhibits a negligibly small solubility in polar solvents. For some dyes, it is possible to use either polar or non-polar solvents. Each solvent exhibits some influence on the absorption and fluorescence spectra. The dyes listed in Table 18.2, plus many others used in laser technology, allow for the covering of the entire visible range, and even beyond toward the ultraviolet and infrared.

Figure 18.12 displays an energy scheme, which is fundamentally adequate for each of the said dyes. The scheme results most likely from $\pi \rightarrow \pi^*$ transitions, where a π bonding electron is excited into an anti-bonding π^* orbital state (see Chapter 7). The scheme exhibits much similarity to that of the helium *He* atom (see Section 5.2.4), comprising two manifolds: A singlet manifold obtained when the excited electron spin is reversed relative to the resultant spin of the remaining molecule electrons, and a triplet manifold when the excited electron spin is co-oriented with the resultant spin of the remaining molecule electrons. Because the considered molecules are very large, complex, and different from one another, the fundamental scheme refrains from indicating neither the electronic configuration nor the Term (see Chapter 6). Instead, the electronic states belonging to the singlet manifold are indicated by the $S^{(p)}$ form, where p is a numeral, indicating the states in an ascending order. The ground-state is a singlet, correspondingly denoted $S^{(0)}$, and it functions as the lower laser level. The first $S^{(1)}$ excited singlet state functions as the upper laser level. States belonging to the triplet manifold are similarly indicated by the $T^{(p)}$ form, and the first excited state is $T^{(1)}$. Each vibrational state is split by other vibrational and rotational motions of the molecule (when dissolved, the rotational motions turn into librational ones). The larger vibrational energies, belonging to the configuration potential of each electronic state, are listed numerically as $\mathrm{v} = 0, 1, 2, 3, \cdots$ at an increasing energy order. Because of the large number of the constituent atoms in each molecule, the vibrational and rotational modes create a dense array of states, which in effect forms a quasi-continuum of allowed energy levels. Nevertheless, as we shall see shortly, both absorption and fluorescence emission spectra exhibit a prominent structure of bands, which reflect the largest vibrational energies. The energy gaps between upper and lower levels in both singlet and triplet manifolds are of the order of 20,000 cm^{-1}; splitting of the major vibrational states is of the order of $1{,}000-1{,}500\ \mathrm{cm}^{-1}$.

As an example, Figure 18.13 presents the measured absorption cross section and fluorescence spectra of BASF-241 dissolved in chloroform $CHCl_3$. The absorption is assigned to the $S^{(0)} \rightarrow S^{(1)}$ electronic

Table 18.2 A short list of common organic dyes, their chemical structures, suitable solvent/s, and fluorescence range. The dye and solvent names are based on their given commercial/accepted shorthand names: EtOH for ethanol, MeOH for methanol, and PMMA for poly(methyl methacrylate) ["Perspex" when polymerized].

Name	Chemical structure	Solvent	Fluorescence range
7-Hydroxy-coumarine	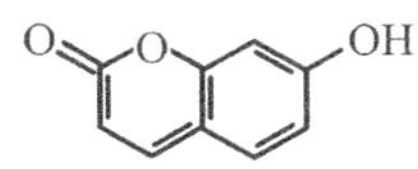	H_2O (pH-9)	450 – 470 nm (blue)
Pyrromethene-556	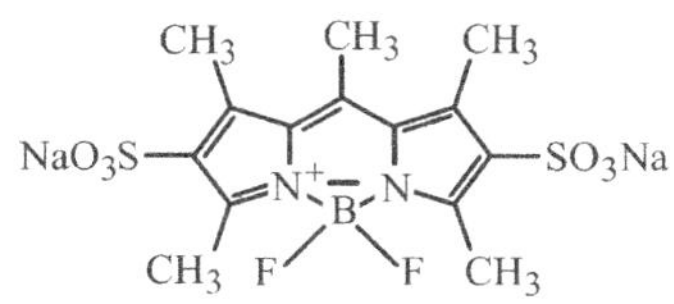	EtOH MeOh H_2O	510 – 580 nm (green)
BASF-241	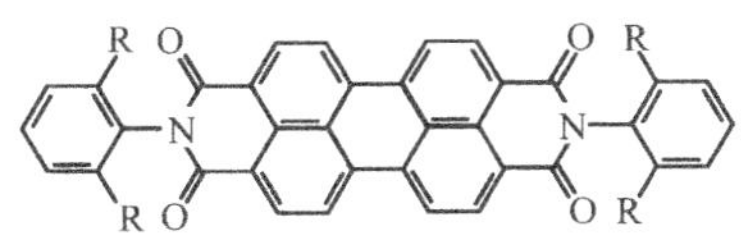 $R \equiv CH(CH_3)_2$	Chloroform	525 – 580 nm (green)
2,7-dichloro-fluorescine	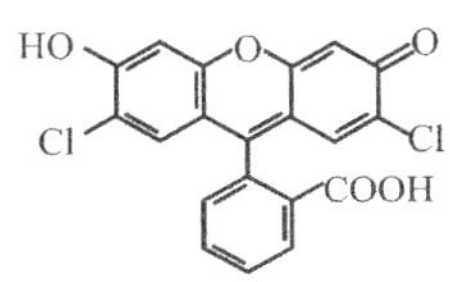	EtOH	530 – 560 nm (green)
Rhodamine-6G	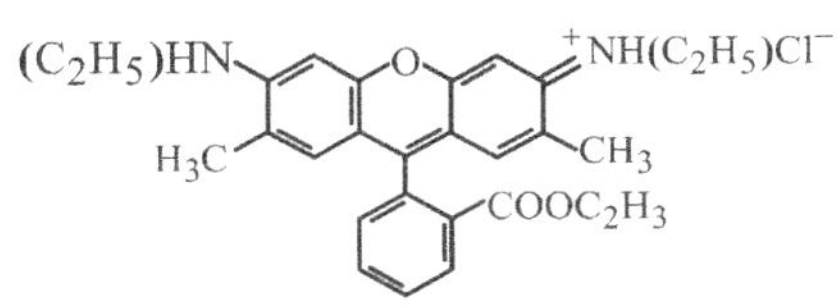	EtOH MeOH H_2O PMMA	570 – 610 nm (yellow)
Rhodamine-B	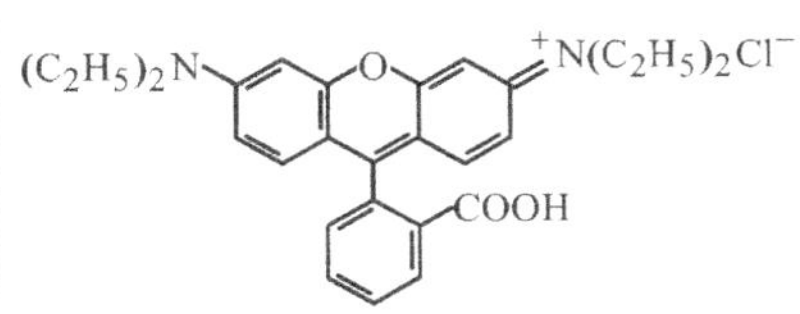	EtOH MeOH PMMA	605 – 635 nm (red)
BASF-1132 (HITCI)	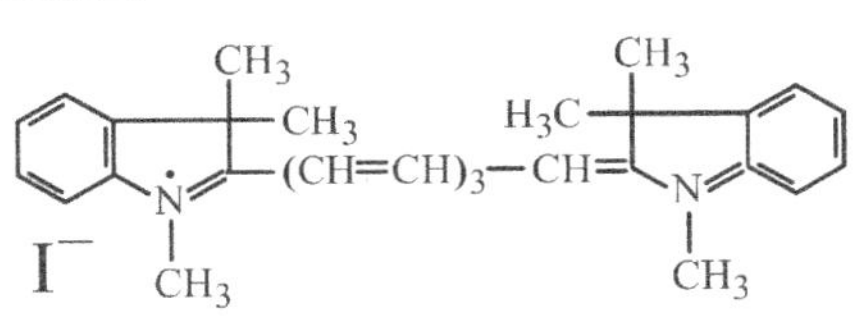	EtOH MeOH acetone	620 – 760 nm (infrared)

transition, and the fluorescence spectrum is assigned to the opposite $S^{(1)} \rightarrow S^{(0)}$ transition. The two reveal prominent structures related to transitions between the large vibrational levels of the different electronic states. Vibrational energies are approximately 1,000 – 1,400 cm^{-1}, and they are assigned to stretching motions of hydrogen atoms bonded by σ-type bonding along the circumference of the molecular skeleton. In noting transitions, vibration ordinal numbers belonging to the $S^{(1)}$ state are added with an asterisk. The ground vibrational orders are noted as "0" and "0*" for $S^{(0)}$ and $S^{(1)}$, respectively. An approximate

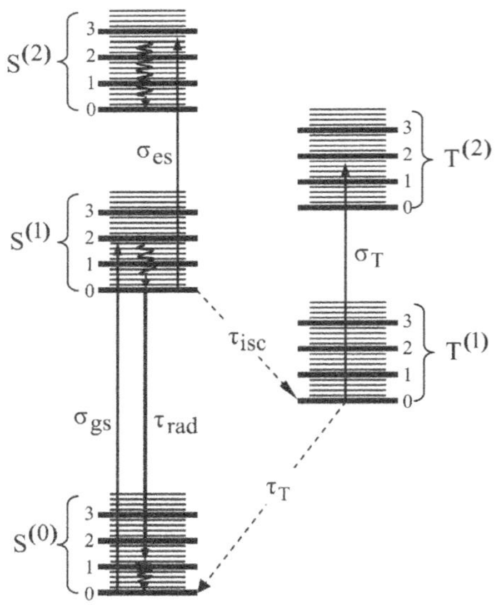

Figure 18.12 Simplistic fundamental energy scheme of organic dye molecules states. Thick horizontal lines indicate vibrational levels belonging to the molecular configuration potential (see Section 8.4); short narrow horizontal lines represent vibrational states of other normal coordinates, including rotational motions. Various radiative transitions are indicated by straight arrows, and non-radiative transitions by wiggled arrows. For discussion, see text.

mirror image is obtained between the absorption and emission spectra. By the Franck-Condon principle (see Sections 7.5.4 and 7.5.5), there also exists a Stokes shift between the $0 \rightarrow 0^*$ and $0^* \rightarrow 0$ transitions; in the present case, it amounts to ~14 nm, equivalent to ~490 cm^{-1}. Importantly, the excited state lifetime belonging to the $S^{(1)} \rightarrow S^{(0)}$ decay transition is $\tau_{rad} = 5.0$ ns, typical of an electric dipole allowed transition; the fluorescence quantum efficiency (ratio between the number of emitted photons to the ones absorbed in the $S^{(0)} \rightarrow S^{(1)}$ transition) is very close to 100%.

Efficient radiative transitions are indicated on the energy scheme by solid arrows, and inefficient ones by dashed arrows. Like in the Helium *He* atom, efficient transitions occur between states belonging to the same manifold (see Figure 5.3). Thus, dye lasers pumping is performed optically using the efficient $S^{(0)} \rightarrow S^{(1)}$ transition. On the other hand, $S^{(1)} \rightarrow S^{(2)}$ excited state transitions cause considerable loss in a dye laser cavity. In addition, inter-manifolds crossing ("isc" – acronym for "intersystem crossing") that are first-order forbidden, occur on continual pumping. The radiative τ_{isc} and τ_T transitions durations (see Figure 18.12) are of the order of 500 ms, but effectively shorten down to ~10 ms by non-radiative transitions (see Chapter 13). They mainly involve phonons belonging to vibrations and rotations of the dye itself, but also by those belonging to the solvent molecules. Despite the involvement of non-radiative transitions, the τ_T lifetime of the lowest triplet state is still very long compared with τ_{rad}. On continual pumping, a large population of accumulated $T^{(1)}$ excited states evolves, contributing a considerable loss by $T^{(1)} \rightarrow T^{(2)}$ light absorptions. For that reason, laser system constructions administer a continual dye solution flow, to bring fresh non-excited dye to the pumped region, and to remove away those molecules that have assumed a $T^{(1)}$ state. Administering a circular flow grants those molecules sufficient time to return to the ground-state, becoming "fresh" again. The solution flowing also assists in dissipating the heat generated during pumping because of various non-radiative transitions. Despite the Stokes shift, lasing will not usually occur by the $0^* - 0$ transition due to the considerable fluorescence overlap with the absorption spectrum (Figure 18.13). Lasing will rather occur by the $0^* - 1$ transition, for which the laser is practically a four-level type system (see Figure 18.1).

Figure 18.14 describes schematically a realization of a dye laser. A transparent glass (usually quartz made) pipe of a rectangular cross section carries the dye solution flow. The two large side-faces are used for light pumping; their typical size is $1 - 2$ cm, and their distance is $2 - 5$ mm. Pumping is performed through one (or two) of the "pump windows". The pump beam cross section is shaped as close as possible to a long rectangle by a combination of regular and cylindrical lenses. The long dimension of the pump beam cross section is inclined at a Brewster angle relative to the "laser windows" (see Figure 14.19). In that way, the beam developing inside the resonance cavity exhibits a linear polarization parallel to the drafting page. The usual cavity length is ~ 15 cm. A reflecting grating at Littrow configuration (see Figure 14.10) provides wavelength tuning. For most dyes, the tuning range is about $15 - 20$ nm. For example, a BASF-241 dye in a similar construction, pumped by ~ 15 ns long pulses of $\lambda_{pump} = 532$ nm originated by a frequency-doubled Nd:YAG laser (see Sections 18.4 and 17.2.3) and ~ 2 mJ/pulse intensity are tunable in the $566 - 585$ nm

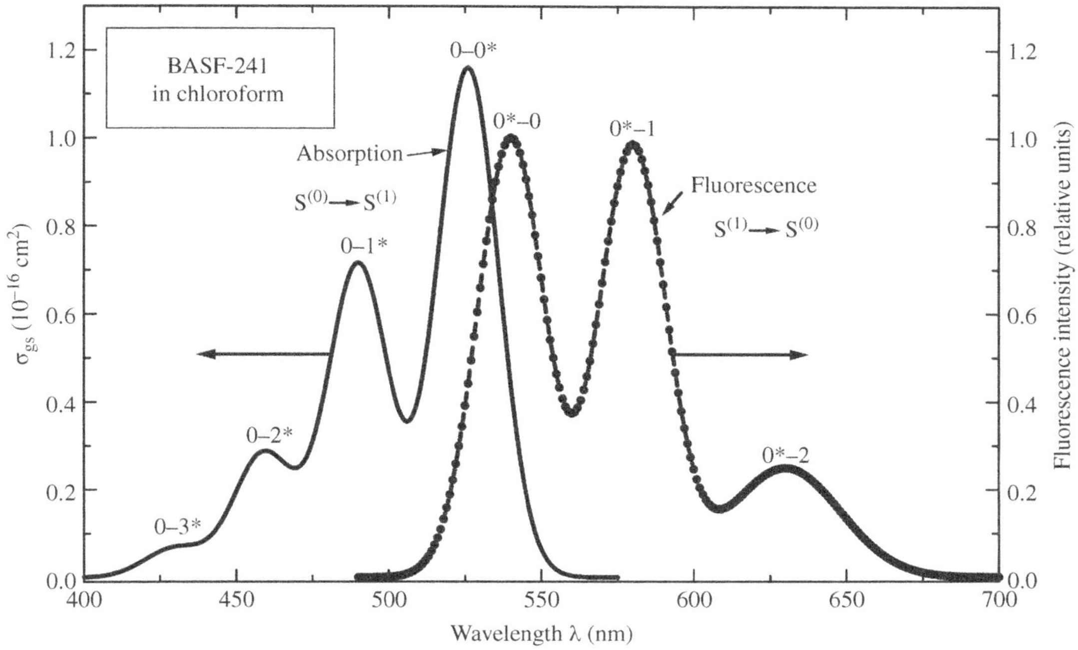

Figure 18.13 Absorption and fluorescence emission spectra of the BASF-241 dye dissolved in chloroform as functions of wavelength. The intensity scales are provided on the left and right axes, respectively. The prominent structures in both spectra are assigned as transitions between vibrational states belonging to different electronic states. For a detailed discussion, see the text.

range. The output is of the order of ~ 100 μJ/pulse over most of the said range. The laser construction shown in Figure 18.14 may also be used to provide a continuous wave (CW) output by administering continuous pumping. Obviously, the pumping wavelength must belong to the dye optical absorption range.

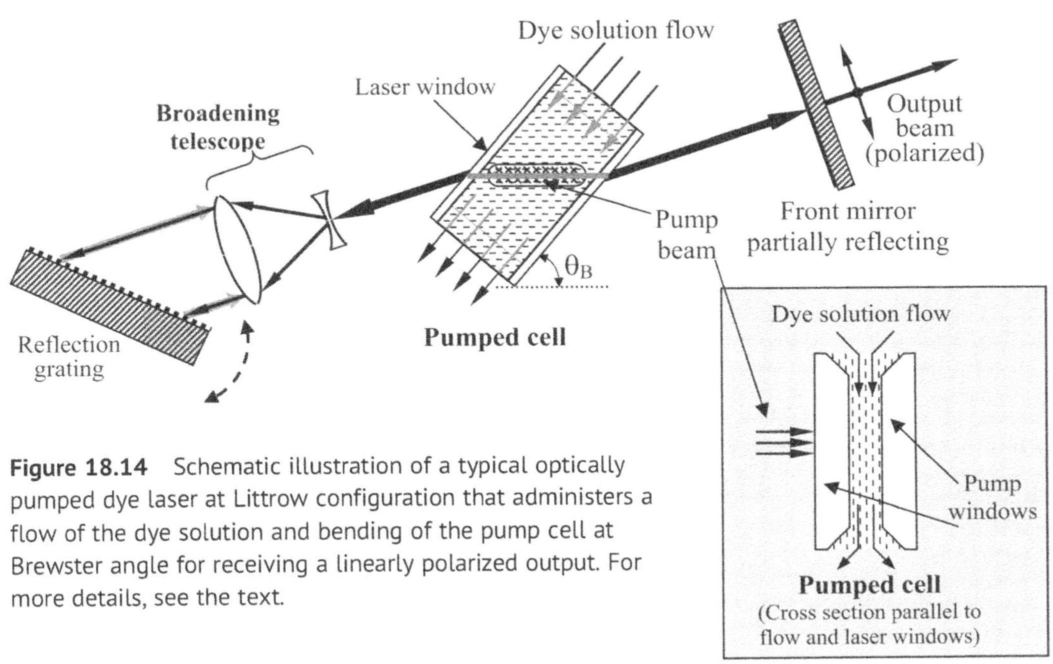

Figure 18.14 Schematic illustration of a typical optically pumped dye laser at Littrow configuration that administers a flow of the dye solution and bending of the pump cell at Brewster angle for receiving a linearly polarized output. For more details, see the text.

18.7 Diode Lasers

The functioning of diode lasers is based on radiation emitted by electron-hole recombination events in semiconductors when minority charge carriers are injected into p-n junctions of semiconductors on application of a forward bias voltage. In the present book, we neither addressed the issue of electrical conduction in general, nor of semiconductors in particular. In the following section, we rely on the background knowledge of readers, and are satisfied with superficial descriptions.

A **p-n junction** is a region of a semiconductor exhibiting transition from p-type into n-type conduction. In p-type conduction, the majority charge carriers are holes populating the valence band top. In n-type conduction, the majority carriers are electrons that populate the conduction band bottom. The conduction band energy is higher than the valence band energy. The energetic difference E_g between the conduction band bottom E_C and the valence band top E_V is called the **energy gap**. Its size in different semiconductors typically ranges between 0.2 eV and 2.0 eV. A p-n junction with two output conducting wires – one connected to the p-region, the other to the n-region – is called a **diode**. On application of a forward voltage across a diode (p-side positive, n-side negative), minority carriers are injected: electrons are injected into the p-region, and holes are injected into the n-region. This effect violates the thermodynamic equilibrium, and initiates recombination between the opposite type carriers.

Recombination involves both radiative (photon) and non-radiative (phonon) emissions. In a direct radiative transition, the photon energy is approximately equal to the E_g energy gap. Photon emissions are dominant in a special type of semiconductors, exhibiting a **direct energy gap**. It exists where the conduction band bottom and the valence band top belong to the same crystalline wave vector (often $\underline{k} = 0$). Therefore there is no need for a simultaneous involvement of phonon transitions (absorption or emission) to assure the conservation of the crystal linear momentum. The oscillator strength is then determined by the electronic states that form the energy bands. Table 18.3 provides a list of several semiconductors that exhibit direct energy gaps, including their relevant basic physical properties: energy gap E_g, and electron and hole mobilities, μ_n and μ_p, respectively. In addition to the listed materials, some of their solid solutions have been found useful, such as $GaAs_xP_{1-x}$, $Ga_xIn_{1-x}As$, or $Pb_{1-x}Sn_xTe$. They preserve the "direct gap" property, while exhibiting intermediate values of other parameters, especially of the energy gap. For example, the $Ga_xIn_{1-x}As$ constitution has an energy gap of 0.35 eV for the $x = 0$ molar fraction, an energy gap of 1.43 eV for the $x = 1$ molar fraction, and a near linear dependence on $x \neq 0$ of the energy gap size between the said two ends.

Majority carrier densities in p-n junctions of diode lasers are usually very large, forming "**degenerate**" conditions, namely such that the Fermi energy is located in the bands themselves (not in the forbidden gap). It results from very high impurity densities in the different crystal zones. At high injection rates of minority carriers, the quasi-Fermi energies describing the injected carriers concentrations may also reside in the corresponding band. Figure 18.15 presents an energy scheme describing bands dispersion curves (energy as function of k wave vector) in a p-type region of a semiconductor under electron injection. A solid arrow indicates a radiative transition (photon emission). As illustrated in the figure, the radiative transition is direct in the crystalline linear momentum, and the emitted photon energies are only slightly larger than the energy gap.

The emission spectrum ranges between E_g as the low energy limit, and $(F_n - F_p)$ as the high-energy limit, where F_p and F_n are the quasi-Fermi energies for holes and electrons, respectively. Notably, the F_n height in the p-region depends on the injection rate, and because of electron-hole recombination, it reduces exponentially as a function of distance from the junction toward the p-region. Except under

Table 18.3 A list of semiconductors exhibiting direct energy gap, and their basic parameters: energy gap E_g, electron mobility μ_n, and hole mobility μ_p.

Material	E_g (eV)		μ_n (cm^2/V·s)	μ_p (cm^2/V·s)
	0 K	300 K		
GaAs	1.52	1.43	8,500	400
InSb	0.23	0.18	77,000	750
InAs	0.36	0.35	33,000	460
InP	1.29	1.35	4,600	150
GaSb	0.81	0.78	4,000	1,400
PbS	0.29	0.36	550	600
PbSe	0.17	0.29	1,020	930
PbTe	0.19	0.3	1,620	750
CdS	2.58	2.42	~200	~100
CdSe	1.84	1.74	650	75
CdTe	1.61	1.45	1,000	90
SnTe	0.3	0.18	?	500

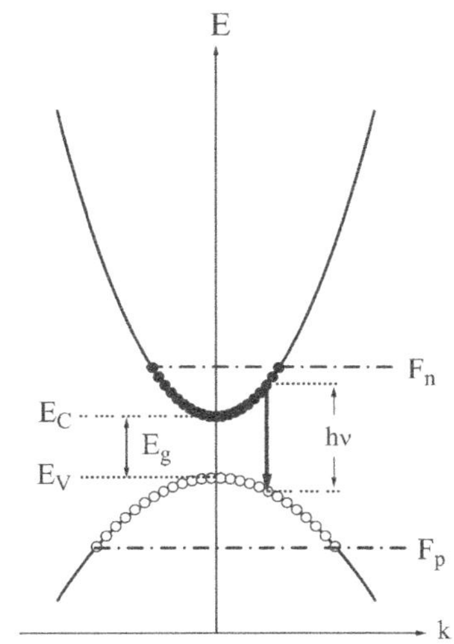

Figure 18.15 Charge carriers concentration distribution over dispersion curves (energy as a function of wave vector) in valence and conduction bands of a semiconductor. Electrons are described as full circles and holes as open circles. A solid arrow indicates a direct recombination transition in the wave-vector (momentum) space for a degenerate p-type semiconductor into which high electron densities were injected.

extreme cases, a relation $(F_n - E_C) < (E_V - F_p)$ always exists. Obviously, symmetric conditions occur on the n-side of the junction under holes injection.

There are, in fact, two types of devices based on electron-hole recombination radiation: **light emitting diode** (LED) and **diode laser**. The latter is an extreme form of the former, where stimulated emission dominates the opposite charge carriers recombination. As an example, we present a diode laser based on the gallium arsenide *GaAs* semiconductor. Figure 18.16 shows a spectrum of the recombination radiation emission in a *GaAs* p-n diode under pre-lasing conditions. It peaks at approximately 840 nm, and exhibits a $\Delta\lambda_{1/2} = 14$ nm full width at half maximum. Under conditions to be described later, and an appropriately increased current, laser emission takes over, accompanied with a drastic reduction in the spectral distribution of the emitted light (see Section 14.1): instead of a 14 nm width, and without tuning devices in the resonance cavity, the spectral width reduces to $\sim$ 0.5 nm, close to the recombination radiation peak.

The effective width w of the zone where recombination radiation occurs near a p-n junction is the sum of diffusion ranges of injected minority carriers: electrons injected into the p-region and holes injected into the n-region, plus the width of the junction transition zone (where electron-hole recombination events also take place). The diffusion ranges sum w_D is given by

$$w_D = \sqrt{D_n\tau_n} + \sqrt{D_p\tau_p} = (k_BT/e)^{1/2}\left(\sqrt{\mu_n\tau_n} + \sqrt{\mu_p\tau_p}\right), \quad (18.2)$$

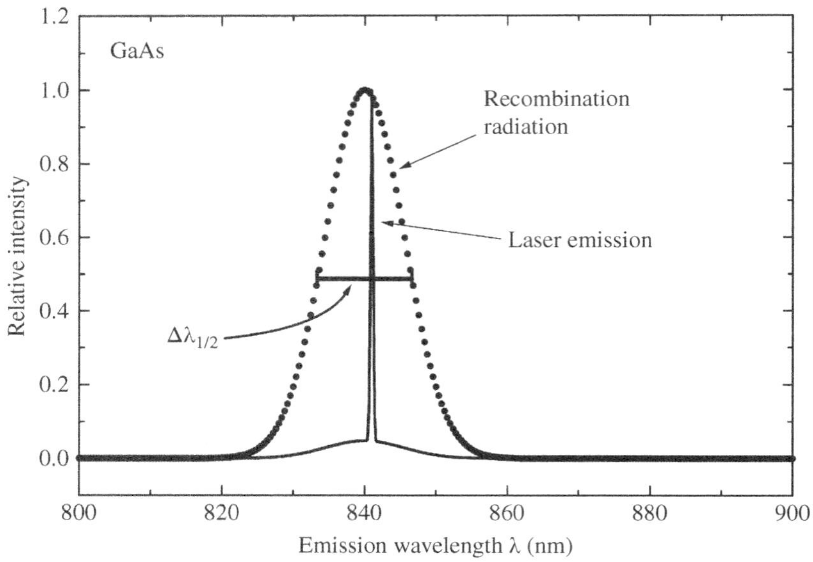

Figure 18.16 Recombination radiation and lasing spectra in a p-n junction of gallium arsenide *GaAs*. The spectra are normalized to a unity at their peaks. The laser peak position departs slightly from the recombination radiation one because of some wavelength dependence of the gain and losses.

where D_n and D_p are the diffusion coefficients of electrons and holes, τ_n and τ_p are their lifetimes as minority carriers, and μ_n and μ_p are their mobility values, respectively. The transition at the right side of Eq. (18.2) is based on Einstein's relation $D = (k_B T/e)\,\mu$ between diffusion coefficient and mobility, k_B being the Boltzmann constant and T the absolute temperature. To demonstrate the effect order of magnitude, we calculate the size of the radiation emitting region around a p-n junction of gallium arsenide *GaAs*. Converting the given mobility values (Table 18.3) into cgs units, they are $\mu_n = 2.55 \times 10^6\ \text{cm}^2/\text{statV·s}$ and $\mu_p = 0.12 \times 10^6\ \text{cm}^2/\text{statV·s}$. Because the material is of a direct energy gap, the lifetime of both type carriers is governed by the radiative recombination lifetime, namely $\tau_n = \tau_p = \tau_{rad} \cong 10^{-9}$ s. Setting the latter in Eq. (18.2), the diffusion ranges at room temperature (300K) are

$$w_D = \left(\frac{1.38 \times 10^{-16} \times 300}{4.8 \times 10^{-10}}\right)^{1/2} \left(\sqrt{2.55 \times 10^6 \times 10^{-9}} + \sqrt{0.12 \times 10^6 \times 10^{-9}}\right) = 5.7 \times 10^{-4}\ \text{cm}\ .$$

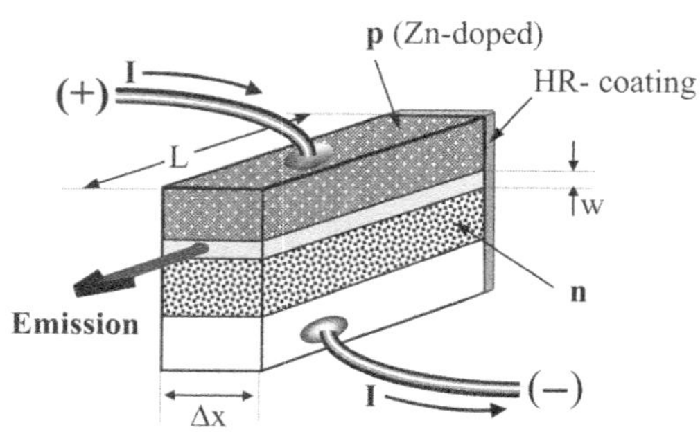

Figure 18.17 Fundamental construction of a diode laser, based on a p-n junction in a gallium arsenide *GaAs* semi-conductor.

The thickness of the transition region in the p-n junction never exceeds about 10^{-4} cm. Thus, the thickness of the light emitting region is dominated by the sum of minority carrier diffusion ranges. Eventually, we may summarize that $w \cong 6 \times 10^{-4}$ cm. In fact, the dominant contribution is provided by the injected electron diffusion.

Figure 18.17 describes the construction of a gallium arsenide *GaAs* diode laser. The junction is formed by inducing in-diffusion of zinc (*Zn*) from the end surface of a rectangular gallium arsenide n-type rod, whose electron concentration ranges between $10^{18}\ \text{cm}^{-3}$ and $10^{20}\ \text{cm}^{-3}$.

The zinc diffusion reverses the in-diffused zone into p-type. The dimensions of a typical device are $L \approx 0.2\,\text{mm}$, the cross-length of the light emitting area is $\Delta x \cong 0.05\,\text{mm}$, and its effective thickness (width) is $w \cong 6\,\mu\text{m}$ (see above). The fundamentally simplest way to form a resonance cavity is to optically polish the surfaces perpendicular to the long device dimension, thus getting a kind of a couple of reflective mirrors resulting from Fresnel reflections at the surfaces (see Eq. [14.31]). In addition, it is possible to provide those surfaces with multilayered dielectric coatings that provide each the adequate reflectance needed for the emitted wavelength spectrum. Use of anti-reflective (AR) coating, external mirrors, and other inside-cavity elements is also possible. Figure 18.17 shows a case where one of the sides is coated for high reflectivity (HR), functioning as a back cavity mirror.

We turn now to calculate the threshold current required for lasing. We summarize the following given parameters: the recombination radiation spectral peak is at $\lambda_{peak} = 840\,\text{nm}$ with $\Delta\lambda_{1/2} = 14\,\text{nm}$ full width at half maximum (see Figure 18.16). The material refractive index is known to be $n = 3.35$. Converting the gain equation (Eq. [2.54]) to exhibit the wavelength λ as variable instead of the frequency ν, we may write the light gain inside the junction as

$$\Gamma\left(\lambda_{peak}\right) = \frac{{\lambda_{peak}}^4 \Delta N}{8\pi n^2 c \tau_{rad}} g\left(\lambda_{peak}\right). \tag{18.3}$$

As a representative quantity for the ΔN population inversion, we choose the ratio $\mathcal{N}/V$ where $\mathcal{N}$ is the *total number* of excess electrons steadily remaining in the junction p-zone during injection; $V = \ell_D A$ is that zone volume, and $A = L \cdot \Delta x$ is the diode junction cross section perpendicularly to the current I orientation. We may assess the steady number of injected electrons $\mathcal{N}$ by the following consideration: under steady state, the electrons' recombination rate equals their injection rate. The recombination rate is given by $\mathcal{N}(1/\tau_{rad} + 1/\tau_{nr})$, where τ_{nr} is the non-radiative recombination lifetime. The injection rate is given by I/e, where I is the junction current and e is the electron elementary charge. The radiative part of the recombination rate $\mathcal{N}/\tau_{rad}$ is obtained by multiplying the current by the radiative recombination quantum efficiency η. Namely,

$$\frac{\mathcal{N}}{\tau_{rad}} = \eta \frac{I}{e}\ ; \tag{18.4}$$

therefore $\mathcal{N} = I\eta\tau_{rad}/e$. For the particular peak wavelength, we may also write $g(\lambda_{peak}) \cong 1/\Delta\lambda_{1/2}$. A further approximation would be to assume that $\ell_D \cong w$. Setting all above approximations into the gain equation, we get

$$\Gamma\left(\lambda_{peak}\right) = \eta\ \frac{{\lambda_{peak}}^4}{8\pi n^2 e c w\, \Delta\lambda_{1/2}}\ J\ , \tag{18.5}$$

where $J = I/A$ is the current density through the junction. Obviously, the junction light amplification is proportional to the current density.

Laser emission occurs when the gain exceeds the summed loses (see Section 14.7). In addition to reflection losses on the resonance cavity (the polished device surfaces – Figure 18.17), the junction exhibits further losses because of scattering, absorption by excitation of electrons from the valence band to the conduction band, and free carrier absorption. An issue that deserves special attention is diffraction losses. Because the developing beam waist w_0 is approximately half the junction width, namely $w_0 \cong 3 \times 10^{-4}\,\text{cm}$

(see Section 15.2), the beam divergence $\Theta = 1.27\lambda/2w_0$ (Eqs. [15.14] and [15.24]) could not be smaller than ~10°, and the Rayleigh range $z_R = \pi w_0^2/\lambda$ could not be longer than about 0.034 mm (compared with an amplifying medium $L = 0.2$ mm – see above). When standing alone, that consideration means that the beam would undergo divergence that would remove most of the beam from the amplifying medium, an effect equal to great losses, called **diffraction losses**. In reality, the described divergence is very little because of a special additional effect existing in the p-n junction. The junction center (transition from an n- to a p-type) is similar to a dielectric medium because both type free carriers concentrations are very small. It is further "sandwiched" between the p- and n-zones, exhibiting very large carrier concentrations. In a sense, it is similar to an insulating medium enclosed between metallic plates performing as mirrors to inner light, thus forming a wave guide, reflecting any beam "attempting to diverge". In the present context, we satisfy ourselves with that qualitative description, and do not elaborate on it quantitatively. The end conclusion is that diffraction losses are rather small. Divergence of the amplified beam expresses itself eventually on exiting the diode junction into free space. Because of the rectangular shape of the beam cross section and the large number of off-axis reflections in the amplifying medium, the beam quality is quite poor, typically $M^2 \approx 100$ (see Section 15.6).

When all loss factors are measured and/or calculated, it totals about 20 cm^{-1}. Quantitatively they are quite different in different diode lasers, and we will not address that issue in the present book. Stable lasing obtains when the gain is greater than the loss. The lasing **threshold** defines a condition where the peak gain $\Gamma(\lambda_{peak})$ (Eq. [(18.5)]) and α_{loss} are equal. Let J_{th} be the threshold current density. Equating between gain and loss, we get

$$J_{th} = \frac{\alpha_{loss} \cdot 8\pi n^2 e c w \, \Delta\lambda_{1/2}}{\eta \lambda_{peak}^4}. \tag{18.6}$$

As an example, we calculate the threshold current density for the following given data:
Given:

$\alpha_{loss} = 20\ \text{cm}^{-1}$ | $w = 6 \times 10^{-4}$ cm

$n = 3.35$ | $\lambda_{peak} = 0.84 \times 10^{-4}$ cm

$e = 4.8 \times 10^{-10}$ esu | $\Delta\lambda_{1/2} = 14\ \text{nm} = 14 \times 10^{-7}$ cm

$c = 3 \times 10^{10}$ cm/s | $\eta \approx 1$

Setting: (Eq. [(18.6)])

$$J_{th} = \frac{20 \times 8 \times 3.14 \times 3.35^2 \times 4.8 \times 10^{-10} \times 3 \times 10^{10} \times 6 \times 10^{-4} \times 1.4 \times 10^{-6}}{0.84^4 \times 10^{-16}};$$

$$J_{th} = 1.37 \times 10^{12} \frac{\text{esu}}{\text{cm}^2 \cdot \text{s}} = 457 \frac{\text{A}}{\text{cm}^2}.$$

(The relation "1 Coulomb = 3×10^9 esu" was used in the latter units conversion step). Experiments show that the above result indeed represents the order of current densities required for achieving lasing, thus justifying the approximations used in development of Eq. (18.6).

Diode lasers technology has undergone a substantial evolution over the last three decades. Most useful diode lasers are based on ternary semiconductor compounds such as gallium-aluminum-arsenide *GaAlAs* and indium-gallium-arsenide *InGaAs*. Changing the fractions between gallium *Ga* and aluminum *Al* in the first noted compound, and between indium *In* and gallium *Ga* in the second noted compound, changes the material energy gap, and allows for the selection of different emission wavelengths.

The said compounds cover spectral regions between 680 nm on the short wavelengths range, and 1,100 nm at the long wavelengths range. Laser diodes based on the indium-gallium-phosphide ternary compound allow laser emissions in the 630 – 690 nm range. There are linear diode arrays comprising several tens of diodes, and two-dimensional arrays comprising hundreds diodes over areas of the order of ~ 1 cm^2. Such arrays may emit tens of watts of a continuous emission, or several hundred Watts of ~ 100 μs long pulses. Efficiencies of electrical power conversion into emitted light are of the order of 30 – 40%. Different micro-lens based devices couple diode laser arrays into optical fibers, allowing the delivery of light over large distances. Research and development efforts on that scientific and technical field are quite intense, aimed at extending the spectral ranges, increasing power conversion efficiencies, and improving beam qualities.

Exercises for Chapter 18

18.1 Use the Russell Saunders method to describe the neon *Ne* energy scheme of Figure 18.2, and determine the states order by Hund's laws (Eqs. [5.48] and [5.64]).

18.2 Assuming that the bonding potential of the hydrogen/deuterium halide molecules listed in Table 18.1 exhibit the Morse form (Eq. [7.11]), calculate each dissociation energy D and β parameter. Use Eq. (7.17) for that purpose. Note that the equation describes the vibrational energies in the Morse potential, whereas the data given in Table 18.1 provide states energies relative to the $\text{v} = 0$ ground state, not relative to the potential minimum energy (Figure 7.1).

18.3 (a) Suggest combinations of orbital angular momentums for the valence electrons in the triply ionized neodymium Nd^{3+} which form the Terms listed in the energy scheme of Figure 18.10.
(b) Present schematically the crystal field splitting of all the electronic states of the Nd^{3+} ion in the YAG crystal that appear in the energy scheme of Figure 18.10. Address solely the orbital angular momentum, and ignore the spin states.

(To address the following exercise, the reader needs to have prior acquaintance with semiconductors electrical properties).

18.4 A p-n junction has been formed using a semiconductor of 0.18 eV energy gap and $n = 4$ infrared refractive index for constructing a diode laser. The conduction and valence bands' effective densities of states is $6 \times 10^{17}\,\text{cm}^{-3}$ and $5 \times 10^{18}\,\text{cm}^{-3}$, respectively, at 80K. The holes density on the p-side of the junction is $8 \times 10^{17}\,\text{cm}^{-3}$, and of electrons on the n-side of the junction is $2 \times 10^{16}\,\text{cm}^{-3}$. Hole mobility is $1,200\ \text{cm}^2/\text{V·s}$, and electron mobility is $80,000\ \text{cm}^2/\text{V·s}$. Excess electrons lifetime on the p-side is 100 ns, and the radiative quantum efficiency is 0.6. The spectral width of the emitted electron-hole recombination radiation on a forward voltage application is approximately 125 nm.
(a) Estimate the peak wavelength of the emitted recombination radiation.
(b) Estimate the thickness of the light emitting zone at 80K and the angular divergence of an emitted beam.
(c) Laser emission obtains for current densities exceeding ~ 500 A/cm^2; estimate the cavity losses during lasing.

Appendix A: Greek alphabet and phonetic names♣

Lower-case	Capital	Name	Lower-case	Capital	Name
α	A	Alfa	ν	N	Nee
β	B	Betta	ξ	Ξ	Ksee
γ	Γ	Gamma	o	O	Ommikron
δ	Δ	Delta	π	Π	Pee
ε	E	Eppsilon	ρ	P	Ro
ζ	Z	Zetta	σ	Σ	Sigma
η	H	Etta	τ	T	Taou
θ, ϑ	Θ	Tetta	υ	Υ	Yppsilon
ι	I	Yotta	ϕ, φ	Φ	Fee
κ	K	Kappa	χ	X	Khee
λ	Λ	Lammda	ψ	Ψ	Psee
μ	M	Mee	ω	Ω	Ommega

♣Indicated phonetic names are aimed at providing the proper Greek pronunciations by using the English alphabet symbols as pronounced by most north-American English speakers. Capital Greek letters in graylighted boxes are not considered in mathematics as Greek, but rather as Roman letters, due to their shape similarity.

Physics, Optics, and Spectroscopy of Materials, First Edition. Zeev Burshtein.

Appendix B: Table of physical constants

Name	Symbol, and size in cgs units
Vacuum speed of light	$c = 2.9979 \times 10^{10}$ cm/s
Elementary charge on electron/proton	$e = 4.8032 \times 10^{-10}$ esu
Planck constant	$h = 6.6262 \times 10^{-27}$ erg·s
Reduced Planck constant	$\hbar \equiv h/2\pi = 1.0546 \times 10^{-27}$ erg·s
Electron rest mass	$m = 9.1096 \times 10^{-28}$ g
Proton rest mass	$M_p = 1.6726 \times 10^{-24}$ g
Electron electrostatic radius♣	$r_0 = 1.409 \times 10^{-13}$ cm
Natural width of electric dipole radiation	$\Delta\lambda_N = 1.1804 \times 10^{-12}$ cm
Bohr radius	$a_0 = 0.52917 \times 10^{-8}$ cm
Rydberg constant	$R_\infty = 109,737$ cm^{-1}
Bohr magneton	$\mu_B = 9.2741 \times 10^{-21}$ erg·gauss^{-1}
Fine-structure constant	$\alpha = 1/137.036$
Avogadro number	$N_A = 6.0222 \times 10^{23}$ mole^{-1}
Boltzmann constant	$k_B = 1.3806 \times 10^{-16}$ erg/**deg**

♣See footnote in Section 1.1.

Appendix C: Dirac δ function

The Dirac $\boldsymbol{\delta}$ function is defined as a functional operator; its operation on a general scalar function $\boldsymbol{F}(x)$ satisfies

$$\int \boldsymbol{F}(x)\boldsymbol{\delta}(x-x_0)dx = \boldsymbol{F}(x_0), \tag{A1}$$

where the integration region contains the x_0 point. A very similar result is obtained when some function, exhibiting a single, very narrow peak at x_0, of a unity integral, is set in Eq. (A1) instead of the $\boldsymbol{\delta}(x - x_0)$ functional, provided the $\boldsymbol{F}(x)$ function is a continuous smooth function exhibiting only very small variation in the x_0 vicinity. The Dirac $\boldsymbol{\delta}$ function may thus be represented as the limit of any single-peak function made narrower while increasing its height to exhibit a unity integral. There are quite a number of analytical functions exhibiting that trait; and they are called **representations** of the $\boldsymbol{\delta}$ function. Several such useful functions are listed below:

$$\left.\begin{aligned}
\delta(x) &= \frac{1}{2\pi}\int_{-\infty}^{\infty} e^{ikx}dk = \frac{1}{2\pi}\int_{-\infty}^{\infty} \boldsymbol{\cos}kx\cdot dk = \lim_{\alpha\to\infty}\frac{\boldsymbol{\sin}\alpha x}{\pi x}\\
&= \lim_{\alpha\to\infty}\frac{1-\boldsymbol{\cos}\alpha x}{\pi\alpha x^2} = \lim_{\alpha\to\infty}\frac{\boldsymbol{\sin}^2\alpha(x/2)}{2\pi\alpha(x/2)^2}\\
&= \lim_{\alpha\to 0}\frac{\alpha}{\pi(\alpha^2+x^2)} = \lim_{\alpha\to 0}\frac{e^{-x^2/4k\alpha}}{\sqrt{4\pi k\alpha}}\ (for\ k>0) = \lim_{\alpha\to 0}\frac{\alpha e^{-\alpha^2x^2}}{\sqrt{\pi}}\\
&= \lim_{\alpha\to 0}\frac{1}{\pi}\int_0^{\infty} e^{-\alpha t}\boldsymbol{\cos}kx\cdot dk
\end{aligned}\right\}. \tag{A2}$$

Some algebraic relations of the $\boldsymbol{\delta}$ function are listed below:

$$\left.\begin{aligned}
&\text{(i)} && \int_{-\infty}^{\infty}\delta(x)dx = 1\\
&\text{(ii)} && \boldsymbol{F}(x)\,\delta(x) = \boldsymbol{F}(0)\,\delta(x)\\
&\text{(iii)} && \delta(x'-x) = \delta(x-x')\\
&\text{(iv)} && x\,\delta(x) = 0\\
&\text{(v)} && \delta(\alpha x) = \frac{1}{|\alpha|}\delta(x)\\
&\text{(vi)} && \delta[(x-x_1)(x-x_2)] = \frac{\delta(x-x_1)+\delta(x-x_2)}{|x_1-x_2|}\\
&\text{(vii)} && \delta(x^2-a^2) = \frac{\delta(x-a)+\delta(x+a)}{2a}\ for\ a>0\\
&\text{(viii)} && \delta(x-x_1)\,\delta(x-x_2) = \delta(x-x_1)\,\delta(x_1-x_2) = \delta(x-x_2)\,\delta(x_1-x_2)
\end{aligned}\right\} \tag{A3}$$

Physics, Optics, and Spectroscopy of Materials, First Edition. Zeev Burshtein.

The nth order derivative of the $\boldsymbol{\delta}$ function, symbolized by $\boldsymbol{\delta}^{(n)}$, is defined by the relation

$$\int \boldsymbol{F}(x)\, \delta^{(n)}(x-x_0)dx = (-1)^n \boldsymbol{F}^{(n)}(x_0). \tag{A4}$$

In the same limits, the nth order derivatives of the functions listed in Eq. (A2) provide representations of $\boldsymbol{\delta}^{(n)}$. The first-order derivative $\boldsymbol{\delta}^{(1)} \equiv \boldsymbol{\delta}'$ satisfies the following listed relations

$$\begin{aligned} &\text{(i)} \quad \delta'(-x) = -\delta'(x) \\ &\text{(ii)} \quad x\,\delta'(x) = -\delta'(x)\,. \\ &\text{(iii)} \quad x^2\,\delta'(x) = 0 \end{aligned} \tag{A5}$$

Appendix D: Literature references for further reading

1. Books

1 L. Pauling, and E. B. Wilson, *Introduction to Quantum Mechanics*, McGraw-Hill, 1935.
2 E. T. Copson, *An Introduction to the Theory of Functions of Complex Variable*, Oxford University Press, 1935.
3 F. Seitz, *The Modern Theory of Solids*, McGraw-Hill, 1940.
4 P. M. Morse, and H. Feshbach, *Methods of Theoretical Physics*, (Vols. I, II) McGraw-Hill, 1952.
5 *International Tables for X-ray crystallography*, Intl. Union of Crystallography, Kynock Press, Birmingham UK, 1952.
6 E. B. Wilson, J. C. Decius, and P. C. Cross, *Molecular Vibrations*, McGraw-Hill, 1955.
7 W. Kauzmann, *Quantum Chemistry*, Academic Press, 1957.
8 A. R. Edmonds, *Angular Momentum in Quantum Mechanics*, Princeton University Press, 1960.
9 9. H. G. Kuhn, *Atomic Spectra*, Longmans, 1961.
10 G. Herzberg, *Spectra of diatomic molecules*, Van Nostrand, 1961.
11 H. Goldstein, *Classical Mechanics*, Addison-Wesley, 1962.
12 A. de-Shalit, and I. Talmi, *Nuclear Shell Theory*, Academic Press, 1963.
13 H. Eyring, J. Walter, and G. E. Kimball, *Quantum Chemistry*, Wiley, 1963.
14 H. C. Allen, and P. C. Cross, *Molecular Vib-Rotors*, Wiley, 1963.
15 C. Kittel. *Quantum Theory of Solids*, Wiley, 1963.
16 E. U. Condon, G. H. Shortley, *The Theory of Atomic Spectra*, Cambridge, 1963.
17 H. Margenau, and G. M. Murphy, *The Mathematics of Physics and Chemistry*, Van Nostrand, 1964.
18 V. M. Agranovich, V. L. Ginzburg, *Spatial Dispersion in Crystal Optics and the Theory of Excitons*, Interscience, 1966.
19 J. J. Sakurai, *Advanced Quantum Mechanics*, Addison-Wesley, 1967.
20 A. Messiah, *Quantum Mechanics*, (Vols. I, II), North-Holland, 1967.
21 L. I. Schiff, *Quantum Mechanics*, McGraw-Hill, 1968.
22 M. Born, K. Huang, *Dynamical Theory of Crystal Lattices*, Oxford, 1968.
23 M. Alonso, and E. J. Finn, *Quantum and Statistical Physics* (Fundamental University Physics vol. III), Addison-Wesley, 1969.
24 E. V. Shpolskii, *Atomic physics*, Iliffe, 1969.
25 S. Sugano, Y. Tanabe, H. Kamimura, *Multiplets of Transition-Metal Ions in Crystals*, Academic Press, 1970.
26 C. Kittel. *Introduction to Solid-State Physics*, Fourth Edition, Wiley, 1971.
27 A. Yariv, *Introduction to Optical Electronics*, Holt-Reihart-Winston, 1971.
28 C. W. Allen, *Astrophysical quantities*, Athlone, 1973.
29 J. Callaway, *Quantum Theory of the Solid State*, Academic Press, 1974.
30 J. D. Jackson, *Classical Electrodynamics*, Wiley, 1975.
31 H. A. Bethe and E. E. Salpeter, *Quantum Mechanics of one- and two-electron atoms*, Plenum, 1977.
32 M. Weissbluth, *Atoms and Molecules*, Academic Press, 1978.
33 J. F. Nye, *Physical Properties of Crystals*, Oxford, 1985.
34 A. Yariv, *Quantum Electronics*, Third Edition, Wiley, 1989.
35 W. Koechner, *Solid State Laser Engineering*, Fifth Edition, Springer, 1999.

Physics, Optics, and Spectroscopy of Materials, First Edition. Zeev Burshtein.

36 V. G. Dmitriev, G. G. Gurzadyan, D. N. Nikogosyan, *Handbook of Nonlinear Optical Crystals*, Third Edition, Springer, 1999.

37 Z. Burshtein, *Light and Matter.* ***part I*** *– Atoms and Molecules;.* ***part II*** *– Dielectric solids, Optics and Lasers*, Kavim, Beer-Sheva, Israel (Hebrew**)**, 2005 and 2009.

2. Journal publications

1 P. M. Morse, *Diatomic molecules according to the wave mechanics. II. Vibrational levels*, Phys. Rev. **34**, 57-64, 1929.

2 R. H. Lyddane, R. G. Sachs, and E. Teller, *On the polar vibrations of alkali halides*, Phys. Rev. **59**, 673-76, 1941.

3 J. Racah, *Theory of complex spectra*, I. Phys. Rev. **61**, 186-97, 1942.
II. Phys. Rev. **62**, 438-62, 1942.
III. Phys. Rev. **63**, 367-82, 1943.

4 W. E. Lamb, Jr., *Anomalous fine structure of hydrogen and singly ionized helium*, Repts. Prog. Phys. **14**, 19-63, 1951.

5 G. F. Koster, *Space groups and their representations*, Solid State Physics (F. Seits and D. Turnbull, eds.) **5**, 173-256, 1957.

6 D. S. McClure, *Electronic spectra of ions in Crystals; Part II: Spectra of ions in crystals*, Solid State Physics (F. Seits and D. Turnbull, eds.) **9**, 400-525, 1959.

7 R. Loudon, *The Raman effect in crystals*, Adv. Phys. **13**, 423-482, 1964.

8 D. M. Roessler, *Kramers-Kronig analysis of non-normal incidence reflection*, Brit. J. Appl. Phys. **16**, 1359-66, 1965.

9 H. Kogelnik and T. Li, *Laser beams and resonators*, Appl. Opt. **5**, 1550-1567, 1966.

10 G. P. Fisher, *The electric dipole moment of a moving magnetic dipole*, Am. J. Phys. **39**, 1528-33, 1971.

11 M. J. Weber, *Multi-phonon relaxation of rare-earth ions in yttrium orthoaluminate*, Phys. Rev. **B**, **8**, 54-64, 1973.

12 E. Silberman and H. W. Morgan, *The use of group theory in the interpretation of infrared and Raman spectra*, Oak-Ridge National Laboratory Report ORNL/TM-**5666**, 1977.

13 D. L. Rousseau et al, *Normal mode determination in crystals*, J. Raman Spect. **10**, 253-90, 1981.

14 S. Rosenwaks and D. Chuchem, *A supersonic multikilohertz pulsed HF chemical laser*, J. Appl. Phys. **54**, 48-49, 1983.

15 S. A. Self, *Focusing of of spherical Gaussian beams*, Appl. Opt. **22**, 658-661, 1983.

16 A. Rose, *The absence of surface dipole contributions to metal work functions*, Sol. Stat. Comm. **45**, 859-64, 1983.

17 J. Ivri, Z. Burshtein, E. Miron, R. Reisfeld, and M. Eyal, *The perylene derivative BASF-241 solution as a new tunable dye laser in the visible*, IEEE J. Quantum Electron, **26**, 1516-20, 1990.

18 Y. V. Orlovskii, K. K. Pukhov, T. T. Basiev, and T. Tsuboi, *Nonlinear mechanism of multiphonon relaxation of the energy of electronic excitation in optical crystals doped with rare-earth ions*, Opt. Mater. **4**, 583-95, 1995.

19 L. D. Deloach R. H. Page, G. D. Wilke, S. A. Payne, and W. F. Krupke, *Transition-metal doped zinc chacogenides: spectroscopy and laser demonstration of a new class of gain media*, IEEE J. Quantum Electron, **32**, 885-95, 1996.

20 Z. Burshtein, P. Blau, Y. Kaliski, Y. Shimony, and M. R. Kokta, *Excited-state absorption studies of Cr^{4+} ions in several garnet host crystals*, IEEE J. Quantum Electron. **34**, 292-99, 1998.

21 G. Muños, *Spin-orbit interaction and the Thomas precession: A comment on the lab frame point of view*, Am. J. Phys. **69**, 554-6, 2001.

22 J. P. Costella et al, *The Thomas rotation*, Am. J. Phys. **69**, 837-47, 2001.

23 M. Brumer et al, *PbSe core, PbSe/PbS, and PbSe/$PbSe_xS_{1-x}$ core/shell nanocrystals as saturable absorbers in passively Q-switched near-infrared lasers*, Appl. Optics **45**, 7488-97, 2006.

24 Z. Burshtein, *Radiative, non-radiative, and mixed-decay transitions of rare-earth ions in dielectric media*, Opt. Eng. **49**, 091005,1-19, 2010.

Index

Physics, Optics, and Spectroscopy of Materials, First Edition. Zeev Burshtein.